Springer Series in Statistics

Springer Series in Statistics

Alho/Spencer: Statistical Demography and Forecasting.
Andersen/Borgan/Gill/Keiding: Statistical Models Based on Counting Processes.
Atkinson/Riani: Robust Diagnostic Regression Analysis.
Atkinson/Riani/Cerioli: Exploring Multivariate Data with the Forward Search.
Berger: Statistical Decision Theory and Bayesian Analysis, 2nd edition.
Borg/Groenen: Modern Multidimensional Scaling: Theory and Applications, 2nd edition.
Brockwell/Davis: Time Series: Theory and Methods, 2nd edition.
Bucklew: Introduction to Rare Event Simulation.
Cappé/Moulines/Rydén: Inference in Hidden Markov Models.
Chan/Tong: Chaos: A Statistical Perspective.
Chen/Shao/Ibrahim: Monte Carlo Methods in Bayesian Computation.
Coles: An Introduction to Statistical Modeling of Extreme Values.
Devroye/Lugosi: Combinatorial Methods in Density Estimation.
Diggle/Ribeiro: Model-based Geostatistics.
Efromovich: Nonparametric Curve Estimation: Methods, Theory, and Applications.
Eggermont/LaRiccia: Maximum Penalized Likelihood Estimation, Volume I: Density Estimation.
Fahrmeir/Tutz: Multivariate Statistical Modeling Based on Generalized Linear Models, 2nd edition.
Fan/Yao: Nonlinear Time Series: Nonparametric and Parametric Methods.
Ferraty/Vieu: Nonparametric Functional Data Analysis: Theory and Practice.
Fienberg/Hoaglin: Selected Papers of Frederick Mosteller.
Frühwirth-Schnatter: Finite Mixture and Markov Switching Models.
Ghosh/Ramamoorthi: Bayesian Nonparametrics.
Glaz/Naus/Wallenstein: Scan Statistics.
Good: Permutation Tests: Parametric and Bootstrap Tests of Hypotheses, 3rd edition.
Gouriéroux: ARCH Models and Financial Applications.
Gu: Smoothing Spline ANOVA Models.
Györfi/Kohler/Krzyżak/Walk: A Distribution-Free Theory of Nonparametric Regression.
Haberman: Advanced Statistics, Volume I: Description of Populations.
Hall: The Bootstrap and Edgeworth Expansion.
Härdle: Smoothing Techniques: With Implementation in S.
Harrell: Regression Modeling Strategies: With Applications to Linear Models, Logistic Regression, and Survival Analysis.
Hart: Nonparametric Smoothing and Lack-of-Fit Tests.
Hastie/Tibshirani/Friedman: The Elements of Statistical Learning: Data Mining, Inference, and Prediction.
Hedayat/Sloane/Stufken: Orthogonal Arrays: Theory and Applications.
Heyde: Quasi-Likelihood and its Application: A General Approach to Optimal Parameter Estimation.
Huet/Bouvier/Poursat/Jolivet: Statistical Tools for Nonlinear Regression: A Practical Guide with S-PLUS and R Examples, 2nd edition.
Ibrahim/Chen/Sinha: Bayesian Survival Analysis.
Jolliffe: Principal Component Analysis, 2nd edition.
Knottnerus: Sample Survey Theory: Some Pythagorean Perspectives.
Küchler/Sørensen: Exponential Families of Stochastic Processes.
Kutoyants: Statistical Inference for Ergodic Diffusion Processes.

(continued after index)

Moshe Shaked
J. George Shanthikumar

Stochastic Orders

Moshe Shaked
Department of Mathematics
University of Arizona
Tucson, AZ 85721
shaked@math.arizona.edu

J. George Shanthikumar
Department of Industrial Engineering and Operations Research
University of California, Berkeley
Berkeley, CA 94720
shanthikumar@ieor.berkeley.edu

Library of Congress Control Number: 2006927724

ISBN-10: 0-387-32915-3
ISBN-13: 978-0387-32915-4

Printed on acid-free paper.

9 8 7 6 5 4 3 2 1

springer.com

To my wife *Edith* and to my children
Tal, Shanna, and *Lila*

M.S.

To my wife *Mellony* and to my children
Devin, *Rajan*, and *Sohan*

J.G.S.

Preface

Stochastic orders and inequalities have been used during the last 40 years, at an accelerated rate, in many diverse areas of probability and statistics. Such areas include reliability theory, queuing theory, survival analysis, biology, economics, insurance, actuarial science, operations research, and management science. The purpose of this book is to collect in one place essentially all that is known about these orders up to the present. In addition, the book illustrates some of the usefulness and applicability of these stochastic orders.

This book is a major extension of the first six chapters in Shaked and Shanthikumar [515]. The idea that led us to write those six chapters arose as follows. In our own research in reliability theory and operations research we have been using, for years, several notions of stochastic orders. Often we would encounter a result that we could easily (or not so easily) prove, but we could not tell whether it was known or new. Even when we were sure that a result was known, we would not know right away where it could be found. Also, sometimes we would prove a result for the purpose of an application, only to realize later that a stronger result (stronger than what we needed) had already been derived elsewhere. We also often have had difficulties giving a reference for *one* source that contained everything about stochastic orders that we needed in a particular paper. In order to avoid such difficulties we wrote the first six chapters in Shaked and Shanthikumar [515].

Since 1994 the theory of stochastic orders has grown significantly. We think that now is the time to put in one place essentially all that is known about these orders. This book is the result of this effort.

The simplest way of comparing two distribution functions is by the comparison of the associated means. However, such a comparison is based on only two single numbers (the means), and therefore it is often not very informative. In addition to this, the means sometimes do not exist. In many instances in applications one has more detailed information, for the purpose of comparison of two distribution functions, than just the two means. Several orders of distribution functions, that take into account various forms of possible knowledge about the two underlying distribution functions, are studied in Chapters 1 and 2.

When one wishes to compare two distribution functions that have the same mean (or that are centered about the same value), one is usually interested in the comparison of the dispersion of these distributions. The simplest way of doing it is by the comparison of the associated standard deviations. However, such a comparison, again, is based on only two single numbers, and therefore it is often not very informative. In addition to this, again, the standard deviations sometimes do not exist. Several orders of distribution functions, which take into account various forms of possible knowledge about the two underlying distribution functions (in addition to the fact that they are centered about the same value), are studied in Chapter 3. Orders that can be used for the joint comparison of both the location and the dispersion of distribution functions are studied in Chapters 4 and 5. The analogous orders for multivariate distribution functions are studied in Chapters 6 and 7.

When one is interested in the comparison of a sequence of distribution functions, associated with the random variables X_i, $i = 1, 2, \ldots$, then one can use, of course, any of the orders described in Chapters 1–7 for the purpose of comparing any two of these distributions. However, the parameter i may now introduce some patterns that connect all the underlying distributions. For example, suppose not only that the random variables X_i, $i = 1, 2, \ldots$, increase stochastically in i, but also that the increase is sharper for larger i's. Then the sequence X_i, $i = 1, 2, \ldots$, is stochastically increasing in a convex sense. Such notions of stochastic convexity and concavity are studied in Chapter 8.

Notions of positive dependence of two random variables X_1 and X_2 have been introduced in the literature in an effort to mathematically describe the property that "large (respectively, small) values of X_1 go together with large (respectively, small) values of X_2." Many of these notions of positive dependence are defined by means of some comparison of the joint distribution of X_1 and X_2 with their distribution under the theoretical assumption that X_1 and X_2 are independent. Often such a comparison can be extended to general pairs of bivariate distributions with given marginals. This fact led researchers to introduce various notions of positive dependence orders. These orders are designed to compare the strength of the positive dependence of the two underlying bivariate distributions. Many of these orders can be further extended to comparisons of general multivariate distributions that have the same marginals. In Chapter 9 we describe these orders.

We have in mind a wide spectrum of readers and users of this book. On one hand, the text can be useful for those who are already familiar with many aspects of stochastic orders, but who are not aware of all the developments in this area. On the other hand, people who are not very familiar with stochastic orders, but who know something about them, can use this book for the purpose of studying or widening their knowledge and understanding of this important area.

We wish to thank Haijun Li, Asok K. Nanda, and Taizhong Hu for critical readings of several drafts of the manuscript. Their comments led to a substantial improvement in the presentation of some of the results in these chapters. We also thank Yigal Gerchak and Marco Scarsini for some illuminating suggestions. We thank our academic advisors John A. Buzacott (of J. G. S.) and Albert W. Marshall (of M. S.) who, years ago, introduced us to some aspects of the area of stochastic orders.

Tucson, Berkeley,
August 16, 2006

Moshe Shaked
J. George Shanthikumar

Contents

Note

Throughout the book "increasing" means "nondecreasing" and "decreasing" means "nonincreasing." Expectations are assumed to exist whenever they are written. The "inverse" of a monotone function (which is not strictly monotone) means the right continuous version of it, unless stated otherwise. For example, if F is a distribution function, then the right continuous version of its inverse is $F^{-1}(u) = \sup\{x : F(x) \le u\}$, $u \in [0,1]$.

The following aging notions will be encountered often throughout the text. Let X be a random variable with distribution function F and survival function $\overline{F} \equiv 1 - F$.

(i) The random variable X (or its distribution) is said to be IFR [increasing failure rate] if $\overline{F}$ is logconcave. It is said to be DFR [decreasing failure rate] if $\overline{F}$ is logconvex.

(ii) The nonnegative random variable X (or its distribution) is said to be IFRA [increasing failure rate average] if $-\log \overline{F}$ is starshaped; that is, if $-\log \overline{F}(t)/t$ is increasing in $t \ge 0$. It is said to be DFRA [decreasing failure rate average] if $-\log \overline{F}$ is antistarshaped; that is, if $-\log \overline{F}(t)/t$ is decreasing in $t \ge 0$.

(iii) The nonnegative random variable X (or its distribution) is said to be NBU [new better than used] if $\overline{F}(s)\overline{F}(t) \ge \overline{F}(s+t)$ for all $s \ge 0$ and $t \ge 0$. It is said to be NWU [new worse than used] if $\overline{F}(s)\overline{F}(t) \le \overline{F}(s+t)$ for all $s \ge 0$ and $t \ge 0$.

(iv) The random variable X (or its distribution) is said to be DMRL [decreasing mean residual life] if $\frac{\int_t^\infty \overline{F}(s)\mathrm{d}s}{\overline{F}(t)}$ is decreasing in t over $\{t : \overline{F}(t) > 0\}$. It is said to be IMRL [increasing mean residual life] if $\frac{\int_t^\infty \overline{F}(s)\mathrm{d}s}{\overline{F}(t)}$ is increasing in t over $\{t : \overline{F}(t) > 0\}$.

(v) The nonnegative random variable X (or its distribution) is said to be NBUE [new better than used in expectation] if $\frac{\int_t^\infty \overline{F}(s)\mathrm{d}s}{\overline{F}(t)} \le EX$ for all

$t \geq 0$. It is said to be NWUE [new worse than used in expectation] if $\frac{\int_t^\infty \overline{F}(s)\mathrm{d}s}{\overline{F}(t)} \geq EX$ for all $t \geq 0$.

The majorization order will be used in some places in the text. Recall from Marshall and Olkin [383] that a vector $\boldsymbol{a} = (a_1, a_2, \dots, a_n)$ is said to be smaller in the majorization order than the vector $\boldsymbol{b} = (b_1, b_2, \dots, b_n)$ (denoted $\boldsymbol{a} \prec \boldsymbol{b}$) if $\sum_{i=1}^n a_i = \sum_{i=1}^n b_i$ and if $\sum_{i=1}^j a_{[i]} \leq \sum_{i=1}^j b_{[i]}$ for $j = 1, 2, \dots, n-1$, where $a_{[i]}$ $[b_{[i]}]$ is the ith largest element of $\boldsymbol{a}$ $[\boldsymbol{b}]$, $i = 1, 2, \dots, n$. An n-dimensional function ϕ is called *Schur convex* [*concave*] if $\boldsymbol{a} \prec \boldsymbol{b} \Longrightarrow \phi(\boldsymbol{a}) \leq [\geq]\ \phi(\boldsymbol{b})$.

The notation $\mathbb{N} \equiv \{\dots, -1, 0, 1, \dots\}$, $\mathbb{N}_+ \equiv \{0, 1, \dots\}$, and $\mathbb{N}_{++} \equiv \{1, 2, \dots\}$ will be used in this text.

1

Univariate Stochastic Orders

In this chapter we study stochastic orders that compare the "location" or the "magnitude" of random variables. The most important and common orders that are considered in this chapter are the usual stochastic order $\le_{\rm st}$, the hazard rate order $\le_{\rm hr}$, and the likelihood ratio order $\le_{\rm lr}$. Some variations of these orders, and some related orders, are also examined in this chapter.

1.A The Usual Stochastic Order

1.A.1 Definition and equivalent conditions

Let X and Y be two random variables such that

$$P\{X > x\} \le P\{Y > x\} \quad \text{for all } x \in (-\infty, \infty). \tag{1.A.1}$$

Then X is said to be *smaller than Y in the usual stochastic order* (denoted by $X \le_{\rm st} Y$). Roughly speaking, (1.A.1) says that X is less likely than Y to take on large values, where "large" means any value greater than x, and that this is the case for all x's. Note that (1.A.1) is the same as

$$P\{X \le x\} \ge P\{Y \le x\} \quad \text{for all } x \in (-\infty, \infty). \tag{1.A.2}$$

It is easy to verify (by noting that every closed interval is an infinite intersection of open intervals) that $X \le_{\rm st} Y$ if, and only if,

$$P\{X \ge x\} \le P\{Y \ge x\} \quad \text{for all } x \in (-\infty, \infty). \tag{1.A.3}$$

In fact, we can recast (1.A.1) and (1.A.3) in a seemingly more general, but actually an equivalent, way as follows:

$$P\{X \in U\} \le P\{Y \in U\} \quad \text{for all upper sets } U \subseteq (-\infty, \infty). \tag{1.A.4}$$

(In the univariate case, that is on the real line, a set U is an upper set if, and only if, it is an open or a closed right half line.) In the univariate case the equivalence of (1.A.4) with (1.A.1) and (1.A.3) is trivial, but in Chapter 6 it will be seen that the generalizations of each of these three conditions to the multivariate case yield different definitions of stochastic orders.

Still another way of rewriting (1.A.1) or (1.A.3) is the following:

$$E[I_U(X)] \le E[I_U(Y)] \quad \text{for all upper sets } U \subseteq (-\infty,\infty), \tag{1.A.5}$$

where I_U denotes the indicator function of U. From (1.A.5) it follows that if $X \le_{\rm st} Y$, then

$$E\Big[\sum_{i=1}^{m} a_i I_{U_i}(X)\Big] - b \le E\Big[\sum_{i=1}^{m} a_i I_{U_i}(Y)\Big] - b \tag{1.A.6}$$

for all $a_i \ge 0$, $i = 1, 2, \ldots, m$, $b \in (-\infty,\infty)$, and $m \ge 0$. Given an increasing function ϕ, it is possible, for each m, to define a sequence of U_i's, a sequence of a_i's, and a b (all of which may depend on m), such that as $m \to \infty$ then (1.A.6) converges to

$$E[\phi(X)] \le E[\phi(Y)], \tag{1.A.7}$$

provided the expectations exist. It follows that $X \le_{\rm st} Y$ *if, and only if,* (1.A.7) *holds for all increasing functions ϕ for which the expectations exist.*

The expressions $\int_x^\infty P\{X > y\}dy$ and $\int_x^\infty P\{Y > y\}dy$ are used extensively in Chapters 2, 3, and 4. It is of interest to note that $X \le_{\rm st} Y$ if, and only if,

$$\int_x^\infty P\{Y > y\}du - \int_x^\infty P\{X > y\}dy \quad \text{is decreasing in } x \in (-\infty,\infty). \tag{1.A.8}$$

If X and Y are discrete random variables taking on values in $\mathbb{N}$, then we have the following. Let $p_i = P\{X = i\}$ and $q_i = P\{Y = i\}$, $i \in \mathbb{N}$. Then $X \le_{\rm st} Y$ if, and only if,

$$\sum_{j=-\infty}^{i} p_j \ge \sum_{j=-\infty}^{i} q_j, \quad i \in \mathbb{N},$$

or, equivalently, $X \le_{\rm st} Y$ if, and only if,

$$\sum_{j=i}^{\infty} p_j \le \sum_{j=i}^{\infty} q_j, \quad i \in \mathbb{N}.$$

1.A.2 A characterization by construction on the same probability space

An important characterization of the usual stochastic order is the following theorem (here $=_{\rm st}$ denotes equality in law).

Theorem 1.A.1. *Two random variables X and Y satisfy $X \leq_{\rm st} Y$ if, and only if, there exist two random variables $\hat{X}$ and $\hat{Y}$, defined on the same probability space, such that*

$$\hat{X} =_{\rm st} X, \tag{1.A.9}$$

$$\hat{Y} =_{\rm st} Y, \tag{1.A.10}$$

and

$$P\{\hat{X} \leq \hat{Y}\} = 1. \tag{1.A.11}$$

Proof. Obviously (1.A.9), (1.A.10), and (1.A.11) imply that $X \leq_{\rm st} Y$. In order to prove the necessity part of Theorem 1.A.1, let F and G be, respectively, the distribution functions of X and Y, and let F^{-1} and G^{-1} be the corresponding right continuous inverses (see Note on page 1). Define $\hat{X} = F^{-1}(U)$ and $\hat{Y} = G^{-1}(U)$ where U is a uniform $[0,1]$ random variable. Then it is easy to see that $\hat{X}$ and $\hat{Y}$ satisfy (1.A.9) and (1.A.10). From (1.A.2) it is seen that (1.A.11) also holds. □

Theorem 1.A.1 is a special case of a more general result that is stated in Section 6.B.2.

From (1.A.2) and Theorem 1.A.1 it follows that the random variables X and Y, with the respective distribution functions F and G, satisfy $X \leq_{\rm st} Y$ if, and only if,

$$F^{-1}(u) \leq G^{-1}(u), \quad \text{for all } u \in (0,1). \tag{1.A.12}$$

Another way of restating Theorem 1.A.1 is the following. We omit the obvious proof of it.

Theorem 1.A.2. *Two random variables X and Y satisfy $X \leq_{\rm st} Y$ if, and only if, there exist a random variable Z and functions ψ_1 and ψ_2 such that $\psi_1(z) \leq \psi_2(z)$ for all z and $X =_{\rm st} \psi_1(Z)$ and $Y =_{\rm st} \psi_2(Z)$.*

In some applications, when the random variables X and Y are such that $X \leq_{\rm st} Y$, one may wish to construct a $\hat{Y}$ $[\hat{X}]$ on the probability space on which X $[Y]$ is defined, such that $\hat{Y} =_{\rm st} Y$ and $P\{X \leq \hat{Y}\} = 1$ $[\hat{X} =_{\rm st} X$ and $P\{\hat{X} \leq Y\} = 1]$. This is always possible. Here we will show how this can be done when the distribution function F $[G]$ of X $[Y]$ is absolutely continuous. When this is the case, $F(X)$ $[G(Y)]$ is uniformly distributed on $[0,1]$, and therefore $\hat{Y} = G^{-1}(F(X))$ $[\hat{X} = F^{-1}(G(Y))]$ is the desired construction $\hat{Y}$ $[\hat{X}]$.

1.A.3 Closure properties

Using (1.A.1) through (1.A.11) it is easy to prove each of the following closure results. The following notation will be used: For any random variable Z and an event A, let $[Z|A]$ denote any random variable that has as its distribution the conditional distribution of Z given A.

Theorem 1.A.3. (a) *If* $X \leq_{\rm st} Y$ *and* g *is any increasing [decreasing] function, then* $g(X) \leq_{\rm st} [\geq_{\rm st}]\ g(Y)$.

(b) *Let* $X_1, X_2, \ldots, X_m$ *be a set of independent random variables and let* $Y_1, Y_2, \ldots, Y_m$ *be another set of independent random variables. If* $X_i \leq_{\rm st} Y_i$ *for* $i = 1, 2, \ldots, m$, *then, for any increasing function* $\psi : \mathbb{R}^m \to \mathbb{R}$, *one has*

$$\psi(X_1, X_2, \ldots, X_m) \leq_{\rm st} \psi(Y_1, Y_2, \ldots, Y_m).$$

In particular,

$$\sum_{j=1}^{m} X_j \leq_{\rm st} \sum_{j=1}^{m} Y_j.$$

That is, the usual stochastic order is closed under convolutions.

(c) *Let* $\{X_j,\ j = 1, 2, \ldots\}$ *and* $\{Y_j,\ j = 1, 2, \ldots\}$ *be two sequences of random variables such that* $X_j \to_{\rm st} X$ *and* $Y_j \to_{\rm st} Y$ *as* $j \to \infty$, *where "*$\to_{\rm st}$*" denotes convergence in distribution. If* $X_j \leq_{\rm st} Y_j$, $j = 1, 2, \ldots$, *then* $X \leq_{\rm st} Y$.

(d) *Let* X, Y, *and* Θ *be random variables such that* $[X \big| \Theta = \theta] \leq_{\rm st} [Y \big| \Theta = \theta]$ *for all* θ *in the support of* Θ. *Then* $X \leq_{\rm st} Y$. *That is, the usual stochastic order is closed under mixtures.*

In the next result and in the sequel we define $\sum_{j=1}^{0} a_j \equiv 0$ for any sequence $\{a_j,\ j = 1, 2, \ldots\}$.

Theorem 1.A.4. *Let* $\{X_j,\ j = 1, 2, \ldots\}$ *be a sequence of nonnegative independent random variables, and let* M *be a nonnegative integer-valued random variable which is independent of the* X_i*'s. Let* $\{Y_j,\ j = 1, 2, \ldots\}$ *be another sequence of nonnegative independent random variables, and let* N *be a nonnegative integer-valued random variable which is independent of the* Y_i*'s. If* $X_i \leq_{\rm st} Y_i$, $i = 1, 2, \ldots$, *and if* $M \leq_{\rm st} N$, *then*

$$\sum_{j=1}^{M} X_j \leq_{\rm st} \sum_{j=1}^{N} Y_j.$$

Another related result is given next.

Theorem 1.A.5. *Let* $\{X_j,\ j = 1, 2, \ldots\}$ *be a sequence of nonnegative independent and identically distributed random variables, and let* M *be a positive integer-valued random variable which is independent of the* X_i*'s. Let* $\{Y_j,\ j = 1, 2, \ldots\}$ *be another sequence of independent and identically distributed random variables, and let* N *be a positive integer-valued random variable which is independent of the* Y_i*'s. Suppose that for some positive integer* K *we have that*

$$\sum_{j=1}^{K} X_j \leq_{\rm st} [\geq_{\rm st}]\ Y_1$$

and

$$M \leq_{\text{st}} [\geq_{\text{st}}] \; KN,$$

then

$$\sum_{j=1}^{M} X_j \leq_{\text{st}} [\geq_{\text{st}}] \sum_{j=1}^{N} Y_j.$$

Proof. The assumptions yield

$$\sum_{i=1}^{M} X_i \leq_{\text{st}} [\geq_{\text{st}}] \sum_{i=1}^{KN} X_i = \sum_{i=1}^{N} \sum_{j=K(i-1)+1}^{Ki} X_j \leq_{\text{st}} [\geq_{\text{st}}] \sum_{i=1}^{N} Y_i. \qquad \square$$

Consider now a family of distribution functions $\{G_\theta,\, \theta \in \mathcal{X}\}$ where $\mathcal{X}$ is a subset of the real line $\mathbb{R}$. Let $X(\theta)$ denote a random variable with distribution function G_θ. For any random variable Θ with support in $\mathcal{X}$, and with distribution function F, let us denote by $X(\Theta)$ a random variable with distribution function H given by

$$H(y) = \int_{\mathcal{X}} G_\theta(y) \mathrm{d}F(\theta), \quad y \in \mathbb{R}.$$

The following result is a generalization of both parts (a) and (c) of Theorem 1.A.3.

Theorem 1.A.6. *Consider a family of distribution functions $\{G_\theta,\, \theta \in \mathcal{X}\}$ as above. Let Θ_1 and Θ_2 be two random variables with supports in $\mathcal{X}$ and distribution functions F_1 and F_2, respectively. Let Y_1 and Y_2 be two random variables such that $Y_i =_{\text{st}} X(\Theta_i)$, $i = 1, 2$; that is, suppose that the distribution function of Y_i is given by*

$$H_i(y) = \int_{\mathcal{X}} G_\theta(y) \mathrm{d}F_i(\theta), \quad y \in \mathbb{R},\; i = 1, 2.$$

If

$$X(\theta) \leq_{\text{st}} X(\theta') \quad \textit{whenever } \theta \leq \theta', \tag{1.A.13}$$

and if

$$\Theta_1 \leq_{\text{st}} \Theta_2, \tag{1.A.14}$$

then

$$Y_1 \leq_{\text{st}} Y_2. \tag{1.A.15}$$

Proof. Note that, by (1.A.13), $P\{X(\theta) > y\}$ is increasing in θ for all y. Thus

$$\begin{aligned} P\{Y_1 > y\} &= \int_{\mathcal{X}} P\{X(\theta) > y\} \mathrm{d}F_1(\theta) \\ &\leq \int_{\mathcal{X}} P\{X(\theta) > y\} \mathrm{d}F_2(\theta) \\ &= P\{Y_2 > y\}, \quad \text{for all } y, \end{aligned}$$

where the inequality follows from (1.A.14) and (1.A.7). Thus (1.A.15) follows from (1.A.1). $\square$

Note that, using the notation that is introduced below before Theorem 1.A.14, (1.A.13) can be rewritten as $\{X(\theta),\ \theta \in \mathcal{X}\} \in \mathrm{SI}$.

The following example shows an application of Theorem 1.A.6 in the area of Bayesian imperfect repair; a related result is given in Example 1.B.16.

Example 1.A.7. Let Θ_1 and Θ_2 be two random variables with supports in $\mathcal{X} = (0,1]$ and distribution functions F_1 and F_2, respectively. For some survival function $\overline{K}$, define

$$\overline{G}_\theta = \overline{K}^{1-\theta}, \quad \theta \in (0,1],$$

and let $X(\theta)$ have the survival function $\overline{K}^{1-\theta}$. Note that (1.A.13) holds because $\overline{K}^{1-\theta}(y) \le \overline{K}^{1-\theta'}(y)$ for all y whenever $0 < \theta \le \theta' \le 1$. Thus, if $\Theta_1 \le_{\mathrm{st}} \Theta_2$ then Y_i, with survival function $\overline{H}_i$ defined by

$$\overline{H}_i(y) = \int_0^1 \overline{K}^{1-\theta}(y)\mathrm{d}F_i(\theta), \quad y \in \mathbb{R},\ i = 1,2,$$

satisfy $Y_1 \le_{\mathrm{st}} Y_2$.

1.A.4 Further characterizations and properties

Clearly, if $X \le_{\mathrm{st}} Y$ then $EX \le EY$. However, as the following result shows, if two random variables are ordered in the usual stochastic order and have the same expected values, they must have the same distribution.

Theorem 1.A.8. *If $X \le_{\mathrm{st}} Y$ and if $E[h(X)] = E[h(Y)]$ for some strictly increasing function h, then $X =_{\mathrm{st}} Y$.*

Proof. First we prove the result when $h(x) = x$. Let $\hat{X}$ and $\hat{Y}$ be as in Theorem 1.A.1. If $P\{\hat{X} < \hat{Y}\} > 0$, then $EX = E\hat{X} < E\hat{Y} = EY$, a contradiction to the assumption $EX = EY$. Therefore $X =_{\mathrm{st}} \hat{X} = \hat{Y} =_{\mathrm{st}} Y$. Now let h be some strictly increasing function. Observe that if $X \le_{\mathrm{st}} Y$, then $h(X) \le_{\mathrm{st}} h(Y)$ and therefore from the above result we have that $h(X) =_{\mathrm{st}} h(Y)$. The strict monotonicity of h yields $X =_{\mathrm{st}} Y$. □

Other results that give conditions, involving stochastic orders, which imply stochastic equalities, are given in Theorems 3.A.43, 3.A.60, 4.A.69, 5.A.15, 6.B.19, 6.G.12, 6.G.13, and 7.A.14–7.A.16.

As was mentioned above, if $X \le_{\mathrm{st}} Y$, then $EX \le EY$. It is easy to find counterexamples which show that the converse is false. However, $X \le_{\mathrm{st}} Y$ implies other moment inequalities (for example, $EX^3 \le EY^3$). Thus one may wonder whether $X \le_{\mathrm{st}} Y$ can be characterized by a collection of moment inequalities. Brockett and Kahane [109, Corollary 1] showed that there exist no finite number of moment inequalities which imply $X \le_{\mathrm{st}} Y$. In fact, they showed it for many other stochastic orders that are studied later in this book.

In order to state the next characterization we define the following class of bivariate functions:

$$\mathcal{G}_{\mathrm{st}} = \{\phi : \mathbb{R}^2 \to \mathbb{R} : \phi(x,y) \text{ is increasing in } x \text{ and decreasing in } y\}.$$

Theorem 1.A.9. *Let X and Y be independent random variables. Then $X \leq_{\rm st} Y$ if, and only if,*

$$\phi(X,Y) \leq_{\rm st} \phi(Y,X) \quad \textit{for all } \phi \in \mathcal{G}_{\rm st}. \tag{1.A.16}$$

Proof. Suppose that (1.A.16) holds. The function ϕ defined by $\phi(x,y) \equiv x$ belongs to $\mathcal{G}_{\rm st}$. Therefore $X \leq_{\rm st} Y$.

In order to prove the "only if" part, suppose that $X \leq_{\rm st} Y$. Let $\phi \in \mathcal{G}_{\rm st}$ and define $\psi(x,y) = \phi(x,-y)$. Then ψ is increasing on $\mathbb{R}^2$. Since X and Y are independent it follows that X and $-Y$ are independent and also that $-X$ and Y are independent. Since $X \leq_{\rm st} Y$ it follows (for example, from Theorem 1.A.1) that $-Y \leq_{\rm st} -X$. Therefore, by Theorem 1.A.3(b), we have

$$\psi(X,-Y) \leq_{\rm st} \psi(Y,-X),$$

that is,

$$\phi(X,Y) \leq_{\rm st} \phi(Y,X). \qquad \square$$

The next result is a similar characterization. In order to state it we need the following notation: Let ϕ_1 and ϕ_2 be two bivariate functions. Denote $\Delta\phi_{21}(x,y) = \phi_2(x,y) - \phi_1(x,y)$. The proof of the following theorem is omitted.

Theorem 1.A.10. *Let X and Y be two independent random variables. Then $X \leq_{\rm st} Y$ if, and only if,*

$$E\phi_1(X,Y) \leq E\phi_2(X,Y)$$

for all ϕ_1 and ϕ_2 which satisfy that, for each y, $\Delta\phi_{21}(x,y)$ decreases in x on $\{x \leq y\}$; for each x, $\Delta\phi_{21}(x,y)$ increases in y on $\{y \geq x\}$; and $\Delta\phi_{21}(x,y) \geq -\Delta\phi_{21}(y,x)$ whenever $x \leq y$.

Another similar characterization is given in Theorem 4.A.36.

Let X and Y be two random variables with distribution functions F and G, respectively. Let $\mathcal{M}(F,G)$ denote the Fréchet class of bivariate distributions with fixed marginals F and G. Abusing notation we write $(\hat{X},\hat{Y}) \in \mathcal{M}(F,G)$ to mean that the jointly distributed random variables $\hat{X}$ and $\hat{Y}$ have the marginal distribution functions F and G, respectively. The Fortret-Mourier-Wasserstein distance between the finite mean random variables X and Y is defined by

$$d(X,Y) = \inf_{(\hat{X},\hat{Y}) \in \mathcal{M}(F,G)} \{E|\hat{Y} - \hat{X}|\}. \tag{1.A.17}$$

Theorem 1.A.11. *Let X and Y be two finite mean random variables such that $EX \leq EY$. Then $X \leq_{\rm st} Y$ if, and only if, $d(X,Y) = EY - EX$.*

Proof. Suppose that $d(X,Y) = EY - EX$. The infimum in (1.A.17) is attained for some $(\hat{X},\hat{Y})$, and we have $E|\hat{Y} - \hat{X}| = E(\hat{Y} - \hat{X})$. Therefore $P\{\hat{X} \leq \hat{Y}\} = 1$, and from Theorem 1.A.1 it follows that $X \leq_{\rm st} Y$.

Conversely, suppose that $X \leq_{\rm st} Y$. Let $\hat{X}$ and $\hat{Y}$ be as in Theorem 1.A.1. Then, for any $(X',Y') \in \mathcal{M}(F,G)$ we have that $E|Y' - X'| \geq |EY' - EX'| = E\hat{Y} - E\hat{X}$. Therefore $d(X,Y) = EY - EX$. $\square$

A simple sufficient condition which implies the usual stochastic order is described next. The following notation will be used. Let $a(x)$ be defined on I, where I is a subset of the real line. The number of sign changes of a in I is defined by

$$S^-(a) = \sup S^-[a(x_1), a(x_2), \dots, a(x_m)], \tag{1.A.18}$$

where $S^-(y_1, y_2, \dots, y_m)$ is the number of sign changes of the indicated sequence, zero terms being discarded, and the supremum in (1.A.18) is extended over all sets $x_1 < x_2 < \cdots < x_m$ such that $x_i \in I$ and $m < \infty$. The proof of the next theorem is simple and therefore it is omitted.

Theorem 1.A.12. *Let X and Y be two random variables with (discrete or continuous) density functions f and g, respectively. If*

$$S^-(g - f) = 1 \quad \textit{and the sign sequence is } -, +,$$

then $X \le_{\rm st} Y$.

Let X_1 be a nonnegative random variable with distribution function F_1 and survival function $\overline{F}_1 \equiv 1 - F_1$. Define the Laplace transform of X_1 by

$$\varphi_{X_1}(\lambda) = \int_0^\infty \mathrm{e}^{-\lambda x} \mathrm{d}F_1(x), \quad \lambda > 0,$$

and denote

$$\overline{a}_\lambda^{X_1}(n) = \frac{(-1)^n}{n!} \frac{\mathrm{d}^n}{\mathrm{d}\lambda^n} \left[\frac{1 - \varphi_{X_1}(\lambda)}{\lambda} \right], \quad n \ge 0,\ \lambda > 0,$$

and

$$\overline{\alpha}_\lambda^{X_1}(n) = \lambda^n \overline{a}_\lambda^{X_1}(n - 1), \quad n \ge 1,\ \lambda > 0.$$

Similarly, for a nonnegative random variable X_2 with distribution function F_2 and survival function $\overline{F}_2 \equiv 1 - F_2$, define $\overline{\alpha}_\lambda^{X_2}(n)$. It can be shown that $\overline{\alpha}_\lambda^{X_1}$ and $\overline{\alpha}_\lambda^{X_2}$ are discrete survival functions (see the proof of the next theorem); denote the corresponding discrete random variables by $N_\lambda(X_1)$ and $N_\lambda(X_2)$. The following result gives a Laplace transform characterization of the order $\le_{\rm st}$.

Theorem 1.A.13. *Let X_1 and X_2 be two nonnegative random variables, and let $N_\lambda(X_1)$ and $N_\lambda(X_2)$ be as described above. Then*

$$X_1 \le_{\rm st} X_2 \Longleftrightarrow N_\lambda(X_1) \le_{\rm st} N_\lambda(X_2) \quad \textit{for all } \lambda > 0.$$

Proof. First suppose that $X_1 \le_{\rm st} X_2$. Select a $\lambda > 0$. Let $Z_1, Z_2, \dots,$ be independent exponential random variables with mean $1/\lambda$. It can be shown that $\overline{\alpha}_\lambda^{X_1}(n) = P\{\sum_{i=1}^n Z_i \le X_1\}$ and that $\overline{\alpha}_\lambda^{X_2}(n) = P\{\sum_{i=1}^n Z_i \le X_2\}$. It thus follows that $N_\lambda(X_1) \le_{\rm st} N_\lambda(X_2)$.

Now suppose that $N_\lambda(X_1) \le_{\rm st} N_\lambda(X_2)$ for all $\lambda > 0$. Select an $x > 0$. Thus $\overline{\alpha}_{n/x}^{X_1}(n) \le \overline{\alpha}_{n/x}^{X_2}(n)$. Letting $n \to \infty$, one obtains $\overline{F}_1(x) \le \overline{F}_2(x)$ for all continuity points x of F_1 and F_2. Therefore, $X_1 \le_{\rm st} X_2$ by (1.A.1). □

The implication $\Longrightarrow$ in Theorem 1.A.13 can be generalized as follows. A family of random variables $\{Z(\theta),\ \theta \in \Theta\}$ (Θ is a subset of the real line) is said to be stochastically increasing in the usual stochastic order (denoted by $\{Z(\theta),\ \theta \in \Theta\} \in \mathrm{SI}$) if $Z(\theta) \le_{\mathrm{st}} Z(\theta')$ whenever $\theta \le \theta'$. Recall from Theorem 1.A.3(a) that if $X_1 \le_{\mathrm{st}} X_2$, then $g(X_1) \le_{\mathrm{st}} g(X_2)$ for any increasing function g. The following result gives a stochastic generalization of this fact.

Theorem 1.A.14. *If $\{Z(\theta),\ \theta \in \Theta\} \in \mathrm{SI}$ and if $X_1 \le_{\mathrm{st}} X_2$, where X_k and $Z(\theta)$ are independent for $k = 1, 2$ and $\theta \in \Theta$, then $Z(X_1) \le_{\mathrm{st}} Z(X_2)$.*

Note that Theorem 1.A.14 is a restatement of Theorem 1.A.6.

Let X be a random variable and denote by $X_{(-\infty,a]}$ the truncation of X at a, that is, $X_{(-\infty,a]}$ has as its distribution the conditional distribution of X given that $X \le a$. $X_{(a,\infty)}$ is similarly defined. It is simple to prove the following result. Results that are stronger than this are contained in Theorems 1.B.20, 1.B.55, and 1.C.27.

Theorem 1.A.15. *Let X be any random variable. Then $X_{(-\infty,a]}$ and $X_{(a,\infty)}$ are increasing in a in the sense of the usual stochastic order.*

An interesting example in which truncated random variables are compared is the following.

Example 1.A.16. Let $X^{(1)}, X^{(2)}, \ldots, X^{(n)}$ be independent and identically distributed random variables. For a fixed t, let $X^{(1)}_{(t,\infty)}, X^{(2)}_{(t,\infty)}, \ldots, X^{(n)}_{(t,\infty)}$ be the corresponding truncations, and assume that they are also independent and identically distributed. Then

$$\big(\max\{X^{(1)}, X^{(2)}, \ldots, X^{(n)}\}\big)_{(t,\infty)} \le_{\mathrm{st}} \max\Big\{X^{(1)}_{(t,\infty)}, X^{(2)}_{(t,\infty)}, \ldots, X^{(n)}_{(t,\infty)}\Big\},$$

where $\big(\max\{X^{(1)}, X^{(2)}, \ldots, X^{(n)}\}\big)_{(t,\infty)}$ denotes the corresponding truncation of $\max\{X^{(1)}, X^{(2)}, \ldots, X^{(n)}\}$. The proof consists of a straightforward verification of (1.A.2) for the compared random variables.

Let ϕ_1 and ϕ_2 be two functions that satisfy $\phi_1(x) \le \phi_2(x)$ for all $x \in \mathbb{R}$, and let X be a random variable. Then, clearly, $\phi_1(X) \le \phi_2(X)$ almost surely. From Theorem 1.A.1 we thus obtain the following result.

Theorem 1.A.17. *Let X be a random variable and let ϕ_1 and ϕ_2 be two functions that satisfy $\phi_1(x) \le \phi_2(x)$ for all $x \in \mathbb{R}$. Then*

$$\phi_1(X) \le_{\mathrm{st}} \phi_2(X).$$

In particular, if ϕ is a function that satisfies $x \le [\ge]\ \phi(x)$ for all $x \in \mathbb{R}$, then $X \le_{\mathrm{st}} [\ge_{\mathrm{st}}]\ \phi(X)$.

Remark 1.A.18. The set of all distribution functions on $\mathbb{R}$ is a lattice with respect to the order $\le_{\mathrm{st}}$. That is, if X and Y are random variables with distributions F and G, then there exist random variables Z and W such that $Z \le_{\mathrm{st}} X$, $Z \le_{\mathrm{st}} Y$, $W \ge_{\mathrm{st}} X$, and $W \ge_{\mathrm{st}} Y$. Explicitly, Z has the survival function $\min\{\overline{F}, \overline{G}\}$ and W has the survival function $\max\{\overline{F}, \overline{G}\}$.

The next four theorems give conditions under which the corresponding spacings are ordered according to the usual stochastic order. Let $X_1, X_2, \ldots, X_m$ be any random variables with the corresponding order statistics $X_{(1)} \le X_{(2)} \le \cdots \le X_{(m)}$. Define the corresponding spacings by $U_{(i)} = X_{(i)} - X_{(i-1)}$, $i = 2, 3, \ldots, m$. When the dependence on m is to be emphasized, we will denote the spacings by $U_{(i:m)}$.

Theorem 1.A.19. *Let $X_1, X_2, \ldots, X_m, X_{m+1}$ be independent and identically distributed* IFR (DFR) *random variables. Then*

$$(m-i+1)U_{(i:m)} \ge_{\rm st} [\le_{\rm st}]\ (m-i)U_{(i+1:m)}, \quad i = 2, 3, \ldots, m-1,$$

and

$$(m-i+2)U_{(i:m+1)} \ge_{\rm st} [\le_{\rm st}]\ (m-i+1)U_{(i:m)}, \quad i = 2, 3, \ldots, m.$$

The proof of Theorem 1.A.19 is not given here. A stronger version of the DFR part of Theorem 1.A.19 is given in Theorem 1.B.31. Some of the conclusions of Theorem 1.A.19 can be obtained under different conditions. These are stated in the next two theorems. Again, the proofs are not given. In the next two theorems we take $X_{(0)} \equiv 0$, and thus $U_{(1)} = X_{(1)}$. For the following theorem recall from page 1 the definition of Schur concavity.

Theorem 1.A.20. *Let $X_1, X_2, \ldots, X_m$ be nonnegative random variables with an absolutely continuous joint distribution function. If the joint density function of $X_1, X_2, \ldots, X_m$ is Schur concave (Schur convex), then*

$$(m-i+1)U_{(i:m)} \ge_{\rm st} [\le_{\rm st}]\ (m-i)U_{(i+1:m)}, \quad i = 1, 2, \ldots, m-1.$$

Theorem 1.A.21. *Let $X_1, X_2, \ldots, X_m$ be independent exponential random variables with possibly different parameters. Then*

$$(m-i+1)U_{(i:m)} \le_{\rm st} (m-i)U_{(i+1:m)}, \quad i = 1, 2, \ldots, m-1.$$

Theorem 1.A.22. *Let $X_1, X_2, \ldots, X_m$ be independent and identically distributed random variables with a finite support, and with an increasing [decreasing] density function over that support. Then*

$$U_{(i:m)} \ge_{\rm st} [\le_{\rm st}]\ U_{(i+1:m)}, \quad i = 2, 3, \ldots, m-1.$$

The proof of Theorem 1.A.22 uses the likelihood ratio order, and therefore it is deferred to Section 1.C, Remark 1.C.3.

Note that any absolutely continuous DFR random variable has a decreasing density function. Thus we see that the assumption in the DFR part of Theorem 1.A.19 is stronger than the assumption in the decreasing part of Theorem 1.A.22, but the conclusion in the DFR part of Theorem 1.A.19 is stronger than the conclusion in the decreasing part of Theorem 1.A.22. It is of interest to compare Theorems 1.A.19–1.A.22 with Theorems 1.B.31 and 1.C.42.

From Theorem 1.A.1 it is obvious that if $X_{(1)} \le X_{(2)} \le \cdots \le X_{(m)}$ are the order statistics corresponding to the random variables $X_1, X_2, \ldots, X_m$, then $X_{(1)} \le_{\rm st} X_{(2)} \le_{\rm st} \cdots \le_{\rm st} X_{(m)}$. Now let $X_{(1)} \le X_{(2)} \le \cdots \le X_{(m)}$ be the order statistics corresponding to the random variables $X_1, X_2, \ldots, X_m$, and let $Y_{(1)} \le Y_{(2)} \le \cdots \le Y_{(m)}$ be the order statistics corresponding to the random variables $Y_1, Y_2, \ldots, Y_m$. As usual, for any distribution function F, we let $\overline{F} \equiv 1 - F$ denote the corresponding survival function.

Theorem 1.A.23. (a) *Let $X_1, X_2, \ldots, X_m$ be independent random variables with distribution functions $F_1, F_2, \ldots, F_m$, respectively. Let $Y_1, Y_2, \ldots, Y_m$ be independent and identically distributed random variables with a common distribution function G. Then $X_{(i)} \le_{\rm st} Y_{(i)}$ for all $i = 1, 2, \ldots, m$ if, and only if,*

$$\prod_{j=1}^{m} F_j(x) \ge G^m(x) \quad \textit{for all } x;$$

that is, if, and only if, $X_{(m)} \le_{\rm st} Y_{(m)}$.

(b) *Let $X_1, X_2, \ldots, X_m$ be independent random variables with survival functions $\overline{F}_1, \overline{F}_2, \ldots, \overline{F}_m$, respectively. Let $Y_1, Y_2, \ldots, Y_m$ be independent and identically distributed random variables with a common survival function $\overline{G}$. Then $X_{(i)} \ge_{\rm st} Y_{(i)}$ for all $i = 1, 2, \ldots, m$ if, and only if,*

$$\prod_{j=1}^{m} \overline{F}_j(x) \ge \overline{G}^m(x) \quad \textit{for all } x;$$

that is, if, and only if, $X_{(1)} \ge_{\rm st} Y_{(1)}$.

The proof of Theorem 1.A.23 is not given here.

More comparisons of order statistics in the usual stochastic order can be found in Theorem 6.B.23 and in Corollary 6.B.24.

The following neat example compares a sum of independent heterogeneous exponential random variables with an Erlang random variable; it is of interest to compare it with Examples 1.B.5 and 1.C.49. We do not give the proof here.

Example 1.A.24. Let X_i be an exponential random variable with mean $\lambda_i^{-1} > 0$, $i = 1, 2, \ldots, m$, and assume that the X_i's are independent. Let Y_i, $i = 1, 2, \ldots, m$, be independent, identically distributed, exponential random variables with mean η^{-1}. Then

$$\sum_{i=1}^{m} X_i \ge_{\rm st} \sum_{i=1}^{m} Y_i \Longleftrightarrow \sqrt[m]{\lambda_1 \lambda_2 \cdots \lambda_m} \le \eta.$$

The next example may be compared with Examples 1.B.6, 1.C.51, and 4.A.45.

Example 1.A.25. Let X_i be a binomial random variable with parameters n_i and p_i, $i = 1, 2, \ldots, m$, and assume that the X_i's are independent. Let Y be a binomial random variable with parameters n and p where $n = \sum_{i=1}^{m} n_i$. Then

$$\sum_{i=1}^{m} X_i \geq_{\text{st}} Y \Longleftrightarrow p \leq \sqrt[n]{p_1^{n_1} p_2^{n_2} \cdots p_m^{n_m}},$$

and

$$\sum_{i=1}^{m} X_i \leq_{\text{st}} Y \Longleftrightarrow 1 - p \leq \sqrt[n]{(1-p_1)^{n_1}(1-p_2)^{n_2} \cdots (1-p_m)^{n_m}}.$$

The following example gives necessary and sufficient conditions for the comparison of normal random variables; it is generalized in Example 6.B.29. See related results in Examples 3.A.51 and 4.A.46.

Example 1.A.26. Let X be a normal random variable with mean μ_X and variance σ_X^2, and let Y be a normal random variable with mean μ_Y and variance σ_Y^2. Then $X \leq_{\text{st}} Y$ if, and only if, $\mu_X \leq \mu_Y$ and $\sigma_X^2 = \sigma_Y^2$.

Example 1.A.27. Let the random variable X have a unimodal density, symmetric about 0. Then

$$(X+a)^2 \leq_{\text{st}} (X+b)^2 \quad \text{whenever } |a| \leq |b|.$$

Example 1.A.28. Let $\boldsymbol{X}$ have a multivariate normal density with mean vector $\boldsymbol{0}$ and variance-covariance matrix $\boldsymbol{\Sigma}_1$. Let $\boldsymbol{Y}$ have a multivariate normal density with mean vector $\boldsymbol{0}$ and variance-covariance matrix $\boldsymbol{\Sigma}_1 + \boldsymbol{\Sigma}_2$, where $\boldsymbol{\Sigma}_2$ is a nonnegative definite matrix. Then

$$\|\boldsymbol{X}\|^2 \leq_{\text{st}} \|\boldsymbol{Y}\|^2,$$

where $\|\cdot\|$ denotes the Euclidean norm.

The next result involves the total time on test (TTT) transform and the observed TTT random variable. Let F be the distribution function of a nonnegative random variable, and suppose, for simplicity, that 0 is the left endpoint of the support of F. The TTT transform associated with F is defined by

$$H_F^{-1}(u) = \int_0^{F^{-1}(u)} \overline{F}(x)\mathrm{d}x, \quad 0 \leq u \leq 1, \tag{1.A.19}$$

where $\overline{F} \equiv 1 - F$ is the survival function associated with F. The inverse, H_F, of the TTT transform is a distribution function. If the mean $\mu = \int_0^\infty x\mathrm{d}F(x) = \int_0^\infty \overline{F}(x)\mathrm{d}x$ is finite, then H_F has support in $[0, \mu]$. If X has the distribution function F, then let X_{ttt} be any random variable that has the distribution H_F. The random variable X_{ttt} is called the *observed total time on test.*

Theorem 1.A.29. *Let X and Y be two nonnegative random variables. Then*

$$X \leq_{\text{st}} Y \Longrightarrow X_{\text{ttt}} \leq_{\text{st}} Y_{\text{ttt}}.$$

See related results in Theorems 3.B.1, 4.A.44, 4.B.8, 4.B.9, and 4.B.29.

1.A.5 Some properties in reliability theory

Recall from page 1 the definitions of the IFR, DFR, NBU, and NWU properties. The next result characterizes random variables that have these properties by means of the usual stochastic order. The statements in the next theorem follow at once from the definitions. Recall from Section 1.A.3 that for any random variable Z and an event A we denote by $[Z|A]$ any random variable that has as its distribution the conditional distribution of Z given A.

Theorem 1.A.30. (a) *The random variable X is* IFR [DFR] *if, and only if,* $[X-t|X>t] \geq_{\text{st}} [\leq_{\text{st}}] [X-t'|X>t']$ *whenever* $t \leq t'$.
(b) *The nonnegative random variable X is* NBU [NWU] *if, and only if,* $X \geq_{\text{st}} [\leq_{\text{st}}] [X-t|X>t]$ *for all* $t>0$.

Note that if X is the lifetime of a device, then $[X-t|X>t]$ is the residual life of such a device with age t. Theorem 1.A.30(a), for example, characterizes IFR and DFR random variables by the monotonicity of their residual lives with respect to the order $\leq_{\text{st}}$. Theorem 1.A.30 should be compared to Theorem 1.B.38, where a similar characterization is given. Some multivariate analogs of Theorem 1.A.30(a) are used in Section 6.B.6 to introduce some multivariate IFR notions.

For a nonnegative random variable X with a finite mean, let A_X denote the corresponding asymptotic equilibrium age. That is, if the distribution function of X is F, then the distribution function F_{e} of A_X is defined by

$$F_{\text{e}}(x) = \frac{1}{EX}\int_0^x \overline{F}(y)\text{d}y, \quad x \geq 0, \tag{1.A.20}$$

where $\overline{F} \equiv 1-F$ is the corresponding survival function. Recall from page 1 the definitions of the NBUE and the NWUE properties. The following result is immediate.

Theorem 1.A.31. *The nonnegative random variable X with finite mean is* NBUE [NWUE] *if, and only if,* $X \geq_{\text{st}} [\leq_{\text{st}}] A_X$.

Another characterization of NBUE random variables is the following. Recall from Section 1.A.4 the definition of the observed total time on test random variable X_{ttt}.

Theorem 1.A.32. *Let X be a nonnegative random variable with finite mean μ. Then X is* NBUE *if, and only if,*

$$X_{\text{ttt}} \geq_{\text{st}} \mathcal{U}(0,\mu),$$

where $\mathcal{U}(0,\mu)$ denotes a uniform random variable on $(0,\mu)$.

Let X be a nonnegative random variable with finite mean and distribution function F, and let A_X be the corresponding asymptotic equilibrium age having the distribution function F_{e} given in (1.A.20). The requirement

$$X \geq_{\rm st} [A_X - t | A_X > t] \quad \text{for all } t \geq 0, \tag{1.A.21}$$

has been used in the literature as a way to define an aging property of the lifetime X. It turns out that this aging property is equivalent to the new better than used in convex ordering (NBUC) notion that is defined in (4.A.31) in Chapter 4.

1.B The Hazard Rate Order

1.B.1 Definition and equivalent conditions

If X is a random variable with an absolutely continuous distribution function F, then the hazard rate of X at t is defined as $r(t) = ({\rm d}/{\rm d}t)(-\log(1 - F(t)))$. The hazard rate can alternatively be expressed as

$$r(t) = \lim_{\Delta t \downarrow 0} \frac{P\{t < X \leq t + \Delta t | X > t\}}{\Delta t} = \frac{f(t)}{\overline{F}(t)}, \quad t \in \mathbb{R}, \tag{1.B.1}$$

where $\overline{F} \equiv 1 - F$ is the survival function and f is the corresponding density function. As can be seen from (1.B.1), the hazard rate $r(t)$ can be thought of as the intensity of failure of a device, with a random lifetime X, at time t. Clearly, the higher the hazard rate is the smaller X should be stochastically. This is the motivation for the order discussed in this section.

Let X and Y be two nonnegative random variables with absolutely continuous distribution functions and with hazard rate functions r and q, respectively, such that

$$r(t) \geq q(t), \quad t \in \mathbb{R}. \tag{1.B.2}$$

Then X is said to be *smaller than* Y *in the hazard rate order* (denoted as $X \leq_{\rm hr} Y$).

Although the hazard rate order is usually applied to random lifetimes (that is, nonnegative random variables), definition (1.B.2) may also be used to compare more general random variables. In fact, even the absolute continuity, which is required in (1.B.2), is not really needed. It is easy to verify that (1.B.2) holds if, and only if,

$$\frac{\overline{G}(t)}{\overline{F}(t)} \quad \text{increases in } t \in (-\infty, \max(u_X, u_Y)) \tag{1.B.3}$$

(here $a/0$ is taken to be equal to ∞ whenever $a > 0$). Here F denotes the distribution function of X and G denotes the distribution function of Y, and u_X and u_Y denote the corresponding right endpoints of the supports of X and of Y. Equivalently, (1.B.3) can be written as

$$\overline{F}(x)\overline{G}(y) \geq \overline{F}(y)\overline{G}(x) \quad \text{for all } x \leq y. \tag{1.B.4}$$

Thus (1.B.3) or (1.B.4) can be used to define the order $X \leq_{\text{hr}} Y$ even if X and/or Y do not have absolutely continuous distributions. A useful further condition, which is equivalent to $X \leq_{\text{hr}} Y$ when X and Y have absolutely continuous distributions with densities f and g, respectively, is the following:

$$\frac{f(x)}{\overline{F}(y)} \geq \frac{g(x)}{\overline{G}(y)} \quad \text{for all } x \leq y. \tag{1.B.5}$$

Rewriting (1.B.4) as

$$\frac{\overline{F}(t+s)}{\overline{F}(t)} \leq \frac{\overline{G}(t+s)}{\overline{G}(t)} \quad \text{for all } s \geq 0 \text{ and all } t,$$

it is seen that $X \leq_{\text{hr}} Y$ if, and only if,

$$P\{X - t > s \big| X > t\} \leq P\{Y - t > s \big| Y > t\} \quad \text{for all } s \geq 0 \text{ and all } t; \tag{1.B.6}$$

that is, if, and only if, the residual lives of X and Y at time t are ordered in the sense $\leq_{\text{st}}$ for all t. Equivalently, (1.B.6) can be written as

$$[X \big| X > t] \leq_{\text{st}} [Y \big| Y > t] \quad \text{for all } t. \tag{1.B.7}$$

Substituting $u = \overline{F}^{-1}(t)$ in (1.B.3) shows that $X \leq_{\text{hr}} Y$ if, and only if,

$$\frac{\overline{G}\,\overline{F}^{-1}(u)}{u} \geq \frac{\overline{G}\,\overline{F}^{-1}(v)}{v} \quad \text{for all } 0 < u \leq v < 1.$$

Simple manipulations show that the latter condition is equivalent to

$$\frac{1 - FG^{-1}(1-u)}{u} \leq \frac{1 - FG^{-1}(1-v)}{v} \quad \text{for all } 0 < u \leq v < 1. \tag{1.B.8}$$

For discrete random variables that take on values in $\mathbb{N}$ the definition of $\leq_{\text{hr}}$ can be written in two different ways. Let X and Y be such random variables. We denote $X \leq_{\text{hr}} Y$ if

$$\frac{P\{X = n\}}{P\{X \geq n\}} \geq \frac{P\{Y = n\}}{P\{Y \geq n\}}, \quad n \in \mathbb{N}. \tag{1.B.9}$$

Equivalently, $X \leq_{\text{hr}} Y$ if

$$\frac{P\{X = n\}}{P\{X > n\}} \geq \frac{P\{Y = n\}}{P\{Y > n\}}, \quad n \in \mathbb{N}.$$

The discrete analog of (1.B.4) is that (1.B.9) holds if, and only if,

$$P\{X \geq n_1\}P\{Y \geq n_2\} \geq P\{X \geq n_2\}P\{Y \geq n_1\} \quad \text{for all } n_1 \leq n_2. \tag{1.B.10}$$

In a similar manner (1.B.3) and (1.B.5) can be modified in the discrete case. Unless stated otherwise, we consider only random variables with absolutely continuous distribution functions in the following sections.

1.B.2 The relation between the hazard rate and the usual stochastic orders

By setting $x = -\infty$ in (1.B.4) (or $n_1 = -\infty$ in (1.B.10)), and then using (1.A.1), we obtain the following result.

Theorem 1.B.1. *If X and Y are two random variables such that $X \le_{\text{hr}} Y$, then $X \le_{\text{st}} Y$.*

1.B.3 Closure properties and some characterizations

Let ϕ be a strictly increasing function with inverse ϕ^{-1}. If X has the survival function $\overline{F}$, then $\phi(X)$ has the survival function $\overline{F}\phi^{-1}$. Similarly, if Y has the survival function $\overline{G}$, then $\phi(Y)$ has the survival function $\overline{G}\phi^{-1}$. If $X \le_{\text{hr}} Y$, then from (1.B.3) it follows that

$$\frac{\overline{G}\phi^{-1}(t)}{\overline{F}\phi^{-1}(t)} \quad \text{increases in } t \text{ over } \{t : \overline{G}\phi^{-1}(t) > 0\}.$$

We have thus shown an important special case of the next theorem. When ϕ is *just* increasing (rather than *strictly* increasing) the result is still true, but the above simple argument is no longer sufficient for its proof.

Theorem 1.B.2. *If $X \le_{\text{hr}} Y$, and if ϕ is any increasing function, then $\phi(X) \le_{\text{hr}} \phi(Y)$.*

In general, if $X_1 \le_{\text{hr}} Y_1$ and $X_2 \le_{\text{hr}} Y_2$, where X_1 and X_2 are independent random variables and Y_1 and Y_2 are also independent random variables, then it is not necessarily true that $X_1 + X_2 \le_{\text{hr}} Y_1 + Y_2$. However, if these random variables are IFR, then it is true. This is shown in Theorem 1.B.4, but first we state and prove the following lemma, which is of independent interest.

Lemma 1.B.3. *If the random variables X and Y are such that $X \le_{\text{hr}} Y$ and if Z is an* IFR *random variable independent of X and Y, then*

$$X + Z \le_{\text{hr}} Y + Z. \tag{1.B.11}$$

Proof. Denote by f_W and $\overline{F}_W$ the density function and the survival function of any random variable W. Note that, for $x \le y$,

$$
\begin{aligned}
&\overline{F}_{X+Z}(x)\overline{F}_{Y+Z}(y) - \overline{F}_{X+Z}(y)\overline{F}_{Y+Z}(x) \\
&= \int_v \int_{u \geq v} \big[f_X(u)\overline{F}_Z(x-u) f_Y(v)\overline{F}_Z(y-v) \\
&\qquad\qquad + f_X(v)\overline{F}_Z(x-v) f_Y(u)\overline{F}_Z(y-u)\big] \mathrm{d}u\mathrm{d}v \\
&- \int_v \int_{u \geq v} \big[f_X(u)\overline{F}_Z(y-u) f_Y(v)\overline{F}_Z(x-v) \\
&\qquad\qquad + f_X(v)\overline{F}_Z(y-v) f_Y(u)\overline{F}_Z(x-u)\big] \mathrm{d}u\mathrm{d}v \\
&= \int_v \int_{u \geq v} \big[\overline{F}_X(u) f_Y(v) - f_X(v)\overline{F}_Y(u)\big] \\
&\qquad\qquad \times \big[\overline{F}_Z(y-v) f_Z(x-u) - f_Z(y-u)\overline{F}_Z(x-v)\big] \mathrm{d}u\mathrm{d}v,
\end{aligned}
$$

where the second equality is obtained by integration by parts with respect to u and by collection of terms. Since $X \leq_{\text{hr}} Y$ it follows from (1.B.5) that the expression within the first set of brackets in the last integral is nonpositive. Since Z is IFR it can be verified that the quantity in the second pair of brackets in the last integral is also nonpositive. Therefore, the integral is nonnegative. This proves (1.B.11). □

The above proof is very similar to the proof that a convolution of two independent IFR random variables is IFR. In fact, this convolution result can be shown to be a consequence of Lemma 1.B.3; see Corollary 1.B.39 in Section 1.B.5.

Theorem 1.B.4. *Let (X_i, Y_i), $i = 1, 2, \ldots, m$, be independent pairs of random variables such that $X_i \leq_{\text{hr}} Y_i$, $i = 1, 2, \ldots, m$. If X_i, Y_i, $i = 1, 2, \ldots, m$, are all* IFR, *then*

$$
\sum_{i=1}^{m} X_i \leq_{\text{hr}} \sum_{i=1}^{m} Y_i.
$$

Proof. Repeated application of (1.B.11), using the closure property of IFR under convolution, yields the desired result. □

The following neat example compares a sum of independent heterogeneous exponential random variables with an Erlang random variable; it is of interest to compare it with Examples 1.A.24 and 1.C.49. We do not give the proof here.

Example 1.B.5. Let X_i be an exponential random variable with mean $\lambda_i^{-1} > 0$, $i = 1, 2, \ldots, m$, and assume that the X_i's are independent. Let Y_i, $i = 1, 2, \ldots, m$, be independent, identically distributed, exponential random variables with mean η^{-1}. Then

$$
\sum_{i=1}^{m} X_i \geq_{\text{hr}} \sum_{i=1}^{m} Y_i \iff \sqrt[m]{\lambda_1 \lambda_2 \cdots \lambda_m} \leq \eta.
$$

The next example may be compared with Examples 1.A.25, 1.C.51, and 4.A.45.

Example 1.B.6. Let X_i be a binomial random variable with parameters n_i and p_i, $i = 1, 2, \ldots, m$, and assume that the X_i's are independent. Let Y be a binomial random variable with parameters n and p where $n = \sum_{i=1}^m n_i$. Then

$$\sum_{i=1}^m X_i \geq_{\text{hr}} Y \Longleftrightarrow p \leq \frac{n}{\sum_{i=1}^m (n_i/p_i)},$$

and

$$\sum_{i=1}^m X_i \leq_{\text{hr}} Y \Longleftrightarrow 1 - p \leq \frac{n}{\sum_{i=1}^m (n_i/(1-p_i))}.$$

A hazard rate order comparison of random sums is given in the following result.

Theorem 1.B.7. *Let $\{X_i,\ i = 1, 2, \ldots\}$ be a sequence of nonnegative* IFR *independent random variables. Let M and N be two discrete positive integer-valued random variables such that $M \leq_{\text{hr}} N$ (in the sense of* (1.B.9) *or* (1.B.10)), *and assume that M and N are independent of the X_i's. Then*

$$\sum_{i=1}^M X_i \leq_{\text{hr}} \sum_{i=1}^N X_i.$$

The hazard rate order (unlike the usual stochastic order; see Theorem 1.A.3(d)) does not have the property of being simply closed under mixtures. However, under quite strong conditions the order $\leq_{\text{hr}}$ is closed under mixtures. This is shown in the next theorem.

Theorem 1.B.8. *Let X, Y, and Θ be random variables such that $[X|\Theta = \theta] \leq_{\text{hr}} [Y|\Theta = \theta']$ for all θ and θ' in the support of Θ. Then $X \leq_{\text{hr}} Y$.*

Proof. Select a θ and a θ' in the support of Θ. Let $\overline{F}(\cdot|\theta)$, $\overline{G}(\cdot|\theta)$, $\overline{F}(\cdot|\theta')$, and $\overline{G}(\cdot|\theta')$ be the survival functions of $[X|\Theta = \theta]$, $[Y|\Theta = \theta]$, $[X|\Theta = \theta']$, and $[Y|\Theta = \theta']$, respectively. For simplicity assume that these random variables have densities which we denote by $f(\cdot|\theta)$, $g(\cdot|\theta)$, $f(\cdot|\theta')$, and $g(\cdot|\theta')$, respectively. It is sufficient to show that for $\alpha \in (0, 1)$ we have

$$\frac{\alpha f(t|\theta) + (1-\alpha) f(t|\theta')}{\alpha \overline{F}(t|\theta) + (1-\alpha)\overline{F}(t|\theta')} \geq \frac{\alpha g(t|\theta) + (1-\alpha) g(t|\theta')}{\alpha \overline{G}(t|\theta) + (1-\alpha)\overline{G}(t|\theta')} \quad \text{for all } t \geq 0. \tag{1.B.12}$$

This is an inequality of the form

$$\frac{a+b}{c+d} \geq \frac{w+x}{y+z},$$

where all eight variables are nonnegative and by the assumptions of the theorem they satisfy

$$\frac{a}{c} \geq \frac{w}{y}, \quad \frac{a}{c} \geq \frac{x}{z}, \quad \frac{b}{d} \geq \frac{w}{y}, \quad \text{and} \quad \frac{b}{d} \geq \frac{x}{z}.$$

It is easy to verify that the latter four inequalities imply the former one, completing the proof of the theorem. □

It should be pointed out, however, that mixtures, of distributions which are ordered by the hazard rate order, are ordered by the usual stochastic order. That is, if X, Y, and Θ are random variables such that $[X|\Theta = \theta] \leq_{\rm hr} [Y|\Theta = \theta]$ for all θ in the support of Θ, then $X \leq_{\rm st} Y$. This follows from a (conditional) application of Theorem 1.B.1, combined with the fact that the usual stochastic order is closed under mixtures (Theorem 1.A.3(d)).

In order to state the next characterization we define the following class of bivariate functions.

$$\mathcal{G}_{\rm hr} = \big\{\phi : \mathbb{R}^2 \to \mathbb{R} : \phi(x,y) \text{ is increasing in } x, \text{ for each } y, \text{ on } \{x \geq y\}, \\ \text{and is decreasing in } y, \text{ for each } x, \text{ on } \{y \geq x\}\big\}.$$

Theorem 1.B.9. *Let X and Y be independent random variables. Then $X \leq_{\rm hr} Y$ if, and only if,*

$$\phi(X,Y) \leq_{\rm st} \phi(Y,X) \quad \textit{for all } \phi \in \mathcal{G}_{\rm hr}. \tag{1.B.13}$$

Proof. Suppose that (1.B.13) holds. Select an x and a y such that $x \geq y$. Let $\phi(u,v) = I_{\{u \geq x, v \geq y\}}$, where I_A denotes the indicator function of the set A. It is easy to see that $\phi(u,v)$ is increasing in u. In addition, for a fixed u and v such that $v \geq u$, we have that $\phi(u,v) = 1$ if $u \geq x$ and $\phi(u,v) = 0$ if $u < x$. Therefore, $\phi \in \mathcal{G}_{\rm hr}$. Hence,

$$\overline{F}(y)\overline{G}(x) = E\phi(Y,X) \geq E\phi(X,Y) = \overline{F}(x)\overline{G}(y) \quad \text{whenever } x \geq y,$$

where $\overline{F}$ and $\overline{G}$ are the survival functions of X and Y, respectively. Therefore, by (1.B.4), $X \leq_{\rm hr} Y$.

Conversely, assume that $X \leq_{\rm hr} Y$. Let $\psi : \mathbb{R} \to \mathbb{R}$ be an increasing function and let $\phi \in \mathcal{G}_{\rm hr}$. Denote $a(x,y) = \psi(\phi(x,y)) - \psi(\phi(y,x))$. For simplicity assume that a is differentiable and that X and Y have densities that we denote by f and g, respectively (otherwise, approximation arguments can be used). Then

$$\begin{aligned} Ea(X,Y) &= \int_{y=-\infty}^{\infty} \int_{x \geq y} a(x,y)[f(x)g(y) - f(y)g(x)] \mathrm{d}x\mathrm{d}y \\ &= \int_{y=-\infty}^{\infty} \int_{x \geq y} \frac{\partial}{\partial x} a(x,y) \left[\overline{F}(x)g(y) - f(y)\overline{G}(x)\right] \mathrm{d}x\mathrm{d}y \leq 0, \end{aligned}$$

where the second equality follows from integration by parts, and the inequality follows from $X \leq_{\rm hr} Y$, the fact that $a(x, y)$ increases in x for all $x \geq y$, and (1.B.5). $\square$

The next result is a similar characterization. It uses the notation of Theorem 1.A.10, and their comparison is of interest. The proof of the following theorem is omitted.

Theorem 1.B.10. *Let X and Y be two independent random variables. Then $X \leq_{\rm hr} Y$ if, and only if,*

$$E\phi_1(X, Y) \leq E\phi_2(X, Y)$$

for all ϕ_1 and ϕ_2 such that, for each x, $\Delta\phi_{21}(x, y)$ increases in y on $\{y \geq x\}$, and such that $\Delta\phi_{21}(x, y) \geq -\Delta\phi_{21}(y, x)$ whenever $x \leq y$.

A further similar characterization is given in Theorem 4.A.36. The next result describes another characterization of the order $\leq_{\rm hr}$.

Theorem 1.B.11. *Let X and Y be two, absolutely continuous or discrete, independent random variables. Then $X \leq_{\rm hr} Y$ if, and only if,*

$$[X \big| \min(X, Y) = z] \leq_{\rm hr} [Y \big| \min(X, Y) = z] \quad \textit{for all } z. \tag{1.B.14}$$

Also, $X \leq_{\rm hr} Y$ if, and only if,

$$[X \big| \min(X, Y) = z] \leq_{\rm st} [Y \big| \min(X, Y) = z] \quad \textit{for all } z. \tag{1.B.15}$$

Proof. First suppose that X and Y are absolutely continuous. Denote the survival functions of X and Y by $\overline{F}$ and $\overline{G}$, respectively, and denote the corresponding density functions by f and g. Then

$$P[X > x \big| \min(X, Y) = z] = \begin{cases} 1, & \text{if } x < z, \\ \frac{\overline{F}(x)g(z)}{f(z)\overline{G}(z)+g(z)\overline{F}(z)}, & \text{if } x \geq z, \end{cases} \tag{1.B.16}$$

and

$$P[Y > x \big| \min(X, Y) = z] = \begin{cases} 1, & \text{if } x < z, \\ \frac{\overline{G}(x)f(z)}{f(z)\overline{G}(z)+g(z)\overline{F}(z)}, & \text{if } x \geq z. \end{cases} \tag{1.B.17}$$

Therefore

$$\frac{P[Y > x | \min(X, Y) = z]}{P[X > x | \min(X, Y) = z]} = \begin{cases} 1, & \text{if } x < z, \\ \frac{\overline{G}(x)}{\overline{F}(x)} \cdot \frac{f(z)}{g(z)}, & \text{if } x \geq z. \end{cases} \tag{1.B.18}$$

If $X \leq_{\rm hr} Y$, then $\frac{\overline{G}(z)}{\overline{F}(z)} \cdot \frac{f(z)}{g(z)} \geq 1$, and $\frac{\overline{G}(x)}{\overline{F}(x)}$ is increasing in x. Thus (1.B.18) is increasing in x, and (1.B.14) follows. Obviously (1.B.15) follows from (1.B.14).

Now suppose that (1.B.15) holds. Then from (1.B.16) and (1.B.17) we get that $\overline{F}(x)g(z) \leq \overline{G}(x)f(z)$ for all $x \geq z$. Therefore $X \leq_{\rm hr} Y$ by (1.B.5).

The proof when X and Y are discrete is similar. $\square$

Some related characterizations are given in the next result.

Theorem 1.B.12. *Let X and Y be two independent random variables. The following conditions are equivalent:*

(a) $X \leq_{\rm hr} Y$.
(b) $E[\alpha(X)]E[\beta(Y)] \leq E[\alpha(Y)]E[\beta(X)]$ *for all functions α and β for which the expectations exist and such that β is nonnegative and α/β and β are increasing.*
(c) *For any two increasing functions a and b such that b is nonnegative, if $E[a(X)b(X)] = 0$, then $E[a(Y)b(Y)] \geq 0$.*

Proof. Assume (a). Let α and β be as in (b). Define $\phi_1(x, y) = \alpha(x)\beta(y)$ and $\phi_2(x, y) = \alpha(y)\beta(x)$. Then $\Delta\phi_{21}(x, y) = \phi_2(x, y) - \phi_1(x, y) = \beta(x)\beta(y) \cdot [\alpha(y)/\beta(y) - \alpha(x)/\beta(x)]$, which is increasing in y. Note that $\Delta\phi_{21}(x, y) + \Delta\phi_{21}(y, x) = 0$. Condition (b) now follows from Theorem 1.B.10.

Assume (b). By taking, for some $u \leq v$, $\alpha(x) = I_{(v,\infty)}(x)$ and $\beta(x) = I_{(u,\infty)}(x)$ in (b) one obtains (1.B.4), from which (a) follows.

Assume (b). Let a and b be two increasing functions such that b is nonnegative and such that $E[a(X)b(X)] = 0$. Define $\beta(x) = b(x)$ and $\alpha(x) = a(x)b(x)$. Substitution in (b) yields $E[a(Y)b(Y)] \geq 0$; that is, (c) holds.

Assume (c). Let α and β be as in (b). Denote $c = E[\alpha(X)]/E[\beta(X)]$. Define $a(x) = \alpha(x)/\beta(x) - c$ and $b(x) = \beta(x)$. Then $E[a(X)b(X)] = 0$. So, by (c), $E[a(Y)b(Y)] \geq 0$. But the latter reduces to $E[\alpha(X)]E[\beta(Y)] \leq E[\alpha(Y)]E[\beta(X)]$, and this establishes (b). □

Example 1.B.13. Let $\{N(t),\ t \geq 0\}$ be a nonhomogeneous Poisson process with mean function Λ (that is, $\Lambda(t) \equiv E[N(t)]$, $t \geq 0$). Let $T_1, T_2, \ldots$ be the successive epoch times, and let $X_n \equiv T_n - T_{n-1}$, $n = 1, 2 \ldots$ (where $T_0 \equiv 0$), be the corresponding inter-epoch times. The survival function of T_n is given by $P\{T_n > t\} = \sum_{i=0}^{n-1} \frac{(\Lambda(t))^i}{i!} \cdot \mathrm{e}^{-\Lambda(t)}$, $t \geq 0$, $n = 1, 2, \ldots$. It is easy to verify that $\frac{P\{T_{n+1}>t\}}{P\{T_n>t\}}$ is increasing in $t \geq 0$, $n = 1, 2, \ldots$, and thus, by (1.B.3),

$$T_n \leq_{\rm hr} T_{n+1}, \quad n = 1, 2, \ldots. \tag{1.B.19}$$

A result that is stronger than (1.B.19) is given in Example 1.C.47.

If we denote by F_n the distribution function of T_n, then

$$\begin{aligned} P\{X_{n+1} > t\} &= \int_0^\infty P\{T_{n+1} - T_n > t \big| T_n = u\} \mathrm{d}F_n(u) \\ &= \int_0^\infty \exp\{-[\Lambda(t + u) - \Lambda(u)]\} \mathrm{d}F_n(u) \\ &= E[\exp\{-[\Lambda(t + T_n) - \Lambda(T_n)]\}, \quad n = 0, 1, \ldots. \end{aligned}$$

Fix $t_1 \leq t_2$ and let $\alpha(x) \equiv \exp\{-[\Lambda(t_2 + x) - \Lambda(x)]\}$ and $\beta(x) \equiv \exp\{-[\Lambda(t_1 + x) - \Lambda(x)]\}$. Note that if Λ is concave, then $\alpha(x)/\beta(x)$ is increasing. Thus, by Theorem 1.B.12(b), if Λ is concave, then

$$\frac{P\{X_{n+1} > t_2\}}{P\{X_n > t_2\}} = \frac{E[\alpha(T_n)]}{E[\alpha(T_{n-1})]} \geq \frac{E[\beta(T_n)]}{E[\beta(T_{n-1})]} = \frac{P\{X_{n+1} > t_1\}}{P\{X_n > t_1\}},$$
$$n = 1, 2, \ldots .$$

It follows, by (1.B.3), that

$$X_n \leq_{\text{hr}} X_{n+1}, \quad n = 1, 2, \ldots .$$

It can be shown in a similar manner that if Λ is convex, then $X_n \geq_{\text{hr}} X_{n+1}$, $n = 1, 2, \ldots$.

As another example of the use of Theorem 1.B.12 consider an increasing convex function H such that $H(0) = 0$. Let X and Y be nonnegative random variables such that $X \leq_{\text{hr}} Y$. Then

$$\frac{E[H(X)]}{E[X]} \leq \frac{E[H(Y)]}{E[Y]}.$$

Rather than using Theorem 1.B.12, one can also obtain the above inequality from (2.B.5) in Chapter 2, and from the fact that the hazard rate order implies the hmrl order (which is discussed there).

Other characterizations of the order $\leq_{\text{hr}}$ can be found in Theorems 2.A.6 and 5.A.22.

Consider now a family of distribution functions $\{G_\theta, \theta \in \mathcal{X}\}$ where $\mathcal{X}$ is a subset of the real line. As in Section 1.A.3 let $X(\theta)$ denote a random variable with distribution function G_θ. For any random variable Θ with support in $\mathcal{X}$, and with distribution function F, let us denote by $X(\Theta)$ a random variable with distribution function H given by

$$H(y) = \int_{\mathcal{X}} G_\theta(y) \mathrm{d}F(\theta), \quad y \in \mathbb{R}.$$

The following result generalizes both Theorems 1.B.2 and 1.B.8, just as Theorem 1.A.6 generalized both parts (a) and (c) of Theorem 1.A.3.

Theorem 1.B.14. *Consider a family of distribution functions $\{G_\theta, \theta \in \mathcal{X}\}$ as above. Let Θ_1 and Θ_2 be two random variables with supports in $\mathcal{X}$ and distribution functions F_1 and F_2, respectively. Let Y_1 and Y_2 be two random variables such that $Y_i =_{\text{st}} X(\Theta_i)$, $i = 1, 2$, that is, suppose that the survival function of Y_i is given by*

$$\overline{H}_i(y) = \int_{\mathcal{X}} \overline{G}_\theta(y) \mathrm{d}F_i(\theta), \quad y \in \mathbb{R}, \ i = 1, 2.$$

If

$$X(\theta) \leq_{\text{hr}} X(\theta') \quad \text{whenever } \theta \leq \theta', \tag{1.B.20}$$

and if

$$\Theta_1 \leq_{\text{hr}} \Theta_2, \tag{1.B.21}$$

then

$$Y_1 \leq_{\text{hr}} Y_2. \tag{1.B.22}$$

Proof. Assumption (1.B.20) means that $\overline{G}_\theta(y)$ is TP_2 (totally positive of order 2) as a function of $\theta \in \mathcal{X}$ and of $y \in \mathbb{R}$ (that is, $\overline{G}_\theta(y)\overline{G}_{\theta'}(y') \geq \overline{G}_\theta(y')\overline{G}_{\theta'}(y)$ whenever $y \leq y'$ and $\theta \leq \theta'$). Assumption (1.B.21) means that $\overline{F}_i(\theta)$, as a function of $i \in \{1,2\}$ and of $\theta \in \mathcal{X}$, is TP_2. Also, from Theorem 1.B.1 it follows that $\overline{G}_\theta(y)$ is increasing in θ. Therefore, by Theorem 2.1 of Joag-Dev, Kochar, and Proschan [259], $\overline{H}_i(y)$ is TP_2 in $i \in \{1,2\}$ and $y \in \mathbb{R}$. That gives (1.B.22). □

The following example shows an interesting and useful application of Theorem 1.B.14

Example 1.B.15. Let $\{X_n^i,\ n \geq 0\}$ be a Markov chain with state space $\{1,2,\ldots,M\}$ (M can be infinity) and transition matrix $\boldsymbol{P}$, which starts from state i; that is, $X_0^i = i$. If $X_1^i \leq_{\mathrm{hr}} X_1^{i'}$ for all $i \leq i'$, then

(a) $I_1 \leq_{\mathrm{hr}} I_2$ implies that $X_n^{I_1} \leq_{\mathrm{hr}} X_n^{I_2}$ for all $n \geq 0$, and
(b) $X_n^1 \leq_{\mathrm{hr}} X_{n'}^1$ whenever $n \leq n'$.

In order to prove (a), first note that the result is trivial for $n = 0$. Suppose that the result is true for $n = k$. Define $Y(i) = X_1^i$. By the Markov property, we have $X_{k+1}^i =_{\mathrm{st}} Y(X_k^i)$ for all i. By induction, $X_k^{I_1} \leq_{\mathrm{hr}} X_k^{I_2}$. In particular, $Y(X_k^i) \leq_{\mathrm{hr}} Y(X_k^{i'})$ for all $i \leq i'$. Therefore, from Theorem 1.B.14 we get $X_{k+1}^{I_1} = Y(X_k^{I_1}) \leq_{\mathrm{hr}} Y(X_k^{I_2}) = X_{k+1}^{I_2}$.

In order to prove (b), note that $X_0^1 = 1 \leq_{\mathrm{hr}} X_1^1$. So, by (a) and the Markov property we have $X_n^1 \leq_{\mathrm{hr}} X_n^{X_1^1} =_{\mathrm{st}} X_{n+1}^1$.

The following example shows an application of Theorem 1.B.14 in the area of Bayesian imperfect repair.

Example 1.B.16. Let Θ_1 and Θ_2 be two random variables as in Example 1.A.7. Let $\overline{G}_\theta$, $X(\theta)$, Y_1, and Y_2 also be as in Example 1.A.7. Note that (1.B.20) holds because $\overline{K}^{1-\theta'}(y)/\overline{K}^{1-\theta}(y)$ is increasing in y whenever $0 < \theta \leq \theta' \leq 1$. Thus, if $\Theta_1 \leq_{\mathrm{hr}} \Theta_2$, then $Y_1 \leq_{\mathrm{hr}} Y_2$.

It is of interest to compare Example 1.B.16 to Example 5.B.13 which deals with random minima and maxima.

The next example deals with the same proportional hazard model as in Example 1.B.16; however, for convenience we change the notation.

Example 1.B.17. Let Θ and X be two nonnegative random variables with distribution function F and G, respectively. Let Y have the survival function $\overline{H}$ defined as

$$\overline{H}(y) = \int_0^\infty \overline{G}^\theta(y)\mathrm{d}F(\theta), \quad y \geq 0.$$

Suppose that G is absolutely continuous with hazard rate function r. Then H is also absolutely continuous, and we denote its hazard rate function by q. We will now show that if $E\Theta \leq 1$, then $X \leq_{\mathrm{hr}} Y$. In order to see it, write $\overline{H}(y) = M(\log \overline{G}(y))$, where M is the moment generating function of Θ. Differentiating $-\log \overline{H}(y)$ we obtain

$$q(y) = -\frac{\mathrm{d}}{\mathrm{d}y} \log \overline{H}(y) = r(y)\frac{M'(\log \overline{G}(y))}{M(\log \overline{G}(y))}$$
$$= r(y)\frac{E\Theta \mathrm{e}^{\Theta \log \overline{G}(y)}}{E\mathrm{e}^{\Theta \log \overline{G}(y)}} \le r(y)\frac{E\Theta E\mathrm{e}^{\Theta \log \overline{G}(y)}}{E\mathrm{e}^{\Theta \log \overline{G}(y)}} = r(y)E\Theta \le r(y),$$

where the first inequality follows from Chebyshev's Inequality (that is, Cov(Θ, $\mathrm{e}^{\Theta \log \overline{G}(y)}) \le 0$), and the second inequality follows from $E\Theta \le 1$. The stated result now follows from (1.B.2).

The following result gives a Laplace transform characterization of the order $\le_{\mathrm{hr}}$. It should be compared with Theorem 1.A.13.

Theorem 1.B.18. *Let X_1 and X_2 be two nonnegative random variables, and let $N_\lambda(X_1)$ and $N_\lambda(X_2)$ be as described in Theorem* 1.A.13. *Then*

$$X_1 \le_{\mathrm{hr}} X_2 \Longleftrightarrow N_\lambda(X_1) \le_{\mathrm{hr}} N_\lambda(X_2) \quad \text{for all } \lambda > 0,$$

where the notation $N_\lambda(X_1) \le_{\mathrm{hr}} N_\lambda(X_2)$ is in the sense of (1.B.9).

Proof. First suppose that $X_1 \le_{\mathrm{hr}} X_2$. Denote

$$\Gamma_\lambda(n, x) = \lambda \mathrm{e}^{-\lambda x}\frac{(\lambda x)^{n-1}}{(n-1)!}, \quad n \ge 1,\ x \ge 0.$$

Let $\overline{\alpha}_\lambda^{X_1}(n) = P\{N_\lambda(X_1) \ge n\}$ and $\overline{\alpha}_\lambda^{X_2}(n) = P\{N_\lambda(X_2) \ge n\}$ be as in the proof of Theorem 1.A.13. Then it can be verified that

$$\overline{\alpha}_\lambda^{X_1}(n) = \int_0^\infty \Gamma_\lambda(n, x)\overline{F}_1(x)\mathrm{d}x \quad \text{and} \quad \overline{\alpha}_\lambda^{X_2}(n) = \int_0^\infty \Gamma_\lambda(n, x)\overline{F}_2(x)\mathrm{d}x,$$

where $\overline{F}_1$ and $\overline{F}_2$ are the survival functions corresponding to X_1 and X_2. For $n_1 \le n_2$, some computation yields

$$\overline{\alpha}_\lambda^{X_1}(n_1)\overline{\alpha}_\lambda^{X_2}(n_2) - \overline{\alpha}_\lambda^{X_1}(n_2)\overline{\alpha}_\lambda^{X_2}(n_1)$$
$$= \int_{y=0}^\infty \int_{x=0}^y [\Gamma_\lambda(n_1, x)\Gamma_\lambda(n_2, y) - \Gamma_\lambda(n_1, y)\Gamma_\lambda(n_2, x)]$$
$$\times \left[\overline{F}_1(x)\overline{F}_2(y) - \overline{F}_1(y)\overline{F}_2(x)\right]\mathrm{d}x\mathrm{d}y.$$

It is not hard to verify that if $x \le y$ and $n_1 \le n_2$, then $[\Gamma_\lambda(n_1, x)\Gamma_\lambda(n_2, y) - \Gamma_\lambda(n_1, y)\Gamma_\lambda(n_2, x)] \ge 0$. Also, using (1.B.4) it is seen that $X_1 \le_{\mathrm{hr}} X_2$ implies $\left[\overline{F}_1(x)\overline{F}_2(y) - \overline{F}_1(y)\overline{F}_2(x)\right] \ge 0$ for $x \le y$. Thus, from (1.B.10) it is seen that $N_\lambda(X_1) \le_{\mathrm{hr}} N_\lambda(X_2)$.

Now suppose that $N_\lambda(X_1) \le_{\rm hr} N_\lambda(X_2)$ for every $\lambda > 0$. Define $c(n,\lambda) = \overline{\alpha}_\lambda^{X_1}(n)/\overline{\alpha}_\lambda^{X_2}(n)$. It can be shown that $c(n,\lambda)$ increases in λ and decreases in n. Thus, $c(n, n/x) \ge c(n, n/y)$ whenever $x \le y$. Letting $n \to \infty$ shows that $\overline{F}_1(x)/\overline{F}_2(x) \ge \overline{F}_1(y)/\overline{F}_2(y)$ for all continuity points x and y of F_1 and F_2 such that $x \le y$. Thus, from (1.B.3) it is seen that $X_1 \le_{\rm hr} X_2$. □

The implication $\Longrightarrow$ in Theorem 1.B.18 can be generalized in the same manner that Theorem 1.A.14 generalizes the implication $\Longrightarrow$ in Theorem 1.A.13. We will not state the result here since it is equivalent to Theorem 1.B.14.

A related result is the following.

Theorem 1.B.19. *Let $X_1, X_2, \dots, X_m$, Θ_1, and Θ_2 be independent nonnegative random variables. Define*

$$N_j(t) = \sum_{i=1}^{n} I_{[\Theta_j X_i]}(t), \quad t \ge 0,\ j = 1,2,$$

where

$$I_{[\Theta_j X_i]}(t) = \begin{cases} 1 & \text{if} \quad \Theta_j X_i > t, \\ 0 & \text{if} \quad \Theta_j X_i \le t. \end{cases}$$

If $\Theta_1 \le_{\rm hr} \Theta_2$ then $N_1(t) \le_{\rm hr} N_2(t)$ in the sense of (1.B.9) *for all $t \ge 0$.*

The following easy-to-prove result strengthens Theorem 1.A.15. An even stronger result appears in Theorem 1.C.27.

Theorem 1.B.20. *Let X be any random variable. Then $X_{(-\infty,a]}$ and $X_{(a,\infty)}$ are increasing in a in the sense of the hazard rate order.*

In Theorem 1.A.17 it was seen that if ϕ is a function which satisfies that $\phi(x) \le x$ for all $x \in \mathbb{R}$, then $\phi(X) \le_{\rm st} X$. The order $\le_{\rm hr}$ does not satisfy such a general property. However, we have the following easy-to-prove result.

Theorem 1.B.21. *Let X be a nonnegative* IFR *random variable, and let $a \le 1$ be a positive constant. Then $aX \le_{\rm hr} X$.*

In fact, a necessary and sufficient condition for a nonnegative random variable X, with survival function $\overline{F}$, to satisfy $aX \le_{\rm hr} X$ for all $0 < a < 1$, is that $\log \overline{F}(\mathrm{e}^x)$ is concave in $x \ge 0$.

In the next result it is shown that a random variable, whose distribution is the mixture of two distributions of hazard rate ordered random variables, is bounded from below and from above, in the hazard rate order sense, by these two random variables.

Theorem 1.B.22. *Let X and Y be two random variables with distribution functions F and G, respectively. Let W be a random variable with the distribution function $pF + (1-p)G$ for some $p \in (0,1)$. If $X \le_{\rm hr} Y$, then $X \le_{\rm hr} W \le_{\rm hr} Y$.*

Proof. Let u_X, u_Y, and u_W denote the right endpoints of the supports of the corresponding random variables, and note that $\max(u_X, u_W) = \max(u_X, u_Y)$. Now, if $X \leq_{\text{hr}} Y$, then

$$\frac{p\overline{F}(t) + (1-p)\overline{G}(t)}{\overline{F}(t)} = p + (1-p)\frac{\overline{G}(t)}{\overline{F}(t)}$$

is increasing in $t \in (-\infty, \max(u_X, u_W))$. Therefore, by (1.B.3), $X \leq_{\text{hr}} W$. The proof that $W \leq_{\text{hr}} Y$ is similar. □

Example 1.B.23. For a nonnegative random variable X with density function f, and for a nonnegative function w such that $E[w(X)]$ exists, define X^w as the random variable with the so-called *weighted* density function f_w given by

$$f_w(x) = \frac{w(x)f(x)}{E[w(X)]}, \quad x \geq 0.$$

Similarly, for another nonnegative random variable Y with density function g, such that $E[w(Y)]$ exists, define Y^w as the random variable with the density function g_w given by

$$g_w(x) = \frac{w(x)g(x)}{E[w(Y)]}, \quad x \geq 0.$$

We will show that if w is increasing, then

$$X \leq_{\text{hr}} Y \Longrightarrow X^w \leq_{\text{hr}} Y^w. \tag{1.B.23}$$

In order to do this, first note that the hazard rate function r_{X^w} of X^w is given by

$$r_{X^w}(x) = \frac{w(x)r_X(x)}{E[w(X)\big|X > x]}, \quad x \geq 0,$$

where r_X is the hazard rate function of X. Similarly, the hazard rate function r_{Y^w} of Y^w is given by

$$r_{Y^w}(x) = \frac{w(x)r_Y(x)}{E[w(Y)\big|Y > x]}, \quad x \geq 0,$$

where r_Y is the hazard rate function of Y. Now, from $X \leq_{\text{hr}} Y$ it follows that $[X\big|X > x] \leq_{\text{hr}} [Y\big|Y > x]$ for all $x \geq 0$. Next, using Theorem 1.B.2 and the monotonicity of w, we get that $[w(X)\big|X > x] \leq_{\text{hr}} [w(Y)\big|Y > x]$, and therefore, by Theorem 1.B.1, $E[w(X)\big|X > x] \leq E[w(Y)\big|Y > x]$. Combining this inequality with $r_X \geq r_Y$, it is seen that $r_{X^w} \geq r_{Y^w}$.

The above random variables are also studied in Example 1.C.59.

In particular, taking w to be the identity function $w(x) = x$, we see from (1.B.23) that the hazard rate ordering of X and Y implies the hazard rate ordering of the corresponding spread (or length-biased) random variables. See Example 8.B.12 for another result involving spreads.

Analogous to the result in Remark 1.A.18, it can be shown that the set of all distribution functions on $\mathbb{R} \cup \{\infty\}$ is a lattice with respect to the order $\leq_{\rm hr}$.

The following example may be compared to Examples 1.C.48, 2.A.22, 3.B.38, 4.B.14, 6.B.41, 6.D.8, 6.E.13, and 7.B.13.

Example 1.B.24. Let X and Y be two absolutely continuous nonnegative random variables with survival functions $\overline{F}$ and $\overline{G}$, respectively. Denote $\Lambda_1 = -\log \overline{F}$, $\Lambda_2 = -\log \overline{G}$, and $\lambda_i = \Lambda_i'$, $i = 1, 2$. Consider two nonhomogeneous Poisson processes $N_1 = \{N_1(t),\ t \geq 0\}$ and $N_2 = \{N_2(t),\ t \geq 0\}$ with mean functions Λ_1 and Λ_2 (see Example 1.B.13), respectively. Let $T_{i,1}, T_{i,2}, \ldots$ be the successive epoch times of process N_i, $i = 1, 2$. Note that $X =_{\rm st} T_{1,1}$ and $Y =_{\rm st} T_{2,1}$.

It turns out that the hazard rate ordering of the first two epoch times implies the hazard rate ordering of all the corresponding later epoch times; that is, it will be shown below that if $X \leq_{\rm hr} Y$, then $T_{1,n} \leq_{\rm hr} T_{2,n}$, $n \geq 1$.

The survival function $\overline{F}_{1,n}$ of $T_{1,n}$ is given by

$$\overline{F}_{1,n}(t) = P(T_{1,n} > t) = \sum_{j=0}^{n-1} \frac{(\Lambda_1(t))^j}{j!} \mathrm{e}^{-\Lambda_1(t)} = \overline{\Gamma}_n(\Lambda_1(t)), \quad t \geq 0, \tag{1.B.24}$$

where $\overline{\Gamma}_n$ is the survival function of the gamma distribution with scale parameter 1 and shape parameter n. The corresponding density function $f_{1,n}$ is given by

$$f_{1,n}(t) = \gamma_n(\Lambda_1(t))\lambda_1(t), \quad t \geq 0,$$

where γ_n is the density function associated with $\overline{\Gamma}_n$. The corresponding hazard rate function $r_{F_{1,n}}$ is given by

$$r_{F_{1,n}}(t) = r_{\Gamma_n}(\Lambda_1(t))\lambda_1(t), \quad t \geq 0,$$

where r_{Γ_n} is the hazard rate function associated with $\overline{\Gamma}_n$. Similarly,

$$r_{F_{2,n}}(t) = r_{\Gamma_n}(\Lambda_2(t))\lambda_2(t), \quad t \geq 0.$$

If $X \leq_{\rm hr} Y$, then

$$r_{F_{1,n}}(t) = r_{\Gamma_n}(\Lambda_1(t))\lambda_1(t) \geq r_{\Gamma_n}(\Lambda_2(t))\lambda_2(t) = r_{F_{2,n}}(t), \quad t \geq 0,$$

where the inequality follows from $\lambda_1(t) \geq \lambda_2(t)$, $\Lambda_1(t) \geq \Lambda_2(t)$, and the fact that the hazard rate function of the gamma distribution described above is increasing.

Now let $X_{i,n} \equiv T_{i,n} - T_{i,n-1}$, $n \geq 1$ (where $T_{i,0} \equiv 0$), be the inter-epoch times of the process N_i, $i = 1, 2$. Again, note that $X =_{\rm st} X_{1,1}$ and $Y =_{\rm st} X_{2,1}$. It turns out that, under some conditions, the hazard rate ordering of the first two inter-epoch times implies the hazard rate ordering of all the corresponding later inter-epoch times. Explicitly, it will be shown below that if $X \leq_{\rm hr} Y$, and if $\overline{F}$ and $\overline{G}$ are logconvex (that is, X and Y are DFR), and if

$$\frac{\lambda_2(t)}{\lambda_1(t)} \quad \text{is increasing in } t \geq 0, \tag{1.B.25}$$

then $X_{1,n} \leq_{\rm hr} X_{2,n}$ for each $n \geq 1$.

For the purpose of this proof let us denote F by F_1, and G by F_2. Let $\overline{G}_{i,n}$ denote the survival function of $X_{i,n}$, $i = 1, 2$. The stated result is obvious for $n = 1$, so let us fix an $n \geq 2$. Then, from (7) in Baxter [62] we obtain

$$\overline{G}_{i,n}(t) = \int_0^\infty \lambda_i(s) \frac{\Lambda_i^{n-2}(s)}{(n-2)!} \overline{F}_i(s+t) \mathrm{d}s, \quad t \geq 0,\ i \in \{1, 2\}. \tag{1.B.26}$$

Condition (1.B.25) means that

$$\lambda_i(t) \text{ is TP}_2 \text{ (totally positive of order 2) in } (i, t).$$

Condition (1.B.25) also implies that $\Lambda_2(t)/\Lambda_1(t)$ is increasing in $t \geq 0$, that is, $\Lambda_i(t)$ is TP_2 in (i, t). Since a product of TP_2 kernels is TP_2 we get that

$$\lambda_i(t) \frac{\Lambda_i^{n-2}(t)}{(n-2)!} \quad \text{is TP}_2 \text{ in } (i, t).$$

The assumption $F_1 \leq_{\rm hr} F_2$ implies that

$$\overline{F}_i(s+t) \text{ is TP}_2 \text{ in } (i, s) \text{ and in } (i, t).$$

Finally, the logconvexity of $\overline{F}_1$ and of $\overline{F}_2$ means that

$$\overline{F}_i(s+t) \text{ is TP}_2 \text{ in } (s, t).$$

Thus, by Theorem 5.1 on page 123 of Karlin [275], we get that $\overline{G}_{i,n}(t)$ is TP_2 in (i, t); that is, $X_{1,n} \leq_{\rm hr} X_{2,n}$.

The inequality $X_{1,n} \leq_{\rm hr} X_{2,n}$, $n \geq 1$, can also be obtained under slightly weaker assumptions, namely, that $X \leq_{\rm hr} Y$, that (1.B.25) holds, and that either X or Y is DFR; see Hu and Zhuang [245].

Example 1.B.25. Let X_1, X_2, Y_1, and Y_2 be independent, nonnegative random variables such that $X_1 =_{\rm st} X_2$ and $Y_1 =_{\rm st} Y_2$. Denote by λ_X and λ_Y the hazard rate functions of X_1 and Y_1, respectively. If $X_1 \leq_{\rm hr} Y_1$, and if λ_Y/λ_X is decreasing on $[0, 1)$, then

$$\min\{\max(X_1, X_2), \max(Y_1, Y_2)\} \leq_{\rm hr} \min\{\max(X_1, Y_1), \max(X_2, Y_2)\}.$$

1.B.4 Comparison of order statistics

Let $X_1, X_2, \ldots, X_m$ be random variables. As usual denote the corresponding order statistics by $X_{(1)} \le X_{(2)} \le \cdots \le X_{(m)}$. When we want to emphasize the dependence on m, we denote the order statistics by $X_{(1:m)} \le X_{(2:m)} \le \cdots \le X_{(m:m)}$. The following three theorems compare the order statistics in the hazard rate order.

Theorem 1.B.26. *If $X_1, X_2, \ldots, X_m$ are independent random variables, then $X_{(k)} \le_{\text{hr}} X_{(k+1)}$ for $k = 1, 2, \ldots, m-1$.*

A relatively simple proof of Theorem 1.B.26 can be obtained using the likelihood ratio order which is discussed in the next section. Therefore the proof of this theorem will be given there in Remark 1.C.40.

Theorem 12.5 in Cramer and Kamps [136] extends Theorem 1.B.26 to the so called sequential order statistics.

Further comparisons of order statistics are given in the next two theorems.

Theorem 1.B.27. *Let $X_1, X_2, \ldots, X_m$ be independent random variables. If $X_j \le_{\text{hr}} X_m$ for all $j = 1, 2, \ldots, m-1$, then $X_{(k-1:m-1)} \le_{\text{hr}} X_{(k:m)}$ for $k = 2, 3, \ldots, m$.*

Theorem 1.B.28. *If $X_1, X_2, \ldots, X_m$ are independent random variables, then $X_{(k:m-1)} \ge_{\text{hr}} X_{(k:m)}$ for $k = 1, 2, \ldots, m-1$.*

From Theorem 1.B.27 it follows that if $X_1, X_2, \ldots, X_m$ are independent random variables, then

$$X_{(1:1)} \ge_{\text{hr}} X_{(1:2)} \ge_{\text{hr}} \cdots \ge_{\text{hr}} X_{(1:m)}. \tag{1.B.27}$$

One may wonder what kind of results of this type hold without the independence assumption. Since $X_{(1:1)} \ge X_{(1:2)} \ge \cdots \ge X_{(1:m)}$ a.s., it follows from Theorem 1.A.1 that $X_{(1:1)} \ge_{\text{st}} X_{(1:2)} \ge_{\text{st}} \cdots \ge_{\text{st}} X_{(1:m)}$ hold without any (independence) assumption. However, a counterexample in the literature shows that (1.B.27) does not always hold. We now describe some conditions under which (1.B.27) holds.

Let $\boldsymbol{X} = (X_1, X_2, \ldots, X_m)$ be a random vector with a partially differentiable survival function $\overline{F}$. The function $R = -\log \overline{F}$ is called the hazard function of $\boldsymbol{X}$, and the vector $\boldsymbol{r_X}$ of partial derivatives, defined by

$$\begin{aligned} \boldsymbol{r_X}(\boldsymbol{x}) &= \left(r_{\boldsymbol{X}}^{(1)}(\boldsymbol{x}), r_{\boldsymbol{X}}^{(2)}(\boldsymbol{x}), \ldots, r_{\boldsymbol{X}}^{(m)}(\boldsymbol{x})\right) \\ &= \left(\frac{\partial}{\partial x_1} R(\boldsymbol{x}), \frac{\partial}{\partial x_2} R(\boldsymbol{x}), \ldots, \frac{\partial}{\partial x_m} R(\boldsymbol{x})\right), \end{aligned} \tag{1.B.28}$$

for all $\boldsymbol{x} \in \{\boldsymbol{x} : \overline{F}(\boldsymbol{x}) > 0\}$, is called the hazard gradient of $\boldsymbol{X}$; see Johnson and Kotz [264] and Marshall [381]. Note that $r_{\boldsymbol{X}}^{(i)}(\boldsymbol{x})$ can be interpreted as the conditional hazard rate of X_i evaluated at x_i, given that $X_j > x_j$ for all $j \ne i$. That is,

$$r_{\boldsymbol{X}}^{(i)}(\boldsymbol{x}) = \frac{f_i(x_i|X_j > x_j, j \neq i)}{\overline{F}_i(x_i|X_j > x_j, j \neq i)},$$

where $f_i(\cdot|X_j > x_j, j \neq i)$ and $\overline{F}_i(\cdot|X_j > x_j, j \neq i)$ are the conditional density and survival functions of X_i, given that $X_j > x_j$ for all $j \neq i$. For convenience, here and below we set $r_{\boldsymbol{X}}^{(i)}(\boldsymbol{x}) = \infty$ for all $\boldsymbol{x} \in \{\boldsymbol{x} : \overline{F}(\boldsymbol{x}) = 0\}$.

For any subset $P \subseteq \{1, 2, \ldots, m\}$ define

$$Y_P = \min_{i \in P} X_i.$$

Denote

$$1_P(i) = \begin{cases} 0 & \text{if } i \notin P, \\ 1 & \text{if } i \in P, \end{cases}$$

$$\mathbf{1}_P = (1_P(1), 1_P(2), \ldots, 1_P(m)), \quad \text{and} \quad \mathbf{1}_{P^c} = \mathbf{1} - \mathbf{1}_P,$$

where $\mathbf{1} = (1, 1, \ldots, 1)$, and P^c denotes the complement of P in $\{1, 2, \ldots, m\}$. Also denote

$$\infty \cdot 1_{P^c}(i) = \begin{cases} 0 & \text{if } i \notin P^c, \\ \infty & \text{if } i \in P^c, \end{cases}$$

and $\infty \cdot \mathbf{1}_{P^c} = (\infty \cdot 1_{P^c}(1), \infty \cdot 1_{P^c}(2), \ldots, \infty \cdot 1_{P^c}(m))$. Then the survival function $\overline{G}_P$ of Y_P can be expressed as

$$\overline{G}_P(t) = \overline{F}(t \cdot \mathbf{1}_P - \infty \cdot \mathbf{1}_{P^c}), \quad t \in \mathbb{R}.$$

Theorem 1.B.29. *Let $(X_1, X_2, \ldots, X_m)$ be a random vector with an absolutely continuous distribution function. Let P and Q be two subsets of $\{1, 2, \ldots, m\}$ such that $P \subset Q$. If*

$$r^{(i)}(t \cdot \mathbf{1}_P - \infty \cdot \mathbf{1}_{P^c}) \leq r^{(i)}(t \cdot \mathbf{1}_Q - \infty \cdot \mathbf{1}_{Q^c}), \quad t \in \mathbb{R}, \ i \in P, \qquad (1.B.29)$$

then

$$Y_P \geq_{\text{hr}} Y_Q.$$

A sufficient condition for (1.B.29) is that

$$r^{(i)}(x_1, x_2, \ldots, x_m) \text{ is increasing in } x_j, \quad j \neq i, \ i = 1, 2, \ldots, m.$$

This is easily seen to be equivalent to the requirement that

$$\frac{\overline{F}(x_1, \ldots, x_{i-1}, x_i', x_{i+1}, \ldots, x_m)}{\overline{F}(x_1, \ldots, x_{i-1}, x_i, x_{i+1}, \ldots, x_m)}$$
$$\text{is decreasing in } x_j, \ j \neq i, \text{ whenever } x_i \leq x_i', \ i = 1, 2, \ldots, m. \qquad (1.B.30)$$

Condition (1.B.30) means that $\overline{F}$ is RR_2 (reverse regular of order 2) in pairs; see Karlin [275]. In particular, it holds when $X_1, X_2, \ldots, X_m$ are independent. Karlin and Rinott [279] showed that some multivariate normal distributions, as well as the Dirichlet distribution, are RR_2 in pairs. So Theorem 1.B.29 applies to these distributions.

When $(X_1, X_2, \ldots, X_m)$ has an exchangeable distribution function, then the corresponding multivariate hazard function R is permutation symmetric. Therefore each $r^{(i)}$ can be expressed by means of $r^{(1)}$ as follows

$$r^{(i)}(x_1, x_2, \ldots, x_{i-1}, x_i, x_{i+1}, \ldots, x_m) \\ = r^{(1)}(x_i, x_2, \ldots, x_{i-1}, x_1, x_{i+1}, \ldots, x_m), \quad i = 2, 3, \ldots, m.$$

Corollary 1.B.30. *Let $(X_1, X_2, \ldots, X_m)$ be a random vector with an absolutely continuous exchangeable distribution function. If*

$$r^{(1)}(\underbrace{t, t, \ldots, t}_{i \text{ times}}, \underbrace{-\infty, -\infty, \ldots, -\infty}_{m-i \text{ times}}) \le r^{(1)}(\underbrace{t, t, \ldots, t}_{i+1 \text{ times}}, \underbrace{-\infty, -\infty, \ldots, -\infty}_{m-i-1 \text{ times}}), \\ t \in \mathbb{R}, \ i = 1, 2, \ldots, m-1, \quad (1.B.31)$$

then

$$X_{(1:1)} \ge_{\mathrm{hr}} X_{(1:2)} \ge_{\mathrm{hr}} \cdots \ge_{\mathrm{hr}} X_{(1:m)}. \quad (1.B.32)$$

If (1.B.31) is not imposed, then (1.B.32) need not be true; this follows from a counterexample in the literature.

The following result strengthens the DFR part of Theorem 1.A.19. Recall that the spacings that correspond to the random variables $X_1, X_2, \ldots, X_m$ are denoted by $U_{(i)} = X_{(i)} - X_{(i-1)}$, $i = 2, 3, \ldots, m$, where the $X_{(i)}$'s are the corresponding order statistics. When the dependence on m is to be emphasized, we will denote the spacings by $U_{(i:m)}$.

Theorem 1.B.31. *Let $X_1, X_2, \ldots, X_m, X_{m+1}$ be independent and identically distributed, absolutely continuous,* DFR *random variables. Then*

$$(m-i+1)U_{(i:m)} \le_{\mathrm{hr}} (m-i)U_{(i+1:m)}, \quad i = 2, 3, \ldots, m-1, \quad (1.B.33)$$

$$(m-i+2)U_{(i:m+1)} \le_{\mathrm{hr}} (m-i+1)U_{(i:m)}, \quad i = 2, 3, \ldots, m, \quad (1.B.34)$$

and

$$U_{(i:m)} \le_{\mathrm{hr}} U_{(i+1:m+1)}, \quad i = 2, 3, \ldots, m. \quad (1.B.35)$$

Note that (1.B.33)–(1.B.35) can be summarized as

$$(m-j+1)U_{(j:m)} \le_{\mathrm{hr}} (n-i+1)U_{(i:n)} \quad \text{whenever } i - j \ge \max\{0, n-m\}.$$

Theorem 1.B.31 is a simple consequence of Theorem 1.C.45 below. It is of interest to compare Theorem 1.B.31 to Theorems 1.A.19 and 1.A.22.

A comparison of such normalized spacings from two different samples is described next. Here $U_{(i:m)}$ denotes, as before, the ith spacing that corresponds to the sample $X_1, X_2, \ldots, X_m$, and $V_{(j:n)}$ denotes the jth spacing that corresponds to the sample $Y_1, Y_2, \ldots, Y_n$. It is of interest to compare the next result with Theorem 1.C.45.

Theorem 1.B.32. *For positive integers m and n, let $X_1, X_2, \ldots, X_m$ be independent identically distributed random variables with an absolutely continuous common distribution function, and let $Y_1, Y_2, \ldots, Y_n$ be independent identically distributed random variables with a possibly different absolutely continuous common distribution function. If $X_1 \le_{\rm hr} Y_1$, and if either X_1 or Y_1 is* DFR, *then*

$$(m-j+1)U_{(j:m)} \le_{\rm st} (n-i+1)V_{(i:n)} \quad \textit{whenever } i-j \ge \max\{0, n-m\}.$$

The hazard rate order is closed under the operation of taking minima, as the next result shows.

Theorem 1.B.33. *Let (X_i, Y_i), $i = 1, 2, \ldots, m$, be independent pairs of random variables such that $X_i \le_{\rm hr} Y_i$, $i = 1, 2, \ldots, m$. Then*

$$\min\{X_1, X_2, \ldots, X_m\} \le_{\rm hr} \min\{Y_1, Y_2, \ldots, Y_m\}.$$

Proof. Clearly, it is enough to show the result when $m = 2$. For simplicity assume that X_1, X_2, Y_1, and Y_2 have hazard rate functions r_1, r_2, q_1, and q_2, respectively. Then it is very easy to see that the hazard rate function of $\min\{X_1, X_2\}$ is $r_1 + r_2$ and the hazard rate function of $\min\{Y_1, Y_2\}$ is $q_1 + q_2$. By the assumptions of the theorem (see (1.B.2)) $r_1(t) \ge q_1(t)$ and $r_2(t) \ge q_2(t)$ for all $t \ge 0$. Therefore $r_1(t) + r_2(t) \ge q_1(t) + q_2(t)$ for all $t \ge 0$, that is, $\min\{X_1, X_2\} \le_{\rm hr} \min\{Y_1, Y_2\}$. $\square$

If the X_i's in Theorem 1.B.33 are identically distributed and if the Y_i's in Theorem 1.B.33 are also identically distributed, then all order statistics (and not just the minima) corresponding to the X_i's and the Y_i's can be compared in the hazard rate order. This is shown in the following result.

Theorem 1.B.34. *Let (X_i, Y_i), $i = 1, 2, \ldots, m$, be independent pairs of absolutely continuous random variables such that $X_i \le_{\rm hr} Y_i$, $i = 1, 2, \ldots, m$. Suppose that the X_i's are identically distributed and that the Y_i's are identically distributed. Then*

$$X_{(k:m)} \le_{\rm hr} Y_{(k:m)}, \quad k = 1, 2, \ldots, m. \tag{1.B.36}$$

If the X_i's or the Y_i's in Theorem 1.B.34 are not identically distributed, then the conclusion (1.B.36) need not hold. However, the following result, from Chapter 16 by Boland and Proschan in [515], gives conditions under which (1.B.36) holds.

Proposition 1.B.35. *Let $X_1, X_2, \ldots, X_m$ [respectively, $Y_1, Y_2, \ldots, Y_m$] be m independent (not necessarily identically distributed) absolutely continuous random variables, all with support (a, b) for some $a < b$. If $X_i \le_{\rm hr} Y_j$ for all i and j, then $X_{(k:m)} \le_{\rm hr} Y_{(k:m)}$, $k = 1, 2, \ldots, m$.*

A result which is stronger than Proposition 1.B.35, but which uses Proposition 1.B.35 in its proof, is the following.

Theorem 1.B.36. *Let $X_1, X_2, \ldots, X_m$ be m independent (not necessarily identically distributed) random variables, and let $Y_1, Y_2, \ldots, Y_n$ be other n independent (not necessarily identically distributed) random variables, all having absolutely continuous distributions with support (a, b) for some $a < b$. If $X_i \le_{\rm hr} Y_j$ for all i and j, then*

$$X_{(j:m)} \le_{\rm hr} Y_{(i:n)} \quad \textit{whenever } i - j \ge \max\{0, n - m\}.$$

The proof of Theorem 1.B.36 uses the likelihood ratio order which is discussed in the next section. Therefore the proof will be given in Remark 1.C.41.

The following example describes an interesting instance in which the two maxima are ordered in the hazard rate order. It may be compared with Example 3.B.32.

Example 1.B.37. Let $Y_1, Y_2, \ldots, Y_m$ be independent exponential random variables with hazard rates $\lambda_1, \lambda_2, \ldots, \lambda_m$, respectively. Let $X_1, X_2, \ldots, X_m$ be independent and identically distributed exponential random variables with hazard rate $\overline{\lambda} = \sum_{i=1}^m \lambda_i/m$. Then

$$X_{(m:m)} \le_{\rm hr} Y_{(m:m)}. \tag{1.B.37}$$

Let $Z_1, Z_2, \ldots, Z_m$ be independent and identically distributed exponential random variables with hazard rate $\tilde{\lambda} = \left(\prod_{i=1}^m \lambda_i\right)^{1/m}$. Then

$$Z_{(m:m)} \le_{\rm hr} Y_{(m:m)}. \tag{1.B.38}$$

In fact, from the arithmetic-geometric mean inequality ($\overline{\lambda} \ge \tilde{\lambda}$) and Proposition 1.B.35, it follows that (1.B.38) implies (1.B.37).

1.B.5 Some properties in reliability theory

The order $\le_{\rm hr}$ can be trivially (but beneficially) used to characterize IFR random variables. The next result lists several such characterizations. Recall from Section 1.A.3 that for any random variable Z and an event A we denote by $[Z|A]$ any random variable that has as its distribution the conditional distribution of Z given A.

Theorem 1.B.38. *The random variable X is* IFR [DFR] *if, and only if, one of the following equivalent conditions holds (when the support of the distribution function of X is bounded, condition* (iii) *does not have a simple* DFR *analog):*

(i) $[X - t|X > t] \ge_{\rm hr} [\le_{\rm hr}]\ [X - t'|X > t']$ *whenever* $t \le t'$.
(ii) $X \ge_{\rm hr} [\le_{\rm hr}]\ [X - t|X > t]$ *for all* $t \ge 0$ *(when X is a nonnegative random variable).*
(iii) $X + t \le_{\rm hr} X + t'$ *whenever* $t \le t'$.

Note that if X is the lifetime of a device, then $[X - t \mid X > t]$ is the residual life of such a device with age t. Theorem 1.B.38(i), for example, characterizes IFR random variables by the monotonicity of their residual lives with respect to the order $\le_{\rm hr}$. Some multivariate analogs of conditions (i) and (ii) of Theorem 1.B.38 are used in Section 6.D.3 to introduce a multivariate IFR notion.

Part (iii) of Theorem 1.B.38 can be used to prove the closure under convolution property of IFR random variables:

Corollary 1.B.39. *Let X and Y be two independent* IFR *random variables. Then $X + Y$ has an* IFR *distribution.*

Proof. From Theorem 1.B.38(iii) it follows that $X + t \le_{\rm hr} X + t'$ whenever $t \le t'$. Also, Y is independent of $X+t$ and of $X+t'$ for all t and t', respectively. From Lemma 1.B.3 it now follows that $X + Y + t \le_{\rm hr} X + Y + t'$ whenever $t \le t'$. Thus, again from Theorem 1.B.38(iii), it follows that $X + Y$ is IFR. □

Recall from (1.A.20) that for a nonnegative random variable X with a finite mean we denote by A_X the corresponding asymptotic equilibrium age. Recall from page 1 the definitions of the DMRL and the IMRL properties. The following result is immediate.

Theorem 1.B.40. *The nonnegative random variable X with finite mean is* DMRL [IMRL] *if, and only if, $X \ge_{\rm hr} [\le_{\rm hr}]\ A_X$.*

1.B.6 The reversed hazard order

If X is a random variable with an absolutely continuous distribution function F, then the reversed hazard rate of X at the point t is defined as $\tilde{r}(t) = ({\rm d}/{\rm d}t)(\log F(t))$. One interpretation of the reversed hazard rate at time t is the following. Suppose that X is nonnegative with distribution function F. Then X can be thought of as the lifetime of some device. Given that the device has already failed by time t, then the probability that it survived up to time $t - \varepsilon$ (for a small $\varepsilon > 0$) is approximately $\varepsilon \cdot \tilde{r}(t)$. Some of the results regarding the hazard rate order have analogs when the hazard rate is replaced by the reversed hazard rate.

Let X and Y be two random variables with absolutely continuous distribution functions and with reversed hazard rate functions $\tilde{r}$ and $\tilde{q}$, respectively, such that

$$\tilde{r}(t) \le \tilde{q}(t), \quad t \in \mathbb{R}. \tag{1.B.39}$$

Then X is said to be *smaller than Y in the reversed hazard rate order* (denoted as $X \le_{\rm rh} Y$).

In fact, the absolute continuity, which is required in (1.B.39), is not really needed. It easy to verify that (1.B.39) holds if, and only if,

$$\frac{G(t)}{F(t)} \quad \text{increases in } t \in (\min(l_X, l_Y), \infty) \tag{1.B.40}$$

(here $a/0$ is taken to be equal to ∞ whenever $a > 0$). Here F denotes the distribution function of X and G denotes the distribution function of Y, and l_X and l_Y denote the corresponding left endpoints of the supports of X and of Y. Equivalently, (1.B.40) can be written as

$$F(x)G(y) \geq F(y)G(x) \quad \text{for all } x \leq y. \tag{1.B.41}$$

Thus (1.B.40) or (1.B.41) can be used to define the order $X \leq_{\text{rh}} Y$ even if X and/or Y do not have absolutely continuous distributions. The analog of (1.B.5) for the reversed hazard order when X and Y have densities f and g, respectively, is that $X \leq_{\text{rh}} Y$ if, and only if,

$$\frac{f(y)}{F(x)} \leq \frac{g(y)}{G(x)} \quad \text{for all } x \leq y. \tag{1.B.42}$$

Another condition that is equivalent to $X \leq_{\text{rh}} Y$ is

$$\frac{GF^{-1}(u)}{u} \leq \frac{GF^{-1}(v)}{v} \quad \text{for all } 0 < u \leq v < 1.$$

Finally, another condition that is equivalent to $X \leq_{\text{rh}} Y$ is

$$P\{X - t \leq -s \big| X \leq t\} \geq P\{Y - t \leq -s \big| Y \leq t\} \quad \text{for all } s \geq 0 \text{ and all } t,$$

or, equivalently,

$$[X \big| X \leq t] \leq_{\text{st}} [Y \big| Y \leq t] \quad \text{for all } t. \tag{1.B.43}$$

For discrete random variables X and Y that take on values in $\mathbb{N}$, we denote $X \leq_{\text{rh}} Y$ if

$$\frac{P\{X = n\}}{P\{X \leq n\}} \leq \frac{P\{Y = n\}}{P\{Y \leq n\}}, \quad n \in \mathbb{N}. \tag{1.B.44}$$

A useful relationship between the hazard rate and the reversed hazard rate orders is described in the following theorem.

Theorem 1.B.41. *Let X and Y be two continuous random variables with supports (l_X, u_X) and (l_Y, u_Y), respectively. Then*

$$X \leq_{\text{hr}} Y \Longrightarrow \phi(X) \geq_{\text{rh}} \phi(Y)$$

for any continuous function ϕ which is strictly decreasing on (l_X, u_Y). Also,

$$X \leq_{\text{rh}} Y \Longrightarrow \phi(X) \geq_{\text{hr}} \phi(Y)$$

for any such function ϕ.

Using Theorem 1.B.41 it is easy to obtain the following analogs of results regarding the order $\leq_{\text{hr}}$.

Theorem 1.B.42. *If X and Y are two random variables such that $X \leq_{\mathrm{rh}} Y$, then $X \leq_{\mathrm{st}} Y$.*

Theorem 1.B.43. *If $X \leq_{\mathrm{rh}} Y$, and if ϕ is any increasing function, then $\phi(X) \leq_{\mathrm{rh}} \phi(Y)$.*

Lemma 1.B.44. *If the random variables X and Y are such that $X \leq_{\mathrm{rh}} Y$, and if Z is a random variable independent of X and Y and has decreasing reversed hazard rate, then*

$$X + Z \leq_{\mathrm{rh}} Y + Z.$$

Theorem 1.B.45. *Let (X_i, Y_i), $i = 1, 2, \ldots, m$, be independent pairs of random variables such that $X_i \leq_{\mathrm{rh}} Y_i$, $i = 1, 2, \ldots, m$. If X_i, Y_i, $i = 1, 2, \ldots, m$, all have decreasing reversed hazard rates, then*

$$\sum_{i=1}^{m} X_i \leq_{\mathrm{rh}} \sum_{i=1}^{m} Y_i.$$

Theorem 1.B.46. *Let X, Y, and Θ be random variables such that $[X \big| \Theta = \theta] \leq_{\mathrm{rh}} [Y \big| \Theta = \theta']$ for all θ and θ' in the support of Θ. Then $X \leq_{\mathrm{rh}} Y$.*

In order to state a bivariate characterization result for the order $\leq_{\mathrm{rh}}$ we define the following class of bivariate functions:

$$\begin{aligned}\mathcal{G}_{\mathrm{rh}} = \{\phi : \mathbb{R}^2 \to \mathbb{R} : \phi(x, y) &\text{ is increasing in } x, \text{ for each } y, \text{ on } \{x \leq y\}, \\ &\text{and is decreasing in } y, \text{ for each } x, \text{ on } \{y \leq x\}\}.\end{aligned}$$

The proof of the next result (Theorem 1.B.47) is similar to the proof of Theorem 1.B.9.

Theorem 1.B.47. *Let X and Y be independent random variables. Then $X \leq_{\mathrm{rh}} Y$ if, and only if,*

$$\phi(X, Y) \leq_{\mathrm{st}} \phi(Y, X) \quad \textit{for all } \phi \in \mathcal{G}_{\mathrm{rh}}.$$

The next result uses the notation of Theorem 1.A.10.

Theorem 1.B.48. *Let X and Y be two independent random variables. Then $X \leq_{\mathrm{rh}} Y$ if, and only if,*

$$E\phi_1(X, Y) \leq E\phi_2(X, Y)$$

for all ϕ_1 and ϕ_2 such that, for each y, $\Delta\phi_{21}(x, y)$ decreases in x on $\{x \leq y\}$, and such that $\Delta\phi_{21}(x, y) \geq -\Delta\phi_{21}(y, x)$ whenever $x \leq y$.

A further similar characterization is given in Theorem 4.A.36.

The following result is an analog of Theorem 1.B.11.

Theorem 1.B.49. *Let X and Y be two independent random variables. Then $X \leq_{\rm rh} Y$ if, and only if,*

$$[X \big| \max(X,Y) = z] \leq_{\rm rh} [Y \big| \max(X,Y) = z] \quad \text{for all } z. \tag{1.B.45}$$

Also, $X \leq_{\rm rh} Y$ if, and only if,

$$[X \big| \max(X,Y) = z] \leq_{\rm st} [Y \big| \max(X,Y) = z] \quad \text{for all } z. \tag{1.B.46}$$

Proof. First suppose that X and Y are absolutely continuous. Denote the distribution functions of X and Y by F and G, respectively, and denote the corresponding density functions by f and g. Then

$$P[X \leq x \big| \max(X,Y) = z] = \begin{cases} \frac{F(x)g(z)}{f(z)G(z)+g(z)F(z)}, & \text{if } x \leq z, \\ 1, & \text{if } x > z, \end{cases} \tag{1.B.47}$$

and

$$P[Y \leq x \big| \max(X,Y) = z] = \begin{cases} \frac{G(x)f(z)}{f(z)G(z)+g(z)F(z)}, & \text{if } x \leq z, \\ 1, & \text{if } x > z. \end{cases} \tag{1.B.48}$$

Therefore

$$\frac{P[Y \leq x | \max(X,Y) = z]}{P[X \leq x | \max(X,Y) = z]} = \begin{cases} \frac{G(x)}{F(x)} \cdot \frac{f(z)}{g(z)}, & \text{if } x \leq z, \\ 1, & \text{if } x > z. \end{cases} \tag{1.B.49}$$

If $X \leq_{\rm rh} Y$, then $\frac{G(x)}{F(x)}$ is increasing in x, and $\frac{G(z)}{F(z)} \cdot \frac{f(z)}{g(z)} \leq 1$. Thus (1.B.49) is increasing in x, and (1.B.45) follows. Obviously (1.B.46) follows from (1.B.45).

Now suppose that (1.B.46) holds. Then from (1.B.47) and (1.B.48) we get that $F(x)g(z) \geq G(x)f(z)$ for all $x \leq z$. Therefore $X \leq_{\rm rh} Y$ by (1.B.42).

The proof when X and Y are discrete is similar. □

The following result is an analog of Theorem 1.B.12.

Theorem 1.B.50. *Let X and Y be two independent random variables. The following conditions are equivalent:*

(a) *$X \leq_{\rm rh} Y$.*
(b) *$E[\alpha(X)]E[\beta(Y)] \geq E[\alpha(Y)]E[\beta(X)]$ for all functions α and β for which the expectations exist and such that β is nonnegative and α/β and β are decreasing.*
(c) *For any increasing function a and a nonnegative decreasing function b, if $E[a(Y)b(Y)] = 0$, then $E[a(X)b(X)] \leq 0$.*

Example 1.B.51. Let X and Y be two random variables with support $[c,d]$, where $c < 0 < d$, and suppose that $E[Y] > 0$. Let u be an increasing differentiable concave function, corresponding to the utility function of a risk-averse individual. Let k_X be a value which maximizes $g_X(k) \equiv E[u(kX)]$, and similarly let k_Y be a value which maximizes $g_Y(k) \equiv E[u(kY)]$. Theorem 1.B.50(c) can be used to prove that if $X \leq_{\rm rh} Y$, then $k_X \leq k_Y$. In order to see it, first note that the result is trivial if $k_X = -\infty$ or if $k_Y = \infty$. Thus, let us assume that k_X and k_Y are finite. Note that then k_X and k_Y satisfy $E[Xu'(k_X X)] = 0$ and $E[Yu'(k_Y Y)] = 0$, where u' denotes the derivative of u. Also note that from the assumption $E[Y] > 0$ it follows that $k_Y > 0$. Without loss of generality let $k_Y = 1$. Thus $E[Yu'(Y)] = 0$, and using the concavity of u the assertion would follow if we show that $E[Xu'(X)] \leq 0$. But this follows from Theorem 1.B.50(c).

Consider now a family of distribution functions $\{G_\theta,\ \theta \in \mathcal{X}\}$ where $\mathcal{X}$ is a subset of the real line. As in Section 1.A.3 let $X(\theta)$ denote a random variable with distribution function G_θ. For any random variable Θ with support in $\mathcal{X}$, and with distribution function F, let us denote by $X(\Theta)$ a random variable with distribution function H given by

$$H(y) = \int_{\mathcal{X}} G_\theta(y)\mathrm{d}F(\theta), \quad y \in \mathbb{R}.$$

The following result generalizes Theorem 1.B.43, just as Theorem 1.A.6 generalized Theorem 1.A.3(a). The proof of the next theorem is similar to the proof of Theorem 1.B.14 and is therefore omitted.

Theorem 1.B.52. *Consider a family of distribution functions* $\{G_\theta,\ \theta \in \mathcal{X}\}$ *as above. Let* Θ_1 *and* Θ_2 *be two random variables with supports in* $\mathcal{X}$ *and distribution functions* F_1 *and* F_2*, respectively. Let* Y_1 *and* Y_2 *be two random variables such that* $Y_i =_{\rm st} X(\Theta_i)$, $i = 1, 2$*; that is, suppose that the distribution function of* Y_i *is given by*

$$H_i(y) = \int_{\mathcal{X}} G_\theta(y)\mathrm{d}F_i(\theta), \quad y \in \mathbb{R},\ i = 1, 2.$$

If

$$X(\theta) \leq_{\rm rh} X(\theta') \quad \textit{whenever } \theta \leq \theta',$$

and if

$$\Theta_1 \leq_{\rm rh} \Theta_2,$$

then

$$Y_1 \leq_{\rm rh} Y_2.$$

The following result, which is the "reversed hazard analog" of Theorem 1.B.18, gives a Laplace transform characterization of the order $\leq_{\rm rh}$.

Theorem 1.B.53. *Let* X_1 *and* X_2 *be two nonnegative random variables, and let* $N_\lambda(X_1)$ *and* $N_\lambda(X_2)$ *be as described in Theorem* 1.A.13*. Then*

$$X_1 \leq_{\rm rh} X_2 \Longleftrightarrow N_\lambda(X_1) \leq_{\rm rh} N_\lambda(X_2) \quad \textit{for all } \lambda > 0,$$

where the notation $N_\lambda(X_1) \leq_{\rm rh} N_\lambda(X_2)$ *is in the sense of* (1.B.44).

The implication $\Longrightarrow$ in Theorem 1.B.53 can be generalized in the same manner that Theorem 1.A.14 generalizes the implication $\Longrightarrow$ in Theorem 1.A.13. We will not state the result here since it is equivalent to Theorem 1.B.52.

The reversed hazard analog of Theorem 1.B.19 is the following.

Theorem 1.B.54. *Let $X_1, X_2, \ldots, X_m$, Θ_1, and Θ_2 be independent nonnegative random variables. Define $N_j(t)$ for $t \geq 0$ and $j = 1, 2$ as in Theorem* 1.B.19. *If $\Theta_1 \leq_{\rm rh} \Theta_2$, then $N_1(t) \leq_{\rm rh} N_2(t)$ in the sense of* (1.B.44) *for all* $t \geq 0$.

The reversed hazard analog of Theorem 1.B.20 is the following.

Theorem 1.B.55. *Let X be any random variable. Then $X_{(-\infty,a]}$ and $X_{(a,\infty)}$ are increasing in a in the sense of the reversed hazard order.*

Analogous to the result in Remark 1.A.18, it can be shown that the set of all distribution functions on $\mathbb{R} \cup \{-\infty\}$ is a lattice with respect to the order $\leq_{\rm rh}$.

The reversed hazard analog of Theorem 1.B.26 is the following.

Theorem 1.B.56. *If $X_1, X_2, \ldots, X_m$ are independent random variables, then $X_{(k)} \leq_{\rm rh} X_{(k+1)}$ for $k = 1, 2, \ldots, m-1$.*

The reversed hazard analog of Theorem 1.B.27 is the following.

Theorem 1.B.57. *Let $X_1, X_2, \ldots, X_m$ be independent random variables. If $X_m \leq_{\rm rh} X_j$ for all $j = 1, 2, \ldots, m-1$, then $X_{(k-1:m-1)} \leq_{\rm rh} X_{(k:m)}$ for $k = 2, 3, \ldots, m$.*

The reversed hazard analog of Theorem 1.B.28 is the following.

Theorem 1.B.58. *If $X_1, X_2, \ldots, X_m$ are independent random variables, then $X_{(k:m-1)} \geq_{\rm rh} X_{(k:m)}$ for $k = 1, 2, \ldots, m-1$.*

The reversed hazard analogs of Theorems 1.B.33, 1.B.34, and 1.B.36 are the following results.

Theorem 1.B.59. *Let (X_i, Y_i), $i = 1, 2, \ldots, m$, be independent pairs of random variables such that $X_i \leq_{\rm rh} Y_i$, $i = 1, 2, \ldots, m$. Then*

$$\max\{X_1, X_2, \ldots, X_m\} \leq_{\rm rh} \max\{Y_1, Y_2, \ldots, Y_m\}.$$

Theorem 1.B.60. *Let (X_i, Y_i), $i = 1, 2, \ldots, m$, be independent pairs of absolutely continuous random variables such that $X_i \leq_{\rm rh} Y_i$, $i = 1, 2, \ldots, m$. Suppose that the X_i's are identically distributed and that the Y_i's are identically distributed. Then*

$$X_{(k:m)} \leq_{\rm rh} Y_{(k:m)}, \quad k = 1, 2, \ldots, m.$$

Theorem 1.B.61. *Let $X_1, X_2, \ldots, X_m$ be m independent (not necessarily identically distributed) random variables, and let $Y_1, Y_2, \ldots, Y_n$ be other n independent (not necessarily identically distributed) random variables, all having absolutely continuous distributions with support (a, b) for some $a < b$. If $X_i \le_{\rm rh} Y_j$ for all i and j, then*

$$X_{(j:m)} \le_{\rm rh} Y_{(i:n)} \quad \textit{whenever } i - j \ge \max\{0, n - m\}.$$

Finally, the reversed hazard analog of Theorem 1.B.38 is the following.

Theorem 1.B.62. *The random variable X with support (a, b), for some $-\infty \le a < b \le \infty$, has decreasing [increasing] reversed hazard rate if, and only if, one of the following equivalent conditions holds:*

(i) $[X - t | X < t] \ge_{\rm rh} [\le_{\rm rh}] \; [X - t' | X < t']$ *whenever* $a < t \le t' < b$.
(ii) $X \le_{\rm rh} [\ge_{\rm rh}] \; [X - t | X < t]$ *for all* $t \in (a, b)$ *(when X is a nonpositive random variable).*
(iii) $X + t \le_{\rm rh} [\ge_{\rm rh}] \; X + t'$ *whenever* $a < t \le t' < b$.

Corollary 1.B.63. *Let X and Y be two independent random variables with decreasing reversed hazard rates. Then $X + Y$ has a decreasing reversed hazard rate.*

1.C The Likelihood Ratio Order

1.C.1 Definition

Let X and Y be continuous [discrete] random variables with densities [discrete densities] f and g, respectively, such that

$$\frac{g(t)}{f(t)} \quad \text{increases in } t \text{ over the union of the supports of } X \text{ and } Y \tag{1.C.1}$$

(here $a/0$ is taken to be equal to ∞ whenever $a > 0$), or, equivalently,

$$f(x)g(y) \ge f(y)g(x) \quad \text{for all } x \le y. \tag{1.C.2}$$

Then X is said to be *smaller than Y in the likelihood ratio order* (denoted by $X \le_{\rm lr} Y$). By integrating (1.C.2) over $x \in A$ and $y \in B$, where A and B are measurable sets in $\mathbb{R}$, it is seen that (1.C.2) is equivalent to

$$P\{X \in A\}P\{Y \in B\} \ge P\{X \in B\}P\{Y \in A\}$$
$$\text{for all measurable sets } A \text{ and } B \text{ such that } A \le B, \tag{1.C.3}$$

where $A \le B$ means that $x \in A$ and $y \in B$ imply that $x \le y$. Note that condition (1.C.3) does not directly involve the underlying densities, and thus it applies uniformly to continuous distributions, or to discrete distributions, or even to mixed distributions.

At a first glance (1.C.1) or (1.C.2) or (1.C.3) seem to be unintuitive technical conditions. However, it turns out that in many situations they are very easy to verify, and this is one of the major reasons for the usefulness and importance of the order $\leq_{\mathrm{lr}}$. It is also easy to verify by a simple differentiation (at least when X and Y have the same support) that

$$X \leq_{\mathrm{lr}} Y \Longleftrightarrow GF^{-1} \text{ is convex.} \tag{1.C.4}$$

Here F and G are the distribution functions of X and Y, respectively.

1.C.2 The relation between the likelihood ratio and the hazard and reversed hazard orders

Note that from (1.C.1) it follows (in the continuous case) that

$$\int_{t=x}^{y} \int_{t'=y}^{\infty} f(t)g(t')\mathrm{d}t'\mathrm{d}t \geq \int_{t=y}^{\infty} \int_{t'=x}^{y} f(t')g(t)\mathrm{d}t'\mathrm{d}t \quad \text{for all } x \leq y,$$

which, in turn, implies that

$$\int_{x}^{\infty} f(t)\mathrm{d}t \int_{y}^{\infty} g(t')\mathrm{d}t' \geq \int_{x}^{\infty} g(t)\mathrm{d}t \int_{y}^{\infty} f(t')\mathrm{d}t' \quad \text{for all } x \leq y,$$

that is, (1.B.4). We thus have shown a part of the following result. The other parts of the next theorem are proven similarly (recall that the discrete versions of the orders $\leq_{\mathrm{hr}}$ and $\leq_{\mathrm{rh}}$ are defined in (1.B.9) and (1.B.44), respectively).

Theorem 1.C.1. *If X and Y are two continuous or discrete random variables such that $X \leq_{\mathrm{lr}} Y$, then $X \leq_{\mathrm{hr}} Y$ and $X \leq_{\mathrm{rh}} Y$ (and therefore $X \leq_{\mathrm{st}} Y$).*

Remark 1.C.2. Neither of the orders $\leq_{\mathrm{hr}}$ and $\leq_{\mathrm{rh}}$ (even if both hold simultaneously) implies the order $\leq_{\mathrm{lr}}$. In order to see it let X be a uniform random variable over the set $\{1, 2, 3, 4\}$ and let Y have the probabilities $P\{Y = 1\} = .1$, $P\{Y = 2\} = .3$, $P\{Y = 3\} = .2$, and $P\{Y = 4\} = .4$. Then it is not true that $X \leq_{\mathrm{lr}} Y$, however, in this case we have that $X \leq_{\mathrm{hr}} Y$ and also that $X \leq_{\mathrm{rh}} Y$.

Remark 1.C.3. Using Theorem 1.C.1 we can now give a proof of Theorem 1.A.22. Let F and f denote, respectively, the distribution function and the density function of X_1. Given $X_{(i-1:m)} = u$ and $X_{(i+1:m)} = v$, the conditional density of $U_{(i:m)}$ at the point w is $\frac{f(u+w)}{F(v)-F(u)}$, $0 \leq w \leq v-u$, and the conditional density of $U_{(i+1:m)}$ at the point w is $\frac{f(v-w)}{F(v)-F(u)}$, $0 \leq w \leq v-u$. Since f is increasing [decreasing] it is seen that, conditionally, $U_{(i:m)} \geq_{\mathrm{lr}} [\leq_{\mathrm{lr}}]\ U_{(i+1:m)}$, and therefore, by Theorem 1.C.1, $U_{(i:m)} \geq_{\mathrm{st}} [\leq_{\mathrm{st}}]\ U_{(i+1:m)}$. Theorem 1.A.22 now follows from Theorem 1.A.3(d).

Although neither of the orders $\leq_{\rm hr}$ and $\leq_{\rm rh}$ implies the order $\leq_{\rm lr}$ (see Remark 1.C.2), the following result gives a simple condition under which this is actually the case. The proof is immediate and is therefore omitted.

Theorem 1.C.4. *Let X and Y be two random variables with distribution functions F and G, (discrete or continuous) hazard rate functions r and q, and (discrete or continuous) reversed hazard rate functions $\tilde{r}$ and $\tilde{q}$, respectively.*

(a) *If $X \leq_{\rm hr} Y$ and if $\frac{q(t)}{r(t)}$ increases in t, then $X \leq_{\rm lr} Y$.*
(b) *If $X \leq_{\rm rh} Y$ and if $\frac{\tilde{q}(t)}{\tilde{r}(t)}$ increases in t, then $X \leq_{\rm lr} Y$.*

1.C.3 Some properties and characterizations

The usual stochastic order has the useful and important constructive property described in Theorem 1.A.1. There is no analogous property associated with the likelihood ratio order. Therefore it is of importance to understand better the relationship between the orders $\leq_{\rm st}$ and $\leq_{\rm lr}$. We already know from Theorems 1.C.1 and 1.B.1 that the likelihood ratio order implies the usual stochastic order. The following result characterizes the likelihood ratio order by means of the order $\leq_{\rm st}$. It says that $X \leq_{\rm lr} Y$ if, and only if, for any interval I, the conditional distribution of X, given that $X \in I$, is stochastically smaller than the conditional distribution of Y, given that $Y \in I$.

As in Section 1.A.3, $[Z|A]$ denotes any random variable that has as its distribution the conditional distribution of Z given A. It is of interest to contrast the next result with (1.B.7) and (1.B.43).

Theorem 1.C.5. *The two random variables X and Y satisfy $X \leq_{\rm lr} Y$ if, and only if,*

$$[X|a \leq X \leq b] \leq_{\rm st} [Y|a \leq Y \leq b] \quad \textit{whenever } a \leq b. \tag{1.C.5}$$

Proof. Suppose that (1.C.5) holds. Select an a and a b such that $a < b$. Then

$$\frac{P\{u \leq X \leq b\}}{P\{a \leq X \leq b\}} \leq \frac{P\{u \leq Y \leq b\}}{P\{a \leq Y \leq b\}} \quad \text{whenever } u \in [a, b].$$

It follows then that

$$\frac{P\{a \leq X < u\}}{P\{u \leq X \leq b\}} \geq \frac{P\{a \leq Y < u\}}{P\{u \leq Y \leq b\}} \quad \text{whenever } u \in [a, b].$$

That is,

$$\frac{P\{a \leq X < u\}}{P\{a \leq Y < u\}} \geq \frac{P\{u \leq X \leq b\}}{P\{u \leq Y \leq b\}} \quad \text{whenever } u \in [a, b].$$

In particular, for $u < b \leq v$,

$$\frac{P\{u \le X < b\}}{P\{u \le Y < b\}} \ge \frac{P\{b \le X \le v\}}{P\{b \le Y \le v\}}.$$

Therefore, when X and Y are continuous random variables,

$$\frac{P\{a \le X < u\}}{P\{a \le Y < u\}} \ge \frac{P\{b \le X \le v\}}{P\{b \le Y \le v\}} \quad \text{whenever } a < u \le b \le v.$$

Now let $a \to u$ and $b \to v$ to obtain (1.C.2). The proof for discrete random variables is similar.

Conversely, suppose that $X \le_{\rm lr} Y$, then clearly, $[X \big| a \le X \le b] \le_{\rm lr} [Y \big| a \le Y \le b]$ whenever $a < b$ (see also Theorem 1.C.6). From Theorems 1.C.1 and 1.B.1 we obtain (1.C.5). $\square$

The likelihood ratio order is preserved under general truncations of the involved random variables. This is stated in the next theorem, the proof of which follows directly from (1.C.2).

Theorem 1.C.6. *If X and Y are two random variables such that $X \le_{\rm lr} Y$, then for any measurable set $A \subseteq \mathbb{R}$ we have $[X \big| X \in A] \le_{\rm lr} [Y \big| Y \in A]$.*

By combining Theorems 1.C.5 and 1.C.6 it is seen that $X \le_{\rm lr} Y$ if, and only if,

$$[X \big| X \in A] \le_{\rm st} [Y \big| Y \in A] \quad \text{for all measurable sets } A \subseteq \mathbb{R}. \tag{1.C.6}$$

In fact, one can take (1.C.6) as the definition of the likelihood ratio order. The advantage of this approach is that it does not directly involve the underlying densities, and thus, similarly to condition (1.C.3), it applies uniformly to continuous distributions, or to discrete distributions, or even to mixed distributions.

Using the characterization (1.C.3), it is not hard to obtain the following result.

Theorem 1.C.7. *Let $\{X_j,\ j = 1, 2, \ldots\}$ and $\{Y_j,\ j = 1, 2, \ldots\}$ be two sequences of random variables such that $X_j \to_{\rm st} X$ and $Y_j \to_{\rm st} Y$ as $j \to \infty$. If $X_j \le_{\rm lr} Y_j$, $j = 1, 2, \ldots$, then $X \le_{\rm lr} Y$.*

Let ψ be a strictly monotone increasing [decreasing] differentiable function with inverse ψ^{-1}. If X has the density function f, then $\psi(X)$ has the density function $(f\psi^{-1})/(\psi'(\psi^{-1}))$. Similarly, if Y has the density function g, then $\psi(Y)$ has the density function $(g\psi^{-1})/(\psi'(\psi^{-1}))$. If $X \le_{\rm lr} Y$, then from (1.C.1) it follows that $\frac{(f\psi^{-1})(u)/(\psi'(\psi^{-1}(u)))}{(g\psi^{-1})(u)/(\psi'(\psi^{-1}(u)))}$ decreases [increases] over the unions of the supports of $\psi(X)$ and $\psi(Y)$. We have thus proved an important special case of Theorem 1.C.8 below. For discrete random variables the result is proven in a similar manner. When ψ is *just* monotone (rather than *strictly* monotone) the result is still true, but the preceding simple argument is no longer sufficient for its proof.

Theorem 1.C.8. *If $X \le_{\rm lr} Y$ and ψ is any increasing [decreasing] function, then $\psi(X) \le_{\rm lr} [\ge_{\rm lr}]\ \psi(Y)$.*

If $X_1 \le_{\rm lr} Y_1$ and $X_2 \le_{\rm lr} Y_2$, where X_1 and X_2 are independent random variables, and Y_1 and Y_2 are also independent random variables, then it is not necessarily true that $X_1+X_2 \le_{\rm lr} Y_1+Y_2$. However, if these random variables have logconcave densities, then it is true. In fact, a slightly stronger result is true:

Theorem 1.C.9. *Let (X_i, Y_i), $i = 1, 2, \ldots, m$, be independent pairs of random variables such that $X_i \le_{\rm lr} Y_i$, $i = 1, 2, \ldots, m$. If X_i, Y_i, $i = 1, 2, \ldots, m$, all have (continuous or discrete) logconcave densities, except possibly one X_l and one Y_k $(l \ne k)$, then*

$$\sum_{i=1}^{m} X_i \le_{\rm lr} \sum_{i=1}^{m} Y_i.$$

Proof. Since a convolution of random variables with logconcave densities has a logconcave density, it is enough to show that if W_1, W_2, and Z are independent random variables such that $W_1 \le_{\rm lr} W_2$, and Z has a logconcave density function, then $W_1 + Z \le_{\rm lr} W_2 + Z$. We will give the proof for the continuous case; the proof for the discrete case is similar. Let f_{W_i}, f_{W_i+Z}, $i = 1, 2$, and f_Z denote the density functions of the indicated random variables. Then

$$f_{W_i+Z}(t) = \int_{-\infty}^{\infty} f_Z(t-w) f_{W_i}(w)\, dw, \quad i = 1, 2,\ t \in \mathbb{R}.$$

The assumption $W_1 \le_{\rm lr} W_2$ means that $f_{W_i}(w)$, as a function of w and of $i \in \{1, 2\}$, is TP_2. The logconcavity of f_Z means that $f_Z(t-w)$, as a function of t and of w, is TP_2. Therefore, by the basic composition formula (Karlin [275]) we see that $f_{W_i+Z}(t)$ is TP_2 in $i \in \{1, 2\}$ and t; that is, $W_1 + Z \le_{\rm lr} W_2 + Z$. □

Example 1.C.10. Consider m independent Bernoulli trials with probability p_i of success in the ith trial. Let $q(k, \boldsymbol{p})$ denote the probability of k successes, $k = 1, 2, \ldots, m$, where $\boldsymbol{p} = (p_1, p_2, \ldots, p_m)$. Then $q(k+1, \boldsymbol{p})/q(k, \boldsymbol{p})$ is increasing in each p_i for $k = 0, 1, \ldots, m-1$. In order to see it, let X_i be a Bernoulli random variable with probability p_i of success, $i = 1, 2, \ldots, m$, and assume that the X_i's are independent. Similarly, let Y_i be a Bernoulli random variable with probability p'_i of success, $i = 1, 2, \ldots, m$, and assume that the Y_i's are independent. Obviously, the discrete density functions of the X_i's and of the Y_i's are logconcave, and if $\boldsymbol{p} \le \boldsymbol{p}'$, then $X_i \le_{\rm lr} Y_i$, $i = 1, 2, \ldots, m$. The stated result thus follows from Theorem 1.C.9.

For nonnegative random variables, Theorem 1.C.9 can be generalized further by having more Y_i's summed than X_i's. Under the assumptions of Theorem 1.C.9, one then obtains, for $m \le n$, that

$$\sum_{i=1}^{m} X_i \le_{\rm lr} \sum_{i=1}^{n} Y_i.$$

Of course, in this case, for $m+1 \le i \le n$, the Y_i's only need to have logconcave densities—they do not have to have corresponding X_i's to which they need to be comparable in the order $\le_{\rm lr}$. One may expect that the latter inequality can be extended to the following one:

$$\sum_{i=1}^{M} X_i \le_{\rm lr} \sum_{i=1}^{N} Y_i,$$

where M and N are two discrete positive integer-valued random variables, independent of the X_i's and of the Y_i's, respectively, such that $M \le_{\rm lr} N$. Indeed this inequality is true under some additional assumptions on the distributions of the X_i's and the Y_i's that will not be stated here. An important special case is the following theorem.

Theorem 1.C.11. *Let $\{X_i,\ i = 1, 2, \dots\}$ be a sequence of nonnegative independent random variables with logconcave densities. Let M and N be two discrete positive integer-valued random variables such that $M \le_{\rm lr} N$, and assume that M and N are independent of the X_i's. Then*

$$\sum_{i=1}^{M} X_i \le_{\rm lr} \sum_{i=1}^{N} X_i.$$

In Pellerey [445] it is claimed that the conclusion of Theorem 1.C.11 holds even under the weaker assumption that $M \le_{\rm hr} N$ (in the sense of (1.B.9) or (1.B.10)). However, there is a mistake in [445] (see Pellerey [446]).

It is of interest to compare Theorem 1.C.11 to the following result, which combines uses of the likelihood ratio and the hazard [reversed hazard] rate orders.

Theorem 1.C.12. *Let $\{X_i,\ i = 1, 2, \dots\}$ be a sequence of nonnegative independent random variables that are* IFR *[have decreasing reversed hazard rates]. Let M and N be two discrete positive integer-valued random variables such that $M \le_{\rm lr} N$, and assume that M and N are independent of the X_i's. Then*

$$\sum_{i=1}^{M} X_i \le_{\rm hr} [\le_{\rm rh}] \sum_{i=1}^{N} X_i.$$

Note that the hazard rate part of Theorem 1.C.12 is weaker than Theorem 1.B.7 because of Theorem 1.C.1.

The hazard rate order can be characterized by means of the likelihood ratio order and the appropriate equilibrium age variables. Recall from (1.A.20) that for nonnegative random variables X and Y with finite means we denote by A_X and A_Y the corresponding asymptotic equilibrium ages. The following result is immediate from (1.B.3) and (1.C.1).

Theorem 1.C.13. *Let X and Y be two nonnegative random variables with finite positive means. Then $X \leq_{\rm hr} Y$ if, and only if, $A_X \leq_{\rm lr} A_Y$.*

In light of Theorem 1.C.13 it is of interest to note that the order $\leq_{\rm lr}$ can also be used to characterize the hazard rate order as is described in the next theorem. Let X and Y be two nonnegative random variables with finite means and suppose that $X \leq_{\rm st} Y$ and that $EX < EY$. Let F and G be the distribution functions of X and of Y, respectively. Define the random variable $Z_{X,Y}$ as the random variable that has the density function h given by

$$h(z) = \frac{\overline{G}(z) - \overline{F}(z)}{EY - EX}, \quad z \geq 0. \tag{1.C.7}$$

Theorem 1.C.14. *Let X and Y be two nonnegative random variables with finite means such that $X \leq_{\rm st} Y$ and such that $EY > EX > 0$. Then*

$$A_X \leq_{\rm lr} Z_{X,Y} \Longleftrightarrow A_Y \leq_{\rm lr} Z_{X,Y} \Longleftrightarrow X \leq_{\rm hr} Y,$$

where $Z_{X,Y}$ has the density function given in (1.C.7).

Proof. Denote by $f_{\rm e}$ the density function of A_Y. Then, using (1.A.20), we obtain

$$\frac{h(x)}{f_{\rm e}(x)} = \frac{EY}{EY - EX}\left(1 - \frac{\overline{F}(x)}{\overline{G}(x)}\right), \quad x \geq 0,$$

and the second stated equivalence follows from (1.C.1) and (1.B.3). The proof of the first equivalence is similar. □

It is of interest to contrast Theorem 1.C.14 with Theorems 2.A.5 and 2.B.3.

The likelihood ratio order enjoys a closure under mixture property which is similar to the closure under mixture property of the hazard rate order stated in Theorem 1.B.8. This is stated next; the proof is similar to the proof of Theorem 1.B.8; we omit the details.

Theorem 1.C.15. *Let X, Y, and Θ be random variables such that $[X|\Theta = \theta] \leq_{\rm lr} [Y|\Theta = \theta']$ for all θ and θ' in the support of Θ. Then $X \leq_{\rm lr} Y$.*

As a corollary of Theorem 1.C.15 we obtain the following result.

Corollary 1.C.16. *Let N be a positive integer-valued random variable, and let X_i, $i = 1, 2, \ldots$, be random variables which are independent of N. Let Y be a random variable such that $X_i \leq_{\rm lr} Y$, $i = 1, 2, \ldots$. Then $X_N \leq_{\rm lr} Y$.*

Consider now a family of (continuous or discrete) density functions $\{g_\theta, \theta \in \mathcal{X}\}$ where $\mathcal{X}$ is a subset of the real line. As in Section 1.A.3 let $X(\theta)$ denote a random variable with density function g_θ. For any random variable Θ with support in $\mathcal{X}$, and with distribution function F, let us denote by $X(\Theta)$ a random variable with density function h given by

$$h(y) = \int_{\mathcal{X}} g_\theta(y) \mathrm{d}F(\theta), \quad y \in \mathbb{R}.$$

The following result generalizes both Theorems 1.C.8 and 1.C.15, just as Theorem 1.A.6 generalized parts (a) and (c) of Theorem 1.A.3.

Theorem 1.C.17. *Consider a family of density functions $\{g_\theta,\ \theta \in \mathcal{X}\}$ as above. Let Θ_1 and Θ_2 be two random variables with supports in $\mathcal{X}$ and distribution functions F_1 and F_2, respectively. Let Y_1 and Y_2 be two random variables such that $Y_i =_{\mathrm{st}} X(\Theta_i)$, $i = 1,2$, that is, suppose that the density function of Y_i is given by*

$$h_i(y) = \int_{\mathcal{X}} g_\theta(y) \mathrm{d}F_i(\theta), \quad y \in \mathbb{R},\ i = 1,2.$$

If

$$X(\theta) \le_{\mathrm{lr}} X(\theta') \quad \textit{whenever } \theta \le \theta', \tag{1.C.8}$$

and if

$$\Theta_1 \le_{\mathrm{lr}} \Theta_2, \tag{1.C.9}$$

then

$$Y_1 \le_{\mathrm{lr}} Y_2. \tag{1.C.10}$$

Proof. We give the proof under the assumption that Θ_1 and Θ_2 are absolutely continuous with density functions f_1 and f_2, respectively. The proof for the discrete case is similar. Assumption (1.C.8) means that $g_\theta(y)$, as a function of θ and of y, is TP_2. Assumption (1.C.9) means that $f_i(\theta)$, as a function of $i \in \{1,2\}$ and of θ, is TP_2. Therefore, by the basic composition formula (Karlin [275]) we see that $h_i(y)$ is TP_2 in $i \in \{1,2\}$ and y. That gives (1.C.10). □

A related result is the following; see also Theorems 1.B.19 and 1.B.54.

Theorem 1.C.18. *Let $X_1, X_2, \ldots, X_m$, Θ_1, and Θ_2 be independent nonnegative random variables. Define $N_j(t)$ for $t \ge 0$ and $j = 1,2$ as in Theorem 1.B.19. If $\Theta_1 \le_{\mathrm{lr}} \Theta_2$, then $N_1(t) \le_{\mathrm{lr}} N_2(t)$ for all $t \ge 0$.*

The following example is an application of Theorem 1.C.17; it may be compared to Examples 1.A.7 and 1.B.16.

Example 1.C.19. Let Θ_1 and Θ_2 be two nonnegative random variables with distribution functions F_1 and F_2, respectively. Let G be some absolutely continuous distribution function, and let g be the corresponding density function. Denote by $X(\theta)$ a random variable with the distribution function G^θ. Define $Y_i = X(\Theta_i)$; that is, the distribution function H_i of Y_i is given by

$$H_i(y) = \int_0^\infty G^\theta(y) \mathrm{d}F_i(\theta), \quad y \in \mathbb{R},\ i = 1,2.$$

Note that the density function k_θ of $X(\theta)$ is given by

$$k_\theta(y) = \theta g(y) G^{\theta-1}(y), \quad y \in \mathbb{R}.$$

It is easy to verify that (1.C.8) holds. Thus, by Theorem 1.C.17, if $\Theta_1 \leq_{\text{lr}} \Theta_2$, then $Y_1 \leq_{\text{lr}} Y_2$.

Now, denote by $\widetilde{X}(\theta)$ a random variable with the survival function $\overline{G}^{\theta}$, where $\overline{G} \equiv 1 - G$. Define $\widetilde{Y}_i = \widetilde{X}(\Theta_i)$; that is, the survival function $\overline{\widetilde{H}}_i$ of $\widetilde{Y}_i$ is given by

$$\overline{\widetilde{H}}_i(y) = \int_0^\infty \overline{G}^{\theta}(y) \mathrm{d}F_i(\theta), \quad y \in \mathbb{R},\ i = 1, 2.$$

Note that the density function $\widetilde{k}_\theta$ of $\widetilde{X}(\theta)$ is given by

$$\widetilde{k}_\theta(y) = \theta g(y) \overline{G}^{\theta-1}(y), \quad y \in \mathbb{R}.$$

It is easy to verify now that $\widetilde{X}(\theta) \geq_{\text{lr}} \widetilde{X}(\theta')$ whenever $\theta \leq \theta'$. Thus, by an obvious modification of Theorem 1.C.17, if $\Theta_1 \leq_{\text{lr}} \Theta_2$, then $Y_1 \geq_{\text{lr}} Y_2$.

In order to state a bivariate characterization result for the order $\leq_{\text{lr}}$ we define the following class of bivariate functions:

$$\mathcal{G}_{\text{lr}} = \{\phi : \mathbb{R}^2 \to \mathbb{R} : \phi(x, y) \leq \phi(y, x) \text{ whenever } x \leq y\}.$$

Theorem 1.C.20. *Let X and Y be independent random variables. Then $X \leq_{\text{lr}} Y$ if, and only if,*

$$\phi(X, Y) \leq_{\text{st}} \phi(Y, X) \quad \textit{for all } \phi \in \mathcal{G}_{\text{lr}}. \tag{1.C.11}$$

Proof. We give the proof for the absolutely continuous case only; the proof for the discrete case is similar. Suppose that (1.C.11) holds. Select u, v, $\Delta u > 0$, and $\Delta v > 0$ such that $u \leq v$. As before, let I_A denote the indicator function of the set A, and define $\phi(x, y) = I_{\{u - \Delta u \leq y \leq u, v \leq x \leq v + \Delta v\}}$. Clearly, $\phi \in \mathcal{G}_{\text{lr}}$. Hence

$$\begin{aligned} P\{v \leq X \leq v + \Delta v, u - \Delta u \leq Y \leq u\} &= E\phi(X, Y) \\ &\leq E\phi(Y, X) = P\{v \leq Y \leq v + \Delta v, u - \Delta u \leq X \leq u\}. \end{aligned}$$

Dividing both sides by $\Delta u \Delta v$ and letting $\Delta u \to 0$ and $\Delta v \to 0$, we obtain (1.C.2), that is, $X \leq_{\text{lr}} Y$.

Conversely, suppose that $X \leq_{\text{lr}} Y$. Let $\phi \in \mathcal{G}_{\text{lr}}$ and let ψ be an increasing function. Then

$$\begin{aligned} &E[\psi(\phi(Y, X)) - \psi(\phi(X, Y))] \\ &= \int_y \int_x [\psi(\phi(y, x)) - \psi(\phi(x, y))] f(x) g(y) \mathrm{d}x \mathrm{d}y \\ &= \int_y \int_{y \geq x} [\psi(\phi(y, x)) - \psi(\phi(x, y))][f(x)g(y) - f(y)g(x)] \mathrm{d}y \mathrm{d}x \geq 0. \square \end{aligned}$$

A typical application of Theorem 1.C.20 is shown in the proof of Theorem 6.B.15 in Chapter 6. Another typical application is the following result.

Theorem 1.C.21. *Let $X_1, X_2, \dots, X_m$ be independent random variables such that $X_1 \le_{\rm lr} X_2 \le_{\rm lr} \cdots \le_{\rm lr} X_m$. Let $a_1, a_2, \dots, a_m$ be constants such that $a_1 \le a_2 \le \cdots \le a_m$. Then*

$$\sum_{i=1}^{m} a_{m-i+1} X_i \le_{\rm st} \sum_{i=1}^{m} a_{\pi_i} X_i \le_{\rm st} \sum_{i=1}^{m} a_i X_i,$$

where $\boldsymbol{\pi} = (\pi_1, \pi_2, \dots, \pi_m)$ denotes any permutation of $(1, 2, \dots, m)$.

Proof. We only give the proof when $m = 2$; the general case then can be obtained by pairwise interchanges. So, suppose that $X_1 \le_{\rm lr} X_2$ and that $a_1 \le a_2$. Define ϕ by $\phi(x, y) = a_1 y + a_2 x$. Then it is easy to verify that $\phi \in \mathcal{G}_{\rm lr}$. Thus, by Theorem 1.C.20, $a_1 X_2 + a_2 X_1 \le_{\rm st} a_1 X_1 + a_2 X_2$. □

The next two results are characterizations similar to the one in Theorem 1.C.20. They use the notation of Theorem 1.A.10, and their comparison is of interest. The proofs of the following two theorems are omitted.

Theorem 1.C.22. *Let X and Y be two independent random variables. Then $X \le_{\rm lr} Y$ if, and only if,*

$$E\phi_1(X, Y) \le E\phi_2(X, Y)$$

for all functions ϕ_1 and ϕ_2 that satisfy $\Delta\phi_{21}(x, y) \ge 0$ whenever $x \le y$, and $\Delta\phi_{21}(x, y) \ge -\Delta\phi_{21}(y, x)$ whenever $x \le y$.

Theorem 1.C.23. *Let X and Y be two independent random variables. Then $X \le_{\rm lr} Y$ if, and only if,*

$$\phi_1(X, Y) \le_{\rm st} \phi_2(X, Y)$$

for all ϕ_1 and ϕ_2 that satisfy $\Delta\phi_{21}(x, y) \ge 0$ whenever $x \le y$, and $\phi_1(x, y) \le \phi_2(y, x)$ for all x and y (then, in particular, $\Delta\phi_{21}(x, y) \ge -\Delta\phi_{21}(y, x)$ whenever $x \le y$).

The next theorem gives a characterization of the likelihood ratio order in the spirit of Theorems 1.B.11 and 1.B.49.

Theorem 1.C.24. *Let X and Y be two independent random variables. Then $X \le_{\rm lr} Y$ if, and only if,*

$$[X \big| \min(X, Y) = z_1, \max(X, Y) = z_2] \\ \le_{\rm lr} [Y \big| \min(X, Y) = z_1, \max(X, Y) = z_2] \quad \text{for all } z_1 \le z_2.$$

Proof. First suppose that X and Y are absolutely continuous with density functions f and g, respectively. Then

$$\begin{aligned} P[X = z_1 \big| \min(X,Y) = z_1, \max(X,Y) = z_2] \\ = 1 - P[X = z_2 \big| \min(X,Y) = z_1, \max(X,Y) = z_2] \\ = P[Y = z_2 \big| \min(X,Y) = z_1, \max(X,Y) = z_2] \\ = 1 - P[Y = z_1 \big| \min(X,Y) = z_1, \max(X,Y) = z_2] \\ = \frac{f(z_1)g(z_2)}{f(z_1)g(z_2) + f(z_2)g(z_1)}, \end{aligned}$$

and the stated result follows.

The proof when X and Y are discrete is similar. □

Another similar characterization is given in Theorem 4.A.36.

The following result gives a Laplace transform characterization of the order $\le_{\text{lr}}$. It should be compared with Theorems 1.A.13, 1.B.18, and 1.B.53. The proof is omitted.

Theorem 1.C.25. *Let X_1 and X_2 be two nonnegative random variables, and let $N_\lambda(X_1)$ and $N_\lambda(X_2)$ be as described in Theorem 1.A.13. Then*

$$X_1 \le_{\text{lr}} X_2 \Longleftrightarrow N_\lambda(X_1) \le_{\text{lr}} N_\lambda(X_2) \quad \textit{for all } \lambda > 0.$$

The implication $\Longrightarrow$ in Theorem 1.C.25 can be generalized in the same manner that Theorem 1.A.14 generalizes the implication $\Longrightarrow$ in Theorem 1.A.13. We will not state the result here since it is equivalent to Theorem 1.C.17.

Some interesting simple implications of the likelihood ratio order are described in the following theorem.

Theorem 1.C.26. *Let X, Y, and Z be independent random variables. If $X \le_{\text{lr}} Y$, then*

$$\begin{aligned} [X \big| X + Y = v] &\le_{\text{lr}} [Y \big| X + Y = v] \quad \textit{for all } v, \\ [X \big| X + Z = v] &\le_{\text{lr}} [Y \big| Y + Z = v] \quad \textit{for all } v, \quad \textit{and} \\ [Z \big| X + Z = v] &\ge_{\text{lr}} [Z \big| Y + Z = v] \quad \textit{for all } v. \end{aligned}$$

Proof. We give only the proof of the first inequality; the proofs of the other two are similar. First suppose that X and Y are absolutely continuous with density functions f and g, respectively. Denote the density function of $X+Y$ by h. Then the density function of $[Y \big| X + Y = v]$ is given by $\frac{f(v-\cdot)g(\cdot)}{h(v)}$, and the density function of $[X \big| X+Y = v]$ is given by $\frac{f(\cdot)g(v-\cdot)}{h(v)}$. It is now seen that the monotonicity of g/f implies the monotonicity of the ratio of the above two density functions.

The proof when X and Y are discrete is similar. □

The next, easily proven, result is stronger than Theorems 1.A.15, 1.B.20, and 1.B.55.

Theorem 1.C.27. *Let X be any random variable. Then $X_{(-\infty,a]}$ and $X_{(a,\infty)}$ are increasing in a in the sense of the likelihood ratio order.*

A similar setting in which the order $\le_{\rm hr}$ gives rise to the order $\le_{\rm lr}$ is described in the following result.

Theorem 1.C.28. *Let X, Y, and T be random variables such that T is independent of (X,Y). If $X \le_{\rm hr} Y$, then*

$$[T \big| T < X] \le_{\rm lr} [T \big| T < Y].$$

Proof. For simplicity assume that T is absolutely continuous with density function f_T. Let $\overline{F}_X$ and $\overline{F}_Y$ be the survival functions of X and Y. The density function of $[T \big| T < X]$ is proportional to $f_T\overline{F}_X$ and the density function of $[T \big| T < Y]$ is proportional to $f_T\overline{F}_Y$. The stated result now follows from (1.B.3). □

An analog of the remark after Theorem 1.B.21 is the following result; its proof is straightforward.

Theorem 1.C.29. *Let X be a nonnegative, absolutely continuous, random variable with the density function f. Then $aX \le_{\rm lr} X$ for all $0 < a < 1$ if, and only if, $\log f(\mathrm{e}^x)$ is concave in $x \ge 0$.*

In the next result it is shown that a random variable, whose distribution is the mixture of two distributions of likelihood ratio ordered random variables, is bounded from below and from above, in the likelihood ratio order sense, by these two random variables.

Theorem 1.C.30. *Let X and Y be two random variables with distribution functions F and G, respectively. Let W be a random variable with the distribution function $pF + (1-p)G$ for some $p \in (0,1)$. If $X \le_{\rm lr} Y$, then $X \le_{\rm lr} W \le_{\rm lr} Y$.*

Proof. Let A and B be two measurable sets such that $A \le B$; see (1.C.3). If $X \le_{\rm lr} Y$, then

$$\begin{aligned} P\{X \in A\}P\{W \in B\} &= P\{X \in A\}(pP\{X \in B\} + (1-p)P\{Y \in B\}) \\ &\ge P\{X \in B\}(pP\{X \in A\} + (1-p)P\{Y \in A\}) \\ &= P\{X \in B\}P\{W \in A\}, \end{aligned}$$

where the inequality follows from (1.C.3). Thus, by (1.C.3), $X \le_{\rm lr} W$. The proof that $W \le_{\rm lr} Y$ is similar. □

Analogous to the result in Remark 1.A.18, it can be shown that some general sets of distribution functions on $\mathbb{R}$ are lattices with respect to the order $\le_{\rm lr}$.

Let $X_1, X_2, \dots, X_m$ be random variables, and let $X_{(k:m)}$ denote the corresponding kth order statistic, $k = 1, 2, \dots, m$.

Theorem 1.C.31. *Let $X_1, X_2, \dots, X_m$ be m independent random variables, all with absolutely continuous distribution functions, all having the same support which is an interval of the real line, and all having differentiable densities.*
(a) *If*

$$X_1 \le_{\rm lr} X_2 \le_{\rm lr} \cdots \le_{\rm lr} X_m,$$

then

$$\begin{aligned} X_{(k-1:m)} &\le_{\rm lr} X_{(k:m)}, \quad 2 \le k \le m, \quad \text{and} \\ X_{(k-1:m-1)} &\le_{\rm lr} X_{(k:m)}, \quad 2 \le k \le m. \end{aligned}$$

(b) *If*

$$X_1 \ge_{\rm lr} X_2 \ge_{\rm lr} \cdots \ge_{\rm lr} X_m,$$

then

$$X_{(k:m)} \le_{\rm lr} X_{(k:m-1)}, \quad 1 \le k \le m-1.$$

A similar result for a finite population is the following. Consider a finite population of size N which is linearly ordered, and suppose, without loss of generality, that it can be represented as $\{1, 2, \dots, N\}$. Here let $X_{(1)} \le X_{(2)} \le \cdots \le X_{(m)}$ denote now the order statistics corresponding to a simple random sample of size m from this population.

Theorem 1.C.32. *Let $X_{(1)} \le X_{(2)} \le \cdots \le X_{(m)}$ be defined as in the preceding paragraph. Then*

$$X_{(1)} \le_{\rm lr} X_{(2)} \le_{\rm lr} \cdots \le_{\rm lr} X_{(m)}.$$

Proof. For each $k \in \{1, 2, \dots, m\}$, let f_k denote the discrete density of $X_{(k)}$. Then

$$f_k(j) = \begin{cases} \frac{\binom{j-1}{k-1}\binom{N-j}{m-k}}{\binom{N}{m}}, & j = k, k+1, \dots, k+N-m; \\ 0, & \text{otherwise.} \end{cases}$$

Therefore, for $k \in \{1, 2, \dots, m-1\}$, we have

$$\frac{f_{k+1}(j)}{f_k(j)} = \begin{cases} 0, & j = k; \\ \frac{(m-k)(j-k)}{k(N-j-m+k+1)}, & j = k+1, k+2, \dots, k+N-m; \\ \infty, & j = k+N-m+1. \end{cases}$$

This is increasing in j, and therefore $X_{(k)} \le_{\rm lr} X_{(k+1)}$. □

Under some conditions the likelihood ratio order is closed under the formation of order statistics. As above, let $X_{(j:m)}$ denote the jth order statistic associated with the random variables $X_1, X_2, \ldots, X_m$, and let $Y_{(i:n)}$ denote similarly the ith order statistic associated with the random variables $Y_1, Y_2, \ldots, Y_n$.

Theorem 1.C.33. *Let $X_1, X_2, \ldots, X_m$ be m independent random variables, and let $Y_1, Y_2, \ldots, Y_n$ be other n independent random variables. If*

$$X_j \le_{\text{lr}} Y_i \quad \textit{for all } 1 \le j \le m \textit{ and } 1 \le i \le n,$$

then

$$X_{(j:m)} \le_{\text{lr}} Y_{(i:n)} \quad \textit{whenever } j \le i \textit{ and } m - j \ge n - i.$$

Proof. First we give the proof when $X_1, X_2, \ldots, X_m$ and $Y_1, Y_2, \ldots, Y_n$ all have absolutely continuous distribution functions. In this proof we use an idea of Chan, Proschan, and Sethuraman [123].

Let f_j, F_j, and $\overline{F}_j \equiv 1 - F_j$ denote the density, distribution, and survival functions of X_j. Similarly, let g_i, G_i, and $\overline{G}_i$ denote the density, distribution, and survival functions of Y_i. The density functions of $X_{(j:m)}$ and $Y_{(i:n)}$ are given by

$$f_{X_{(j:m)}}(t) = \sum_{\boldsymbol{\pi}} f_{\pi_1}(t) F_{\pi_2}(t) \cdots F_{\pi_j}(t) \overline{F}_{\pi_{j+1}}(t) \cdots \overline{F}_{\pi_m}(t),$$

and

$$g_{Y_{(i:n)}}(t) = \sum_{\boldsymbol{\sigma}} g_{\sigma_1}(t) G_{\sigma_2}(t) \cdots G_{\sigma_i}(t) \overline{G}_{\sigma_{i+1}}(t) \cdots \overline{G}_{\sigma_n}(t),$$

where $\sum_{\boldsymbol{\pi}}$ signifies the sum over all permutations $\boldsymbol{\pi} = (\pi_1, \pi_2, \ldots, \pi_m)$ of $(1, 2, \ldots, m)$, and $\sum_{\boldsymbol{\sigma}}$ similarly denotes the sum over all permutations $\boldsymbol{\sigma} = (\sigma_1, \sigma_2, \ldots, \sigma_n)$ of $(1, 2, \ldots, n)$. Write

$$\frac{g_{Y_{(i:n)}}(t)}{f_{X_{(j:m)}}(t)} = \frac{\sum_{\boldsymbol{\sigma}} g_{\sigma_1}(t) G_{\sigma_2}(t) \cdots G_{\sigma_i}(t) \overline{G}_{\sigma_{i+1}}(t) \cdots \overline{G}_{\sigma_n}(t)}{\sum_{\boldsymbol{\pi}} f_{\pi_1}(t) F_{\pi_2}(t) \cdots F_{\pi_j}(t) \overline{F}_{\pi_{j+1}}(t) \cdots \overline{F}_{\pi_m}(t)}. \tag{1.C.12}$$

Now, for any choice of a permutation $\boldsymbol{\pi}$ of $(1, 2, \ldots, m)$ and a permutation $\boldsymbol{\sigma}$ of $(1, 2, \ldots, n)$ we have

$$\begin{aligned}
&\frac{g_{\sigma_1}(t) G_{\sigma_2}(t) \cdots G_{\sigma_i}(t) \overline{G}_{\sigma_{i+1}}(t) \cdots \overline{G}_{\sigma_n}(t)}{f_{\pi_1}(t) F_{\pi_2}(t) \cdots F_{\pi_j}(t) \overline{F}_{\pi_{j+1}}(t) \cdots \overline{F}_{\pi_m}(t)} \\
&\qquad = \frac{g_{\sigma_1}(t)}{f_{\pi_1}(t)} \times \frac{G_{\sigma_2}(t) \cdots G_{\sigma_j}(t)}{F_{\pi_2}(t) \cdots F_{\pi_j}(t)} \times \frac{\overline{G}_{\sigma_{i+1}}(t) \cdots \overline{G}_{\sigma_n}(t)}{\overline{F}_{\pi_{m-n+i+1}}(t) \cdots \overline{F}_{\pi_m}(t)} \\
&\qquad\qquad \times \frac{G_{\sigma_{j+1}}(t) \cdots G_{\sigma_i}(t)}{\overline{F}_{\pi_{j+1}}(t) \cdots \overline{F}_{\pi_{m-n+i}}(t)}.
\end{aligned}$$

Since $X_{\pi_1} \le_{\rm lr} Y_{\sigma_1}$ we see from (1.C.1) that the first fraction above is increasing in t. From $X_{\pi_k} \le_{\rm lr} Y_{\sigma_k}$ and Theorem 1.C.1 it follows that $X_{\pi_k} \le_{\rm rh} Y_{\sigma_k}$; but that means that $G_{\sigma_k}(t)/F_{\pi_k}(t)$ is increasing in t, $k = 2, \ldots, j$, and therefore the second fraction above is increasing in t. Similarly, from $X_{\pi_{k+m-n}} \le_{\rm lr} Y_{\sigma_k}$ and Theorem 1.C.1 it also follows that $X_{\pi_{k+m-n}} \le_{\rm hr} Y_{\sigma_k}$; but that means that $\overline{G}_{\sigma_k}(t)/\overline{F}_{\pi_{k+m-n}}(t)$ is increasing in t, $k = i+1, \ldots, n$, and therefore the third fraction above is increasing in t. The fourth fraction above obviously increases in t too, and thus the whole product increases in t.

Note that if $a_1, a_2, \ldots, a_m$ and $b_1, b_2, \ldots, b_n$ are all nonnegative univariate functions, such that $a_j(t)/b_i(t)$ is increasing in t for all $1 \le j \le m$ and $1 \le i \le n$, then $\sum_{j=1}^m a_j(t)/\sum_{i=1}^n b_i(t)$ is also increasing in t. It follows from this fact, and from (1.C.12), that $g_{Y_{(i:n)}}(t)/f_{X_{(j:m)}}(t)$ is increasing in t, and from (1.C.1) we obtain the stated result.

The result for the case when the random variables do not necessarily have absolutely continuous distribution functions follows from the above proof and the closure of the likelihood ratio order under weak convergence (Theorem 1.C.7). □

Some of the results that are described in the following pages are stated in the literature (see Section 1.E) only for random variables with absolutely continuous distribution functions. However, by the closure of the likelihood ratio order under weak convergence (Theorem 1.C.7) these results are true also for random variables that do not necessarily have absolutely continuous distribution functions.

As a corollary of Theorem 1.C.33 we obtain the following result.

Corollary 1.C.34. *Let $X_1, X_2, \ldots, X_m$ be m independent random variables and let $Y_1, Y_2, \ldots, Y_m$ be other m independent random variables. If $X_j \le_{\rm lr} Y_i$, for all choices of i and j, then $X_{(k)} \le_{\rm lr} Y_{(k)}$, $k = 1, 2, \ldots, m$.*

Example 1.C.35. Let X and Y be two independent random variables. If $X \le_{\rm lr} Y$, then $\min\{X, Y\} \le_{\rm lr} Y$ and $X \le_{\rm lr} \max\{X, Y\}$.

Example 1.C.36. Let X, Y, and Z be three independent random variables. If $X \le_{\rm lr} Y \le_{\rm lr} Z$, then $\min\{X, Y\} \le_{\rm lr} \min\{Y, Z\}$ and $\max\{X, Y\} \le_{\rm lr} \max\{Y, Z\}$.

By letting all the X_j's and Y_i's in Theorem 1.C.33 be identically distributed we obtain the following result.

Theorem 1.C.37. *For positive integers m and n, let $X_1, X_2, \ldots, X_{\max\{m,n\}}$ be independent identically distributed random variables. Then*

$$X_{(j:m)} \le_{\rm lr} X_{(i:n)} \quad \textit{whenever } j \le i \textit{ and } m - j \ge n - i.$$

In particular, it follows from Theorem 1.C.37 that

$$X_1 \le_{\rm lr} X_{(m:m)}, \quad m = 2, 3, \ldots \tag{1.C.13}$$

and

$$X_1 \geq_{\rm lr} X_{(1:m)}, \quad m = 2, 3, \ldots. \tag{1.C.14}$$

Note that (1.C.13) and (1.C.14) can also be obtained by induction from Example 1.C.35.

The following two corollaries of Theorem 1.C.37 can be compared to Theorems 1.B.27 and 1.B.28.

Corollary 1.C.38. *Let $X_1, X_2, \ldots, X_m$ be independent identically distributed random variables. Then $X_{(k-1:m-1)} \leq_{\rm lr} X_{(k:m)}$ for $k = 2, 3, \ldots, m$.*

Corollary 1.C.39. *Let $X_1, X_2, \ldots, X_m$ be independent identically distributed random variables. Then $X_{(k:m-1)} \geq_{\rm lr} X_{(k:m)}$ for $k = 1, 2, \ldots, m-1$.*

Remark 1.C.40. The likelihood ratio order can be used to provide a proof of Theorem 1.B.26. Let $X_1, X_2, \ldots, X_m$ be independent nonnegative random variables, and let $X_{(1)} \leq X_{(2)} \leq \cdots \leq X_{(m)}$ denote the corresponding order statistics. Fix s and t such that $0 \leq s \leq t$. For $j = 1, 2, \ldots, m$, define $M_j = 1$ if $X_j \leq s$, and $M_j = 0$ if $X_j > s$, and also define $N_j = 1$ if $X_j \leq t$, and $N_j = 0$ if $X_j > t$. Denote $M = \sum_{j=1}^m M_j$ and $N = \sum_{j=1}^m N_j$. Note that, for $j = 1, 2, \ldots, m$, we have

$$\begin{aligned} P\{M < j\} &= P\{X_{(j)} > s\}, \quad \text{and} \\ P\{N < j\} &= P\{X_{(j)} > t\}. \end{aligned}$$

Since $P\{M_j = 1\} = P\{X_j \leq s\} \leq P\{X_j \leq t\} = P\{N_j = 1\}$ it is easily seen that $M_j \leq_{\rm lr} N_j$, $j = 1, 2, \ldots, m$. Also, obviously, M_j and N_j have logconcave discrete density functions. Thus, from Theorem 1.C.9 it is seen that $M \leq_{\rm lr} N$. Therefore, by Theorem 1.C.1, $M \leq_{\rm rh} N$. Thus, from (1.B.44), we get that

$$\frac{P\{N < j\}}{P\{M < j\}} \quad \text{is increasing in } j \geq 1.$$

Therefore, for k such that $1 \leq k \leq m-1$ we have

$$\frac{P\{X_{(k)} > t\}}{P\{X_{(k)} > s\}} = \frac{P\{N < k\}}{P\{M < k\}} \leq \frac{P\{N < k+1\}}{P\{M < k+1\}} = \frac{P\{X_{(k+1)} > t\}}{P\{X_{(k+1)} > s\}}.$$

From (1.B.3) it thus follows that $X_{(k)} \leq_{\rm hr} X_{(k+1)}$.

Remark 1.C.41. The likelihood ratio order can be used to provide a proof of Theorem 1.B.36. Let the X_i's and the Y_j's be as in that theorem. Assume that $X_i \leq_{\rm hr} Y_j$ for all i, j. We first show that there exists a random variable Z with support (a, b) such that $X_i \leq_{\rm hr} Z \leq_{\rm hr} Y_j$ for all i, j. Let r_{X_i} and r_{Y_j} denote the hazard rate functions of the indicated random variables. From the assumption that $X_i \leq_{\rm hr} Y_j$ for all i, j it follows by (1.B.2) that

$$\min\{r_{X_1}(t), r_{X_2}(t), \ldots, r_{X_m}(t)\} \geq \max\{r_{Y_1}(t), r_{Y_2}(t), \ldots, r_{Y_n}(t)\}, \quad t \in (a, b).$$

Let q be a function which satisfies

$$\min\{r_{X_1}(t), r_{X_2}(t), \dots, r_{X_m}(t)\} \geq q(t) \geq \max\{r_{Y_1}(t), r_{Y_2}(t), \dots, r_{Y_n}(t)\}, \quad t \in (a, b);$$

for example, let $q(t) = \min\{r_{X_1}(t), r_{X_2}(t), \dots, r_{X_m}(t)\}$. It can be shown that q is indeed a hazard rate function. Let Z be a random variable with the hazard rate function q. Then indeed $X_i \leq_{\rm hr} Z \leq_{\rm hr} Y_j$ for all i, j.

Now, let $Z_1, Z_2, \dots, Z_{\max\{m,n\}}$ be independent random variables which are distributed as Z. Then, for $j \leq i$ and $m - j \geq n - i$ we have

$$\begin{aligned} X_{(i:m)} &\leq_{\rm hr} Z_{(i:m)} && \text{(by Proposition 1.B.35)} \\ &\leq_{\rm lr} Z_{(j:n)} && \text{(by Theorem 1.C.37)} \\ &\leq_{\rm hr} Y_{(j:n)} && \text{(by Proposition 1.B.35)}, \end{aligned}$$

and Theorem 1.B.36 follows from the fact that the likelihood ratio order implies the hazard rate order.

Recall that for a collection $X_1, X_2, \dots, X_m$ of nonnegative random variables, the spacings are defined by $U_{(i)} \equiv X_{(i)} - X_{(i-1)}$, $i = 1, 2, \dots, m$, where $X_{(0)} \equiv 0$. The following result may be compared with Theorems 1.A.19, 1.A.21, and 1.B.31.

Theorem 1.C.42. *Let $X_1, X_2, \dots, X_m$ be independent exponential random variables with possibly different parameters. Then*

$$U_{(1)} \leq_{\rm lr} \frac{m - i + 1}{m} \cdot U_{(i)}, \quad i = 1, 2, \dots, m.$$

It is worth mentioning that Kochar and Kirmani [313] claimed that if $X_1, X_2, \dots, X_m$ are independent and identically distributed random variables with a common logconvex density, then $U_{(i)} \leq_{\rm lr} ((m - i)/(m - i + 1))U_{(i+1)}$ for $i = 1, 2, \dots, m - 1$. However, Misra and van der Meulen [396] showed via a counterexample that this is not correct.

For spacings that are not "normalized" we have the following results. We denote by $U_{(i:m)} = X_{(i:m)} - X_{(i-1:m)}$ the ith spacing that corresponds to a sample $X_1, X_2, \dots, X_m$ of size m.

Theorem 1.C.43. *Let $X_1, X_2, \dots, X_m, X_{m+1}$ be independent, identically distributed, nonnegative random variables with a common logconvex density. Then*

$$\begin{aligned} U_{(i:m)} &\leq_{\rm lr} U_{(i+1:m)}, && 1 \leq i \leq m - 1, \\ U_{(i:m+1)} &\leq_{\rm lr} U_{(i:m)}, && 1 \leq i \leq m, \end{aligned}$$

and

$$U_{(i:m)} \leq_{\rm lr} U_{(i+1:m+1)}, \quad 1 \leq i \leq m.$$

Note that the three statements of the above theorem can be summarized as

$$U_{(j:m)} \le_{\rm lr} U_{(i:n)} \quad \text{whenever } i - j \ge \max\{0, n - m\}.$$

We also have the following result.

Theorem 1.C.44. *Let $X_1, X_2, \ldots, X_m, X_{m+1}$ be independent, identically distributed, nonnegative random variables with a common logconcave density. Then*

$$U_{(i:m)} \ge_{\rm lr} U_{(i+1:m+1)}, \quad 1 \le i \le m.$$

A comparison of spacings from two different samples, that is similar to Theorem 1.B.32, is described next. In fact, it will be argued after the next theorem that the next result strengthens Theorem 1.B.31. Here $U_{(i:m)} = X_{(i:m)} - X_{(i-1:m)}$ denotes, as before, the ith spacing that corresponds to the sample $X_1, X_2, \ldots, X_m$, and $V_{(j:n)}$ denotes, similarly, the jth spacing that corresponds to the sample $Y_1, Y_2, \ldots, Y_n$. Other results which give related comparisons can be found in Theorem 4.B.17 and in Examples 6.B.25 and 6.E.15.

Theorem 1.C.45. *For positive integers m and n, let $X_1, X_2, \ldots, X_m$ be independent identically distributed random variables with an absolutely continuous common distribution function, and let $Y_1, Y_2, \ldots, Y_n$ be independent identically distributed random variables with a possibly different absolutely continuous common distribution function. If $X_1 \le_{\rm lr} Y_1$, and if either X_1 or Y_1 is* DFR, *then*

$$(m - j + 1)U_{(j:m)} \le_{\rm hr} (n - i + 1)V_{(i:n)} \quad \textit{whenever } i - j \ge \max\{0, n - m\}.$$

Taking $X_1 =_{\rm st} Y_1$ in Theorem 1.C.45 it is seen that Theorem 1.B.31 is a consequence of Theorem 1.C.45.

In the following example it is shown that, under the proper conditions, random minima and maxima are ordered in the likelihood ratio order sense; see related results in Examples 3.B.39, 4.B.16, 5.A.24 and 5.B.13.

Example 1.C.46. Let $X_1, X_2, \ldots$ be a sequence of absolutely continuous nonnegative independent and identically distributed random variables with a common distribution function F_{X_1} and a common density function f_{X_1}. Let N_1 and N_2 be two positive integer-valued random variables which are independent of the X_i's. Denote $X_{(1:N_j)} \equiv \min\{X_1, X_2, \ldots, X_{N_j}\}$ and $X_{(N_j:N_j)} \equiv \max\{X_1, X_2, \ldots, X_{N_j}\}$, $j = 1, 2$. Then the density function of $X_{(N_j:N_j)}$ is given by

$$f_{X_{(N_j:N_j)}}(x) = \sum_{n=1}^{\infty} nF_{X_1}^{n-1}(x) f_{X_1}(x) P\{N_j = n\}, \quad x \ge 0,\ j = 1, 2.$$

If $N_1 \le_{\rm lr} N_2$, then $P\{N_j = n\}$ is TP$_2$ in $n \ge 1$ and $j \in \{1, 2\}$. Also, $nF_{X_1}^{n-1}(x) f_{X_1}(x)$ is TP$_2$ in $n \ge 1$ and $x \ge 0$. Therefore, by the Basic Composition Formula (Karlin [275]) it follows that $f_{X_{(N_j:N_j)}}(x)$ is TP$_2$ in $x \ge 0$

and $j \in \{1,2\}$. That is, $X_{(N_1:N_1)} \le_{\rm lr} X_{(N_2:N_2)}$. In a similar fashion it can be shown also that $X_{(1:N_1)} \ge_{\rm lr} X_{(1:N_2)}$.

Example 1.C.47. Let $\{N(t),\ t \ge 0\}$ be a nonhomogeneous Poisson process with mean function Λ (that is, $\Lambda(t) \equiv E[N(t)]$, $t \ge 0$), and let $T_1, T_2, \ldots$ be the successive epoch times. The survival function of T_n is given by $P\{T_n > t\} = \sum_{i=0}^{n-1} \frac{(\Lambda(t))^i}{i!} \mathrm{e}^{-\Lambda(t)}$, $t \ge 0$, and the density function of T_n is given by $f_n(t) = \lambda(t)\frac{(\Lambda(t))^{(n-1)}}{(n-1)!}\mathrm{e}^{-\Lambda(t)}$, $t \ge 0$, where $\lambda(t) \equiv \frac{\mathrm{d}}{\mathrm{d}t}\Lambda(t)$, $n = 1, 2, \ldots$. It is easy to verify that $\frac{f_{n+1}(t)}{f_n(t)}$ is increasing in $t \ge 0$, $n = 1, 2, \ldots$, and therefore

$$T_n \le_{\rm lr} T_{n+1}, \quad n = 1, 2, \ldots.$$

Theorem 2.6 on page 182 of Kamps [273] extends Example 1.C.47 (as it extends Theorem 1.C.45) to the so called generalized order statistics. A further extension is described in Franco, Ruiz, and Ruiz [205].

The following example may be compared to Examples 1.B.24, 2.A.22, 3.B.38, 4.B.14, 6.B.41, 6.D.8, 6.E.13, and 7.B.13.

Example 1.C.48. Let X and Y be two absolutely continuous nonnegative random variables with survival functions $\overline{F}$ and $\overline{G}$ and density functions f and g, respectively. Denote $\Lambda_1 = -\log \overline{F}$, $\Lambda_2 = -\log \overline{G}$, and $\lambda_i = \Lambda_i'$, $i = 1, 2$. Consider two nonhomogeneous Poisson processes $N_1 = \{N_1(t),\ t \ge 0\}$ and $N_2 = \{N_2(t),\ t \ge 0\}$ with mean functions Λ_1 and Λ_2 (see Example 1.B.13), respectively. Let $T_{i,1}, T_{i,2}, \ldots$ be the successive epoch times of process N_i, $i = 1, 2$. Note that $X =_{\rm st} T_{1,1}$ and $Y =_{\rm st} T_{2,1}$.

It turns out that, under some conditions, the likelihood ratio ordering of the first two epoch times implies the likelihood ratio ordering of all the corresponding later epoch times. Explicitly, it will be shown below that if $X \le_{\rm lr} Y$, and if

$$\frac{\Lambda_2(t)}{\Lambda_1(t)} \quad \text{is increasing in } t \ge 0, \tag{1.C.15}$$

then $T_{1,n} \le_{\rm lr} T_{2,n}$, $n \ge 1$.

From (1.B.24) it is easy to see that the density function $f_{1,n}$ of $T_{1,n}$ is given by

$$f_{1,n}(t) = f(t)\frac{(\Lambda_1(t))^{n-1}}{(n-1)!}, \quad t \ge 0,\ n \ge 1,$$

and that the density function $f_{2,n}$ of $T_{2,n}$ is given by

$$f_{2,n}(t) = g(t)\frac{(\Lambda_2(t))^{n-1}}{(n-1)!}, \quad t \ge 0,\ n \ge 1.$$

Thus,

$$\frac{f_{2,n}(t)}{f_{1,n}(t)} = \frac{g(t)}{f(t)}\left(\frac{\Lambda_2(t)}{\Lambda_1(t)}\right)^{n-1}.$$

Now, if $X \le_{\rm lr} Y$ and (1.C.15) holds, then $f_{2,n}/f_{1,n}$ is increasing and we obtain $T_{1,n} \le_{\rm lr} T_{2,n}$.

Now let $X_{i,n} \equiv T_{i,n} - T_{i,n-1}$, $n \ge 1$ (where $T_{i,0} \equiv 0$), be the inter-epoch times of the process N_i, $i = 1,2$. Again, note that $X =_{\rm st} X_{1,1}$ and $Y =_{\rm st} X_{2,1}$. It turns out that, under some conditions, the likelihood ratio ordering of the first two inter-epoch times implies the likelihood ratio ordering of all the corresponding later inter-epoch times. Explicitly, it will be shown below that if $X \le_{\rm hr} Y$, if f and g are logconvex, and if (1.B.25) holds, then $X_{1,n} \le_{\rm lr} X_{2,n}$ for each $n \ge 1$.

First note that by Theorem 1.C.4 we have $X \le_{\rm lr} Y$. For the purpose of the following proof we denote f by f_1 and g by f_2. Let $g_{i,n}$ denote the density function of $X_{i,n}$, $i = 1,2$. The stated result is obvious for $n = 1$, so let us fix an $n \ge 2$. From (1.B.26) we obtain

$$g_{i,n}(t) = \int_0^\infty \lambda_i(s)\frac{\Lambda_i^{n-2}(s)}{(n-2)!} f_i(s+t)\,ds, \quad t \ge 0,\ i = 1,2.$$

As in Example 1.B.24, we have that

$$\lambda_i(t)\frac{\Lambda_i^{n-2}(t)}{(n-2)!} \quad \text{is TP}_2 \text{ in } (i,t).$$

The assumption $F_1 \le_{\rm lr} F_2$ implies that

$$f_i(s+t) \text{ is TP}_2 \text{ in } (i,s) \text{ and in } (i,t).$$

Finally, the logconvexity of f_1 and of f_2 means that

$$f_i(s+t) \text{ is TP}_2 \text{ in } (s,t).$$

Thus, by Theorem 5.1 on page 123 of Karlin [275], we get that $g_{i,n}(t)$ is TP$_2$ in (i,t); that is, $X_{1,n} \le_{\rm lr} X_{2,n}$.

The following neat example compares a sum of independent heterogeneous exponential random variables with an Erlang random variable; it is of interest to compare it with Examples 1.A.24 and 1.B.5. We do not give the proof here.

Example 1.C.49. Let X_i be an exponential random variable with mean $\lambda_i^{-1} > 0$, $i = 1,2,\ldots,m$, and assume that the X_i's are independent. Let Y_i, $i = 1,2,\ldots,m$, be independent, identically distributed, exponential random variables with mean η^{-1}. Then

$$\sum_{i=1}^m X_i \ge_{\rm lr} \sum_{i=1}^m Y_i \Longleftrightarrow \frac{\lambda_1 + \lambda_2 + \cdots + \lambda_m}{m} \le \eta.$$

A related example is the following. Recall from page 2 the definition of the majorization order $\prec$ among n-dimensional vectors. It is of interest to compare the example below with Example 3.B.34.

Example 1.C.50. Let X_i be an exponential random variable with mean $\lambda_i^{-1} > 0$, $i = 1, 2, \ldots, m$, and let Y_i be an exponential random variable with mean $\eta_i^{-1} > 0$, $i = 1, 2, \ldots, m$. If $(\lambda_1, \lambda_2, \ldots, \lambda_m) \succ (\eta_1, \eta_2, \ldots, \eta_m)$, then

$$\sum_{i=1}^{m} X_i \geq_{\text{lr}} \sum_{i=1}^{m} Y_i.$$

The next example may be compared with Examples 1.A.25, 1.B.6, and 4.A.45.

Example 1.C.51. Let X_i be a binomial random variable with parameters n_i and p_i, $i = 1, 2, \ldots, m$, and assume that the X_i's are independent. Let Y be a binomial random variable with parameters n and p where $n = \sum_{i=1}^{m} n_i$. Then

$$\sum_{i=1}^{m} X_i \geq_{\text{lr}} Y \Longleftrightarrow p \leq \frac{n}{\sum_{i=1}^{m}(n_i/p_i)},$$

and

$$\sum_{i=1}^{m} X_i \leq_{\text{lr}} Y \Longleftrightarrow 1 - p \leq \frac{n}{\sum_{i=1}^{m}(n_i/(1-p_i))}.$$

The order $\leq_{\text{lr}}$ can be used to characterize random variables with logconcave densities. The next result lists several such characterizations. It shows that logconcavity can be interpreted as an aging notion in reliability theory by a correct use of the likelihood ratio ordering. This theorem may be compared to Theorem 1.B.38.

Theorem 1.C.52. *The random variable X has a logconcave density (that is, a Polya frequency of order 2 (PF$_2$)) if, and only if, one of the following equivalent conditions holds:*

(i) $[X - t \mid X > t] \geq_{\text{lr}} [X - t' \mid X > t']$ *whenever* $t \leq t'$.
(ii) $X \geq_{\text{lr}} [X - t \mid X > t]$ *for all* $t \geq 0$ *(when X is a nonnegative random variable).*
(iii) $X + t \leq_{\text{lr}} X + t'$ *whenever* $t \leq t'$.

Random variables that satisfy (i) in Theorem 1.C.52 (and hence any of the conditions of that theorem) are said to have the ILR (increasing likelihood ratio) property; see Section 13.D.2 by Righter in [515].

A multivariate extension of parts (i) and (ii) of Theorem 1.C.52 is given in Section 6.E.3.

Another connection between logconcavity and the likelihood ratio order is illustrated in the next result. It is worthwhile to compare the following result with Theorem 6.B.9 in Section 6.B.3.

Theorem 1.C.53. *Let $X_1, X_2, \ldots, X_m$ be independent random variables having logconcave density functions. Then*

$$\left[X_i \middle| \sum_{j=1}^{m} X_j = s\right] \le_{\rm lr} \left[X_i \middle| \sum_{j=1}^{m} X_j = s'\right] \quad \textit{whenever } s \le s',\ i = 1, 2, \ldots, m.$$

Proof. Since the convolution of logconcave density functions is logconcave, it is sufficient to prove the result for $m = 2$ and $i = 1$. Let f_1 and f_2 denote the density functions of X_1 and X_2, respectively. The conditional density of X_1, given $X_1 + X_2 = s$, is

$$f_{X_1|X_1+X_2=s}(x_1) = \frac{f_1(x_1)f_2(s - x_1)}{\int f_1(u)f_2(s - u)\mathrm{d}u}.$$

Thus,

$$\frac{f_{X_1|X_1+X_2=s'}(x_1)}{f_{X_1|X_1+X_2=s}(x_1)} = \frac{f_2(s' - x_1)\int f_1(u)f_2(s - u)\mathrm{d}u}{f_2(s - x_1)\int f_1(u)f_2(s' - u)\mathrm{d}u}. \tag{1.C.16}$$

The logconcavity of f_2 implies that the expression in (1.C.16)) increases in x_1, whenever $s \le s'$. By (1.C.1) the proof is complete. □

Theorems 1.C.52 and 1.C.53 have straightforward discrete analogs, which we do not state here. A few other properties of the order $\le_{\rm lr}$ can be found in Lemma 13.D.1 in Chapter 13 by Righter, and in (14.B.7) in Chapter 14 by Shanthikumar and Yao, in [515].

An interesting closure property of logconcave density functions is described in the following result.

Theorem 1.C.54. *Let $X_1, X_2, \ldots, X_m$ be independent, identically distributed random variables with a common logconcave density function. Then the ith order statistic $X_{(i:m)}$ also has a logconcave density function, $1 \le i \le m$.*

Proof. Let f, F, and $\overline{F}$ denote, respectively, the density, distribution, and survival function of X_1. Then the density function of $X_{(i:m)}$ is given by

$$f_{(i:m)}(x) = m\binom{m-1}{i-1}F^{i-1}(x)f(x)\overline{F}^{m-i}(x).$$

Since the logconcavity of f implies the logconcavity of F and of $\overline{F}$, it follows that $f_{(i:m)}$ is logconcave. □

Misra and van der Meulen [396] showed the preservation of logconcavity and logconvexity from the parent density to the density of the corresponding spacings.

The likelihood ratio order can be used to characterize some aging notions in reliability theory. Recall from (1.A.20) that for a nonnegative random variable X with a finite mean we denote by A_X the corresponding asymptotic equilibrium age. Recall from page 1 the definitions of the IFR and the DFR properties. The following result is immediate. It is of interest to contrast it with Theorems 1.A.31 and 1.B.40

Theorem 1.C.55. *The nonnegative random variable X with finite mean is* IFR [DFR] *if, and only if,* $X \geq_{\rm lr} [\leq_{\rm lr}] A_X$.

An interesting comparison of asymptotic equilibrium ages is described in the next example. Recall from page 1 the definitions of the DMRL property.

Example 1.C.56. Let X and Y be two independent nonnegative DMRL random variables with survival functions $\overline{F}$ and $\overline{G}$, density functions f and g, and asymptotic equilibrium ages A_X and A_Y, respectively. Let $A_{\min\{X,Y\}}$ denote the asymptotic equilibrium age of $\min\{X,Y\}$. Then

$$\min\{A_X, A_Y\} \leq_{\rm lr} A_{\min\{X,Y\}}.$$

In order to see this, assume, for simplicity, that the supports of X and of Y are $(0,\infty)$. Note that the density function of $\min\{A_X, A_Y\}$ is given by

$$f_{\min\{A_X,A_Y\}}(t) = (EXEY)^{-1}\Big(\overline{F}(t)\int_t^\infty \overline{G}(x)\,dx+\overline{G}(t)\int_t^\infty \overline{F}(x)\mathrm{d}x\Big), \quad t \geq 0,$$

and the density function of $A_{\min\{X,Y\}}$ is given by

$$f_{A_{\min\{X,Y\}}}(t) = \big(E[\min\{X,Y\}]\big)^{-1}\overline{F}(t)\overline{G}(t), \quad t \geq 0.$$

Therefore

$$\frac{f_{A_{\min\{X,Y\}}}(t)}{f_{\min\{A_X,A_Y\}}(t)} = \frac{EXEY}{E[\min\{X,Y\}]}\big(m(t)+l(t)\big)^{-1}, \quad t \geq 0,$$

where m and l are the mean residual life functions of X and of Y, given by $m(t) = E[X-t\,|\,X>t]$ and $l(t) = E[Y-t\,|\,Y>t]$, $t \geq 0$. The functions m and l are decreasing by the DMRL assumptions, and therefore $\min\{A_X, A_Y\} \leq_{\rm lr} A_{\min\{X,Y\}}$ by (1.C.1).

In the following example it is shown that if X is increasing in Θ in the likelihood ratio sense, then the posterior distribution of Θ is increasing in X in the same sense.

Example 1.C.57. Let X be a random variable whose distribution function depends on the real parameter Θ. Denote the prior density function of Θ by π, and denote the posterior density function of Θ, given $X = x$, by $\pi^*(\cdot|x)$. Also, denote the conditional density of X, given $\Theta = \theta$ by $f(\cdot|\theta)$, and denote the marginal density of X by g. If X is increasing in Θ in the likelihood ratio sense (that is, if $[X|\Theta=\theta] \leq_{\rm lr} [X|\Theta=\theta']$ whenever $\theta \leq \theta'$), then Θ is increasing in X in the likelihood ratio sense (that is, $[\Theta|X=x] \leq_{\rm lr} [\Theta|X=x']$ whenever $x \leq x'$). The proof of this statement is easy by noting that

$$\pi^*(\theta|x) = \frac{f(x|\theta)\pi(\theta)}{g(x)}.$$

An extension of Example 1.C.57 to the multivariate likelihood ratio order is given in Example 6.E.16.

Example 1.C.58. Let X be a random variable whose distribution function depends on the random parameter Θ_1 or, in other circumstances, on the random parameter Θ_2. Denote the prior density functions, of Θ_1 and Θ_2, by π_1 and π_2, respectively, and denote the posterior density functions of Θ_1 and Θ_2, given $X = x$, by $\pi_1^*(\cdot|x)$ and $\pi_2^*(\cdot|x)$, respectively. Also, denote the conditional density of X, given $\Theta_1 = \theta$ or $\Theta_2 = \theta$, by $f(\cdot|\theta)$, and denote the marginal density of X by g_1 or by g_2, according to whether X depends on Θ_1 or on Θ_2. Then, for any x, we have that

$$\Theta_1 \leq_{\rm lr} \Theta_2 \Longrightarrow [\Theta_1|X = x] \leq_{\rm lr} [\Theta_2|X = x].$$

The proof of this statement is easy by noting that

$$\pi_i^*(\theta|x) = \frac{f(x|\theta)\pi_i(\theta)}{g_i(x)}, \quad i = 1, 2.$$

Example 1.C.59. Recall from Example 1.B.23 that for a nonnegative random variable X with density function f, and for a nonnegative function w such that $E[w(X)]$ exists, we denote by X^w the random variable with the weighted density function f_w given by

$$f_w(x) = \frac{w(x)f(x)}{E[w(X)]}, \quad x \geq 0. \tag{1.C.17}$$

Similarly, for another nonnegative random variable Y with density function g, such that $E[w(Y)]$ exists, we denote by Y^w the random variable with the density function g_w given by

$$g_w(x) = \frac{w(x)g(x)}{E[w(Y)]}, \quad x \geq 0. \tag{1.C.18}$$

It is then obvious that $X \leq_{\rm lr} Y \Longrightarrow X^w \leq_{\rm lr} Y^w$.

Example 1.C.60. Let X be a nonnegative random variable with density function f, and for a nonnegative function w such that $E[w(X)]$ exists, let X^w be the random variable with the weighted density function f_w given in (1.C.17). It is then obvious that if w is increasing [decreasing], then $X \leq_{\rm lr} [\geq_{\rm lr}] X^w$. In particular, the inequality $X \leq_{\rm lr} X^w$ holds when X^w is the length-biased version of X; that is, when $w(x) = x$, $x \geq 0$.

Example 1.C.61. Let the random variable X have a generalized skew normal distribution with parameters n and λ, that is, suppose that its density function is given by

$$f(x; n, \lambda) = \frac{\Phi^n(\lambda x)\phi(x)}{C(n, \lambda)}, \quad x \in \mathbb{R},$$

where ϕ and Φ are, respectively, the density and distribution functions of a standard normal random variable, and $C(n,\lambda)$ is given by

$$C(n,\lambda) = \int_{-\infty}^{\infty} \Phi^n(\lambda x)\phi(x)\mathrm{d}x.$$

Let Y have a generalized skew normal distribution with parameters n_1 and λ. It is easy to see that if $\lambda > [<]\ 0$ and $n \le n_1$, then $X \le_{\mathrm{lr}} [\ge_{\mathrm{lr}}]\ Y$.

1.C.4 Shifted likelihood ratio orders

In this subsection we consider only random variables with absolutely continuous distribution functions and interval supports (although it is possible to state and prove analogs of many of the results here also for discrete random variables). So let X and Y be such random variables. Let l_X and u_X be the left and the right endpoints of the support of X. Similarly define l_Y and u_Y. The values l_X, u_X, l_Y, and u_Y may be infinite. Let f and g denote the density functions of X and Y, respectively. Suppose that

$$X - x \le_{\mathrm{lr}} Y \quad \text{for each } x \ge 0. \tag{1.C.19}$$

Then X is said to be *smaller than Y in the up shifted likelihood ratio order* (denoted as $X \le_{\mathrm{lr}\uparrow} Y$). Rewriting (1.C.19) using (1.C.1) it is seen that $X \le_{\mathrm{lr}\uparrow} Y$ if, and only if, for each $x \ge 0$ we have

$$\frac{g(t)}{f(t+x)} \quad \text{is increasing in } t \in (l_X - x, u_X - x) \cup (l_Y, u_Y). \tag{1.C.20}$$

It is readily apparent that

$$X \le_{\mathrm{lr}\uparrow} Y \Longrightarrow X \le_{\mathrm{lr}} Y.$$

The up shifted likelihood ratio order satisfies some closure properties given in the next theorem.

Theorem 1.C.62. (a) *Let $X_1, X_2, \ldots, X_m$ be a set of independent random variables and let $Y_1, Y_2, \ldots, Y_m$ be another set of independent random variables. If $X_i \le_{\mathrm{lr}\uparrow} Y_i$ for $i = 1, 2, \ldots, m$, then*

$$\sum_{j=1}^{m} X_j \le_{\mathrm{lr}\uparrow} \sum_{j=1}^{m} Y_j.$$

That is, the up likelihood ratio order is closed under convolutions.

(b) *Let $\{X_j,\ j = 1, 2, \ldots\}$ and $\{Y_j,\ j = 1, 2, \ldots\}$ be two sequences of random variables such that $X_j \to_{\mathrm{st}} X$ and $Y_j \to_{\mathrm{st}} Y$ as $j \to \infty$. If $X_j \le_{\mathrm{lr}\uparrow} Y_j$, $j = 1, 2, \ldots$, then $X \le_{\mathrm{lr}\uparrow} Y$.*

Shanthikumar and Yao [530] proved Theorem 1.C.62(a) by establishing a stochastic monotonicity property of birth and death processes. Hu and Zhu [242] provided a straightforward analytic proof of this result. This result is generalized in Hu, Nanda, Xie, and Zhu [237].

From Theorem 1.C.15 we obtain the following result.

Theorem 1.C.63. *Let X, Y, and Θ be random variables such that $[X|\Theta = \theta] \le_{\text{lr}\uparrow} [Y|\Theta = \theta']$ for all θ and θ' in the support of Θ. Then $X \le_{\text{lr}\uparrow} Y$.*

Some further properties of the up shifted likelihood ratio order are listed in the following theorems.

Theorem 1.C.64. *Let X and Y be two absolutely continuous random variables with interval supports. If X or Y or both have logconcave densities, and if $X \le_{\text{lr}} Y$, then $X \le_{\text{lr}\uparrow} Y$.*

Theorem 1.C.65. *Let X and Y be two absolutely continuous random variables with differentiable densities on the respective interval supports. Then $X \le_{\text{lr}\uparrow} Y$ if, and only if, there exists a random variable Z with a logconcave density such that $X \le_{\text{lr}} Z \le_{\text{lr}} Y$.*

Theorem 1.C.66. *Let X be an absolutely continuous random variable with an interval support. Then $X \le_{\text{lr}\uparrow} X$ if, and only if, f is logconcave on $(-\infty, \infty)$.*

Example 1.C.67. Let X be a random variable with a density function h. For each $\theta \in (-\infty, \infty)$, let X_θ be a random variable with density function f_θ defined by

$$f_\theta(x) = h(x - \theta), \quad x \in (-\infty, \infty).$$

Then it is easy to see that $X_{\theta_1} \le_{\text{lr}\uparrow} X_{\theta_2}$ whenever $\theta_1 \le \theta_2$ if, and only if, $X \le_{\text{lr}\uparrow} X$; that is, by Theorem 1.C.66, if, and only if, h is logconcave on $(-\infty, \infty)$.

A preservation result of the order $\le_{\text{lr}\uparrow}$ is described next.

Theorem 1.C.68. *Let X and Y be two absolutely continuous random variables with interval supports. If $X \le_{\text{lr}\uparrow} Y$ and if the density function of X is increasing [respectively, decreasing] on (l_Y, u_X), then $\phi(X) \le_{\text{lr}\uparrow} \phi(Y)$ for any strictly increasing twice differentiable convex [respectively, concave] function ϕ (with first and second derivatives ϕ' and ϕ'') such that $\phi''(x)/(\phi'(x))^2$ is increasing.*

A characterization of the relation $X \le_{\text{lr}\uparrow} Y$ for nonnegative random variables is given next.

Theorem 1.C.69. *Let X and Y be two nonnegative absolutely continuous random variables with interval supports; that is, assume that $l_X \ge 0$ and $l_Y \ge 0$. Then $X \le_{\text{lr}\uparrow} Y$ if, and only if,*

$$[X - x|X > x] \le_{\text{lr}} Y \quad \textit{for all } x \in (l_X, u_X).$$

Another shifted likelihood ratio stochastic order is defined next. Let X and Y be two absolutely continuous random variables with support $[0,\infty)$. Suppose that

$$X \leq_{\mathrm{lr}} [Y - x \big| Y > x] \quad \text{for all } x \geq 0.$$

Then X is said to be *smaller than* Y *in the down shifted likelihood ratio order* (denoted as $X \leq_{\mathrm{lr}\downarrow} Y$).

Note that in the above definition only nonnegative random variables are compared. This is because for the down shifted likelihood ratio order it is not possible to take an analog of (1.C.19), such as $X \leq_{\mathrm{lr}} Y - x$, as a definition. The reason is that here, by taking x very large, it is seen that practically no random variables would satisfy such an order relation. Note that in the definition above, the right-hand side $[Y - x \big| Y > x]$ can take on (as x varies) any value in the right neighborhood of 0. Therefore the support of the compared random variables is restricted here to be $[0,\infty)$.

Let f and g denote the density functions of X and Y, respectively. An analog of (1.C.20) is the following:

$$X \leq_{\mathrm{lr}\downarrow} Y \Longleftrightarrow \frac{g(t+x)}{f(t)} \text{ is increasing in } t \geq 0 \text{ for all } x \geq 0. \tag{1.C.21}$$

(A discrete version of the down shifted likelihood ratio order is defined and used in Section 6.B.3.)

It is readily apparent that for nonnegative random variables with support $[0,\infty)$ we have

$$X \leq_{\mathrm{lr}\downarrow} Y \Longrightarrow X \leq_{\mathrm{lr}} Y.$$

We describe now some further properties of the down shifted likelihood ratio order.

Theorem 1.C.70. *Let $\{X_j,\ j = 1,2,\dots\}$ and $\{Y_j,\ j = 1,2,\dots\}$ be two sequences of random variables, with support $[0,\infty)$, such that $X_j \to_{\mathrm{st}} X$ and $Y_j \to_{\mathrm{st}} Y$ as $j \to \infty$. If $X_j \leq_{\mathrm{lr}\downarrow} Y_j$, $j = 1,2,\dots$, then $X \leq_{\mathrm{lr}\downarrow} Y$.*

The following result is an analog of Theorem 1.C.63, however, it does not follow at once from Theorem 1.C.15. Its proof can be found in Lillo, Nanda, and Shaked [361].

Theorem 1.C.71. *Let X, Y, and Θ be random variables such that $[X \big| \Theta = \theta]$ and $[Y \big| \Theta = \theta]$ are absolutely continuous and have the support $[0,\infty)$ for all θ in the support of Θ. If $[X \big| \Theta = \theta] \leq_{\mathrm{lr}\downarrow} [Y \big| \Theta = \theta']$ for all θ and θ' in the support of Θ, then $X \leq_{\mathrm{lr}\downarrow} Y$.*

More properties are listed next.

Theorem 1.C.72. *Let X and Y be two absolutely continuous random variables with support $[0,\infty)$. If X or Y or both have logconvex densities on $[0,\infty)$, and if $X \leq_{\mathrm{lr}} Y$, then $X \leq_{\mathrm{lr}\downarrow} Y$.*

Theorem 1.C.73. *Let X and Y be two absolutely continuous random variables with differentiable densities on their support $[0,\infty)$. Then $X \le_{\rm lr\downarrow} Y$ if, and only if, there exists a random variable Z with a logconvex density on $[0,\infty)$ such that $X \le_{\rm lr} Z \le_{\rm lr} Y$.*

Theorem 1.C.74. *Let X be an absolutely continuous random variable with support $[0,\infty)$. Then $X \le_{\rm lr\downarrow} X$ if, and only if, f is logconvex on $[0,\infty)$.*

Theorem 1.C.75. *Let X and Y be two absolutely continuous random variables with support $[0,\infty)$. If $X \le_{\rm lr\downarrow} Y$ and if Y has a decreasing density function on $[0,\infty)$, then $\phi(X) \le_{\rm lr\downarrow} \phi(Y)$ for any strictly increasing twice differentiable convex function $\phi : [0,\infty) \to [0,\infty)$ (with first and second derivatives ϕ' and ϕ'') such that $\phi''(x)/(\phi'(x))^2$ is decreasing.*

Example 1.C.76. An interesting family of distribution functions, with associated random variables that are ordered in the down shifted likelihood ratio order, is the Pareto family. Explicitly, for $\theta \in (0,\infty)$, let X_θ be a random variable with density function f_θ defined by

$$f_\theta(x) = \theta/(1+x)^{\theta+1}, \quad x \ge 0.$$

Then, by verifying (1.C.21), it is easy to see that $X_{\theta_1} \le_{\rm lr\downarrow} X_{\theta_2}$ whenever $\theta_1 \ge \theta_2 > 0$.

Some results that compare order statistics in the shifted likelihood ratio orders are described next. Again, $X_{(j:m)}$ denotes the jth order statistic associated with the random variables $X_1, X_2, \ldots, X_m$, and $Y_{(i:n)}$ denotes the ith order statistic associated with the random variables $Y_1, Y_2, \ldots, Y_n$. An analog of Theorem 1.C.33 for the order $\le_{\rm lr\uparrow}$ is the following result. Note that in the following theorem the assumption is stronger than the assumption in Theorem 1.C.33, but so is the conclusion.

Theorem 1.C.77. *Let $X_1, X_2, \ldots, X_m$ be m independent random variables, and let $Y_1, Y_2, \ldots, Y_n$ be other n independent random variables, all having absolutely continuous distributions. If $X_j \le_{\rm lr\uparrow} Y_i$ for all $1 \le j \le m$ and $1 \le i \le n$, then*

$$X_{(j:m)} \le_{\rm lr\uparrow} Y_{(i:n)} \quad \text{whenever } j \le i \text{ and } m-j \ge n-i.$$

Proof. Fix an $x \ge 0$ and denote by $(X-x)_{(j:m)}$ the jth order statistic among the random variables $X_1 - x, X_2 - x, \ldots, X_m - x$. By assumption we have $X_j - x \le_{\rm lr\uparrow} Y_i$ for all $1 \le j \le m$ and $1 \le i \le n$. Therefore from Theorem 1.C.33 we get $(X-x)_{(j:m)} \le_{\rm lr} Y_{(i:n)}$ whenever $j \le i$ and $m-j \ge n-i$. The stated result follows from the fact that $(X-x)_{(j:m)} = X_{(j:m)} - x$. □

For the down shifted likelihood ratio order, the method of proof used in the proof of Theorem 1.C.33 only yields comparisons of minima as described in the following result.

Theorem 1.C.78. *Let $X_1, X_2, \dots, X_m$ be m independent random variables, and let $Y_1, Y_2, \dots, Y_n$ be other n independent random variables, all having absolutely continuous distributions with support $[0, \infty)$. If $X_j \le_{\mathrm{lr}\downarrow} Y_i$ for all $1 \le j \le m$ and $1 \le i \le n$, then*

$$X_{(1:m)} \le_{\mathrm{lr}\downarrow} Y_{(1:n)} \quad \textit{whenever } m \ge n.$$

Now let $X_1, X_2, \dots$ be independent and identically distributed random variables. Taking $Y_i =_{\mathrm{st}} X_j$ for all i and j in Theorems 1.C.77 and 1.C.78, and using Theorems 1.C.66 and 1.C.74, we obtain the following analogs of Theorem 1.C.37. Note that in the next theorem (unlike in Theorem 1.C.37) we assume logconcavity or logconvexity of the underlying density function, but the conclusion in part (a) of the next theorem is stronger than the conclusion in Theorem 1.C.37.

Theorem 1.C.79. (a) *Let $X_1, X_2, \dots$ be independent and identically distributed absolutely continuous random variables with an interval support. If the common density function is logconcave, then*

$$X_{(j:m)} \le_{\mathrm{lr}\uparrow} X_{(i:n)} \quad \textit{whenever } j \le i \textit{ and } m - j \ge n - i.$$

(b) *Let $X_1, X_2, \dots$ be independent and identically distributed absolutely continuous random variables with support $[0, \infty)$. If the common density function is logconvex on $[0, \infty)$, then*

$$X_{(1:m)} \le_{\mathrm{lr}\downarrow} X_{(1:n)} \quad \textit{whenever } m \ge n.$$

1.D The Convolution Order

Let X and Y be two random variables such that

$$Y =_{\mathrm{st}} X + U, \tag{1.D.1}$$

where U is a nonnegative random variable, independent of X. Then X is said to be *smaller than Y in the convolution order* (denoted as $X \le_{\mathrm{conv}} Y$). Obviously, the convolution order is a partial order. It is equivalent to the information order which is defined for statistical experiments when the underlying parameter is a location parameter.

The convolution order is obviously closed under increasing linear transformations. That is, for any $a \in \mathbb{R}$ and $b \ge 0$ we have

$$X \le_{\mathrm{conv}} Y \Longrightarrow a + bX \le_{\mathrm{conv}} a + bY.$$

The convolution order is obviously also closed under convolutions. That is, let $X_1, X_2, \dots, X_n$ be a set of independent random variables, and let $Y_1, Y_2, \dots, Y_n$ be another set of independent random variables. Then

$$\left(X_j \le_{\text{conv}} Y_j,\ j = 1, 2, \ldots, n\right) \Longrightarrow \sum_{i=1}^{n} X_i \le_{\text{conv}} \sum_{i=1}^{n} Y_i.$$

It is obvious from Theorem 1.A.2 and (1.D.1) that

$$X \le_{\text{conv}} Y \Longrightarrow X \le_{\text{st}} Y.$$

For any nonnegative random variable X we denote by L_X its classical Laplace transform, that is,

$$L_X(s) = E[\mathrm{e}^{-sX}], \quad s \ge 0.$$

Recall that a nonnegative function ϕ is a Laplace transform of a nonnegative measure on $(0, \infty)$ if, and only if, ϕ is completely monotone, that is, all the derivatives $\phi^{(n)}$ of ϕ exist, and they satisfy $(-1)^n \phi^{(n)}(x) \ge 0$ for all $x \ge 0$ and $n = 1, 2, \ldots$. It follows that for nonnegative random variables we have

$$X \le_{\text{conv}} Y \Longleftrightarrow \frac{L_Y(s)}{L_X(s)} \text{ is a completely monotone function in } s \ge 0. \quad (1.D.2)$$

Example 1.D.1. Let X_i be an exponential random variable with mean $1/\lambda_i$, $i = 1, 2$. If $\lambda_1 > \lambda_2$, then $X_1 \le_{\text{conv}} X_2$. To see this, note that the ratio of the Laplace transforms of X_2 and X_1 at s is equal to $(\lambda_2/\lambda_1)((s+\lambda_1)/(s+\lambda_2))$, and it is easy to verify that this ratio is completely monotone. The result thus follows from (1.D.2).

Example 1.D.2. Let $X_1, X_2, \ldots, X_n$ be independent and identically distributed exponential random variables with mean $1/\lambda$ for some $\lambda > 0$. Denote the corresponding order statistics by $X_{(1)} \le X_{(2)} \le \cdots \le X_{(n)}$. Then

$$X_{(i)} \le_{\text{conv}} X_{(j)} \quad \text{whenever } 1 \le i < j \le n.$$

To see this, note that

$$X_{(k+1)} =_{\text{st}} X_{(k)} + Z_k,$$

where Z_k is an exponential random variable with mean $((n-k)\lambda)^{-1}$, $k = 1, 2, \ldots, n-1$, and use the transitivity property of the order $\le_{\text{conv}}$.

1.E Complements

Section 1.A: The usual stochastic order is being used in many areas of applications, but there is no single source where many of the basic results can all be found. Some standard references are the books of Lehmann [342], Marshall and Olkin [383], Ross [475], and Müller and Stoyan [419], where most of the results described in Section 1.A can be found. For example, Theorem 1.A.2 can be found in Marshall and Olkin [383]. The characterization of the usual stochastic order by the monotonicity described in

(1.A.8) is taken from Müller [407], whereas the characterization given in (1.A.12) can be found in Fellman [193]. The comparison of the random sums in Theorem 1.A.5 is motivated by ideas in Pellerey and Shaked [455]; it was communicated to us by Pellerey [444]. The application of the order $\leq_{\rm st}$ in Bayesian imperfect repair (Example 1.A.7) is taken from Lim, Lu, and Park [364]. The result which gives conditions for stochastic equality (Theorem 1.A.8) can be found in Baccelli and Makowski [27] and in Scarsini and Shaked [494]. Lemma 2.1 of Costantini and Pasqualucci [135] with $n = 1$ is an interesting variation of Theorem 1.A.8. The bivariate characterizations in Theorems 1.A.9 and 1.A.10 are taken from Shanthikumar and Yao [532] and from Righter and Shanthikumar [466], respectively. The characterization of the order $\leq_{\rm st}$ by means of the Fortret-Mourier-Wasserstein distance (Theorem 1.A.11) is taken from Adell and de la Cal [3]. The Laplace transform characterization of the order $\leq_{\rm st}$ (Theorem 1.A.13) can be found in Kebir [281] and in Kan and Yi [274]. An extension of Theorem 1.A.13 to more general orders can be found in Nanda [422]. The closure of the order $\leq_{\rm st}$ under a stochastically increasing family of random variables (Theorem 1.A.14) is taken from Shaked and Wong [524]. The condition for the usual stochastic order, given in Theorem 1.A.17, has been communicated to us by Gerchak and He [210]. The comparison of truncated maximum with truncations maximum (Example 1.A.16) can be found in Pellerey and Petakos [453]. The lattice property of the order $\leq_{\rm st}$ (Remark 1.A.18) is given in Müller and Scarsini [418]. The four results that give the stochastic orderings of the spacings, Theorems 1.A.19–1.A.22, can be found in Barlow and Proschan [35], Ebrahimi and Spizzichino [178], Pledger and Proschan [458], and Joag-Dev [258], respectively. The stochastic comparison of order statistics of independent random variables with the order statistics of independent and identically distributed random variables (Theorem 1.A.23) is taken from Ma [371]; it generalizes some previous results in the literature. The stochastic comparison of a sum of independent heterogeneous exponential random variables with a proper Erlang random variable (Example 1.A.24) is taken from Bon and Păltănea [105], where more refined comparisons can also be found. The stochastic comparison of a sum of independent heterogeneous binomial random variables with a proper binomial random variable (Example 1.A.25) is taken from Boland, Singh, and Cukic [102]. The necessary and sufficient conditions for the comparison of normal random variables (Example 1.A.26) are taken from Müller [413]; an extension of this result to Kotz-type distributions is given in Ding and Zhang [168]. The stochastic comparisons of norms, in Examples 1.A.27 and 1.A.28, are taken from Lapidoth and Moser [333]. The TTT transform (1.A.19) is introduced in Barlow, Bartholomew, Bremner, and Brunk [32], and is further studied in Barlow and Doksum [34] and in Barlow and Campo [33]. The observed total time on test random variable $X_{\rm ttt}$ is defined and studied in Li and Shaked [356], where the implication in Theorem 1.A.29 can be found. The characterizations of the NBUE and the NWUE aging notions by means of the usual stochastic order (Theorem 1.A.31) can be found in Whitt [565] and in Fagiuoli and Pellerey [187]. The other characterization, by means of the random variable X_{ttt} (Theorem 1.A.32), is taken from Li and Shaked [356]. The aging notion that is described in (1.A.21) is studied in Mugdadi and Ahmad [402].

Boland, Singh, and Cukic [103] studied an order, called the *stochastic precedence order*, according to which the random variable X is smaller than the random variable Y if $P\{X < Y\} \geq P\{Y < X\}$. If X and Y are independent, then $X \leq_{\text{st}} Y$ implies that X is smaller than Y in the stochastic precedence order.

Section 1.B: Many of the basic results regarding the hazard rate order can be found in Ross [475] and in Müller and Stoyan [419]. The characterization (1.B.8) can be found in Lehmann and Rojo [345]. The results regarding the preservation of the orders $\leq_{\text{hr}}$ and $\leq_{\text{rh}}$ under monotone increasing transformations (Theorems 1.B.2 and 1.B.43) can be found in Keilson and Sumita [283]. The closure under convolutions result (Theorem 1.B.4) and the bivariate characterization result (Theorem 1.B.9) are taken from Kijima [291] and Shanthikumar and Yao [532]. A special case of Lemma 1.B.3 can be found in Mukherjee and Chatterjee [403]. The hazard rate order comparison of a sum of independent heterogeneous exponential random variables with a proper Erlang random variable (Example 1.B.5) is taken from Bon and Păltănea [105], where more refined comparisons can also be found. The hazard rate order comparison of a sum of independent heterogeneous binomial random variables with a proper binomial random variable (Example 1.B.6) is taken from Boland, Singh, and Cukic [102]. The hazard rate order comparison of random sums (Theorem 1.B.7) can be found in Pellerey [445]; some related results are Theorem 7.2 of Kijima [291] and Proposition 2.2 of Kebir [282]. The closure under mixtures result (Theorem 1.B.8) can be found in Boland, El-Neweihi, and Proschan [97]; a generalization of it is contained in Nanda, Jain, and Singh [424]. The bivariate characterizations in Theorems 1.B.10 and 1.B.11 are taken from Righter and Shanthikumar [466] and from Cheng and Righter [128], respectively. The characterizations given in Theorem 1.B.12 can be found in Capéraà [118] and in Joag-Dev, Kochar, and Proschan [259]. The hazard rate ordering result regarding the inter-epoch times of a nonhomogeneous Poisson process (Example 1.B.13) is taken from Kochar [309] where other applications of Theorem 1.B.12 can also be found. The hazard rate ordering of the epoch times of a nonhomogeneous Poisson process (1.B.19) can be found in Baxter [62]. The closure property of the order $\leq_{\text{hr}}$ under hazard rate ordered mixtures (Theorem 1.B.14) is taken from Shaked and Wong [524]; a related result is Proposition 4.1 of Kebir [282]. The preservation of the order $\leq_{\text{hr}}$ under the formation of a proper Markov chain (Example 1.B.15) can essentially be found in Ross, Shanthikumar, and Zhu [478]; they gave a version of this preservation result for the order $\leq_{\rm rh}$. The application of the order $\leq_{\rm hr}$ in Bayesian imperfect repair (Example 1.B.16) is inspired by Lim, Lu, and Park [364], but the result given here is stronger than their Theorem 4.1(iii). The hazard rate order comparison of a proportional hazard mixture with its parent distribution (Example 1.B.17) is taken from Gupta and Gupta [214]. The Laplace transform characterization of the order $\leq_{\rm hr}$ (Theorem 1.B.18) can be found in Kebir [281] and in Kan and Yi [274]. An extension of Theorem 1.B.18 to more general orders can be found in Nanda [422]. The result about the inheritance of the order $\leq_{\rm hr}$, from the mixing scales to the underlying counting processes (Theorem 1.B.19), is essentially taken from Ma [374]. The closure property which is given in Theorem 1.B.21 can be found in Kochar [305]; the necessary and sufficient condition, given after Theorem 1.B.21, is taken from Ma [374]. The result involving the hazard rate comparison of weighted random variables (Example 1.B.23) is taken from Nanda and Jain [423]; see also Bartoszewicz and Skolimowska [51]. The hazard rate comparison of epoch times of nonhomogeneous Poisson processes in Example 1.B.24 can be found in Ahmadi and Arghami [6] and in Belzunce, Lillo, Ruiz, and Shaked [69]; in the latter paper the result is extended to nonhomogeneous pure birth processes. The hazard rate order comparison of inter-epoch times of nonhomogeneous Poisson processes in Example 1.B.24 is taken from Belzunce, Lillo, Ruiz, and Shaked [69], who also obtained a similar result for the more general nonhomogeneous pure birth processes. The hazard rate order comparison of series systems of parallel systems (Example 1.B.25) can be found in Valdés and Zequeira [553]. The proof of Theorem 1.B.26 (given in Remark 1.C.40) is taken from Boland, Shaked, and Shanthikumar [101]. The hazard rate order comparisons of order statistics described in Theorems 1.B.27 and 1.B.28 can be found in Korwar [321]. The conditions that lead to the hazard rate ordering of minima (Theorem 1.B.29 and Corollary 1.B.30) are taken from Navarro and Shaked [430]. The two results that give the hazard rate orderings of the spacings (Theorem 1.B.31) can be found in Kochar and Kirmani [313] and in Khaledi and Kochar [285], whereas the comparison of spacings from two different samples (Theorem 1.B.32) is taken from Khaledi and Kochar [285]; further results can be found in Hu and Wei [240] and in Misra and van der Meulen [396]. The closure property under formations of order statistics (Theorem 1.B.34) is taken from Singh and Vijayasree [537]; see also Lynch, Mimmack, and Proschan [369]. Boland, El-Neweihi, and Proschan [97] show, by a counterexample, that the conclusion of Theorem 1.B.34 need not hold when the X_i's or the Y_i's are not identically distributed. Extensions of Theorem 1.B.34 can be found in Shaked and Shanthikumar [516], in Belzunce, Mercader, and Ruiz [70], and in Hu and Zhuang [247]. The general comparison result, given in Theorem 1.B.36, is taken from Boland, Hu, Shaked, and Shanthikumar [99]; see related results in Franco, Ruiz, and Ruiz [205] and in Hu and Zhuang [247]. The hazard rate order comparisons of maxima of heterogeneous exponential random variables (Example 1.B.37) are taken from Dykstra, Kochar, and Rojo [174] and from Khaledi and Kochar [287]. The closure under convolution property of IFR random variables (Corollary 1.B.39) can be found, for example, in Barlow and Proschan [36, page 100]). The characterizations of the DMRL and the IMRL aging notions by means of the hazard rate order (Theorem 1.B.40) can be found in Brown [111, page 229], in Whitt [565], and in Fagiuoli and Pellerey [187]. The observation that essentially reduces the study of the reversed hazard rate order into the study of the hazard rate order (Theorem 1.B.41) is taken from Nanda and Shaked [428]. The bivariate characterization results for the reversed hazard order (Theorems 1.B.47 and 1.B.49) can be found in Shanthikumar, Yamazaki, and Sakasegawa [529] and in Cheng and Righter [128]. The application of the reversed hazard order in economics, described in Example 1.B.51, is taken from Eeckhoudt and Gollier [180]; further results in this vein can be found in Kijima and Ohnishi [293]. The closure property of the order $\leq_{\text{rh}}$ under reversed hazard rate ordered mixtures (Theorem 1.B.52) is taken from Shaked and Wong [524]; a related result is Proposition 4.1 of Kebir [282]. The Laplace transform characterization of the order $\leq_{\text{rh}}$ (Theorem 1.B.53) is taken from Kebir [281]. The result about the inheritance of the order $\leq_{\text{rh}}$, from the mixing scales to the underlying counting processes (Theorem 1.B.54), is essentially taken from Ma [374]. The results about the reversed hazard rate ordering of order statistics (Theorems 1.B.56 and 1.B.57), and the characterizations of the reversed hazard rate order given in Theorem 1.B.62, can be found in Block, Savits, and Singh [96], whereas the result described in Theorem 1.B.58 is taken from Hu and He [232]. The preservation of the order statistics in the sense of the order $\leq_{\text{rh}}$ (Theorem 1.B.60) can be found in Nanda, Jain, and Singh [426].

An order among nonnegative random variables, which is defined by stipulating the monotonicity of the ratio of the hazard rate functions (when they exist), is studied in Kalashnikov and Rachev [271], Sengupta and Deshpande [500], and Rowell and Siegrist [479]. Equivalently, if $\overline{F}$ and $\overline{G}$ are survival functions, and we denote $R_F = -\log \overline{F}$ and $R_G = -\log \overline{G}$, then the order mentioned above can be defined by requiring that the composition $R_F \circ R_G^{-1}$ be convex on $[0, \infty)$. The notion of the monotonicity of the ratio of hazard rate functions is used in Examples 1.B.24 (see (1.B.25)) and 1.B.25, as well as in Theorem 1.C.4. Sengupta and Deshpande [500] and Rowell and Siegrist [479] also studied the orders defined by stipulating that $R_F \circ R_G^{-1}$ be starshaped or superadditive.

Brown and Shanthikumar [112], Lillo, Nanda, and Shaked [361], Hu and Zhu [242], Di Crescenzo and Longobardi [165], and Belzunce, Ruiz, and Ruiz [74] have introduced and studied various shifted hazard and reversed hazard rate orders. Similar orders which extend the likelihood ratio order are studied in Section 1.C.4.

Section 1.C: Again, many of the basic results regarding the likelihood ratio order can be found in Ross [475] and in Müller and Stoyan [419]. Condition (1.C.3) is implicit in Block, Savits, and Shaked [95], and it is explicit in Müller [408]. The relation (1.C.4) is mentioned in Chan, Proschan, and Sethuraman [123]. The sufficient conditions for $X \leq_{\rm lr} Y$, given in Theorem 1.C.4, have been noted in Belzunce, Lillo, Ruiz, and Shaked [69]. The closure property of the likelihood ratio order under conditioning (Theorem 1.C.5) is observed in Whitt [561]. Many variations of Theorem 1.C.5 with respect to general sample spaces can be found in Whitt [561] and in Rüschendorf [485]. The closure under limits property of the order $\leq_{\rm lr}$ (Theorem 1.C.7) is taken from Müller [408]. The result regarding the preservation of the order $\leq_{\rm lr}$ under monotone increasing transformations (Theorem 1.C.8) can be found in Keilson and Sumita [283]. The several closure under convolution results (Theorems 1.C.9, 1.C.11, and 1.C.12) as well as the bivariate characterization result (Theorem 1.C.20) are taken from Shanthikumar and Yao [532]; a related result is Proposition 2.4 of Kebir [282]. A special case of Theorem 1.C.9 can be found in Mukherjee and Chatterjee [403]. The result about the number of successes in independent trials (Example 1.C.10) is statement (7) in Samuels [488], who attributed it to Ghurye and Wallace. The characterization of the order $\leq_{\rm hr}$ by means of the order $\leq_{\rm lr}$, given in Theorem 1.C.14, is taken from Di Crescenzo [164]; a density of the form (1.C.7) can be found in Adell and Lekuona [4, page 773]. The likelihood ratio order comparison of a random random variable with a fixed random variable (Corollary 1.C.16) is a slight generalization of Problem B in Szekli [544, page 22]. The closure property of the order $\leq_{\rm lr}$ under likelihood ratio ordered mixtures (Theorem 1.C.17) is an extension of a result in Kebir [282]. The result about the inheritance of the order $\leq_{\rm lr}$, from the mixing scales to the underlying counting processes (Theorem 1.C.18), is taken from Ma [374]. Example 1.C.19 is inspired by Theorem 4.12 of Asadi and Shanbhag [23], but Example 1.C.19 has weaker assumptions (Θ_1 and Θ_2 need not be degenerate) and stronger conclusions (Y_1 and Y_2 are ordered in the likelihood ratio order, rather than in the hazard rate order) than the result of Asadi and Shanbhag [23]. The result in Theorem 1.C.21 is a special case of a result in Ross [475]. The bivariate characterizations in Theorems 1.C.22, 1.C.23, and 1.C.24 are taken from Righter and Shanthikumar [466] and from Chapter 13 by Righter in [515]. The Laplace transform characterization of the order $\leq_{\rm lr}$ (Theorem 1.C.25) can be found in Kebir [281]. An extension of Theorem 1.C.25 to more general orders can be found in Nanda [422]. The conditional likelihood ratio orderings, described in Theorem 1.C.26, can be found in Ku and Niu [324] and in Chapter 14 by Shanthikumar and Yao in [515]. The setting in which the order $\leq_{\rm hr}$ gives rise to the order $\leq_{\rm lr}$, as described in Theorem 1.C.28, is essentially taken from Ross, Shanthikumar, and Zhu [478]; they gave a version of this result for the order $\leq_{\rm rh}$. The necessary and sufficient condition for $aX \leq_{\rm lr} X$ (Theorem 1.C.29) can be found in

Hu, Nanda, Xie, and Zhu [237]. The likelihood ratio order comparisons of the order statistics given in Theorem 1.C.31 are taken from Bapat and Kochar [31] and from Hu, Zhu, and Wei [243]; an extension of the first part of Theorem 1.C.31(a) can be found in Ma [373]. The result about the likelihood ratio order comparison of order statistics of a simple random sample from a finite population (Theorem 1.C.32) can be found in Kochar and Korwar [315]. The general result which compares order statistics from two samples of different size (Theorem 1.C.33) is taken from Lillo, Nanda, and Shaked [362]; see related results in Franco, Ruiz, and Ruiz [205] and in Hu and Zhuang [247]. Belzunce and Shaked [78] extended Theorem 1.C.33 to comparison of lifetimes of coherent systems in reliability theory; see also Belzunce, Franco, Ruiz, and Ruiz [66]. The closure property under formation of order statistics (Corollary 1.C.34) can be found in Chan, Proschan, and Sethuraman [123]; a special case of this result can be found in Singh and Vijayasree [537]. The likelihood ratio order comparison of the order statistics given in Theorem 1.C.37 is taken from Raqab and Amin [465]. Theorem 2.6 in Kamps [273, page 182] extends Theorem 1.C.37 to the so called generalized order statistics; see also Korwar [322] and Hu and Zhuang [247]. The special case of Theorem 1.C.37 when $j = i$, is extended in Nanda, Misra, Paul, and Singh [427] to the case when the sample sizes m and n are random. Nanda, Misra, Paul, and Singh [427] also extend the special case of Theorem 1.C.37 when $m = n$, to the case when the common sample size is random. The likelihood ratio order comparison of normalized spacings (Theorem 1.C.42) can be found in Kochar and Korwar [314], whereas the comparisons for nonnormalized spacings (Theorem 1.C.43) are special cases of results in Misra and van der Meulen [396] and in Hu and Zhuang [246, 248]. The comparison of spacings that correspond to random variables with logconcave density (Theorem 1.C.44) is a special case of a result of Hu and Zhuang [246, 248]. The comparison of spacings from two different samples (Theorem 1.C.45) is taken from Khaledi and Kochar [285]; an extension of this result can be found in Franco, Ruiz, and Ruiz [205], and a related result can be found in Belzunce, Mercader, and Ruiz [70]. The results about the likelihood ratio order comparisons of random minima and maxima (Example 1.C.46) are taken from Shaked and Wong [526]; see a related result in Bartoszewicz [49]. The result about the likelihood ratio comparison of the successive epochs of a nonhomogeneous Poisson process (Example 1.C.47) is given in Kochar [307, 309], where it is also shown that it implies the likelihood order comparison of successive record values of a sequence of independent and identically distributed random variables. The likelihood ratio comparisons of epoch and inter-epoch times of nonhomogeneous Poisson processes (Example 1.C.48) are taken from Belzunce, Lillo, Ruiz, and Shaked [69], who also extended them to comparisons of epoch and inter-epoch times of nonhomogeneous pure birth processes. The likelihood ratio order comparison of a sum of independent heterogeneous exponential random variables with a proper Erlang random variable (Example 1.C.49) is a combination of results from Boland, El-Neweihi, and Proschan [98] and from Bon and Păltănea [105], where more refined comparisons can also be found. For instance, the comparison in Example 1.C.50 is given in Boland, El-Neweihi, and Proschan [98]. The likelihood ratio order comparison of a sum of independent heterogeneous binomial random variables with a proper binomial random variable (Example 1.C.51) is taken from Boland, Singh, and Cukic [102]. An interpretation of logconcavity and logconvexity as aging notions can be found in Shaked and Shanthikumar [506], where the proof of parts (i) and (ii) of Theorem 1.C.52 can be found. A proof of (1.C.13) can also be found there. The likelihood ratio ordering of random variables conditioned on their sum (Theorem 1.C.53) is essentially Example 12 of Lehmann [343]. The closure property of logconcave densities under order statistics (Theorem 1.C.54) is a generalization of an observation in Li and Lu [355]. The characterizations of the IFR and the DFR aging notions by means of the likelihood ratio order (Theorem 1.C.55) can be found in Whitt [565]. The likelihood ratio order comparison of the asymptotic equilibrium ages, given in Example 1.C.56, is a special case of a result of Bon and Illayk [104]. The likelihood ratio monotonicity of the parameter in the observation, given the likelihood ratio monotonicity of the observation in the parameter (Example 1.C.57), can be found in Whitt [560], whereas the preservation of the likelihood ratio order of the priors by the posteriors (Example 1.C.58) is given as Remark 3.14 in Spizzichino [539]. The comparison of the weighted random variables (Example 1.C.59) can be found in Bartoszewicz and Skolimowska [51]. An extension of the implication in Example 1.C.59, when X^w and Y^w are the length-biased versions of X and of Y, respectively, is given in Hu and Zhuang [244]. An extension of the implication in Example 1.C.59 to multivariate weighted distributions can be found in Jain and Nanda [253]. The result in Example 1.C.60 is taken from Bartoszewicz and Skolimowska [51]; extensions of the inequality $X \le_{\rm lr} X^w$, when X^w is the length-biased version of X, are given in Ross [476]. The ordering of generalized skew normal random variables (Example 1.C.61) is taken from Gupta and Gupta [215]. The up shifted likelihood ratio order is introduced in Shanthikumar and Yao [530]. The results described in Section 1.C.4 can mostly be found in Lillo, Nanda, and Shaked [361, 362]. An extension of Theorem 1.C.77 is given in Belzunce, Ruiz, and Ruiz [74]; see also Belzunce and Shaked [78]. Ramos Romero and Sordo Díaz [464] defined an order that is reminiscent of the order $\le_{\rm lr\uparrow}$ as defined in (1.C.19). According to their definition, the nonnegative random variable X is said to be smaller than the nonnegative random variable Y if $aX \le_{\rm lr} Y$ for every $0 < a < 1$.

Lehmann and Rojo [345] used the characterization (1.C.4) in order to define stochastic orders that are stronger than $\le_{\rm lr}$. For example, let X and Y be two random variables with distribution functions F and G, respectively, and consider the stipulation that, for a fixed k,

$$\frac{\mathrm{d}^n}{\mathrm{d}u^n} GF^{-1}(u) \geq 0 \quad \text{for all } 0 < u < 1 \text{ and all } n = 1, 2, \ldots, k.$$

If $k \geq 3$, then X is stochastically smaller than Y in a sense that is stronger than $\leq_{\mathrm{lr}}$. The order $\leq_{\mathrm{lr}}$ is obtained when $k = 2$. Lehmann and Rojo [345] showed, for example, that if $X_1, X_2, \ldots, X_m$ are independent, identically distributed, then X_1 is smaller than $\max\{X_1, X_2, \ldots, X_m\}$, in the above sense, with $k = m$.

Chang [126] considered four exponential random variables X_1, X_2, Y_1, and Y_2, with the corresponding rates λ_1, λ_2, μ_1, and μ_2, where X_1 and X_2 are independent, and Y_1 and Y_2 are independent. He obtained the necessary and sufficient conditions on λ_1, λ_2, μ_1, and μ_2, for each of the following results: (i) $X_1 + X_2 \leq_{\mathrm{lr}} Y_1 + Y_2$, (ii) $X_1 + X_2 \geq_{\mathrm{lr}} Y_1 + Y_2$, and (iii) $X_1 + X_2$ and $Y_1 + Y_2$ are not comparable in the likelihood ratio order.

Section 1.D: The discussion in this section follows Shaked and Suarez-Llorens [520].

Fagiuoli and Pellerey [185] have introduced an approach that describes a unified point of view regarding some of the orders studied in this chapter and some of the orders studied in Chapters 2, 3, and 4. This approach led Fagiuoli and Pellerey to introduce some families of new orders. Several properties of these orders were studied in Fagiuoli and Pellerey [185], in Nanda, Jain, and Singh [424, 425], and in Hu, Kundu, and Nanda [236]; see also Hesselager [221]. Another general approach that unifies some of the orders studied in this chapter and in Chapter 2 was introduced in Hu, Nanda, Xie, and Zhu [237].

Other orders that are related to the orders $\leq_{\mathrm{st}}$ and $\leq_{\mathrm{lr}}$ have been introduced and studied in Di Crescenzo [163]. Yanagimoto and Sibuya [571], Zijlstra and de Kroon [577], and Shanthikumar and Yao [532], extended the definitions of $X \leq_{\mathrm{st}} Y$, $X \leq_{\mathrm{hr}} Y$, and $X \leq_{\mathrm{lr}} Y$, to jointly distributed random variables X and Y; see also Arcones, Kvam, and Samaniego [15]. Ebrahimi and Pellerey [177] have introduced a stochastic order based on a notion of uncertainty and studied its relationship to some of the orders studied in this chapter.

2

Mean Residual Life Orders

In this chapter we study two orders that are based on comparisons of functionals of mean residual lives. Like the orders in Chapter 1, the purpose of the orders here is to compare the "location" or the "magnitude" of random variables. Among other things, the relationship between the orders of Chapter 1 and the orders in this chapter will be analyzed.

2.A The Mean Residual Life Order

2.A.1 Definition

If X is a random variable with a survival function $\overline{F}$ and a finite mean μ, the mean residual life of X at t is defined as

$$m(t) = \begin{cases} E[X - t | X > t], & \text{for } t < t^*; \\ 0, & \text{otherwise}, \end{cases} \tag{2.A.1}$$

where $t^* = \sup\{t : \overline{F}(t) > 0\}$. Note that if X is an almost surely positive random variable, then $m(0) = \mu$. By the finiteness of μ we have that $m(t) < \infty$ for all $t < \infty$. However, it is possible that $m(\infty) \equiv \lim_{t \to \infty} m(t) = \infty$. A useful observation is that $m(t) = (\int_t^\infty \overline{F}(x)\mathrm{d}x)/\overline{F}(t)$ when $t^* = \infty$.

Although in (2.A.1) there is no restriction on the support of X, the mean residual life function is usually of interest when X is a nonnegative random variable. In that case X can be thought of as a lifetime of a device and $m(t)$ then expresses the conditional expected residual life of the device at time t given that the device is still alive at time t. Clearly, $m(t) \geq 0$, but not every nonnegative function is a *mean residual life* (mrl) function corresponding to some random variable. In fact, a function m is an mrl function of some nonnegative random variable with an absolutely continuous distribution function if, and only if, m satisfies the following properties:

(i) $0 \leq m(t) < \infty$ for all $t \geq 0$,

(ii) $m(0) > 0$,
(iii) m is continuous,
(iv) $m(t) + t$ is increasing on $[0, \infty]$, and
(v) when there exists a t_0 such that $m(t_0) = 0$, then $m(t) = 0$ for all $t \geq t_0$. Otherwise, when there does not exist such a t_0 with $m(t_0) = 0$, then

$$\int_0^\infty \frac{1}{m(t)} \mathrm{d}t = \infty.$$

Clearly, the smaller the mrl function is, the smaller X should be in some stochastic sense. This is the motivation for the order discussed in this section.

Let X and Y be two random variables with mrl functions m and l, respectively, such that

$$m(t) \leq l(t) \quad \text{for all } t. \tag{2.A.2}$$

Then X is said to be *smaller than* Y *in the mean residual life order* (denoted as $X \leq_{\mathrm{mrl}} Y$).

Analogously to (1.B.3), it can be shown that $X \leq_{\mathrm{mrl}} Y$ if, and only if,

$$\frac{\int_t^\infty \overline{G}(u)\mathrm{d}u}{\int_t^\infty \overline{F}(u)\mathrm{d}u} \quad \text{increases in } t \text{ over } \{t : \int_t^\infty \overline{F}(u)\mathrm{d}u > 0\}, \tag{2.A.3}$$

or equivalently, if, and only if,

$$\overline{G}(t) \int_t^\infty \overline{F}(u)\mathrm{d}u \leq \overline{F}(t) \int_t^\infty \overline{G}(u)\mathrm{d}u \quad \text{for all } t, \tag{2.A.4}$$

or equivalently, if, and only if,

$$\frac{E[(Y-t)_+]}{E[(X-t)_+]} \quad \text{increases in } t \text{ over } \{t : E[(X-t)_+] > 0\}, \tag{2.A.5}$$

where, for any real number a, we let a_+ denote the positive part of a; that is, $a_+ = a$ if $a \geq 0$ and $a_+ = 0$ if $a < 0$.

Analogously to (1.B.5), we also have that $X \leq_{\mathrm{mrl}} Y$ if, and only if,

$$\frac{\overline{F}(s)}{\int_t^\infty \overline{F}(u)\mathrm{d}u} \geq \frac{\overline{G}(s)}{\int_t^\infty \overline{G}(u)\mathrm{d}u} \quad \text{for all } s \leq t \tag{2.A.6}$$

such that the denominators are positive.

It is worthwhile to note that Condition (2.A.5) uses the expectations $E[(X-t)_+]$ and $E[(Y-t)_+]$ as (3.A.5) in Chapter 3 and (4.A.4) in Chapter 4 do.

For discrete random variables that take on values in $\mathbb{N}_+$ the definition of $\leq_{\mathrm{mrl}}$ should be modified. Let X be such a random variable with a finite mean μ. The mrl function of X at n is defined as

$$m(n) = \begin{cases} E[X-n | X \geq n], & \text{for } n \leq n^*; \\ 0, & \text{otherwise,} \end{cases}$$

where $n^* = \max\{n : P\{X \geq n\} > 0\}$. Note that for such a random variable $m(0) = \mu$. By the finiteness of μ we have that $m(n) < \infty$ for $n < \infty$. Let X and Y be two such random variables with mrl functions m and l, respectively. We denote $X \leq_{\rm mrl} Y$ if

$$m(n) \leq l(n) \quad \text{for all } n \geq 0. \tag{2.A.7}$$

The discrete analog of (2.A.3) is that (2.A.7) holds if, and only if,

$$\frac{\sum_{j=n}^{\infty} P\{Y \geq j\}}{\sum_{j=n}^{\infty} P\{X \geq j\}} \quad \text{increases in } n \text{ over } \mathbb{N}_+ \cap \{n : \sum_{j=n}^{\infty} P\{X \geq j\} > 0\}.$$

The discrete analog of (2.A.4) is that (2.A.7) holds if, and only if,

$$P\{Y \geq n\} \sum_{j=n+1}^{\infty} P\{X \geq j\} \leq P\{X \geq n\} \sum_{j=n+1}^{\infty} P\{Y \geq j\} \quad \text{for all } n \geq 0.$$

The discrete analog of (2.A.6) is that $X \leq_{\rm mrl} Y$ if, and only if,

$$\frac{P\{X \geq m\}}{\sum_{j=n+1}^{\infty} P\{X \geq j\}} \geq \frac{P\{Y \geq m\}}{\sum_{j=n+1}^{\infty} P\{Y \geq j\}} \quad \text{for all } m \leq n$$

such that the denominators are positive.

2.A.2 The relation between the mean residual life and some other stochastic orders

If X is a random variable with mrl function m and hazard rate function r, it is not hard to verify that

$$m(t) = \int_t^{t^*} \exp\Big\{ - \int_t^x r(u) \mathrm{d}u \Big\} \mathrm{d}x, \quad \text{for } t < t^*. \tag{2.A.8}$$

Therefore, if Y is another random variable with mrl function l and hazard rate function q and (1.B.2) is satisfied, that is, $X \leq_{\rm hr} Y$, then $X \leq_{\rm mrl} Y$. We thus have proved the following result.

Theorem 2.A.1. *If X and Y are two random variables such that $X \leq_{\rm hr} Y$, then $X \leq_{\rm mrl} Y$.*

Neither of the orders $\leq_{\rm st}$ and $\leq_{\rm mrl}$ implies the other; counterexamples can be found in the literature. The next result, however, gives a condition under which $X \leq_{\rm mrl} Y$ if, and only if, $X \leq_{\rm hr} Y$. Therefore, in particular, under that condition, $X \leq_{\rm mrl} Y \Longrightarrow X \leq_{\rm st} Y$.

Theorem 2.A.2. *Let X and Y be two random variables with mrl functions m and l, respectively. Suppose that $\frac{m(t)}{l(t)}$ increases in t. Then, if $X \leq_{\rm mrl} Y$, then $X \leq_{\rm hr} Y$.*

Proof. It is not hard to verify that m is differentiable over $\{t : P\{X > t\} > 0\}$ and that if X has the hazard rate function r, then

$$r(t) = \frac{m'(t) + 1}{m(t)},$$

where m' denotes the derivative of m. Similarly, if Y has the hazard rate function q, then

$$q(t) = \frac{l'(t) + 1}{l(t)}.$$

The monotonicity of $m(t)/l(t)$, together with (2.A.2), implies that

$$r(t) = \frac{m'(t)}{m(t)} + \frac{1}{m(t)} \geq \frac{l'(t)}{l(t)} + \frac{1}{l(t)} = q(t),$$

that is, $X \leq_{\text{hr}} Y$. □

Under a condition that is weaker than the one in Theorem 2.A.2 one merely obtains that $X \leq_{\text{mrl}} Y$ implies that $X \leq_{\text{st}} Y$. This is shown in the next result.

Theorem 2.A.3. *Let X and Y be two nonnegative random variables with mrl functions m and l, respectively. Suppose that $\frac{m(t)}{l(t)} \geq \frac{m(0)}{l(0)}$ (that is, $\frac{m(t)}{l(t)} \geq \frac{EX}{EY}$ when X and Y are almost surely positive), $t \geq 0$. If $X \leq_{\text{mrl}} Y$, then $X \leq_{\text{st}} Y$.*

Proof. Let $\overline{F}$ be the survival function of X. It is not hard to verify that

$$\overline{F}(t) = \frac{EX}{m(t)} \exp\left\{ - \int_0^t \frac{1}{m(x)} \mathrm{d}x \right\} \quad \text{over } \{t : P\{X > t\} > 0\}.$$

Similarly, the survival function of Y can be expressed as

$$\overline{G}(t) = \frac{EY}{l(t)} \exp\left\{ - \int_0^t \frac{1}{l(x)} \mathrm{d}x \right\} \quad \text{over } \{t : P\{Y > t\} > 0\}.$$

Therefore, under the assumptions of the theorem, it is seen that $\frac{\overline{G}(t)}{\overline{F}(t)} \geq 1$. □

The mean residual life order can be characterized by means of the hazard rate order and the appropriate equilibrium age variables. Recall from (1.A.20) that for nonnegative random variables X and Y with finite means we denote by A_X and A_Y the corresponding asymptotic equilibrium ages. The following result follows at once from (1.B.3) and (2.A.3). It may be contrasted with Theorem 1.C.13.

Theorem 2.A.4. *For nonnegative random variables X and Y with finite means we have $X \leq_{\text{mrl}} Y$ if, and only if, $A_X \leq_{\text{hr}} A_Y$.*

In the next theorem the order $\leq_{\rm mrl}$ is characterized by ordering two related random variables in the sense of the hazard rate order. Let X and Y be two nonnegative random variables with finite means and suppose that $X \leq_{\rm st} Y$ and that $EX < EY$. Let F and G be the distribution functions of X and of Y, respectively. Define the random variable $Z_{X,Y}$ as the random variable that has the density function h given by (1.C.7), as in Theorem 1.C.14; see also Theorem 2.B.3.

Theorem 2.A.5. *Let X and Y be two nonnegative random variables with finite means such that $X \leq_{\rm st} Y$ and such that $EY > EX > 0$. Then*

$$X \leq_{\rm mrl} Y \Longleftrightarrow A_Y \leq_{\rm hr} Z_{X,Y} \Longleftrightarrow A_X \leq_{\rm hr} Z_{X,Y},$$

where $Z_{X,Y}$ has the density function given in (1.C.7).

Proof. Denote by $\overline{G}_{\rm e}$ and $\overline{H}$ the survival functions of A_Y and $Z_{X,Y}$, respectively. Using (1.A.20) and (1.C.7) we compute

$$\frac{\overline{H}(x)}{\overline{G}_{\rm e}(x)} = \frac{EY}{EY - EX}\left(1 - \frac{\int_x^\infty \overline{F}(u)\mathrm{d}u}{\int_x^\infty \overline{G}(u)\mathrm{d}u}\right), \quad x \geq 0,$$

and the first stated equivalence follows from (2.A.3) and (1.B.3). The second equivalence is proven similarly. □

Some characterizations of the hazard rate order by means of the order $\leq_{\rm mrl}$ are given below. We denote by $\mathrm{Exp}(\mu)$ any exponential random variable with mean μ.

Theorem 2.A.6. *Let X and Y be two continuous nonnegative random variables. Then $X \leq_{\rm hr} Y$ if, and only if,*

$$\min\{X, \mathrm{Exp}(\mu)\} \leq_{\rm mrl} \min\{Y, \mathrm{Exp}(\mu)\} \quad \textit{for all } \mu > 0.$$

The proof of Theorem 2.A.6 uses the Laplace transform order which is discussed in Chapter 5, and it will be given in Remark 5.A.23.

Note that from Theorem 2.A.6 it follows, for continuous nonnegative random variables, that $X \leq_{\rm hr} Y$ if, and only if,

$$\min\{X, Z\} \leq_{\rm mrl} \min\{Y, Z\}$$

for any nonnegative random variable Z which is independent of X and of Y. This is so because $X \leq_{\rm hr} Y$ implies $\min\{X, Z\} \leq_{\rm hr} \min\{Y, Z\}$ by Theorem 1.B.33, and the latter implies the above inequality by Theorem 2.A.1.

The proof of the next result is not given here.

Theorem 2.A.7. *Let X and Y be two continuous nonnegative random variables. Then $X \leq_{\rm hr} Y$ if, and only if,*

$$1 - \mathrm{e}^{-sX} \leq_{\rm mrl} 1 - \mathrm{e}^{-sY} \quad \textit{for all } s > 0.$$

A characterization of the order $\leq_{\rm mrl}$, by means of the increasing convex order, is given in Theorem 4.A.24.

2.A.3 Some closure properties

In general, if $X_1 \le_{\rm mrl} Y_1$ and $X_2 \le_{\rm mrl} Y_2$, where X_1 and X_2 are independent random variables and Y_1 and Y_2 are also independent random variables, then it is not necessarily true that $X_1+X_2 \le_{\rm mrl} Y_1+Y_2$. However, if these random variables are IFR, then it is true. This is shown in Theorem 2.A.9, but first we state and prove the following lemma, which is of independent interest.

Lemma 2.A.8. *If the random variables X and Y are such that $X \le_{\rm mrl} Y$ and if Z is an* IFR *random variable which is independent of X and Y, then*

$$X+Z \le_{\rm mrl} Y+Z. \tag{2.A.9}$$

Proof. Denote by f_W and $\overline{F}_W$ the density function and the survival function of any random variable W. Note that

$$\int_{x=s}^{\infty} \overline{F}_{X+Z}(x)\mathrm{d}x = \int_{-\infty}^{\infty} \overline{F}_X(u)\overline{F}_Z(s-u)\mathrm{d}u \quad \text{for all } s.$$

Now, for $s \le t$, compute

$$\begin{aligned}
&\int_{x=s}^{\infty} \overline{F}_{X+Z}(x)\mathrm{d}x \int_{y=t}^{\infty} \overline{F}_{Y+Z}(y)\mathrm{d}y - \int_{x=t}^{\infty} \overline{F}_{X+Z}(x)\mathrm{d}x \int_{y=s}^{\infty} \overline{F}_{Y+Z}(y)\mathrm{d}y \\
&= \int_v \int_{u \ge v} \Big[\overline{F}_X(u)\overline{F}_Z(s-u)\overline{F}_Y(v)\overline{F}_Z(t-v) \\
&\qquad\qquad + \overline{F}_X(v)\overline{F}_Z(s-v)\overline{F}_Y(u)\overline{F}_Z(t-u)\Big]\mathrm{d}u\mathrm{d}v \\
&- \int_v \int_{u \ge v} \Big[\overline{F}_X(u)\overline{F}_Z(t-u)\overline{F}_Y(v)\overline{F}_Z(t-v) \\
&\qquad\qquad + \overline{F}_X(v)\overline{F}_Z(t-v)\overline{F}_Y(u)\overline{F}_Z(s-u)\Big]\mathrm{d}u\mathrm{d}v \\
&= \int_v \int_{u \ge v} \Big[\int_{x=u}^{\infty} \overline{F}_X(x)\,dx \cdot \overline{F}_Y(v) - \int_{x=u}^{\infty} \overline{F}_Y(x)\,dx \cdot \overline{F}_X(v)\Big] \\
&\qquad\qquad \times [f_Z(s-u)\overline{F}_Z(t-v) - f_Z(t-u)\overline{F}_Z(s-v)]\mathrm{d}u\mathrm{d}v,
\end{aligned}$$

where the second equality is obtained by integration of parts and by collection of terms. Since $X \le_{\rm mrl} Y$ it follows from (2.A.4) that the expression within the first set of brackets in the last integral is nonpositive. Since Z is IFR it can be verified that the quantity in the second pair of brackets in the last integral is also nonpositive. Therefore the integral is nonnegative. This proves (2.A.9). □

Theorem 2.A.9. *Let (X_i, Y_i), $i = 1, 2, \dots, m$, be independent pairs of random variables such that $X_i \le_{\rm mrl} Y_i$, $i = 1, 2, \dots, m$. If X_i, Y_i, $i = 1, 2, \dots, m$, are all* IFR, *then*

$$\sum_{i=1}^{m} X_i \le_{\rm mrl} \sum_{i=1}^{m} Y_i.$$

Proof. Repeated application of (2.A.9), using the closure property of IFR under convolution, yields the desired result. □

Another interesting lemma is stated next. Recall that a random variable X is said to be (or to have) *decreasing mean residual life* (DMRL) if $m(t)$ is decreasing in t.

Lemma 2.A.10. *If the random variables X and Y are such that $X \leq_{\rm hr} Y$ and if Z is a* DMRL *random variable independent of X and Y, then*

$$X + Z \leq_{\rm mrl} Y + Z.$$

Proof. Integrating the identity in the proof of Lemma 1.B.3, we obtain that, for $s \leq t$, one has

$$\int_{x=s}^{\infty} \overline{F}_{X+Z}(x)\mathrm{d}x \int_{y=t}^{\infty} \overline{F}_{Y+Z}(y)\mathrm{d}y - \int_{x=t}^{\infty} \overline{F}_{X+Z}(x)\mathrm{d}x \int_{y=s}^{\infty} \overline{F}_{Y+Z}(y)\mathrm{d}y$$
$$= \int_v \int_{u \geq v} \left[\overline{F}_X(u) f_Y(v) - f_X(v)\overline{F}_Y(u)\right]$$
$$\times \left[\int_{y=t}^{\infty} \overline{F}_Z(y-v)\mathrm{d}y \cdot \overline{F}_Z(s-u) - \int_{x=s}^{\infty} \overline{F}_Z(x-v)\mathrm{d}x \cdot \overline{F}_Z(t-u)\right] \mathrm{d}u\mathrm{d}v.$$

The result now follows from the assumptions. □

It should be pointed out that a theorem such as Theorem 2.A.9 cannot be obtained from Lemma 2.A.10. The reason is that the inductive argument used to prove Theorem 2.A.9 does not have an analog based on Lemma 2.A.10.

Theorem 2.A.11. *Let X be a* DMRL *random variable, and let Z be a non-negative random variable independent of X. Then*

$$X \leq_{\rm mrl} X + Z.$$

Proof. Let F_X, F_Z, and F_{X+Z} denote the distribution functions of the corresponding random variables, and let $\overline{F}_X$ and $\overline{F}_{X+Z}$ denote the corresponding survival functions. Then, for any $t \in \mathbb{R}$ we have

$$\begin{aligned}
\overline{F}_X(t) \int_t^{\infty} \overline{F}_{X+Z}(u)\mathrm{d}u &= \overline{F}_X(t) \int_t^{\infty} \int_0^{\infty} \overline{F}_X(u-z)\mathrm{d}F_Z(z)\mathrm{d}u \\
&= \overline{F}_X(t) \int_0^{\infty} \int_t^{\infty} \overline{F}_X(u-z)\mathrm{d}u\mathrm{d}F_Z(z) \\
&= \int_0^{\infty} \overline{F}_X(t) \int_{t-z}^{\infty} \overline{F}_X(u)\mathrm{d}u\mathrm{d}F_Z(z) \\
&\geq \int_0^{\infty} \overline{F}_X(t-z) \int_t^{\infty} \overline{F}_X(u)\mathrm{d}u\mathrm{d}F_Z(z) \\
&= \overline{F}_{X+Z}(t) \int_t^{\infty} \overline{F}_X(u)\mathrm{d}u,
\end{aligned}$$

where the inequality follows from the assumption that X is DMRL. The stated result now follows from (2.A.4). □

A mean residual life order comparison of random sums is given in the following result.

Theorem 2.A.12. *Let $\{X_i,\ i = 1, 2, \ldots\}$ be a sequence of independent and identically distributed nonnegative* IFR *random variables. Let M and N be two discrete positive integer-valued random variables such that $M \le_{\rm mrl} N$ (in the sense of* (2.A.7)*), and assume that M and N are independent of the X_i's. Then*

$$\sum_{i=1}^{M} X_i \le_{\rm mrl} \sum_{i=1}^{N} X_i.$$

The mean residual life order does not have the property of being simply closed under mixtures. However, under quite strong conditions the order $\le_{\rm mrl}$ is closed under mixtures. This is shown in the next theorem which may be compared with Theorem 1.B.8.

Theorem 2.A.13. *Let X, Y, and Θ be random variables such that $[X|\Theta = \theta] \le_{\rm mrl} [Y|\Theta = \theta']$ for all θ and θ' in the support of Θ. Then $X \le_{\rm mrl} Y$.*

Proof. The proof is similar to the proof of Theorem 1.B.8. Select a θ and a θ' in the support of Θ. Let $\overline{F}(\cdot|\theta)$, $\overline{G}(\cdot|\theta)$, $\overline{F}(\cdot|\theta')$, and $\overline{G}(\cdot|\theta')$ be the survival functions of $[X|\Theta = \theta]$, $[Y|\Theta = \theta]$, $[X|\Theta = \theta']$, and $[Y|\Theta = \theta']$, respectively. It is sufficient to show that for $\alpha \in (0, 1)$ we have

$$\frac{\alpha \int_t^\infty \overline{F}(u|\theta)\mathrm{d}u + (1-\alpha)\int_t^\infty \overline{F}(u|\theta')\mathrm{d}u}{\alpha\overline{F}(t|\theta) + (1-\alpha)\overline{F}(t|\theta')} \le \frac{\alpha \int_t^\infty \overline{G}(u|\theta)\mathrm{d}u + (1-\alpha)\int_t^\infty \overline{G}(u|\theta')\mathrm{d}u}{\alpha\overline{G}(t|\theta) + (1-\alpha)\overline{G}(t|\theta')} \quad \text{for all } t \ge 0.$$

The proof of this inequality is similar to the proof of (1.B.12). □

An analog of Theorem 1.B.12 exists for the order $\le_{\rm mrl}$. This is stated next.

Theorem 2.A.14. *Let X and Y be two nonnegative independent random variables. Then $X \le_{\rm mrl} Y$ if, and only if, for all functions α and β such that β is nonnegative and α/β and β are increasing, one has*

$$E[\alpha^*(X)]E[\beta^*(Y)] \le E[\alpha^*(Y)]E[\beta^*(X)],$$

provided the expectations exist, where

$$\alpha^*(x) = \int_0^x \alpha(u)\mathrm{d}u \quad \textit{and} \quad \beta^*(x) = \int_0^x \beta(u)\mathrm{d}u.$$

In particular, if $X \le_{\rm mrl} Y$, then

$$\frac{E[Y^n]}{E[X^n]} \quad \text{is increasing in } n. \tag{2.A.10}$$

Consider now a family of distribution functions $\{G_\theta,\ \theta \in \mathcal{X}\}$ where $\mathcal{X}$ is a subset of the real line. As in Sections 1.A.3 and 1.C.3 let $X(\theta)$ denote a random variable with distribution function G_θ. For any random variable Θ with support in $\mathcal{X}$, and with distribution function F, let us denote by $X(\Theta)$ a random variable with distribution function H given by

$$H(y) = \int_{\mathcal{X}} G_\theta(y) \mathrm{d}F(\theta), \quad y \in \mathbb{R}.$$

The following result is comparable to Theorems 1.A.6, 1.B.14, 1.B.52, and 1.C.17.

Theorem 2.A.15. *Consider a family of distribution functions $\{G_\theta,\ \theta \in \mathcal{X}\}$ as above. Let Θ_1 and Θ_2 be two random variables with supports in $\mathcal{X}$ and distribution functions F_1 and F_2, respectively. Let Y_1 and Y_2 be two random variables such that $Y_i =_{\text{st}} X(\Theta_i)$, $i = 1, 2$, that is, suppose that the distribution function of Y_i is given by*

$$H_i(y) = \int_{\mathcal{X}} G_\theta(y) \mathrm{d}F_i(\theta), \quad y \in \mathbb{R},\ i = 1, 2.$$

If

$$X(\theta) \le_{\text{mrl}} X(\theta') \quad \textit{whenever } \theta \le \theta', \tag{2.A.11}$$

and if

$$\Theta_1 \le_{\text{hr}} \Theta_2, \tag{2.A.12}$$

then

$$Y_1 \le_{\text{mrl}} Y_2. \tag{2.A.13}$$

The proof of Theorem 2.A.15 uses the increasing convex order, and is therefore given in Remark 4.A.29 in Chapter 4.

A Laplace transform characterization of the order $\le_{\text{mrl}}$ is given next; it may be compared to Theorems 1.A.13, 1.B.18, 1.B.53, and 1.C.25.

Theorem 2.A.16. *Let X_1 and X_2 be two nonnegative random variables, and let $N_\lambda(X_1)$ and $N_\lambda(X_2)$ be as described in Theorem 1.A.13. Then*

$$X_1 \le_{\text{mrl}} X_2 \Longleftrightarrow N_\lambda(X_1) \le_{\text{mrl}} N_\lambda(X_2) \quad \textit{for all } \lambda > 0,$$

where the notation $N_\lambda(X_1) \le_{\text{mrl}} N_\lambda(X_2)$ is in the sense of (2.A.7).

Proof. We use the notation of Theorem 1.A.13. Denote the distribution and survival functions of X_k by F_k and $\overline{F}_k$, $k = 1, 2$. For $k = 1, 2$, note that $\overline{\alpha}_\lambda^{X_k}(n)$ can be written as

$$\begin{aligned} \overline{\alpha}_\lambda^{X_k}(n) &= \int_0^\infty \sum_{i=n}^\infty \mathrm{e}^{-\lambda x} \frac{(\lambda x)^i}{i!} \mathrm{d}F_k(x) \\ &= \begin{cases} 1, & n = 0, \\ \int_0^\infty \lambda \mathrm{e}^{-\lambda x} \frac{(\lambda x)^{n-1}}{(n-1)!} \overline{F}_k(x) \mathrm{d}x, & n = 1, 2, \ldots . \end{cases} \end{aligned} \tag{2.A.14}$$

Therefore

$$P[X_k = n] = \int_0^\infty e^{-\lambda x} \frac{(\lambda x)^n}{n!} dF_k(x), \quad n = 0, 1, 2, \ldots. \tag{2.A.15}$$

From (2.A.15) it is seen that

$$E[N_\lambda(X_k)] = \lambda E[X_k], \quad k = 1, 2, \tag{2.A.16}$$

provided the expectations exist.

First assume that $X_1 \le_{\mathrm{mrl}} X_2$. For the sake of this proof replace temporarily the notation $\overline{\alpha}_\lambda^{X_1}(n)$ and $\overline{\alpha}_\lambda^{X_2}(n)$, by $\overline{\alpha}_{\lambda,1}(n)$ and $\overline{\alpha}_{\lambda,2}(n)$, respectively. We also denote $E[X_1]$ and $E[X_2]$ by μ_1 and μ_2, respectively. The proof of $N_\lambda(X_1) \le_{\mathrm{mrl}} N_\lambda(X_2)$ will consist of showing the following three inequalities:

$$\frac{\sum_{n=0}^\infty \overline{\alpha}_{\lambda,2}(n)}{\sum_{n=0}^\infty \overline{\alpha}_{\lambda,1}(n)} \le \frac{\sum_{n=1}^\infty \overline{\alpha}_{\lambda,2}(n)}{\sum_{n=1}^\infty \overline{\alpha}_{\lambda,1}(n)}, \tag{2.A.17}$$

$$\frac{\sum_{n=1}^\infty \overline{\alpha}_{\lambda,2}(n)}{\sum_{n=1}^\infty \overline{\alpha}_{\lambda,1}(n)} \le \frac{\sum_{n=2}^\infty \overline{\alpha}_{\lambda,2}(n)}{\sum_{n=2}^\infty \overline{\alpha}_{\lambda,1}(n)}, \tag{2.A.18}$$

and

$$\sum_{n=m}^\infty \overline{\alpha}_{\lambda,k}(n) \text{ is TP}_2 \text{ in } k \in \{1,2\} \text{ and } m \ge 2. \tag{2.A.19}$$

In order to prove (2.A.17) note that from (2.A.16) it follows that

$$\sum_{n=0}^\infty \overline{\alpha}_{\lambda,k}(n) = 1 + \lambda\mu_k \quad k = 1, 2, \quad \text{and}$$

$$\sum_{n=1}^\infty \overline{\alpha}_{\lambda,k}(n) = \mu_k, \quad k = 1, 2. \tag{2.A.20}$$

But since $X_1 \le_{\mathrm{mrl}} X_2$ implies that $\mu_1 \le \mu_2$ it follows that

$$\frac{1 + \lambda\mu_2}{1 + \lambda\mu_1} \le \frac{\lambda\mu_2}{\lambda\mu_1},$$

and (2.A.17) is obtained.

Next notice that (2.A.18) is equivalent to

$$\frac{\overline{\alpha}_{\lambda,2}(1)}{\overline{\alpha}_{\lambda,1}(1)} \le \frac{\sum_{n=1}^\infty \overline{\alpha}_{\lambda,2}(n)}{\sum_{n=1}^\infty \overline{\alpha}_{\lambda,1}(n)}. \tag{2.A.21}$$

Since $\sum_{n=1}^\infty \overline{\alpha}_{\lambda,k}(n) = \lambda\mu_k$, $k = 1, 2$, and

$$\overline{\alpha}_{\lambda,k}(1) = \int_0^\infty \lambda e^{-\lambda x} \overline{F}_k(x) dx = \lambda \Big[\mu_k - \int_0^\infty \lambda e^{-\lambda x} \int_x^\infty \overline{F}_k(u) du dx\Big],$$
$$k = 1, 2,$$

it follows that (2.A.21) is the same as

$$\mu_1 \int_0^\infty \lambda e^{-\lambda x} \int_x^\infty \overline{F}_2(u)\mathrm{d}u\mathrm{d}x - \mu_2 \int_0^\infty \lambda e^{-\lambda x} \int_x^\infty \overline{F}_1(u)\mathrm{d}u\mathrm{d}x \geq 0. \quad (2.A.22)$$

Rewriting the left-hand side of (2.A.22) we see that

$$\begin{aligned}
&\int_0^\infty \lambda e^{-\lambda x}\Big[\mu_1 \int_x^\infty \overline{F}_2(u)\mathrm{d}u - \mu_2 \int_x^\infty \overline{F}_1(u)\mathrm{d}u\Big]\mathrm{d}x \\
&= \int_0^\infty \lambda e^{-\lambda x}\left[\int_0^\infty \overline{F}_1(u)\mathrm{d}u \int_x^\infty \overline{F}_2(u)\mathrm{d}u - \int_0^\infty \overline{F}_2(u)\mathrm{d}u \int_x^\infty \overline{F}_1(u)\mathrm{d}u\right]\mathrm{d}x \\
&\geq 0,
\end{aligned}$$

where the inequality follows from the TP_2-ness of $\int_x^\infty \overline{F}_k(u)\mathrm{d}u$ in $k = 1, 2$, and $x \geq 0$ (see (2.A.3)). This proves (2.A.22), and hence (2.A.18).

Finally, in order to prove (2.A.19), notice, using a straightforward computation, that, for $m \geq 2$,

$$\sum_{n=m}^\infty \overline{\alpha}_{\lambda,k}(n) = \int_0^\infty \lambda^2 e^{-\lambda x}\frac{(\lambda x)^{m-2}}{(m-2)!}\int_x^\infty \overline{F}_k(u)\mathrm{d}u\mathrm{d}x. \quad (2.A.23)$$

By assumption, $\int_x^\infty \overline{F}_k(u)\mathrm{d}u$ is TP_2 in $k \in \{1, 2\}$ and $x \geq 0$. Furthermore, $\lambda^2 e^{-\lambda x}\frac{(\lambda x)^{m-2}}{(m-2)!}$ is TP_2 in $m \geq 2$ and $x \geq 0$. Thus, it follows that $\sum_{n=m}^\infty \overline{\alpha}_{\lambda,k}(n)$ is TP_2 in $k \in \{1, 2\}$ and $m \geq 2$, and this establishes (2.A.19).

Now suppose that $N_\lambda(X_1) \leq_{\mathrm{mrl}} N_\lambda(X_2)$ for all $\lambda > 0$. Then

$$\frac{\sum_{n=m}^\infty \overline{\alpha}_{\lambda,1}(n)}{\overline{\alpha}_{\lambda,1}(m)} \leq \frac{\sum_{n=m}^\infty \overline{\alpha}_{\lambda,2}(n)}{\overline{\alpha}_{\lambda,2}(m)}, \quad m = 0, 1, 2, \ldots.$$

For $m \geq 2$, by (2.A.23) and (2.A.14),

$$\frac{\int_0^\infty \lambda e^{-\lambda u}\frac{(\lambda u)^{m-2}}{(m-2)!}\left[\int_u^\infty \overline{F}_1(x)\mathrm{d}x\right]\mathrm{d}u}{\int_0^\infty \lambda e^{-\lambda u}\frac{(\lambda u)^{m-1}}{(m-1)!}\overline{F}_1(u)\mathrm{d}u} \leq \frac{\int_0^\infty \lambda e^{-\lambda u}\frac{(\lambda u)^{m-2}}{(m-2)!}\left[\int_u^\infty \overline{F}_2(x)\mathrm{d}x\right]\mathrm{d}u}{\int_0^\infty \lambda e^{-\lambda u}\frac{(\lambda u)^{m-1}}{(m-1)!}\overline{F}_2(u)\mathrm{d}u}. \quad (2.A.24)$$

For a fixed $y > 0$, define $\lambda = (m-1)/y$. Letting $m \to \infty$ ($\lambda \to \infty$), we have

$$\int_0^\infty \lambda e^{-\lambda u}\frac{(\lambda u)^{m-2}}{(m-2)!}\left[\int_u^\infty \overline{F}_k(x)\mathrm{d}x\right]\mathrm{d}u \to \int_y^\infty \overline{F}_k(x)\mathrm{d}x,$$

and

$$\int_0^\infty \lambda e^{-\lambda u}\frac{(\lambda u)^{m-1}}{(m-1)!}\overline{F}_k(u)\mathrm{d}u \to \overline{F}_k(y), \quad k = 1, 2,$$

as long as y is a continuity point of $\overline{F}_1(x)$ and $\overline{F}_2(x)$. For such y's, (2.A.24) gives us

$$\frac{\int_y^\infty \overline{F}_1(x)\mathrm{d}x}{\overline{F}_1(y)} \le \frac{\int_y^\infty \overline{F}_2(x)\mathrm{d}x}{\overline{F}_2(y)}.$$

It follows that $X_1 \le_{\mathrm{mrl}} X_2$ since the set of continuity points of $\overline{F}_1(x)$ and $\overline{F}_2(x)$ is dense in the set of positive real numbers. □

An analog of Theorem 1.B.21 is the following result.

Theorem 2.A.17. *Let X be a nonnegative* DMRL *random variable, and let $a \le 1$ be a positive constant. Then $aX \le_{\mathrm{mrl}} X$.*

Proof. It is easy to verify that the mean residual life function of aX is given by $am(\frac{t}{a})$, for all t, where m is the mean residual life function of X. Now

$$am(\frac{t}{a}) \le m(\frac{t}{a}) \le m(t) \quad \text{for all } t,$$

where the first inequality follows from $a \in [0,1]$ and the second inequality follows from the assumption that X is DMRL. The proof now follows from (2.A.2). □

In the next result it is shown that a random variable, whose distribution is the mixture of two distributions of mean residual life ordered random variables, is bounded from below and from above, in the mean residual life order sense, by these two random variables.

Theorem 2.A.18. *Let X and Y be two random variables with distribution functions F and G, respectively. Let W be a random variable with the distribution function $pF + (1-p)G$ for some $p \in (0,1)$. If $X \le_{\mathrm{mrl}} Y$, then $X \le_{\mathrm{mrl}} W \le_{\mathrm{mrl}} Y$.*

The proof of Theorem 2.A.18 is similar to the proof of Theorem 1.B.22, but it uses (2.A.3) instead of (1.B.3). We omit the details.

The following result is proven in Remark 4.A.25 of Section 4.A.3.

Theorem 2.A.19. *Let X and Y be two random variables. If $X \le_{\mathrm{mrl}} Y$, then $\phi(X) \le_{\mathrm{mrl}} \phi(Y)$ for every increasing convex function ϕ.*

Analogous to the result in Remark 1.A.18, it can be shown that the set of all distribution functions on $\mathbb{R}_+$ with finite means is a lattice with respect to the order $\le_{\mathrm{mrl}}$.

Let $X_1, X_2, \ldots, X_m$ be random variables, and let $X_{(k:m)}$ denote the corresponding kth order statistic, $k = 1, 2, \ldots, m$.

Theorem 2.A.20. *Let $X_1, X_2, \ldots, X_m$ be m independent random variables. If*

$$X_i \le_{\mathrm{mrl}} X_m, \quad i = 1, 2, \ldots, m-1,$$

then

$$X_{(m-1:m-1)} \le_{\mathrm{mrl}} X_{(m:m)}.$$

Let $X_1, X_2, \ldots, X_m$ be nonnegative random variables and let $U_{(i:m)} = X_{(i:m)} - X_{(i-1:m)}$ denote the corresponding spacings, $i = 1, 2, \ldots, m$ (where $U_{(1:m)} = X_{(1:m)}$). Similarly, let $Y_1, Y_2, \ldots, Y_n$ be nonnegative random variables and let $V_{(i:n)}$ denote the corresponding spacings, $i = 1, 2, \ldots, n$.

Theorem 2.A.21. *For positive integers m and n, let $X_1, X_2, \ldots, X_m$ be independent identically distributed nonnegative random variables, and let $Y_1, Y_2, \ldots, Y_n$ be other independent identically distributed nonnegative random variables. If $X_1 \le_{\text{mrl}} Y_1$, and if X_1 is* IMRL *and Y_1 is* DMRL, *then*

$$(m - j + 1)U_{(j:m)} \le_{\text{mrl}} (n - i + 1)V_{(i:n)} \quad \textit{for } j \le m \textit{ and } i \le n.$$

The following example may be compared to Examples 1.B.24, 1.C.48, 3.B.38, 4.B.14, 6.B.41, 6.D.8, 6.E.13, and 7.B.13.

Example 2.A.22. Let X and Y be two absolutely continuous nonnegative random variables with survival functions $\overline{F}$ and $\overline{G}$ and density functions f and g, respectively. Denote $\Lambda_1 = -\log \overline{F}$, $\Lambda_2 = -\log \overline{G}$, and $\lambda_i = \Lambda_i'$, $i = 1, 2$. Consider two nonhomogeneous Poisson processes $N_1 = \{N_1(t),\, t \ge 0\}$ and $N_2 = \{N_2(t),\, t \ge 0\}$ with mean functions Λ_1 and Λ_2 (see Example 1.B.13), respectively. Let $T_{i,1}, T_{i,2}, \ldots$ be the successive epoch times of process N_i, and let $X_{i,n} \equiv T_{i,n} - T_{i,n-1}$, $n \ge 1$ (where $T_{i,0} \equiv 0$), be the inter-epoch times of the process N_i, $i = 1, 2$. Note that $X =_{\text{st}} X_{1,1}$ and $Y =_{\text{st}} X_{2,1}$. It turns out that, under some conditions, the mean residual life ordering of the first two inter-epoch times implies the mean residual life ordering of all the corresponding later inter-epoch times. Explicitly, it will be shown below that if $X \le_{\text{mrl}} Y$, if X and Y are IMRL, and if (1.B.25) holds, then $X_{1,n} \le_{\text{mrl}} X_{2,n}$ for each $n \ge 1$.

For the purpose of this proof we denote $\overline{F}$ by $\overline{F}_1$ and $\overline{G}$ by $\overline{F}_2$. The stated result is obvious for $n = 1$. So let us fix $n \ge 2$. The survival function $\overline{G}_{i,n}$ of $X_{i,n}$, $i = 1, 2$, is given in (1.B.26). From (2.A.3) it is seen that the stated result is equivalent to

$$\int_t^\infty \overline{G}_{i,n}(x)\mathrm{d}x \quad \text{is TP}_2 \text{ in } (i, t);$$

that is, to

$$\int_{s=0}^\infty \lambda_i(s)\frac{\Lambda_i^{n-2}(s)}{(n-2)!}\int_{u=s+t}^\infty \overline{F}_i(u)\mathrm{d}u\mathrm{d}s \quad \text{is TP}_2 \text{ in } (i, t). \tag{2.A.25}$$

Now, from Example 1.B.24 we know that (1.B.25) implies that $\lambda_i(s)\frac{\Lambda_i^{n-2}(s)}{(n-2)!}$ is TP$_2$ in (i, s). The assumption $F_1 \le_{\text{mrl}} F_2$ means that

$$\int_{u=s+t}^\infty \overline{F}_i(u)\mathrm{d}u \quad \text{is TP}_2 \text{ in } (i, s) \text{ and in } (i, t).$$

Finally, the assumption that F_i is IMRL means that

$$\int_{u=s+t}^{\infty} \overline{F}_i(u)\mathrm{d}u \quad \text{is TP}_2 \text{ in } (s,t).$$

Thus (2.A.25) follows from Theorem 5.1 on page 123 of Karlin [275].

2.A.4 A property in reliability theory

The order $\leq_{\rm mrl}$ can be used to characterize DMRL random variables. As in Section 1.A.3, $[Z|A]$ denotes any random variable that has as its distribution the conditional distribution of Z given A.

Theorem 2.A.23. *The random variable X is* DMRL *if, and only if, any one of the following equivalent conditions holds:*

(i) $[X-t|X>t] \geq_{\rm mrl} [X-t'|X>t']$ *whenever* $t \leq t'$.
(ii) $X \geq_{\rm mrl} [X-t|X>t]$ *for all* $t \geq 0$ *(when X is a nonnegative random variable).*
(iii) $X+t \leq_{\rm mrl} X+t'$ *whenever* $t \leq t'$.

The proofs of all these statements are trivial and are thus omitted.

Other characterizations of DMRL and IMRL random variables, by means of other stochastic orders, can be found in Theorems 2.B.17, 3.A.56, 3.C.13, and 4.A.51.

A multivariate extension of parts (i) and (ii) of Theorem 2.A.23 is given in Section 6.F.3.

An interesting application of part (iii) of Theorem 2.A.23 is the following corollary. Its proof consists of a combination of Theorem 2.A.23(iii) with Lemma 2.A.8 (or, alternatively, a combination of Theorem 2.A.23(iii), Theorem 1.B.38(iii), and Lemma 2.A.10).

Corollary 2.A.24. *Let X be a* DMRL *random variable and let Y be an* IFR *random variable. If X and Y are independent, then $X+Y$ is* DMRL.

2.B The Harmonic Mean Residual Life Order

2.B.1 Definition

Let X and Y be two nonnegative random variables with mrl functions m and l, respectively, and suppose that the harmonic averages of m and l are comparable as follows:

$$\left[\frac{1}{x}\int_0^x \frac{1}{m(u)}\mathrm{d}u\right]^{-1} \leq \left[\frac{1}{x}\int_0^x \frac{1}{l(u)}\mathrm{d}u\right]^{-1} \quad \text{for all } x>0. \tag{2.B.1}$$

Then X is said to be *smaller than Y in the harmonic mean residual life order* (denoted as $X \leq_{\rm hmrl} Y$).

Notice that

$$\frac{1}{m(u)} = \frac{\overline{F}(u)}{\int_u^\infty \overline{F}(v)\mathrm{d}v} = -\frac{\mathrm{d}}{\mathrm{d}u}\log\Big(\int_u^\infty \overline{F}(v)\mathrm{d}v\Big).$$

Therefore

$$\int_0^x \frac{1}{m(u)}\mathrm{d}u = \log\left(\frac{EX}{\int_x^\infty \overline{F}(u)\mathrm{d}u}\right).$$

Similarly

$$\int_0^x \frac{1}{l(u)}\mathrm{d}u = \log\left(\frac{EY}{\int_x^\infty \overline{G}(u)\mathrm{d}u}\right).$$

Thus it is seen that (2.B.1) holds if, and only if,

$$\frac{\int_x^\infty \overline{F}(u)\mathrm{d}u}{EX} \le \frac{\int_x^\infty \overline{G}(u)\mathrm{d}u}{EY} \quad \text{for all } x \ge 0. \tag{2.B.2}$$

For discrete random variables that take on values in $\mathbb{N}_+$ the definition of $\le_{\text{hmrl}}$ should be modified. Let X and Y be two such random variables. We denote $X \le_{\text{hmrl}} Y$ if

$$\frac{\sum_{j=n}^\infty P\{X \ge j\}}{E[X]} \le \frac{\sum_{j=n}^\infty P\{Y \ge j\}}{E[Y]}, \quad n = 1, 2, \ldots. \tag{2.B.3}$$

2.B.2 The relation between the harmonic mean residual life and some other stochastic orders

Since the harmonic averages of m and l are increasing functionals of m and l, respectively, it follows that

$$X \le_{\text{mrl}} Y \Longrightarrow X \le_{\text{hmrl}} Y.$$

The order $\le_{\text{hmrl}}$ is closely related to the order $\le_{\text{icx}}$ which is studied in Section 4.A. The reader may find it helpful to browse over that section now, since some of the ideas that are explained there are used below.

Note that both (2.B.2) and (2.B.3) are equivalent to

$$\frac{E[(X-t)_+]}{E[X]} \le \frac{E[(Y-t)_+]}{E[Y]} \quad \text{for all } t \ge 0, \tag{2.B.4}$$

and from (2.B.4) it follows that $X \le_{\text{hmrl}} Y$ if, and only if,

$$\frac{E[\phi(X)]}{E[X]} \le \frac{E[\phi(Y)]}{E[Y]} \quad \text{for all increasing convex functions } \phi : [0,\infty) \to \mathbb{R}, \tag{2.B.5}$$

such that the expectations exist. It is worthwhile to note that condition (2.B.4) uses the expectations $E[(X-t)_+]$ and $E[(Y-t)_+]$ as (2.A.5) and as (3.A.5) in Chapter 3 and (4.A.4) in Chapter 4 do. In Chapter 4, where the order $\le_{\rm icx}$ is studied, we will use (2.B.4) in order to derive a relationship between the orders $\le_{\rm hmrl}$ and $\le_{\rm icx}$ (see Theorem 4.A.28).

Neither of the orders $\le_{\rm st}$ and $\le_{\rm hmrl}$ implies the other; counterexamples can be found in the literature.

Letting $x \to 0$ in (2.B.1) we obtain $m(0) \le l(0)$, that is,

$$X \le_{\rm hmrl} Y \Longrightarrow E[X|X>0] \le E[Y|Y>0].$$

Thus, when X and Y are positive almost surely, then

$$X \le_{\rm hmrl} Y \Longrightarrow EX \le EY. \tag{2.B.6}$$

If $EX = EY$, then

$$X \le_{\rm hmrl} Y \Longleftrightarrow X \le_{\rm cx} Y, \tag{2.B.7}$$

where the order $\le_{\rm cx}$ is studied in Section 3.A (see (3.A.7)). Thus, from (3.A.4) it follows that if $X \le_{\rm hmrl} Y$ and $EX = EY$, then $\text{Var}[X] \le \text{Var}[Y]$. Under the proper condition, even if X and Y do not have the same mean, one can still get the variance inequality; this is shown in the next result.

Theorem 2.B.1. *Let X and Y be two almost surely positive random variables with finite second moments. If $X \le_{\rm hmrl} Y$, and if Y is* NWUE, *then* $\text{Var}[X] \le \text{Var}[Y]$.

Proof. From (2.B.5) we get

$$\frac{E[X^2]}{E[X]} \le \frac{E[Y^2]}{E[Y]}. \tag{2.B.8}$$

From Barlow and Proschan [36, page 187] it is seen that $\text{Var}[Y] \ge \{E[Y]\}^2$, since Y is NWUE. Thus, using (2.B.6), we see that $\text{Var}[Y] \ge E[Y]E[X]$. Therefore

$$\begin{aligned}
\text{Var}[Y] &\ge \frac{E[X]}{E[Y]}\text{Var}[Y] + \{E[Y]-E[X]\}E[X] \\
&= \frac{E[X]}{E[Y]} \cdot E[Y^2] - \{E[X]\}^2 \\
&\ge E[X^2] - \{E[X]\}^2 \\
&= \text{Var}[X],
\end{aligned}$$

where the last inequality follows from (2.B.8). □

The harmonic mean residual life order can be characterized by means of the usual stochastic order and the appropriate equilibrium age variables. Recall from (1.A.20) that for nonnegative random variables X and Y with finite means we denote by A_X and A_Y the corresponding asymptotic equilibrium ages. The following result follows at once from (1.A.1) and (2.B.2). It may be contrasted with Theorems 1.C.13 and 2.A.4.

Theorem 2.B.2. *For nonnegative random variables* X *and* Y *with finite means we have* $X \leq_{\rm hmrl} Y$ *if, and only if,* $A_X \leq_{\rm st} A_Y$.

In the next theorem the order $\leq_{\rm hmrl}$ is characterized by ordering two related random variables in the sense of the usual stochastic order. Let X and Y be two nonnegative random variables with finite means and suppose that $X \leq_{\rm st} Y$ and that $EX < EY$. Let F and G be the distribution functions of X and of Y, respectively. Define the random variable $Z_{X,Y}$ as the random variable that has the density function h given by (1.C.7), as in Theorem 1.C.14; see also Theorem 2.A.5.

Theorem 2.B.3. *Let* X *and* Y *be two nonnegative random variables with finite means such that* $X \leq_{\rm st} Y$ *and such that* $EY > EX > 0$. *Then*

$$X \leq_{\rm hmrl} Y \Longleftrightarrow A_Y \leq_{\rm st} Z_{X,Y} \Longleftrightarrow A_X \leq_{\rm st} Z_{X,Y},$$

where $Z_{X,Y}$ *has the density function given in* (1.C.7).

Proof. It is easy to see that (here $\overline{H}$ is the survival function of Z, $\overline{G}_{\rm e}$ is the survival function of A_Y, and $\overline{F}_{\rm e}$ is as in (1.A.20))

$$\overline{H}(x) - \overline{G}_{\rm e}(x) = \frac{EX}{EY - EX}\Big[\overline{G}_{\rm e}(x) - \overline{F}_{\rm e}(x)\Big], \quad x \geq 0.$$

Thus the first stated equivalence follows from Theorem 2.B.2. The proof of the second equivalence is similar. □

The order $\leq_{\rm hmrl}$ can characterize the order $\leq_{\rm mrl}$ as follows.

Theorem 2.B.4. *Let* X *and* Y *be two nonnegative random variables with finite means. Then* $X \leq_{\rm mrl} Y$ *if, and only if,* $[X - t|X > t] \leq_{\rm hmrl} [Y - t|Y > t]$ *for all* $t \geq 0$.

The proof of Theorem 2.B.4 consists of applying (2.B.2) to $[X - t|X > t]$ and $[Y - t|Y > t]$, for each $t \geq 0$, and then showing that the resulting inequality is equivalent to (2.A.3). We omit the details.

2.B.3 Some closure properties

Under the proper conditions, the order $\leq_{\rm hmrl}$ is closed under the operation of convolution. First we prove the following lemma. Recall that a nonnegative random variable X with a finite mean is called NBUE (new better than used in expectation) if $E[X - t|X > t] \leq E[X]$ for all $t > 0$. Note that a nonnegative NBUE random variable must be almost surely positive.

Lemma 2.B.5. *If the two almost surely positive random variables* X *and* Y *are such that* $X \leq_{\rm hmrl} Y$, *and if* Z *is an* NBUE *nonnegative random variable independent of* X *and* Y, *then*

$$X + Z \leq_{\rm hmrl} Y + Z.$$

Proof. Let F, G, and H [$\overline{F}$, $\overline{G}$, and $\overline{H}$] be the distribution [survival] functions corresponding to X, Y, and Z, respectively. The corresponding equilibrium age distribution [survival] functions will be denoted by $F_{\rm e}$, $G_{\rm e}$, and $H_{\rm e}$ [$\overline{F}_{\rm e}$, $\overline{G}_{\rm e}$, and $\overline{H}_{\rm e}$]. Let A_X, A_Y, A_Z, A_{X+Z}, and A_{Y+Z} denote the asymptotic equilibrium ages corresponding to X, Y, Z, $X+Z$, and $Y+Z$, respectively. Now compute

$$\begin{aligned}
P\{A_{X+Z}>t\} &= \frac{1}{E[X+Z]}\int_{v=t}^{\infty} P\{X+Z>v\}{\rm d}v \\
&= \frac{1}{EX+EZ}\int_{v=t}^{\infty}\int_{u=0}^{\infty}\overline{F}(v-u){\rm d}H(u){\rm d}v \\
&= \frac{1}{EX+EZ}\int_{u=0}^{\infty}\int_{v=t}^{\infty}\overline{F}(v-u){\rm d}v{\rm d}H(u) \\
&= \frac{1}{EX+EZ}\int_{u=0}^{\infty}\int_{v=t-u}^{\infty}\overline{F}(v){\rm d}v{\rm d}H(u) \\
&= \frac{1}{EX+EZ}\left[\int_{u=0}^{t}\int_{v=t-u}^{\infty}\overline{F}(v){\rm d}v{\rm d}H(u)\right. \\
&\qquad \left. +\int_{u=t}^{\infty}\int_{v=0}^{\infty}\overline{F}(v){\rm d}v{\rm d}H(u)+\int_{u=t}^{\infty}\int_{v=t-u}^{0}{\rm d}v{\rm d}H(u)\right] \\
&= \frac{1}{EX+EZ}\left[EX\int_0^t\overline{F}_{\rm e}(t-u){\rm d}H(u)\right. \\
&\qquad \left. +EX\cdot\overline{H}(t)+\int_t^{\infty}\overline{H}(u){\rm d}u\right] \\
&= \frac{1}{EX+EZ}\left[EX\cdot P\{A_X+Z>t\}+EZ\cdot\overline{H}_{\rm e}(t)\right],
\end{aligned}$$

where A_X and Z are taken to be independent in the above expression. Now, since Z is NBUE we have that $Z \geq_{\rm st} A_Z$. Therefore

$$P\{A_X+Z>t\}\geq P\{A_X+A_Z>t\}\geq P\{A_Z>t\}=\overline{H}_{\rm e}(t). \tag{2.B.9}$$

Now notice that

$$\begin{aligned}
P\{A_{X+Z}>t\} &= \frac{1}{EX+EZ}\left[EX\cdot P\{A_X+Z>t\}+EZ\cdot\overline{H}_{\rm e}(t)\right] \\
&\leq \frac{1}{EY+EZ}\left[EY\cdot P\{A_X+Z>t\}+EZ\cdot\overline{H}_{\rm e}(t)\right] \\
&\leq \frac{1}{EY+EZ}\left[EY\cdot P\{A_Y+Z>t\}+EZ\cdot\overline{H}_{\rm e}(t)\right] \\
&= P\{A_{Y+Z}>t\}
\end{aligned}$$

(A_Y and Z are taken to be independent in the above), where the first inequality follows from (2.B.6) and (2.B.9), and the second inequality follows from Theorem 2.B.2. The result now follows from Theorem 2.B.2. □

Repeated application of Lemma 2.B.5, using the closure property of NBUE under convolution, and noting that every NBUE random variable is almost surely positive, yields the following result.

Theorem 2.B.6. *Let (X_i, Y_i), $i = 1, 2, \ldots, m$, be independent pairs of nonnegative random variables such that $X_i \le_{\mathrm{hmrl}} Y_i$, $i = 1, 2, \ldots, m$. If X_i, Y_i, $i = 1, 2, \ldots, m$, are all* NBUE, *then*

$$\sum_{i=1}^{m} X_i \le_{\mathrm{hmrl}} \sum_{i=1}^{m} Y_i.$$

Using Theorem 2.B.6 we can prove the following result.

Theorem 2.B.7. *Let $X_1, X_2, \ldots$ and $Y_1, Y_2, \ldots$ each be a sequence of* NBUE *nonnegative independent and identically distributed random variables such that $X_i \le_{\mathrm{hmrl}} Y_i$, $i = 1, 2, \ldots$. Let M and N be integer-valued positive random variables that are independent of the $\{X_i\}$ and the $\{Y_i\}$ sequences, respectively, such that $M \le_{\mathrm{hmrl}} N$. Then*

$$\sum_{j=1}^{M} X_j \le_{\mathrm{hmrl}} \sum_{j=1}^{N} Y_j.$$

Proof. The proof here is similar to the proof of Theorem 4.A.9. The reader may wish to look at that proof before continuing to read the present proof.

From Theorem 2.B.6 and (2.B.4) it is seen that $\frac{1}{mE[X_1]} E\big[\big(\sum_{i=1}^{m} X_i - u\big)_+\big] \le \frac{1}{mE[Y_1]} E\big[\big(\sum_{i=1}^{m} Y_i - u\big)_+\big]$ (all the X_i's have the same mean, and also all the Y_i's have the same mean). Therefore

$$\frac{E\big[\big(\sum_{i=1}^{m} X_i - u\big)_+\big]}{E[X_1]} \le \frac{E\big[\big(\sum_{i=1}^{m} Y_i - u\big)_+\big]}{E[Y_1]} \quad \text{for all } u \ge 0,\ m = 1, 2, \ldots.$$

Thus

$$\begin{aligned}
\frac{E\big[\big(\sum_{i=1}^{M} X_i - u\big)_+\big]}{E\big[\sum_{i=1}^{M} X_i\big]} &= \frac{\sum_{m=1}^{\infty} E\big[\big(\sum_{i=1}^{m} X_i - u\big)_+\big] P\{M = m\}}{E[M]E[X_1]} \\
&\le \frac{\sum_{m=1}^{\infty} E\big[\big(\sum_{i=1}^{m} Y_i - u\big)_+\big] P\{M = m\}}{E[M]E[Y_1]} \\
&= \frac{E\big[\big(\sum_{i=1}^{M} Y_i - u\big)_+\big]}{E\big[\sum_{i=1}^{M} Y_i\big]}.
\end{aligned}$$

Therefore (again by (2.B.4)) we have

$$\sum_{i=1}^{M} X_i \le_{\mathrm{hmrl}} \sum_{i=1}^{M} Y_i. \tag{2.B.10}$$

Now let ϕ be an increasing convex function and denote $g(n) \equiv E[\phi(Y_1 + Y_2 + \cdots + Y_n)]$. In the proof of Theorem 4.A.9 it is shown that $g(n)$ is increasing and convex in n. Therefore, since $M \leq_{\text{hmrl}} N$, we have that $\frac{E\left[\phi\left(\sum_{i=1}^{M} Y_i\right)\right]}{E[M]} \leq \frac{E\left[\phi\left(\sum_{i=1}^{N} Y_i\right)\right]}{E[N]}$, and since the Y_i's have the same mean we have that

$$\frac{E\left[\phi\left(\sum_{i=1}^{M} Y_i\right)\right]}{E\left[\sum_{i=1}^{M} Y_i\right]} = \frac{E\left[\phi\left(\sum_{i=1}^{M} Y_i\right)\right]}{E[M]E[Y_1]} \leq \frac{E\left[\phi\left(\sum_{i=1}^{N} Y_i\right)\right]}{E[N]E[Y_1]} = \frac{E\left[\phi\left(\sum_{i=1}^{N} Y_i\right)\right]}{E\left[\sum_{i=1}^{N} Y_i\right]}.$$

Thus we have that

$$\sum_{i=1}^{M} Y_i \leq_{\text{hmrl}} \sum_{i=1}^{N} Y_i. \tag{2.B.11}$$

The inequalities (2.B.10) and (2.B.11) yield the stated result. □

A result that is related to Theorem 2.B.7 is given next. It is of interest to compare it to Theorem 1.A.5.

Theorem 2.B.8. *Let $\{X_j,\ j = 1, 2, \ldots\}$ be a sequence of nonnegative independent and identically distributed* NBUE *random variables, and let M be a positive integer-valued random variable which is independent of the X_i's. Let $\{Y_j,\ j = 1, 2, \ldots\}$ be another sequence of nonnegative independent and identically distributed* NBUE *random variables, and let N be a positive integer-valued random variable which is independent of the Y_i's. Suppose that for some positive integer K we have*

$$\sum_{i=1}^{K} X_i \leq_{\text{hmrl}} [\geq_{\text{hmrl}}]\ Y_1,$$

and

$$M \leq_{\text{hmrl}} [\geq_{\text{hmrl}}]\ KN.$$

Then

$$\sum_{j=1}^{M} X_j \leq_{\text{hmrl}} [\geq_{\text{hmrl}}] \sum_{j=1}^{N} Y_j.$$

We do not give a detailed proof of Theorem 2.B.8 here since it is similar to the proof of Theorem 4.A.12 in Section 4.A.1. In order to construct a proof of Theorem 2.B.8 from the proof of Theorem 4.A.12 one just uses the equivalence (2.B.7) and one replaces the application of Theorem 4.A.9 by an application of Theorem 2.B.7.

Two other similar theorems are the following. Their proofs are similar to the proofs of Theorems 4.A.13 and 4.A.14 in Section 4.A.1.

Theorem 2.B.9. *Let $\{X_j,\ j = 1, 2, \ldots\}$ be a sequence of nonnegative independent and identically distributed* NBUE *random variables, and let M*

be a positive integer-valued random variable which is independent of the X_i*'s. Let* $\{Y_j,\ j=1,2,\dots\}$ *be another sequence of nonnegative independent and identically distributed* NBUE *random variables, and let* N *be a positive integer-valued random variable which is independent of the* Y_i*'s. Also, let* $\{N_j,\ j=1,2,\dots\}$ *be a sequence of independent random variables that are distributed as* N*. If for some positive integer* K *we have*

$$\sum_{i=1}^{K} X_i \le_{\mathrm{hmrl}} Y_1 \quad \text{and} \quad M \le_{\mathrm{hmrl}} \sum_{i=1}^{K} N_i,$$

or if we have

$$KX_1 \le_{\mathrm{hmrl}} Y_1 \quad \text{and} \quad M \le_{\mathrm{hmrl}} KN,$$

or if we have

$$KX_1 \le_{\mathrm{hmrl}} Y_1 \quad \text{and} \quad M \le_{\mathrm{hmrl}} \sum_{i=1}^{K} N_i,$$

then

$$\sum_{j=1}^{M} X_j \le_{\mathrm{hmrl}} \sum_{j=1}^{N} Y_j.$$

Theorem 2.B.10. *Let* $\{X_j,\ j=1,2,\dots\}$ *be a sequence of nonnegative independent and identically distributed* NBUE *random variables, and let* M *be a positive integer-valued random variable which is independent of the* X_i*'s. Let* $\{Y_j,\ j=1,2,\dots\}$ *be another sequence of nonnegative independent and identically distributed* NBUE *random variables, and let* N *be a positive integer-valued random variable which is independent of the* Y_i*'s. If for some positive integers* K_1 *and* K_2*, such that* $K_1 \le K_2$*, we have*

$$\sum_{i=1}^{K_1} X_i \le_{\mathrm{hmrl}} \frac{K_1}{K_2} Y_1 \quad \text{and} \quad M \le_{\mathrm{hmrl}} K_2 N,$$

then

$$\sum_{j=1}^{M} X_j \le_{\mathrm{hmrl}} \sum_{j=1}^{N} Y_j.$$

The harmonic mean residual life order does not have the property of being simply closed under mixtures. However, under quite strong conditions the order $\le_{\mathrm{hmrl}}$ is closed under mixtures. This is shown in the next theorem which may be compared with Theorems 1.B.8 and 2.A.13.

Theorem 2.B.11. *Let* X *and* Y *be nonnegative random variables, and let* Θ *be another random variable, such that* $[X \mid \Theta = \theta] \le_{\mathrm{hmrl}} [Y \mid \Theta = \theta']$ *for all* θ *and* θ' *in the support of* Θ*. Then* $X \le_{\mathrm{hmrl}} Y$*.*

Proof. The proof is similar to the proof of Theorem 1.B.8. Select a θ and a θ' in the support of Θ. Let $\overline{F}(\cdot|\theta)$, $\overline{G}(\cdot|\theta)$, $\overline{F}(\cdot|\theta')$, and $\overline{G}(\cdot|\theta')$ be the survival functions of $[X|\Theta=\theta]$, $[Y|\Theta=\theta]$, $[X|\Theta=\theta']$, and $[Y|\Theta=\theta']$, respectively. Let $E[X|\theta]$, $E[Y|\theta]$, $E[X|\theta']$, and $E[Y|\theta']$ be the corresponding expectations. By (2.B.2) it is sufficient to show that for $\alpha \in (0,1)$ we have

$$\frac{\alpha\int_t^\infty \overline{F}(u|\theta)\mathrm{d}u + (1-\alpha)\int_t^\infty \overline{F}(u|\theta')\mathrm{d}u}{\alpha E[X|\theta] + (1-\alpha)E[X|\theta']} \leq \frac{\alpha\int_t^\infty \overline{G}(u|\theta)\mathrm{d}u + (1-\alpha)\int_t^\infty \overline{G}(u|\theta')\mathrm{d}u}{\alpha E[Y|\theta] + (1-\alpha)E[Y|\theta']} \quad \text{for all } t \geq 0. \tag{2.B.12}$$

The proof of this inequality is similar to the proof of (1.B.12). □

Another condition under which the order $\leq_{\text{hmrl}}$ is closed under mixtures is given in the following theorem.

Theorem 2.B.12. *Let X and Y be nonnegative random variables, and let Θ be another random variable, such that $[X|\Theta=\theta] \leq_{\text{hmrl}} [Y|\Theta=\theta]$ for all θ in the support of Θ. Furthermore, assume that*

$$\frac{E[Y|\Theta=\theta]}{E[X|\Theta=\theta]} = k \quad (\textit{independent of } \theta). \tag{2.B.13}$$

Then $X \leq_{\text{hmrl}} Y$.

Proof. As in the proof of Theorem 2.B.11, select a θ and a θ' in the support of Θ. Let $\overline{F}(\cdot|\theta)$, $\overline{G}(\cdot|\theta)$, $\overline{F}(\cdot|\theta')$, and $\overline{G}(\cdot|\theta')$ be the survival functions of $[X|\Theta=\theta]$, $[Y|\Theta=\theta]$, $[X|\Theta=\theta']$, and $[Y|\Theta=\theta']$, respectively. Let $E[X|\theta]$, $E[Y|\theta]$, $E[X|\theta']$, and $E[Y|\theta']$ be the corresponding expectations.

Let $\alpha \in (0,1)$. Note that from (2.B.13) we obtain

$$\frac{\alpha E[Y|\theta] + (1-\alpha)E[Y|\theta']}{\alpha E[X|\theta] + (1-\alpha)E[X|\theta']} = k. \tag{2.B.14}$$

Also, from $[X|\Theta=\theta] \leq_{\text{hmrl}} [Y|\Theta=\theta]$, $[X|\Theta=\theta'] \leq_{\text{hmrl}} [Y|\Theta=\theta']$, and (2.B.13), we get, for $t \geq 0$, that

$$k\int_t^\infty \overline{F}(u|\theta)\mathrm{d}u \leq \int_t^\infty \overline{G}(u|\theta)\mathrm{d}u \quad \text{and} \quad k\int_t^\infty \overline{F}(u|\theta')\mathrm{d}u \leq \int_t^\infty \overline{G}(u|\theta')\mathrm{d}u,$$

and hence

$$k\Big[\alpha\int_t^\infty \overline{F}(u|\theta)\mathrm{d}u + (1-\alpha)\int_t^\infty \overline{F}(u|\theta')\mathrm{d}u\Big] \leq \alpha\int_t^\infty \overline{G}(u|\theta)\mathrm{d}u + (1-\alpha)\int_t^\infty \overline{G}(u|\theta')\mathrm{d}u.$$

From this inequality and (2.B.14) we obtain (2.B.12), and this completes the proof. □

Consider now a family of distribution functions $\{G_\theta,\ \theta \in \mathcal{X}\}$ where $\mathcal{X}$ is a subset of the real line. As in Sections 1.A.3 and 1.C.3 let $X(\theta)$ denote a random variable with distribution function G_θ. For any random variable Θ with support in $\mathcal{X}$, and with distribution function F, let us denote by $X(\Theta)$ a random variable with distribution function H given by

$$H(y) = \int_{\mathcal{X}} G_\theta(y)\mathrm{d}F(\theta), \quad y \in \mathbb{R}.$$

The following result is comparable to Theorems 1.A.6, 1.B.14, 1.B.52, 1.C.17 and 2.A.15.

Theorem 2.B.13. *Consider a family of distribution functions $\{G_\theta,\ \theta \in \mathcal{X}\}$ as above. Let Θ_1 and Θ_2 be two random variables with supports in $\mathcal{X}$ and distribution functions F_1 and F_2, respectively. Let Y_1 and Y_2 be two random variables such that $Y_i =_{\text{st}} X(\Theta_i)$, $i = 1, 2$, that is, suppose that the distribution function of Y_i is given by*

$$H_i(y) = \int_{\mathcal{X}} G_\theta(y)\mathrm{d}F_i(\theta), \quad y \in \mathbb{R},\ i = 1, 2.$$

If

$$X(\theta) \le_{\text{hmrl}} X(\theta') \quad \textit{whenever } \theta \le \theta', \tag{2.B.15}$$

and if

$$\Theta_1 \le_{\text{hr}} \Theta_2, \tag{2.B.16}$$

then

$$Y_1 \le_{\text{hmrl}} Y_2. \tag{2.B.17}$$

The proof of Theorem 2.B.13 uses the increasing convex order, and is therefore given in Remark 4.A.29 in Chapter 4.

A Laplace transform characterization of the order $\le_{\text{hmrl}}$ is given next; it may be compared to Theorems 1.A.13, 1.B.18, 1.B.53, 1.C.25, and 2.A.16.

Theorem 2.B.14. *Let X_1 and X_2 be two nonnegative random variables, and let $N_\lambda(X_1)$ and $N_\lambda(X_2)$ be as described in Theorem* 1.A.13. *Then*

$$X_1 \le_{\text{hmrl}} X_2 \Longleftrightarrow N_\lambda(X_1) \le_{\text{hmrl}} N_\lambda(X_2) \quad \textit{for all } \lambda > 0,$$

where the notation $N_\lambda(X_1) \le_{\text{hmrl}} N_\lambda(X_2)$ is in the sense of (2.B.3).

Proof. First assume that $X \le_{\text{hmrl}} Y$. As in the proof of Theorem 2.A.16 we temporarily replace the notation $\overline{\alpha}_\lambda^{X_1}(n)$ and $\overline{\alpha}_\lambda^{X_2}(n)$, by $\overline{\alpha}_{\lambda,1}(n)$ and $\overline{\alpha}_{\lambda,2}(n)$, respectively. We also denote the survival function and the mean of X_k by $\overline{F}_k$ and μ_k, respectively, $k = 1, 2$. Let $m \ge 2$. Using (2.A.23) we have

$$\mu_1 \sum_{n=m}^{\infty} \overline{P}_2(n) - \mu_2 \sum_{n=m}^{\infty} \overline{P}_1(n)$$
$$= \int_0^\infty \lambda^2 \mathrm{e}^{-\lambda x} \frac{(\lambda x)^{m-2}}{(m-2)!} \Big[\mu_1 \int_x^\infty \overline{F}_2(u)\mathrm{d}u - \mu_2 \int_x^\infty \overline{F}_1(u)\mathrm{d}u\Big]\mathrm{d}x.$$

The integrand is nonnegative by the assumption of the theorem, and one direction of the proof is complete.

The proof of the converse statement is similar to the proof of the converse of Theorem 2.A.16. □

The following result gives necessary and sufficient conditions for two random variables to be equal in the sense of the order $\leq_{\text{hmrl}}$.

Theorem 2.B.15. *Let X and Y be two nonnegative random variables with positive expectations, such that $EX \leq EY$. Then $X =_{\text{hmrl}} Y$ if, and only if, $X =_{\text{st}} BY$ for some Bernoulli random variable B, independent of Y.*

Proof. First assume that $X =_{\text{st}} BY$ for some Bernoulli random variable B, independent of Y. Then

$$\frac{E[(X-t)_+]}{E[X]} = \frac{E[(BY-t)_+]}{E[BY]} = \frac{E[(Y-t)_+]P\{B=1\}}{E[Y]P\{B=1\}} = \frac{E[(Y-t)_+]}{E[Y]} \quad \text{for all } t \geq 0,$$

and thus $X =_{\text{hmrl}} Y$ follows from (2.B.4).

Conversely, suppose that $X =_{\text{hmrl}} Y$. By (2.B.2) this means that

$$\frac{\int_t^\infty P\{X>u\}du}{EX} = \frac{\int_t^\infty P\{Y>u\}du}{EY} \quad \text{for all } t \geq 0,$$

which yields

$$P\{X>t\} = \frac{EX}{EY} \cdot P\{Y>t\}, \quad t \geq 0.$$

That is, $X =_{\text{st}} BY$, where B is a Bernoulli random variable such that $P\{B = 1\} = EX/EY$. □

From the proof of Theorem 2.B.15 it is seen, in contrast to (2.B.6), that if $X \leq_{\text{hmrl}} Y$, then it does not necessarily follow that $EX \leq EY$ (unless X and Y are positive almost surely).

In the next result it is shown that a random variable, whose distribution is the mixture of two distributions of harmonic mean residual life ordered random variables, is bounded from below and from above, in the harmonic mean residual life order sense, by these two random variables.

Theorem 2.B.16. *Let X and Y be two nonnegative random variables with distribution functions F and G, respectively. Let W be a random variable with the distribution function $pF + (1-p)G$ for some $p \in (0,1)$. If $X \leq_{\text{hmrl}} Y$, then $X \leq_{\text{hmrl}} W \leq_{\text{hmrl}} Y$.*

Proof. By assumption, (2.B.2) holds. Therefore

$$\frac{\int_x^\infty \overline{F}(u)\mathrm{d}u}{EX} \le \frac{p\int_x^\infty \overline{F}(u)\mathrm{d}u + (1-p)\int_x^\infty \overline{G}(u)\mathrm{d}u}{pEX + (1-p)EY} \le \frac{\int_x^\infty \overline{G}(u)\mathrm{d}u}{EY} \quad \text{for all } x \ge 0,$$

and the stated result follows from (2.B.2). □

2.B.4 Properties in reliability theory

The order $\le_{\text{hmrl}}$ can be used to characterize DMRL random variables. As in Section 1.A.3, $[Z|A]$ denotes any random variable that has as its distribution the conditional distribution of Z given A.

Theorem 2.B.17. *The nonnegative random variable X is* DMRL *if, and only if, $[X-t|X>t] \ge_{\text{hmrl}} [X-t'|X>t']$ whenever $t' \ge t \ge 0$.*

The proof is simple and thus omitted.

Other characterizations of DMRL and IMRL random variables, by means of other stochastic orders, can be found in Theorems 2.A.23, 3.A.56, 3.C.13, and 4.A.51.

The order $\le_{\text{hmrl}}$ can also be used to characterize NBUE random variables as follows.

Theorem 2.B.18. *Let X be a nonnegative random variable with a finite positive mean. Then the following assertions are equivalent:*

(i) *$X \le_{\text{hmrl}} X+Y$ for any nonnegative random variable Y with a finite positive mean, which is independent of X.*
(ii) *X is* NBUE.
(iii) *$X+Y_1 \le_{\text{hmrl}} X+Y_2$ whenever Y_1 and Y_2 are almost surely positive random variables with finite means, which are independent of X, such that $Y_1 \le_{\text{hmrl}} Y_2$.*

Proof. Suppose that (i) holds. Then, taking $Y =_{\text{a.s.}} y$ for some $y>0$, we get from (2.B.4) that

$$\frac{E[(X-t)_+]}{E[X]} \le \frac{E[(X+y-t)_+]}{E[X]+y}, \quad t \ge 0.$$

Upon rearrangement this gives

$$yE[(X-t)_+] \le E[X]\big\{E[(X+y-t)_+] - E[(X-t)_+]\big\}, \quad t \ge 0;$$

that is,

$$E[(X-t)_+] \le \frac{E[X]}{y}\int_{t-y}^{t} P\{X>u\}\mathrm{d}u, \quad t \ge 0.$$

Letting $y \to 0$ we obtain

$$E[(X-t)_+] \le E[X]P\{X>t\}, \quad t \ge 0,$$

that is, X is NBUE.

The statement (ii)$\Longrightarrow$(iii) is Lemma 2.B.5.

Now assume that (iii) holds. Let $Y_1 =_{\text{a.s}} a$ and $Y_2 =_{\text{a.s}} y$, where $0 < a < y$. It is easy to verify (for instance, using (2.B.4)) that $Y_1 \leq_{\text{hmrl}} Y_2$. That is,

$$(E[X] + y)E[(X + a - t)_+] \leq (E[X] + a)E[(X + y - t)_+], \quad t \geq 0.$$

Letting $a \to 0$ we obtain

$$(E[X] + y)E[(X - t)_+] \leq E[X]E[(X + y - t)_+], \quad t \geq 0, \ y \geq 0.$$

Integrating both sides of the above inequality with respect to the distribution of Y (Y is any random variable as described in (i)) we obtain

$$(E[X] + E[Y])E[(X - t)_+] \leq E[X]E[(X + Y - t)_+], \quad t \geq 0,$$

that is, by (2.B.4), we have $X \leq_{\text{hmrl}} X + Y$. $\square$

Another characterization of NBUE random variables by means of the usual stochastic order is given in Theorem 1.A.31.

2.C Complements

Section 2.A: Basic properties of the mrl function (which is also called the *biometric function*) can be found in Yang [572] and references therein. Some properties of the mrl functions are summarized in Shaked and Shanthikumar [513], where further references can be found. The counterexamples mentioned after Theorem 2.A.1 can also be found in that paper and further counterexamples can be found in Gupta and Kirmani [216] and in Alzaid [12]. The conditions under which the $\leq_{\text{mrl}}$ order implies the $\leq_{\text{hr}}$ and the $\leq_{\text{st}}$ orders (Theorems 2.A.2 and 2.A.3) are taken from Gupta and Kirmani [216]. The equivalence of the order $\leq_{\text{mrl}}$ and (2.A.3) can be found, for example, in Singh [536]. The characterization of the order $\leq_{\text{mrl}}$ which is given in Theorem 2.A.5 is taken from Di Crescenzo [164]. The characterizations of the order $\leq_{\text{hr}}$ by means of the order $\leq_{\text{mrl}}$, given in Theorems 2.A.6 and 2.A.7, can be found in Belzunce, Gao, Hu, and Pellerey [67]. The closure under convolution results of the order $\leq_{\text{mrl}}$ in Section 2.A.3 were communicated to us by Pellerey [444]. A special case of Lemma 2.A.8 can be found in Mukherjee and Chatterjee [403]. Theorem 2.A.9 can be found in Pellerey [448] and Theorem 2.A.12 can be found in Fagiuoli and Pellerey [186]. The fact that a DMRL random variable increases in the order $\leq_{\text{mrl}}$ when a nonnegative random variable is added to it (Theorem 2.A.11) is a result that is slightly stronger than a result in Frostig [207]. The closure under mixtures result (Theorem 2.A.13) is taken from Nanda, Jain, and Singh [424]. The characterization of the mrl order that is given in Theorem 2.A.14 can be found in Joag-Dev, Kochar, and Proschan [259], whereas its special case given in (2.A.10) is taken from Fagiuoli and Pellerey [187]. Fagiuoli and Pellerey [187] have extended (2.A.10) to sums of mrl ordered random variables. The closure under mixtures property of the order $\leq_{\rm mrl}$ (Theorem 2.A.15) is a special case of a result of Hu, Kundu, and Nanda [236], and it can also be found in Hu, Nanda, Xie, and Zhu [237]; see also Theorem 3.4 in Ahmed [7]. The Laplace transform characterization of the order $\leq_{\rm mrl}$ (Theorem 2.A.16) is taken from Shaked and Wong [524]; see also Kan and Yi [274]. An extension of Theorem 2.A.16 to more general orders can be found in Nanda [422]. The mean residual life order comparisons of order statistics (Theorems 2.A.20 and 2.A.21) can be found in Hu, Zhu, and Wei [243] and in Hu and Wei [240]. The comparison of inter-epoch times of two nonhomogeneous Poisson processes in the sense of the mean residual life order (Example 2.A.22) is taken from Belzunce, Lillo, Ruiz, and Shaked [69]. The result that a convolution of an IFR and a DMRL random variables is DMRL (Corollary 2.A.24) can be found in Kopocinska and Kopocinski [320].

Nanda, Singh, Misra, and Paul [429] studied a notion of reversed residual lifetime, and introduced and studied a stochastic order based on it.

An order which is related to the mean residual life order is introduced in Ebrahimi and Zahedi [179]. If m and l are the mrl functions of X and Y, respectively, then the order is defined by requiring $\frac{\rm d}{{\rm d}t}(l(t) - m(t))$ to be monotone in t. Ebrahimi and Zahedi [179] show that this order implies the mean residual life order.

In Kirmani [297] it is claimed that the spacings, from a sample of independent and identically distributed IMRL random variables, are ordered in the mean residual life order. However, the proof of Kirmani is erroneous; see Kirmani [298].

Section 2.B: The order $\leq_{\rm hmrl}$ is studied, for example, in Deshpande, Singh, Bagai, and Jain [161] and in Heilmann and Schröter [219]. Baccelli and Makowski [28] call it the *forward recurrence times stochastic order* (see an additional comment on the paper of Baccelli and Makowski [28] in Section 4.C). The counterexamples mentioned after (2.B.5) can be found, for example, in Mi [394]. In fact, Gerchak and Golani [209] have noticed that the example given on page 489 of Wolff [567] shows that it is possible for both $X \leq_{\rm st} Y$ and $Y \leq_{\rm hmrl} X$ to hold simultaneously in the strict sense. The comparison of the expectations of $\leq_{\rm hmrl}$ ordered random variables, described in (2.B.6), is a special case of a result of Nanda, Jain, and Singh [425]. The variance inequality (Theorem 2.B.1) can be found in Kirmani [297]. The characterization of the order $\leq_{\rm hmrl}$ which is given in Theorem 2.B.3 is taken from Di Crescenzo [164]. The characterization of the order $\leq_{\rm mrl}$ by means of the order $\leq_{\rm hmrl}$ (Theorem 2.B.4) can be found in Hu, Kundu, and Nanda [236]. The preservation under convolution property of the order $\leq_{\rm hmrl}$ (Theorem 2.B.6) is taken from Pellerey [448, 449] (the latter is a correction note), and the closure under random summations property of the order $\leq_{\rm hmrl}$ (Theorem 2.B.7) is also taken from Pellerey [448, 449], though it is alluded to in Heilmann and Schröter [219]. These results (Theorems 2.B.6 and 2.B.7) can also be found in Baccelli and Makowski [28]. A slight extension of Theorem 2.B.6 is given in Lefèvre and Utev [340]. Theorems 2.B.8–2.B.10 have been communicated to us by Pellerey [447]. The closure under mixtures properties of the order $\leq_{\rm hmrl}$ (Theorems 2.B.11 and 2.B.12) are taken from Nanda, Jain, and Singh [424] and from Lefèvre and Utev [340], respectively, whereas Theorem 2.B.13 is inspired by Ahmed, Soliman, and Khider [9]. The Laplace transform characterization of the order $\leq_{\rm hmrl}$ (Theorem 2.B.14) is taken from Shaked and Wong [524]. An extension of Theorem 2.B.14 to more general orders can be found in Nanda [422]. The conditions under which $X =_{\rm hmrl} Y$ (Theorem 2.B.15) can be found in Lefèvre and Utev [340]. The NBUE characterization, given in Theorem 2.B.18, is taken from Lefèvre and Utev [340].

3

Univariate Variability Orders

In this chapter we study stochastic orders that compare the "variability" or the "dispersion" of random variables. The most important and common orders that are studied in this chapter are the convex and the dispersive orders. We also study in this chapter the excess wealth order (which is also called the right spread order) which is found to be useful in an increasing number of applications. Various related orders are also examined in this chapter.

3.A The Convex Order

3.A.1 Definition and equivalent conditions

Let X and Y be two random variables such that

$$E[\phi(X)] \le E[\phi(Y)] \quad \text{for all convex functions } \phi : \mathbb{R} \to \mathbb{R}, \tag{3.A.1}$$

provided the expectations exist. Then X is said to be *smaller than Y in the convex order* (denoted as $X \le_{\rm cx} Y$). Roughly speaking, convex functions are functions that take on their (relatively) larger values over regions of the form $(-\infty, a) \cup (b, \infty)$ for $a < b$. Therefore, if (3.A.1) holds, then Y is more likely to take on "extreme" values than X. That is, Y is "more variable" than X. It should be mentioned here that in (3.A.1) it is sufficient to consider only functions ϕ that are convex on the union of the supports of X and Y rather than over the whole real line; we will not keep repeating this point throughout this chapter.

One can also define a *concave order* by requiring (3.A.1) to hold for all concave functions ϕ (denoted as $X \le_{\rm cv} Y$). However, $X \le_{\rm cv} Y$ if, and only if, $Y \le_{\rm cx} X$. Therefore, it is not necessary to have a separate discussion for the concave order.

Note that the functions ϕ_1 and ϕ_2, defined by $\phi_1(x) = x$ and $\phi_2(x) = -x$, are both convex. Therefore, from (3.A.1) it easily follows that

$$X \leq_{\rm cx} Y \Longrightarrow E[X] = E[Y], \tag{3.A.2}$$

provided the expectations exist. Later it will be helpful to observe that if $E[X] = E[Y]$, then

$$\int_{-\infty}^{\infty} \Big[F(u) - G(u)\Big] du = \int_{-\infty}^{\infty} \Big[\overline{F}(u) - \overline{G}(u)\Big] du = 0, \tag{3.A.3}$$

provided the integrals exist, where $\overline{F}$ [F] and $\overline{G}$ [G] are the survival [distribution] functions of X and Y, respectively. The function ϕ, defined by $\phi(x) = x^2$, is convex. Therefore, from (3.A.1) and (3.A.2), it follows that

$$X \leq_{\rm cx} Y \Longrightarrow {\rm Var}[X] \leq {\rm Var}[Y], \tag{3.A.4}$$

whenever $\mathrm{Var}(Y) < \infty$.

For a fixed a, the function ϕ_a, defined by $\phi_a(x) = (x-a)_+$, and the function φ_a, defined by $\varphi_a = (a-x)_+$, are both convex. (The reader is encouraged to draw a sketch of ϕ_a and φ_a since they are very handy in the analysis of the order $\leq_{\rm cx}$ as well as in the analysis of the monotone convex and the monotone concave orders discussed in Chapter 4.) Therefore, if $X \leq_{\rm cx} Y$, then

$$E[(X-a)_+] \leq E[(Y-a)_+] \quad \text{for all } a \tag{3.A.5}$$

and

$$E[(a-X)_+] \leq E[(a-Y)_+] \quad \text{for all } a, \tag{3.A.6}$$

provided the expectations exist. Alternatively, using a simple integration by parts, it is seen that (3.A.5) and (3.A.6) can be rewritten as

$$\int_x^{\infty} \overline{F}(u)du \leq \int_x^{\infty} \overline{G}(u)du \quad \text{for all } x \tag{3.A.7}$$

and

$$\int_{-\infty}^{x} F(u)du \leq \int_{-\infty}^{x} G(u)du \quad \text{for all } x, \tag{3.A.8}$$

provided the integrals exist.

In fact, when $E[X] = E[Y]$, (3.A.7) is equivalent to $X \leq_{\rm cx} Y$. To see this equivalence, note that every convex function can be approximated by (that is, is a limit of) positive linear combinations of the functions ϕ_a's, for various choices of a's, and of the function $\phi(x) = -x$. By (3.A.7), $E[\phi_a(X)] \leq E[\phi_a(Y)]$ for all a's, and this fact, together with the equality of the means of X and Y, implies (3.A.1). We thus have proved the first part of the following result. The other part is proven similarly.

Theorem 3.A.1. *Let X and Y be two random variables such that $E[X] = E[Y]$. Then*

(a) *$X \leq_{\rm cx} Y$ if, and only if, (3.A.7) holds.*

(b) $X \leq_{\text{cx}} Y$ *if, and only if,* (3.A.8) *holds.*

By adding a to both sides of the inequality in (3.A.5), it is seen that (3.A.5) can be rewritten as

$$\max\{X, a\}] \leq E[\max\{Y, a\}] \quad \text{for all } a. \tag{3.A.9}$$

Thus, when $E[X] = E[Y]$, then (3.A.9) is equivalent to $X \leq_{\text{cx}} Y$. In a similar manner (3.A.6) can be rewritten.

The following theorem provides another characterization of the convex order.

Theorem 3.A.2. *Let X and Y be two random variables such that $E[X] = E[Y]$. Then $X \leq_{\text{cx}} Y$ if, and only if,*

$$E|X - a| \leq E|Y - a| \quad \textit{for all } a \in \mathbb{R}. \tag{3.A.10}$$

Proof. Clearly, if $X \leq_{\text{cx}} Y$, then (3.A.10) holds. So suppose that (3.A.10) holds. Without loss of generality it can be assumed that $EX = EY = 0$. A straightforward computation gives

$$E|X - a| = a + 2\int_a^{\infty} \overline{F}(u)du = -a + 2\int_{-\infty}^{a} F(u)du. \tag{3.A.11}$$

The result now follows from (3.A.7) or (3.A.8). □

The function $-E|X - \cdot|$ is called the potential of the probability measure of X. Similarly, $-E|Y - \cdot|$ is the potential of the probability measure of Y. Thus, (3.A.10) can be written as $-E|X - \cdot| \geq -E|Y - \cdot|$ pointwise. Using this observation, we obtain from Chacon and Walsh [122] the following characterization.

Theorem 3.A.3. *Let X and Y be two random variables such that $E[X] = E[Y] = 0$. Then $X \leq_{\text{cx}} Y$ if, and only if, for a standard Brownian motion from 0, $\{B(t),\ t \geq 0\}$, there exist two stopping times T_1 and T_2, such that $T_1 \leq T_2$ almost surely, and $X =_{\text{st}} B(T_1)$ and $Y =_{\text{st}} B(T_2)$.*

An immediate consequence of (3.A.5) is shown next. Denote the supports of X and Y by $\text{supp}(X)$ and $\text{supp}(Y)$. Let $l_X = \inf\{x : x \in \text{supp}(X)\}$ and $u_X = \sup\{x : x \in \text{supp}(X)\}$. Define l_Y and u_Y similarly. Then we have that if $X \leq_{\text{cx}} Y$, then $l_Y \leq l_X$ and $u_Y \geq u_X$. As proof, suppose, for example, that $u_Y < u_X$. Let a be such that $u_Y < a < u_X$. Then $E[(Y - a)_+] = 0 < E[(X - a)_+]$, in contradiction to (3.A.5). Therefore we must have $u_Y \geq u_X$. Similarly, using (3.A.6), it can be shown that $l_Y \leq l_X$. As a consequence we have that if X and Y are random variables whose supports are intervals, then

$$X \leq_{\text{cx}} Y \Longrightarrow \text{supp}(X) \subseteq \text{supp}(Y). \tag{3.A.12}$$

An important characterization of the convex order by construction on the same probability space is stated in the next theorem.

Theorem 3.A.4. *The random variables X and Y satisfy $X \leq_{\text{cx}} Y$ if, and only if, there exist two random variables $\hat{X}$ and $\hat{Y}$, defined on the same probability space, such that*

$$\hat{X} =_{\text{st}} X,$$
$$\hat{Y} =_{\text{st}} Y,$$

and $\{\hat{X}, \hat{Y}\}$ is a martingale, that is,

$$E[\hat{Y}|\hat{X}] = \hat{X} \quad \text{a.s.} \tag{3.A.13}$$

Furthermore, the random variables $\hat{X}$ and $\hat{Y}$ can be selected such that $[\hat{Y}|\hat{X} = x]$ is increasing in x in the usual stochastic order $\leq_{\text{st}}$.

It is not easy to prove the constructive part of Theorem 3.A.4. However, it is easy to prove that if random variables $\hat{X}$ and $\hat{Y}$ as described in the theorem exist, then $X \leq_{\text{cx}} Y$. Just note that if ϕ is a convex function, then by Jensen's Inequality,

$$E[\phi(X)] = E[\phi(\hat{X})] = E\phi(E[\hat{Y}|\hat{X}]) \leq E\{E[\phi(\hat{Y})|\hat{X}]\} = E[\phi(\hat{Y})] = E[\phi(Y)],$$

which is (3.A.1).

Other characterizations of the convex order are described in the next theorem.

Theorem 3.A.5. *Let X and Y be two random variables with distribution functions F and G, respectively, and with equal finite means. Then each of the following two statements is a necessary and sufficient condition for $X \leq_{\text{cx}} Y$:*

$$\int_0^p F^{-1}(u)\mathrm{d}u \geq \int_0^p G^{-1}(u)\mathrm{d}u \quad \textit{for all } p \in [0,1]; \tag{3.A.14}$$

and

$$\int_p^1 F^{-1}(u)\mathrm{d}u \leq \int_p^1 G^{-1}(u)\mathrm{d}u \quad \textit{for all } p \in [0,1]. \tag{3.A.15}$$

Proof. Since $EX = \int_0^1 F^{-1}(u)\mathrm{d}u$ and $EY = \int_0^1 G^{-1}(u)\mathrm{d}u$, and since $EX = EY$, it follows that for any $p \in [0,1]$ the inequality

$$\int_p^1 F^{-1}(u)\mathrm{d}u \leq \int_p^1 G^{-1}(u)\mathrm{d}u \tag{3.A.16}$$

is equivalent to the inequality

$$\int_0^p F^{-1}(u)\mathrm{d}u \geq \int_0^p G^{-1}(u)\mathrm{d}u. \tag{3.A.17}$$

It follows that (3.A.14) and (3.A.15) are equivalent. Thus, we just need to show that $X \leq_{\text{cx}} Y$ is equivalent to (3.A.14).

We only give the proof for the case when the distribution functions F and G of X and Y are continuous; the proof for the general case is similar, though notationally more complex. Without loss of generality, suppose that F and G are not identical. Since $EX = EY$, it follows that F and G must cross each other at least once. If either (3.A.7) or (3.A.14) hold, then, if there is a first time that F crosses G, it must cross it there from below. Similarly, if there is a last time that F crosses G, it also must cross it there from below. (Thus, if there is a finite number of crossings, then it must be odd.)

Let (y_0, p_0), (y_1, p_1), and (y_2, p_2) be three consecutive crossing points as depicted in Figure 3.A.1. Note that (y_0, p_0) may be $(-\infty, 0)$ (we then adopt the convention that $0 \cdot (-\infty) \equiv 0$), and that (y_2, p_2) may be $(\infty, 1)$ (we then adopt the convention that $0 \cdot \infty \equiv 0$). Note that by the continuity assumption we have

$$p_i = F(y_i) = G(y_i), \quad i = 0, 1, 2.$$

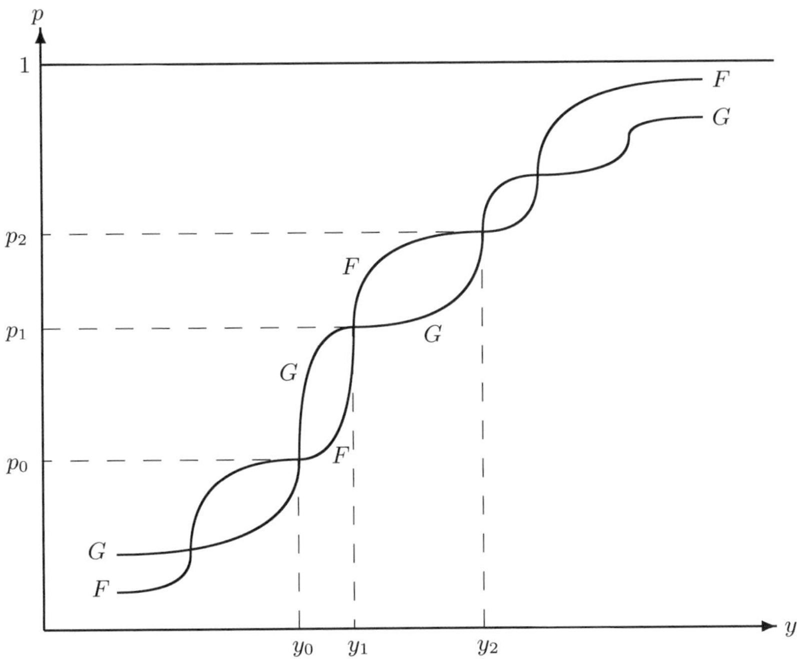

Fig. 3.A.1. Typical segments of F and G when $X \le_{\rm cx} Y$

Assume that $X \le_{\rm cx} Y$. Then

$$\int_{y_2}^{\infty} \overline{F}(x)\mathrm{d}x \le \int_{y_2}^{\infty} \overline{G}(x)\mathrm{d}x. \tag{3.A.18}$$

Thus

$$\begin{aligned}\int_{p_2}^{1} F^{-1}(u)\mathrm{d}u &= y_2(1-p_2) + \int_{y_2}^{\infty} \overline{F}(x)\mathrm{d}x \\ &\le y_2(1-p_2) + \int_{y_2}^{\infty} \overline{G}(x)\mathrm{d}x \qquad \text{(by (3.A.18))} \qquad (3.A.19) \\ &= \int_{p_2}^{1} G^{-1}(u)\mathrm{d}u.\end{aligned}$$

Now, for $u \in [p_1, p_2]$ we have that $F^{-1}(u) - G^{-1}(u) \le 0$ (see Figure 3.A.1). Thus $\int_p^1 (F^{-1}(u) - G^{-1}(u))\mathrm{d}u$ is increasing in $p \in [p_1, p_2]$. Therefore, from (3.A.19) we get that

$$\int_p^1 F^{-1}(u)\mathrm{d}u \le \int_p^1 G^{-1}(u)\mathrm{d}u \quad \text{for } p \in [p_1, p_2]. \tag{3.A.20}$$

From $X \le_{\mathrm{cx}} Y$ we also have

$$\int_{-\infty}^{y_0} F(x)\mathrm{d}x \le \int_{-\infty}^{y_0} G(x)\mathrm{d}x. \tag{3.A.21}$$

Thus

$$\begin{aligned}\int_0^{p_0} F^{-1}(u)\mathrm{d}u &= y_0 p_0 - \int_{-\infty}^{y_0} F(x)\mathrm{d}x \\ &\ge y_0 p_0 - \int_{-\infty}^{y_0} G(x)\mathrm{d}x \qquad \text{(by (3.A.21))} \qquad (3.A.22) \\ &= \int_0^{p_0} G^{-1}(u)\mathrm{d}u.\end{aligned}$$

Now, for $u \in [p_0, p_1]$ we have that $F^{-1}(u) - G^{-1}(u) \ge 0$ (see Figure 3.A.1). Thus $\int_0^p (F^{-1}(u) - G^{-1}(u))\mathrm{d}u$ is increasing in $p \in [p_0, p_1]$. Therefore, from (3.A.22) we get that

$$\int_0^p F^{-1}(u)\mathrm{d}u \ge \int_0^p G^{-1}(u)\mathrm{d}u \quad \text{for } p \in [p_0, p_1]. \tag{3.A.23}$$

Thus we see from (3.A.20) and (3.A.23) that for each $p \in [0, 1]$ either (3.A.16) or (3.A.17) hold. Therefore, (3.A.14) (or, equivalently, (3.A.15)) holds.

Conversely, assume that (3.A.14) (or, equivalently, (3.A.15)) holds. Then

$$\int_{p_2}^1 F^{-1}(u)\mathrm{d}u \le \int_{p_2}^1 G^{-1}(u)\mathrm{d}u. \tag{3.A.24}$$

Thus

$$\begin{aligned}\int_{y_2}^{\infty} \overline{F}(x)\mathrm{d}x &= \int_{p_2}^{1} F^{-1}(u)\mathrm{d}u - y_2(1-p_2) \\ &\le \int_{p_2}^{1} G^{-1}(u)\mathrm{d}u - y_2(1-p_2) \qquad \text{(by (3.A.24))} \\ &= \int_{y_2}^{\infty} \overline{G}(x)\mathrm{d}x.\end{aligned} \tag{3.A.25}$$

Now, for $x \in [y_1, y_2]$ we have that $\overline{F}(x) - \overline{G}(x) \le 0$ (see Figure 3.A.1). Thus $\int_y^{\infty}(\overline{F}(x) - \overline{G}(x))\mathrm{d}x$ is increasing in $y \in [y_1, y_2]$. Therefore, from (3.A.25) we get that

$$\int_{y}^{\infty} \overline{F}(x)\mathrm{d}x \le \int_{y}^{\infty} \overline{G}(x)\mathrm{d}x \quad \text{for } y \in [y_1, y_2]. \tag{3.A.26}$$

From (3.A.14) we also have

$$\int_{0}^{p_0} F^{-1}(u)\mathrm{d}u \ge \int_{0}^{p_0} G^{-1}(u)\mathrm{d}u. \tag{3.A.27}$$

Thus

$$\begin{aligned}\int_{-\infty}^{y_0} F(x)\mathrm{d}x &= y_0 p_0 - \int_{0}^{p_0} F^{-1}(u)\mathrm{d}u \\ &\le y_0 p_0 - \int_{0}^{p_0} G^{-1}(u)\mathrm{d}u \qquad \text{(by (3.A.27))} \\ &= \int_{-\infty}^{y_0} G(x)\mathrm{d}x.\end{aligned} \tag{3.A.28}$$

Now, for $x \in [y_0, y_1]$ we have that $F(x) - G(x) \le 0$ (see Figure 3.A.1). Thus $\int_{-\infty}^{y}(F(x) - G(x))\mathrm{d}x$ is decreasing in $y \in [y_0, y_1]$. Therefore, from (3.A.28) we get that

$$\int_{-\infty}^{y} F(x)\mathrm{d}x \le \int_{-\infty}^{y} G(x)\mathrm{d}x \quad \text{for } y \in [y_0, y_1]. \tag{3.A.29}$$

Thus we see from (3.A.26) and (3.A.29) that for each $y \in \mathbb{R}$ either (3.A.7) or (3.A.8) hold. Therefore $X \le_{\text{cx}} Y$. □

We now give a bivariate characterization result for the order $\le_{\text{cx}}$ that is similar to the characterizations given in Theorems 1.A.9, 1.B.9, 1.B.47, and 1.C.20, for the orders $\le_{\text{st}}$, $\le_{\text{hr}}$, $\le_{\text{rh}}$, and $\le_{\text{lr}}$, respectively. We define the following class of bivariate functions:

$$\mathcal{G}_{\text{cx}} = \{\phi : \mathbb{R}^2 \to \mathbb{R} : \phi(x, y) - \phi(y, x) \text{ is convex in } x \text{ for all } y\}.$$

Theorem 3.A.6. *Let X and Y be independent random variables. Then $X \le_{\text{cx}} Y$ if, and only if,*

$$E[\phi(X, Y)] \le E[\phi(Y, X)] \quad \textit{for all } \phi \in \mathcal{G}_{\text{cx}}. \tag{3.A.30}$$

Proof. Suppose that (3.A.30) holds. Let ψ be a univariate convex function. Define $\phi(x,y) = \psi(x)$. Then $\phi \in \mathcal{G}_{\text{cx}}$ and from (3.A.30) we see that $X \leq_{\text{cx}} Y$.

Conversely, suppose that $X \leq_{\text{cx}} Y$. Let $\phi \in \mathcal{G}_{\text{cx}}$ and let $\hat{Y}$ be another random variable, independent of X and Y, such that $\hat{Y} =_{\text{st}} Y$. Define ψ by $\psi(x) \equiv E[\phi(x,\hat{Y}) - \phi(\hat{Y},x)]$. From the independence of X and $\hat{Y}$ it follows that ψ is convex. Therefore, since $X \leq_{\text{cx}} Y$, it follows that

$$E[\phi(X,Y)] - E[\phi(Y,X)] = E[\psi(X)] \leq E[\psi(Y)] = 0. \qquad \square$$

Another characterization of the convex order, by means of the number of sign changes of two distribution functions, is given in Theorem 3.A.45 in Section 3.A.3.

Let X be a random variable with survival function $\overline{F}$, and let $h : [0,1] \to [0,1]$ be an increasing function that satisfies $h(0) = 0$ and $h(1) = 1$. Such a function h is called a *probability transformation function*. Consider the functional

$$V_h(X) = -\int_{-\infty}^{\infty} x \mathrm{d}h(\overline{F}(x)); \tag{3.A.31}$$

this functional is called the *Yaari functional* and it is of interest in economics.

Theorem 3.A.7. *Let X and Y be two random variables with the same finite means. Then $X \leq_{\text{cx}} Y$ if, and only if,*

$$V_h(X) \leq V_h(Y) \quad \textit{for every convex probability transformation function } h.$$

As can be seen from (3.A.2), only random variables that have the same means can be compared by the order $\leq_{\text{cx}}$. Often, however, we do not want a variability order to depend on the location of the involved distributions. Several ideas for using the order $\leq_{\text{cx}}$ to define a variability order that is independent of the locations of the underlying random variables X and Y have been suggested in the literature. When X and Y have finite means, one idea is to say that X is less variable than Y if

$$[X - EX] \leq_{\text{cx}} [Y - EY]. \tag{3.A.32}$$

This is sometimes called the *dilation order*. When the random variables X and Y satisfy (3.A.32), we denote $X \leq_{\text{dil}} Y$.

For nonnegative random variables X and Y with finite means one can define X as less variable than Y if

$$\frac{X}{EX} \leq_{\text{cx}} \frac{Y}{EY}. \tag{3.A.33}$$

This is sometimes called the *Lorenz order*. When the nonnegative random variables X and Y satisfy (3.A.33), we denote $X \leq_{\text{Lorenz}} Y$. Bhattacharjee and Sethuraman [88] introduced a stochastic order, for nonnegative random variables with finite means, denoted by $\leq_{\text{hnbue}}$. Kochar [306] showed that the orders $\leq_{\text{hnbue}}$ and $\leq_{\text{Lorenz}}$ are equivalent.

The dilation order can be characterized as follows.

Theorem 3.A.8. *Let X and Y be two random variables with distribution functions F and G, respectively, and with finite expectations. Then $X \leq_{\rm dil} Y$ if, and only if,*

$$\frac{1}{1-p}\int_p^1 [F^{-1}(u) - G^{-1}(u)]\mathrm{d}u \leq \int_0^1 [F^{-1}(u) - G^{-1}(u)]\mathrm{d}u \quad \textit{for all } p \in [0,1). \tag{3.A.34}$$

Proof. Denote $\Delta = EX - EY$. Then the stochastic inequality $X \leq_{\rm dil} Y$ can be rewritten as $X - \Delta \leq_{\rm cx} Y$. Denote by F_Δ the distribution function of $X - \Delta$, and note that from Theorem 3.A.5 we have that $X - \Delta \leq_{\rm cx} Y$ if, and only if,

$$\int_p^1 F_\Delta^{-1}(u)\mathrm{d}u \leq \int_p^1 G^{-1}(u)\mathrm{d}u \quad \text{for all } p \in [0,1]. \tag{3.A.35}$$

Since $F_\Delta(x) = F(x + \Delta)$ for all $x \in \mathbb{R}$ it follows that $F_\Delta^{-1}(u) = F^{-1}(u) - \Delta$ for all $u \in [0,1]$. Therefore (3.A.35) is equivalent to

$$\int_p^1 [F^{-1}(u) - \Delta]\mathrm{d}u \leq \int_p^1 G^{-1}(u)\mathrm{d}u \quad \text{for all } p \in [0,1];$$

that is,

$$\int_p^1 [F^{-1}(u) - G^{-1}(u)]\mathrm{d}u \leq \int_p^1 [EX - EY]\mathrm{d}u \quad \text{for all } p \in [0,1];$$

that is,

$$\frac{1}{1-p}\int_p^1 [F^{-1}(u) - G^{-1}(u)]\mathrm{d}u \leq EX - EY \quad \text{for all } p \in [0,1). \tag{3.A.36}$$

Now, since $EX = \int_0^1 F^{-1}(u)\mathrm{d}u$ and $EY = \int_0^1 G^{-1}(u)\mathrm{d}u$ it is seen that (3.A.36) is equivalent to (3.A.34). □

For each $p \in (0,1)$, the quantity $\int_0^1 [F^{-1}(u) - G^{-1}(u)]\mathrm{d}u$ on the right-hand side of (3.A.34) is a weighted average of $\frac{1}{1-p}\int_p^1 [F^{-1}(u) - G^{-1}(u)]\mathrm{d}u$ and of $\frac{1}{p}\int_0^p [F^{-1}(u) - G^{-1}(u)]\mathrm{d}u$. Thus, from Theorem 3.A.8 we obtain that $X \leq_{\rm dil} Y$ if, and only if,

$$\int_0^1 [F^{-1}(u) - G^{-1}(u)]\mathrm{d}u \leq \frac{1}{p}\int_0^p [F^{-1}(u) - G^{-1}(u)]\mathrm{d}u \quad \text{for all } p \in (0,1].$$

Also, $X \leq_{\rm dil} Y$ if, and only if,

$$\frac{1}{1-p}\int_p^1 [F^{-1}(u) - G^{-1}(u)]\mathrm{d}u \leq \frac{1}{p}\int_0^p [F^{-1}(u) - G^{-1}(u)]\mathrm{d}u \quad \text{for all } p \in (0,1).$$

For $p \in [0,1]$, let us denote the p-quantiles of X and of Y by $x(p) = F^{-1}(p)$ and $y(p) = G^{-1}(p)$, respectively. As in Jewitt [256], we observe that $\int_0^p F^{-1}(u)\mathrm{d}u = \int_0^{F^{-1}(p)} x\mathrm{d}F(x) = pE[X|X \le x(p)]$. Similarly, $\int_p^1 F^{-1}(u)\mathrm{d}u = (1-p)E[X|X \ge x(p)]$, $\int_0^p G^{-1}(u)\mathrm{d}u = pE[Y|Y \le y(p)]$, and $\int_p^1 G^{-1}(u)\mathrm{d}u = (1-p)E[Y|Y \ge y(p)]$. Thus we see that each of the following three statements is a necessary and sufficient condition for $X \le_{\mathrm{dil}} Y$:

$$E[X|X \ge x(p)] - E[Y|Y \ge y(p)] \le EX - EY \quad \text{for all } p \in [0,1), \tag{3.A.37}$$

$$E[X|X \le x(p)] - E[Y|Y \le y(p)] \ge EX - EY \quad \text{for all } p \in (0,1], \tag{3.A.38}$$

and

$$E[X|X \ge x(p)] - E[Y|Y \ge y(p)] \le E[X|X \le x(p)] - E[Y|Y \le y(p)] \quad \text{for all } p \in (0,1).$$

Rewriting (3.A.37) and (3.A.38) we see that under the conditions of Theorem 3.A.8 we have that $X \le_{\mathrm{dil}} Y$ if, and only if,

$$E[X - EX|X \ge x(p)] \le E[Y - EY|Y \ge y(p)] \quad \text{for all } p \in [0,1). \tag{3.A.39}$$

Also, $X \le_{\mathrm{dil}} Y$ if, and only if,

$$E[X - EX|X \le x(p)] \ge E[Y - EY|Y \le y(p)] \quad \text{for all } p \in (0,1]. \tag{3.A.40}$$

When $EX = EY$ we have that $X \le_{\mathrm{dil}} Y \Longleftrightarrow X \le_{\mathrm{cx}} Y$. Therefore, when $EX = EY$, the convex order can be characterized by noting that $X \le_{\mathrm{cx}} Y$ if, and only if,

$$E[X|X \ge x(p)] \le E[Y|Y \ge y(p)] \quad \text{for all } p \in [0,1). \tag{3.A.41}$$

Also then, $X \le_{\mathrm{cx}} Y$ if, and only if,

$$E[X|X \le x(p)] \ge E[Y|Y \le y(p)] \quad \text{for all } p \in (0,1]. \tag{3.A.42}$$

Another characterization of the dilation order is given next.

Theorem 3.A.9. *Let X and Y be two random variables with finite means, and let the corresponding distribution functions be F and G, respectively. Then $X \le_{\mathrm{dil}} Y$ if, and only if,*

$$\int_0^1 \phi(p)[F^{-1}(p) - EX]\mathrm{d}p \le \int_0^1 \phi(p)[G^{-1}(p) - EY]\mathrm{d}p,$$

for any increasing function ϕ on $[0,1]$ for which the integrals above are well-defined.

The Lorenz order is closely connected to the so-called Lorenz curve defined as follows. Let X be a nonnegative random variable with distribution function F. The Lorenz curve L_X, corresponding to X, is defined as

$$L_X(p) = \frac{\int_0^p F^{-1}(u)\mathrm{d}u}{\int_0^1 F^{-1}(u)\mathrm{d}u}, \quad p \in [0,1]. \tag{3.A.43}$$

The Lorenz curve is used in economics to measure the inequality of incomes. Let Y be another nonnegative random variable with distribution function G. The Lorenz curve L_Y, corresponding to Y, is defined analogously. The next theorem, which follows from Theorem 3.A.5, highlights the connection between the Lorenz curve and the Lorenz order.

Theorem 3.A.10. *Let X and Y be two nonnegative random variables with equal means. Then $X \le_{\text{Lorenz}} Y$ (or, equivalently, $X \le_{\text{cx}} Y$) if, and only if,*

$$L_X(p) \ge L_Y(p) \quad \textit{for all } p \in [0,1].$$

Another related characterization of the Lorenz order is described next. Let Ψ be the set of all measurable mappings from $\mathbb{R}_+$ to $[0,1]$. For any nonnegative random variable X with a finite mean define the Lorenz zonoid in $\mathbb{R}_+^2$ by

$$L(X) = \left\{ \left(\int_0^\infty \psi(x)\mathrm{d}F(x), \frac{1}{EX}\int_0^\infty x\psi(x)\mathrm{d}F(x) \right) : \psi \in \Psi \right\},$$

where F denotes the distribution function of X.

Theorem 3.A.11. *Let X and Y be two nonnegative random variables with finite means. Then*

$$X \le_{\text{Lorenz}} Y \Longleftrightarrow L(X) \subseteq L(Y).$$

Ramos and Sordo [463] defined what they called a "second-order absolute Lorenz order" by requiring two random variables X and Y, with finite means and with distribution functions F and G, respectively, to satisfy

$$\int_p^1 \int_0^u [F^{-1}(v) - EX]\mathrm{d}v\mathrm{d}u \ge \int_p^1 \int_0^u [G^{-1}(v) - EY]\mathrm{d}v\mathrm{d}u \quad \text{for all } p \in [0,1].$$

3.A.2 Closure and other properties

Using (3.A.1) through (3.A.13) it is easy to prove each of the closure results in the first two parts of the following theorem. (Recall from Section 1.A.3 that for any random variable Z and any event A we denote by $[Z|A]$ any random variable whose distribution is the conditional distribution of Z given A.)

Theorem 3.A.12. (a) *Let X and Y be two random variables. Then*

$$X \le_{\text{cx}} Y \Longleftrightarrow -X \le_{\text{cx}} -Y.$$

(b) *Let X, Y, and Θ be random variables such that $[X|\Theta = \theta] \leq_{\rm cx} [Y|\Theta = \theta]$ for all θ in the support of Θ. Then $X \leq_{\rm cx} Y$. That is, the convex order is closed under mixtures.*

(c) *Let $\{X_j,\ j = 1, 2, \ldots\}$ and $\{Y_j,\ j = 1, 2, \ldots\}$ be two sequences of random variables such that $X_j \to_{\rm st} X$ and $Y_j \to_{\rm st} Y$ as $j \to \infty$. Assume that*

$$E|X_j| \to E|X| \quad \text{and} \quad E|Y_j| \to E|Y| \quad \text{as } j \to \infty. \tag{3.A.44}$$

If $X_j \leq_{\rm cx} Y_j$, $j = 1, 2, \ldots$, then $X \leq_{\rm cx} Y$.

(d) *Let $X_1, X_2, \ldots, X_m$ be a set of independent random variables and let $Y_1, Y_2, \ldots, Y_m$ be another set of independent random variables. If $X_i \leq_{\rm cx} Y_i$ for $i = 1, 2, \ldots, m$, then*

$$\sum_{j=1}^{m} X_j \leq_{\rm cx} \sum_{j=1}^{m} Y_j.$$

That is, the convex order is closed under convolutions.

In order to prove part (c) of Theorem 3.A.12 we will use the characterization of the convex order given in Theorem 3.A.2. Without loss of generality it can be assumed that $EX_j = EY_j = EX = EY = 0$ for all j. From (3.A.11) we have that $E|X_j - a| = -a + 2\int_{-\infty}^{a} F_j(u)du$ for all a, where F_j denotes the distribution function of X_j. In particular, when $a = 0$, it is seen that $E|X_j| = 2\int_{-\infty}^{0} F_j(u)du$. Therefore $E|X_j - a| = E|X_j| - a + 2\int_0^a F_j(u)du$. Using (3.A.44) it is seen that, as $j \to \infty$, the latter expression converges to $E|X| - a + 2\int_0^a F(u)du = E|X - a|$, where F is the distribution function of X. That is, for all a, $E|X_j - a| \to E|X - a|$, as $j \to \infty$. Similarly, $E|Y_j - a| \to E|Y - a|$, as $j \to \infty$. The result now follows from Theorem 3.A.2.

One way of proving part (d) of Theorem 3.A.12 is the following. Note that part (b) of Theorem 3.A.12 can be rephrased as follows: Let Z_1, Z_2, and Θ be independent random variables and let g be a bivariate function such that

$$g(Z_1, \theta) \leq_{\rm cx} g(Z_2, \theta) \quad \text{for all } \theta \text{ in the support of } \Theta. \tag{3.A.45}$$

Then

$$g(Z_1, \Theta) \leq_{\rm cx} g(Z_2, \Theta).$$

If Z_1 and Z_2 satisfy $Z_1 \leq_{\rm cx} Z_2$, then the function g, defined by $g(z, \theta) = z + \theta$, satisfies (3.A.45), since the order $\leq_{\rm cx}$ is closed under shifts. Thus we have shown that if $Z_1 \leq_{\rm cx} Z_2$ and Θ is any random variable independent of Z_1 and Z_2, then

$$Z_1 + \Theta \leq_{\rm cx} Z_2 + \Theta. \tag{3.A.46}$$

Repeated applications of (3.A.46) yield part (d) of Theorem 3.A.12.

It should be pointed out, in contrast to part (a) of Theorem 3.A.12, that if X and Y are such that $X \leq_{\rm cx} Y$, it is not necessarily true that $X \leq_{\rm cx} -Y$ also, even when $EX = EY = 0$. This can be seen easily from (3.A.12).

Without condition (3.A.44) the conclusion of part (c) of Theorem 3.A.12 need not be true. For example, let the X_j's be all uniformly distributed on $[.5, 1.5]$. And let the Y_j's be such that $P\{Y_j = 0\} = (j-1)/j$ and $P\{Y_j = j\} = 1/j$, $j \ge 2$. Note that the distributions of the Y_j's converge to a distribution that is degenerate at 0. Here $X_j \le_{\rm cx} Y_j$, $j = 2, 3, \ldots$, but it is not true that $X \le_{\rm cx} Y$.

For nonnegative random variables, a "random sums" analog of Theorem 3.A.12(d) follows. We omit the proof (however, in Theorem 8.A.13 of Chapter 8 we give a proof of a special case of the following theorem).

Theorem 3.A.13. *Let $X_1, X_2, \ldots$ and $Y_1, Y_2, \ldots$ each be a sequence of nonnegative independent random variables such that $X_i \le_{\rm cx} Y_i$, $i = 1, 2, \ldots$. Let M and N be integer-valued positive random variables that are independent of the $\{X_i\}$ and $\{Y_i\}$ sequences, respectively, such that $M \le_{\rm cx} N$. If the X_i's or the Y_i's are increasing in i in the convex order, then*

$$\sum_{j=1}^{M} X_j \le_{\rm cx} \sum_{j=1}^{N} Y_j.$$

A result that is related to Theorem 3.A.13 is given next. It is of interest to compare it to Theorems 1.A.5 and 2.B.8.

Theorem 3.A.14. *Let $\{X_j,\ j = 1, 2, \ldots\}$ be a sequence of nonnegative independent and identically distributed random variables, and let M be a positive integer-valued random variable which is independent of the X_i's. Let $\{Y_j,\ j = 1, 2, \ldots\}$ be another sequence of independent and identically distributed random variables, and let N be a positive integer-valued random variable which is independent of the Y_i's. Suppose that for some positive integer K we have*

$$\sum_{i=1}^{K} X_i \le_{\rm cx} [\ge_{\rm cx}]\ Y_1,$$

and

$$M \le_{\rm cx} [\ge_{\rm cx}]\ KN.$$

Then

$$\sum_{j=1}^{M} X_j \le_{\rm cx} [\ge_{\rm cx}] \sum_{j=1}^{N} Y_j.$$

We do not give a detailed proof of Theorem 3.A.14 here since it is similar to the proof of Theorem 4.A.12 in Section 4.A.1.

Two other similar theorems are the following. Their proofs are similar to the proofs of Theorems 4.A.13 and 4.A.14 in Section 4.A.1.

Theorem 3.A.15. *Let $\{X_j,\ j = 1, 2, \ldots\}$ be a sequence of nonnegative independent and identically distributed random variables, and let M be a positive integer-valued random variable which is independent of the X_i's. Let*

$\{Y_j,\ j = 1, 2, \dots\}$ be another sequence of independent and identically distributed random variables, and let N be a positive integer-valued random variable which is independent of the Y_i's. Also, let $\{N_j, j = 1, 2, \dots\}$ be a sequence of independent random variables that are distributed as N. If for some positive integer K we have

$$\sum_{i=1}^{K} X_i \le_{\rm cx} Y_1 \quad \text{and} \quad M \le_{\rm cx} \sum_{i=1}^{K} N_i,$$

or if we have

$$KX_1 \le_{\rm cx} Y_1 \quad \text{and} \quad M \le_{\rm cx} KN,$$

or if we have

$$KX_1 \le_{\rm cx} Y_1 \quad \text{and} \quad M \le_{\rm cx} \sum_{i=1}^{K} N_i,$$

then

$$\sum_{j=1}^{M} X_j \le_{\rm cx} \sum_{j=1}^{N} Y_j.$$

Theorem 3.A.16. *Let $\{X_j,\ j = 1, 2, \dots\}$ be a sequence of nonnegative independent and identically distributed random variables, and let M be a positive integer-valued random variable which is independent of the X_i's. Let $\{Y_j,\ j = 1, 2, \dots\}$ be another sequence of independent and identically distributed random variables, and let N be a positive integer-valued random variable which is independent of the Y_i's. If for some positive integers K_1 and K_2, such that $K_1 \le K_2$, we have*

$$\sum_{i=1}^{K_1} X_i \le_{\rm cx} \frac{K_1}{K_2} Y_1 \quad \text{and} \quad M \le_{\rm cx} K_2 N,$$

then

$$\sum_{j=1}^{M} X_j \le_{\rm cx} \sum_{j=1}^{N} Y_j.$$

Another result which involves a comparison of random sums, with respect to the convex order, is given in Example 9.A.19.

Theorem 3.A.12(d) can be generalized to situations in which the X_j's or the Y_j's are not necessarily independent. For example, the result (7.A.13) in Chapter 7 is a generalization of Theorem 3.A.12(d). The next result is a trivial illustration of a case in which one of the independence assumptions is dropped.

Theorem 3.A.17. *Let X be a random variable with a finite mean. Then*

$$X + EX \leq_{\rm cx} 2X.$$

Proof. Let X' be an independent copy of X. Then, for any convex function ϕ for which the expectations below exist, one has

$$\begin{aligned} E\phi(X + EX) &= E\phi(E(X + X'|X)) \\ &\leq E\phi(X + X') \\ &\leq E\phi(2X), \end{aligned}$$

where the first inequality follows from Jensen's Inequality and the second inequality follows from Example 3.A.29 below (with $n = 2$). □

Theorem 3.A.17 can also be easily proven using Theorem 3.A.4.

The following result provides a generalization of Theorem 3.A.17; see a comment after Theorem 3.A.18. Recall from (3.A.32) the definition of the dilation order.

Theorem 3.A.18. *Let X be a random variable with a finite mean. Then*

$$X \leq_{\rm dil} aX \quad \textit{whenever } a \geq 1.$$

Proof. Without loss of generality assume that $EX = 0$. Let ϕ be a convex function which, without loss of generality, can be assumed to satisfy $\phi(0) = 0$. Then, for $k \geq 1$ we have

$$E\phi(X) \leq E[k\phi(X)] \leq E\phi(kX).$$ □

From Theorem 3.A.18 it follows that

$$X + (k-1)EX \leq_{\rm cx} kX \quad \text{whenever } k \geq 1,$$

which is, indeed, a generalization of Theorem 3.A.17.

From Theorem 3.A.12(a) it is not hard to see that

$$X \leq_{\rm dil} Y \Longleftrightarrow -X \leq_{\rm dil} -Y.$$

Another property of the dilation and of the convex orders is described in the following theorem.

Theorem 3.A.19. *Let X_1 and X_2 (Y_1 and Y_2) be two independent copies of X (Y), where X and Y have finite means. If $X \leq_{\rm dil} Y$, then $X_1 - X_2 \leq_{\rm dil} Y_1 - Y_2$. If $X \leq_{\rm cx} Y$, then $X_1 - X_2 \leq_{\rm cx} Y_1 - Y_2$.*

Proof. Using the fact that $X \leq_{\rm dil} Y$ if, and only if, $-X \leq_{\rm dil} -Y$, and the fact that the dilation order is closed under convolutions (see Theorem 3.A.12(d)), the stated result follows. The proof of $X \leq_{\rm cx} Y \Longrightarrow X_1 - X_2 \leq_{\rm cx} Y_1 - Y_2$ is similar (using Theorem 3.A.12(a) and (d)). □

An interesting comparison of sums of random variables in the convex order is the following result.

Theorem 3.A.20. *Let $X_1, X_2, \ldots, X_n$, and Z be random variables. Then*

$$X_1 + X_2 + \cdots + X_n \geq_{\rm cx} E[X_1 | Z] + E[X_2 | Z] + \cdots + E[X_n | Z],$$

provided the conditional expectations above exist.

Proof. Let ϕ be a convex function. By Jensen's Inequality we have

$$\begin{aligned} E\phi(X_1 + X_2 + \cdots + X_n) &= E\big[E[\phi(X_1 + X_2 + \cdots + X_n) | Z]\big] \\ &\geq E\big[\phi(E[X_1 | Z] + E[X_2 | Z] + \cdots + E[X_n | Z])\big], \end{aligned}$$

and the stated result follows. □

Consider now a family of distribution functions $\{G_\theta,\ \theta \in \mathcal{X}\}$ where $\mathcal{X}$ is a convex subset (that is, an interval) of the real line or of $\mathbb{N}$. As in Section 1.A.3 let $X(\theta)$ denote a random variable with distribution function G_θ. For any random variable Θ with support in $\mathcal{X}$, and with distribution function F, let us denote by $X(\Theta)$ a random variable with distribution function H given by

$$H(y) = \int_{\mathcal{X}} G_\theta(y) \mathrm{d}F(\theta), \quad y \in \mathbb{R}.$$

Theorem 3.A.21. *Consider a family of distribution functions $\{G_\theta,\ \theta \in \mathcal{X}\}$ as above. Let Θ_1 and Θ_2 be two random variables with supports in $\mathcal{X}$ and distribution functions F_1 and F_2, respectively. Let Y_1 and Y_2 be two random variables such that $Y_i =_{\rm st} X(\Theta_i)$, $i = 1, 2$; that is, suppose that the distribution function of Y_i is given by*

$$H_i(y) = \int_{\mathcal{X}} G_\theta(y) \mathrm{d}F_i(\theta), \quad y \in \mathbb{R},\ i = 1, 2.$$

If for every convex function ϕ

$$E[\phi(X(\theta))] \quad \textit{is convex in } \theta, \tag{3.A.47}$$

and if

$$\Theta_1 \leq_{\rm cx} \Theta_2,$$

then

$$Y_1 \leq_{\rm cx} Y_2.$$

The proof of Theorem 3.A.21 is similar to the proof of Theorem 4.A.18 below, and therefore we omit it.

It is worth mentioning that condition (3.A.47) is the same as the condition $\{X(\theta),\ \theta \in \mathcal{X}\} \in \mathrm{SCX}$ which is studied in Section 8.A of Chapter 8.

The following corollary of Theorem 3.A.21 shows that the convex order is closed under products of nonnegative random variables. A variation of this corollary is given in Example 4.A.19.

Corollary 3.A.22. *Let X_1 and X_2 be a pair of independent random variables, and let Y_1 and Y_2 be another pair of independent random variables. If $X_i \leq_{\text{cx}} Y_i$, $i = 1, 2$, then $X_1 X_2 \leq_{\text{cx}} Y_1 Y_2$.*

Proof. Using Theorem 3.A.21 twice we see that

$$X_1 X_2 \leq_{\text{cx}} Y_1 X_2 \leq_{\text{cx}} Y_1 Y_2,$$

and the stated result follows from the transitivity property of the convex order. □

An interesting variation of Theorem 3.A.21 is the following. Again, we omit the proof because it is similar to the proof of Theorem 4.A.18.

Theorem 3.A.23. *Consider a family of distribution functions $\{G_\theta,\ \theta \in \mathcal{X}\}$ as described before Theorem 3.A.21. Let Θ_1 and Θ_2 be two random variables with supports in $\mathcal{X}$ and distribution functions F_1 and F_2, respectively. Let Y_1 and Y_2 be two random variables such that $Y_i =_{\text{st}} X(\Theta_i)$, $i = 1, 2$; that is, suppose that the distribution function of Y_i is given by*

$$H_i(y) = \int_{\mathcal{X}} G_\theta(y) \mathrm{d}F_i(\theta), \quad y \in \mathbb{R},\ i = 1, 2.$$

If for every convex function ϕ

$$E[\phi(X(\theta))] \quad \text{is increasing in } \theta,$$

and if

$$\Theta_1 \leq_{\text{st}} \Theta_2,$$

then

$$Y_1 \leq_{\text{cx}} Y_2.$$

The next result indicates the "minimal" and the "maximal" random variables, with respect to the order $\leq_{\text{cx}}$, when the support and the mean are given. The proof, using (3.A.7) or (3.A.8) for example, is trivial and is thus omitted.

Theorem 3.A.24. *Let X be a random variable with mean EX. Denote the left [right] endpoint of the support of X by l_X [u_X] (see the paragraph preceding (3.A.12) for the exact definition of l_X and u_X). Let Z be a random variable such that $P\{Z = l_X\} = (u_X - EX)/(u_X - l_X)$ and $P\{Z = u_X\} = (EX - l_X)/(u_X - l_X)$. Then*

$$EX \leq_{\text{cx}} X \leq_{\text{cx}} Z, \tag{3.A.48}$$

where in (3.A.48) (and in (3.A.49)) EX denotes a random variable that takes on the value EX with probability 1.

Another result that indicates the "minimal" random variable, with respect to the order $\leq_{\text{cx}}$, for some rich families of random variables when the mean is given is Theorem 3.A.46.

It follows from the first inequality of (3.A.48) and from the fact that for any two random variables U and V one has $U \leq_{\text{cx}} V \Longleftrightarrow V \leq_{\text{cv}} U$, that if X is a random variable with mean EX, then

$$X \leq_{\text{cv}} EX. \tag{3.A.49}$$

In analogy to Theorem 1.A.17 we have the following results. We omit the proof of Theorem 3.A.25; however, the necessity part of Theorem 3.A.25 is a special case of Theorem 3.A.26. In the next three theorems we assume that all the random variables that are considered have finite means.

Theorem 3.A.25. *Let X be a nonnegative random variable that is not degenerate at 0 and let g be a nonnegative function defined on $[0,\infty)$. If $g(x) > 0$ for all $x > 0$, and if g is increasing on $[0,\infty)$, and if $g(x)/x$ is decreasing [increasing] on $(0,\infty)$, then*

$$g(X) \leq_{\text{Lorenz}} [\geq_{\text{Lorenz}}]\ X.$$

For example, if X is a nonnegative random variable, then

$$X + a \leq_{\text{Lorenz}} X \quad \text{whenever } a > 0.$$

The proof of the next theorem follows from results in Chapter 4 (see Theorem 4.B.5 and the first part of the proof of Theorem 4.B.4).

Theorem 3.A.26. *Let X be a nonnegative random variable that is not degenerate at 0, and let g and h be nonnegative increasing functions, defined on $[0,\infty)$, such that $g(x) > 0$ and $h(x) > 0$ for all $x > 0$. If $h(x)/g(x)$ is increasing in $x \in (0,\infty)$, then*

$$g(X) \leq_{\text{Lorenz}} h(X).$$

Using Theorem 3.A.25 it is not too hard to prove the following result.

Theorem 3.A.27. *Let X and Z be two independent nonnegative random variables that are not degenerate at 0 and let g be a nonnegative function defined on $[0,\infty)^2$ such that $g(Z,X)$ is not degenerate at 0. If $g(z,x)/x$ is increasing in x for every z, and if $g(z,x)$ is increasing in x for every z, then*

$$X \leq_{\text{Lorenz}} g(Z,X).$$

The Lorenz order often implies the harmonic mean residual life order, as the following result shows.

Theorem 3.A.28. *Let X and Y be two nonnegative random variables with positive expectations. If $X \leq_{\text{Lorenz}} Y$ and if $EX \leq EY$, then $X \leq_{\text{hmrl}} Y$.*

Proof. For $t \geq 0$ we have

$$\frac{E[(X-t)_+]}{EX} \leq \frac{E\left[\left(\frac{EX}{EY} \cdot Y - t\right)_+\right]}{EX} = E\left[\left(\frac{Y}{EY} - \frac{t}{EX}\right)_+\right]$$
$$= \frac{E\left[\left(Y - \frac{EY}{EX} \cdot t\right)_+\right]}{EY} \leq \frac{E[(Y-t)_+]}{EY},$$

where the first inequality follows from $X \leq_{\text{Lorenz}} Y$ (that is, $X \leq_{\text{cx}} \left(\frac{EX}{EY}\right)Y$), and the second inequality follows from $EX \leq EY$. The stated result now follows from (2.B.4). □

Let us now return to the characterization of the convex order given in Theorem 3.A.4. This characterization is sometimes useful for establishing the relation $\leq_{\text{cx}}$ between two random variables. The next example is a fine illustration of this procedure.

Example 3.A.29. Let $X_1, X_2, \ldots$ be independent and identically distributed random variables. Denote by $\overline{X}_n$ the sample mean of $X_1, X_2, \ldots, X_n$. That is, $\overline{X}_n = (X_1 + X_2 + \cdots + X_n)/n$. It is well known that if the variances exist, then for every $n \geq 2$ one has $\text{Var}(\overline{X}_n) \leq \text{Var}(\overline{X}_{n-1})$. But more than that is true. In fact, if the expectation of X_1 exists, then for each $n \geq 2$ one has

$$\overline{X}_n \leq_{\text{cx}} \overline{X}_{n-1}.$$

In order to see it note that from the exchangeability of $X_1, X_2, \ldots, X_n$ it follows that $E[X_i | \overline{X}_n] = \overline{X}_n$ for all $i \leq n$. Therefore $E[\overline{X}_{n-1} | \overline{X}_n] = \overline{X}_n$. That is, $\{\overline{X}_n, \overline{X}_{n-1}\}$ is a martingale. The result now follows from Theorem 3.A.4.

An extension of Example 3.A.29 to the multivariate case is given in Example 7.A.11.

A result that is similar to Example 3.A.29 is the following (actually it is a generalization of Example 3.A.29 as will be argued below).

Theorem 3.A.30. *Let $X_1, X_2, \ldots, X_n$ be independent and identically distributed random variables. Let $\phi_1, \phi_2, \ldots, \phi_n$ be measurable real functions. Denote $\overline{\phi} = \frac{1}{n}\sum_{i=1}^n \phi_i$. Then*

$$\sum_{i=1}^n \overline{\phi}(X_i) \leq_{\text{cx}} \sum_{i=1}^n \phi_i(X_i).$$

The proof of Theorem 3.A.30 consists of verifying, using the exchangeability of the X_i's, that

$$E\left[\frac{1}{n!}\sum_{\pi}\sum_{i=1}^n \phi_{\pi_i}(X_i) \middle| \sum_{i=1}^n \overline{\phi}(X_i)\right] = \sum_{i=1}^n \overline{\phi}(X_i)$$

and that

$$\frac{1}{n!}\sum_{\pi}\sum_{i=1}^{n}\phi_{\pi_i}(X_i) =_{\rm st} \sum_{i=1}^{n}\phi_i(X_i).$$

The desired result then follows from Theorem 3.A.4.

Corollary 3.A.31. *Let $X_1, X_2, \ldots, X_n$ be independent and identically distributed random variables. Let $a_1, a_2, \ldots, a_n$ be real constants. Denote $\overline{a} = \frac{1}{n}\sum_{i=1}^{n} a_i$. Then*

$$\overline{a}\sum_{i=1}^{n} X_i \le_{\rm cx} \sum_{i=1}^{n} a_i X_i.$$

By taking $a_i = 1/(n-1)$ for $i = 1, 2, \ldots, n-1$, and $a_n = 0$, it is easily seen that Example 3.A.29 is a special case of Corollary 3.A.31.

Example 3.A.32. Let $m \le m'$ be two positive integers, and let M and N be two Poisson random variables with means $m\lambda$ and $m'\lambda$, respectively, for some $\lambda > 0$. Define $X = mN$ and $Y = m'M$. Then, using Example 3.A.29, it can be shown that $X \le_{\rm cx} Y$. This result can be extended to the case where m and m' are not integers, by approximating m/m' with rational numbers.

Two other simple results that follow from Theorem 3.A.4 are the following theorems.

Theorem 3.A.33. *Let X and Y be independent random variables with finite means and suppose that $EY = a$. Then*

$$aX \le_{\rm cx} YX.$$

Proof. Clearly, $E[YX|X] = aX$ and the result now follows from Theorem 3.A.4. This result is also an immediate consequence of Corollary 3.A.22 if one takes there $X_1 = a$ almost surely, $X_2 = X$, and $Y_1 = Y$ and $Y_2 = X$. □

Theorem 3.A.34. *Let X and Y be independent random variables with finite means and suppose that $EY = 0$. Then*

$$X \le_{\rm cx} X + Y.$$

Proof. Clearly, $E[X+Y|X] = X$ and the result follows from Theorem 3.A.4. Another way of proving this result is to use Theorem 3.A.12(d). □

Recall from (3.A.32) the definition of the dilation order. From Theorem 3.A.34 it follows that if X and Y are independent random variables with finite means, then

$$X \le_{\rm dil} X + Y. \tag{3.A.50}$$

Recall from page 2 the definition of the majorization order $\boldsymbol{a} \prec \boldsymbol{b}$ among n-dimensional vectors. The next result strengthens Corollary 3.A.31.

Theorem 3.A.35. *Let $X_1, X_2, \ldots, X_n$ be exchangeable random variables. Let $\boldsymbol{a} = (a_1, a_2, \ldots, a_n)$ and $\boldsymbol{b} = (b_1, b_2, \ldots, b_n)$ be two vectors of constants. If $\boldsymbol{a} \prec \boldsymbol{b}$, then*

$$\sum_{i=1}^{n} a_i X_i \le_{\mathrm{cx}} \sum_{i=1}^{n} b_i X_i. \tag{3.A.51}$$

Proof. Below, for any constants a, b, c, and d the notation $a \le [b, c]$ stands for $a \le \min\{b, c\}$, and the notation $[b, c] \le d$ stands for $\max\{b, c\} \le d$. By a well-known property of the majorization order it suffices to prove the result only for $n = 2$. Let X_1 and X_2 be exchangeable random variables, and let a_1, a_2, b_1, and b_2 be four constants such that $b_1 \le a_1 \le a_2 \le b_2$ and $a_1 + a_2 = b_1 + b_2$. Denote $X_{(1)} = \min\{X_1, X_2\}$ and $X_{(2)} = \max\{X_1, X_2\}$. Then, almost surely,

$$b_1 X_{(2)} + b_2 X_{(1)} \le [a_1 X_{(1)} + a_2 X_{(2)}, a_1 X_{(2)} + a_2 X_{(1)}] \le b_1 X_{(1)} + b_2 X_{(2)}$$

and

$$a_1 X_{(2)} + a_2 X_{(1)} + a_1 X_{(1)} + a_2 X_{(2)} = b_1 X_{(2)} + b_2 X_{(1)} + b_1 X_{(1)} + b_2 X_{(2)}.$$

Hence for any convex function ϕ we have, almost surely,

$$\begin{aligned} &\phi(a_1 X_{(2)} + a_2 X_{(1)}) + \phi(a_1 X_{(1)} + a_2 X_{(2)}) \\ &\qquad\qquad \le \phi(b_1 X_{(2)} + b_2 X_{(1)}) + \phi(b_1 X_{(1)} + b_2 X_{(2)}). \end{aligned}$$

Therefore,

$$\begin{aligned} 2E\phi(a_1 X_1 + a_2 X_2) &= E[\phi(a_1 X_{(2)} + a_2 X_{(1)}) + \phi(a_1 X_{(1)} + a_2 X_{(2)})] \\ &\le E[\phi(b_1 X_{(2)} + b_2 X_{(1)}) + \phi(b_1 X_{(1)} + b_2 X_{(2)})] = 2E\phi(b_1 X_1 + b_2 X_2), \end{aligned}$$

and the stated result follows. □

A result that is related to Theorem 3.A.35 is Theorem 4.A.39. Another result that is related to Theorem 3.A.35 is Theorem 7.B.8 in Chapter 7 by Tong in [515]; the latter compares $\sum_{i=1}^{n} b_i X_i$ and $\sum_{i=1}^{n} a_i X_i$ in the sense of the peakedness order of Section 3.D, rather than in the sense of the order $\le_{\mathrm{cx}}$.

From Theorem 3.A.35 it follows that if the X_i's are exchangeable (in particular, if they are identically distributed), if $a_i \ge 0$, $i = 1, 2, \ldots, n$, and $\sum_{i=1}^{n} a_i = 1$, and if $X_1 \le_{\mathrm{cx}} Y$ for some random variable Y, then

$$\sum_{i=1}^{n} a_i X_i \le_{\mathrm{cx}} Y. \tag{3.A.52}$$

The next result shows that (3.A.52) is true even if the X_i's are not exchangeable, but have any joint distribution.

Theorem 3.A.36. *Let $X_1, X_2, \ldots, X_n$ and Y be $n+1$ random variables. If $X_i \le_{\rm cx} Y$, $i = 1, 2, \ldots, n$, then*

$$\sum_{i=1}^{n} a_i X_i \le_{\rm cx} Y,$$

whenever $a_i \ge 0$, $i = 1, 2, \ldots, n$, and $\sum_{i=1}^{n} a_i = 1$.

Proof. Let ϕ be any convex function for which the expectations below exist. Then

$$E\Big[\phi\Big(\sum_{i=1}^{n} a_i X_i\Big)\Big] \le E\Big[\sum_{i=1}^{n} a_i \phi(X_i)\Big] = \sum_{i=1}^{n} a_i E[\phi(X_i)]$$
$$\le \sum_{i=1}^{n} a_i E[\phi(Y)] = E[\phi(Y)],$$

where the first inequality follows from the convexity of ϕ, and the second inequality from $X_i \le_{\rm cx} Y$, $i = 1, 2, \ldots, n$. $\square$

Similar results are described in Theorems 5.A.14, 5.C.8, and 5.C.18.

An interesting result in which the coefficients in (3.A.51) are replaced by Bernoulli random variables is described next. Let I_p denote a Bernoulli random variable with probability of success p, that is, $P\{I_p = 1\} = 1 - P\{I_p = 0\} = p$.

Theorem 3.A.37. *Let $X_1, X_2, \ldots, X_n$ be nonnegative exchangeable random variables, and let $I_{p_1}, I_{p_2}, \ldots, I_{p_n}$ and $I_{q_1}, I_{q_2}, \ldots, I_{q_n}$ be independent Bernoulli random variables that are independent of $X_1, X_2, \ldots, X_n$. If $\mathbf{p} \prec \mathbf{q}$, then*

$$\sum_{i=1}^{n} I_{p_i} X_i \ge_{\rm cx} \sum_{i=1}^{n} I_{q_i} X_i.$$

A result that is related to Theorem 3.A.37 is Theorem 4.A.38.

Example 3.A.38. If the X_i's in Theorem 3.A.37 are all identically equal to 1, then we get that $\boldsymbol{p} \prec \boldsymbol{q}$ implies that

$$\sum_{i=1}^{n} I_{p_i} \ge_{\rm cx} \sum_{i=1}^{n} I_{q_i}.$$

In particular,

$$\sum_{i=1}^{n} I_{q_i} \le_{\rm cx} Y,$$

where Y is a binomial random variable having the parameters n and $\overline{q} = \big(\sum_{i=1}^{n} q_i\big)/n$.

Conceptually it can be expected that if the random variables $X_1, X_2, \ldots,$ X_n are "more positively [negatively] associated" than the random variables $Y_1, Y_2, \ldots, Y_n$ in some sense, but otherwise $X_i =_{\text{st}} Y_i$ for each i, then $\sum_{i=1}^n X_i \geq_{\text{cx}} [\leq_{\text{cx}}] \sum_{i=1}^n Y_i$. The following result is a formalization of this idea. Recall that random variables $X_1, X_2, \ldots, X_n$ are said to be *positively associated* if

$$\text{Cov}(h_1(X_1, X_2, \ldots, X_n), h_2(X_1, X_2, \ldots, X_n)) \geq 0 \tag{3.A.53}$$

for all increasing functions h_1 and h_2 for which the above covariance is defined. Similarly, $X_1, X_2, \ldots, X_n$ are said to be *negatively associated* if

$$\text{Cov}(h_1(X_{i_1}, X_{i_2}, \ldots, X_{i_k}), h_2(X_{j_1}, X_{j_2}, \ldots, X_{j_{n-k}})) \leq 0 \tag{3.A.54}$$

for all choices of disjoint subsets $\{i_1, i_2, \ldots, i_k\}$ and $\{j_1, j_2, \ldots, j_{n-k}\}$ of $\{1, 2, \ldots, n\}$, and for all increasing functions h_1 and h_2 for which the above covariance is defined.

Theorem 3.A.39. *Let $X_1, X_2, \ldots, X_n$ be positively [negatively] associated random variables, and let $Y_1, Y_2, \ldots, Y_n$ be independent random variables such that $X_i =_{\text{st}} Y_i$, $i = 1, 2, \ldots, n$. Then*

$$\sum_{i=1}^n X_i \geq_{\text{cx}} [\leq_{\text{cx}}] \sum_{i=1}^n Y_i.$$

Theorem 3.A.39 follows from Theorem 9.A.23 in Chapter 9; see a comment there after that theorem.

A Laplace transform characterization of the order $\leq_{\text{cx}}$ is stated next; it may be compared to Theorems 1.A.13, 1.B.18, 1.B.53, 1.C.25, 2.A.16, and 2.B.14. We do not give the proof of this characterization here since it follows easily from Theorem 4.A.21 in Chapter 4.

Theorem 3.A.40. *Let X_1 and X_2 be two nonnegative random variables, and let $N_\lambda(X_1)$ and $N_\lambda(X_2)$ be as described in Theorem* 1.A.13. *Then*

$$X_1 \leq_{\text{cx}} X_2 \iff N_\lambda(X_1) \leq_{\text{cx}} N_\lambda(X_2) \quad \textit{for all } \lambda > 0.$$

Example 3.A.41. Let Θ be a random variable whose realization, θ, is a parameter of interest. In the context of statistical inference the distribution function of Θ is called a prior distribution. Let X and Y be two random variables whose distribution functions depend on θ, that is, the conditional distribution of X given $\Theta = \theta$ is, say, F_θ, and the conditional distribution of Y given $\Theta = \theta$ is, say, G_θ. Let $L(a, \theta)$ be the loss incurred when $\Theta = \theta$ and when the action a has been taken (a is a number in the action space A which is a compact subset of $\mathbb{R}$).

In the following discussion, every expected value that is mentioned is assumed to exist.

If $X = x$ is observed, and action a is taken, then the expected loss is

$$E\big[L(a, \Theta)\big|X = x\big].$$

The minimal expected loss, given that $X = x$ has been observed, is then

$$\min_{a \in A} E\big[L(a, \Theta)\big|X = x\big].$$

Therefore the expected minimal expected loss, for an experiment in which X is used for inference on θ, is

$$E\big\{\min_{a \in A} E\big[L(a, \Theta)\big|X\big]\big\}.$$

Similarly, the expected minimal expected loss, for an experiment in which Y is used for inference on θ, is

$$E\big\{\min_{a \in A} E\big[L(a, \Theta)\big|Y\big]\big\}.$$

We say that Y is *more informative* than X for Θ if

$$E\big\{\min_{a \in A} E\big[L(a, \Theta)\big|X\big]\big\} \geq E\big\{\min_{a \in A} E\big[L(a, \Theta)\big|Y\big]\big\} \tag{3.A.55}$$

for any loss function L, and any action space A, for which the minima and the expected values above are well defined.

Let $U = E[\Theta|X]$ and $V = E[\Theta|Y]$ be the posterior means in the corresponding experiments. Obviously,

$$EU = E[E[\Theta|X]] = E\Theta = E[E[\Theta|Y]] = EV. \tag{3.A.56}$$

Take $A = [0, 1]$ and consider the loss function

$$L_c(a, \theta) = a(\theta - c),$$

where c is some constant. Then

$$\min_{a \in A} E\big[L_c(a, \Theta)\big|X\big] = \min_{a \in A} a[E[\Theta - c|X]] = \min\{0, U - c\} = -(c - U)_+,$$

and, similarly,

$$\min_{a \in A} E\big[L_c(a, \Theta)\big|Y\big] = -(c - V)_+.$$

From (3.A.55) we get that $E[(c - U)_+] \leq E[(c - V)_+]$ for all c. Therefore, from (3.A.6) and (3.A.56) it follows that

$$E[\Theta|X] \leq_{\text{cx}} E[\Theta|Y].$$

The following result is an analog of Theorem 1.A.8; similar results are Theorems 3.A.59, 4.A.48, 4.A.69, 5.A.15, 6.B.19, 6.G.12, 6.G.13, and 7.A.14–7.A.16.

Theorem 3.A.42. *Let X and Y be two random variables. Suppose that $X \le_{\rm cx} Y$ $[X \le_{\rm cv} Y]$ and that $E[X^2] = E[Y^2]$, provided the expectations exist. Then $X =_{\rm st} Y$.*

Proof. Denote the distribution functions of X and Y by F and G, respectively. Then

$$E[Y^2] - E[X^2] = 2\int_{u=-\infty}^{0}\left[\int_{v=-\infty}^{u}(G(v)-F(v))\mathrm{d}v\right]\mathrm{d}u + 2\int_{u=0}^{\infty}\left[\int_{v=u}^{\infty}(\overline{G}(v)-\overline{F}(v))\mathrm{d}v\right]\mathrm{d}u.$$

By Theorem 3.A.1 both inner integrals are nonnegative. From $E[X^2] = E[Y^2]$ we thus obtain $\int_{v=-\infty}^{u} F(v)\mathrm{d}v = \int_{v=-\infty}^{u} G(v)\mathrm{d}v$ for $u \le 0$, and $\int_{v=u}^{\infty} \overline{F}(v)\mathrm{d}v = \int_{v=u}^{\infty} \overline{G}(v)\mathrm{d}v$ for $u \ge 0$. Differentiating these equalities we obtain $F = G$. □

Theorem 3.A.42 can be strengthened as follows; we do not detail the proof here.

Theorem 3.A.43. *Let X and Y be two random variables. Suppose that $X \le_{\rm cx} Y$ $[X \le_{\rm cv} Y]$ and that for some strictly convex function ϕ we have that $E[\phi(X)] = E[\phi(Y)]$, provided the expectations exist. Then $X =_{\rm st} Y$.*

Theorem 3.A.60 below is a generalization of Theorem 3.A.43.

3.A.3 Conditions that lead to the convex order

Once the relation $X \le_{\rm cx} Y$ has been established between the two random variables X and Y it can be of great use. However, given the two random variables and their distributions it is sometimes not clear how to verify that $X \le_{\rm cx} Y$. In this section we point out several simple conditions that imply the convex order. Recall the notation $S^-(a)$ (defined in (1.A.18)) for the number of sign changes of the function a.

Theorem 3.A.44. *Let X and Y be two random variables with equal means, density functions f and g, distribution functions F and G, and survival functions $\overline{F}$ and $\overline{G}$, respectively. Then $X \le_{\rm cx} Y$ if any of the following conditions hold:*

$$S^-(g-f) = 2 \quad \text{and the sign sequence is } +,-,+, \tag{3.A.57}$$

$$S^-(\overline{F}-\overline{G}) = 1 \quad \text{and the sign sequence is } +,-, \tag{3.A.58}$$

$$S^-(G-F) = 1 \quad \text{and the sign sequence is } +,-. \tag{3.A.59}$$

Proof. We will prove the result for the continuous case; the proof in the discrete case is similar. Suppose that $S^-(g-f) = 2$ and that the sign sequence is $+,-,+$. Let a and b $(a < b)$ be two of the crossing points, where the definition of a crossing point is self-explanatory. Denote $I_1 = (-\infty, a]$, $I_2 = (a, b]$, and $I_3 = (b, \infty)$. Then $g(x) - f(x) \geq 0$ on I_1, $g(x) - f(x) \leq 0$ on I_2, and $g(x) - f(x) \geq 0$ on I_3. Therefore

$$G(x) - F(x) = \int_{-\infty}^{x} [g(u) - f(u)]\mathrm{d}u$$

is increasing on I_1, decreasing on I_2, and increasing on I_3. It is also clear that $\lim_{x\to-\infty}[G(x) - F(x)] = \lim_{x\to\infty}[G(x) - F(x)] = 0$. Combining all these observations shows that $S^-(G - F) = 1$ and that the sign sequence is $+, -$. Now suppose that $S^-(G-F) = 1$ and that the sign sequence is $+, -$. Let c be a crossing point. Denote $J_1 = (-\infty, c]$ and $J_2 = (c, \infty)$. Then $G(x) - F(x) \geq 0$ on J_1 and $G(x) - F(x) \leq 0$ on J_2. Clearly

$$\lim_{x\to-\infty} \int_{-\infty}^{x} [G(u) - F(u)]\mathrm{d}u = 0$$

and from the equality of the means (see (3.A.3)) it follows that

$$\lim_{x\to\infty} \int_{-\infty}^{x} [G(u) - F(u)]\mathrm{d}u = 0.$$

Combining these observations shows that (3.A.8) holds. This proves that (3.A.57) and (3.A.59) imply $X \leq_{\mathrm{cx}} Y$. Note that $S^-(\overline{F} - \overline{G}) = S^-(G - F)$ with the same sign sequence. This observation, together with (3.A.59), shows that (3.A.58) implies $X \leq_{\mathrm{cx}} Y$. $\square$

The condition (3.A.58) (or, equivalently, (3.A.59)) is not only sufficient for $X \leq_{\mathrm{cx}} Y$, but, for nonnegative random variables, it can also characterize the convex order as the following theorem shows.

Theorem 3.A.45. *Let X and Y be two nonnegative random variables with equal means. Then $X \leq_{\mathrm{cx}} Y$ if, and only if, there exist random variables $Z_1, Z_2, \ldots$, with distribution functions $F_1, F_2, \ldots$, such that $Z_1 =_{\mathrm{st}} X$, $EZ_j = EY$, $j = 1, 2, \ldots$, $Z_j \to_{\mathrm{st}} Y$ as $j \to \infty$, and $S^-(\overline{F}_j - \overline{F}_{j+1}) = 1$ and the sign sequence is $+, -$, $j = 1, 2, \ldots$.*

If the random variables in Theorem 3.A.45 are not nonnegative then the sufficiency part of that theorem is not correct. This can be seen by noting that Example 1 of Müller [410] describes a sequence of distribution functions (say of the random variables $Z_1, Z_2, \ldots$), and two other distribution functions (say of the random variables X and Y, which are not nonnegative), which satisfy all the conditions in Theorem 3.A.45, but such that $X \not\leq_{\mathrm{cx}} Y$. We thank Taizhong Hu for pointing out this fact to us.

In Theorem 3.A.24 we obtained the "minimal" random variable with respect to the order $\leq_{\mathrm{cx}}$ when the support and the mean are given. Now, with the aid of Theorem 3.A.44, we can obtain the "minimal" random variables with respect to the order $\leq_{\mathrm{cx}}$ for some rich families of random variables when the mean is given. This is shown in the next result.

Theorem 3.A.46. *Let X be a nonnegative random variable with mean μ.*

(a) *Suppose that X has a density function that is decreasing on $[0,\infty)$. Let Y be uniformly distributed over the interval $[0,2\mu]$ (so that $EY=\mu$). Then*

$$Y \le_{\rm cx} X.$$

(b) *Suppose that X has a density function that is decreasing and convex on $[0,\infty)$. Let Z have the triangular distribution over the interval $[0,3\mu]$ with density function*

$$f_Z(x) = \begin{cases} \frac{2}{3\mu} - \frac{2}{9\mu^2}x, & \text{if } 0 \le x \le 3\mu, \\ 0, & \text{otherwise} \end{cases}$$

(so that $EZ=\mu$). Then

$$Z \le_{\rm cx} X.$$

Proof. In order to prove (a) let f_X and f_Y denote the density functions of X and Y, respectively. It is easy to see, using the fact that $EX = EY$, that $S^-(f_X - f_Y) = 2$ and that the sign sequence is $+,-,+$. The result now follows from Theorem 3.A.44. The proof of (b) is similar. $\square$

Some illustrations of the applicability of Theorem 3.A.44 are shown in the following examples.

Example 3.A.47. The following statements can be proven by verifying, using the method in Shaked [502], that in each one of them the two random variables have the same mean, and that their densities satisfy (3.A.57).

(a) Let X and Y have, respectively, the Poisson and the Pascal distributions with the discrete densities f and g given by

$$f(x) = e^{-\lambda/\alpha}\frac{(\lambda/\alpha)^x}{x!}, \qquad x = 0,1,\ldots,$$
$$g(x) = \Big(\frac{\alpha}{1+\alpha}\Big)^{\lambda}\Big(\frac{1}{1+\alpha}\Big)^{x}\frac{\Gamma(x+\lambda)}{\Gamma(\lambda)x!}, \quad x = 0,1,\ldots,$$

where $\alpha > 0$ and $\lambda > 0$. Then $X \le_{\rm cx} Y$.

(b) Let X and Y have, respectively, the exponential and the power distributions with the densities f and g given by

$$f(x) = (\gamma-1)\delta^{-1}\exp\{-(\gamma-1)\delta^{-1}x\}, \quad x \ge 0,$$
$$g(x) = (\gamma/\delta)(1+x/\gamma)^{-\gamma-1}, \qquad x \ge 0,$$

where $\gamma > 1$ and $\delta > 0$. Then $X \le_{\rm cx} Y$.

(c) Let X and Y have, respectively, the binomial and the Polya distributions with the discrete densities f and g given by

$$f(x) = \binom{n}{x}\left(\frac{\alpha}{\alpha+\beta}\right)^x\left(\frac{\beta}{\alpha+\beta}\right)^{n-x}, \qquad x = 0,1,\ldots,n,$$
$$g(x) = \binom{n}{x}\frac{\Gamma(\alpha+\beta)\Gamma(\alpha+x)\Gamma(\beta+n-x)}{\Gamma(\alpha)\Gamma(\beta)\Gamma(\alpha+\beta+n)}, \quad x = 0,1,\ldots,n,$$

where $\alpha > 0$ and $\beta > 0$. Then $X \leq_{\rm cx} Y$.

(d) Let X and Y have, respectively, the discrete densities f and g given by

$$f(x) = \left(\frac{\alpha}{\alpha+\beta-1}\right)^\lambda\left(\frac{\beta-1}{\alpha+\beta-1}\right)^x\frac{\Gamma(x+\lambda)}{\Gamma(\lambda)x!}, \quad x = 0,1,\ldots,$$
$$g(x) = \frac{\Gamma(\alpha+\beta)\Gamma(\beta+\lambda)\Gamma(\lambda+x)\Gamma(\alpha+x)}{\Gamma(\alpha)\Gamma(\beta)\Gamma(\lambda)\Gamma(\alpha+\beta+\lambda+x)x!}, \quad x = 0,1,\ldots,$$

where $\alpha > 0$, $\beta > 1$, and $\lambda > 0$. Then $X \leq_{\rm cx} Y$.

Example 3.A.48. Let X and Y be Bernoulli random variables with parameters p and q, respectively, where $0 < p \leq q \leq 1$. Then

$$\frac{X}{p} \geq_{\rm cx} \frac{Y}{q}.$$

This can be seen by easily verifying (3.A.59), where F and G there are the distribution functions of Y and X, respectively.

A further illustration of the applicability of Theorem 3.A.44 is shown in the following example.

Example 3.A.49. Let $U_{(i:n)}$ be the ith order statistic from a sample of n uniform $[0,1]$ random variables. By examination of the density functions of the normalized variables $\frac{n+1}{i}U_{(i:n)}$ it is possible to verify (3.A.57) and obtain the following results (see also Example 4.B.13):

$$U_{(i+1:n)} \leq_{\rm Lorenz} U_{(i:n)}, \qquad \text{for all } i \leq n-1,$$
$$U_{(i:n)} \leq_{\rm Lorenz} U_{(i:n+1)}, \qquad \text{for all } i \leq n+1,$$
$$U_{(n-i+1:n+1)} \leq_{\rm Lorenz} U_{(n-i:n)}, \qquad \text{for all } i \leq n,$$

and

$$U_{(n+2:2n+3)} \leq_{\rm Lorenz} U_{(n+1:2n+1)}, \quad \text{for all } n.$$

The last inequality may be described as "sample medians exhibit less variability as sample size increases." Arnold and Villasenor [21], who derived the above results, give many other Lorenz order inequalities for order statistics and record values associated with various parametric families; see also Wilfling [566] and Kleiber [304].

Example 3.A.50. Let $X_{(i:n)}$ denote the ith order statistic in a sample of n independent and identically distributed random variables having the common distribution F, survival function $\overline{F}$, and density function f. Recall that a function $\phi : [0, \infty) \to [0, \infty)$ is said to be regularly varying at ∞ with index $\rho \in \mathbb{R}$ if

$$\lim_{x\to\infty} \frac{\phi(tx)}{\phi(x)} = t^{\rho}, \quad \text{for all } t \in [0, \infty).$$

The function ϕ is said to be regularly varying at $-\infty$ with index ρ if $\phi(-x)$ is regularly varying at ∞ with index ρ. Finally, the function ϕ is said to be regularly varying at 0 with index ρ if $\phi(x^{-1})$ is regularly varying at ∞ with index ρ.

For F with support $(-\infty, \infty)$ Kleiber [303] showed:

(a) If F is regularly varying at $-\infty$ with index $\alpha < 0$, and if f is monotone on $(-\infty, c]$ for some c, then $X_{(j:m)} \le_{\text{dil}} X_{(i:n)}$ implies $i \le j$.
(b) If $\overline{F}$ is regularly varying at ∞ with index $\alpha < 0$, and if f is monotone on $[c, \infty)$ for some c, then $X_{(j:m)} \le_{\text{dil}} X_{(i:n)}$ implies $n - i \le m - j$.

For F with support $[0, \infty)$ Kleiber [303] also showed:

(c) If F is regularly varying at 0 with index $\alpha < 0$, and if f is monotone on $(0, c]$ for some $c > 0$, then $X_{(j:m)} \le_{\text{Lorenz}} X_{(i:n)}$ implies $i \le j$.
(d) If $\overline{F}$ is regularly varying at ∞ with index $\alpha < 0$, and if f is monotone on $[c, \infty)$ for some c, then $X_{(j:m)} \le_{\text{Lorenz}} X_{(i:n)}$ implies $n - i \le m - j$.

The following example gives necessary and sufficient conditions for the comparison of normal random variables; it is generalized in Example 7.A.13. See related results in Examples 1.A.26 and 4.A.46.

Example 3.A.51. Let X be a normal random variable with mean μ_X and variance σ_X^2, and let Y be a normal random variable with mean μ_Y and variance σ_Y^2. Then $X \le_{\text{cx}} Y$ if, and only if, $\mu_X = \mu_Y$ and $\sigma_X^2 \le \sigma_Y^2$.

Analogous to the result in Remark 1.A.18, it can be shown that the set of all distribution functions on $\mathbb{R}$, with any fixed finite mean, is a lattice with respect to the order $\le_{\text{cx}}$.

Let X and Y be two random variables with densities f and g, respectively. Recall that supp(X) and supp(Y) denote the respective supports. We say that X is *uniformly less variable than* Y (denoted as $X \le_{\text{uv}} Y$) if supp(X) $\subseteq$ supp(Y) and the ratio $f(x)/g(x)$ is unimodal over supp(Y), where the mode is a supremum, but X and Y are not ordered by the usual stochastic order (see definition in Section 1.A).

The relation $\le_{\text{uv}}$ is not a transitive order. It is possible to have $X \le_{\text{uv}} Y$ and $Y \le_{\text{uv}} Z$ but not $X \le_{\text{uv}} Z$. However, it is useful as a simple condition which implies (3.A.57). The next theorem points out this relationship. The proof of the theorem is easy and is therefore omitted.

Theorem 3.A.52. *Let X and Y be two random variables with densities f and g, respectively, such that* $\text{supp}(X) \subseteq \text{supp}(Y)$. *Then $X \le_{\text{uv}} Y$ if, and only if,*

$$S^-(g - cf) \le 2 \textit{ whenever } c > 0, \quad \textit{and in case of equality the sign sequence is } +, -, +. \tag{3.A.60}$$

From (3.A.60) and (3.A.57) we see that the order $\le_{\text{uv}}$ is a sufficient condition for the order $\le_{\text{cx}}$ provided the underlying random variables have equal means. This is formally stated in the next theorem.

Theorem 3.A.53. *Let X and Y be two random variables with absolutely continuous distributions and equal means such that* $\text{supp}(X) \subseteq \text{supp}(Y)$. *If $X \le_{\text{uv}} Y$, then $X \le_{\text{cx}} Y$.*

A relation that is even stronger than $\le_{\text{uv}}$ is defined next. Its usefulness is that it gives a simple sufficient condition for the order $\le_{\text{uv}}$ and therefore for the order $\le_{\text{cx}}$. Again, let X and Y be two random variables with densities f and g, respectively. We say that X is *logconcave relative to* Y (denoted by $X \le_{\text{lc}} Y$) if f/g is logconcave. The relation $\le_{\text{lc}}$, unlike the relation $\le_{\text{uv}}$, is transitive, and it implies the relation $\le_{\text{uv}}$ as the next result shows. Again, the proof is trivial and hence it is omitted.

Theorem 3.A.54. *Let X and Y be two random variables with densities f and g, respectively, such that* $\text{supp}(X) \subseteq \text{supp}(Y)$ *and $S^-(g - f) = 2$. Then $X \le_{\text{lc}} Y \Longrightarrow X \le_{\text{uv}} Y$.*

3.A.4 Some properties in reliability theory

Recall from page 1 the definitions of NBUE and NWUE random variables. Such random variables are of interest in reliability theory. The next result shows that NBUE [NWUE] random variables are smaller [larger] than exponential random variables with the same means with respect to the convex order. We denote by $\text{Exp}(\mu)$ an exponential random variable with mean μ.

Theorem 3.A.55. *If X is an* NBUE [NWUE] *random variable with mean μ, then*

$$X \le_{\text{cx}} [\ge_{\text{cx}}] \ \text{Exp}(\mu), \tag{3.A.61}$$

or, equivalently,

$$X \ge_{\text{cv}} [\le_{\text{cv}}] \ \text{Exp}(\mu). \tag{3.A.62}$$

The proof consists of showing that if $\overline{F}$ is the survival function of X, then

$$\int_x^\infty \overline{F}(u) \mathrm{d}u \le [\ge] \ \mu \mathrm{e}^{-x/\mu}, \quad x \ge 0,$$

and the result then follows from (3.A.7). We omit the details.

Random variables that satisfy (3.A.61) are called harmonic new better [worse] than used in expectation (HNBUE [HNWUE]). Sometimes such random variables are defined by $X \le_{\text{hmrl}} [\ge_{\text{hmrl}}] \text{Exp}(\mu)$ rather than by (3.A.61), but by (2.B.7) these two definitions are the same.

Recall from page 1 the definition of IMRL and DMRL random variables. The following result characterizes such random variables by means of the dilation order defined in (3.A.32). Other characterizations of DMRL and IMRL random variables, by means of other stochastic orders, can be found in Theorems 2.A.23, 2.B.17, 3.C.13, and 4.A.51.

Theorem 3.A.56. *The nonnegative random variable X is* DMRL [IMRL] *if, and only if,*

$$[X|X > t] \ge_{\text{dil}} [\le_{\text{dil}}] [X|X > t'] \quad \text{whenever } t \le t'.$$

Two related results are stated next without proofs.

Theorem 3.A.57. *Let X and Y be two random variables that have a common support of the form $(0, \infty)$, and that have finite means. If X and/or Y is* IMRL, *and if $X \le_{\text{mrl}} Y$, then $X \le_{\text{dil}} Y$.*

Theorem 3.A.58. *Let X and Y be two random variables that have a common support of the form $(0, \infty)$, and that have finite means. If X is* NBUE *and Y is* NWUE, *then*

$$X \le_{\text{mrl}} Y \iff X \le_{\text{dil}} Y \iff EX \le EY.$$

3.A.5 The m-convex orders

Let $\mathcal{S}$ be a subinterval of the real line. The subinterval $\mathcal{S}$ may be open, half-open, or closed, finite or infinite. Fix a positive integer m, and consider the class $\mathcal{M}^{\mathcal{S}}_{m\text{-cx}}$ of all functions $\phi : \mathcal{S} \to \mathbb{R}$ whose mth derivative $\phi^{(m)}$ exists and satisfies $\phi^{(m)}(x) \ge 0$, for all $x \in \mathcal{S}$, or which are limits of sequences of functions whose mth derivative is continuous and nonnegative on $\mathcal{S}$.

Let X and Y be two random variables that take on values in $\mathcal{S}$ such that

$$E[\phi(X)] \le E[\phi(Y)] \quad \text{for all functions } \phi \in \mathcal{M}^{\mathcal{S}}_{m\text{-cx}}, \tag{3.A.63}$$

provided the expectations exist. Then X is said to be *smaller than Y in the m-convex order* (denoted as $X \le^{\mathcal{S}}_{m\text{-cx}} Y$). For random variables X and Y that take on values in $\mathbb{N}_{++}$ the definition of the m-cx order is similar — it can be found in Denuit and Lefèvre [146].

In a similar manner one can define the m-concave order and observe that

$$X \le^{\mathcal{S}}_{m\text{-cx}} Y \iff \begin{cases} X \le^{\mathcal{S}}_{m\text{-cv}} Y & \text{when } m \text{ is odd,} \\ Y \le^{\mathcal{S}}_{m\text{-cv}} X & \text{when } m \text{ is even.} \end{cases}$$

It can be shown that

$$X \le^{\mathcal{S}}_{m\text{-cx}} Y \Longleftrightarrow \begin{cases} EX^k = EY^k, \quad k = 1, 2, \ldots, m-1, \quad \text{and} \\ E(X-t)_+^{m-1} \le E(Y-t)_+^{m-1} \quad \text{for all } t \in \mathcal{S}, \end{cases} \tag{3.A.64}$$

and also that

$$X \le^{\mathcal{S}}_{m\text{-cx}} Y \Longleftrightarrow \begin{cases} EX^k = EY^k, \quad k = 1, 2, \ldots, m-1, \quad \text{and} \\ (-1)^m \big[E(t-Y)_+^{m-1} - E(t-X)_+^{m-1}\big] \ge 0 \text{ for all } t \in \mathcal{S}. \end{cases} \tag{3.A.65}$$

Note that the order $\le^{\mathcal{S}}_{1\text{-cx}}$ is just the order $\le_{\text{st}}$, and that the order $\le^{\mathcal{S}}_{2\text{-cx}}$ is the order $\le_{\text{cx}}$. Menezes, Geiss, and Tressler [390] gave the following interpretation to the order $\le^{\mathcal{S}}_{3\text{-cx}}$: if $X \le^{\mathcal{S}}_{3\text{-cx}} Y$, then, of course, X and Y have the same mean and variance, but X then has smaller rightside risk than Y.

Let F and G be the distribution functions of X and Y, respectively. Denote $F^{[0]}(t) = F(t)$, and, for $k \ge 1$, denote $F^{[k]}(t) = \int_{-\infty}^t F^{[k-1]}(x)\mathrm{d}x$. Similarly, denote $\overline{F}^{[0]}(t) = \overline{F}(t)$, and, for $k \ge 1$, denote $\overline{F}^{[k]}(t) = \int_t^{\infty} \overline{F}^{[k-1]}(x)\mathrm{d}x$. Define $G^{[k]}$ and $\overline{G}^{[k]}$ in a similar manner. Using the identities

$$F_Y^{[m-1]}(t) - F_X^{[m-1]}(t) = \frac{E(t-Y)_+^{m-1} - E(t-X)_+^{m-1}}{(m-1)!}$$

and

$$\overline{F}_Y^{[m-1]}(t) - \overline{F}_X^{[m-1]}(t) = \frac{E(Y-t)_+^{m-1} - E(X-t)_+^{m-1}}{(m-1)!}$$

(which are easily proven by induction and Fubini's Theorem) we obtain from (3.A.64) and (3.A.65) that

$$X \le^{\mathcal{S}}_{m\text{-cx}} Y \Longleftrightarrow \begin{cases} EX^k = EY^k, \quad k = 1, 2, \ldots, m-1, \quad \text{and} \\ (-1)^m \left[F_Y^{[m-1]}(t) - F_X^{[m-1]}(t)\right] \ge 0 \quad \text{for all } t \in \mathbb{R}, \end{cases} \tag{3.A.66}$$

and also that

$$X \le^{\mathcal{S}}_{m\text{-cx}} Y \Longleftrightarrow \begin{cases} EX^k = EY^k, \quad k = 1, 2, \ldots, m-1, \quad \text{and} \\ \overline{F}_Y^{[m-1]}(t) - \overline{F}_X^{[m-1]}(t) \ge 0 \quad \text{for all } t \in \mathbb{R}. \end{cases} \tag{3.A.67}$$

Using the identities

$$m! \int_t^{\infty} \left[\overline{F}_Y^{[m-1]}(x) - \overline{F}_X^{[m-1]}(x)\right]\mathrm{d}x = E(Y-t)_+^m - E(X-t)_+^m$$

and

$$m! \int_{-\infty}^t \left[F_Y^{[m-1]}(x) - F_X^{[m-1]}(x)\right]\mathrm{d}x = E(t-Y)_+^m - E(t-X)_+^m,$$

we obtain from (3.A.66) and (3.A.67) that

$$X \leq_{m\text{-cx}}^{\mathcal{S}} Y \Longleftrightarrow \begin{cases} EX^k = EY^k, \quad k = 1, 2, \ldots, m-1, \quad \text{and} \\ E(Y-t)_+^m - E(X-t)_+^m \quad \text{is decreasing in } t \in \mathbb{R}, \end{cases}$$

and also that

$$X \leq_{m\text{-cx}}^{\mathcal{S}} Y \Longleftrightarrow \begin{cases} EX^k = EY^k, \quad k = 1, 2, \ldots, m-1, \quad \text{and} \\ (-1)^m \left[E(t-Y)_+^m - E(t-X)_+^m\right] \text{ is increasing in } t \in \mathbb{R}. \end{cases}$$

Fishburn [203] has reported some attempts at obtaining an analog of Theorem 3.A.4 for the 3-cx order.

From (3.A.63) it is seen that if $X \leq_{m\text{-cx}}^{\mathcal{S}} Y$, then

$$EX^k \leq EY^k \quad \text{for } k \geq m \text{ such that } k - m \text{ is even.}$$

If, moreover, X and Y are nonnegative, then

$$EX^k \leq EY^k \quad \text{for } k \geq m.$$

Motivated by Theorem 3.A.42 (see also Theorems 1.A.8, 4.A.48, 4.A.69, 5.A.15, 6.B.19, 6.G.12, 6.G.13, and 7.A.14–7.A.16) we have the following result.

Theorem 3.A.59. *Let X and Y be two random variables that take on values in $\mathcal{S}$. If $X \leq_{m\text{-cx}}^{\mathcal{S}} Y$, and if $E[X^m] = E[Y^m]$, then $X =_{\text{st}} Y$.*

Theorem 3.A.59 can be strengthened to the following result in a way that is analogous to the way in which Theorem 3.A.43 strengthened Theorem 3.A.42; we do not detail the proof here.

Theorem 3.A.60. *Let X and Y be two random variables that take on values in $\mathcal{S}$. If $X \leq_{m\text{-cx}}^{\mathcal{S}} Y$ and if $E[\phi(X)] = E[\phi(Y)]$ for some $\phi \in \mathcal{M}_{m\text{-cx}}^{\mathcal{S}}$ which satisfies $\phi^{(m)}(x) > 0$ for all $x \in \mathcal{S}$, then $X =_{\text{st}} Y$.*

Note that Theorems 1.A.8, 3.A.43, and 3.A.59 are all special cases of Theorem 3.A.60.

A generalization of (3.A.12) is given in the next theorem. The notations l_X, u_X, l_Y, and u_Y are described before (3.A.12).

Theorem 3.A.61. *Let X and Y be two random variables that take on values in $\mathcal{S}$. If $X \leq_{m\text{-cx}}^{\mathcal{S}} Y$, then $u_X \leq u_Y$. Also, if m is even, then $l_X \geq l_Y$, and if m is odd, then $l_X \leq l_Y$.*

Some closure properties of the order $\leq_{m\text{-cx}}^{\mathcal{S}}$ are given in the next theorem.

Theorem 3.A.62. (a) *Let X and Y be two random variables that take on values in $\mathcal{S}$. Then*

$$X \leq_{m\text{-cx}}^{\mathcal{S}} Y \Longleftrightarrow \begin{cases} -X \leq_{m\text{-cx}}^{-\mathcal{S}} -Y & \text{when } s \text{ is even,} \\ -Y \leq_{m\text{-cx}}^{-\mathcal{S}} -X & \text{when } s \text{ is odd.} \end{cases}$$

(b) *Let* X, Y, *and* Θ *be random variables such that* $[X|\Theta=\theta] \le_{m\text{-cx}}^{\mathcal{S}} [Y|\Theta=\theta]$ *for all* θ *in the support of* Θ. *Then* $X \le_{m\text{-cx}}^{\mathcal{S}} Y$. *That is, the* m*-convex order is closed under mixtures.*

(c) *If* $X \le_{m\text{-cx}}^{\mathcal{S}} Y$, *then* $cX \le_{m\text{-cx}}^{c\mathcal{S}} cY$ *whenever* $c > 0$, *where* $c\mathcal{S} = \{x \in \mathbb{R} : x/c \in \mathcal{S}\}$.

(d) *If* $X \le_{m\text{-cx}}^{\mathcal{S}} Y$, *then* $cX \le_{m\text{-cx}}^{c\mathcal{S}} cY$ *whenever* $c < 0$ *and* m *is even, and* $cY \le_{m\text{-cx}}^{c\mathcal{S}} cX$ *whenever* $c < 0$ *and* m *is odd.*

(e) *If* $X \le_{m\text{-cx}}^{\mathcal{S}} Y$, *then* $X + d \le_{m\text{-cx}}^{\mathcal{S}+d} Y + d$ *for all* $d \in \mathbb{R}$, *where* $\mathcal{S} + d = \{x \in \mathbb{R} : x - d \in \mathcal{S}\}$; *that is, the* m*-convex order is shift-invariant.*

(f) *Let* $\{X_j,\ j = 1, 2, \dots\}$ *and* $\{Y_j,\ j = 1, 2, \dots\}$ *be two sequences of random variables that take on values in* $\mathcal{S}$, *such that* $X_j \to_{\text{st}} X$ *and* $Y_j \to_{\text{st}} Y$ *as* $j \to \infty$. *Assume that* $E(X)_+^{m-1}$ *and* $E(Y)_+^{m-1}$ *are finite and that* $E(X_j)_+^{m-1} \to E(X)_+^{m-1}$ *and that* $E(Y_j)_+^{m-1} \to E(Y)_+^{m-1}$ *as* $j \to \infty$. *If* $X_i \le_{m\text{-cx}}^{\mathcal{S}} Y_i$ *for all integers* i, *then* $X \le_{m\text{-cx}}^{\mathcal{S}} Y$. *That is, the* m*-convex order is closed under limits.*

(g) *Let* $X_1, X_2, \dots, X_n$ *be a set of independent random variables and let* $Y_1, Y_2, \dots, Y_n$ *be another set of independent random variables, all taking on values in* $\mathcal{S}$. *If* $X_i \le_{m\text{-cx}}^{\mathcal{S}} Y_i$ *for* $i = 1, 2, \dots, n$, *then*

$$\sum_{j=1}^{n} X_j \le_{m\text{-cx}}^{\mathcal{R}} \sum_{j=1}^{n} Y_j,$$

where $\mathcal{R}$ *denotes the union of the supports of the distribution functions of the two sums. That is, the* m*-convex order is closed under convolutions.*

(h) *Let* $X_1, X_2, \dots$ *be a set of independent random variables and let* $Y_1, Y_2, \dots$ *be another set of independent random variables, all taking on values in* $\mathcal{S}$. *If* $X_i \le_{m\text{-cx}}^{\mathcal{S}} Y_i$ *for* $i = 1, 2, \dots$, *then, for any positive integer-valued random variable* N *which is independent of the* X_i*'s and of the* Y_j*'s, one has*

$$\sum_{j=1}^{N} X_j \le_{m\text{-cx}}^{\tilde{\mathcal{R}}} \sum_{j=1}^{N} Y_j,$$

where $\tilde{\mathcal{R}}$ *denotes the union of the supports of the distribution functions of the two compound sums.*

Theorem 3.A.62(h) can be extended as follows.

Theorem 3.A.63. *Let* $X_1, X_2, \dots$ *be a set of independent random variables and let* $Y_1, Y_2, \dots$ *be another set of independent random variables, all taking on values in* $\mathcal{S}$. *Let* N_1 *be an integer-valued random variable that is independent of the* X_i*'s, and let* N_2 *be an integer-valued random variable that is independent of the* Y_i*'s, both taking on values in* $\mathcal{Q}$. *If* $X_i \le_{m\text{-cx}}^{\mathcal{S}} Y_i$ *for* $i = 1, 2, \dots$, *and if* $N_1 \le_{m\text{-cx}}^{\mathcal{Q}} N_2$, *then*

$$\sum_{j=1}^{N_1} X_j \le_{m\text{-cx}}^{\tilde{\mathcal{R}}} \sum_{j=1}^{N_2} Y_j,$$

where $\tilde{\mathcal{R}}$ denotes the union of the supports of the distribution functions of the two compound sums.

Theorem 3.A.19 can be extended as follows.

Theorem 3.A.64. *Let X_1 and X_2 (Y_1 and Y_2) be two independent copies of X (Y). If $X \le_{2m\text{-cx}}^{\mathcal{S}} Y$, then $X_1 - X_2 \le_{2m\text{-cx}}^{\mathcal{R}} Y_1 - Y_2$, where $\mathcal{R}$ denotes the union of the supports of the distribution functions of the two differences.*

The proof of Theorem 3.A.64 is similar to the proof of Theorem 3.A.19 (using Theorem 3.A.62(a) and (g)).

Recall from (1.A.20) that for nonnegative random variables X and Y with finite means, we denote by A_X and A_Y the corresponding asymptotic equilibrium ages.

Theorem 3.A.65. *Let X and Y be two nonnegative random variables. Then, for $m \ge 2$ we have*

$$X \le_{m\text{-cx}}^{[0,\infty)} Y \Longleftrightarrow A_X \le_{(m-1)\text{-cx}}^{[0,\infty)} A_Y.$$

In particular,

$$X \le_{\text{cx}} Y \Longleftrightarrow A_X \le_{\text{st}} A_Y.$$

We now describe a generalization of Theorem 3.A.44. Let $\mathcal{B}_m(\mathcal{S}; \mu_1, \mu_2, \dots, \mu_{m-1})$ denote the class of all the random variables X whose distribution functions have support in $\mathcal{S}$ and which have the first $m-1$ moments $EX^k = \mu_k$, $k = 1, 2, \dots, m-1$.

Theorem 3.A.66. *Let X and Y be two random variables in $\mathcal{B}_m(\mathcal{S}; \mu_1, \mu_2, \dots, \mu_{m-1})$ with distribution functions F and G, respectively, and with density functions f and g, respectively.*

(a) *If $S^-(F-G) \le m-1$ and if the last sign of $F-G$ is a $+$, then $X \le_{m\text{-cx}}^{\mathcal{S}} Y$.*
(b) *If $S^-(f-g) \le m$ and if the last sign of $g-f$ is a $+$, then $X \le_{m\text{-cx}}^{\mathcal{S}} Y$.*

The following example describes typical applications of Theorem 3.A.66.

Example 3.A.67. Let X have the Gamma density given by

$$f_X(x) = \frac{\beta^\alpha}{\Gamma(\alpha)} x^{\alpha-1} e^{-\beta x}, \quad x > 0,$$

where $\alpha > 0$ and $\beta > 0$ are constants, and let Y have the inverse Gaussian density given by

$$f_Y(x) = \frac{\alpha x^{-3/2}}{\sqrt{2\pi\beta}} \exp\left\{ -\frac{(\alpha - \beta x)^2}{2\beta x} \right\}, \quad x > 0,$$

where also here $\alpha > 0$ and $\beta > 0$ are constants. Note that X and Y have the same mean α/β and the same second moment $\alpha(\alpha+1)/\beta^2$. We claim that

$X \le_{\text{3-cx}}^{[0,\infty)} Y$. In order to see it, first note that without loss of generality we can take the means to be equal to 1, that is, $\beta = \alpha$. Now, a straightforward computation yields

$$\log \frac{f_X(x)}{f_Y(x)} = C + \left(\alpha + \frac{1}{2}\right) \log x + \frac{\alpha}{2x} - \frac{\alpha x}{2}, \qquad x > 0,$$

where C is some constant. The first derivative of the above expression is a quadratic form in $1/x$, which cannot have more than two zeroes, so the expression itself has no more than three sign changes. In addition, the above expression tends to $-\infty$ as $x \to \infty$. The stated result now follows from Theorem 3.A.66(b).

Let Z have the lognormal density given by

$$f_Z(x) = \frac{1}{x\tau\sqrt{2\pi}} \exp\left\{-\frac{(\log x - \nu)^2}{2\tau^2}\right\}, \qquad x > 0,$$

where $\tau > 0$ and $\nu > 0$ are constants. With the choice $\tau^2 = \log(1 + \frac{1}{\alpha})$ and $\nu = \frac{1}{2} \log \frac{\alpha^3}{(\alpha+1)\beta^2}$ we have that X and Z have the same mean α/β and the same second moment $\alpha(\alpha+1)/\beta^2$. We now claim that $X \le_{\text{3-cx}}^{[0,\infty)} Z$. In order to see it, again note that without loss of generality we can take the means to be equal to 1, that is, $\beta = \alpha$. Now, a straightforward computation yields

$$\log \frac{f_X(x)}{f_Z(x)} = C + \left(\alpha + \frac{1}{2}\right) \log x - \alpha x + \frac{\log^2 x}{2\tau^2}, \qquad x > 0,$$

where C is some constant. Substituting $u = \log x$, the above expression is seen to be the difference of a quadratic form in u and an exponential function, which cannot have more than three sign changes. In addition, the above expression tends to $-\infty$ as $x \to \infty$. The stated result again follows from Theorem 3.A.66(b).

Theorem 3.A.24 can be viewed as a result that gives the "minimal" and the "maximal" random variables with respect to the order $\le_{\text{2-cx}}^{\mathcal{S}}$ when the (bounded) support and the mean are given. The following theorem gives the extrema with respect to the order $\le_{\text{3-cx}}^{\mathcal{S}}$ when the first two moments are given. Here we take $\mathcal{S} = [a, b]$ for some finite a and b.

Theorem 3.A.68. *Let $X \in \mathcal{B}_3([a,b]; \mu_1, \mu_2)$. Consider the random variables $X_{\min}^{(3)}$ and $X_{\max}^{(3)}$ in $\mathcal{B}_3([a,b]; \mu_1, \mu_2)$ defined by*

$$X_{\min}^{(3)} = \begin{cases} a & \textit{with probability } \frac{\mu_2 - \mu_1^2}{(a-\mu_1)^2 + \mu_2 - \mu_1^2}, \\ \mu_1 + \frac{\mu_2 - \mu_1^2}{\mu_1 - a} & \textit{with probability } \frac{(a-\mu_1)^2}{(a-\mu_1)^2 + \mu_2 - \mu_1^2}, \end{cases}$$

and

$$X_{\max}^{(3)} = \begin{cases} \mu_1 - \frac{\mu_2-\mu_1^2}{b-\mu_1} & \textit{with probability} \quad \frac{(b-\mu_1)^2}{(b-\mu_1)^2+\mu_2-\mu_1^2}, \\ b & \textit{with probability} \quad \frac{\mu_2-\mu_1^2}{(b-\mu_1)^2+\mu_2-\mu_1^2}. \end{cases}$$

Then $X_{\min}^{(3)} \leq_{\text{3-cx}}^{[a,b]} X \leq_{\text{3-cx}}^{[a,b]} X_{\max}^{(3)}$.

An effective method for deriving the support points and the associated probabilities of the stochastic extrema in general (that is, for m's other than 3) will be described next. For the purpose of somewhat simplifying the expressions below we take $a = 0$. Thus we describe how to obtain the support points and the associated probabilities of $X_{\min}^{(m)}$ and $X_{\max}^{(m)}$ in $\mathcal{B}_m([0,b];\mu_1,\mu_2,\ldots,\mu_{m-1})$.

If m is even, $m = 2k$, say, then the support of $X_{\min}^{(2k)}$ in $\mathcal{B}_{2k}([0,b];\mu_1,\mu_2,\ldots,\mu_{2k-1})$ consists of k interior points $x_1,x_2,\ldots,x_k$, $0 < x_1 < x_2 < \cdots < x_k < b$, which are the k distinct roots of the equation (denoting $\mu_0 = 1$)

$$\begin{vmatrix} 1 & x & x^2 & \cdots & x^k \\ \mu_0 & \mu_1 & \mu_2 & \cdots & \mu_k \\ \mu_1 & \mu_2 & \mu_3 & \cdots & \mu_{k+1} \\ \vdots & \vdots & \vdots & \ddots & \vdots \\ \mu_{k-1} & \mu_k & \mu_{k+1} & \cdots & \mu_{2k-1} \end{vmatrix} = 0;$$

the corresponding probabilities $p_1,p_2,\ldots,p_k$ are now found by solving

$$p_1x_1^j + p_2x_2^j + \cdots + p_kx_k^j = \mu_j, \quad j = 0,1,\ldots,k-1.$$

The support of $X_{\max}^{(2k)}$ in $\mathcal{B}_{2k}([0,b];\mu_1,\mu_2,\ldots,\mu_{2k-1})$ consists of the points 0, b, and $k-1$ interior points $x_2,x_3,\ldots,x_k$, $0 < x_2 < x_3 < \cdots < x_k < b$, which are the $k-1$ distinct roots of the equation

$$\begin{vmatrix} 1 & x & x^2 & \cdots & x^{k-1} \\ \mu_2 - b\mu_1 & \mu_3 - b\mu_2 & \mu_4 - b\mu_3 & \cdots & \mu_{k+1} - b\mu_k \\ \mu_3 - b\mu_2 & \mu_4 - b\mu_3 & \mu_5 - b\mu_4 & \cdots & \mu_{k+2} - b\mu_{k+1} \\ \vdots & \vdots & \vdots & \ddots & \vdots \\ \mu_k - b\mu_{k-1} & \mu_{k+1} - b\mu_k & \mu_{k+2} - b\mu_{k+1} & \cdots & \mu_{2k-1} - b\mu_{2k-2} \end{vmatrix} = 0;$$

the corresponding probabilities $p_1,p_2,\ldots,p_{k+1}$ are now found by solving the Vandermonde system

$$\begin{cases} p_1 + p_2 + \cdots + p_{k+1} = 1, \\ p_2x_2^j + p_3x_3^j + \cdots + p_kx_k^j + p_{k+1}b^j = \mu_j, \quad j = 1,2,\ldots,k. \end{cases}$$

When m is odd, $m = 2k+1$, say, then the support of $X_{\min}^{(2k+1)}$ in $\mathcal{B}_{2k+1}([0,b];\mu_1,\mu_2,\ldots,\mu_{2k})$ consists of 0 and k interior points $x_2,x_3,\ldots,x_{k+1}$, $0 < x_2 < x_3 < \cdots < x_{k+1} < b$, which are the k distinct roots of the equation

$$\begin{vmatrix} 1 & x & x^2 & \cdots & x^k \\ \mu_1 & \mu_2 & \mu_3 & \cdots & \mu_{k+1} \\ \mu_2 & \mu_3 & \mu_4 & \cdots & \mu_{k+2} \\ \vdots & \vdots & \vdots & \ddots & \vdots \\ \mu_k & \mu_{k+1} & \mu_{k+2} & \cdots & \mu_{2k} \end{vmatrix} = 0;$$

the corresponding probabilities $p_1, p_2, \ldots, p_{k+1}$ are now found by solving

$$\begin{cases} p_1 + p_2 + \cdots + p_{k+1} = 1, \\ p_2 x_2^j + p_3 x_3^j + \cdots + p_{k+1} x_{k+1}^j = \mu_j, \quad j = 1, 2, \ldots, k. \end{cases}$$

The support of $X_{\max}^{(2k+1)}$ in $\mathcal{B}_{2k+1}([0,b];\mu_1,\mu_2,\ldots,\mu_{2k})$ consists of the points b and k interior points $x_1, x_2, \ldots, x_k$, $0 < x_1 < x_2 < \cdots < x_k < b$, which are the k distinct roots of the equation

$$\begin{vmatrix} 1 & x & x^2 & \cdots & x^k \\ \mu_1 - b & \mu_2 - b\mu_1 & \mu_3 - b\mu_2 & \cdots & \mu_{k+1} - b\mu_k \\ \mu_2 - b\mu_1 & \mu_3 - b\mu_2 & \mu_4 - b\mu_3 & \cdots & \mu_{k+2} - b\mu_{k+1} \\ \vdots & \vdots & \vdots & \ddots & \vdots \\ \mu_k - b\mu_{k-1} & \mu_{k+1} - b\mu_k & \mu_{k+2} - b\mu_{k+1} & \cdots & \mu_{2k} - b\mu_{2k-1} \end{vmatrix} = 0;$$

the corresponding probabilities $p_1, p_2, \ldots, p_{k+1}$ are now found by solving the Vandermonde system

$$p_1 x_1^j + p_2 x_2^j + \cdots + p_{k+1} x_{k+1}^j + p_{k+1} b^j = \mu_j, \quad j = 0, 1, \ldots, k.$$

Explicit descriptions for the distribution functions of $X_{\min}^{(m)}$ and $X_{\max}^{(m)}$, for values of m up to 5, are given in Tables 3.A.1 and 3.A.2, where in Table 3.A.2 we use the notation

$$\begin{aligned} \varrho = \big((\mu_1 - b)(\mu_4 - b\mu_3) - (\mu_2 - b\mu_1)(\mu_3 - b\mu_2)\big)^2 \\ - 4\big((\mu_1 - b)(\mu_3 - b\mu_2) - (\mu_2 - b\mu_1)^2\big) \\ \times \big((\mu_2 - b\mu_1)(\mu_4 - b\mu_3) - (\mu_3 - b\mu_2)^2\big). \end{aligned}$$

Denuit, De Vylder, and Lefèvre [142] obtained also the extrema with respect to the order $\leq_{m\text{-cx}}^{\mathcal{S}}$ when not only the first $m-1$ moments and the support are given, but also when the density function of X is known to be unimodal with a known mode. Tables that are similar to Tables 3.A.1 and 3.A.2, but when the mode is known, are available in Denuit, Lefèvre, and Shaked [153, 154].

3.B The Dispersive Order

3.B.1 Definition and equivalent conditions

Let X and Y be two random variables with distribution functions F and G, respectively. Let F^{-1} and G^{-1} be the right continuous inverses of F and G,

Table 3.A.1. Probability distribution of $X_{\min}^{(m)} \in \mathcal{B}_m([0,b];\mu_1,\mu_2,\ldots,\mu_{m-1})$

m	Support point	Probability mass
1	0	1
2	μ_1	1
3	0	$\frac{\mu_2-\mu_1^2}{\mu_2}$
	$\frac{\mu_2}{\mu_1}$	$\frac{\mu_1^2}{\mu_2}$
4	$r_+ = \frac{\mu_3-\mu_1\mu_2+\sqrt{(\mu_3-\mu_1\mu_2)^2-4(\mu_2-\mu_1^2)(\mu_1\mu_3-\mu_2^2)}}{2(\mu_2-\mu_1^2)}$	$\frac{\mu_1-r_-}{r_+-r_-}$
	$r_- = \frac{\mu_3-\mu_1\mu_2-\sqrt{(\mu_3-\mu_1\mu_2)^2-4(\mu_2-\mu_1^2)(\mu_1\mu_3-\mu_2^2)}}{2(\mu_2-\mu_1^2)}$	$1-\frac{\mu_1-r_-}{r_+-r_-}$
5	0	$1-p_+-p_-$
	$t_+ = \frac{\mu_1\mu_4-\mu_2\mu_3+\sqrt{(\mu_1\mu_4-\mu_2\mu_3)^2-4(\mu_1\mu_3-\mu_2^2)(\mu_2\mu_4-\mu_3^2)}}{2(\mu_1\mu_3-\mu_2^2)}$	$p_+ = \frac{\mu_2-t_-\mu_1}{t_+(t_+-t_-)}$
	$t_- = \frac{\mu_1\mu_4-\mu_2\mu_3-\sqrt{(\mu_1\mu_4-\mu_2\mu_3)^2-4(\mu_1\mu_3-\mu_2^2)(\mu_2\mu_4-\mu_3^2)}}{2(\mu_1\mu_3-\mu_2^2)}$	$p_- = \frac{\mu_2-t_+\mu_1}{t_-(t_--t_+)}$

Table 3.A.2. Probability distribution of $X_{\max}^{(m)} \in \mathcal{B}_m([0,b];\mu_1,\mu_2,\ldots,\mu_{m-1})$

m	Support point	Probability mass
1	b	1
2	0	$\frac{b-\mu_1}{b}$
	b	$\frac{\mu_1}{b}$
3	$\frac{b\mu_1-\mu_2}{b-\mu_1}$	$\frac{(b-\mu_1)^2}{(b-\mu_1)^2+\mu_2-\mu_1^2}$
	b	$\frac{\mu_2-\mu_1^2}{(b-\mu_1)^2+\mu_2-\mu_1^2}$
4	0	$1-p_1-p_2$
	$\frac{\mu_3-b\mu_2}{\mu_2-b\mu_1}$	$p_1 = \frac{(\mu_2-b\mu_1)^3}{(\mu_3-b\mu_2)(\mu_3-2b\mu_2+b^2\mu_1)}$
	b	$p_2 = \frac{\mu_1\mu_3-\mu_2^2}{b(\mu_3-2b\mu_2+b^2\mu_1)}$
5	$z_+ = \frac{(\mu_1-b)(\mu_4-b\mu_3)-(\mu_2-b\mu_1)(\mu_3-b\mu_2)+\sqrt{\varrho}}{2((\mu_1-b)(\mu_3-b\mu_2)-(\mu_2-b\mu_1)^2)}$	$q_+ = \frac{\mu_2-(b+z_-)\mu_1+bz_-}{(z_+-z_-)(z_+-b)}$
	$z_- = \frac{(\mu_1-b)(\mu_4-b\mu_3)-(\mu_2-b\mu_1)(\mu_3-b\mu_2)-\sqrt{\varrho}}{2((\mu_1-b)(\mu_3-b\mu_2)-(\mu_2-b\mu_1)^2)}$	$q_- = \frac{\mu_2-(b+z_+)\mu_1+bz_+}{(z_--z_+)(z_--b)}$
	b	$1-q_+-q_-$

respectively, and assume that

$$F^{-1}(\beta) - F^{-1}(\alpha) \le G^{-1}(\beta) - G^{-1}(\alpha) \quad \text{whenever } 0 < \alpha \le \beta < 1. \tag{3.B.1}$$

Then X is said to be *smaller than* Y *in the dispersive order* (denoted as $X \le_{\text{disp}} Y$). It is conceptually clear that the order $\le_{\text{disp}}$ indeed corresponds to a comparison of X and Y by variability because it requires the difference between any two quantiles of X to be smaller than the corresponding quantiles of Y.

It is clear from (3.B.1) that the order $\le_{\text{disp}}$ is location-free. That is,

$$X \le_{\text{disp}} Y \Longleftrightarrow X + c \le_{\text{disp}} Y \quad \text{for any real } c. \tag{3.B.2}$$

For a fixed α, one can find a c such that the inverse of the distribution of $X+c$, which is $F^{-1}(\cdot)+c$, satisfies $F^{-1}(\alpha)+c = G^{-1}(\alpha) = x_0$, say. It follows then from (3.B.2) that $F(x-c) \ge G(x)$ for all $x \ge x_0$. Similarly, it can be seen that $F(x-c) \le G(x)$ for all $x \le x_0$. This is true for every α (c and x_0 are determined by α). By varying α one can obtain any desired c of the form $G^{-1}(\alpha) - F^{-1}(\alpha)$. In fact, it can be shown that $X \le_{\text{disp}} Y$ if, and only if,

$$S^{-}(F(\cdot - c) - G(\cdot)) \le 1 \quad \text{for all } c, \text{ with the sign sequence being } -,+ \text{ in the case of equality.} \tag{3.B.3}$$

It is not hard to prove that condition (3.B.3) is equivalent to the following condition:

$$G(G^{-1}(\alpha) + c) \le F(F^{-1}(\alpha) + c) \quad \text{for all } \alpha \in (0,1) \text{ and } c > 0, \tag{3.B.4}$$

or, equivalently,

$$G(G^{-1}(\alpha) - c) \ge F(F^{-1}(\alpha) - c) \quad \text{for all } \alpha \in (0,1) \text{ and } c > 0. \tag{3.B.5}$$

Alternatively, (3.B.4) and (3.B.5) can be written as

$$(X - F^{-1}(\alpha))_+ \le_{\text{st}} (Y - G^{-1}(\alpha))_+, \quad \alpha \in (0,1). \tag{3.B.6}$$

From (3.B.1) it is clear that $X \le_{\text{disp}} Y$ if, and only if,

$$G^{-1}(\alpha) - F^{-1}(\alpha) \quad \text{increases in } \alpha \in (0,1), \tag{3.B.7}$$

or, equivalently, if, and only if,

$$\overline{G}^{-1}(\alpha) - \overline{F}^{-1}(\alpha) \quad \text{decreases in } \alpha \in (0,1), \tag{3.B.8}$$

where $\overline{F} \equiv 1 - F$ and $\overline{G} \equiv 1 - G$ are the survival functions associated with X and Y, respectively. Let $R \equiv -\log \overline{F}$ and $Q \equiv -\log \overline{G}$ denote the cumulative hazard functions of X and Y, respectively. Note that $R^{-1}(z) = \overline{F}^{-1}(\mathrm{e}^{-z})$ and

$Q^{-1}(z) = \overline{G}^{-1}(e^{-z})$. Thus from (3.B.8) we obtain that $X \leq_{\text{disp}} Y$ if, and only if,

$$Q^{-1}(z) - R^{-1}(z) \quad \text{increases in } z \geq 0. \tag{3.B.9}$$

Substituting $\alpha = F(x)$ in (3.B.7) we obtain that $X \leq_{\text{disp}} Y$ if, and only if,

$$G^{-1}(F(x)) - x \quad \text{increases in } x. \tag{3.B.10}$$

When X and Y have densities f and g, respectively, then $X \leq_{\text{disp}} Y$ if, and only if,

$$g(G^{-1}(\alpha)) \leq f(F^{-1}(\alpha)) \quad \text{for all } \alpha \in (0,1); \tag{3.B.11}$$

this can be obtained at once by differentiation of (3.B.10) and a simple substitution. When X and Y have hazard rate functions r and q, then (3.B.11) can alternatively be recast as

$$q(G^{-1}(\alpha)) \leq r(F^{-1}(\alpha)) \quad \text{for all } \alpha \in (0,1). \tag{3.B.12}$$

The dispersive order can be characterized also by comparing transformations of the random variables X and Y. For example, for continuous random variables X and Y we have that $X \leq_{\text{disp}} Y$ if, and only if,

$$Y =_{\text{st}} \phi(X) \quad \text{for some } \phi \text{ which satisfies}$$
$$\phi(x') - \phi(x) \geq x' - x \text{ whenever } x \leq x'. \tag{3.B.13}$$

In order to prove it just let ϕ be $G^{-1}F$. When the ϕ in (3.B.13) is differentiable, the condition on ϕ there is the same as $\phi' \geq 1$, where ϕ' denotes the derivative of ϕ. An equivalent way of recasting (3.B.13) for continuous random variables X and Y is the following:

$$Y =_{\text{st}} X + \psi(X) \quad \text{for some increasing function } \psi. \tag{3.B.14}$$

Condition (3.B.13) can also be rewritten as

$$X =_{\text{st}} \varphi(Y) \quad \text{for some increasing } \varphi \text{ which satisfies}$$
$$\varphi(x') - \varphi(x) \leq x' - x \text{ whenever } x \leq x'. \tag{3.B.15}$$

In fact, (3.B.15) characterizes $X \leq_{\text{disp}} Y$ even if X and Y are not continuous random variables.

The next characterization of the dispersive order that we describe is by means of observed total time on test random variables (see Section 1.A.4). Let F be an absolutely continuous distribution function of a nonnegative random variable X, and suppose, for simplicity, that 0 is the left endpoint of the support of F. Let H_F^{-1} be as defined in (1.A.19), and let X_{ttt} have the distribution function H_F. Denote by h_F the density function associated with H_F. Then it is easy to see that

$$h_F\big(H_F^{-1}(u)\big) = \frac{f(F^{-1}(u))}{1-u}, \quad 0 \le u < 1, \tag{3.B.16}$$

where f is the density function associated with F. Similarly, if G is another absolutely continuous distribution function, then the density h_G, of the inverse of the TTT transform H_G that is associated with G, satisfies

$$h_G\big(H_G^{-1}(u)\big) = \frac{g(G^{-1}(u))}{1-u}, \quad 0 \le u < 1, \tag{3.B.17}$$

where g is the density function associated with G.

Let Y and Y_{ttt} have the distribution functions G and H_G, respectively. From (3.B.11), (3.B.16), and (3.B.17) we obtain the following result.

Theorem 3.B.1. *Let X and Y be two nonnegative random variables with absolutely continuous distribution functions having 0 as the left endpoint of their supports. Then*

$$X \le_{\text{disp}} Y \Longleftrightarrow X_{\text{ttt}} \le_{\text{disp}} Y_{\text{ttt}}.$$

See related results in Theorems 1.A.29, 4.A.44, 4.B.8, 4.B.9, and 4.B.29.

Next we mention a characterization by means of the so-called Q-addition (*quantiles-addition*). The random variable Y with distribution function G is said to be the Q-addition of the random variables X and Z, with corresponding distribution functions F and H, if $G^{-1}(\alpha) = F^{-1}(\alpha) + H^{-1}(\alpha)$ for all $\alpha \in (0,1)$. If X and Y have distribution functions F and G, respectively, then by (3.B.1), $X \le_{\text{disp}} Y$ if, and only if,

$$H^{-1}(\alpha) \equiv G^{-1}(\alpha) - F^{-1}(\alpha) \quad \text{is increasing in } \alpha \in (0,1).$$

That means that H^{-1} is an inverse of a distribution function of a random variable Z, say. Thus we see that $X \le_{\text{disp}} Y$ if, and only if,

$$Y \text{ is a Q-addition of } X \text{ and } Z \text{ for some random variable } Z.$$

Another characterization of the order $\le_{\text{disp}}$ is given in the following theorem.

Theorem 3.B.2. *Let X and Y be two random variables. Then $X \le_{\text{disp}} Y$ if, and only if, for every increasing function ϕ and increasing concave function h such that ϕ and $\psi(\cdot) \equiv h(\phi(\cdot))$ are integrable with respect to the distribution of Y, and for every real number c, we have that*

$$E\phi(X-c) \ge E\phi(Y) \Longrightarrow E\psi(X-c) \ge E\psi(Y).$$

It is worthwhile to mention that two twice differentiable functions ϕ and ψ satisfy $\psi(\cdot) \equiv h(\phi(\cdot))$ for some increasing concave function h if, and only if, $\phi''/\phi' \ge \psi''/\psi'$ (see Pratt [459] or Arrow [22]).

Like the convex order (see Theorem 3.A.7), the dispersive order can be characterized by means of Yaari functionals V_h defined in (3.A.31).

Theorem 3.B.3. *Let X and Y be two random variables with the same finite means. Then $X \leq_{\text{disp}} Y$ if, and only if,*

$$V_h(X) \leq V_h(Y) \quad \textit{for every probability transformation function } h \leq 1.$$

Before leaving this subsection we should mention an alternative way of comparing by dispersion random variables that are *symmetric* about 0. In such a case one may say (as an alternative to (3.B.1)) that X is less dispersed than Y if $F^{-1}(\alpha) - F^{-1}(1/2) \leq [\geq]\ G^{-1}(\alpha) - G^{-1}(1/2)$ whenever $\alpha \geq [\leq]\ 1/2$. If X and/or Y are not necessarily symmetric, then one can define an order that is weaker than $\leq_{\text{disp}}$ by requiring

$$F^{-1}(\alpha) - F^{-1}(1-\alpha) \leq G^{-1}(\alpha) - G^{-1}(1-\alpha), \quad \alpha \in [1/2, 1];$$

see Townsend and Colonius [552].

If X and Y are *positive* random variables, then, as an alternative to (3.B.1), one can say that X is less dispersed than Y if $\log X \leq_{\text{disp}} \log Y$. The latter condition is equivalent to $\log X \leq_* \log Y$, where the order $\leq_*$ is defined in Section 4.B (see Theorem 4.B.1).

3.B.2 Properties

The dispersive order satisfies some desirable closure properties but does not satisfy some other desirable properties. For example, it is very easy to verify the following result (compare it to Theorem 3.A.18).

Theorem 3.B.4. *Let X be a random variable. Then*

$$X \leq_{\text{disp}} aX \quad \textit{whenever } a \geq 1.$$

Theorem 3.B.4 can be generalized as follows. For two functions ϕ and ψ let us denote $\phi \leq_{\text{disp}} \psi$ if

$$\phi(y) - \phi(x) \leq \psi(y) - \psi(x) \quad \text{whenever } x \leq y. \tag{3.B.18}$$

Note that if ϕ and ψ are differentiable then $\phi \leq_{\text{disp}} \psi$ if, and only if, $\phi' \leq \psi'$, where ϕ' and ψ' are the derivatives of ϕ and ψ, respectively. Now let X be a random variable. Write $\psi(X) = \phi(X) + (\psi(X) - \phi(X))$. From (3.B.14) we obtain the following result.

Theorem 3.B.5. *Let X be a random variable. Then*

$$\phi(X) \leq_{\text{disp}} \psi(X) \quad \textit{whenever } \phi \leq_{\text{disp}} \psi.$$

Another simple desirable property that is easily verified is the following theorem.

Theorem 3.B.6. *Let X and Y be two random variables. Then*

$$X \leq_{\text{disp}} Y \iff -X \leq_{\text{disp}} -Y. \tag{3.B.19}$$

However, the dispersive order is not closed under convolutions. In fact, it is not even true in general that for any two independent random variables X and Y we have that $X \le_{\rm disp} X + Y$. This observation follows from the next theorem, the proof of which is omitted.

Theorem 3.B.7. *The random variable X satisfies*

$$X \le_{\rm disp} X + Y \quad \text{for any random variable } Y \text{ independent of } X$$

if, and only if, X has a logconcave density.

A random variable Z is said to be *dispersive* if $X+Z \le_{\rm disp} Y+Z$ whenever $X \le_{\rm disp} Y$ and Z is independent of X and Y. From Theorem 3.B.7 it follows that every dispersive random variable must be strongly unimodal (that is, have a logconcave density). It turns out that strong unimodality is also a sufficient condition for dispersivity, as the next result shows. Again the proof is omitted.

Theorem 3.B.8. *The random variable X is dispersive if, and only if, X has a logconcave density.*

Other characterizations of random variables with logconcave densities are given in Theorem 1.C.52.

From Theorem 3.B.8 we obtain, by iteration, the following result.

Theorem 3.B.9. *Let $X_1, X_2, \ldots, X_n$ be a set of independent random variables, and let $Y_1, Y_2, \ldots, Y_n$ be another set of independent random variables. If the X_i's and the Y_i's have logconcave densities, and if $X_i \le_{\rm disp} Y_i$, $i = 1, 2, \ldots, n$, then*

$$\sum_{i=1}^{n} X_i \le_{\rm disp} \sum_{i=1}^{n} Y_i.$$

The dispersive order is closed under increasing convex and decreasing concave transformations when the underlying random variables are ordered in the usual stochastic order. We have the following result.

Theorem 3.B.10. *Let X and Y be two random variables such that $X \le_{\rm st} Y$.*

(a) *If $X \le_{\rm disp} Y$, then*

$$\phi(X) \le_{\rm disp} \phi(Y) \quad \text{for all increasing convex and all decreasing concave functions } \phi. \tag{3.B.20}$$

(b) *If $X \le_{\rm disp} Y$, then*

$$\phi(X) \ge_{\rm disp} \phi(Y) \quad \text{for all decreasing convex and all increasing concave functions } \phi. \tag{3.B.21}$$

Proof. First we prove (3.B.20) when ϕ is increasing and convex. Let F and G denote the distribution functions of X and Y, respectively, and let F^{-1} and G^{-1} be the respective inverses. For simplicity suppose that F, G, and ϕ are differentiable with derivatives f, g, and ϕ', respectively. The condition $X \leq_{\text{st}} Y$ implies that (see (1.A.12))

$$F^{-1}(\alpha) \leq G^{-1}(\alpha) \quad \text{for all } \alpha \in (0,1).$$

Since ϕ is convex it follows that ϕ' is increasing. Therefore

$$\phi'(F^{-1}(\alpha)) \leq \phi'(G^{-1}(\alpha)) \quad \text{for all } \alpha \in (0,1). \tag{3.B.22}$$

The condition $X \leq_{\text{disp}} Y$ implies that (see (3.B.11))

$$g(G^{-1}(\alpha)) \leq f(F^{-1}(\alpha)) \quad \text{for all } \alpha \in (0,1). \tag{3.B.23}$$

Since ϕ is increasing it follows that $\phi' \geq 0$. Therefore, combining (3.B.22) and (3.B.23), we see that

$$g(G^{-1}(\alpha))\phi'(F^{-1}(\alpha)) \leq f(F^{-1}(\alpha))\phi'(G^{-1}(\alpha)) \quad \text{for all } \alpha \in (0,1),$$

and, again from (3.B.11), it is seen that the latter inequality is equivalent to $\phi(X) \leq_{\text{disp}} \phi(Y)$.

If ϕ is decreasing and concave, then $-\phi$ is increasing and convex. Therefore, from what we just proved it follows that $-\phi(X) \leq_{\text{disp}} -\phi(Y)$. From Theorem 3.B.6 we obtain that $\phi(X) \leq_{\text{disp}} \phi(Y)$. The proof of (3.B.21) is similar. □

Theorem 3.B.10 can be generalized in several ways. Here are two generalizations of the increasing convex part of (3.B.20).

Theorem 3.B.11. *Let X and Y be two random variables such that $X \leq_{\text{st}} Y$.*

(a) *If $X \leq_{\text{disp}} Y$, then $\phi(X) \leq_{\text{disp}} \psi(Y)$ whenever $\phi \leq_{\text{disp}} \psi$ (in the sense of (3.B.18)) and ϕ or ψ is an increasing convex function.*
(b) *If $X \leq_{\text{disp}} Y$, then $\phi(X) \leq_{\text{disp}} \psi(Y)$ whenever ϕ and ψ are differentiable and their derivatives, ϕ' and ψ', respectively, satisfy $\phi'(x) \leq \psi'(y)$ for all $x \leq y$.*

A relation similar to (3.B.20) can be used as a sufficient condition for $X \leq_{\text{disp}} Y$. The next result states such a condition. Note that in (3.B.24) the directions of the monotonicity in the convex and the concave cases are interchanged.

Theorem 3.B.12. *Let X and Y be two random variables such that $X \leq_{\text{st}} Y$. If*

$$\phi(X) \leq_{\text{disp}} \phi(Y) \quad \textit{for some decreasing convex or increasing concave function } \phi, \tag{3.B.24}$$

then $X \leq_{\text{disp}} Y$.

The proof of Theorem 3.B.12 uses Theorem 3.B.10. If ϕ in (3.B.24) is increasing and concave, then ϕ^{-1} is increasing and convex. Since $X \leq_{\rm st} Y$ it follows that $\phi(X) \leq_{\rm st} \phi(Y)$. Therefore, by Theorem 3.B.10, $X = \phi^{-1}(\phi(X)) \leq_{\rm disp} \phi^{-1}(\phi(Y)) = Y$. The proof for a decreasing and convex ϕ is similar.

For random variables with equal left-end support points the assumption in Theorems 3.B.10 and 3.B.11 of the comparison of X and Y in the usual stochastic order need not be stated. This is because of the following observation. Here, for random variables X and Y, we denote the corresponding endpoints of their supports by l_X, u_X, l_Y, and u_Y as defined before (3.A.12).

Theorem 3.B.13. (a) *If X and Y are random variables such that $l_X = l_Y > -\infty$, then*

$$X \leq_{\rm disp} Y \Longrightarrow X \leq_{\rm st} Y.$$

(b) *If X and Y are random variables such that $u_X = u_Y < \infty$, then*

$$X \leq_{\rm disp} Y \Longrightarrow X \geq_{\rm st} Y.$$

For example, if X and Y are nonnegative random variables such that $l_X = l_Y = 0$, then Theorem 3.B.13(a) applies. A stronger version of this fact is described in Remark 4.B.35.

The proof of Theorem 3.B.13(a) is based on the fact that if F and G are the distribution functions of X and Y, respectively, then $F^{-1}(0) = l_X = l_Y = G^{-1}(0)$. Therefore, from (3.B.1) one obtains that $F^{-1}(\beta) \leq G^{-1}(\beta)$ for all $\beta \in (0, 1)$, that is, $X \leq_{\rm st} Y$ by (1.A.12). The proof of Theorem 3.B.13(b) is similar. The following result can be shown using the same kind of argument.

Theorem 3.B.14. *If X and Y are random variables having the same finite support and satisfying $X \leq_{\rm disp} Y$, then they must have the same distribution.*

The next result is an analog of (3.A.12). We omit the proof.

Theorem 3.B.15. *Let X and Y be random variables whose supports are intervals. Then*

$$X \leq_{\rm disp} Y \Longrightarrow \mu\{\text{supp}(X)\} \leq \mu\{\text{supp}(Y)\},$$

where μ denotes the Lebesgue measure.

Suppose that X and Y are two random variables with distributions F and G, respectively, such that $X \leq_{\rm disp} Y$. Then by taking $c = 0$ in (3.B.3) we see that (3.A.59) holds for the random variables $X - EX$ and $Y - EY$. We thus have proved the following implication.

Theorem 3.B.16. *Let X and Y be two random variables with finite means. Then*

$$X \leq_{\rm disp} Y \Longrightarrow X \leq_{\rm dil} Y.$$

A more refined result can be obtained by combining (3.C.7) and (3.C.9) in Section 3.C below.

From Theorem 3.B.16, (3.A.32), and (3.A.4) it follows that if $X \le_{\text{disp}} Y$, then

$$\mathrm{Var}(X) \le \mathrm{Var}(Y), \tag{3.B.25}$$

whenever $\mathrm{Var}(Y) < \infty$.

From Theorem 3.B.7 it follows that

$$\big(X \le_{\text{conv}} Y, \text{ and } X \text{ has a logconcave density}\big) \Longrightarrow X \le_{\text{disp}} Y. \tag{3.B.26}$$

In contrast to (3.B.19), if $X \le_{\text{disp}} Y$, it does not necessarily follow that $X \le_{\text{disp}} -Y$. In order to see it, let X be an exponential random variable with mean 1. Clearly $X \le_{\text{disp}} X$ (in fact, this is the case for any random variable X). The distribution function of X is concave on $[0, \infty)$, and the distribution function of $-X$ is convex on $(-\infty, 0)$. Since the order $\le_{\text{disp}}$ is preserved under shifts, it follows that $X \not\le_{\text{disp}} -X$.

Using an argument as in the proof of Theorem 3.A.44, we obtain the following sufficient condition for the dispersive order.

Theorem 3.B.17. *Let X and Y be random variables with respective densities f and g. If*

$$\begin{aligned} &S^-(f(\cdot - c) - g(\cdot)) \le 2 \textit{ for all } c, \\ &\qquad \textit{with the sign sequence being } -,+,- \textit{ in the case of equality,} \end{aligned} \tag{3.B.27}$$

then $X \le_{\text{disp}} Y$.

Another sufficient condition for $X \le_{\text{disp}} Y$ is given next.

Theorem 3.B.18. *Let X and Y be two absolutely continuous random variables with hazard rate functions (see (1.B.1)) r and q, respectively. If*

$$r(u) \ge q(u + x) \quad \textit{for all } u \textit{ and } x \ge 0, \tag{3.B.28}$$

then $X \le_{\text{disp}} Y$.

Proof. Let F and G denote the distribution functions of X and Y, respectively. Condition (3.B.28) implies that $r(u) \ge q(u)$; that is, $X \le_{\text{hr}} Y$. This, in turn, implies $X \le_{\text{st}} Y$, which, in turn, implies (1.A.12).

Now, (3.B.28) therefore gives $r(F^{-1}(\alpha)) \ge q(G^{-1}(\alpha))$ for all $\alpha \in (0, 1)$, which is equivalent to $X \le_{\text{disp}} Y$ by (3.B.12). □

Let X be a random variable and denote by $X_{(-\infty,a]}$ the truncation of X at a as defined in Section 1.A.4. One would expect $X_{(-\infty,a]}$ to increase in a in the sense of the dispersion order. This is not always the case, but it is the case if the distribution function F of X is logconcave; that is, if X has decreasing reverse hazard (see Section 1.B.6). This is shown in the next result, which is an analog of Theorem 1.A.15 for the dispersion order. The proof of the first part of the theorem consists of verifying that for $\alpha \le \beta$ the quantity $F^{-1}(\beta F(a)) - F^{-1}(\alpha F(a))$ increases in a when F is logconcave. The other parts of the theorem are proven similarly. The notation $X_{(a,b)}$ for $a < b$ is self-explanatory.

Theorem 3.B.19. *Let X be a random variable with distribution function F and density f.*

(a) *If F is logconcave, then $X_{(-\infty,a]}$ increases in a in the sense of the dispersion order.*
(b) *If $\overline{F}$ is logconcave (that is, if X is* IFR*), then $X_{(a,\infty)}$ decreases in a in the sense of the dispersion order.*
(c) *If f is logconcave, then $X_{(a,b)}$ decreases in a $(< b)$ and increases in b $(> a)$ in the sense of the dispersion order.*

Recall from page 1 the definitions of the IFR, DFR, NBU, NWU, DMRL and IMRL properties. The following theorems list some relations between the dispersion order and some other orders. The proofs are mostly straightforward and are not detailed here.

Theorem 3.B.20. *Let X and Y be two nonnegative random variables.*

(a) *If $X \le_{\text{hr}} Y$ and X or Y is* DFR*, then $X \le_{\text{disp}} Y$.*
(b) *If $X \le_{\text{disp}} Y$ and X or Y is* IFR*, then $X \le_{\text{hr}} Y$.*
(c) *If X is* NBU *and Y is* NWU*, then $X \le_{\text{disp}} Y \iff X \le_{\text{hr}} Y$.*

A version of parts (a) and (b) of Theorem 3.B.19, where $\le_{\text{hr}}$ is replaced by $\le_{\text{rh}}$, and DFR and IFR are replaced by monotonicity conditions on the reversed hazard rate function, can be found in Bartoszewicz [44].

Recall from (1.A.20) that for nonnegative random variables X and Y with finite means, we denote by A_X and A_Y the corresponding asymptotic equilibrium ages. The following result may be contrasted with Theorem 2.A.4.

Theorem 3.B.21. *Let X and Y be two nonnegative random variables.*

(a) *If $X \le_{\text{mrl}} Y$ and X or Y is* IMRL*, then $A_X \le_{\text{disp}} A_Y$.*
(b) *If $A_X \le_{\text{disp}} A_Y$ and X or Y is* DMRL*, then $X \le_{\text{mrl}} Y$.*
(c) *If $X \le_{\text{disp}} Y$ and X is* DMRL *and Y is* IMRL*, then $X \le_{\text{mrl}} Y$.*

Example 3.B.22. Let $X_1, X_2, \ldots, X_n$ be independent DFR random variables, and let $X_{(1)} \le X_{(2)} \le \cdots \le X_{(n)}$ be the corresponding order statistics. Then $X_{(1)}$ is DFR (since its hazard rate function is the sum of the hazard rate functions of the X_i's). From Theorem 1.B.26 we see that $X_{(1)} \le_{\text{hr}} X_{(i)}$, $i = 2, 3, \ldots, n$. Therefore, by Theorem 3.B.20(a), we have that

$$X_{(1)} \le_{\text{disp}} X_{(i)}, \quad i = 2, 3, \ldots, n.$$

Example 3.B.23. Let $X_1, X_2, \ldots, X_m, X_{m+1}$ be independent and identically distributed DFR random variables and let the corresponding spacings be denoted by $U_{(i:m)}$ as in Theorem 1.B.31. It is easy to see then that the spacings are DFR random variables (see Barlow and Proschan [35]). Then, from Theorems 1.B.31 and 3.B.20(a) we get

$$(m-i+1)U_{(i:m)} \le_{\text{disp}} (m-i)U_{(i+1:m)}, \quad i = 2,3,\ldots,m-1, \tag{3.B.29}$$

$$(m-i+2)U_{(i:m+1)} \le_{\text{disp}} (m-i+1)U_{(i:m)}, \quad i = 2,3,\ldots,m, \tag{3.B.30}$$

and

$$U_{(i:m)} \le_{\text{disp}} U_{(i+1:m+1)}, \quad i = 2,3,\ldots,m. \tag{3.B.31}$$

Note that (3.B.29)–(3.B.31) can be summarized as

$$(m-j+1)U_{(j:m)} \le_{\text{disp}} (n-i+1)U_{(i:n)} \quad \text{whenever } i-j \ge \max\{0, n-m\}.$$

The dispersive order can be used to characterize IFR and DFR random variables as the following result shows.

Theorem 3.B.24. *Let X be a nonnegative random variable. Then X is* IFR [DFR] *if, and only if,* $[X-t \mid X>t] \ge_{\text{disp}} [\le_{\text{disp}}]\ [X-t' \mid X>t']$ *whenever* $t \le t'$.

Proof. If X is IFR, then, by Theorem 3.B.19(b), $[X \mid X>t]$ is decreasing in t in the sense of the dispersive order. Since the dispersive order is preserved under shifts, it is seen that $[X-t \mid X>t]$ is decreasing in t in the sense of the dispersive order. The proof of the DFR case is similar, though one first needs to prove a DFR version of Theorem 3.B.19(b). The converses of the above statements are consequences of Theorems 1.A.30(a) and 3.B.13(a). □

Under some regularity conditions on the distribution function of X and on its support, but without the assumption of nonnegativity of X, we have a related characterization of the IFR and the DFR properties. We do not give the proof of this result here.

Theorem 3.B.25. *Let X be a random variable with a continuous distribution function, and with support of the form (a, ∞), where $a \ge -\infty$ [respectively, $a > -\infty$]. Then X is* IFR [DFR] *if, and only if,* $X \ge_{\text{disp}} [\le_{\text{disp}}]\ [X-t \mid X>t]$ *for all $t > a$.*

The next result states a preservation property of the order $\le_{\text{disp}}$ which is useful in reliability theory as well as in nonparametric statistics. Let X and Y be two random variables. Let $X_{(1:n)} \le X_{(2:n)} \le \cdots \le X_{(n:n)}$ denote the order statistics from a sample $X_1, X_2, \ldots, X_n$ of independent and identically distributed random variables that have the same distribution as X. Similarly, let $Y_{(1:n)} \le Y_{(2:n)} \le \cdots \le Y_{(n:n)}$ denote the order statistics from another sample $Y_1, Y_2, \ldots, Y_n$ of independent and identically distributed random variables that have the same distribution as Y.

Theorem 3.B.26. *Let X and Y be two random variables. If $X \leq_{\text{disp}} Y$, then $X_{(j:n)} \leq_{\text{disp}} Y_{(j:n)}$ for $j = 1, 2, \ldots, n$.*

The proof follows at once from (3.B.10) and the fact that

$$G_{j:n}^{-1} F_{j:n} = G^{-1} F \quad \text{for } j = 1, 2, \ldots, n,$$

where F, $F_{j:n}$, G, and $G_{j:n}$ are the distribution functions of X, $X_{(j:n)}$, Y, and $Y_{(j:n)}$, respectively.

For the next result about comparison of order statistics we will need the following lemma.

Lemma 3.B.27. *Let $E_{(j:m)}$ and $E_{(i:n)}$ denote the jth and the ith order statistics of samples from the exponential distribution with rate $\lambda > 0$ of sizes m and n, respectively. Then*

$$E_{(j:m)} \leq_{\text{disp}} E_{(i:n)} \quad \textit{whenever } i - j \geq \max\{0, n - m\}.$$

Proof. Write $E_{(j:m)} =_{\text{st}} \sum_{k=1}^{j} E_{m-j+k}$, where E_{m-j+k} is an exponential random variable with rate $(m - j + k)\lambda$, $k = 1, 2, \ldots, j$, and the E_{m-j+k}'s are independent. Similarly, write $E_{(i:n)} =_{\text{st}} \sum_{k=1}^{i} E'_{n-i+k}$, where E'_{n-i+k} is an exponential random variable with rate $(n - i + k)\lambda$, $k = 1, 2, \ldots, i$, and the E'_{n-i+k}'s are independent. It is easy to check, for instance using Theorem 3.B.4, that $E_{m-j+k} \leq_{\text{disp}} E'_{n-i+k}$ because $m - j \geq n - i$. Since exponential random variables have logconcave densities, we obtain from Theorems 3.B.9 and 3.B.7, respectively, that

$$E_{(j:m)} =_{\text{st}} \sum_{k=1}^{j} E_{m-j+k} \leq_{\text{disp}} \sum_{k=1}^{j} E'_{n-i+k} \leq_{\text{disp}} \sum_{k=1}^{i} E'_{n-i+k} =_{\text{st}} E_{(i:n)}$$

because $j \leq i$. □

Theorem 3.B.28. *Let $X_{(j:m)}$ and $X_{(i:n)}$ denote the jth and the ith order statistics of samples from a* DFR *distribution F of sizes m and n, respectively. Then*

$$X_{(j:m)} \leq_{\text{disp}} X_{(i:n)} \quad \textit{whenever } i - j \geq \max\{0, n - m\}.$$

Proof. The distribution $F_{j:m}$ of $X_{(j:m)}$ can be expressed as $F_{j:m} = B_{j:m} F$, where $B_{j:m}$ is the beta distribution with parameters j and $m - j + 1$. Similarly, the distribution $F_{i:n}$ of $X_{(i:n)}$ can be expressed as $F_{i:n} = B_{i:n} F$. Now write

$$F_{j:m} = B_{j:m} G G^{-1} F = H_{j:m} G^{-1} F,$$

where G denotes the distribution function of an exponential random variable with mean 1, and $H_{j:m} = B_{j:m} G$. Note that $H_{j:m}$ is the distribution function of $E_{(j:m)}$ in Lemma 3.B.27. Similarly, write

$$F_{i:n} = H_{i:n} G^{-1} F,$$

and finally notice that

$$F_{i:n}^{-1} F_{j:m} = \psi H_{i:n}^{-1} H_{j:m} \psi^{-1},$$

where $\psi = F^{-1}G$. From Lemma 3.B.27 and (3.B.10) we see that $H_{i:n}^{-1}H_{j:m}(x) - x$ is increasing in x. The function ψ is strictly convex because F is DFR, and it satisfies $\psi(0) = 0$. Therefore, by a result of Bartoszewicz [40] it follows that $F_{i:n}^{-1}F_{j:m}(x) - x$ is increasing in x. The stated result now follows from (3.B.10). □

As a corollary of Theorems 3.B.26 and 3.B.28 we get the following result.

Theorem 3.B.29. *Let $X_{(j:m)}$ and $Y_{(i:n)}$ denote the jth and the ith order statistics of samples from the distribution F and G of sizes m and n, respectively. If F or G is* DFR, *and if $X \le_{\rm disp} Y$, then*

$$X_{(j:m)} \le_{\rm disp} Y_{(i:n)} \quad \textit{whenever } i - j \ge \max\{0, n - m\}.$$

Proof. If F is DFR, then $X_{(j:m)} \le_{\rm disp} X_{(i:n)} \le_{\rm disp} Y_{(i:n)}$ by Theorems 3.B.28 and 3.B.26, respectively. If G is DFR, then $X_{(j:m)} \le_{\rm disp} Y_{(j:m)} \le_{\rm disp} Y_{(i:n)}$ by Theorems 3.B.26 and 3.B.28, respectively. □

It is of interest to compare Theorem 3.B.29 to the following example (which follows from Example 3.A.50 and Theorem 3.B.16).

Example 3.B.30. Let $X_{(i:n)}$ denote the ith order statistic in a sample of n independent and identically distributed random variables having the common distribution function F, survival function $\overline{F}$, and density function f. Recall the definition of regular variation from Example 3.A.50. For F with support $(-\infty, \infty)$ we have:

(a) If F is regularly varying at $-\infty$ with index $\alpha < 0$, and if f is monotone on $(-\infty, c]$ for some c, then $X_{(j:m)} \le_{\rm disp} X_{(i:n)}$ implies $i \le j$.
(b) If $\overline{F}$ is regularly varying at ∞ with index $\alpha < 0$, and if f is monotone on $[c, \infty)$ for some c, then $X_{(j:m)} \le_{\rm disp} X_{(i:n)}$ implies $n - i \le m - j$.

The dispersive order between X and Y implies the usual stochastic order between the corresponding spacings as the next result shows. In order to state it we use the following notation. Let $X_{(1:n)} \le X_{(2:n)} \le \cdots \le X_{(n:n)}$ and $Y_{(1:n)} \le Y_{(2:n)} \le \cdots \le Y_{(n:n)}$ be the order statistics as above. The corresponding spacings are defined by $U_{(i:n)} \equiv X_{(i:n)} - X_{(i-1:n)}$ and $V_{(i:n)} \equiv Y_{(i:n)} - Y_{(i-1:n)}$, $i = 2, 3, \ldots, n$. The proof of the next theorem is given in Example 6.B.25 in Chapter 6.

Theorem 3.B.31. *Let X and Y be two random variables. If $X \le_{\rm disp} Y$, then $U_{(i:n)} \le_{\rm st} V_{(i:n)}$ for $i = 2, 3, \ldots, n$.*

Theorem 2.7 on page 182 of Kamps [273] extends Theorem 3.B.31 to the spacings of the so called generalized order statistics.

The following example describes an interesting instance in which the two maxima are ordered in the dispersive order. It may be compared with Example 1.B.37.

Example 3.B.32. Let $Y_1, Y_2, \ldots, Y_n$ be independent exponential random variables with hazard rates $\lambda_1, \lambda_2, \ldots, \lambda_n$, respectively. Let $X_1, X_2, \ldots, X_n$ be independent and identically distributed exponential random variables with hazard rate $\overline{\lambda} = \sum_{i=1}^n \lambda_i / n$. Then

$$X_{(n:n)} \le_{\text{disp}} Y_{(n:n)}. \tag{3.B.32}$$

Let $Z_1, Z_2, \ldots, Z_n$ be independent and identically distributed exponential random variables with hazard rate $\tilde{\lambda} = (\prod_{i=1}^n \lambda_i)^{1/n}$. Then

$$Z_{(n:n)} \le_{\text{disp}} Y_{(n:n)}. \tag{3.B.33}$$

Note that from the arithmetic-geometric mean inequality ($\overline{\lambda} \ge \tilde{\lambda}$) it follows that $X_1 \le_{\text{hr}} Z_1$. Therefore, by Theorem 3.B.20(a), $X_1 \le_{\text{disp}} Z_1$. Alternatively, we can see that $X_1 \le_{\text{disp}} Z_1$ from Example 1.D.1 and (3.B.26). Hence, by Theorem 3.B.26, $X_{(n:n)} \le_{\text{disp}} Z_{(n:n)}$. That is, actually (3.B.33) is a stronger result than (3.B.32).

Example 3.B.33. Let $Y_1, Y_2, \ldots, Y_n$ and $X_1, X_2, \ldots, X_n$ be as in Example 3.B.32. Denote the corresponding spacings by $U_{(i:n)} \equiv X_{(i:n)} - X_{(i-1:n)}$ and $V_{(i:n)} \equiv Y_{(i:n)} - Y_{(i-1:n)}$, $i = 2, 3, \ldots, n$. Then

$$U_{(i:n)} \le_{\text{disp}} V_{(i:n)}, \quad i = 2, 3, \ldots, n.$$

A related example is the following. Recall from page 2 the definition of the majorization order $\prec$ among n-dimensional vectors. It is of interest to compare the example below with Example 1.C.50.

Example 3.B.34. Let X_i be an exponential random variable with mean $\lambda_i^{-1} > 0$, $i = 1, 2, \ldots, m$, and let Y_i be an exponential random variable with mean $\eta_i^{-1} > 0$, $i = 1, 2, \ldots, m$. If $(\lambda_1, \lambda_2, \ldots, \lambda_m) \succ (\eta_1, \eta_2, \ldots, \eta_m)$, then

$$\sum_{i=1}^m X_i \ge_{\text{disp}} \sum_{i=1}^m Y_i.$$

Similar examples are the following.

Example 3.B.35. Let X_i be a uniform random variable on $[0, \lambda_i^{-1}]$, $i = 1, 2, \ldots, m$, and let Y_i be a uniform random variable on $[0, \eta_i^{-1}]$, $i = 1, 2, \ldots, m$. If $(\lambda_1, \lambda_2, \ldots, \lambda_m) \succ (\eta_1, \eta_2, \ldots, \eta_m)$, then

$$\sum_{i=1}^m X_i \ge_{\text{disp}} \sum_{i=1}^m Y_i.$$

Example 3.B.36. Let X_i be a Gamma random variable with density function $(1/\Gamma(\alpha))\lambda_i^\alpha x^{\alpha-1} \mathrm{e}^{-\lambda_i x}$, $x > 0$, $i = 1, 2, \ldots, m$, and let Y_i be a Gamma random variable with density function $(1/\Gamma(\alpha))\eta_i^\alpha x^{\alpha-1} \mathrm{e}^{-\eta_i x}$, $x > 0$, $i =$

$1, 2, \ldots, m$. Here $\alpha \geq 1$, and the λ_i's and the η_i's are positive parameters. If $(\lambda_1, \lambda_2, \ldots, \lambda_m) \succ (\eta_1, \eta_2, \ldots, \eta_m)$, then

$$\sum_{i=1}^{m} X_i \geq_{\text{disp}} \sum_{i=1}^{m} Y_i.$$

The proof of the next example is omitted.

Example 3.B.37. Let $\{N(t),\ t \geq 0\}$ be a nonhomogeneous Poisson process with mean function Λ (that is, $\Lambda(t) \equiv E[N(t)],\ t \geq 0$), and let $T_1, T_2, \ldots$ be the successive epoch times. If Λ is strictly increasing and concave, then

$$T_n \leq_{\text{disp}} T_{n+1}, \quad n = 1, 2, \ldots.$$

In the following example the idea of the proof of Theorem 3.B.26 is used. This example may be compared with Examples 1.B.24, 1.C.48, 2.A.22, 4.B.14, 6.B.41, 6.D.8, 6.E.13, and 7.B.13.

Example 3.B.38. Let X and Y be two absolutely continuous nonnegative random variables with survival functions $\overline{F}$ and $\overline{G}$, respectively. Denote $\Lambda_1 = -\log \overline{F}$ and $\Lambda_2 = -\log \overline{G}$, $i = 1, 2$. Consider two nonhomogeneous Poisson processes $N_1 = \{N_1(t),\ t \geq 0\}$ and $N_2 = \{N_2(t),\ t \geq 0\}$ with mean functions Λ_1 and Λ_2 (see Example 3.B.37), respectively. Let $T_{i,1}, T_{i,2}, \ldots$ be the successive epoch times of process N_i, $i = 1, 2$. Note that $X =_{\text{st}} T_{1,1}$ and $Y =_{\text{st}} T_{2,1}$.

It turns out that the dispersive ordering of the first two epoch times implies the dispersive ordering of all the corresponding later epoch times; that is, it will be shown below that if $X \leq_{\text{disp}} Y$, then $T_{1,n} \leq_{\text{disp}} T_{2,n}$, $n \geq 1$.

In order to see it, fix an $n \geq 1$, and denote by $F_{1,n}$ and $F_{2,n}$ the distribution functions of $T_{1,n}$ and $T_{2,n}$, respectively. Note from (1.B.24) that

$$F_{1,n}(t) = \psi_n(F(t)) \quad \text{and} \quad F_{2,n}(t) = \psi_n(G(t)),$$

where $\psi_n(u) \equiv \Gamma_n(-\log(1-u))$, $u \in [0, 1]$. Therefore,

$$F_{2,n}^{-1}(F_{1,n}(t)) - t = (\psi_n(G))^{-1}(\psi_n(F(t))) - t = G^{-1}(F(t)) - t, \quad t \geq 0.$$

Thus, from (3.B.10) it is seen that $X \leq_{\text{disp}} Y$ if, and only if, $T_{1,n} \leq_{\text{disp}} T_{2,n}$.

In the following example it is shown that, under the proper conditions, random minima and maxima are ordered in the dispersive order sense; see related results in Examples 1.C.46, 4.B.16, 5.A.24, and 5.B.13.

Example 3.B.39. Let $X_1, X_2, \ldots$, and $Y_1, Y_2, \ldots$, each be a sequence of independent and identically distributed random variables with common distribution functions F_{X_1} and F_{Y_1}, respectively, and common survival functions $\overline{F}_{X_1}$ and $\overline{F}_{Y_1}$, respectively. Let N be a positive integer-valued random variable, independent of the X_i's and of the Y_i's, with a Laplace transform L_N.

Denote $X_{(1,N)} = \min\{X_1, X_2, \ldots, X_N\}$, $X_{(N,N)} = \max\{X_1, X_2, \ldots, X_N\}$, $Y_{(1,N)} = \min\{Y_1, Y_2, \ldots, Y_N\}$, and $Y_{(N,N)} = \max\{Y_1, Y_2, \ldots, Y_N\}$. The distribution functions of $X_{(N,N)}$ and $Y_{(N,N)}$ are given by

$$F_{X_{(N,N)}}(x) = L_N(-\log F_{X_1}(x)), \quad x \geq 0, \ j = 1, 2,$$

and

$$F_{Y_{(N,N)}}(x) = L_N(-\log F_{Y_1}(x)), \quad x \geq 0, \ j = 1, 2.$$

If $X_1 \leq_{\text{disp}} Y_1$, then, for $0 < \alpha \leq \beta < 1$ we compute

$$\begin{aligned} F^{-1}_{X_{(N:N)}}(\beta) - F^{-1}_{X_{(N:N)}}(\alpha) &= F^{-1}_{X_1}\big(\mathrm{e}^{-L_N^{-1}(\beta)}\big) - F^{-1}_{X_1}\big(\mathrm{e}^{-L_N^{-1}(\alpha)}\big) \\ &\leq F^{-1}_{Y_1}\big(\mathrm{e}^{-L_N^{-1}(\beta)}\big) - F^{-1}_{Y_1}\big(\mathrm{e}^{-L_N^{-1}(\alpha)}\big) \\ &= F^{-1}_{Y_{(N:N)}}(\beta) - F^{-1}_{Y_{(N:N)}}(\alpha). \end{aligned}$$

Therefore $X_{(N:N)} \leq_{\text{disp}} Y_{(N:N)}$. Similarly it can be shown that if $X_1 \leq_{\text{disp}} Y_1$, then $X_{(1:N)} \leq_{\text{disp}} Y_{(1:N)}$.

Example 3.B.40. Let X (respectively, Y) have the central t-distribution with ν_X (respectively, ν_Y) degrees of freedom. If $\nu_X \leq \nu_Y$, then $X \geq_{\text{disp}} Y$.

Example 3.B.41. As in Example 1.C.59, for nonnegative absolutely continuous random variables X and Y, let X^w and Y^w be the random variables with the weighted density functions f_w and g_w given in (1.C.17) and (1.C.18). Suppose that $X \leq_{\text{disp}} Y$. If X is DFR, if Y is IFR, and if w is decreasing and convex, then $X^w \leq_{\text{disp}} Y^w$.

Analogous to the result in Remark 1.A.18, it can be shown that a certain quotient set of all distribution functions on $\mathbb{R}$ is a lattice with respect to the order $\leq_{\text{disp}}$.

A consequence of the order $\leq_{\text{disp}}$ is given in the next theorem. It is a motivation for a multivariate dispersion order that is described in Chapter 7.

Theorem 3.B.42. *Let X and X' be two independent and identically distributed random variables and let Y and Y' be two other independent and identically distributed random variables. If $X \leq_{\text{disp}} Y$, then $|X - X'| \leq_{\text{st}} |Y - Y'|$, that is,*

$$P\{|X - X'| > z\} \leq P\{|Y - Y'| > z\} \quad \textit{for all } z \geq 0. \tag{3.B.34}$$

Proof. Denote the common distribution function of X and X' [respectively, Y and Y'] by F [G]. Select a $z \geq 0$. Then

$$\begin{aligned} P\{|X - X'| \leq z\} &= \int_{-\infty}^{\infty} [F(x+z) - F(x-z)]\mathrm{d}F(x) \\ &= \int_0^1 \{F[F^{-1}(u) + z] - F[F^{-1}(u) - z]\}\mathrm{d}u \end{aligned}$$

$$\ge \int_0^1 \{G[G^{-1}(u)+z] - G[G^{-1}(u)-z]\}\mathrm{d}u$$
$$= P\{|Y - Y'| \le z\},$$

where the inequality is a consequence of (3.B.4) and (3.B.5). This proves (3.B.34). □

3.C The Excess Wealth Order

3.C.1 Motivation and definition

Let X be a nonnegative random variable with distribution function F and with a finite mean. Recall from (3.A.43) the definition of the Lorenz curve. The nonstandardized (or the generalized) Lorenz curve $\tilde{L}_X$, corresponding to X, is defined as

$$\tilde{L}_X(p) = \int_0^p F^{-1}(u)\mathrm{d}u, \quad p \in [0,1].$$

Note that the requirement that X is nonnegative is not needed in order for $\tilde{L}_X$ to be well defined. Thus, in this section we will not assume the nonnegativity of the discussed random variables, unless stated otherwise.

For a nonnegative random variable X with a finite mean, a transform that is closely related to the nonstandardized Lorenz curve is the transform T_X defined as

$$T_X(p) = \int_0^{F^{-1}(p)} \overline{F}(x)\mathrm{d}x, \quad p \in [0,1].$$

The transform T_X is called the TTT transform, and is denoted by H_F^{-1} in (1.A.19).

A third transform, that is related to the nonstandardized Lorenz curve and to the TTT transform, and which will be heavily used in this section, is the excess wealth transform W_X defined as

$$W_X(p) = \int_{F^{-1}(p)}^{\infty} \overline{F}(x)\mathrm{d}x, \quad p \in (0,1].$$

Note that it is not necessary for the random variable X to be nonnegative in order for W_X to be well defined; it is only required that X has a finite mean. This useful property of the excess wealth transform is one of the main reasons for its applicability as a tool that defines a stochastic order.

The transforms $\tilde{L}_X$, T_X, and W_X, when X is nonnegative with a finite mean, are depicted in Figure 3.C.1. For $p \in (0,1)$ the value $\tilde{L}_X(p)$ is depicted as the area of the region A in the figure. The value $T_X(p)$ is the area of $A \cup B$, and the value $W_X(p)$ is the area of C. Note that the area of $A \cup B \cup C$ is EX.

The order which is determined by the pointwise comparison of the excess wealth transforms of two random variables is of interest in this section. Let X

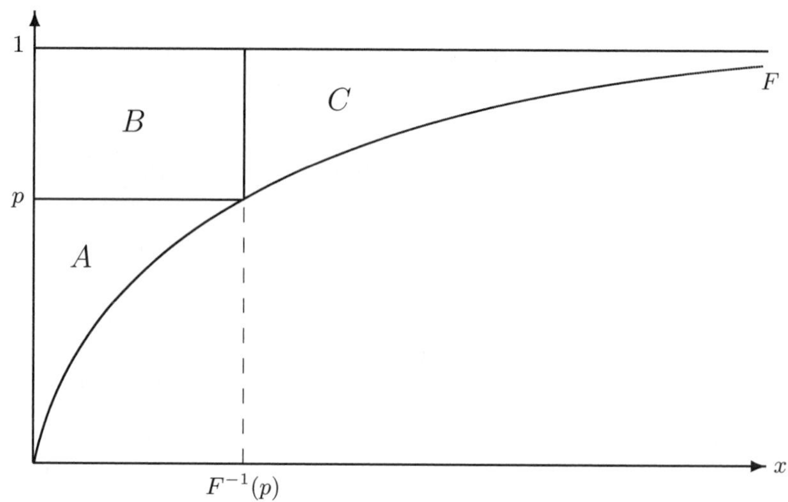

Fig. 3.C.1. Depiction of $\tilde{L}_X(p)$, $T_X(p)$, and $W_X(p)$.

and Y be two random variables with distribution functions F and G. Assume that

$$W_X(p) \equiv \int_{F^{-1}(p)}^{\infty} \overline{F}(x)\mathrm{d}x \le \int_{G^{-1}(p)}^{\infty} \overline{G}(x)\mathrm{d}x \equiv W_Y(p) \quad \text{for all } p \in (0,1). \tag{3.C.1}$$

Then X is said to be *smaller than* Y *in the excess wealth order* (denoted as $X \le_{\mathrm{ew}} Y$).

Note that since $F^{-1}(p) = \overline{F}^{-1}(1-p)$ and $G^{-1}(p) = \overline{G}^{-1}(1-p)$, we see that $X \le_{\mathrm{ew}} Y$ if, and only if,

$$\int_{\overline{F}^{-1}(p)}^{\infty} \overline{F}(x)\mathrm{d}x \le \int_{\overline{G}^{-1}(p)}^{\infty} \overline{G}(x)\mathrm{d}x \quad \text{for all } p \in (0,1).$$

If we define $\Psi_X(y) = \int_y^\infty \overline{F}(x)\mathrm{d}x$ and $\Psi_Y(y) = \int_y^\infty \overline{G}(x)\mathrm{d}x$, $x \in \mathbb{R}$, then $X \le_{\mathrm{ew}} Y$ if, and only if,

$$\Psi_Y^{-1}(z) - \Psi_X^{-1}(z) \quad \text{is decreasing in } z \ge 0. \tag{3.C.2}$$

In order to obtain another characterization of the $\le_{\mathrm{ew}}$ order, rewrite (3.C.1) as

$$\int_p^1 \big(F^{-1}(u) - F^{-1}(p)\big)\mathrm{d}u \le \int_p^1 \big(G^{-1}(u) - G^{-1}(p)\big)\mathrm{d}u \tag{3.C.3}$$

(this can be formally verified by Fubini's Theorem or, informally, by rewriting the area of the region C in Figure 3.C.1 as the left-hand side above). It is thus seen that $X \le_{\mathrm{ew}} Y$ if, and only if,

$$G^{-1}(p) - F^{-1}(p) \le \frac{1}{1-p}\int_p^1 \big(G^{-1}(u) - F^{-1}(u)\big)\mathrm{d}u, \quad p \in (0,1).$$

By a straightforward differentiation it can be verified that the latter is equivalent to

$$\frac{1}{1-p}\int_p^1 \big(G^{-1}(u) - F^{-1}(u)\big)\mathrm{d}u \quad \text{is increasing in } p \in (0,1). \qquad (3.C.4)$$

Thus, $X \le_{\rm ew} Y$ if, and only if, (3.C.4) holds.

Let m_X and m_Y, defined by $m_X(t) \equiv \frac{\int_t^\infty \overline{F}(x)\mathrm{d}x}{\overline{F}(t)}$ and $m_Y(t) \equiv \frac{\int_t^\infty \overline{G}(x)\mathrm{d}x}{\overline{G}(t)}$ (for t's for which the denominators are not 0), denote the mean residual life functions associated with X and Y (see (2.A.1)). Then it is seen that $X \le_{\rm ew} Y$ if, and only if,

$$m_X(F^{-1}(p)) \le m_Y(G^{-1}(p)), \quad p \in (0,1). \qquad (3.C.5)$$

Also, $X \le_{\rm ew} Y$ if, and only if,

$$m_X(\overline{F}^{-1}(p)) \le m_Y(\overline{G}^{-1}(p)), \quad p \in (0,1). \qquad (3.C.6)$$

Another characterization of the excess wealth order is given in Theorem 4.A.43.

Like the convex and the dispersive orders (see Theorems 3.A.7 and 3.B.3), the excess wealth order can be characterized by means of Yaari functionals V_h defined in (3.A.31). Recall that an increasing function $h : [0,1] \to [0,1]$ is starshaped if $h(t)/t$ is increasing on $[0,1]$.

Theorem 3.C.1. *Let X and Y be two random variables with the same finite means. Then $X \le_{\rm ew} Y$ if, and only if,*

$V_h(X) \le V_h(Y)$ *for every starshaped probability transformation function h.*

Jewitt [256] considered an order, called the *location independent riskier* order that can be denoted by $\le_{\rm lir}$. It is shown in Fagiuoli, Pellerey, and Shaked [188] that $X \le_{\rm lir} Y \Longleftrightarrow -X \le_{\rm ew} -Y$. Thus every result that holds for the order $\le_{\rm ew}$ can be reworded by means of the order $\le_{\rm lir}$.

3.C.2 Properties

It is easy to verify that the excess wealth order is location-independent. That is,

$$X \le_{\rm ew} Y \Longrightarrow X + a \le_{\rm ew} Y \quad \text{for any } a \in \mathbb{R}.$$

From (3.C.4) and Theorem 3.A.8 we see that

$$X \le_{\rm ew} Y \Longrightarrow X \le_{\rm dil} Y. \qquad (3.C.7)$$

It follows that if $EX = EY$, then

$$X \leq_{\rm ew} Y \Longrightarrow X \leq_{\rm cx} Y. \tag{3.C.8}$$

Shaked and Shanthikumar [518] showed that if $X \leq_{\rm cx} Y$, then it does not necessarily follow that $X \leq_{\rm ew} Y$.

From (3.C.7), (3.A.32), and (3.A.4) it follows that

$$X \leq_{\rm ew} Y \Longrightarrow \text{Var}(X) \leq \text{Var}(Y),$$

provided $\text{Var}(Y) < \infty$.

From (3.C.3) and (3.B.1) we see that for random variables with finite means, we have

$$X \leq_{\rm disp} Y \Longrightarrow X \leq_{\rm ew} Y. \tag{3.C.9}$$

A characterization of the excess wealth order, which is similar to the characterization of the dispersive order, given in Theorem 3.B.2, is given next.

Theorem 3.C.2. *Let X and Y be two random variables. Then $X \leq_{\rm ew} Y$ if, and only if, for all increasing convex functions ϕ and h such that ϕ and $\psi(\cdot) \equiv h(\phi(\cdot))$ are integrable with respect to the distribution of Y, and for every real number c, we have that*

$$E\phi(X - c) \leq E\phi(Y) \Longrightarrow E\psi(X - c) \leq E\psi(Y).$$

It is worthwhile to mention that two twice differentiable functions ϕ and ψ satisfy $\psi(\cdot) \equiv h(\phi(\cdot))$ for some increasing convex function h if, and only if, $\phi''/\phi' \leq \psi''/\psi'$.

Another characterization of the excess wealth order is described in the following theorem. It is similar to the characterization of the convex order in Theorem 3.A.45. Below, for any random variable Z, the function Ψ_Z is as defined before (3.C.2).

Theorem 3.C.3. *Let X and Y be two random variables with equal means. Then $X \leq_{\rm ew} Y$ if, and only if, there exist random variables $Z_1, Z_2, \ldots,$ with distribution functions $F_1, F_2, \ldots,$ such that $Z_1 =_{\rm st} X$, $EZ_j = EY$, $j = 1, 2, \ldots,$ $\Psi_{Z_j}(x) \to \Psi_Y(x)$ as $j \to \infty$ for all $x \in \mathbb{R}$, and, for any $c \geq 0$, it holds that $S^-\big(\overline{F}_j(\cdot) - \overline{F}_{j+1}(\cdot - c)\big) = 1$ and the sign sequence is $+, -$, $j = 1, 2, \ldots$.*

An important closure property of the excess wealth order is given next.

Theorem 3.C.4. *Let X and Y be two continuous random variables with finite means. Then, for any increasing convex function ϕ, we have*

$$X \leq_{\rm ew} Y \Longrightarrow \phi(X) \leq_{\rm ew} \phi(Y).$$

In the next two results we describe some relationships between the orders $\leq_{\rm ew}$ and $\leq_{\rm mrl}$. We denote the left endpoint of the support of a random variable X by l_X.

Theorem 3.C.5. *Let X and Y be two random variables with distribution functions F and G, respectively, with finite means, and with finite left endpoints l_X and l_Y such that $l_X \le l_Y$. If $X \le_{\rm ew} Y$, and if either X or Y or both are* DMRL, *then $X \le_{\rm mrl} Y$.*

Proof. We only give the proof for the case when the distribution functions F and G of X and Y are continuous; the proof for the general case is similar, though notationally more complex. Let (y_0, p_0), (y_1, p_1), and (y_2, p_2) be three consecutive points of crossing as in the proof of Theorem 3.A.5 (see Figure 3.A.1). Note that by the continuity assumption we have $p_i = F(y_i) = G(y_i)$, $i = 0, 1, 2$.

Suppose that Y is DMRL. For $p \in [p_1, p_2]$ we have $F^{-1}(p) \le G^{-1}(p)$, and therefore, for such a p we have

$$m_X(F^{-1}(p)) \le m_Y(G^{-1}(p)) \le m_Y(F^{-1}(p)),$$

where the first inequality follows from (3.C.5), and the second from the assumption that Y is DMRL. Thus,

$$m_X(y) \le m_Y(y) \quad \text{for } y \in [y_1, y_2]. \tag{3.C.10}$$

If X (rather than Y) is DMRL, then (3.C.10) follows from

$$m_X(G^{-1}(p)) \le m_X(F^{-1}(p)) \le m_Y(G^{-1}(p)), \quad p \in [p_1, p_2],$$

where the first inequality follows from the assumption that X is DMRL, and the second from (3.C.5).

Since $y_0 = F^{-1}(p_0) = G^{-1}(p_0)$, from $X \le_{\rm ew} Y$ we also have that

$$\int_{y_0}^{\infty} \overline{F}(x)\mathrm{d}x \le \int_{y_0}^{\infty} \overline{G}(x)\mathrm{d}x. \tag{3.C.11}$$

Now let $y \in (y_0, y_1)$. For $x \in (y_0, y)$ we have $\overline{F}(x) \ge \overline{G}(x)$. Therefore

$$\int_{y_0}^{y} \overline{F}(x)\mathrm{d}x \ge \int_{y_0}^{y} \overline{G}(x)\mathrm{d}x. \tag{3.C.12}$$

Hence

$$\begin{aligned}
\int_{y}^{\infty} \overline{F}(x)\mathrm{d}x &= \int_{y_0}^{\infty} \overline{F}(x)\mathrm{d}x - \int_{y_0}^{y} \overline{F}(x)\mathrm{d}x \\
&\le \int_{y_0}^{\infty} \overline{G}(x)\mathrm{d}x - \int_{y_0}^{y} \overline{F}(x)\mathrm{d}x \qquad \text{[by (3.C.11)]} \\
&\le \int_{y_0}^{\infty} \overline{G}(x)\mathrm{d}x - \int_{y_0}^{y} \overline{G}(x)\mathrm{d}x \qquad \text{[by (3.C.12)]} \\
&= \int_{y}^{\infty} \overline{G}(x)\mathrm{d}x.
\end{aligned}$$

Therefore
$$\int_y^\infty \overline{F}(x)\mathrm{d}x \le \int_y^\infty \overline{G}(x)\mathrm{d}x, \quad \text{for all } y \in [y_0, y_1].$$
But since $\overline{F}(y) \ge \overline{G}(y)$ for $y \in [y_0, y_1]$, we see that
$$\frac{\int_y^\infty \overline{F}(x)\mathrm{d}x}{\overline{F}(y)} \le \frac{\int_y^\infty \overline{G}(x)\mathrm{d}x}{\overline{F}(y)} \le \frac{\int_y^\infty \overline{G}(x)\mathrm{d}x}{\overline{G}(y)}.$$
So
$$m_X(y) \le m_Y(y) \quad \text{for } y \in [y_0, y_1]. \tag{3.C.13}$$
That is, from (3.C.10) and (3.C.13) we have
$$m_X(y) \le m_Y(y) \quad \text{for } y \in [y_0, y_2].$$

In order to complete the proof we need to show that the interval $[l_X, \infty)$ is a union of segments $[y_0, y_2)$ as above. Suppose that a last point of crossing of F and G exists, and denote it by (y_l, p_l). Denote $(y_0'', p_0'') = (y_{l-1}, p_{l-1})$, $(y_1'', p_1'') = (y_l, p_l)$, and $(y_2'', p_2'') = (\infty, 1)$, where (y_{l-1}, p_{l-1}) is the point of the next to the last crossing of F and G. From the facts that $F^{-1}(p_1'') = G^{-1}(p_1'')$, and that $X \le_{\text{ew}} Y$ implies $\int_{F^{-1}(p_1'')}^\infty \overline{F}(x)\mathrm{d}x \le \int_{G^{-1}(p_1'')}^\infty \overline{G}(x)\mathrm{d}x$, it follows that F crosses G from below at (y_1'', p_1''), and therefore the interval $[y_0'', \infty)$ is of the type described above.

Now suppose that a first point of crossing of F and G exists, and denote it by (y_f, p_f). If $l_X < l_Y$, then at the first point of crossing, F crosses G from above. Thus, from the above proof it follows that $m_X(y) \le m_Y(y)$ for all $y \ge y_f$. The proof that $m_X(y) \le m_Y(y)$ also for $y < y_f$ is similar to the proof of (3.C.10).

If $l_X = l_Y$, then consider two possible cases: (a) in the first point of crossing, F crosses G from above, and (b) in the first point of crossing, F crosses G from below. In case (a) we obtain $m_X(y) \le m_Y(y)$ for all y, as we obtained it above when $l_X < l_Y$. In case (b) denote $y_0' = \sup\{y : F(y) = G(y)\}$, $p_0' = F(y_0')$, $(y_1', p_1') = (y_f, p_f)$, and $(y_2', p_2') = (y_{f+1}, p_{f+1})$, where (y_{f+1}, p_{f+1}) is the point of the second crossing of F and G. The interval $[y_0', y_2')$ is of the kind described above, and therefore $m_X(y) \le m_Y(y)$ for $y \in [y_0', y_2')$, and from it it also follows that $m_X(y) \le m_Y(y)$ for $y < y_0'$. □

Theorem 3.C.6. *Let X and Y be two random variables with distribution functions F and G, respectively, with finite means, and with finite left endpoints l_X and l_Y such that $l_X \le l_Y$. If $X \le_{\text{mrl}} Y$, and if either X or Y or both are* IMRL, *then $X \le_{\text{ew}} Y$.*

Proof. Again, we only give the proof for the case when the distribution functions F and G of X and Y are continuous; the proof for the general case is similar, though notationally more complex. Let (y_0, p_0), (y_1, p_1), and (y_2, p_2) be three consecutive points of crossing as in the proof of Theorem 3.A.5 (see Figure 3.A.1).

Suppose that Y is IMRL. For $p \in [p_1, p_2]$ we have $F^{-1}(p) \le G^{-1}(p)$, and therefore, for such a p we have

$$m_X(F^{-1}(p)) \le m_Y(F^{-1}(p)) \le m_Y(G^{-1}(p)),$$

where the first inequality follows from $X \le_{\text{mrl}} Y$, and the second from the assumption that Y is IMRL. Thus,

$$m_X(F^{-1}(p)) \le m_Y(G^{-1}(p)) \quad \text{for } p \in [p_1, p_2]. \tag{3.C.14}$$

If X (rather than Y) is IMRL, then (3.C.14) follows from

$$m_X(F^{-1}(p)) \le m_X(G^{-1}(p)) \le m_Y(G^{-1}(p)), \quad p \in [p_1, p_2],$$

where the first inequality follows from the assumption that X is IMRL, and the second from $X \le_{\text{mrl}} Y$.

Since $y_0 = F^{-1}(p_0) = G^{-1}(p_0)$, from $X \le_{\text{mrl}} Y$ we also have that

$$m_X(F^{-1}(p_0)) \le m_Y(G^{-1}(p_0)). \tag{3.C.15}$$

Now let $p \in (p_0, p_1)$. Since $\overline{F}(x) \ge \overline{G}(x)$ for $x \in [y_0, y_1]$ we see that

$$\int_{y_0}^{F^{-1}(p)} \overline{F}(x)\mathrm{d}x \ge \int_{y_0}^{F^{-1}(p)} \overline{G}(x)\mathrm{d}x \ge \int_{y_0}^{G^{-1}(p)} \overline{G}(x)\mathrm{d}x, \tag{3.C.16}$$

where the second inequality follows from $F^{-1}(p) \ge G^{-1}(p)$. Therefore

$$\begin{aligned} m_X(F^{-1}(p)) &= \frac{\int_{F^{-1}(p)}^{\infty} \overline{F}(x)\mathrm{d}x}{1-p} \\ &= \frac{\int_{y_0}^{\infty} \overline{F}(x)\mathrm{d}x - \int_{y_0}^{F^{-1}(p)} \overline{F}(x)\mathrm{d}x}{1-p} \\ &\le \frac{\int_{y_0}^{\infty} \overline{G}(x)\mathrm{d}x - \int_{y_0}^{G^{-1}(p)} \overline{G}(x)\mathrm{d}x}{1-p} \qquad \text{[by (3.C.15) and (3.C.16)]} \\ &= m_Y(G^{-1}(p)), \quad \text{for } p \in [p_0, p_1]. \end{aligned}$$

So, from the preceding inequality and from (3.C.14) we obtain

$$m_X(F^{-1}(p)) \le m_Y(G^{-1}(p)) \quad \text{for } p \in [p_0, p_2].$$

In order to complete the proof we need to show that the interval $[l_X, \infty)$ is a union of segments $[y_0, y_2)$ as above. Suppose that a last point of crossing of F and G exists, and denote it by (y_l, p_l). Denote $(y_0'', p_0'') = (y_{l-1}, p_{l-1})$, $(y_1'', p_1'') = (y_l, p_l)$, and $(y_2'', p_2'') = (\infty, 1)$, where (y_{l-1}, p_{l-1}) is the point of the next to the last crossing of F and G. From the facts that $\overline{F}(y_1'') = \overline{G}(y_1'')$, and that $X \le_{\text{mrl}} Y$ implies $\int_{y_1''}^{\infty} \overline{F}(x)\mathrm{d}x \le \int_{y_1''}^{\infty} \overline{G}(x)\mathrm{d}x$, it follows that F crosses

G from below at (y_1'', p_1''), and therefore the interval $[y_0'', \infty)$ is of the type described above.

Now suppose that a first point of crossing of F and G exists, and denote it by (y_f, p_f). If $l_X < l_Y$, then at the first point of crossing, F crosses G from above. Thus, from the above proof it follows that $m_X(F^{-1}(p)) \le m_Y(G^{-1}(p))$ for all $p \ge p_f$. The proof that $m_X(F^{-1}(p)) \le m_Y(G^{-1}(p))$ also for $p < p_f$ is similar to the proof of (3.C.14).

If $l_X = l_Y$, then consider two possible cases: (a) in the first point of crossing, F crosses G from above, and (b) in the first point of crossing, F crosses G from below. In case (a) we obtain $m_X(F^{-1}(p)) \le m_Y(G^{-1}(p))$ for all p, as we obtained it above when $l_X < l_Y$. In case (b) denote $y_0' = \sup\{y : F(y) = G(y)\}$, $p_0' = F(y_0')$, $(y_1', p_1') = (y_f, p_f)$, and $(y_2', p_2') = (y_{f+1}, p_{f+1})$, where (y_{f+1}, p_{f+1}) is the point of the second crossing of F and G. The interval $[y_0', y_2')$ is of the kind described above, and therefore $m_X(F^{-1}(p)) \le m_Y(G^{-1}(p))$ for $p \in [p_0', p_2')$, and from it it also follows that $m_X(F^{-1}(p)) \le m_Y(G^{-1}(p))$ for $p \le p_0'$.

In summary, we have shown that $m_X(F^{-1}(p)) \le m_Y(G^{-1}(p))$ for all $p \in (0, 1)$. Therefore $X \le_{\rm ew} Y$ by (3.C.5). □

The following few results give conditions under which the order $\le_{\rm ew}$ is closed under convolutions.

Theorem 3.C.7. *Let (X_i, Y_i), $i = 1, 2, \ldots, m$, be independent pairs of random variables such that $X_i \le_{\rm ew} Y_i$, $i = 1, 2, \ldots, m$. If X_i, Y_i, $i = 1, 2, \ldots, m$, all have (continuous or discrete) logconcave densities, except possibly one X_l and one Y_k $(l \neq k)$, then*

$$\sum_{i=1}^{m} X_i \le_{\rm ew} \sum_{i=1}^{m} Y_i.$$

In order to prove Theorem 3.C.7 one first proves that if X and Y are two random variables such that $X \le_{\rm ew} Y$, and if Z is a random variable with logconcave density that is independent of X and of Y, then $X + Z \le_{\rm ew} Y + Z$. The statement of the theorem can then be derived from the fact that a convolution of random variables with logconcave densities has a logconcave density. We do not give the details here.

The next two results are analogs of Theorems 3.B.7 and 3.B.8.

Theorem 3.C.8. *The random variable X satisfies*

$$X \le_{\rm ew} X + Y \quad \text{for any random variable } Y \text{ independent of } X$$

if, and only if, X is IFR.

Theorem 3.C.9. *Let Z be a random variable. Then*

$$X + Z \le_{\rm ew} Y + Z \quad \text{whenever } X \le_{\rm disp} Y \text{ and } Z \text{ is independent of } X \text{ and } Y$$

if, and only if, Z is IFR.

Since a convolution of IFR random variables is IFR (see Corollary 1.B.39), repeated application of Theorem 3.C.9 yields the following result, which is an analog of Theorem 3.B.9.

Theorem 3.C.10. *Let $X_1, X_2, \ldots, X_n$ be a set of independent random variables, and let $Y_1, Y_2, \ldots, Y_n$ be another set of independent random variables. If the X_i's and the Y_i's are all* IFR, *and if $X_i \le_{\text{disp}} Y_i$, $i = 1, 2, \ldots, n$, then*

$$\sum_{i=1}^{n} X_i \le_{\text{ew}} \sum_{i=1}^{n} Y_i.$$

An interesting closure property of the order $\le_{\text{ew}}$ is given next.

Theorem 3.C.11. *Let $X_1, X_2, \ldots$ be a collection of independent and identically distributed random variables, and let $Y_1, Y_2, \ldots$ be another collection of independent and identically distributed random variables. Also, let N be a positive, integer-valued, random variable, independent of the X_i's and of the Y_i's. If $X_1 \le_{\text{ew}} Y_1$, then* $\max\{X_1, X_2, \ldots, X_N\} \le_{\text{ew}} \max\{Y_1, Y_2, \ldots, Y_N\}$.

The following result may be compared to Theorem 3.B.31. By (3.C.9), we assume below less than is assumed in Theorem 3.B.31, but the conclusion is weaker. We use below the notation for spacings that was used in Theorem 3.B.31.

Theorem 3.C.12. *Let X and Y be two random variables. If $X \le_{\text{ew}} Y$, then $EU_{(n-1:n)} \le EV_{(n-1:n)}$ for $n = 2, 3, \ldots$.*

The order $\le_{\text{ew}}$ can be used to characterize DMRL and IMRL random variables. The following result may be compared with Theorems 2.A.23, 2.B.17, 3.A.56, and 4.A.51. As in Section 1.A.3, $[Z|A]$ denotes any random variable that has as its distribution the conditional distribution of Z given A.

Theorem 3.C.13. *Let X be a continuous random variable with a finite left endpoint of its support l_X. Then X is* DMRL [IMRL] *if, and only if, any one of the following equivalent conditions holds:*

(i) $[X - t|X > t] \ge_{\text{ew}} [\le_{\text{ew}}]\ [X - t'|X > t']$ *whenever* $t' \ge t \ge l_X$.
(ii) $X \ge_{\text{ew}} [\le_{\text{ew}}]\ [X - t|X > t]$ *for all* $t \ge l_X$ *(when* $l_X = 0$*).*

The proof of this result is omitted.

3.D The Peakedness Order

3.D.1 Definition

In this section we discuss a variability order that applies to random variables with *symmetric* distribution functions. This is one of the oldest (if not *the* oldest) variability notions that can be found in the literature. It stochastically compares random variables according to their distance from their center of symmetry.

Let X be a random variable with a distribution function that is symmetric about μ, and let Y be another random variable with a distribution function that is symmetric about ν. Suppose that

$$|X-\mu| \leq_{\rm st} |Y-\nu|.$$

Then X is said to be *smaller than Y in the peakedness order* (denoted by $X \leq_{\rm peak} Y$). Note that, in the literature, often X is said to be *more peaked* about μ than Y about ν if $X \leq_{\rm peak} Y$.

The following result is easy to prove.

Theorem 3.D.1. *Let X and Y be two random variables with different distribution functions, but with the same mean. Suppose that the distribution functions F and G, of X and Y, respectively, are symmetric about the common mean. Then $X \leq_{\rm peak} Y$ if, and only if,*

$$S^-(G-F) = 1 \text{ and the sign sequence is } +,-,$$

where S^- is defined in (1.A.18).

3.D.2 Some properties

The peakedness order satisfies some desirable closure properties. For example, it is easy to verify the following result.

Theorem 3.D.2. *Let X be a random variable with a symmetric distribution function. Then*

$$X \leq_{\rm peak} aX \quad \text{whenever } a \geq 1.$$

The closure results in the next theorem can also be easily verified.

Theorem 3.D.3. (a) *Let X, Y, and Θ be random variables such that the distribution functions of $[X|\Theta=\theta]$ are symmetric about some μ (which is independent of θ) and the distribution functions of $[Y|\Theta=\theta]$ are symmetric about some ν (which is also independent of θ) and such that $[X|\Theta=\theta] \leq_{\rm peak} [Y|\Theta=\theta]$ for all θ in the support of Θ. Then $X \leq_{\rm peak} Y$. That is, the peakedness order is closed under mixtures.*

(b) *Let $\{X_j,\ j=1,2,\dots\}$ and $\{Y_j,\ j=1,2,\dots\}$ be two sequences of random variables with symmetric distribution functions such that $X_j \to_{\rm st} X$ and $Y_j \to_{\rm st} Y$ as $j\to\infty$. If $X_j \leq_{\rm peak} Y_j$, $j=1,2,\dots$, then $X \leq_{\rm peak} Y$.*

The peakedness order is also closed under convolutions of random variables that have unimodal symmetric distribution functions (that is, with mode at their center of symmetry). This is shown next.

Theorem 3.D.4. *Let $X_1, X_2, \ldots, X_n$ and $Y_1, Y_2, \ldots, Y_n$ be two sets of independent random variables, all having distribution functions that are symmetric about possibly different centers, and all having unimodal densities with possibly some probability mass at their respective centers. If $X_i \le_{\text{peak}} Y_i$ for $i = 1, 2, \ldots, n$, then*

$$\sum_{i=1}^{n} X_i \le_{\text{peak}} \sum_{i=1}^{n} Y_i.$$

In particular, $\overline{X} \le_{\text{peak}} \overline{Y}$, where $\overline{X}$ and $\overline{Y}$ denote the corresponding sample means.

Proof. Without loss of generality we may assume that all the centers of the X_i's and of the Y_i's are 0. First we prove the result for $n = 2$. Let F_1, F_2, G_1, and G_2 denote the distribution functions of X_1, X_2, Y_1, and Y_2, respectively. Select an $a > 0$. Then

$$\begin{aligned}
P\{|X_1 + X_2| \le a\} &= 2\int_0^\infty [F_1(x+a) - F_1(x-a)]\mathrm{d}F_2(x) \\
&\ge 2\int_0^\infty [F_1(x+a) - F_1(x-a)]\mathrm{d}G_2(x) \\
&= 2\int_0^\infty [G_2(x+a) - G_2(x-a)]\mathrm{d}F_1(x) \\
&\ge 2\int_0^\infty [G_2(x+a) - G_2(x-a)]\mathrm{d}G_1(x) \\
&= P\{|Y_1 + Y_2| \le a\},
\end{aligned}$$

where the first inequality follows from the unimodality of X_1 (therefore, the integrand is decreasing in $x \ge 0$) and from $X_2 \le_{\text{peak}} Y_2$, and the second inequality follows from the unimodality of Y_2 and from $X_1 \le_{\text{peak}} Y_1$. This proves the result for $n = 2$. The general result can be obtained by a simple induction together with the observation that a sum of independent random variables, all having distribution functions that are symmetric about 0 and all having unimodal densities, also has a unimodal density symmetric about 0. □

If $X_1, X_2, \ldots$ are independent and identically distributed random variables, then, for each n, we denote by $\overline{X}_n$ the sample mean of $X_1, X_2, \ldots, X_n$. That is, $\overline{X}_n = (X_1 + X_2 + \cdots + X_n)/n$. In Example 3.A.29 it is shown that $\overline{X}_n \le_{\text{cx}} \overline{X}_{n-1}$. The following result shows that a similar property holds for the peakedness order under an additional condition.

Theorem 3.D.5. *If $X_1, X_2, \ldots$ are independent and identically distributed random variables, having a common logconcave density function that is symmetric about a common value, then for each $n \ge 2$ one has*

$$\overline{X}_n \le_{\text{peak}} \overline{X}_{n-1}.$$

A relationship between the dispersive and the peakedness orders is described next.

Theorem 3.D.6. *Let X and Y be two random variables having distribution functions that are symmetric about possibly different centers. If $X \leq_{\text{disp}} Y$, then $X \leq_{\text{peak}} Y$.*

3.E Complements

Section 3.A: For historical reasons, the convex order is sometimes referred to as "dilation." However, in recent literature the order defined in (3.A.32) is often called the dilation order. Some standard references about the convex order are Ross [475] and Müller and Stoyan [419], where many of the results described in Section 3.A can be found. Another monograph that studies the convex order (under the mask of the Lorenz order) is Arnold [19], and many of the results in this section that deal directly with the Lorenz order can be found there. The proof of Theorem 3.A.2 is taken from Muñoz-Perez and Sanches-Gomez [421]; an alternative proof, using ideas from the area of comparison of experiments, can be found in Torgersen [551, page 369]. Result (3.A.12) is taken from Hickey [223]. The present version of the characterization of the convex order given in Theorem 3.A.4 is taken from Müller and Rüschendorf [415]. The characterization of the convex order in Theorem 3.A.5 can be found in Fagiuoli, Pellerey, and Shaked [188]; see also Levy and Kroll [346] and Ramos and Sordo [463]. The characterization of the convex order by means of Yaari functionals (Theorem 3.A.7) can be found in Chateauneuf, Cohen, and Meilijson [127]. The characterization of the dilation order, given in Theorem 3.A.8, is taken from Fagiuoli, Pellerey, and Shaked [188]. The characterization given in Theorem 3.A.9 can be found in Ramos and Sordo [463]. The characterization of the Lorenz order by means of the Lorenz zonoids (Theorem 3.A.11) is taken from Arnold [20]. The result about the convex ordering of random sums (Theorem 3.A.13) is a special case of a result of Jean-Marie and Liu [254]; the extensions of it when the underlying random variables are identically distributed (Theorems 3.A.14–3.A.16) are taken from Pellerey [450]. Theorem 3.A.17 and some related results can be found in Berger [79]. The property of the increase in the dilation order with an increase in the scale (Theorem 3.A.18) is taken from Hickey [223]. The result about the dilation ordering of two differences (Theorem 3.A.19) can be found in Kochar and Carrière [312]. The convex order lower bound on $\sum_{i=1}^{n} X_i$, given in Theorem 3.A.20, is taken from Vyncke, Goovaerts, De Schepper, Kaas, and Dhaene [557]. The property of inheritance of the convex order from the mixing random variables to the mixed ones (Theorem 3.A.21) can be found in Schweder [499]; its variation, Theorem 3.A.23, is taken from Kottas and Gelfand [323]. The property of the preservation of the convex order under products of nonnegative random variables (Corollary 3.A.22) can be found in Whitt [562]. The Lorenz order comparison of $g(X)$ and $h(X)$ (Theorem 3.A.26) can be found in Wilfling [566]. The relationship between the orders $\leq_{\text{Lorenz}}$ and $\leq_{\text{hmrl}}$, given in Theorem 3.A.28, is taken from Lefèvre and Utev [340]. The result about the convex ordering of the sample means (Example 3.A.29) can be found in Marshall and Olkin [383, page 288]. Its generalizations (Theorem 3.A.30 and Corollary 3.A.31) are taken from Denuit and Vermandele [158] and from O'Cinneide [439]. The convex order comparison of scaled Poisson random variables (Example 3.A.32) is inspired by a result at the end of page 1078 in Bäuerle [60]. The convex order comparison in Theorem 3.A.33 can be found in O'Cinneide [439]. The closure property (3.A.50) of the dilation order can be found in Muñoz-Perez and Sanches-Gomez [421]. The majorization result (Theorem 3.A.35) is a special case of a result of Marshall and Proschan [384]; related results can be found in Ma [375]. The preservation of the convex order under linear convex combinations (Theorem 3.A.36) is taken from Pellerey [452]. The convex order comparison of sums of random variables with random coefficients (Theorem 3.A.37) can be found in Ma [375]. The particular case of it, given in Example 3.A.38, is a result of Karlin and Novikoff [277]; Marshall and Olkin [383, Section 15.E] obtained a generalization of this special case which is different from the result in Theorem 3.A.37. The convex order comparison of sums of positively [respectively, negatively] associated random variables, and independent random variables, given in Theorem 3.A.39, can be found in Denuit, Dhaene, and Ribas [143] [respectively, Shao [535]]; see also Boutsikas and Vaggelatou [107]. The Laplace transform characterization of the order $\leq_{\text{cx}}$ (Theorem 3.A.40) is taken from Shaked and Wong [524]; see also Kan and Yi [274]. The convex order comparison of posterior means, in the context of statistical experiments (Example 3.A.41), is essentially taken from Baker [30]. The condition for stochastic equality of $\leq_{\text{cx}}$-ordered random variables (Theorem 3.A.42) is a special case of a result by Denuit, Lefèvre, and Shaked [151], whereas its generalization (Theorem 3.A.43) has been motivated by a result in Bhattacharjee and Bhattacharya [87]; see also Huang and Lin [249]. The result that gives sufficient conditions for the convex order by means of the number of crossings of the underlying densities or distribution functions (Theorem 3.A.44) is taken from Shaked [502], but its origins may be found in Karlin and Novikoff [277], if not before. A proof of the characterization of the convex order by means of the number of crossings of two distribution functions (Theorem 3.A.45) can be found in Müller [407]; similar results are given in Borglin and Keiding [106]. The convex order comparison of normalized Bernoulli random variables (Example 3.A.48) can be found in Makowski [379]. The necessary and sufficient conditions for the comparison of normal random variables (Example 3.A.51) are taken from Müller [413]. The relations $\leq_{\text{uv}}$ and $\leq_{\text{lc}}$ were introduced in Whitt [564] as means to identify the order $\leq_{\text{cx}}$. The characterization of DMRL and IMRL random variables by means of the convex order (Theorem 3.A.56) is taken from Belzunce, Candel, and Ruiz [64]. The relationships between the orders $\le_{\text{dil}}$ and $\le_{\text{mrl}}$ that are described in Theorems 3.A.57 and 3.A.58 can be found in Belzunce, Pellerey, Ruiz, and Shaked [72] where further related results can also be found.

The results on the m-convex order (Section 3.A.5) are mostly taken from Denuit, Lefèvre, and Shaked [151]. The condition that implies the stochastic equality of $\le_{m\text{-cx}}^{\mathcal{S}}$-ordered random variables (Theorem 3.A.60) is taken from Denuit, Lefèvre, and Shaked [152]. The method for deriving the distributions of the stochastic extrema in $\mathcal{B}_m([0,b];\mu_1,\mu_2,\ldots,\mu_{m-1})$ is taken from Denuit, De Vylder, and Lefèvre [142]. The stochastic comparisons of the Gamma, inverse Gaussian, and lognormal random variables (Example 3.A.67) are taken from Kaas and Hesselager [270]. Tables 3.A.1 and 3.A.2 can be found in Denuit, Lefèvre, and Shaked [153, 154]. Theorem 3.A.63 can be found in Denuit, Lefèvre, and Utev [155]. The result about the $2m$-cx ordering of two differences (Theorem 3.A.64) is taken from Bassan, Denuit, and Scarsini [52]. Denuit and Lefèvre [146], Denuit, Lefèvre, and Utev [156], and Denuit, Lefèvre, and Mesfioui [149] studied discrete analogs of the m-convex order; in particular they obtained some analogs of the results in Section 3.A.5 for arithmetic random variables, as well as some specific results for the discrete case. Denuit, Lefèvre, and Utev [155] extended the m-convex order to Tchebycheff-type orders; see also Lynch [367].

Bhattacharjee [85] studied the order $\le_{\text{cx}}$ under the restriction that the compared random variables are discrete.

Metzger and Rüschendorf [393] studied variability orderings, which are related to $\le_{\text{uv}}$ and $\le_{\text{lc}}$, defined by requiring the ratio of the distribution functions F/G or of the survival functions $\overline{F}/\overline{G}$ to be unimodal. For example, they showed that if X and Y are two random variables with distribution functions F and G, respectively, such that $\text{supp}(X) \subseteq \text{supp}(Y)$, and if $X \le_{\text{uv}} Y$, then F/G is unimodal. They also considered the order defined by requiring the ratio of a shifted density to another density $f(\cdot + a)/g(\cdot)$ to be unimodal for all a. This order is to be compared with the order $\le_{\text{uv}}$ and also with the order $\le_{\text{lr}\uparrow}$ studied in Section 1.C.4.

Müller [412] considered an order that is defined by requiring (3.A.1) to hold for all so-called (a,b)-concave functions. Other related stochastic orders can be found in Müller [412] as well.

An order which is related to the Lorenz order is studied in Zenga [576].

Section 3.B: Doksum [169] studied some properties of the dispersive order by stipulating (3.B.10) and calling it the "tail-order" (see Deshpande and Kochar [159] for further early references in which this order is studied). A basic paper on the dispersive order is Shaked [503] where many of the equivalent conditions described in Section 3.B.1 can be found. The conditions (3.B.3), (3.B.4), and (3.B.6) are taken, respectively, from Saunders [489], Hickey [223], and Muñoz-Perez [420]. Another characterization of the order $\leq_{\text{disp}}$, which is related to (3.B.3), is given in Burger [115]. The observation (3.B.15) has been noted in Müller and Stoyan [419]. The characterization of the dispersive order by means of the observed total time on test random variables (Theorem 3.B.1) can be found in Bartoszewicz [42]; other related results can be found in Bartoszewicz [39, 42]. The notion of Q-addition was introduced in Muñoz-Perez [420]. The characterization of the dispersive order given in Theorem 3.B.2 is taken from Landsberger and Meilijson [330]. The characterization of the dispersive order by means of Yaari functionals (Theorem 3.B.3) can be found in Chateauneuf, Cohen, and Meilijson [127]. The properties described in Section 3.B.2 have been collected from many sources. The result of Theorem 3.B.7 can be found in Droste and Wefelmeyer [171]. Several versions of Theorem 3.B.8 can be found in Lewis and Thompson [347] and in Lynch, Mimmack, and Proschan [368]. Some versions of Theorem 3.B.10 can be found in Bartoszewicz [37] and in Rojo and He [472]. Some related results appear in Hickey [223]; for example, his Theorem 4 can be obtained from (3.B.21) applied to the decreasing convex case. Theorems 3.B.14 and 3.B.15 are also taken from that paper. The relationship between the orders $\leq_{\text{disp}}$ and $\leq_{\text{conv}}$, given in (3.B.26), was noted in Shaked and Suarez-Llorens [520]. The sufficient condition for the dispersive order by means of comparison of shifted hazard rate functions (Theorem 3.B.18) can be found in Belzunce, Lillo, Ruiz, and Shaked [69]. Theorem 3.B.19 has been proved in Mailhot [377], whereas Theorem 3.B.20 combines results from Bartoszewicz [38, 40] and Bagai and Kochar [29]. The relationships between the orders $\leq_{\text{disp}}$ and $\leq_{\text{mrl}}$, given in Theorem 3.B.21, can be found in Bartoszewicz [44]. The result about the dispersive ordering of order statistics of DFR random variables (Example 3.B.22) is taken from Kochar [308]; some other related results can also be found there. The results about the dispersive ordering of the spacings of DFR random variables (Example 3.B.23) are taken from Kochar and Kirmani [313] and from Khaledi and Kochar [285]; an extension of these results can be found in Belzunce, Hu, and Khaledi [68]. The characterizations of IFR and DFR random variables by means of the dispersive order (Theorems 3.B.24 and 3.B.25) have been derived by Belzunce, Candel, and Ruiz [64], and by Pellerey and Shaked [456]. The results on the dispersive order comparisons of order statistics and spacings (Theorems 3.B.26, 3.B.28, 3.B.29, and 3.B.31) can be found in Bartoszewicz [39], in Khaledi and Kochar [286], and in Oja [440], whereas Example 3.B.30 is mentioned in Kleiber [303]; related results can be found in Alzaid and Proschan [14], in Belzunce, Hu, and Khaledi [68], in Belzunce, Mercader, and Ruiz [70], and in Hu and Zhuang [247]. An extension of Theorem 3.B.26 to order statistics from samples with random size can be found in Nanda, Misra, Paul, and Singh [427].

The dispersive order comparisons of maxima of heterogeneous exponential random variables (Example 3.B.32) are taken from Dykstra, Kochar, and Rojo [174] and from Khaledi and Kochar [287], whereas the comparison of the spacings (Example 3.B.33) is taken from Kochar and Korwar [314]. The comparison of sums of heterogeneous exponential random variables (Example 3.B.34) can be found in Kochar and Ma [317]. The comparisons of sums of uniform and Gamma random variables (Examples 3.B.35 and 3.B.36) are slightly weaker than results that are given in Khaledi and Kochar [288, 289]. The result about the dispersive order comparison of the successive epochs of a nonhomogeneous Poisson process (Example 3.B.37) is given in Kochar [310], though it is stated by means of the dispersive order comparison of successive record values of a sequence of independent and identically distributed random variables with a common DFR distribution function. The dispersive order comparison of epoch times of nonhomogeneous Poisson processes (Example 3.B.38) can be found in Belzunce, Lillo, Ruiz, and Shaked [69] and in Yue and Cao [575]. The results about the dispersive order comparisons of random minima and maxima (Example 3.B.39) are taken from Shaked and Wong [526]; a simple proof of these results is given in Bartoszewicz [49]. The comparison of t-distributed random variables (Example 3.B.40) can be found in Arias-Nicolás, Fernández-Ponce, Luque-Calvo, and Suárez-Llorens [17], whereas the comparison of weighted random variables (Example 3.B.41) can be found in Bartoszewicz and Skolimowska [51]. Finally, the result of Theorem 3.B.42 has been derived by Giovagnoli and Wynn [211] in order to motivate a definition of multivariate dispersive order (see Section 7.B); Theorem 3.B.42 was also obtained by Kusum, Kochar, and Deshpande [327] who actually derived it for logarithms of positive random variables.

Fernández-Ponce and Suárez-Llorens [197] introduced a "weakly dispersive" order by requiring that, corresponding to every interval of length ε in the support of the "larger" variable, there exists an interval of the same length in the support of the "smaller" variable, such that the probability mass of the latter with respect to the distribution of the "smaller" variable is at least as large as the probability mass of the former with respect to the distribution of the "larger" variable.

Belzunce, Hu, and Khaledi [68] studied an order, which they denoted by $\leq_{\text{disp-hr}}$, that is stronger than the order $\leq_{\text{disp}}$.

Condition (3.B.1) can be written as

$$\frac{F^{-1}(\beta) - F^{-1}(\alpha)}{G^{-1}(\beta) - G^{-1}(\alpha)} \leq M \quad \text{whenever } 0 < \alpha < \beta < 1,$$

where $M = 1$. Lehmann [344] considered this condition for other possible values of M in order to compare the tails of F and G. Burger [115] studied, among other things, the above condition (with $M = 1$), but only for α

and β such that $0 < \alpha < G^{-1}(\mu) < \beta < 1$, where μ is some constant. Rojo [471] studied the above condition with $M = \infty$ in the sense

$$\limsup_{u \to 1} \frac{F^{-1}(u)}{G^{-1}(u)} < \infty,$$

and Bartoszewicz [43] obtained comparison results, with respect to the latter order, for the observed total time on test random variables X_{ttt} and Y_{ttt}, with distribution functions as defined in (1.A.19).

Section 3.C: Most of the results, about the excess wealth order, that are described in this section are taken from Shaked and Shanthikumar [518], Fagiuoli, Pellerey, and Shaked [188], and Kochar, Li, and Shaked [316]. Fernandez-Ponce, Kochar, and Muñoz-Perez [195] also studied the excess wealth order by the name of the *right spread order.* The characterization of the excess wealth order given in (3.C.2) is taken from Chateauneuf, Cohen, and Meilijson [127]; the characterization of the excess wealth order by means of Yaari functionals (Theorem 3.C.1) can be found in that paper as well. The characterization of the excess wealth order given in Theorem 3.C.2 is a translation of the definition of the order $\leq_{\text{lir}}$ into the order $\leq_{\text{ew}}$, which can be done by virtue of Lemma 3.1 of Fagiuoli, Pellerey, and Shaked [188]. The characterization of the excess wealth order by means of the number of crossings of two distribution functions (Theorem 3.C.3) can be obtained in a similar manner from a correction by Müller [410] of Theorem 1 in Landsberger and Meilijson [330]. The conditions for the preservation of the order $\leq_{\text{ew}}$ under convolutions (Theorems 3.C.7–3.C.10) can essentially all be found in Hu, Chen, and Yao [231]. The result about the preservation of the excess wealth order under random maxima (Theorem 3.C.11) is taken from Li and Zuo [358], and the result that compares the expected values of the extreme spacings (Theorem 3.C.12) is a special case of a result of Li [353]. The characterization of DMRL and IMRL random variables by the order $\leq_{\text{ew}}$ (Theorem 3.C.13) is taken from Belzunce [63].

Belzunce, Hu, and Khaledi [68] studied a stochastic order, denoted by $\leq_{\text{disp-mrl}}$, which is stronger than the order $\leq_{\text{ew}}$.

Section 3.D: The peakedness order was introduced by Birnbaum [90]. The characterization of this order, given in Theorem 3.D.1, was observed in Kottas and Gelfand [323]. Theorem 3.D.4 was essentially proven by Birnbaum [90]; the proof given here is adopted from Bickel and Lehmann [89]. The result about the monotonicity of the sample means in the sense of the peakedness order (Theorem 3.D.5) is given in Proschan [461]; an extension of Theorem 3.D.5 can be found in Ma [372]. The relationship between the dispersive and the peakedness orders, given in Theorem 3.D.6, was observed in Shaked [503].

4

Univariate Monotone Convex and Related Orders

In Chapter 1 we studied orders that compare random variables according to their "magnitude". In Chapter 3 the studied orders compare random variables according to their "variability". The orders that are discussed in this chapter compare random variables according to both their "location" and their "spread". The most important and common orders that are studied in this chapter are the increasing convex and the increasing concave orders. Also the transform orders that are studied here, that is, the convex, the star, and the superadditive orders, are of interest in many theoretical and practical applications. In addition, some other related orders are investigated in this chapter as well.

4.A The Monotone Convex and Monotone Concave Orders

4.A.1 Definitions and equivalent conditions

Let X and Y be two random variables such that

$$E[\phi(X)] \leq E[\phi(Y)] \quad \text{for all increasing convex [concave] functions } \phi : \mathbb{R} \to \mathbb{R}, \tag{4.A.1}$$

provided the expectations exist. Then X is said to be *smaller than* Y *in the increasing convex [concave] order* (denoted by $X \leq_{\text{icx}} Y$ [$X \leq_{\text{icv}} Y$]). Roughly speaking, if $X \leq_{\text{icx}} Y$, then X is both "smaller" and "less variable" than Y in some stochastic sense. Similarly, if $X \leq_{\text{icv}} Y$, then X is both "smaller" and "more variable" than Y in some stochastic sense.

One can also define a *decreasing convex [concave] order* by requiring (4.A.1) to hold for all decreasing convex [concave] functions ϕ (denoted by $X \leq_{\text{dcx}}$ [$\leq_{\text{dcv}}$] Y). The terms "decreasing convex" and "decreasing concave" are counterintuitive in the sense that if X is smaller than Y in the sense of either of these two orders, then X is "larger" than Y in some stochastic sense. These orders can be easily characterized using the orders $\le_{\text{icx}}$ and $\le_{\text{icv}}$. Therefore, it is not necessary to have a separate discussion for these orders.

In analogy with Theorem 3.A.12(a), the orders $\le_{\text{icx}}$ and $\le_{\text{icv}}$ are related to each other as follows.

Theorem 4.A.1. *Let X and Y be two random variables. Then*

$$X \le_{\text{icx}} [\le_{\text{icv}}]\, Y \Longleftrightarrow -X \ge_{\text{icv}} [\ge_{\text{icx}}] -Y.$$

The proof of Theorem 4.A.1 is based on the fact that a function ϕ satisfies that $\phi(x)$ is increasing and convex in x if, and only if, $-\phi(-x)$ is increasing and concave in x. We omit the straightforward details.

Note that the function ϕ, defined by $\phi(x) = x$, is increasing and is both convex and concave. Therefore, from (4.A.1) it follows that

$$X \le_{\text{icx}} Y \Longrightarrow E[X] \le E[Y] \tag{4.A.2}$$

and that

$$X \le_{\text{icv}} Y \Longrightarrow E[X] \le E[Y], \tag{4.A.3}$$

provided the expectations exist.

Let $\overline{F}$ $[F]$ and $\overline{G}$ $[G]$ be the survival [distribution] functions of X and Y, respectively. For a fixed a, the function ϕ_a, defined by $\phi_a(x) = (x-a)_+$, is increasing and convex. Therefore, if $X \le_{\text{icx}} Y$, then

$$E[(X-a)_+] \le E[(Y-a)_+] \quad \text{for all } a, \tag{4.A.4}$$

provided the expectations exist. Alternatively, using a simple integration by parts, it is seen that (4.A.4) can be rewritten as

$$\int_x^\infty \overline{F}(u)du \le \int_x^\infty \overline{G}(u)du \quad \text{for all } x, \tag{4.A.5}$$

provided the integrals exist. For any real number a let a_- denote the negative part of a, that is, $a_- = a$ if $a \le 0$ and $a_- = 0$ if $a > 0$. For a fixed a, the function ζ_a, defined by $\zeta_a(x) = (x-a)_-$, is increasing and concave. Therefore, if $X \le_{\text{icv}} Y$, then

$$E[(X-a)_-] \le E[(Y-a)_-] \quad \text{for all } a, \tag{4.A.6}$$

provided the expectations exist. Alternatively, again using a simple integration by parts, it is seen that (4.A.6) can be rewritten as

$$\int_{-\infty}^x F(u)du \ge \int_{-\infty}^x G(u)du \quad \text{for all } x, \tag{4.A.7}$$

provided the integrals exist.

In fact (4.A.5) [(4.A.7)] is equivalent to $X \leq_{\text{icx}} Y$ [$X \leq_{\text{icv}} Y$]. To see it, note that every increasing convex [concave] function can be approximated by (that is, is a limit of) positive linear combinations of the functions ϕ_a's [ζ_a's], for various choices of a's. By (4.A.5), $E[\phi_a(X)] \leq E[\phi_a(Y)]$ for all a, and this fact implies (4.A.1) in the increasing convex case. Similarly, by (4.A.7), $E[\zeta_a(X)] \leq E[\zeta_a(Y)]$ for all a, and this fact implies (4.A.1) in the increasing concave case. We thus have proved the following result.

Theorem 4.A.2. *Let X and Y be two random variables. Then $X \leq_{\text{icx}} Y$ [$X \leq_{\text{icv}} Y$] if, and only if,* (4.A.5) [(4.A.7)] *holds.*

The next two results give further characterizations of the order $\leq_{\text{icx}}$. The first one is an analog of Theorem 3.A.5.

Theorem 4.A.3. *Let X and Y be two random variables with distribution functions F and G, respectively. Then $X \leq_{\text{icx}} Y$ if, and only if,*

$$\int_p^1 F^{-1}(u)\mathrm{d}u \leq \int_p^1 G^{-1}(u)\mathrm{d}u \quad \textit{for all } p \in [0,1].$$

Theorem 4.A.4. *Let X and Y be two random variables with distribution functions F and G, respectively. Then $X \leq_{\text{icx}} Y$ if, and only if,*

$$\int_0^1 F^{-1}(u)\mathrm{d}\phi(u) \leq \int_0^1 G^{-1}(u)\mathrm{d}\phi(u) \quad \textit{for all increasing convex functions } \phi : [0,1] \to \mathbb{R}.$$

Another necessary and sufficient condition for $X \leq_{\text{icx}} Y$ is the following:

$$F^{-1}(p) + \frac{1}{1-p}\int_{F^{-1}(p)}^{\infty} \overline{F}(x)\mathrm{d}x \leq G^{-1}(p) + \frac{1}{1-p}\int_{G^{-1}(p)}^{\infty} \overline{G}(x)\mathrm{d}x, \quad p \in (0,1). \tag{4.A.8}$$

Condition (4.A.8) may be compared with (3.C.1); see also Corollary 4.A.32.

An important characterization of the increasing convex and the increasing concave orders by construction on the same probability space is stated next.

Theorem 4.A.5. *Two random variables X and Y satisfy $X \leq_{\text{icx}} Y$ [$X \leq_{\text{icv}} Y$] if, and only if, there exist two random variables $\hat{X}$ and $\hat{Y}$, defined on the same probability space, such that*

$$\hat{X} =_{\text{st}} X,$$
$$\hat{Y} =_{\text{st}} Y,$$

and $\{\hat{X}, \hat{Y}\}$ is a submartingale [$\{\hat{Y}, \hat{X}\}$ is a supermartingale], that is,

$$E[\hat{Y}|\hat{X}] \geq \hat{X} \quad [E[\hat{X}|\hat{Y}] \leq \hat{Y}] \quad \textit{almost surely.} \tag{4.A.9}$$

Furthermore, the random variables $\hat{X}$ *and* $\hat{Y}$ *can be selected such that* $[\hat{Y}|\hat{X} = x]$ $[[\hat{X}|\hat{Y} = x]]$ *is increasing in* x *in the usual stochastic order* $\leq_{\mathrm{st}}$.

The proof of this theorem is similar to the proof of Theorem 3.A.4. It is not easy to prove the constructive part of Theorem 4.A.5. However, it is easy to prove that if random variables $\hat{X}$ and $\hat{Y}$ as described in the theorem exist, then $X \leq_{\mathrm{icx}} Y$ $[X \leq_{\mathrm{icv}} Y]$. For example, if the first inequality in (4.A.9) holds and if ϕ is an increasing convex function, then, using Jensen's Inequality,

$$\begin{aligned} E[\phi(X)] = E[\phi(\hat{X})] &\leq E\{\phi(E[\hat{Y}|\hat{X}])\} \\ &\leq E\{E[\phi(\hat{Y})|\hat{X}]\} = E[\phi(\hat{Y})] = E[\phi(Y)], \end{aligned}$$

which is (4.A.1).

Theorem 4.A.6. (a) *Two random variables* X *and* Y *satisfy* $X \leq_{\mathrm{icx}} Y$ *if, and only if, there exists a random variable* Z *such that*

$$X \leq_{\mathrm{st}} Z \leq_{\mathrm{cx}} Y.$$

(b) *Two random variables* X *and* Y *satisfy* $X \leq_{\mathrm{icx}} Y$ *if, and only if, there exists a random variable* Z *such that*

$$X \leq_{\mathrm{cx}} Z \leq_{\mathrm{st}} Y.$$

(c) *Two random variables* X *and* Y *satisfy* $X \leq_{\mathrm{icv}} Y$ *if, and only if, there exists a random variable* Z *such that*

$$X \leq_{\mathrm{cv}} Z \leq_{\mathrm{st}} Y.$$

(d) *Two random variables* X *and* Y *satisfy* $X \leq_{\mathrm{icv}} Y$ *if, and only if, there exists a random variable* Z *such that*

$$X \leq_{\mathrm{st}} Z \leq_{\mathrm{cv}} Y.$$

Proof. First we prove part (a). It is obvious (see, for example, Theorem 4.A.34 below) that $X \leq_{\mathrm{st}} Z \leq_{\mathrm{cx}} Y \Longrightarrow X \leq_{\mathrm{icx}} Y$. So suppose that $X \leq_{\mathrm{icx}} Y$. Let $\hat{X}$ and $\hat{Y}$ be defined on the same probability space, as in Theorem 4.A.5. Define $\hat{Z} = E[\hat{Y}|\hat{X}]$. It is seen that $E[\hat{Y}|\hat{Z}] = E[\hat{Y}|\hat{X}] = \hat{Z}$. Thus, by Theorem 3.A.4, $\hat{Z} \leq_{\mathrm{cx}} \hat{Y}$. Also, by Theorem 4.A.5, $\hat{X} \leq \hat{Z}$, and therefore, by Theorem 1.A.1, $\hat{X} \leq_{\mathrm{st}} \hat{Z}$. Letting Z have the same distribution as $\hat{Z}$, we obtain the stated result.

Now we prove part (b). Again it is obvious that $X \leq_{\mathrm{cx}} Z \leq_{\mathrm{st}} Y \Longrightarrow X \leq_{\mathrm{icx}} Y$. So suppose that $X \leq_{\mathrm{icx}} Y$. Let $\hat{X}$ and $\hat{Y}$ be defined on the same probability space, as in Theorem 4.A.5. Let $\hat{Z} = \hat{Y} + \hat{X} - E[\hat{Y}|\hat{X}]$. Then, by Theorem 4.A.5, $\hat{Z} \leq \hat{Y}$, and therefore, by Theorem 1.A.1, $\hat{Z} \leq_{\mathrm{st}} \hat{Y}$. Also,

$E[\hat{Z}|\hat{X}] = \hat{X}$, and thus, by Theorem 3.A.4, $\hat{X} \le_{\rm cx} \hat{Z}$. Letting Z have the same distribution as $\hat{Z}$, we obtain the stated result.

Parts (c) and (d) can be proven similarly. Alternatively, using Theorem 4.A.1, part (c) can be obtained from part (a), and part (d) can be obtained from part (b). □

The following bivariate characterization of the orders $\le_{\rm icx}$ and $\le_{\rm icv}$ is analogous to Theorem 3.A.6. Its proof is similar to the proof of Theorem 3.A.6 and is therefore omitted. Define the following classes of bivariate functions:

$$\mathcal{G}_{\rm icx} = \{\phi : \mathbb{R}^2 \to \mathbb{R} : \phi(x,y) - \phi(y,x) \text{ is increasing and convex in } x \text{ for all } y\}$$

and

$$\mathcal{G}_{\rm icv} = \{\phi : \mathbb{R}^2 \to \mathbb{R} : \phi(x,y) - \phi(y,x) \text{ is increasing and concave in } x \text{ for all } y\}.$$

Theorem 4.A.7. *Let X and Y be independent random variables. Then $X \le_{\rm icx} Y$ $[X \le_{\rm icv} Y]$ if, and only if,*

$$E[\phi(X,Y)] \le E[\phi(Y,X)] \quad \textit{for all } \phi \in \mathcal{G}_{\rm icx} \ [\mathcal{G}_{\rm icv}].$$

Another characterization of the increasing convex order, by means of the number of sign changes of two distribution functions, is given in Theorem 4.A.23 below.

4.A.2 Closure properties and some characterizations

Using (4.A.1) through (4.A.9) it is easy to prove each of the closure results in the first two parts of the following theorem. The last two parts can be proven as in Theorem 3.A.12. (Recall from Section 1.A.3 that for any random variable Z and any event A we denote by $[Z|A]$ any random variable whose distribution is the conditional distribution of Z given A.)

Theorem 4.A.8. (a) *If $X \le_{\rm icx} Y$ $[X \le_{\rm icv} Y]$ and g is any increasing and convex [concave] function, then $g(X) \le_{\rm icx} [\le_{\rm icv}]\ g(Y)$.*

(b) *Let X, Y, and Θ be random variables such that $[X|\Theta = \theta] \le_{\rm icx} [\le_{\rm icv}]$ $[Y|\Theta = \theta]$ for all θ in the support of Θ. Then $X \le_{\rm icx} [\le_{\rm icv}]\ Y$. That is, the increasing convex [concave] order is closed under mixtures.*

(c) *Let $\{X_j,\ j = 1, 2, \dots\}$ and $\{Y_j,\ j = 1, 2, \dots\}$ be two sequences of random variables such that $X_j \to_{\rm st} X$ and $Y_j \to_{\rm st} Y$ as $j \to \infty$. Assume that EX_+ $[EX_-]$ and EY_+ $[EY_-]$ are finite and that*

$$E(X_j)_+ \to EX_+ \ [E(X_j)_- \to EX_-] \quad \textit{and} \quad E(Y_j)_+ \to EY_+ \ [E(Y_j)_- \to EY_-] \quad \textit{as } j \to \infty. \tag{4.A.10}$$

If $X_j \le_{\rm icx} [\le_{\rm icv}]\ Y_j$, $j = 1, 2, \dots$, then $X \le_{\rm icx} [\le_{\rm icv}]\ Y$.

(d) *Let $X_1, X_2, \ldots, X_m$ be a set of independent random variables and let $Y_1, Y_2, \ldots, Y_m$ be another set of independent random variables. If $X_i \le_{\text{icx}}$ $[\le_{\text{icv}}]$ Y_i for $i = 1, 2, \ldots, m$, then*

$$\sum_{j=1}^{m} X_j \le_{\text{icx}} [\le_{\text{icv}}] \sum_{j=1}^{m} Y_j.$$

That is, the increasing convex [concave] order is closed under convolutions.

In part (c), as in Theorem 3.A.12, the condition (4.A.10) is necessary — without it the conclusion of part (c) may not hold.

Part (d) of Theorem 4.A.8 can be strengthened as follows.

Theorem 4.A.9. *Let $X_1, X_2, \ldots$ and $Y_1, Y_2, \ldots$ each be a sequence of nonnegative independent and identically distributed random variables such that $X_i \le_{\text{icx}} [\le_{\text{icv}}] Y_i$, $i = 1, 2, \ldots$. Let M and N be positive integer-valued random variables that are independent of the $\{X_i\}$ and the $\{Y_i\}$ sequences, respectively, such that $M \le_{\text{icx}} [\le_{\text{icv}}] N$. Then*

$$\sum_{j=1}^{M} X_j \le_{\text{icx}} [\le_{\text{icv}}] \sum_{j=1}^{N} Y_j.$$

Proof. Let ϕ be an increasing convex [concave] function and denote $g(n) \equiv E[\phi(X_1 + X_2 + \cdots + X_n)]$. Clearly $g(n)$ increases in n. Denote $S_n = X_1 + X_2 + \cdots + X_n$ for $n \ge 1$. Now, $E[\phi(S_n + X_{n+1}) - \phi(S_n) \big| S_n = s] = E[\phi(s + X_{n+1}) - \phi(s)] = h(s)$, say. Since ϕ is convex [concave] it follows that $h(s)$ is increasing [decreasing] in s. Since S_n is increasing in n in the usual stochastic order, it follows that $g(n+1) - g(n) = E[h(S_n)]$ is increasing [decreasing] in n. That is, $g(n)$ is increasing and convex [concave] in n. Therefore

$$E\Big[\phi\Big(\sum_{i=1}^{M} X_i\Big)\Big] \le E\Big[\phi\Big(\sum_{i=1}^{N} X_i\Big)\Big],$$

that is,

$$\sum_{i=1}^{M} X_i \le_{\text{icx}} [\le_{\text{icv}}] \sum_{i=1}^{N} X_i. \tag{4.A.11}$$

From Theorem 4.A.8 (b) and (d) it follows that

$$\sum_{i=1}^{N} X_i \le_{\text{icx}} [\le_{\text{icv}}] \sum_{i=1}^{N} Y_i,$$

and the proof is complete by the transitivity property of the order $\le_{\text{icx}}$ $[\le_{\text{icv}}]$. □

A special case of Theorem 4.A.9 is stated, and proven in a different manner, in Chapter 8 (see Theorem 8.A.13).

Remark 4.A.10. If in Theorem 4.A.9 the X_i's are only assumed to be increasing [decreasing] in i in the increasing convex [concave] order (rather than being identically distributed), or if the same is assumed about the Y_i's, then the conclusion of the theorem is still true.

As a special case of the result mentioned in Remark 4.A.10 we obtain the following theorem.

Theorem 4.A.11. *Let $\{X_i,\ i=1,2,\dots\}$ be a sequence of nonnegative independent random variables such that $X_i \le_{\rm st} X_{i+1}$, $i=1,2,\dots$. Let M and N be two discrete positive integer-valued random variables such that $M \le_{\rm icx} N$, and assume that M and N are independent of the X_i's. Then*

$$\sum_{i=1}^{M} X_i \le_{\rm icx} \sum_{i=1}^{N} X_i.$$

The following result follows easily from Theorem 4.A.9. It is of interest to compare it to Theorems 1.A.5, 2.B.8, and 3.A.14.

Theorem 4.A.12. *Let $\{X_j,\ j=1,2,\dots\}$ be a sequence of nonnegative independent and identically distributed random variables, and let M be a positive integer-valued random variable which is independent of the X_i's. Let $\{Y_j,\ j=1,2,\dots\}$ be another sequence of independent and identically distributed random variables, and let N be a positive integer-valued random variable which is independent of the Y_i's. Suppose that for some positive integer K we have*

$$\sum_{i=1}^{K} X_i \le_{\rm icx} [\ge_{\rm icx}, \le_{\rm icv}, \ge_{\rm icv}]\ Y_1,$$

and

$$M \le_{\rm icx} [\ge_{\rm icx}, \le_{\rm icv}, \ge_{\rm icv}]\ KN.$$

Then

$$\sum_{j=1}^{M} X_j \le_{\rm icx} [\ge_{\rm icx}, \le_{\rm icv}, \ge_{\rm icv}] \sum_{j=1}^{N} Y_j.$$

Proof. The assumptions yield

$$\begin{aligned}\sum_{i=1}^{M} X_i &\le_{\rm icx} [\ge_{\rm icx}, \le_{\rm icv}, \ge_{\rm icv}] \sum_{i=1}^{KN} X_i\\ &= \sum_{i=1}^{N} \sum_{j=K(i-1)+1}^{Ki} X_j \le_{\rm icx} [\ge_{\rm icx}, \le_{\rm icv}, \ge_{\rm icv}] \sum_{i=1}^{N} Y_i,\end{aligned}$$

where the inequalities follow from Theorem 4.A.9. This gives the stated result. □

Some results that are related to Theorem 4.A.12 are given in the next theorem.

Theorem 4.A.13. *Let $\{X_j,\ j = 1, 2, \ldots\}$ be a sequence of nonnegative independent and identically distributed random variables, and let M be a positive integer-valued random variable which is independent of the X_i's. Let $\{Y_j,\ j = 1, 2, \ldots\}$ be another sequence of independent and identically distributed random variables, and let N be a positive integer-valued random variable which is independent of the Y_i's. Also, let $\{N_j, j = 1, 2, \ldots\}$ be a sequence of independent random variables that are distributed as N. If for some positive integer K we have*

$$\sum_{i=1}^{K} X_i \le_{\text{icx}} Y_1 \quad \textit{and} \quad M \le_{\text{icx}} \sum_{i=1}^{K} N_i, \tag{4.A.12}$$

or if we have

$$KX_1 \le_{\text{icx}} Y_1 \quad \textit{and} \quad M \le_{\text{icx}} KN, \tag{4.A.13}$$

or if we have

$$KX_1 \le_{\text{icx}} Y_1 \quad \textit{and} \quad M \le_{\text{icx}} \sum_{i=1}^{K} N_i, \tag{4.A.14}$$

then

$$\sum_{j=1}^{M} X_j \le_{\text{icx}} \sum_{j=1}^{N} Y_j. \tag{4.A.15}$$

Proof. Assume that (4.A.13) holds. Then

$$\sum_{i=1}^{M} X_i \le_{\text{icx}} \sum_{i=1}^{KN} X_i = \sum_{i=1}^{N} \sum_{j=K(i-1)+1}^{Ki} X_j \le_{\text{cx}} \sum_{i=1}^{N} KX_i \le_{\text{icx}} \sum_{i=1}^{N} Y_i,$$

where the first and the third inequalities follow from Theorem 4.A.9, and the second inequality follows from Theorem 3.A.13 and Example 3.A.29. This gives (4.A.15).

Next note, using Example 3.A.29, that $\sum_{i=1}^{K} N_i \le_{\text{icx}} KN$. Thus, by Theorem 4.A.12, the conditions in (4.A.12) imply (4.A.15), and, by (4.A.13), the conditions in (4.A.14) imply (4.A.15). □

A slight generalization of the conditions in (4.A.12) is given in the next theorem.

Theorem 4.A.14. *Let $\{X_j,\ j=1,2,\dots\}$ be a sequence of nonnegative independent and identically distributed random variables, and let M be a positive integer-valued random variable which is independent of the X_i's. Let $\{Y_j,\ j=1,2,\dots\}$ be another sequence of independent and identically distributed random variables, and let N be a positive integer-valued random variable which is independent of the Y_i's. If for some positive integers K_1 and K_2, such that $K_1 \le K_2$, we have*

$$\sum_{i=1}^{K_1} X_i \le_{\rm icx} \frac{K_1}{K_2} Y_1 \quad \text{and} \quad M \le_{\rm icx} K_2 N,$$

then

$$\sum_{j=1}^{M} X_j \le_{\rm icx} \sum_{j=1}^{N} Y_j.$$

Proof. The first assumption and Example 3.A.29 yield

$$K_1 \cdot \frac{\sum_{i=1}^{K_2} X_i}{K_2} \le_{\rm cx} K_1 \cdot \frac{\sum_{i=1}^{K_1} X_i}{K_1} \le_{\rm icx} \frac{K_1}{K_2} Y_1;$$

that is, $\sum_{i=1}^{K_2} X_i \le_{\rm icx} Y_1$. The result now follows from Theorem 4.A.12. □

Parts (a) and (d) of Theorem 4.A.8 can be generalized as follows.

Theorem 4.A.15. *Let $X_1, X_2, \dots, X_m$ be a set of independent random variables and let $Y_1, Y_2, \dots, Y_m$ be another set of independent random variables. If $X_i \le_{\rm icx} Y_i$ for $i = 1, 2, \dots, m$, then*

$$g(X_1, X_2, \dots, X_m) \le_{\rm icx} g(Y_1, Y_2, \dots, Y_m) \tag{4.A.16}$$

for every increasing and componentwise convex function g.

Proof. Without loss of generality we can assume that all the $2m$ random variables are independent because such an assumption does not affect the distributions of $g(X_1, X_2, \dots, X_m)$ and $g(Y_1, Y_2, \dots, Y_m)$. The proof is by induction on m. For $m = 1$ the result is just Theorem 4.A.8(a). Assume that (4.A.16) is true for vectors of size $m-1$. Let g and ϕ be increasing and componentwise convex functions. Then

$$\begin{aligned} E[\phi(g(X_1, X_2, \dots, X_m))\big|X_1 = x] &= E[\phi(g(x, X_2, \dots, X_m))] \\ &\le E[\phi(g(x, Y_2, \dots, Y_m))] \\ &= E[\phi(g(X_1, Y_2, \dots, Y_m))\big|X_1 = x], \end{aligned}$$

where the equalities above follow from the independence assumption and the inequality follows from the induction hypothesis. Taking expectations with respect to X_1, we obtain

$$E[\phi(g(X_1, X_2, \ldots, X_m))] \le E[\phi(g(X_1, Y_2, \ldots, Y_m))].$$

Repeating the argument, but now conditioning on $Y_2, \ldots, Y_m$ and using (4.A.16) with $m = 1$, we see that

$$E[\phi(g(X_1, Y_2, \ldots, Y_m))] \le E[\phi(g(Y_1, Y_2, \ldots, Y_m))],$$

and this proves the result. □

From Theorem 4.A.15 we obtain the following corollary.

Corollary 4.A.16. *Let $X_1, X_2, \ldots, X_m$ be a set of independent random variables and let $Y_1, Y_2, \ldots, Y_m$ be another set of independent random variables. If $X_i \le_{\rm icx} Y_i$ for $i = 1, 2, \ldots, m$, then*

$$\max\{X_1, X_2, \ldots, X_m\} \le_{\rm icx} \max\{Y_1, Y_2, \ldots, Y_m\}.$$

From Corollary 4.A.16 and Theorem 4.A.1 it is easy to see that if $X_1, X_2, \ldots, X_m$ are independent random variables, and if $Y_1, Y_2, \ldots, Y_m$ are independent random variables, and if $X_i \le_{\rm icv} Y_i$ for $i = 1, 2, \ldots, m$, then

$$\min\{X_1, X_2, \ldots, X_m\} \le_{\rm icv} \min\{Y_1, Y_2, \ldots, Y_m\}.$$

A comparison of maxima of two partial sums in the increasing convex order is given next. Recall from (3.A.54) the definition of negatively associated random variables.

Theorem 4.A.17. *Let $X_1, X_2, \ldots, X_n$ be negatively associated random variables, and let $Y_1, Y_2, \ldots, Y_n$ be independent random variables such that $X_i =_{\rm st} Y_i$, $i = 1, 2, \ldots, n$. Then*

$$\max_{1 \le k \le n} \sum_{i=1}^{k} X_i \le_{\rm icx} \max_{1 \le k \le n} \sum_{i=1}^{k} Y_i.$$

Theorem 4.A.17 follows from Theorem 9.A.23 in Chapter 9; see a comment there after that theorem.

Consider now a family of distribution functions $\{G_\theta,\ \theta \in \mathcal{X}\}$ where $\mathcal{X}$ is a convex subset (that is, an interval) of the real line or of $\mathbb{N}$. As in Section 1.A.3 let $X(\theta)$ denote a random variable with distribution function G_θ. For any random variable Θ with support in $\mathcal{X}$, and with distribution function F, let us denote by $X(\Theta)$ a random variable with distribution function H given by

$$H(y) = \int_{\mathcal{X}} G_\theta(y) \mathrm{d}F(\theta), \quad y \in \mathbb{R}.$$

The following result generalizes Theorem 4.A.8(a), just as Theorem 1.A.6 generalized Theorem 1.A.3(a).

Theorem 4.A.18. *Consider a family of distribution functions $\{G_\theta,\ \theta \in \mathcal{X}\}$ as above. Let Θ_1 and Θ_2 be two random variables with supports in $\mathcal{X}$ and distribution functions F_1 and F_2, respectively. Let Y_1 and Y_2 be two random variables such that $Y_i =_{\rm st} X(\Theta_i)$, $i = 1, 2$; that is, suppose that the distribution function of Y_i is given by*

$$H_i(y) = \int_{\mathcal{X}} G_\theta(y) \mathrm{d}F_i(\theta), \quad y \in \mathbb{R},\ i = 1, 2.$$

If for every increasing convex [concave] function ϕ

$$E[\phi(X(\theta))] \quad \text{is increasing and convex [concave] in } \theta, \tag{4.A.17}$$

and if

$$\Theta_1 \le_{\rm icx} [\le_{\rm icv}]\ \Theta_2, \tag{4.A.18}$$

then

$$Y_1 \le_{\rm icx} [\le_{\rm icv}]\ Y_2. \tag{4.A.19}$$

Proof. Select an increasing convex [concave] function ϕ for which the expectations below exist, denote

$$\psi(\theta) = E[\phi(X(\theta))], \quad \theta \in \mathcal{X},$$

and notice that ψ is increasing and convex [concave] by (4.A.17). Then

$$E[\phi(Y_1)] = E[\psi(\Theta_1)] \le E[\psi(\Theta_2)] = [E[\phi(Y_2)],$$

where the inequality follows from (4.A.18). This gives (4.A.19). □

Note that (4.A.11) can be easily obtained from the result above. It is worth mentioning also that condition (4.A.17) is weaker than the condition $\{X(\theta),\ \theta \in \mathcal{X}\} \in$ SICX [SICV] which is studied in Section 8.A of Chapter 8. An extension of Theorem 4.A.18 is given as Theorem 4.A.65 below.

The following example illustrates the use of Theorem 4.A.18. It may be compared to Corollary 3.A.22.

Example 4.A.19. Let U, Θ_1, and Θ_2 be independent positive random variables. Define

$$Y_1 = \frac{U}{\Theta_1} \quad \text{and} \quad Y_2 = \frac{U}{\Theta_2}.$$

If $\Theta_1 \le_{\rm icv} [\le_{\rm icx}]\ \Theta_2$, then $Y_1 \ge_{\rm icx} [\ge_{\rm icv}]\ Y_2$. This can be proven by a simple application of Theorems 4.A.18 and 4.A.1.

An interesting variation of Theorem 4.A.18 is the following. Its proof is similar to the proof of Theorem 4.A.18 and is therefore omitted.

Theorem 4.A.20. *Consider a family of distribution functions $\{G_\theta,\ \theta \in \mathcal{X}\}$ as described before Theorem 4.A.18. Let Θ_1 and Θ_2 be two random variables with supports in $\mathcal{X}$ and distribution functions F_1 and F_2, respectively. Let Y_1 and Y_2 be two random variables such that $Y_i =_{\rm st} X(\Theta_i)$, $i = 1, 2$; that is, suppose that the distribution function of Y_i is given by*

$$H_i(y) = \int_{\mathcal{X}} G_\theta(y) \mathrm{d}F_i(\theta), \quad y \in \mathbb{R},\ i = 1, 2.$$

If for every increasing convex [concave] function ϕ

$$E[\phi(X(\theta))] \quad \textit{is increasing in } \theta,$$

and if

$$\Theta_1 \le_{\rm st} \Theta_2,$$

then

$$Y_1 \le_{\rm icx} [\le_{\rm icv}]\ Y_2.$$

A Laplace transform characterization of the orders $\le_{\rm icx}$ and $\le_{\rm icv}$ is given next; it may be compared to Theorems 1.A.13, 1.B.18, 1.B.53, 1.C.25, 2.A.16, and 2.B.14.

Theorem 4.A.21. *Let X_1 and X_2 be two nonnegative random variables, and let $N_\lambda(X_1)$ and $N_\lambda(X_2)$ be as described in Theorem 1.A.13. Then*

$$X_1 \le_{\rm icx} [\le_{\rm icv}]\ X_2 \Longleftrightarrow N_\lambda(X_1) \le_{\rm icx} [\le_{\rm icv}]\ N_\lambda(X_2) \quad \textit{for all } \lambda > 0.$$

Proof. First assume that $X_1 \le_{\rm icx} [\le_{\rm icv}]\ X_2$. For $k = 1, 2$, denote the distribution function of X_k by F_k. Let ϕ be an increasing convex [concave] function. Without loss of generality assume that $\phi(0) = 0$. Then, from (2.A.16) we have that

$$E[\phi(X_k)] = \int_0^\infty \sum_{n=1}^\infty \phi(n) \mathrm{e}^{-\lambda x} \frac{(\lambda x)^n}{n!} \mathrm{d}F_k(x),$$

and therefore it is seen that it suffices to show that

$$g(x) \equiv \sum_{n=1}^\infty \phi(n) \mathrm{e}^{-\lambda x} \frac{(\lambda x)^n}{n!}$$

is increasing and convex [concave] in x. Now compute

$$\begin{aligned} g'(x) &= \sum_{n=1}^\infty \phi(n) \lambda \mathrm{e}^{-\lambda x} \left[\frac{(\lambda x)^{n-1}}{(n-1)!} - \frac{(\lambda x)^n}{n!} \right] \\ &= \lambda \sum_{n=0}^\infty [\phi(n+1) - \phi(n)] \mathrm{e}^{-\lambda x} \frac{(\lambda x)^n}{n!}. \end{aligned}$$

If we denote $\Delta_\phi(n) \equiv \phi(n+1) - \phi(n)$, then it is seen that

$$g'(x) = \lambda E\{\Delta_\phi[N(x)]\},$$

where $\{N(x), x \geq 0\}$ is a Poisson process with rate λ. Since $\Delta_\phi(n) \geq 0$, by the monotonicity of ϕ, it follows that $g'(x) \geq 0$. Also, since $\Delta_\phi(n) \uparrow [\downarrow]\ n$ by the convexity [concavity] of ϕ, and since $N(x) \uparrow_{\rm st} x$, it follows that $g'(x) \uparrow [\downarrow]\ x$. Therefore g is increasing and convex [concave].

Now suppose that $N_\lambda(X_1) \leq_{\rm icx} N_\lambda(X_2)$ for all $\lambda > 0$, that is, using the notation of the proof of Theorem 2.A.16,

$$\sum_{n=m}^{\infty} \overline{\alpha}_{\lambda,1}(n) \leq \sum_{n=m}^{\infty} \overline{\alpha}_{\lambda,2}(n), \quad m = 0, 1, 2, \ldots.$$

Then for $m \geq 2$, (2.A.23) yields

$$\int_0^\infty \lambda e^{-\lambda u} \frac{(\lambda u)^{m-2}}{(m-2)!} \left[\int_u^\infty \overline{F}_1(x) dx \right] du \leq \int_0^\infty \lambda e^{-\lambda u} \frac{(\lambda u)^{m-2}}{(m-2)!} \left[\int_u^\infty \overline{F}_2(x) dx \right] du.$$

For any fixed $y > 0$ set $\lambda = (m-1)/y$. It follows that as $m \to \infty$ (then $\lambda \to \infty$),

$$\int_0^\infty \lambda e^{-\lambda u} \frac{(\lambda u)^{m-2}}{(m-2)!} \left[\int_u^\infty \overline{F}_k(x) dx \right] du \to \int_y^\infty \overline{F}_k(x) dx, \quad k = 1, 2.$$

Therefore we obtain

$$\int_y^\infty \overline{F}_1(x) dx \leq \int_y^\infty \overline{F}_2(x) dx, \quad y > 0,$$

that is $X_1 \leq_{\rm icx} X_2$ (see (4.A.5)). The proof of the converse for the $\leq_{\rm icv}$ order is similar. □

The implication $\Longrightarrow$ in Theorem 4.A.21 can be generalized in the same manner that Theorem 1.A.14 generalizes the implication $\Longrightarrow$ in Theorem 1.A.13. We will not state the result here since it is equivalent to Theorem 4.A.18.

4.A.3 Conditions that lead to the increasing convex and increasing concave orders

Once the relation $X \leq_{\rm icx} Y$ or the relation $X \leq_{\rm icv} Y$ has been established between the two random variables X and Y, it can be of great use. However, given the two random variables and their distribution functions it is sometimes not clear how to verify that $X \leq_{\rm icx} Y$ or that $X \leq_{\rm icv} Y$. Parallel to the analysis in Section 3.A.3 we point out here some simple conditions that imply the increasing convex and the increasing concave orders.

Theorem 4.A.22. *Let X and Y be two random variables with distribution functions F and G and survival functions $\overline{F}$ and $\overline{G}$, respectively, and with finite means such that $EX \le EY$.*

(a) *If $S^-(\overline{F}-\overline{G}) \le 1$ and the sign sequence is $+,-$ $[-,+]$ when equality holds, then $X \le_{\rm icx} Y$ $[X \le_{\rm icv} Y]$.*
(b) *If $S^-(G-F) \le 1$ and the sign sequence is $+,-$ $[-,+]$ when equality holds, then $X \le_{\rm icx} Y$ $[X \le_{\rm icv} Y]$.*

The proof of this theorem is similar to the proof of Theorem 3.A.44 and is not detailed here.

The condition in part (a) (or, equivalently, in part (b)) of Theorem 4.A.22 is not only sufficient for $X \le_{\rm icx} Y$, but, for nonnegative random variable, it can also characterize the increasing convex order in a similar manner in which (3.A.58) (or, equivalently, (3.A.59)) characterizes the convex order in Theorem 3.A.45. This is stated next.

Theorem 4.A.23. *Let X and Y be two nonnegative random variables such that $EX \le EY$. Then $X \le_{\rm icx}$ $[\le_{\rm icv}]$ Y if, and only if, there exist random variables $Z_1, Z_2, \ldots,$ with distribution functions $F_1, F_2, \ldots,$ such that $Z_1 =_{\rm st} X$, $EZ_j \le EY$, $j = 1, 2, \ldots,$ $Z_j \to_{\rm st} Y$ as $j \to \infty$, $EZ_j \to EY$ as $j \to \infty$, and $S^-(\overline{F}_j - \overline{F}_{j+1}) = 1$ and the sign sequence is $+,-$ $[-,+]$, $j = 1, 2, \ldots$.*

If the random variables in Theorem 4.A.23 are not nonnegative, then the sufficiency part of that theorem is not correct. This follows from the remark after Theorem 3.A.45.

An interesting characterization of the mean residual life order by means of the increasing convex order is the following result.

Theorem 4.A.24. *Let X and Y be two random variables. Then $X \le_{\rm mrl} Y$ if, and only if,*

$$[X - s \,|\, X > s] \le_{\rm icx} [Y - s \,|\, Y > s] \quad \text{for all } s. \tag{4.A.20}$$

Proof. Let $\overline{F}$ and $\overline{G}$ be the survival functions of X and Y, respectively. Condition (4.A.20) can be written as

$$\frac{\int_t^\infty \overline{F}(s+u)\mathrm{d}u}{\overline{F}(s)} \le \frac{\int_t^\infty \overline{G}(s+u)\mathrm{d}u}{\overline{G}(s)} \quad \text{for all } s \text{ and all } t \ge 0,$$

which is equivalent to $X \le_{\rm mrl} Y$ by (2.A.6). □

Remark 4.A.25. Let ϕ be an increasing convex function. For any s let s' be selected such that $\phi(s') = s$. Note that if (4.A.20) holds, then $[X \,|\, X > s'] \le_{\rm icx} [Y \,|\, Y > s']$. Therefore $E[\phi(X) \,|\, X > s'] \le E[\phi(Y) \,|\, Y > s']$, and therefore $E[\phi(X) - s \,|\, \phi(X) > s] \le E[\phi(Y) - s \,|\, \phi(Y) > s]$. Thus we have proven that if $X \le_{\rm mrl} Y$, then $\phi(X) \le_{\rm mrl} \phi(Y)$ for every increasing convex function ϕ.

From Theorem 4.A.24 we see that if $X \le_{\rm mrl} Y$, then $[X|X > s] \le_{\rm icx} [Y|Y > s]$ for all s. Letting $s \to -\infty$ we obtain from Theorem 4.A.8(c) the following result.

Theorem 4.A.26. *Let X and Y be two random variables with finite means. If $X \le_{\rm mrl} Y$, then $X \le_{\rm icx} Y$.*

An analog of Theorem 4.A.26 for the increasing concave order is the following result.

Theorem 4.A.27. *Let X and Y be two random variables with finite means. If*

$$E[X|X \le x] \le E[Y|Y \le x] \quad \text{for all } x \in \mathbb{R},$$

then $X \le_{\rm icv} Y$.

For positive random variables we have a result that is stronger than Theorem 4.A.26:

Theorem 4.A.28. *Let X and Y be two almost surely positive random variables with finite means. If $X \le_{\rm hmrl} Y$, then $X \le_{\rm icx} Y$.*

Proof. Let $\overline{F}$ and $\overline{G}$ be the survival functions of X and Y, respectively. From (2.B.4) (or, equivalently, from (2.B.2)) it follows that

$$\frac{\int_t^\infty \overline{F}(u)\mathrm{d}u}{EX} \le \frac{\int_t^\infty \overline{G}(u)\mathrm{d}u}{EY} \quad \text{for all } t \ge 0. \tag{4.A.21}$$

Since, for almost surely positive random variables, $X \le_{\rm hmrl} Y$ implies that $EX \le EY$ (see (2.B.6)), it follows that (4.A.5) holds. □

Remark 4.A.29. With the help of Theorem 4.A.28 we can now provide proofs for Theorems 2.A.15 and 2.B.13.

First we prove Theorem 2.A.15. From (2.A.3) it is seen that assumption (2.A.11) means that $\int_y^\infty \overline{G}_\theta(u)\mathrm{d}u$, as a function of θ and of y, is TP_2, where $\overline{G}_\theta$ is the survival function associated with G_θ. Assumption (2.A.12) means that $\overline{F}_i(\theta)$, as a function of $i \in \{1,2\}$ and of θ, is TP_2. From Theorem 4.A.28 and (4.A.5) it follows that $\int_y^\infty \overline{G}_\theta(u)\mathrm{d}u$ is increasing in θ. Therefore, by Theorem 2.1(i) of Lynch, Mimmack, and Proschan [369], $\int_{\mathcal{X}} \int_y^\infty \overline{G}_\theta(u)\mathrm{d}u\, \mathrm{d}F_i(\theta)$ is TP_2 in $i \in \{1,2\}$ and y. But $\int_{\mathcal{X}} \int_y^\infty \overline{G}_\theta(u)\mathrm{d}u\, \mathrm{d}F_i(\theta) = \int_y^\infty \big[\int_{\mathcal{X}} \overline{G}_\theta(u)\mathrm{d}F_i(\theta)\big]\mathrm{d}u$, and that, by (2.A.3), gives (2.A.13).

Next we prove Theorem 2.B.13. Fix an $x > 0$. From (2.B.2) it is seen that assumption (2.B.15) implies that $\int_y^\infty \overline{G}_\theta(u)\mathrm{d}u$ is TP_2 in $y \in \{0,x\}$ and θ, where $\overline{G}_\theta$ is the survival function associated with G_θ. Assumption (2.B.16) means that $\overline{F}_i(\theta)$, as a function of $i \in \{1,2\}$ and of θ, is TP_2. From Theorem 4.A.28 and (4.A.5) it follows that $\int_y^\infty \overline{G}_\theta(u)\mathrm{d}u$ is increasing in θ. Therefore, by Theorem 2.1(i) of Lynch, Mimmack, and Proschan [369], $\int_{\mathcal{X}} \int_y^\infty \overline{G}_\theta(u)\mathrm{d}u\, \mathrm{d}F_i(\theta)$ is TP_2 in $i \in \{1,2\}$ and $y \in \{0,x\}$. But $\int_{\mathcal{X}} \int_y^\infty \overline{G}_\theta(u)\mathrm{d}u\, \mathrm{d}F_i(\theta) = \int_y^\infty \big[\int_{\mathcal{X}} \overline{G}_\theta(u)\mathrm{d}F_i(\theta)\big]\mathrm{d}u$ and this expression is TP_2 in $i \in \{1,2\}$ and $y \in \{0,x\}$ for all $x > 0$. Thus, by (2.B.2), we obtain (2.B.17).

Under quite weak conditions the order $\leq_{\rm dil}$ implies the order $\leq_{\rm icx}$. This is shown in the next theorem. For any random variable Z, let l_Z denote the left endpoint of the support of Z.

Theorem 4.A.30. *Let X and Y be two random variables with finite means. If*

$$l_X \leq l_Y \tag{4.A.22}$$

and if $X \leq_{\rm dil} Y$, then $X \leq_{\rm icx} Y$.

Proof. Suppose that $X \leq_{\rm dil} Y$. Then

$$[X - EX] \leq_{\rm cx} [Y - EY]. \tag{4.A.23}$$

Therefore, by (3.A.12) we get that $\text{supp}(X - EX) \subseteq \text{supp}(Y - EY)$. Thus $l_Y - EY \leq l_X - EX$. Hence,

$$EY - EX \geq l_Y - l_X. \tag{4.A.24}$$

Combining (4.A.22) with (4.A.24) it is seen that

$$EX \leq EY. \tag{4.A.25}$$

From (4.A.23) it follows that

$$X \leq_{\rm cx} Y - (EY - EX), \tag{4.A.26}$$

and from (4.A.25) it follows that

$$Y - (EY - EX) \leq_{\rm st} Y. \tag{4.A.27}$$

Using Theorem 4.A.6(b) it is seen that, from (4.A.26) and (4.A.27), we obtain $X \leq_{\rm icx} Y$. It is also easy to obtain $X \leq_{\rm icx} Y$ from (4.A.26) and (4.A.27) by noticing that the usual stochastic order and the convex order both imply the increasing convex order. □

As a corollary of Theorem 4.A.30 we obtain the following result.

Corollary 4.A.31. *Let X and Y be two nonnegative random variables with finite means, such that X has the support $[0, \infty)$. If $X \leq_{\rm dil} Y$, then $X \leq_{\rm icx} Y$.*

A corollary of Theorem 4.A.30 and of (3.C.7) is the following result.

Corollary 4.A.32. *Let X and Y be two random variables with finite means. If $l_X \leq l_Y$ and if $X \leq_{\rm ew} Y$, then $X \leq_{\rm icx} Y$.*

The next result gives a simple condition that implies the increasing convex order between a given random variable and a scale transformation of another random variable. Let $X_1, X_2, \ldots$ be a sequence of independent and identically distributed nonnegative random variables with a common distribution function F, and let $Y_1, Y_2, \ldots$ be another sequence of independent and identically distributed nonnegative random variables with a common distribution function G. Let $X_{(n)} \equiv \max\{X_1, X_2, \ldots, X_n\}$ be the nth order statistic of a sample of size n from the distribution F, $n = 1, 2, \ldots$. Let $Y_{(n)}$ be similarly defined for $n = 1, 2, \ldots$. Note that from Corollary 4.A.16 it follows that if $X_1 \le_{\rm icx} Y_1$, then $X_{(n)} \le_{\rm icx} Y_{(n)}$ for all $n = 1, 2, \ldots$. The following theorem is a weak converse of this observation. The proof is not given here.

Theorem 4.A.33. *Let $X_1, X_2, \ldots$ be a sequence of independent and identically distributed nonnegative random variables and let $Y_1, Y_2, \ldots$ be another sequence of independent and identically distributed nonnegative random variables. If $E[X_{(n)}] \le E[Y_{(n)}]$ for all $n = 1, 2, \ldots$, then $X_1 \le_{\rm icx} \kappa Y_1$ for some constant $\kappa \ge 1$ that is independent of the distributions of X_1 and Y_1. The constant κ can be taken to be equal to $2(1 - {\rm e}^{-1})^{-1}$.*

4.A.4 Further properties

Let X and Y be two random variables. If $E[\phi(X)] \le E[\phi(Y)]$ for all increasing functions ϕ, then (4.A.1) definitely holds. If $E[\phi(X)] \le E[\phi(Y)]$ for all convex [concave] functions ϕ, then (4.A.1) also holds. From (1.A.7) and (3.A.1) we thus obtain the following result. Note that in the conclusion of the second part of (b) in the next theorem the random variables X and Y are interchanged.

Theorem 4.A.34. *Let X and Y be two random variables.*

(a) *If $X \le_{\rm st} Y$, then $X \le_{\rm icx} Y$ and $X \le_{\rm icv} Y$.*
(b) *If $X \le_{\rm cx} Y$, then $X \le_{\rm icx} Y$ and $Y \le_{\rm icv} X$.*

Thus we see that indeed the increasing convex [concave] order has both properties of ordering by size and ordering by variability. One indication of the ordering by size property is (4.A.2) [(4.A.3)], that is, the ordering of the expected values (when they exist) that follows from the increasing convex [concave] order. It turns out that the ordering of the expected values is actually the only indication of the ordering by size property. If the two means are equal, then the monotone convex and the monotone concave orders reduce to the convex order of Section 3.A. This is stated formally in the following theorem.

Theorem 4.A.35. *Let X and Y be two random variables with finite means.*

(a) *If $X \le_{\rm icx} Y$ and $EX = EY$, then $X \le_{\rm cx} Y$.*
(b) *If $X \le_{\rm icv} Y$ and $EX = EY$, then $Y \le_{\rm cx} X$.*

Proof. If $X \le_{\rm icx} Y$, then (4.A.5) (which is the same as (3.A.7)) holds. Part (a) now follows from Theorem 3.A.1(a). Part (b) is proven similarly using (4.A.7), (3.A.8), and Theorem 3.A.1(b). □

The order $\leq_{\rm icx}$ can be used to yield bivariate characterizations of the orders $\leq_{\rm st}$, $\leq_{\rm hr}$, $\leq_{\rm rh}$, and $\leq_{\rm lr}$ (compare the following result to Theorems 1.A.10, 1.B.10, 1.B.48, 1.C.22, and 1.C.23). Let ϕ_1 and ϕ_2 be two bivariate functions and let $\Delta\phi_{21}(x,y) = \phi_2(x,y) - \phi_1(x,y)$. Consider the following set of conditions on ϕ_1 and ϕ_2:

(a) $\Delta\phi_{21}(x,y) \geq -\Delta\phi_{21}(y,x)$ whenever $x \leq y$.
(b) $\Delta\phi_{21}(x,y) \geq 0$ whenever $x \leq y$.
(c) $\phi_1(y,x) \leq \phi_2(x,y)$ whenever $x \leq y$.
(d) For each x, $\phi_2(x,y)$ increases in y on $\{y \geq x\}$.
(e) For each y, $\phi_2(x,y)$ decreases in x on $\{x \leq y\}$.
(f) For each x, $\Delta\phi_{21}(x,y)$ increases in y on $\{y \geq x\}$.
(g) For each y, $\Delta\phi_{21}(x,y)$ decreases in x on $\{x \leq y\}$.

The proof of the next theorem is omitted.

Theorem 4.A.36. *Let X and Y be two independent random variables. Then*

(i) $X \leq_{\rm st} Y$ *if, and only if,*

$$\phi_1(X,Y) \leq_{\rm icx} \phi_2(X,Y) \tag{4.A.28}$$

for all ϕ_1 and ϕ_2 satisfying (a), (b), (c), (d), (e), (f), *and* (g).

(ii) $X \leq_{\rm hr} Y$ *if, and only if,* (4.A.28) *holds for all ϕ_1 and ϕ_2 satisfying* (a), (b), (c), (d), *and* (f).

(iii) $X \leq_{\rm rh} Y$ *if, and only if,* (4.A.28) *holds for all ϕ_1 and ϕ_2 satisfying* (a), (b), (c), (e), *and* (g).

(iv) $X \leq_{\rm lr} Y$ *if, and only if,* (4.A.28) *holds for all ϕ_1 and ϕ_2 satisfying* (a), (b), *and* (c).

A typical application of Theorem 4.A.36 is the following result (compare it to Theorem 1.C.21).

Theorem 4.A.37. *Let $X_1, X_2, \ldots, X_m$ be independent random variables such that $X_1 \leq_{\rm rh} X_2 \leq_{\rm rh} \cdots \leq_{\rm rh} X_m$. Let $a_1, a_2, \ldots, a_m$ be constants such that $a_1 \leq a_2 \leq \cdots \leq a_m$. Then*

$$\sum_{i=1}^m a_{m-i+1}X_i \leq_{\rm icv} \sum_{i=1}^m a_{\pi_i}X_i \leq_{\rm icv} \sum_{i=1}^m a_iX_i,$$

where $\boldsymbol{\pi} = (\pi_1, \pi_2, \ldots, \pi_m)$ denotes any permutation of $(1, 2, \ldots, m)$.

Proof. We only give the proof when $m = 2$; the general case then can be obtained by pairwise interchanges. So, suppose that $X_1 \leq_{\rm rh} X_2$ and that $a_1 \leq a_2$. Define ϕ_1 and ϕ_2 by $\phi_1(x,y) = -a_1x - a_2y$ and $\phi_2(x,y) = -a_1y - a_2x$. Then it is easy to verify that (a), (b), (c), (e), and (g) above hold. Thus, by Theorem 4.A.36(iii), $-a_1X_1 - a_2X_2 \leq_{\rm icx} -a_1X_2 - a_2X_1$. By Theorem 4.A.1 this means $a_1X_2 + a_2X_1 \leq_{\rm icv} a_1X_1 + a_2X_2$. □

In the next few results we denote by I_p a Bernoulli random variable with probability of success p, that is, $P\{I_p = 1\} = 1 - P\{I_p = 0\} = p$. Recall from page 2 the definition of the majorization order $\prec$ among n-dimensional vectors. It is shown after the next theorem that it partially extends Theorem 3.A.37.

Theorem 4.A.38. *Let $X_1, X_2, \ldots, X_n$ be independent nonnegative random variables, and let $I_{p_1}, I_{p_2}, \ldots, I_{p_n}$ and $I_{q_1}, I_{q_2}, \ldots, I_{q_n}$ be independent Bernoulli random variables that are independent of $X_1, X_2, \ldots, X_n$. Suppose that*

(i) $1 \geq p_1 \geq p_2 \geq \cdots \geq p_n$ *and* $1 \geq q_1 \geq q_2 \geq \cdots \geq q_n$,
(ii) $X_n \leq_{\text{st}} X_{n-1} \leq_{\text{st}} \cdots \leq_{\text{st}} X_1$, *and*
(iii) $\boldsymbol{p} \prec \boldsymbol{q}$.

Then

$$\sum_{i=1}^{n} I_{p_i} X_i \leq_{\text{icv}} \sum_{i=1}^{n} I_{q_i} X_i.$$

If $X_1, X_2, \ldots, X_n$ in Theorem 4.A.38 are identically distributed, then $E\big(\sum_{i=1}^{n} I_{p_i} X_i\big) = E\big(\sum_{i=1}^{n} I_{q_i} X_i\big)$ and therefore the conclusion in this theorem is $\sum_{i=1}^{n} I_{p_i} X_i \leq_{\text{cv}} \sum_{i=1}^{n} I_{q_i} X_i$; that is, $\sum_{i=1}^{n} I_{p_i} X_i \geq_{\text{cx}} \sum_{i=1}^{n} I_{q_i} X_i$. This is the same as the conclusion of Theorem 3.A.37.

The following result partially extends Theorem 3.A.35.

Theorem 4.A.39. *Let $X_1, X_2, \ldots, X_n$ be independent and identically distributed nonnegative random variables, and let $I_{p_1}, I_{p_2}, \ldots, I_{p_n}$ be independent Bernoulli random variables that are independent of $X_1, X_2, \ldots, X_n$. Let $\boldsymbol{a} = (a_1, a_2, \ldots, a_n)$ and $\boldsymbol{b} = (b_1, b_2, \ldots, b_n)$ be two vectors of constants. Suppose that*

(i) $1 \geq p_1 \geq p_2 \geq \cdots \geq p_n$,
(ii) $a_1 \geq a_2 \geq \cdots \geq a_n$ *and* $b_1 \geq b_2 \geq \cdots \geq b_n$, *and*
(iii) $\boldsymbol{a} \prec \boldsymbol{b}$.

Then

$$\sum_{i=1}^{n} I_{p_i} a_i X_i \leq_{\text{icx}} \sum_{i=1}^{n} I_{p_i} b_i X_i.$$

A family of nonnegative random variables $\{X(\theta),\ \theta > 0\}$ is said to have the semigroup property if, for all $\theta_1 > 0$ and $\theta_2 > 0$, one has $X(\theta_1 + \theta_2) =_{\text{st}} X(\theta_1) + X(\theta_2)$, where $X(\theta_1)$ and $X(\theta_2)$ are independent. As a corollary of Theorem 4.A.39 we obtain the following result.

Corollary 4.A.40. *Let $\{X(\theta),\ \theta > 0\}$ be a family of random variables with the semigroup property, and let $I_{p_1}, I_{p_2}, \ldots, I_{p_n}$ be independent Bernoulli random variables that are independent of $\{X(\theta),\ \theta > 0\}$. Let $\boldsymbol{\theta} = (\theta_1, \theta_2, \ldots, \theta_n)$ and $\boldsymbol{\gamma} = (\gamma_1, \gamma_2, \ldots, \gamma_n)$ be two vectors of constants. Suppose that*

(i) $1 \geq p_1 \geq p_2 \geq \cdots \geq p_n$,

(ii) $\theta_1 \geq \theta_2 \geq \cdots \geq \theta_n$ *and* $\gamma_1 \geq \gamma_2 \geq \cdots \geq \gamma_n$, *and*
(iii) $\boldsymbol{\theta} \prec \boldsymbol{\gamma}$.

Then

$$\sum_{i=1}^{n} I_{p_i} X(\theta_i) \leq_{\text{icx}} \sum_{i=1}^{n} I_{p_i} X(\gamma_i).$$

The following characterizations of the dilation order, by means of the order $\leq_{\text{icx}}$, are similar to characterizations (3.A.39) and (3.A.40).

Theorem 4.A.41. *Let X and Y be two random variables with distribution functions F and G, respectively, and with finite expectations. Then $X \leq_{\text{dil}} Y$ if, and only if, any of the following two statements hold:*

$$[X - EX \big| X \geq F^{-1}(p)] \leq_{\text{icx}} [Y - EY \big| Y \geq G^{-1}(p)] \quad \textit{for all } p \in [0,1),$$

and

$$[X - EX \big| X \leq F^{-1}(p)] \geq_{\text{icx}} [Y - EY \big| Y \leq G^{-1}(p)] \quad \textit{for all } p \in [0,1).$$

The following characterizations of the convex order, by means of the order $\leq_{\text{icx}}$, are similar to characterizations (3.A.41) and (3.A.42). These characterizations follow at once from Theorem 4.A.41 and from (3.A.32).

Theorem 4.A.42. *Let X and Y be two random variables with distribution functions F and G, respectively, and with equal finite means. Then $X \leq_{\text{cx}} Y$ if, and only if, any of the following two statements hold:*

$$[X \big| X \geq F^{-1}(p)] \leq_{\text{icx}} [Y \big| Y \geq G^{-1}(p)] \quad \textit{for all } p \in [0,1),$$

and

$$[X \big| X \leq F^{-1}(p)] \geq_{\text{icx}} [Y \big| Y \leq G^{-1}(p)] \quad \textit{for all } p \in [0,1).$$

In a manner similar to the characterization (3.B.6) of the dispersive order by the usual stochastic order, the increasing convex order can characterize the excess wealth order as follows.

Theorem 4.A.43. *Let X and Y be two continuous random variables with distribution functions F and G, respectively. Then $X \leq_{\text{ew}} Y$ if, and only if,*

$$(X - F^{-1}(\alpha))_+ \leq_{\text{icx}} (Y - G^{-1}(\alpha))_+, \quad \alpha \in (0,1). \tag{4.A.29}$$

Proof. We give the proof under the assumption that F and G are strictly increasing; the more general proof can be found in the literature.

First assume that (4.A.29) holds. Then, by (4.A.2) we get

$$E[(X - F^{-1}(\alpha))_+] \leq E[(Y - G^{-1}(\alpha))_+], \quad \alpha \in (0,1).$$

The latter inequality is easily seen to be equivalent to (3.C.5), and therefore $X \leq_{\text{ew}} Y$.

In order to obtain the converse note that (4.A.29) is equivalent to

$$H(t,\alpha) \equiv \int_{t+G^{-1}(\alpha)}^{\infty} \overline{G}(x)\mathrm{d}x - \int_{t+F^{-1}(\alpha)}^{\infty} \overline{F}(x)\mathrm{d}x \geq 0, \quad (t,\alpha) \in [0,\infty) \times (0,1).$$

Select an $\alpha \in (0,1)$. Note that $\lim_{t\to\infty} H(t,\alpha) = 0$. If $H(\cdot,\alpha)$ attains a minimum at t^*, since $H(\cdot,\alpha)$ is continuous and differentiable, t^* should satisfy $\frac{\partial H(t,\alpha)}{\partial t}\big|_{t=t^*} = 0$. This equality holds if, and only if,

$$F(t^* + F^{-1}(\alpha)) = F(t^* + G^{-1}(\alpha)) = \beta, \text{ say.}$$

Since F and G are strictly increasing it is seen that $F^{-1}(\beta) = t^* + F^{-1}(\alpha)$ and $G^{-1}(\beta) = t^* + G^{-1}(\alpha)$. Therefore

$$H(t^*,\alpha) = \int_{G^{-1}(\beta)}^{\infty} \overline{G}(x)\mathrm{d}x - \int_{F^{-1}(\beta)}^{\infty} \overline{F}(x)\mathrm{d}x \geq 0,$$

where the inequality follows from $X \leq_{\mathrm{ew}} Y$. □

Let X and Y be two nonnegative random variables with respective distribution functions F and G. Let H_F^{-1} and H_G^{-1} be the TTT transforms associated with F and G, respectively (see (1.A.19)), and let H_F and H_G be the respective inverses. Let X_{ttt} and Y_{ttt} be random variables with distribution functions H_F and H_G (see Section 1.A.4).

Theorem 4.A.44. *Let X and Y be two nonnegative random variables. Then*

$$X \leq_{\mathrm{icv}} Y \Longrightarrow X_{\mathrm{ttt}} \leq_{\mathrm{icv}} Y_{\mathrm{ttt}}.$$

See related results in Theorems 1.A.29, 3.B.1, 4.B.8, 4.B.9, and 4.B.29.

The next example may be compared with Examples 1.A.25, 1.B.6, and 1.C.51.

Example 4.A.45. Let X_i be a binomial random variable with parameters n_i and p_i, $i = 1, 2, \ldots, m$, and assume that the X_i's are independent. Let Y be a binomial random variable with parameters n and p where $n = \sum_{i=1}^{m} n_i$. Then

$$\sum_{i=1}^{m} X_i \geq_{\mathrm{icx}} Y \Longleftrightarrow p \leq \sqrt[n]{p_1^{n_1} p_2^{n_2} \cdots p_m^{n_m}},$$

and

$$\sum_{i=1}^{m} X_i \leq_{\mathrm{icx}} Y \Longleftrightarrow p \geq \frac{\sum_{i=1}^{m} n_i p_i}{n}.$$

The following example gives necessary and sufficient conditions for the comparison of normal random variables; it is generalized in Example 7.A.13. See related results in Examples 1.A.26 and 3.A.51.

Example 4.A.46. Let X be a normal random variable with mean μ_X and variance σ_X^2, and let Y be a normal random variable with mean μ_Y and variance σ_Y^2. Then $X \le_{\rm icx} Y$ if, and only if, $\mu_X \le \mu_Y$ and $\sigma_X^2 \le \sigma_Y^2$.

Example 4.A.47. Let $X_1, X_2, \dots, X_n$ be independent exponential random variables with distinct hazard rates $\lambda_1 > \lambda_2 > \dots > \lambda_n > 0$. Then $\frac{1}{n}\sum_{i=1}^n X_i \le_{\rm icx} X_n$.

Conditions for stochastic equality, for random variables that are $\le_{\rm icx}$- or $\le_{\rm icv}$-ordered, are given in the following result. This result may be compared to Theorems 1.A.8, 3.A.43, 3.A.60, 4.A.69, 5.A.15, 6.B.19, 6.G.12, 6.G.13, and 7.A.14–7.A.16.

Theorem 4.A.48. *Let X and Y be two nonnegative random variables. Suppose that $X \le_{\rm icx} Y$ [$X \le_{\rm icv} Y$] and that $E[X^r] = E[Y^r]$ for some $r \in (1, \infty)$ [$r \in (0, 1)$], provided the expectations exist. Then $X =_{\rm st} Y$.*

This result is a corollary of Theorem 4.A.69 below with $p = 1$.

In fact, the following stronger result, which is an analog of Theorem 3.A.43, holds for the orders $\le_{\rm icx}$ and $\le_{\rm icv}$.

Theorem 4.A.49. *Let X and Y be two random variables. Suppose that $X \le_{\rm icx}$ [$\le_{\rm icv}$] Y and that for some increasing strictly convex [concave] function ϕ we have that $E[\phi(X)] = E[\phi(Y)]$, provided the expectations exist. Then $X =_{\rm st} Y$.*

Of course, in Theorem 4.A.49 we can replace "increasing strictly convex [concave] function" by "decreasing strictly concave [convex] function."

Theorem 4.A.50. *Let $X_1, X_2, \dots, X_n$ and $Y_1, Y_2, \dots, Y_n$ ($n \ge 2$) be two collections of independent and identically distributed random variables. If $X_1 \le_{\rm icx} Y_1$ and if $E[\max\{X_1, X_2, \dots, X_n\}] = E[\max\{Y_1, Y_2, \dots, Y_n\}]$, then $X_1 =_{\rm st} Y_1$.*

Analogous to the result in Remark 1.A.18, it can be shown that the set of all distribution functions on $\mathbb{R}$ with finite means is a lattice with respect to the order $\le_{\rm icx}$.

Meilijson and Nádas [389] have proved the following result which, for the sake of simplicity, we describe informally. Let X be a random variable with mean residual life function m (see, for example, (2.A.1)). Define H by $H(x) = m(x) + x = E[X \big| X > x]$, for all x, and note that H is increasing. Denote $\tilde{X} = H(X)$. Then $\tilde{X} \ge_{\rm st} Y$ for every random variable Y which satisfies $Y \le_{\rm icx} X$. In fact, Meilijson and Nádas [389] proved that $\tilde{X}$ is the least stochastic majorant in the sense that if another random variable Z also satisfies $Z \ge_{\rm st} Y$ for every Y such that $Y \le_{\rm st} X$, then $\tilde{X} \le_{\rm st} Z$.

4.A.5 Some properties in reliability theory

We have seen in Theorem 1.A.30 that a nonnegative random variable is IFR [DFR] if, and only if, $[X - t|X > t] \geq_{\text{st}} [\leq_{\text{st}}] [X - t'|X > t']$ whenever $t \leq t'$. A question of interest then is what does one get if in the above condition one replaces the order $\geq_{\text{st}}$ by the order $\geq_{\text{icx}}$. It turns out that the order $\geq_{\text{icx}}$ can characterize another familiar aging notion in reliability theory. Recall from page 1 the definitions of DMRL and IMRL random variables. A combination of Theorems 2.A.23 and 4.A.24 provides a proof of the DMRL part of the next theorem. The proof of the IMRL part is similar.

Theorem 4.A.51. *The nonnegative random variable X is* DMRL [IMRL] *if, and only if,* $[X - t|X > t] \geq_{\text{icx}} [\leq_{\text{icx}}] [X - t'|X > t']$ *whenever* $t \leq t'$.

Other characterizations of DMRL and IMRL random variables, by means of other stochastic orders, can be found in Theorems 2.A.23, 2.B.17, 3.A.56, and 3.C.13.

We will now describe a generalization of the sufficiency part of Theorem 4.A.51. For two independent random variables X and T, let X_T denote a random variable that has the distribution of $[X - T|X > T]$. Note that X_T is *not* the residual life of X given T.

Theorem 4.A.52. *Let X, T_1, and T_2 be independent random variables. If $T_1 \leq_{\text{rh}} T_2$, and if X is* DMRL [IMRL], *then* $X_{T_1} \geq_{\text{icx}} [\leq_{\text{icx}}] X_{T_2}$.

Proof. We will prove the DMRL part only. The proof of the IMRL part is similar. Let $\overline{F}$ denote the survival function of X, and let $\overline{G}_i$ denote the survival function of X_{T_i}, $i = 1, 2$. Then, for any fixed x we have

$$\int_x^\infty \overline{G}_2(y)\mathrm{d}y - \int_x^\infty \overline{G}_1(y)\mathrm{d}y$$
$$= \frac{E\Big[\overline{F}(T_1)\Big]E\Big[\int_x^\infty \overline{F}(T_2 + y)\mathrm{d}y\Big] - E\Big[\overline{F}(T_2)\Big]E\Big[\int_x^\infty \overline{F}(T_1 + y)\mathrm{d}y\Big]}{E\Big[\overline{F}(T_1)\Big]E\Big[\overline{F}(T_2)\Big]}. \tag{4.A.30}$$

Define the functions α and β by $\alpha(t) = \int_x^\infty \overline{F}(t + y)\mathrm{d}y$ and $\beta(t) = \overline{F}(t)$. Note that β is nonnegative and decreasing, and that α/β is decreasing because X is DMRL. Therefore, by Theorem 1.B.50(b), we see that the numerator in (4.A.30) is nonpositive for any x. It follows, by (4.A.5), that $X_{T_1} \geq_{\text{icx}} X_{T_2}$. □

Note that if the nonnegative random variable X is DMRL [IMRL], then, from Theorem 4.A.51 it follows that

$$X \geq_{\text{icx}} [\leq_{\text{icx}}] [X - t|X > t] \quad \text{for all } t \geq 0. \tag{4.A.31}$$

Nonnegative random variables that satisfy (4.A.31) are called new better [worse] than used in convex ordering (NBUC [NWUC]) or new better [worse] than used in mean (NBUM [NWUM]). An equivalent definition of the NBUC notion, by means of the usual stochastic order, is given in (1.A.21).

It is of interest to note that a nonnegative random variable X with survival function $\overline{F}$ is NBUC if, and only if,

$$\int_{x+t}^{\infty} \overline{F}(y)\mathrm{d}y \leq \frac{\overline{F}(t)}{1-\overline{F}(t)} \int_{x}^{x+t} \overline{F}(y)\mathrm{d}y \quad \text{for all } t \geq 0 \text{ and } x \geq 0. \tag{4.A.32}$$

It is worthwhile to point out that a nonnegative random variable X that satisfies (4.A.31), but with the increasing concave (rather than the increasing convex) order, is said to be NBU(2) [NWU(2)].

If a nonnegative random variable X satisfies

$$[X-t|X>t] \geq_{\text{icv}} [\leq_{\text{icv}}] \; [X-t'|X>t'] \quad \text{whenever } t \leq t', \tag{4.A.33}$$

then, in some places in the literature, the random variable X is said to have the IFR(2) [DFR(2)] property. However, Belzunce, Hu, and Khaledi [68] proved that the IFR(2) [DFR(2)] property is the same as the IFR [DFR] property. Thus they obtained the following characterization of the IFR [DFR] property.

Theorem 4.A.53. *The nonnegative random variable X is* IFR [DFR] *if, and only if,* (4.A.33) *holds.*

4.A.6 The starshaped order

A function $\phi : [0, \infty) \rightarrow [0, \infty)$, which satisfies $\phi(0) = 0$, is called starshaped if $\phi(x)/x$ is increasing in x on $(0, \infty)$ (here we use the convention $a/\infty = 0$ for $a > 0$). Note that such a function is increasing. Note also that every increasing convex function ϕ on $[0, \infty)$, such that $\phi(0) = 0$, is starshaped.

Let X and Y be two nonnegative random variables such that

$$E[\phi(X)] \leq E[\phi(Y)] \quad \text{for all starshaped functions } \phi : [0, \infty) \rightarrow [0, \infty), \tag{4.A.34}$$

provided the expectations exist. Then X is said to be *smaller than Y in the starshaped order* (denoted by $X \leq_{\text{ss}} Y$).

Theorem 4.A.54. *Let X and Y be two nonnegative random variables with distribution functions F and G, respectively. Then $X \leq_{\text{ss}} Y$ if, and only if,*

$$\int_{y}^{\infty} x\mathrm{d}F(x) \leq \int_{y}^{\infty} x\mathrm{d}G(x), \quad y \geq 0. \tag{4.A.35}$$

Proof. The function ϕ_y, defined by

$$\phi_y(x) = \begin{cases} 0, & x \leq y, \\ x, & x > y, \end{cases}$$

is starshaped. Thus, (4.A.34) $\Longrightarrow$ (4.A.35). Conversely, let ϕ be a starshaped function. Then $h(x) = \phi(x)/x$ is increasing in x on $(0, \infty)$. Approximate h by a sequence of increasing step functions h_n. Then (4.A.35) yields

$$\int_0^\infty xh_n(x)\mathrm{d}F(x) \le \int_0^\infty xh_n(x)\mathrm{d}G(x).$$

Letting $n \to \infty$, we obtain (4.A.34). $\square$

Theorem 4.A.54 shows that when the compared random variables have the same mean, then the starshaped order is equivalent to the usual stochastic ordering of the corresponding length-biased (or spread) random variables. Such random variables are studied in Examples 1.B.23, 1.C.59, 1.C.60, and 8.B.12.

Theorem 4.A.55. *Let X and Y be two nonnegative random variables. Then*

$$X \le_{\rm st} Y \Longrightarrow X \le_{\rm ss} Y \Longrightarrow X \le_{\rm icx} Y.$$

Proof. The first implication follows from the fact that a starshaped function ϕ, such that $\phi(0) = 0$, is increasing. In order to prove the second implication, let ϕ be an increasing convex function. First suppose that $\phi(0) = 0$. Then ϕ is starshaped and the inequality in (4.A.1) follows from $X \le_{\rm ss} Y$. If $\phi(0) = a \ne 0$, then define $\tilde{\phi}(x) = \phi(x) - a$, $x \ge 0$. The function $\tilde{\phi}$ is increasing convex, and it satisfies $\tilde{\phi}(0) = 0$. Thus, by the previous argument $E[\tilde{\phi}(X)] \le E[\tilde{\phi}(Y)]$; that is, $E[\phi(X)] - a \le E[\phi(Y)] - a$, and the inequality in (4.A.1) follows. $\square$

Some closure properties of the starshaped order are given in the next theorem.

Theorem 4.A.56. (a) *If the nonnegative random variables X and Y are such that $X \le_{\rm ss} Y$, and g is any starshaped function with $g(0) = 0$, then $g(X) \le_{\rm ss} g(Y)$. In particular, $cX \le_{\rm ss} cY$ for any $c > 0$.*

(b) *Let X, Y, and Θ be random variables such that $[X|\Theta = \theta] \le_{\rm ss} [Y|\Theta = \theta]$ for all θ in the support of Θ. Then $X \le_{\rm ss} Y$. That is, the starshaped order is closed under mixtures.*

(c) *Let $\{X_j,\ j = 1, 2, \dots\}$ and $\{Y_j,\ j = 1, 2, \dots\}$ be two sequences of nonnegative random variables such that $X_j \to_{\rm st} X$ and $Y_j \to_{\rm st} Y$ as $j \to \infty$. Assume that EX^2 and EY^2 are finite and that*

$$\frac{EX_j^2}{EX_j} \to \frac{EX^2}{EX} \quad \textit{and} \quad \frac{EY_j^2}{EY_j} \to \frac{EY^2}{EY} \quad \textit{as } j \to \infty.$$

If $X_j \le_{\rm ss} Y_j$, $j = 1, 2, \dots$, then $X \le_{\rm ss} Y$.

Theorem 4.A.57. *Let X be a nonnegative random variable. Then $I_{[a,\infty)}(X) \le_{\rm ss} I_{[b,\infty)}(X)$ whenever $b \ge a \ge 0$, where $I_{[a,\infty)}$ and $I_{[b,\infty)}$ are the indicator functions of the indicated intervals.*

The proof of Theorem 4.A.57 consists of verifying (4.A.35) in each of the cases $y \le a$, $a < y \le b$, and $y > b$.

4.A.7 Some related orders

Let X and Y be two random variables with survival function $\overline{F}$ and $\overline{G}$, and distribution functions F and G, respectively. Let $F^{[k]}$, $\overline{F}^{[k]}$, $G^{[k]}$, and $\overline{G}^{[k]}$ be defined as in (3.A.66) and (3.A.67). The inequalities (4.A.5) and (4.A.7) can be generalized as follows: For a positive integer m suppose that

$$\overline{F}^{[m-1]}(x) \le \overline{G}^{[m-1]}(x) \quad \text{for all } x, \tag{4.A.36}$$

or that

$$F^{[m-1]}(x) \ge G^{[m-1]}(x) \quad \text{for all } x, \tag{4.A.37}$$

provided these integrals are finite (the integrals are finite if F and G have finite $(m-1)$st moments). If (4.A.36) holds, then X is said to be *smaller than Y in the m-icx order* (denoted by $X \le_{m\text{-icx}} Y$). If it is known that X and Y take on values in $\mathbb{N}_{++}$, then the definition of the m-icx order can be modified, exploiting the special structure of $\mathbb{N}_{++}$; see Denuit and Lefèvre [146]. If (4.A.37) holds, then X is said to be *smaller than Y in the m-icv order* (denoted by $X \le_{m\text{-icv}} Y$).

It is seen from the definition that the orders $\le_{1\text{-icx}}$ and $\le_{1\text{-icv}}$ are equivalent to the order $\le_{\text{st}}$, the order $\le_{2\text{-icx}}$ is equivalent to the order $\le_{\text{icx}}$, and the order $\le_{2\text{-icv}}$ is equivalent to the order $\le_{\text{icv}}$.

The orders $\le_{m\text{-icx}}$ and $\le_{m\text{-icv}}$ have some properties that are similar to the properties of the orders $\le_{\text{icx}}$ and $\le_{\text{icv}}$. For example, the extension of (4.A.4) is that $X \le_{m\text{-icx}} Y$ if, and only if,

$$E[(X-a)_+]^{m-1} \le E[(Y-a)_+]^{m-1} \quad \text{for all } a. \tag{4.A.38}$$

The extension of (4.A.6) is that $X \le_{m\text{-icv}} Y$ if, and only if,

$$E[(X-a)_-]^{m-1} \le E[(Y-a)_-]^{m-1} \quad \text{for all } a. \tag{4.A.39}$$

The characterization (4.A.1) of the orders $\le_{\text{icx}}$ and $\le_{\text{icv}}$ has an analog for the orders $\le_{m\text{-icx}}$ and $\le_{m\text{-icv}}$. We will not give the technical details here (see Section 4.C for a reference), but we just mention the following results. For $m = 1, 2, \ldots$, let $\mathcal{M}_{m\text{-icx}}$ be the set of all functions $\phi : \mathbb{R} \to \mathbb{R}$ such that $\lim_{x\to-\infty} \phi(x)$ is finite, and whose first $m-1$ derivatives, $\phi^{(1)}, \phi^{(2)}, \ldots, \phi^{(m-1)}$, exist, and are such that $\lim_{x\to-\infty} \phi^{(j)}(x) = 0$, $j = 1, 2, \ldots, m-1$, and $\phi^{(m-1)}$ is increasing. Let $\overline{\mathcal{M}}_{m\text{-icx}}$ be the closure of $\mathcal{M}_{m\text{-icx}}$ in the topology of weak convergence (that is, pointwise convergence in each continuity point of the limit). Let X and Y be two random variables and suppose that the support of each of them contains an interval of the form $(-\infty, a)$ for some a. Then $X \le_{m\text{-icx}} Y$ if, and only if,

$$E[\phi(X)] \le E[\phi(Y)] \quad \text{for all functions } \phi \in \overline{\mathcal{M}}_{m\text{-icx}}, \tag{4.A.40}$$

provided the expectations exist.

Next, for $m = 1, 2, \ldots$, let $\mathcal{M}_{m\text{-icv}}$ be the set of all functions $\phi : \mathbb{R} \to \mathbb{R}$ such that $\lim_{x\to\infty} \phi(x)$ is finite, whose first $m-1$ derivatives, $\phi^{(1)}, \phi^{(2)}, \ldots, \phi^{(m-1)}$, exist, and are such that $\lim_{x\to\infty} \phi^{(j)}(x) = 0$, $j = 1, 2, \ldots, m-1$, and $(-1)^{m-1}\phi^{(m-1)}$ is increasing. Let $\overline{\mathcal{M}}_{m\text{-icv}}$ be the closure of $\mathcal{M}_{m\text{-icv}}$ in the topology of weak convergence. Let X and Y be two random variables and suppose that the support of each of them contains an interval of the form (a, ∞) for some a. Then $X \leq_{m\text{-icv}} Y$ if, and only if,

$$E[\phi(X)] \leq E[\phi(Y)] \quad \text{for all functions } \phi \in \overline{\mathcal{M}}_{m\text{-icv}}, \tag{4.A.41}$$

provided the expectations exist.

Let us denote $X \leq_{\infty\text{-icx}} [\leq_{\infty\text{-icv}}]\ Y$ if

$$X \leq_{m\text{-icx}} [\leq_{m\text{-icv}}]\ Y \quad \text{for all positive integers } m. \tag{4.A.42}$$

A characterization of the order $\leq_{\infty\text{-icv}}$ is given in Theorem 5.A.17.

It can be shown that if X and Y have finite $(m-1)$st moments, then

$$X \leq_{m\text{-icx}} Y \Longrightarrow E[X] \leq E[Y]$$

and

$$X \leq_{m\text{-icv}} Y \Longrightarrow E[X] \leq E[Y],$$

provided the expectations exist. In fact we have the following more general result.

Theorem 4.A.58. *Let X and Y be two random variables with finite first $m-1$ moments. If $X \leq_{m\text{-icx}} Y$ $[X \leq_{m\text{-icv}} Y]$, then $EX^k < EY^k$ $[(-1)^{k+1}EX^k < (-1)^{k+1}EY^k]$ for the smallest k for which $EX^k \neq EY^k$.*

Some closure properties of the orders $\leq_{m\text{-icx}}$ and $\leq_{m\text{-icv}}$ are stated next. We omit the proof of the following theorem. Note, however, that parts (b) and (c) of the next theorem are easy to prove. The proof of part (a) uses the fact that if $\phi \in \mathcal{M}_{m\text{-icx}}$ $[\mathcal{M}_{m\text{-icv}}]$ then $\phi^{(j)}$ $[(-1)^j\phi^{(j)}]$ is nonnegative and increasing [decreasing] for all $j \in \{1, 2, \ldots, m-1\}$, and therefore $\mathcal{M}_{m\text{-icx}}$ and $\mathcal{M}_{m\text{-icv}}$ are closed under compositions.

Theorem 4.A.59. (a) *Let X and Y be two random variables and suppose that the support of each of them contains an interval of the form $(-\infty, a)$ $[(a, \infty)]$ for some a. If $X \leq_{m\text{-icx}} [\leq_{m\text{-icv}}]\ Y$ and if g is any function in $\overline{\mathcal{M}}_{m\text{-icx}}$ $[\overline{\mathcal{M}}_{m\text{-icv}}]$, then $g(X) \leq_{m\text{-icx}} [\leq_{m\text{-icv}}]\ g(Y)$.*

(b) *Let X, Y, and Θ be random variables such that, for all θ in the support of Θ, we have that $[X \big| \Theta = \theta] \leq_{m\text{-icx}} [\leq_{m\text{-icv}}]\ [Y \big| \Theta = \theta]$. Then $X \leq_{m\text{-icx}}$ $[\leq_{m\text{-icv}}]\ Y$. That is, the m-icx [m-icv] order is closed under mixtures.*

(c) *Let $\{X_j,\ j = 1, 2, \ldots\}$ and $\{Y_j,\ j = 1, 2, \ldots\}$ be two sequences of random variables such that $X_j \to_{\text{st}} X$ and $Y_j \to_{\text{st}} Y$ as $j \to \infty$. Assume that $E(X_+)^{m-1}$ and $E(Y_+)^{m-1}$ are finite and that*

$$E(X_j)_+^{m-1} \to EX_+^{m-1} \; [E(X_j)_-^{m-1} \to EX_-^{m-1}] \quad \textit{and}$$
$$E(Y_j)_+^{m-1} \to EY_+^{m-1} \; [E(Y_j)_-^{m-1} \to EY_-^{m-1}] \quad \textit{as } j \to \infty. \quad (4.A.43)$$

If $X_j \leq_{m\text{-icx}} [\leq_{m\text{-icv}}] \, Y_j$, $j = 1, 2, \ldots$, *then* $X \leq_{m\text{-icx}} [\leq_{m\text{-icv}}] \, Y$.

(d) *Let* $X_1, X_2, \ldots, X_l$ *be a set of independent random variables and let* $Y_1, Y_2, \ldots, Y_l$ *be another set of independent random variables. If* $X_i \leq_{m\text{-icx}} [\leq_{m\text{-icv}}] \, Y_i$ *for* $i = 1, 2, \ldots, l$, *then*

$$\sum_{j=1}^{l} X_j \leq_{m\text{-icx}} [\leq_{m\text{-icv}}] \sum_{j=1}^{l} Y_j.$$

That is, the m-icx [m-icv] order is closed under convolutions.

In part (c), as in Theorem 3.A.12, the condition (4.A.43) is necessary — without it the conclusion of part (c) may not hold.

The following result, which extends the m-icx part of Theorem 4.A.59(d), is essentially the same as Theorem 8.A.29.

Theorem 4.A.60. *Let* $X_1, X_2, \ldots$ *be a set of independent random variables and let* $Y_1, Y_2, \ldots$ *be another set of independent random variables. Let* N_1 *be an integer-valued random variable that is independent of the* X_i*'s, and let* N_2 *be an integer-valued random variable that is independent of the* Y_i*'s. If* $X_i \leq_{m\text{-icx}} Y_i$ *for* $i = 1, 2, \ldots$, *and if* $N_1 \leq_{m\text{-icx}} N_2$, *then*

$$\sum_{j=1}^{N_1} X_j \leq_{m\text{-icx}} \sum_{j=1}^{N_2} Y_j.$$

For the orders $\leq_{m\text{-icx}}$ and $\leq_{m\text{-icv}}$, the analog of Theorem 3.A.12(a) is the following.

Theorem 4.A.61. *Let* X *and* Y *be two random variables. Then*

$$X \leq_{m\text{-icx}} [\leq_{m\text{-icv}}] \, Y \iff -X \geq_{m\text{-icv}} [\geq_{m\text{-icx}}] -Y.$$

The proof of Theorem 4.A.61 easily follows from (4.A.36) and (4.A.37).

It is not hard to verify the next statement.

Theorem 4.A.62. *Consider two random variables* X *and* Y. *If* $X \leq_{m_1\text{-icx}} [\leq_{m_1\text{-icv}}] \, Y$, *then* $X \leq_{m_2\text{-icx}} [\leq_{m_2\text{-icv}}] \, Y$ *for all* $m_2 \geq m_1$.

Since the order $\leq_{1\text{-icx}}$ is the same as the order $\leq_{\text{st}}$ we see that

$$X \leq_{\text{st}} Y \Longrightarrow X \leq_{m\text{-icx}} Y$$

and that

$$X \leq_{\text{st}} Y \Longrightarrow X \leq_{m\text{-icv}} Y.$$

The following obvious relationships hold between the orders of Section 3.A.5 and the present orders:

$$X \leq^S_{m\text{-cx}} Y \Longrightarrow X \leq_{m\text{-icx}} Y,$$

and

$$X \leq^S_{m\text{-cv}} Y \Longrightarrow X \leq_{m\text{-icv}} Y.$$

Sufficient conditions for $X \leq_{m\text{-icv}} Y$ and $X \leq_{m\text{-icv}} Y$ are given in the next result, which is related to Theorem 4.A.22. It is of interest to compare the next result with Theorem 3.A.66.

Theorem 4.A.63. *Let X and Y be two nonnegative random variables with distribution functions F and G, respectively, and with density functions f and g, respectively, such that $E[X^i] = E[Y^i]$, $i = 1, 2, \ldots, m-2$, and $E[X^{m-1}] \leq E[Y^{m-1}]$.*

(a) *If $S^-(F-G) \leq m-1$ and if the last sign of $F-G$ is a $+$, then $X \leq_{m\text{-icx}} Y$.*
(b) *If $S^-(f-g) \leq m$ and if the last sign of $g-f$ is a $+$, then $X \leq_{m\text{-icx}} Y$.*

The following example describes a typical application of Theorem 4.A.63.

Example 4.A.64. Let the inverse Gaussian random variable Y, and the log-normal random variable Z, be as in Example 3.A.67; in particular they both have the mean α/β and the second moment $\alpha(\alpha+1)/\beta^2$. We claim that $Y \leq_{4\text{-icx}} Z$. In order to see it, first note, as in Example 3.A.67, that without loss of generality we can take the means to be equal to 1, that is, $\beta = \alpha$. Now, a straightforward computation yields

$$\log \frac{f_Y(x)}{f_X(x)} = C + \frac{\log^2 x}{2\tau^2} - \frac{\alpha x}{2} - \frac{\alpha}{2x}, \quad x > 0,$$

where C is some constant. Substituting $u = \log x$, the second derivative of the above expression is seen to have two sign changes. Therefore the expression itself has at most four sign changes. We also have here, by a lengthy computation (see Kaas and Hesselager [270]), that $E[Y^3] < E[Z^3]$. The stated result now follows from Theorem 4.A.63(b).

In fact, it can be shown that if X, Y, and Z, are, respectively, Gamma, inverse Gaussian, and lognormal random variables (with parameters that are different from the ones in Example 3.A.67), such that $E[X] = E[Y] = E[Z]$ and $E[X^2] \leq E[Y^2] \leq E[Z^2]$, then $X \leq_{3\text{-icx}} Y$, $X \leq_{3\text{-icx}} Z$, and $Y \leq_{4\text{-icx}} Z$.

Some comparisons of Gamma, inverse Gaussian, lognormal, and Birnbaum-Saunders random variables in the $\leq_{3\text{-icv}}$ sense were derived by Klar [300].

Consider now a family of distribution functions $\{G_\theta, \theta \in \mathbb{R}\}$. As in Section 1.A.3 let $X(\theta)$ denote a random variable with distribution function G_θ. For any random variable Θ with support $\mathbb{R}$, and with distribution function F, let us denote by $X(\Theta)$ a random variable with distribution function H given by

$$H(y) = \int_{\mathcal{X}} G_\theta(y) \mathrm{d}F(\theta), \quad y \in \mathbb{R}.$$

The following result generalizes Theorem 4.A.8(a), just as Theorem 1.A.6 generalized Theorem 1.A.3(a). Its proof is similar to the proof of Theorem 4.A.18, using the fact that $\mathcal{M}_{m\text{-icx}}$ and $\mathcal{M}_{m\text{-icv}}$ are closed under compositions. We omit the details.

Theorem 4.A.65. *Consider a family of distribution functions $\{G_\theta, \theta \in \mathbb{R}\}$ as above. Let Θ_1 and Θ_2 be two random variables with support $\mathbb{R}$ and distribution functions F_1 and F_2, respectively. Let Y_1 and Y_2 be two random variables such that $Y_i =_{\text{st}} X(\Theta_i)$, $i = 1, 2$; that is, suppose that the distribution function of Y_i is given by*

$$H_i(y) = \int_{\mathcal{X}} G_\theta(y) \mathrm{d}F_i(\theta), \quad y \in \mathbb{R}, \ i = 1, 2.$$

If ψ_ϕ, defined by $\psi_\phi(\theta) \equiv E[\phi(X(\theta))]$, is in $\mathcal{M}_{m\text{-icx}}$ $[\mathcal{M}_{m\text{-icv}}]$ whenever $\phi \in \mathcal{M}_{m\text{-icx}}$ $[\phi \in \mathcal{M}_{m\text{-icv}}]$, and if

$$\Theta_1 \leq_{m\text{-icx}} [\leq_{m\text{-icv}}] \, \Theta_2,$$

then

$$Y_1 \leq_{m\text{-icx}} [\leq_{m\text{-icv}}] \, Y_2.$$

For example, the family $\{G_\theta, \ \theta \geq 0\}$ of the Poisson distributions (or, in fact, every family of distribution functions whose associated density functions $\{g_\theta, \ \theta \in \mathbb{R}\}$ satisfy that $g_\theta(x)$ is totally positive of order m; see Karlin [275]) satisfies the condition in Theorem 4.A.65 that ψ_ϕ is in $\mathcal{M}_{m\text{-icx}}$ $[\mathcal{M}_{m\text{-icv}}]$ whenever $\phi \in \mathcal{M}_{m\text{-icx}}$ $[\phi \in \mathcal{M}_{m\text{-icv}}]$.

A Laplace transform characterization of the orders $\leq_{m\text{-icx}}$ and $\leq_{m\text{-icv}}$ is given next; it may be compared to Theorems 1.A.13, 1.B.18, 1.B.53, 1.C.25, 2.A.16, 2.B.14, and 4.A.21. Before stating it we make a few observations. First, note that the random variables X_1 and X_2 in the theorem below have the support $[0, \infty)$. Then the characterizations (4.A.40) and (4.A.41) are still valid provided the test functions ϕ in (4.A.40) satisfy that $\phi^{(j)}(0) = 0$ (rather than $\lim_{x \to -\infty} \phi^{(j)}(x) = 0$), $j = 1, 2, \ldots, m-1$. Next, note that the random variables $N_\lambda(X_1)$ and $N_\lambda(X_2)$ in the theorem below are discrete with support $\mathbb{N}_+$. There are several ways of defining the orders $\leq_{m\text{-icx}}$ and $\leq_{m\text{-icv}}$ for such random variables. One possible way is by the requirement (4.A.36) or (4.A.37) (or, equivalently, by (4.A.38) or (4.A.39)). Another possible way is by replacing the integrals in (4.A.36) or (4.A.37) by sums. In the theorem below we adopt a definition that is a discrete analog of (4.A.40) and (4.A.41). For $m = 1, 2, \ldots$, let $\mathcal{K}_{m\text{-icx}}$ be the set of functions $\phi : \mathbb{N}_+ \to \mathbb{R}$ such that $\Delta_\phi^{(j)}(0) = 0$, $j = 0, 1, \ldots, m-1$ (where $\Delta_\phi^{(0)}(n) \equiv \phi(n)$ and $\Delta_\phi^{(j)}(n) = \Delta_\phi^{(j-1)}(n+1) - \Delta_\phi^{(j-1)}(n)$, $j = 1, 2, \ldots$), and such that $\Delta_\phi^{(m-1)}(n)$ is increasing on $\mathbb{N}_+$. For the discrete random variables M_1 and M_2 denote $M_1 \leq'_{m\text{-icx}} M_2$ if $E[\phi(M_1)] \leq E[\phi(M_2)]$ for all functions $\phi \in \mathcal{K}_{m\text{-icx}}$. Similarly, let $\mathcal{K}_{m\text{-icv}}$ be the set of functions $\phi : \mathbb{N}_+ \to \mathbb{R}$ such that $\lim_{n \to \infty} \Delta_\phi^{(j)}(n) = 0$, $j = 0, 1, \ldots, m-1$, and such that

$(-1)^{m-1}\Delta_\phi^{(m-1)}(n)$ is increasing on $\mathbb{N}_+$. For the discrete random variables M_1 and M_2 denote $M_1 \le'_{m\text{-icv}} M_2$ if $E[\phi(M_1)] \le E[\phi(M_2)]$ for all functions $\phi \in \mathcal{K}_{m\text{-icv}}$.

Theorem 4.A.66. *Let X_1 and X_2 be two nonnegative random variables, and let $N_\lambda(X_1)$ and $N_\lambda(X_2)$ be as described in Theorem* 1.A.13. *Then*

$$X_1 \le_{m\text{-icx}} [\le_{m\text{-icv}}]\ X_2 \Longleftrightarrow N_\lambda(X_1) \le'_{m\text{-icx}} [\le'_{m\text{-icv}}]\ N_\lambda(X_2) \quad \textit{for all } \lambda > 0.$$

The proof of this theorem is similar to the proof of Theorem 4.A.21 and is therefore omitted.

Another family of orders that are related to the $\le_{\rm cx}$, $\le_{\rm icx}$, and $\le_{\rm icv}$ orders can be defined by a generalization of (4.A.5) and (4.A.7) that is different from the generalization that is described in (4.A.36) and (4.A.37). Let X and Y be two random nonnegative variables with distribution functions F and G, and survival functions $\overline{F}$ and $\overline{G}$, respectively. Let $p > 0$ and suppose that $E[X^p]$ and $E[Y^p]$ exist. If

$$\int_x^\infty u^{p-1}\overline{F}(u)du \le \int_x^\infty u^{p-1}\overline{G}(u)du \text{ for all } x, \quad \text{and} \quad E[X^p] = E[Y^p],$$

then X is said to be *smaller than Y in pth order* (denoted by $X \le_p Y$). If

$$\int_x^\infty u^{p-1}\overline{F}(u)du \le \int_x^\infty u^{p-1}\overline{G}(u)du \quad \text{for all } x,$$

then X is said to be *smaller than Y in $p+$ order* (denoted by $X \le_{p+} Y$). Finally, if

$$\int_0^x u^{p-1}F(u)du \ge \int_0^x u^{p-1}G(u)du \quad \text{for all } x,$$

then X is said to be *smaller than Y in $p-$ order* (denoted by $X \le_{p-} Y$).

It is not hard to verify that for nonnegative random variables X and Y we have

$$X \le_p Y \Longleftrightarrow X^p \le_{\rm cx} Y^p, \tag{4.A.44}$$

$$X \le_{p+} Y \Longleftrightarrow X^p \le_{\rm icx} Y^p,$$

and

$$X \le_{p-} Y \Longleftrightarrow X^p \le_{\rm icv} Y^p. \tag{4.A.45}$$

It is seen at once that

$$X \le_p Y \Longrightarrow X \le_{p+} Y,$$

and that

$$X \le_p Y \Longrightarrow Y \le_{p-} X.$$

Notice that, for $p = m$, the order $\le_{p+}$ $[\le_{p-}]$ is not the same as the order $\le_{m\text{-icx}}$ $[\le_{m\text{-icv}}]$. In fact, $X \le_{m+} Y$ if, and only if,

$$E[(X^m - a)_+] \leq E[(Y^m - a)_+] \quad \text{for all } a$$

(compare this to (4.A.38)), and $X \leq_{m-} Y$ if, and only if,

$$E[(X^m - a)_-] \leq E[(Y^m - a)_-] \quad \text{for all } a$$

(compare this to (4.A.39)).

It is easy to verify that the orders $\leq_p$, $\leq_{p+}$ and $\leq_{p-}$ are closed under mixtures. They are also closed under limits in distribution provided a condition on convergence of moments, which is an obvious modification of (4.A.10) (similar to (4.A.43)), holds. The following result points out some interrelationships among these orders.

Theorem 4.A.67. *Let X and Y be two nonnegative random variables. If $X \leq_{p+} [\leq_{p-}] Y$, then $X \leq_{q+} [\leq_{q-}] Y$ whenever $q \geq p$ $[q \leq p]$.*

A relationship to the order $\leq_*$ is given next (the order $\leq_*$ is defined in Section 4.B below).

Theorem 4.A.68. *Let X and Y be two nonnegative random variables that have finite pth moments and that are not degenerate at 0. If $X \leq_* Y$ and if $E[X^p] = E[Y^p]$, then $X \leq_p Y$.*

A simple proof of Theorem 4.A.68 will be given in Remark 4.B.24.

Motivated by the result of Theorem 1.A.8 (see also Theorems 3.A.43, 3.A.60, 4.A.48, 5.A.15, 6.B.19, 6.G.12, 6.G.13, and 7.A.14–7.A.16), the following results have been derived.

Theorem 4.A.69. *Let X and Y be two nonnegative random variables. Suppose that $X \leq_{p+} Y$ $[X \geq_{p-} Y]$ and that $E[X^r] = E[Y^r]$ for some $r \in (p, \infty)$ $[r \in (0, p)]$, provided the expectations exist. Then $X =_{\text{st}} Y$.*

Theorem 4.A.70. *Let X and Y be two nonnegative random variables with finite means and distribution functions F and G, respectively. If $X \leq_p Y$ and if*

$$\int_0^1 \left[F^{-1}(t)\right]^r \mathrm{d}\phi(t) = \int_0^1 \left[G^{-1}(t)\right]^r \mathrm{d}\phi(t)$$

for some $r \geq p$ and some increasing and strictly convex function $\phi : [0,1] \to \mathbb{R}$, then $X =_{\text{st}} Y$.

We end this section by mentioning still another sequence of orders that is based on iterated integrals. If F is a distribution function, then let F^{-1} denote the inverse of F (see page 1). Denote recursively

$$F_1^{-1}(p) = F^{-1}(p), \quad p \in [0,1],$$

and

$$F_n^{-1}(p) = \int_p^1 F_{n-1}^{-1}(u)\mathrm{d}u, \quad p \in [0,1], \tag{4.A.46}$$

for $n = 2, 3, \ldots$. Similarly define G_n^{-1} for a distribution function G. For any positive integer m, if the distribution functions F and G, of the random variables X and Y, satisfy

$$F_m^{-1}(p) \le G_m^{-1}(p) \quad \text{for } p \in [0, 1],$$

then we denote $X \le_m^{-1} Y$. It is easy to see that

$$X \le_1^{-1} Y \Longleftrightarrow X \le_{\rm st} Y.$$

Also, if $EX = EY$, then, by Theorem 3.A.5 we see that

$$X \le_2^{-1} Y \Longleftrightarrow X \le_{\rm cx} Y.$$

From (4.A.46) we obtain at once the following result

Theorem 4.A.71. *Let X and Y be two random variables. If $X \le_{m_1}^{-1} Y$, then $X \le_{m_2}^{-1} Y$ for all $m_2 \ge m_1$.*

A necessary condition for $X \le_m^{-1} Y$ is given in the next result.

Theorem 4.A.72. *Let X and Y be two random variables. If $X \le_m^{-1} Y$, then*

$$E[\max\{X_1, X_2, \ldots, X_k\}] \le E[\max\{Y_1, Y_2, \ldots, Y_k\}], \quad k \ge m - 1,$$

where the X_i's [Y_i's] are independent random variables, all distributed according to the distribution of X [Y].

Proof. Let F and G denote the distribution functions of X and Y, respectively. A straightforward computation yields

$$F_m^{-1}(0) = E[\max\{X_1, X_2, \ldots, X_{m-1}\}]$$

and

$$G_m^{-1}(0) = E[\max\{Y_1, Y_2, \ldots, Y_{m-1}\}].$$

Therefore $E[\max\{X_1, X_2, \ldots, X_{m-1}\}] \le E[\max\{Y_1, Y_2, \ldots, Y_{m-1}\}]$. The inequality for $k > m - 1$ now follows from Theorem 4.A.71. □

4.B Transform Orders: The Convex, Star, and Superadditive Orders

4.B.1 Definitions

Let X and Y be two nonnegative random variables with distribution functions F and G, respectively. Suppose that the support of X is an interval (finite or infinite).

We say that X is *smaller than Y in the convex transform order* (denoted as $X \le_{\rm c} Y$) if $G^{-1}F(x)$ is convex in x on the support of F.

We say that X is *smaller than Y in the star order* (denoted by $X \le_* Y$) if $G^{-1}F(x)$ is starshaped in x (that is, if $G^{-1}F(x)/x$ increases in $x \ge 0$). It is easily seen that $X \le_* Y$ if, and only if,

$$\frac{G^{-1}(u)}{F^{-1}(u)} \quad \text{is increasing in } u \in (0,1). \tag{4.B.1}$$

Also, recalling the definition of the number of sign changes in (1.A.18), it is easily seen that $X \le_* Y$ if, and only if, for all $b > 0$ we have that

$$S^-(F(\cdot) - G(b\cdot)) \le 1, \tag{4.B.2}$$

and the sign sequence is $-, +$ if a crossing occurs.

We say that X is *smaller than Y in the superadditive order* (denoted by $X \le_{\rm su} Y$) if $G^{-1}F(x)$ is superadditive in x (that is, if $G^{-1}F(x+y) \ge G^{-1}F(x) + G^{-1}F(y)$ for all $x \ge 0$ and $y \ge 0$).

4.B.2 Some properties

Every nonnegative function that vanishes at 0, and that is increasing and convex on $[0,\infty)$, is also starshaped on $[0,\infty)$. Furthermore, every nonnegative function that vanishes at 0, and that is increasing and starshaped on $[0,\infty)$, is also superadditive on $[0,\infty)$. Therefore, for any two nonnegative random variables X and Y we have

$$X \le_{\rm c} Y \Longrightarrow X \le_* Y, \tag{4.B.3}$$

and

$$X \le_* Y \Longrightarrow X \le_{\rm su} Y.$$

The star order is related to the dispersion order as follows:

Theorem 4.B.1. *Let X and Y be two nonnegative random variables. Then*

$$X \le_* Y \Longleftrightarrow \log X \le_{\rm disp} \log Y. \tag{4.B.4}$$

Proof. The relation $X \le_* Y$ holds if, and only if, $G^{-1}F(x)/x$ is increasing in $x \ge 0$; that is, if, and only if, $\log G^{-1}F(x) - \log x = \log G^{-1}F(e^{\log x}) - \log x$ is increasing in x. The result now follows from (3.B.10). $\square$

An equivalent way of writing (4.B.4) is the following. For any two nonnegative random variables X and Y,

$$X \le_{\rm disp} Y \Longleftrightarrow e^X \le_* e^Y.$$

Under an obvious restriction, the superadditive (and hence also the star and the convex transform) order implies the dispersion order as is shown in the next theorem.

Theorem 4.B.2. *Let X and Y be two nonnegative random variables such that $X \leq_{\rm st} Y$. If $X \leq_{\rm su} Y$, then $X \leq_{\rm disp} Y$.*

Proof. Let F and G denote the distribution functions of X and Y, respectively, and let S_F denote the support of F. Let x and y be two values in S_F. Then

$$\begin{aligned} G^{-1}F(x+y) - (x+y) &\geq G^{-1}F(x) + G^{-1}F(y) - (x+y) \\ &\geq G^{-1}F(y) - y, \end{aligned}$$

where the first inequality follows from $X \leq_{\rm su} Y$ and the second inequality follows from $F(x) \geq G(x)$. Thus $G^{-1}F(x) - x$ is increasing in x. Now, from (3.B.10), we obtain $X \leq_{\rm disp} Y$. □

The condition $X \leq_{\rm st} Y$ is clearly needed because without it it is impossible that $X \leq_{\rm disp} Y$ (see Theorem 3.B.13). The condition $X \leq_{\rm su} Y$ by itself (in fact, even the condition $X \leq_* Y$) does not necessarily imply that $X \leq_{\rm st} Y$.

Theorem 4.B.2, together with (4.B.3), implies that if X and Y are two nonnegative random variables with finite means such that $X \leq_{\rm st} Y$ and if $X \leq_{\rm su} Y$ (and therefore if $X \leq_{\rm c} Y$ or if $X \leq_* Y$), then (see Theorem 3.B.16) $[X - EX] \leq_{\rm cx} [Y - EY]$, and in particular,

$$\mathrm{Var}(X) \leq \mathrm{Var}(Y).$$

Another condition, under which $X \leq_{\rm su} Y$ implies $X \leq_{\rm disp} Y$, is given in the next theorem.

Theorem 4.B.3. *Let X and Y be two nonnegative random variables with distributions F and G, respectively, such that $\lim_{x \to 0}(G^{-1}F(x)/x) \geq 1$. If $X \leq_{\rm su} Y$, then $X \leq_{\rm disp} Y$.*

In particular, if F and G are absolutely continuous with $F(0) = G(0) = 0$ and their corresponding density functions f and g are such that $f(0) \geq g(0) > 0$, then $X \leq_{\rm su} Y$ implies $X \leq_{\rm disp} Y$.

The relationship between the orders $\leq_*$ and $\leq_{\rm icx}$ is described in the next theorem.

Theorem 4.B.4. *Let X and Y be two nonnegative random variables such that $EX \leq EY$. If $X \leq_* Y$, then $X \leq_{\rm icx} Y$.*

Proof. First we show that $X \leq_* Y \Longrightarrow X \leq_{\rm Lorenz} Y$. For this end we can assume temporarily, without loss of generality, since both orders are scale invariant, that $EX = EY = 1$. Let F and G denote the distribution functions of X and of Y, respectively. If $F \equiv G$, then the result is trivial. Thus assume $F \not\equiv G$. From (4.B.2) (with $b = 1$), and from the fact that $EX = EY$, it follows that $S^-(G - F) = 1$, and that the sign sequence is $+, -$. Thus, from (3.A.59) we obtain $X \leq_{\rm Lorenz} Y$. (Another proof of $X \leq_* Y \Longrightarrow X \leq_{\rm Lorenz} Y$ can be found in Section 4.B.3.)

Now suppose that $X \leq_{\rm Lorenz} Y$ and that $EX \leq EY$. Then

$$X \leq_{\text{cx}} \frac{EX}{EY} \cdot Y \leq_{\text{st}} Y.$$

Thus we see from Theorem 4.A.6(b) that $X \leq_{\text{icx}} Y$. $\square$

The following theorem describes a star order comparison of two functions of the same random variable.

Theorem 4.B.5. *Let X be a nonnegative random variable that is not degenerate at 0, and let g and h be nonnegative increasing functions, defined on $[0,\infty)$, such that $g(x) > 0$ and $h(x) > 0$ for all $x > 0$. If $h(x)/g(x)$ is increasing in $x \in (0,\infty)$, then*

$$g(X) \leq_* h(X).$$

Proof. Denote by F the distribution function of X. From the assumption that $h(x)/g(x)$ is increasing in $x \in (0,\infty)$ it follows that

$$\frac{h(F^{-1}(u))}{g(F^{-1}(u))} \quad \text{is increasing in } u \in (0,1).$$

Therefore, denoting by F_g and F_h the distribution functions of $g(X)$ and of $h(X)$, we have that

$$\frac{F_h^{-1}(u)}{F_g^{-1}(u)} \quad \text{is increasing in } u \in (0,1).$$

Thus $g(X) \leq_* h(X)$ by (4.B.1). $\square$

For example, if X is a nonnegative random variable, then

$$X + a \leq_* X \quad \text{whenever } a > 0.$$

An interesting property of the order $\leq_*$ is given in the next theorem.

Theorem 4.B.6. *Let X and Y be positive random variables. If $X \leq_* Y$, then $X^p \leq_* Y^p$ for any $p \neq 0$. In particular, $1/X \leq_* 1/Y$.*

Proof. Let F and G be the distribution functions of X and Y, respectively. First consider the case where $p > 0$. Then the distribution functions $\widetilde{F}$ and $\widetilde{G}$ of X^p and Y^p, respectively, are given by

$$\widetilde{F}(x) = F(x^{1/p}) \quad \text{and} \quad \widetilde{G}(x) = G(x^{1/p}), \quad x \geq 0.$$

Noting that $\widetilde{G}^{-1}(\widetilde{F}(x)) = (G^{-1}(F(x^{1/p})))^p$ we compute

$$\frac{\widetilde{G}^{-1}(\widetilde{F}(x))}{x} = \frac{(G^{-1}(F(x^{1/p})))^p}{x} = \Big(\frac{G^{-1}(F(y))}{y}\Big)^p,$$

where $y = x^{1/p}$. From the assumption $X \le_* Y$ it is seen that the right-hand side of the above equation is increasing in $y \ge 0$, and therefore the left-hand side of that equation is increasing in $x \ge 0$.

Now, in order to complete the proof it is only necessary to prove that $1/X \le_* 1/Y$. Let now $\widetilde{F}$ and $\widetilde{G}$ denote the distribution functions of $1/X$ and $1/Y$, respectively. These are given by

$$\widetilde{F}(x) = \overline{F}(1/x) \quad \text{and} \quad \widetilde{G}(x) = \overline{G}(1/x), \quad x \ge 0,$$

where $\overline{F} \equiv 1 - F$ and $\overline{G} \equiv 1 - G$. Noting that $\widetilde{G}^{-1} = 1/\overline{G}^{-1}$ and that $\overline{G}^{-1}\overline{F} = G^{-1}F$, we compute

$$\frac{\widetilde{G}^{-1}(\widetilde{F}(x))}{x} = \frac{1}{\overline{G}^{-1}(\overline{F}(1/x))x} = \frac{1/x}{G^{-1}(F(1/x))}.$$

From the assumption $X \le_* Y$ it is seen that the latter expression is increasing in $x \ge 0$. □

Example 4.B.7. Let X and Y be two positive random variables, and let E_1 be a mean 1 exponential random variable which is independent of both X and Y. Define $\widetilde{X} = E_1/X$ and $\widetilde{Y} = E_1/Y$; that is, the distributions of both $\widetilde{X}$ and $\widetilde{Y}$ are scale mixtures of exponential distributions. Then

$$X \le_* Y \Longrightarrow \widetilde{X} \le_* \widetilde{Y}.$$

The proof is obtained by showing that $X \le_* Y \Longrightarrow 1/\widetilde{X} \le_* 1/\widetilde{Y}$, and then using Theorem 4.B.6. We omit the details.

See Remarks 5.A.2 and 5.B.1 for similar results.

A characterization of the order $\le_{\rm c}$ by means of the observed total time on test random variables (see Section 1.A.4) is given next. Let X and Y be two random variables with absolutely continuous distribution functions F and G, respectively. Suppose that 0 is the left endpoint of the supports of X and Y. Let H_F^{-1} and H_G^{-1} be the TTT transforms associated with F and G, respectively (see (1.A.19)), and let H_F and H_G be the respective inverses. Let $X_{\rm ttt}$ and $Y_{\rm ttt}$ be random variables with distribution functions H_F and H_G.

Theorem 4.B.8. *Let X and Y be two nonnegative random variables with absolutely continuous distribution functions having 0 as the left endpoint of their supports. Then*

$$X \le_{\rm c} Y \Longleftrightarrow X_{\rm ttt} \le_{\rm c} Y_{\rm ttt}.$$

Proof. Note that $X \le_{\rm c} Y$ if, and only if,

$$\frac{f(F^{-1}(u))}{g(G^{-1}(u))} \quad \text{is increasing in } u \in [0,1], \tag{4.B.5}$$

where f and g are the densities associated with F and G. From (4.B.5), (3.B.16), and (3.B.17) it is seen that $X \le_{\mathrm{c}} Y$ if, and only if, the ratio $h_F(H_F^{-1}(u))/h_G(H_G^{-1}(u))$ is increasing in $u \in [0,1]$ where h_F and h_G are the density functions associated with H_F and H_G, respectively. Thus, again by (4.B.5), we obtain the stated result. □

A related result is the following.

Theorem 4.B.9. *Let X and Y be two nonnegative random variables with absolutely continuous distribution functions having 0 as the left endpoint of their supports. If $X \le_* Y$, then $X_{\mathrm{ttt}} \le_* Y_{\mathrm{ttt}}$.*

See related results in Theorems 1.A.29, 3.B.1, 4.A.44, and 4.B.29.

The following characterization of the order $\le_*$ is similar to the characterization of the order $\le_{\mathrm{hr}}$ in Theorem 1.B.12.

Theorem 4.B.10. *Let X and Y be two random variables with continuous distribution functions F and G, respectively, with common support $[0,\infty)$. The following conditions are equivalent:*

(a) $X \le_* Y$.

(b) *For all functions α and β, such that α is nonnegative and α and α/β are decreasing, and such that $\int_0^1 \alpha(u)\mathrm{d}F^{-1}(u) < \infty$, $\int_0^1 \alpha(u)\mathrm{d}G^{-1}(u) < \infty$, $0 \ne \int_0^1 \beta(u)\mathrm{d}F^{-1}(u) < \infty$, and $0 \ne \int_0^1 \beta(u)\mathrm{d}G^{-1}(u) < \infty$, we have*

$$\frac{\int_0^1 \alpha(u)\mathrm{d}G^{-1}(u)}{\int_0^1 \beta(u)\mathrm{d}G^{-1}(u)} \le \frac{\int_0^1 \alpha(u)\mathrm{d}F^{-1}(u)}{\int_0^1 \beta(u)\mathrm{d}F^{-1}(u)}.$$

(c) *For any two increasing functions a and b such that b is nonnegative, if $\int_0^1 a(u)b(u)\mathrm{d}F^{-1}(u) = 0$, then $\int_0^1 a(u)b(u)\mathrm{d}G^{-1}(u) \le 0$.*

The orders $\le_{\mathrm{c}}$, $\le_*$, and $\le_{\mathrm{su}}$ can be used to characterize, respectively, IFR, IFRA, and NBU random variables as follows.

Theorem 4.B.11. *Let* Exp *denote any exponential random variable (no matter what its mean is). Let X be a nonnegative random variable. Then*

$$\begin{aligned} X \text{ is IFR} &\Longleftrightarrow X \le_{\mathrm{c}} \mathrm{Exp}, \\ X \text{ is IFRA} &\Longleftrightarrow X \le_* \mathrm{Exp}, \quad \text{and} \\ X \text{ is NBU} &\Longleftrightarrow X \le_{\mathrm{su}} \mathrm{Exp}. \end{aligned}$$

The theorem follows at once from the definitions and the observation that a random variable is IFR [IFRA, NBU] if, and only if, the negative of the logarithm of its survival function is convex [starshaped, superadditive] on $(0,\infty)$.

The claim in the next example is easy to prove.

Example 4.B.12. Let X be a nonnegative random variable with an absolutely continuous distribution function. Then X has a decreasing density if, and only if, $U \le_{\mathrm{c}} X$, where U is a uniform$[0,1]$ random variable.

Example 4.B.13. Let $U_{(j:m)}$ and $U_{(i:n)}$ denote the jth and the ith order statistics of samples from the uniform distribution on $[0,1]$ of sizes m and n, respectively. Then

$$U_{(j:m)} \le_* U_{(i:n)} \quad \text{whenever } i-j \ge \max\{0, n-m\}.$$

This follows from Lemma 3.B.27 and (4.B.4), and from the fact that if U is a uniform random variable on $[0,1]$, then $-\log(1-U)$ is a standard exponential random variable.

It is worthwhile to mention that the above inequality, together with Theorem 4.B.4, yields the first three inequalities in Example 3.A.49.

The following example may be compared with Examples 1.B.24, 1.C.48, 2.A.22, 3.B.38, 6.B.41, 6.D.8, and 6.E.13.

Example 4.B.14. Let X and Y be two absolutely continuous nonnegative random variables with survival functions $\overline{F}$ and $\overline{G}$, respectively. Denote $\Lambda_1 = -\log \overline{F}$ and $\Lambda_2 = -\log \overline{G}$, $i = 1,2$. Consider two nonhomogeneous Poisson processes $N_1 = \{N_1(t),\ t \ge 0\}$ and $N_2 = \{N_2(t),\ t \ge 0\}$ with mean functions Λ_1 and Λ_2 (see Example 3.B.37), respectively. Let $T_{i,1}, T_{i,2}, \ldots$ be the successive epoch times of process N_i, $i = 1,2$. Note that $X =_{\text{st}} T_{1,1}$ and $Y =_{\text{st}} T_{2,1}$.

It turns out that any of the three transform orderings of the first two epoch times implies the same ordering of all the corresponding later epoch times; that is, if $X \le_{\text{c}} [\le_*, \le_{\text{su}}]\ Y$, then $T_{1,n} \le_{\text{c}} [\le_*, \le_{\text{su}}]\ T_{2,n}$, $n \ge 1$. The proof of this fact is similar to the proof in Example 3.B.38, and is therefore omitted.

Similar to the orders $\le_{\text{st}}$, $\le_{\text{hr}}$, and $\le_{\text{lr}}$ (see Theorems 1.B.34, 1.C.33, and 6.B.23), the orders $\le_{\text{c}}$, $\le_*$, and $\le_{\text{su}}$ are also preserved under the formation of orders statistics. This is shown in the next result.

Theorem 4.B.15. *Let (X_i, Y_i), $i = 1,2,\ldots,m$, be independent pairs of random variables such that $X_i \le_{\text{c}} [\le_*, \le_{\text{su}}]\ Y_i$, $i = 1,2,\ldots,m$. Denote the corresponding order statistics by $X_{(1)} \le X_{(2)} \le \cdots \le X_{(m)}$ and $Y_{(1)} \le Y_{(2)} \le \cdots \le Y_{(m)}$. Suppose that the X_i's are identically distributed and that the Y_i's are identically distributed. Then*

$$X_{(k)} \le_{\text{c}} [\le_*, \le_{\text{su}}]\ Y_{(k)}, \quad k = 1,2,\ldots,m. \tag{4.B.6}$$

Proof. Let F $[G]$ denote the common distribution function of the X_i's $[Y_i$'s$]$ and let $F_{(k)}$ $[G_{(k)}]$ denote the distribution function of $X_{(k)}$ $[Y_{(k)}]$. Then it is well known that

$$F_{(k)}(x) = \frac{m!}{(k-1)!(m-k)!} \int_0^{F(x)} u^{k-1}(1-u)^{m-k} du$$

and, similarly,

$$G_{(k)}(x) = \frac{m!}{(k-1)!(m-k)!} \int_0^{G(x)} u^{k-1}(1-u)^{m-k} du.$$

Thus, $G_{(k)}^{-1} F_{(k)} = G^{-1}F$, and (4.B.6) follows from the assumptions of the theorem. □

In the following example it is shown that, under the proper conditions, random minima and maxima are ordered in the convex transform, star, and superadditive order senses; see related results in Examples 1.C.46, 3.B.39, 5.A.24, and 5.B.13.

Example 4.B.16. Let $X_1, X_2, \ldots,$ and $Y_1, Y_2, \ldots,$ each be a sequence of independent and identically distributed random variables. Let N be a positive integer-valued random variable, independent of the X_i's and of the Y_i's. Denote $X_{(1,N)} = \min\{X_1, X_2, \ldots, X_N\}$, $X_{(N,N)} = \max\{X_1, X_2, \ldots, X_N\}$, $Y_{(1,N)} = \min\{Y_1, Y_2, \ldots, Y_N\}$, and $Y_{(N,N)} = \max\{Y_1, Y_2, \ldots, Y_N\}$. It can be shown that if $X_1 \le_{\rm c} [\le_*, \le_{\rm su}]\ Y_1$, then $X_{(1:N)} \le_{\rm c} [\le_*, \le_{\rm su}]\ Y_{(1:N)}$ and $X_{(N:N)} \le_{\rm c} [\le_*, \le_{\rm su}]\ Y_{(N:N)}$.

The convex transform order between X and Y implies the usual stochastic order between ratios of the corresponding spacings as the next result shows; related results can be found in Theorem 1.C.45, and in Examples 6.B.25 and 6.E.15. In the next result we use the following notation. Let $X_{(1:n)} \le X_{(2:n)} \le \cdots \le X_{(n:n)}$ and $Y_{(1:n)} \le Y_{(2:n)} \le \cdots \le Y_{(n:n)}$ be the order statistics corresponding to samples $X_1, X_2, \ldots, X_n$ and $Y_1, Y_2, \ldots, Y_n$; each consists of independent, identically distributed random variables, where the X_i's have the same distribution as X, and the Y_i's have the same distribution as Y. The corresponding spacings are defined by $U_{(i:n)} \equiv X_{(i:n)} - X_{(i-1:n)}$ and $V_{(i:n)} \equiv Y_{(i:n)} - Y_{(i-1:n)}$, $i = 2, 3, \ldots, n$.

Theorem 4.B.17. *Let X and Y be two random variables. If $X \le_{\rm c} Y$, then*

$$\frac{U_{(j:n)}}{U_{(i:n)}} \le_{\rm st} \frac{V_{(j:n)}}{V_{(i:n)}} \quad \text{for } 2 \le i \le j \le n.$$

Proof. First note that from the convexity of $G^{-1}F$ we obtain

$$\frac{G^{-1}F(x_2) - G^{-1}F(x_1)}{x_2 - x_1} \le \frac{G^{-1}F(x_4) - G^{-1}F(x_3)}{x_4 - x_3} \quad \text{whenever } x_1 \le [x_2, x_3] \le x_4,$$

where $x_1 \le [x_2, x_3] \le x_4$ denotes $x_1 \le x_2 \le x_4$ and $x_1 \le x_3 \le x_4$. Thus, for $2 \le i \le j \le n$,

$$P\left\{\frac{U_{(j:n)}}{U_{(i:n)}} > z\right\} = P\left\{\frac{X_{(j:n)} - X_{(j-1:n)}}{X_{(i:n)} - X_{(i-1:n)}} > z\right\}$$
$$\leq P\left\{\frac{G^{-1}F(X_{(j:n)}) - G^{-1}F(X_{(j-1:n)})}{G^{-1}F(X_{(i:n)}) - G^{-1}F(X_{(i-1:n)})} > z\right\}$$
$$= P\left\{\frac{V_{(j:n)}}{V_{(i:n)}} > z\right\},$$

where the last equality follows from the observation that the joint distribution of $G^{-1}F(X_{(i:n)})$, $G^{-1}F(X_{(i-1:n)})$, $G^{-1}F(X_{(j:n)})$, and $G^{-1}F(X_{(j-1:n)})$ is the same as the joint distribution of $Y_{(i:n)}$, $Y_{(i-1:n)}$, $Y_{(j:n)}$, and $Y_{(j-1:n)}$. □

Under a weaker assumption than the one in Theorem 4.B.17 we have the following results.

Theorem 4.B.18. *Let X and Y be two random variables with distribution functions F and G, respectively, such that $F(0) = G(0) = 0$. Let $0 \leq p \leq q$. If $X \leq_* Y$, then*

(a) $E[X^q_{(i:n)}]/E[Y^p_{(i:n)}]$ *is decreasing in i,*
(b) $E[X^q_{(i:n)}]/E[Y^p_{(i:n)}]$ *is increasing in n, and*
(c) $E[X^q_{(n-i:n)}]/E[Y^p_{(n-i:n)}]$ *is decreasing in n,*

provided the expectations exist.

The notation in Theorem 4.B.17 is used in the next result.

Theorem 4.B.19. *Let X and Y be two nonnegative random variables. If $X \leq_* Y$, then $E[U_{(i:n)}] \leq E[V_{(i:n)}]$, $i = 2, 3, \ldots, n$.*

4.B.3 Some related orders

In this subsection we consider random variables X and Y with distribution functions F and G, respectively, and with supports of the form $[0, a)$, $a > 0$ (a can be infinity). We assume throughout this subsection that X and Y have finite means. Denote the mrl functions (see (2.A.1)) that are associated with X and Y, by m and l, respectively.

The random variable X is said to be *smaller than Y in the* DMRL *order* (denoted by $X \leq_{\text{dmrl}} Y$) if

$$\frac{l(G^{-1}(u))}{m(F^{-1}(u))} \quad \text{is increasing in } u \in [0, 1]. \tag{4.B.7}$$

Note that (4.B.7) is the same as the condition

$$\frac{\frac{1}{EY}\int_{G^{-1}(u)}^{\infty} \overline{G}(x)\mathrm{d}x}{\frac{1}{EX}\int_{F^{-1}(u)}^{\infty} \overline{F}(x)\mathrm{d}x} \quad \text{is increasing in } u \in [0, 1], \tag{4.B.8}$$

where $\overline{F}$ and $\overline{G}$ are the survival functions associated with F and G, respectively. Condition (4.B.8) can be written equivalently as

$$\frac{EY - H_G^{-1}(u)}{EX - H_F^{-1}(u)} \quad \text{is increasing in } u \in [0,1],$$

where H_F^{-1} and H_G^{-1} are the TTT transforms (see (1.A.19)) that are associated with F and G, respectively.

Theorem 4.B.20. *Let X and Y be two random variables, each with support of the form $[0, a)$. If $X \le_{\mathrm{c}} Y$, then $X \le_{\mathrm{dmrl}} Y$.*

Proof. Let the equilibrium survival functions associated with F and G be defined as

$$\overline{F}_{\mathrm{e}}(x) = \int_x^\infty \frac{\overline{F}(t)}{EX}\mathrm{d}t \quad \text{and} \quad \overline{G}_{\mathrm{e}}(x) = \int_x^\infty \frac{\overline{G}(t)}{EY}\mathrm{d}t.$$

Let $\alpha(x) \equiv G^{-1}(F(x)) = \overline{G}^{-1}(\overline{F}(x))$ and let

$$\gamma(u) \equiv \overline{F}_{\mathrm{e}}\big(\alpha^{-1}(\overline{G}_{\mathrm{e}}^{-1}(u))\big), \quad u \in [0,1].$$

For simplicity suppose that α and γ are differentiable. A lengthy straightforward computation gives

$$\gamma'(u) = \frac{EX}{EY} \cdot \frac{\mathrm{d}}{\mathrm{d}x}\alpha^{-1}(x)\bigg|_{\overline{G}_{\mathrm{e}}^{-1}(u)}. \tag{4.B.9}$$

By assumption, α is convex. It follows from (4.B.9) that γ is convex, and therefore γ is starshaped. That is,

$$\frac{\overline{F}_{\mathrm{e}}\big(F^{-1}(G(\overline{G}_{\mathrm{e}}^{-1}(u)))\big)}{u} \quad \text{is increasing in } u \in [0,1].$$

Equivalently,

$$\frac{\overline{F}_{\mathrm{e}}\big(F^{-1}(u)\big)}{\overline{G}_{\mathrm{e}}\big(G^{-1}(u)\big)} \quad \text{is decreasing in } u \in [0,1],$$

and (4.B.8) is obtained. □

The random variable X is said to be *smaller than Y in the* NBUE *order* (denoted by $X \le_{\mathrm{nbue}} Y$) if

$$\frac{m\big(F^{-1}(u)\big)}{l\big(G^{-1}(u)\big)} \le \frac{EX}{EY} \quad \text{for all } u \in [0,1]. \tag{4.B.10}$$

Note that (4.B.10) is the same as the condition

$$\frac{1}{EX}\int_{F^{-1}(u)}^{\infty}\overline{F}(x)\mathrm{d}x \le \frac{1}{EY}\int_{G^{-1}(u)}^{\infty}\overline{G}(x)\mathrm{d}x \quad \text{for all } u \in [0,1]. \tag{4.B.11}$$

Condition (4.B.11) can be written equivalently as

$$\frac{H_F^{-1}(u)}{EX} \ge \frac{H_G^{-1}(u)}{EY} \quad \text{for all } u \in [0,1].$$

From (4.B.11) and (3.C.1) it follows that if $EX = EY$, then $X \le_{\rm nbue} Y \Longleftrightarrow X \le_{\rm ew} Y$. In other words, for nonnegative random variables X and Y we have

$$X \le_{\rm nbue} Y \Longleftrightarrow \frac{X}{EX} \le_{\rm ew} \frac{Y}{EY}. \tag{4.B.12}$$

Without the condition that $EX = EY$ the orders $\le_{\rm nbue}$ and $\le_{\rm ew}$ are distinct (see Kochar, Li, and Shaked [316]).

The following result is immediate from (4.B.7) and (4.B.10).

Theorem 4.B.21. *Let X and Y be two random variables, each with support of the form $[0,a)$. If $X \le_{\rm dmrl} Y$, then $X \le_{\rm nbue} Y$.*

In the following two theorems some further relationships among some orders are proven.

Theorem 4.B.22. *Let X and Y be two random variables, each with support of the form $[0,a)$. If $X \le_* Y$, then $X \le_{\rm nbue} Y$.*

Proof. If $X \le_* Y$, then, from Theorem 4.B.9, we have that $\frac{H_G^{-1}(u)}{H_F^{-1}(u)}$ is increasing in $u \in [0,1]$ (see (4.B.1)). Therefore, $\frac{H_G^{-1}(u)}{H_F^{-1}(u)} \le \frac{H_G^{-1}(1)}{H_F^{-1}(1)} = \frac{EY}{EX}$. □

Recall from Theorem 3.B.16 that for random variables with the same means the dispersion order implies the convex order. Thus, from Theorem 4.B.2 it follows that for nonnegative random variables X and Y with finite means, such that $X(EX)^{-1} \le_{\rm st} Y(EY)^{-1}$, we have that the star order implies the Lorenz order. However, a stronger result is true — one can obtain the Lorenz order without assuming any usual stochastic comparison associated with X and Y. This follows from Theorem 4.B.22 and the next result.

Theorem 4.B.23. *Let X and Y be two nonnegative random variables. If $X \le_{\rm nbue} Y$, then $X \le_{\rm Lorenz} Y$.*

Proof. The proof follows at once from (4.B.11) and (3.C.8). □

A summary of the implications among orders that were mentioned so far in this section is given in the following chart.

$$\begin{array}{ccccc}
X \le_{\rm c} Y & \Rightarrow & X \le_* Y & \Rightarrow & X \le_{\rm su} Y \\
\Downarrow & & \Downarrow & & \\
X \le_{\rm dmrl} Y & \Rightarrow & X \le_{\rm nbue} Y & \Rightarrow & X \le_{\rm Lorenz} Y
\end{array}$$

Remark 4.B.24. Using the above facts, we provide here a simple proof of Theorem 4.A.68. Recall from Theorem 4.B.6 that for any $p > 0$ we have that $X \le_* Y$ if, and only if, $X^p \le_* Y^p$. Thus, from (4.A.44) and from Theorems 4.B.22 and 4.B.23 it is seen that if $X \le_* Y$, then $X^p \le_{\text{Lorenz}} Y^p$. This observation, again with the aid of (4.A.44), proves Theorem 4.A.68.

The orders $\le_{\text{dmrl}}$ and $\le_{\text{nbue}}$ can be used to characterize, respectively, DMRL and NBUE random variables as follows.

Theorem 4.B.25. *Let* Exp *denote any exponential random variable (no matter what its mean is). Let X be a nonnegative random variable. Then*

$$X \textit{ is } \text{DMRL} \Longleftrightarrow X \le_{\text{dmrl}} \text{Exp}, \textit{ and}$$
$$X \textit{ is } \text{NBUE} \Longleftrightarrow X \le_{\text{nbue}} \text{Exp}.$$

The theorem follows at once from the definitions and the observation that the mrl function of an exponential random variable is a constant.

Recall from (1.A.19) the definition of the TTT transform. We will now introduce and discuss an order that is defined through a comparison of TTT transforms. Let X and Y be two nonnegative random variables with distribution functions F and G, respectively. If

$$\int_0^{F^{-1}(u)} \overline{F}(x)\mathrm{d}x \le \int_0^{G^{-1}(u)} \overline{G}(x)\mathrm{d}x, \quad \text{for all } u \in (0,1) \tag{4.B.13}$$

then X is said to be *smaller than Y in the* TTT *order* (denoted by $X \le_{\text{ttt}} Y$).

A simple sufficient condition for the order $\le_{\text{ttt}}$ is the usual stochastic order:

$$X \le_{\text{st}} Y \Longrightarrow X \le_{\text{ttt}} Y. \tag{4.B.14}$$

In order to verify (4.B.14) one may just notice that if $X \le_{\text{st}} Y$, then $F^{-1}(u) \le G^{-1}(u)$ for all $u \in (0,1)$ (see (1.A.12)).

By letting $u \to 1$ in (4.B.13) it is seen that

$$X \le_{\text{ttt}} Y \Longrightarrow EX \le EY. \tag{4.B.15}$$

From (4.B.11) and (4.B.13) it follows that if $EX = EY$, then $X \le_{\text{ttt}} Y \Longleftrightarrow X \ge_{\text{nbue}} Y$. In other words, for nonnegative random variables X and Y we have

$$X \ge_{\text{nbue}} Y \Longleftrightarrow \frac{X}{EX} \le_{\text{ttt}} \frac{Y}{EY};$$

see a similar relation in (4.B.12).

It is easy to see that for any two nonnegative random variables X and Y we have

$$X \le_{\text{ttt}} Y \Longrightarrow aX \le_{\text{ttt}} aY \quad \text{for any } a > 0.$$

An important closure property of the order $\le_{\text{ttt}}$, analogous to Theorem 3.C.4, is given next.

Theorem 4.B.26. *Let X and Y be two finite mean continuous nonnegative random variables with interval supports, and with 0 being the common left endpoint of the supports. Then, for any increasing concave function ϕ, such that $\phi(0) = 0$, we have*

$$X \le_{\text{ttt}} Y \Longrightarrow \phi(X) \le_{\text{ttt}} \phi(Y).$$

As a corollary we obtain an analog of (3.C.8):

Corollary 4.B.27. *Let X and Y be two finite mean continuous nonnegative random variables with interval supports, and with 0 being the common left endpoint of the supports. Then*

$$X \le_{\text{ttt}} Y \Longrightarrow X \le_{\text{icv}} Y.$$

Proof. Suppose that $X \le_{\text{ttt}} Y$. Let ϕ be an increasing concave function defined on $[0, \infty)$. Define $\tilde{\phi}(\cdot) = \phi(\cdot) - \phi(0)$, so that $\tilde{\phi}(0) = 0$. From Theorem 4.B.26 we obtain $\tilde{\phi}(X) \le_{\text{ttt}} \tilde{\phi}(Y)$. Hence from (4.B.15) we get $E[\tilde{\phi}(X)] \le E[\tilde{\phi}(Y)]$, and this reduces to $E[\phi(X)] \le E[\phi(Y)]$, provided the expectations exist. □

An interesting closure property of the order $\le_{\text{ttt}}$, analogous to Theorem 3.C.11, is given next.

Theorem 4.B.28. *Let $X_1, X_2, \dots$ be a collection of independent and identically distributed random variables, and let $Y_1, Y_2, \dots$ be another collection of independent and identically distributed random variables. Also, let N be a positive, integer-valued, random variable, independent of the X_i's and of the Y_i's. If X_1 and Y_1 are nonnegative, and if $X_1 \le_{\text{ttt}} Y_1$, then* $\min\{X_1, X_2, \dots, X_N\} \le_{\text{ttt}} \min\{Y_1, Y_2, \dots, Y_N\}$.

Some interesting connections between the order $\le_{\text{ttt}}$ and observed total time on test random variables are given in the next theorem. Let X and Y be two nonnegative random variables. Recall from Section 1.A.4 the definition of the observed total time on test random variables X_{ttt} and Y_{ttt}.

Theorem 4.B.29. *Let X and Y be two nonnegative random variables. Then*

$$X_{\text{ttt}} \le_{\text{st}} Y_{\text{ttt}} \Longleftrightarrow X \le_{\text{ttt}} Y$$

and

$$X \le_{\text{ttt}} Y \Longrightarrow X_{\text{ttt}} \le_{\text{ttt}} Y_{\text{ttt}}.$$

Some related results can be found in Theorems 1.A.29, 3.B.1, 4.A.44, 4.B.8, and 4.B.9.

The following example describes comparisons of random variables that arise in the model of imperfect repair, and as the lifetimes of series systems.

Example 4.B.30. Let X be a nonnegative random variable with survival function $\overline{F}$. For $\theta > 0$, let $X(\theta)$ denote a random variable with the survival function $(\overline{F})^\theta$. Similarly, if Y is a nonnegative random variable with the survival function $\overline{G}$, then denote by $Y(\theta)$ a random variable with survival function $(\overline{G})^\theta$. Suppose that both X and Y have 0 as the left endpoint of their supports.

(a) If $\theta > 1$, then $X \le_{\text{ttt}} Y \Longrightarrow X(\theta) \le_{\text{ttt}} Y(\theta)$.
(b) If $\theta < 1$, then $X(\theta) \le_{\text{ttt}} Y(\theta) \Longrightarrow X \le_{\text{ttt}} Y$.

A generalization of the TTT order is described next. This generalization contains as special cases the orders $\le_{\text{st}}$, $\le_{\text{lir}}$, and $\le_{\text{ttt}}$. Let $\mathfrak{H}$ denote the set of all functions h such that $h(u) > 0$ for $u \in (0,1)$, and $h(u) = 0$ for $u \notin [0,1]$. For $h \in \mathfrak{H}$, if

$$\int_{-\infty}^{F^{-1}(p)} h(F(x))\mathrm{d}x \le \int_{-\infty}^{G^{-1}(p)} h(G(x))\mathrm{d}x, \quad p \in (0,1),$$

then we say that X is smaller than Y in the generalized total time on test transform order with respect to h. We denote this by $X \le_{\text{ttt}}^{(h)} Y$.

Example 4.B.31. Let X and Y be random variables with the same left endpoint of support $a > -\infty$. Let h be a constant function on $[0,1]$; that is, $h(u) = c$, $u \in [0,1]$, for some $c > 0$, and $h(u) = 0$ otherwise. Then $X \le_{\text{ttt}}^{(h)} Y$ if, and only if,

$$F^{-1}(p) \le G^{-1}(p), \quad p \in (0,1);$$

that is (by (1.A.12)), if, and only if, $X \le_{\text{st}} Y$.

Example 4.B.32. Let $h(u) = u$, $u \in [0,1]$, and $h(u) = 0$ otherwise. Then $X \le_{\text{ttt}}^{(h)} Y$ if, and only if,

$$\int_{-\infty}^{F^{-1}(p)} F(x)\mathrm{d}x \le \int_{-\infty}^{G^{-1}(p)} G(x)\mathrm{d}x, \quad p \in (0,1);$$

that is, if, and only if, $X \le_{\text{lir}} Y$; the order $\le_{\text{lir}}$ is defined in Section 3.C.1.

Example 4.B.33. Let X and Y be nonnegative random variables with 0 being the left endpoint of their supports. Let $h(u) = 1 - u$, $u \in [0,1]$, and $h(u) = 0$ otherwise. Then $X \le_{\text{ttt}}^{(h)} Y$ if, and only if,

$$\int_{0}^{F^{-1}(p)} \overline{F}(x)\mathrm{d}x \le \int_{0}^{G^{-1}(p)} \overline{G}(x)\mathrm{d}x, \quad p \in (0,1);$$

that is, if, and only if, $X \le_{\text{ttt}} Y$.

The next result describes a relationship among the orders $\le_{\text{ttt}}^{(h)}$ for different h's.

Theorem 4.B.34. *Let X and Y be two random variables with continuous distribution functions, having 0 as the left endpoint of their supports. Let $h_1, h_2 \in \mathfrak{H}$. Suppose that*

$$h_2(u)/h_1(u) \text{ is decreasing on } (0,1).$$

Then

$$X \leq_{\text{ttt}}^{(h_1)} Y \Longrightarrow X \leq_{\text{ttt}}^{(h_2)} Y.$$

Remark 4.B.35. In Theorem 4.B.34 let $h_1(u) = u$ and $h_2(u) = c$ for some constant $c > 0$, $u \in [0,1]$. Then by Theorem 4.B.34, $X \leq_{\text{ttt}}^{(h_1)} Y \Longrightarrow X \leq_{\text{ttt}}^{(h_2)} Y$; that is, by Examples 4.B.31 and 4.B.32,

$$X \leq_{\text{lir}} Y \Longrightarrow X \leq_{\text{st}} Y \tag{4.B.16}$$

when X and Y are two random variables with continuous distribution functions, having 0 as the left endpoint of their supports.

Recall from Theorem 3.B.13(a) that if X and Y have 0 as the left endpoint of their supports, then $X \leq_{\text{disp}} Y \Longrightarrow X \leq_{\text{st}} Y$. It is not hard to see that $X \leq_{\text{disp}} Y \Longrightarrow X \leq_{\text{lir}} Y$. Thus (4.B.16) strengthens Theorem 3.B.13(a) when X and Y have 0 as the left endpoint of their supports.

Some relationships between the usual stochastic order $\leq_{\text{st}}$ and the orders $\leq_{\text{ttt}}^{(h)}$ are given next.

Theorem 4.B.36. *Let X and Y be two nonnegative random variables with continuous distribution functions, having 0 as the left endpoint of their supports. Let $h \in \mathfrak{H}$.*

(a) *If h is decreasing on $[0,1]$, then $X \leq_{\text{st}} Y \Longrightarrow X \leq_{\text{ttt}}^{(h)} Y$.*
(b) *If h is increasing on $[0,1]$, then $X \leq_{\text{ttt}}^{(h)} Y \Longrightarrow X \leq_{\text{st}} Y$.*

A relationship between the order $\leq_{\text{icv}}$ and some orders $\leq_{\text{ttt}}^{(h)}$ is described next.

Theorem 4.B.37. *Let X and Y be two random variables with continuous distribution functions, and supports $[0,a)$ and $[0,b)$, respectively, for some finite or infinite constants a and b. Let $h \in \mathfrak{H}$ be decreasing on $[0,1]$. Then*

$$X \leq_{\text{ttt}}^{(h)} Y \Longrightarrow X \leq_{\text{icv}} Y.$$

4.C Complements

Section 4.A: Some standard references for the monotone convex and concave orders are Ross [475] and Müller and Stoyan [419], where many of the results that are described in Section 4.A can be found. The characterizations of the order $\leq_{\text{icx}}$ by means of the quantile functions (Theorems

4.A.3 and 4.A.4) are taken from Sordo and Ramos [538]. The condition (4.A.8) is studied in Hürlimann [251]; there it is called the RaC (risk-adjusted capital) order. The present version of the characterizations of the orders $\leq_{\rm icx}$ and $\leq_{\rm icv}$, given in Theorem 4.A.5, is taken from Müller and Rüschendorf [415]. The two characterizations of the order $\leq_{\rm icx}$, given in Theorem 4.A.6, can be found in Makowski [378]; an alternative proof of these results is given in Müller [407]. The result that gives the closure under random convolutions property of the monotone convex and concave orders (Theorem 4.A.9) and its proof are taken from Ross and Schechner [477]. Extensions of Theorem 4.A.9 are given in Jean-Marie and Liu [254]; for example, the results mentioned in Remark 4.A.10 can be found there. Theorem 4.A.11 can be found in Fagiuoli and Pellerey [186]. The comparisons of the random sums in Theorems 4.A.12–4.A.14 are motivated by ideas in Pellerey and Shaked [455]; they can be found in Pellerey [450]. The result that gives the closure under general convex increasing transformations property of the increasing convex order (Theorem 4.A.15) and its proof can be found in Ross [475]. The ordering of the maxima in the sense of $\leq_{\rm icx}$ (Corollary 4.A.16) is implicit in Theorem 9 of Li, Li, and Jing [354]. The increasing convex order comparison of maxima of partial sums (Theorem 4.A.17) is taken from Shao [535]; see also Bulinski and Suquet [114]. The icx and icv comparisons of ratios (Example 4.A.19) are restatements of results of Pellerey and Semeraro [454]. The result that gives the closure under mixtures property of the increasing convex and concave orders (Theorem 4.A.20) has been motivated by a result of Ahmed, Soliman, and Khider [10]. The Laplace transform characterization of the orders $\leq_{\rm icx}$ and $\leq_{\rm icv}$ (Theorem 4.A.21) is essentially taken from Ross and Schechner [477] and from Shaked and Wong [524]. A proof of the characterization of the increasing convex order by means of the number of crossings of two distribution functions (Theorem 4.A.23) can be found in Müller [407]. The characterization of the order $\leq_{\rm mrl}$ by the order $\leq_{\rm icx}$ (Theorem 4.A.24) is taken from Brown and Shanthikumar [112]. The closure property of the order $\leq_{\rm mrl}$ given in Remark 4.A.25 is also taken from Brown and Shanthikumar [112]. The sufficient condition for the increasing concave order in Theorem 4.A.27 is given on page 484 of Landsberger and Meilijson [329]. The fact that the order $\leq_{\rm hmrl}$ implies the order $\leq_{\rm icx}$ (Theorem 4.A.28) can be found in Fagiuoli and Pellerey [185]. The relationship between the orders $\leq_{\rm dil}$ and $\leq_{\rm icx}$ that is described in Theorem 4.A.30 and in Corollary 4.A.31 can be found in Belzunce, Pellerey, Ruiz, and Shaked [72]. The relationship between the orders $\leq_{\rm ew}$ and $\leq_{\rm icx}$ that is described in Corollary 4.A.32 can be found in Fagiuoli, Pellerey, and Shaked [188]; in Kochar, Li, and Shaked [316] it is shown that Corollary 4.A.32 can be easily obtained from Theorem 3.C.4. The result about the expected values of the extremes and the increasing convex order (Theorem 4.A.33) is taken from Downey and Maier [170]. The bivariate characterizations of the orders $\leq_{\rm st}$, $\leq_{\rm hr}$, and $\leq_{\rm lr}$ in Theorem 4.A.36

are taken from Righter and Shanthikumar [466]; its application (Theorem 4.A.37) is taken from Kijima and Ohnishi [292]. The increasing convex and concave comparisons of linear functions of random variables with random coefficients, whose parameters are comparable in the majorization order (Theorems 4.A.38 and 4.A.39 and Corollary 4.A.40), are taken from Denuit and Frostig [144]; further results of this type can be found there. The characterizations of the dilation and the icx orders by means of the increasing convex order (Theorems 4.A.41 and 4.A.42) are taken from Sordo and Ramos [538]. The characterization of the excess wealth order by means of the increasing convex order (Theorem 4.A.43) can be found in Belzunce [63]. The inheritance of the icv order by the observed total time on test random variables (Theorem 4.A.44) is given in Li and Shaked [356]. The icx order comparisons of a sum of independent heterogeneous binomial random variables with a proper binomial random variable (Example 4.A.45) is taken from Boland, Singh, and Cukic [102]. The necessary and sufficient conditions for the comparison of normal random variables (Example 4.A.46) are taken from Müller [413]. The icx comparison of average of exponential random variables with the largest among them (Example 4.A.47) can be found in Argon and Andradóttir [16]. The condition for stochastic equality in the icx case of Theorem 4.A.48 can be found Bhattacharjee and Bhattacharya [87]; the condition for stochastic equality in the icv case of Theorem 4.A.48 follows from the above condition and from Theorem 4.A.1. The condition for stochastic equality in Theorem 4.A.50 is taken from Sordo and Ramos [538]. The characterization of the DMRL and IMRL aging notions by means of the increasing convex order (Theorem 4.A.51) can be found in Cao and Wang [117], who also defined and studied the classes of NBUC and NWUC random variables. The terminology of NBUM and NWUM is due to Bergmann [81]. The characterization of NBUC random variables, given in (4.A.32), is taken from Belzunce, Ortega, and Ruiz [71]. The notions of NBU(2) and NWU(2) are defined in Deshpande, Kochar, and Singh [160]. The extension of the sufficiency condition in Theorem 4.A.51, given in Theorem 4.A.52, is taken from Li and Zuo [359]. Most of the results about the starshaped order (Section 4.A.6) can be found in Alzaid [13]. Most of the results on the orders $\leq_{m\text{-icx}}$ and $\leq_{m\text{-icv}}$ (Section 4.A.7) are taken from Rolski [473]; see also Mukherjee and Chatterjee [404], Fishburn and Lavalle [204], Wang and Young [558], Cheng and Pai [129], and references therein. Lefèvre and Utev [339] studied some stochastic orders among discrete random variables by replacing the integrals in (4.A.36) and in (4.A.37) by summations. Fishburn and Lavalle [204] also studied discrete analogs of the $\leq_{m\text{-icv}}$ orders. Thorlund-Petersen [549] characterized the $\leq_{3\text{-icv}}$ comparison of arithmetic random variables. The definition of the order $\leq_{\infty\text{-icv}}$ can be found in Thistle [548] or in Fishburn and Lavalle [204] and in other references that are given in the latter paper. The moment inequalities that are given in Theorem 4.A.58 are also taken from Fishburn and Lavalle [204]; see further references there, and see also Carletti and Pellerey [121]. Theorem 4.A.60 can be found in Denuit, Lefèvre, and Utev [155]. The sufficient conditions for the m-icx order, in terms of sign changes (Theorem 4.A.63), are taken from Kaas and Hesselager [270]; the stochastic comparisons of the Gamma, inverse Gaussian, and lognormal random variables (Example 4.A.64) can also be found there. A variation of Theorem 4.A.65 can be found in Hesselager [222]. Some results that are related to Theorem 4.A.66 have been derived in Denuit [140]. Fishburn [201, 202] and Stoyan [540, page 22] extended the orders $\leq_{m\text{-icx}}$ and $\leq_{m\text{-icv}}$ by allowing m to be any positive number (that is, not necessarily an integer). They did it by letting the m in (4.A.38) and in (4.A.39) be any number greater than 0. Shaked and Wong [524] considered orders defined by requiring the test functions ϕ in (4.A.40) [respectively, (4.A.41)] to satisfy that $\phi^{(j)}$ [respectively, $(-1)^j\phi^{(j)}$] is increasing, $j = 0, 1, \ldots, m-1$. Denuit, Lefèvre, and Shaked [151] studied the orders defined by requiring (4.A.38) and (4.A.39) to hold as well as $E(X-a)^i \leq E(Y-a)^i$, $i = 1, 2, \ldots, m-1$, where a is the left endpoint of the support of the underlying random variables, and a is assumed to be finite. The results about the orders $\leq_p$, $\leq_{p+}$, and $\leq_{p-}$ (Theorems 4.A.67–4.A.69) are taken from Bhattacharjee and Sethuraman [88], Bhattacharjee [83], Li and Zhu [351], and Jun [265]. Note that the order that we denote by $\leq_{p-}$ is not the same as, but is a modification of, an order discussed by these authors. Some generalizations of Theorem 4.A.69 can be found in Cai and Wu [116]. The condition for stochastic equality in Theorem 4.A.70 is taken from Sordo and Ramos [538]. The discussion involving the orders $\leq_m^{-1}$ is motivated by Muliere and Scarsini [406]; extensions of these orders are developed in Wang and Young [558] and in Maccheroni, Muliere, and Zoli [376].

Bhattacharjee [85] studied the order $\leq_{\text{icx}}$ under the restriction that the compared random variables are discrete.

Baccelli and Makowski [28] denote $X \leq_{\text{FR-st}} Y$ whenever (4.A.21) holds (that is, $X \leq_{\text{FR-st}} Y \Longleftrightarrow X \leq_{\text{hmrl}} Y$). They also define the orders $\leq_{\text{FR-cx}}$ and $\leq_{\text{FR-icx}}$ in a similar manner, and they study many closure properties of the orders $\leq_{\text{FR-st}}$, $\leq_{\text{FR-cx}}$, and $\leq_{\text{FR-icx}}$. The order $\leq_{\text{FR-icx}}$ is a "hybrid" of the orders $\leq_{\text{hmrl}}$ (see (2.B.2)) and $\leq_{\text{3-icx}}$ (see (4.A.36)). It is defined by saying that the nonnegative random variables X and Y satisfy $X \leq_{\text{FR-icx}} Y$ if (here $\overline{F}$ and $\overline{G}$ denote the survival functions of X and Y, respectively)

$$\frac{\int_x^\infty \int_{x_2}^\infty \overline{F}(x_1)\mathrm{d}_1\mathrm{d}x_2}{EX} \leq \frac{\int_x^\infty \int_{x_2}^\infty \overline{G}(x_1)\mathrm{d}x_1\mathrm{d}x_2}{EY} \quad \text{for all } x \geq 0.$$

Clearly, if $EX = EY$, then $X \leq_{\text{FR-icx}} Y$ if, and only if, $X \leq_{\text{2-icx}} Y$. The order $\leq_{\text{FR-cx}}$ is defined by saying that the nonnegative random variables X and Y satisfy $X \leq_{\text{FR-cx}} Y$ if $X \leq_{\text{FR-icx}} Y$ and if $E[X^2]/E[X] = E[Y^2]/E[Y]$.

Section 4.B: A good reference about the convex transform, star, and superadditive orders is Barlow and Proschan [36], where further references can be found. Many of the results given in this section can be found there. Another basic reference about the convex transform order is van Zwet [578]. The result about the relation of the star order and the dispersive order (Theorem 4.B.1) is implicit in Shaked [503], whereas the results about the relation of the superadditive order and the dispersive order (Theorems 4.B.2 and 4.B.3) can be found in Ahmed, Alzaid, Bartoszewicz, and Kochar [8]. The relationship between the star order and the icx order (Theorem 4.B.4) is taken from Szekli [544, page 23]; the idea of the first part of the proof of Theorem 4.B.4 is adopted from Arnold and Villasenor [21]. The property of the star order given in Theorem 4.B.6, when $p = -1$, can be found in Taillie [546]; Rivest [469] has obtained it for a general $p \neq 0$. The comparison of the exponential mixtures with respect to the order $\leq_*$, given in Example 4.B.7, is taken from Bartoszewicz [50]. The characterization of the order $\leq_{\rm c}$ by means of observed total time on test random variables (Theorem 4.B.8) can be found in Barlow and Doksum [34]. The proof of the implication that is given in Theorem 4.B.9 can be found in Bartoszewicz [42, 45]. An interesting study of the relationship between the convex transform, star, and superadditive orders and some variability orders can be found in Metzger and Rüschendorf [393]. A characterization of the star order, by means of the monotonicity in k of the ratio of the quantile functions of the corresponding order statistics $X_{(k)}$ and $Y_{(k)}$ (see (4.B.6)), is given in Bartoszewicz [41]. The characterization of the star order given in Theorem 4.B.10 is taken from Bartoszewicz [45]. The star ordering of order statistics from uniform distribution (Example 4.B.13) can be found in Jeon, Kochar, and Park [255]. The three transform orderings of the epoch times of two nonhomogeneous Poisson processes (Example 4.B.14) are given in Gupta and Kirmani [217]. The result about the preservation of the convex transform, star, and superadditive orders under formation of order statistics (Theorem 4.B.15) is a special case of a result in Belzunce, Mercader, and Ruiz [70]. The results about the convex transform, star, and superadditive order comparisons of random minima and maxima (Example 4.B.16) are taken from Bartoszewicz [49]. An extension of Theorem 4.B.15 to order statistics from samples with a random size can be found in Nanda, Misra, Paul, and Singh [427]. This extension of Nanda, Misra, Paul, and Singh [427] also extends the results in Example 4.B.16. The fact that the convex transform order implies the usual stochastic order among ratios of spacings (Theorem 4.B.17) can be found in Oja [440]. The result about the monotonicity of the ratios of expected values of the order statistics which is implied by the order $\leq_*$ (Theorem 4.B.18) is given in Bartoszewicz [45]; see also Barlow and Proschan [35]. The inequalities between the expected values of spacings from different samples (Theorem 4.B.19) are taken from Paul and Gutierrez [443]. The discussion of the DMRL and the NBUE orders in

Section 4.B.3 follows the work of Kochar and Wiens [319] and of Kochar [306], although some of the proofs here are different; see also Belzunce, Candel, and Ruiz [65] and Fernandez-Ponce, Kochar, and Muñoz-Perez [195]. The discussion of the TTT order in Section 4.B.3 follows the work of Kochar, Li, and Shaked [316]. The result about the preservation of the TTT order under random minima (Theorem 4.B.28) is taken from Li and Zuo [358]. The connections between the order $\le_{\rm ttt}$ and observed total time on test random variables (Theorem 4.B.29) can be found in Li and Shaked [356]. The comparisons of random variables of interest in reliability theory, given in Example 4.B.30, are taken from Li and Shaked [357]. The generalization $\le_{\rm ttt}^{(h)}$ of the TTT order has been introduced and studied in Li and Shaked [357].

The definitions of the orders $\le_{\rm c}$, $\le_*$, and $\le_{\rm su}$, given in Section 4.B.1, are proper when the comparisons apply to distributions of nonnegative random variables. Van Zwet [578], Lawrence [334], and Loh [365] study modifications of these orders which apply to symmetric distributions.

5

The Laplace Transform and Related Orders

The most important common order that is studied in this chapter is the Laplace transform order. Like the orders that were discussed in Chapter 4, the Laplace transform order compares random variables according to both their "location" and their "spread". Two other useful orders, based on ratios of Laplace transforms, are also discussed in this chapter. In addition, some other related orders are investigated in this chapter as well.

5.A The Laplace Transform Order

5.A.1 Definitions and equivalent conditions

The relations $X \leq_{\text{st}} Y$, $X \leq_{\text{cx}} Y$, $X \leq_{\text{icx}} Y$, and $X \leq_{\text{icv}} Y$, as well as many others, are defined by requiring $E[\phi(X)] \leq E[\phi(Y)]$ to hold for all functions ϕ in some class of functions. For example, the class of functions which corresponds to the usual stochastic order is the class of all increasing functions. The order that is discussed in this section corresponds to the class of functions ϕ of the form $\phi(x) = -\mathrm{e}^{-sx}$ where s is a positive number.

More explicitly, let X and Y be two nonnegative random variables such that

$$E[\exp\{-sX\}] \geq E[\exp\{-sY\}] \quad \text{for all } s > 0. \tag{5.A.1}$$

Then X is said to be *smaller than* Y *in the Laplace transform order* (denoted by $X \leq_{\text{Lt}} Y$). Throughout this section we consider only nonnegative random variables.

For a nonnegative random variable X with distribution function F and survival function $\overline{F} \equiv 1 - F$, denote by

$$f^*(s) = \int_0^\infty \mathrm{e}^{-sx} \mathrm{d}F(x) \quad \text{and} \quad \overline{F}^*(s) = \int_0^\infty \mathrm{e}^{-sx} \overline{F}(x) \mathrm{d}x$$

the Laplace-Stieltjes transform of F (or the Laplace transform of X) and the Laplace transform of $\overline{F}$, respectively. Then it is easy to verify that

$$\overline{F}^*(s) = s^{-1}(1 - f^*(s)) \quad \text{for all } s > 0. \tag{5.A.2}$$

Using (5.A.2), the following result is easy to verify.

Theorem 5.A.1. *Let X and Y be two nonnegative random variables with survival functions $\overline{F}$ and $\overline{G}$, respectively. Then $X \le_{\rm Lt} Y$ if, and only if,*

$$\int_0^\infty \mathrm{e}^{-sx}\overline{F}(x)\mathrm{d}x \le \int_0^\infty \mathrm{e}^{-sx}\overline{G}(x)\mathrm{d}x \quad \textit{for all } s > 0. \tag{5.A.3}$$

Note that (5.A.3) can be written as

$$E\min\{X, E_s\} \le E\min\{Y, E_s\} \quad \text{for all } s > 0, \tag{5.A.4}$$

where E_s is an exponential random variable with mean $1/s$, which is independent of X and of Y.

Using (5.A.2) it is also easy to verify the statement that is given in the following remark.

Remark 5.A.2. Let X and Y be two positive random variables, and let E_1 be a mean 1 exponential random variable which is independent of both X and Y. Define $\widetilde{X} = E_1/X$ and $\widetilde{Y} = E_1/Y$; that is, the distributions of both $\widetilde{X}$ and $\widetilde{Y}$ are scale mixtures of exponential distributions. Then

$$X \le_{\rm Lt} Y \Longleftrightarrow \widetilde{Y} \le_{\rm st} \widetilde{X}.$$

See similar results in Example 4.B.7 and in Remark 5.B.1.

If $X \le_{\rm Lt} Y$, then $(1 - E[\exp\{-sX\}])/s \le (1 - E[\exp\{-sY\}])/s$ for all $s > 0$. Letting $s \downarrow 0$ it is seen that

$$X \le_{\rm Lt} Y \Longrightarrow EX \le EY, \tag{5.A.5}$$

provided the expectations exist.

A function $\phi : [0, \infty) \to \mathbb{R}$ is said to be completely monotone if all its derivatives $\phi^{(n)}$ exist and satisfy $\phi^{(0)}(x) \equiv \phi(x) \ge 0$, $\phi^{(1)}(x) \le 0$, $\phi^{(2)}(x) \ge 0, \ldots$; that is, ϕ is completely monotone if $(-1)^n\phi^{(n)}(x) \ge 0$ for all $x > 0$ and $n = 0, 1, 2, \ldots$. It is well known that ϕ is completely monotone if, and only if, there exists a measure μ on $(0, \infty)$ such that

$$\phi(x) = \int_0^\infty \mathrm{e}^{-xu}\mu(\mathrm{d}u).$$

Therefore, if $X \le_{\rm Lt} Y$ and ϕ is completely monotone, then

$$\begin{aligned} E[\phi(X)] = E\Big[\int_0^\infty \mathrm{e}^{-Xu}\mu(\mathrm{d}u)\Big] &= \int_0^\infty E[\mathrm{e}^{-Xu}]\mu(\mathrm{d}u) \\ &\ge \int_0^\infty E[\mathrm{e}^{-Yu}]\mu(\mathrm{d}u) = E[\phi(Y)], \end{aligned}$$

provided the expectations exist. The function ϕ, which is defined by $\phi(x) = \exp\{-sx\}$, is completely monotone for each $s > 0$. We thus have proven the following characterization of the order $\le_{\rm Lt}$.

Theorem 5.A.3. *Let X and Y be two nonnegative random variables. Then $X \leq_{\rm Lt} Y$ if, and only if,*

$$E[\phi(X)] \geq E[\phi(Y)] \tag{5.A.6}$$

for all completely monotone functions ϕ, provided the expectations exist.

A similar result is the following.

Theorem 5.A.4. *Let X and Y be two nonnegative random variables. Then $X \leq_{\rm Lt} Y$ if, and only if,*

$$E[\phi(X)] \leq E[\phi(Y)]$$

for all differentiable functions ϕ on $[0, \infty)$ with a completely monotone derivative, provided the expectations exist.

Next we characterize the order $\leq_{\rm Lt}$ by a function of the respective moments. In order to do that we notice that if X is a nonnegative random variable with survival function $\overline{F}$ such that all its moments exist, then

$$\int_0^\infty \mathrm{e}^{-sx}\overline{F}(x)\mathrm{d}x = \sum_{i=0}^\infty \frac{(-s)^i}{i!}\int_0^\infty x^i\overline{F}(x)\mathrm{d}x = \sum_{i=0}^\infty \frac{(-s)^i}{i!}\frac{EX^{i+1}}{i+1}.$$

Using this fact and Theorem 5.A.1, the proof of the next theorem is apparent.

Theorem 5.A.5. *Let X and Y be nonnegative random variables that possess moments μ_i and ν_i, respectively, $i = 1, 2, \ldots$. Then $X \leq_{\rm Lt} Y$ if, and only if,*

$$\sum_{i=0}^\infty \frac{(-s)^i}{(i+1)!}\mu_{i+1} \leq \sum_{i=0}^\infty \frac{(-s)^i}{(i+1)!}\nu_{i+1} \quad \text{for all } s > 0.$$

A Laplace transform characterization of the order $\leq_{\rm Lt}$ is stated next. It may be compared to Theorems 1.A.13, 1.B.18, 1.B.53, 1.C.25, 2.A.16, 2.B.14, and 4.A.21. We omit its proof.

Theorem 5.A.6. *Let X_1 and X_2 be two nonnegative random variables, and let $N_\lambda(X_1)$ and $N_\lambda(X_2)$ be as described in Theorem 1.A.13. Then*

$$X_1 \leq_{\rm Lt} X_2 \Longleftrightarrow N_\lambda(X_1) \leq_{\rm Lt} N_\lambda(X_2) \quad \text{for all } \lambda > 0.$$

5.A.2 Closure and other properties

Using (5.A.1) and (5.A.6) it is easy to prove each of the closure results in the following theorem. The first part of the theorem follows from the observation that if ϕ is a completely monotone function and g is a positive function with a completely monotone derivative, then $\phi(g)$ is completely monotone. Comments about the proof of the last part are given after the statement of the theorem. (Recall from Section 1.A.3 that for any random variable Z and any event A we denote by $[Z|A]$ any random variable whose distribution is the conditional distribution of Z given A.)

Theorem 5.A.7. (a) *If $X \le_{\rm Lt} Y$ and g is any positive function with a completely monotone derivative, then $g(X) \le_{\rm Lt} g(Y)$.*

(b) *Let X, Y, and Θ be random variables such that $[X|\Theta = \theta] \le_{\rm Lt} [Y|\Theta = \theta]$ for all θ in the support of Θ. Then $X \le_{\rm Lt} Y$. That is, the Laplace transform order is closed under mixtures.*

(c) *Let $\{X_j,\ j = 1,2,\dots\}$ and $\{Y_j,\ j = 1,2,\dots\}$ be two sequences of random variables such that $X_j \to_{\rm st} X$ and $Y_j \to_{\rm st} Y$ as $j \to \infty$. If $X_j \le_{\rm Lt} Y_j$, $j = 1,2,\dots$, then $X \le_{\rm Lt} Y$.*

(d) *Let $X_1, X_2, \dots, X_m$ be a set of independent random variables and let $Y_1, Y_2, \dots, Y_m$ be another set of independent random variables. If $X_i \le_{\rm Lt} Y_i$ for $i = 1,2,\dots,m$, then*

$$g(X_1, X_2, \dots, X_m) \le_{\rm Lt} g(Y_1, Y_2, \dots, Y_m)$$

for all nonnegative functions g on $[0,\infty)^n$ such that $(\partial/\partial x_i)g(x_1, x_2, \dots, x_n)$ is completely monotone in x_i, $i = 1,2,\dots,m$. In particular, the Laplace transform order is closed under convolutions.

The proof of Theorem 5.A.7(d) is very similar to the proof of Theorem 4.A.15. The basic difference is that one should use Theorem 5.A.7(a) rather than Theorem 4.A.8(a) in the first step of the inductive argument.

Another closure property of the order $\le_{\rm Lt}$ is described in the following theorem.

Theorem 5.A.8. *Let $X_1, X_2, \dots$ and $Y_1, Y_2, \dots$ each be a sequence of nonnegative independent random variables, and let M and N be integer-valued positive random variables that are independent of the $\{X_i\}$ and the $\{Y_i\}$ sequences, respectively. Suppose that there exists a nonnegative random variable Z such that $X_i \le_{\rm Lt} Z \le_{\rm Lt} Y_j$ for all i and j. If $M \le_{\rm Lt} N$, then*

$$\sum_{j=1}^{M} X_j \le_{\rm Lt} \sum_{j=1}^{N} Y_j.$$

Proof. Note that for all $s > 0$ we have

$$\begin{aligned}
E\Big[\exp\Big\{-s\sum_{j=1}^{M} X_j\Big\}\Big] &= \sum_{n=1}^{\infty} P\{M = n\} \prod_{j=1}^{n} E[\exp\{-sX_j\}] \\
&\ge \sum_{n=1}^{\infty} P\{M = n\} (E[\exp\{-sZ\}])^n \\
&= \sum_{n=1}^{\infty} P\{M = n\} \exp\{-n(-\log E[\exp\{-sZ\}])\} \\
&\ge \sum_{n=1}^{\infty} P\{N = n\} \exp\{-n(-\log E[\exp\{-sZ\}])\}
\end{aligned}$$

$$
\begin{aligned}
&= \sum_{n=1}^{\infty} P\{N = n\}(E[\exp\{-sZ\}])^n \\
&\ge \sum_{n=1}^{\infty} P\{N = n\} \prod_{j=1}^{n} E[\exp\{-sY_j\}] \\
&= E\Big[\exp\Big\{-s\sum_{j=1}^{N} Y_j\Big\}\Big],
\end{aligned}
$$

where the first and the last equalities follow from the independence of M and N of the $\{X_i\}$ and the $\{Y_i\}$ sequences, the first and the last inequalities follow from $X_i \le_{\rm Lt} Z \le_{\rm Lt} Y_j$ for all i and j, and the middle inequality follows from $M \le_{\rm Lt} N$. The stated result now follows. □

As a corollary of Theorem 5.A.8 we obtain the next result, which is an analog of Theorem 4.A.9. It is worthwhile to point out that Theorem 7.D.7, which is proven in Section 7.D.1, is a more general result than the following theorem.

Theorem 5.A.9. *Let $X_1, X_2, \ldots$ and $Y_1, Y_2, \ldots$ each be a sequence of nonnegative independent and identically distributed random variables such that $X_i \le_{\rm Lt} Y_i$, $i = 1, 2, \ldots$. Let M and N be integer-valued positive random variables that are independent of the $\{X_i\}$ and the $\{Y_i\}$ sequences, respectively, such that $M \le_{\rm Lt} N$. Then*

$$
\sum_{j=1}^{M} X_j \le_{\rm Lt} \sum_{j=1}^{N} Y_j.
$$

A result that is related to Theorem 5.A.9 is given next. It is of interest to compare it to Theorems 1.A.5, 2.B.8, 3.A.14, and 4.A.12.

Theorem 5.A.10. *Let $\{X_j,\ j = 1, 2, \ldots\}$ be a sequence of nonnegative independent and identically distributed random variables, and let M be a positive integer-valued random variable which is independent of the X_i's. Let $\{Y_j,\ j = 1, 2, \ldots\}$ be another sequence of independent and identically distributed random variables, and let N be a positive integer-valued random variable which is independent of the Y_i's. Suppose that for some positive integer K we have*

$$
\sum_{i=1}^{K} X_i \le_{\rm Lt} [\ge_{\rm Lt}]\ Y_1,
$$

and

$$
M \le_{\rm Lt} [\ge_{\rm Lt}]\ KN.
$$

Then

$$
\sum_{j=1}^{M} X_j \le_{\rm Lt} [\ge_{\rm Lt}] \sum_{j=1}^{N} Y_j.
$$

We do not give a detailed proof of Theorem 5.A.10 here since it is similar to the proof of Theorem 4.A.12 in Section 4.A.1.

Two other similar theorems are the following. Their proofs are similar to the proofs of Theorems 4.A.13 and 4.A.14 in Section 4.A.1.

Theorem 5.A.11. *Let $\{X_j,\ j = 1, 2, \dots\}$ be a sequence of nonnegative independent and identically distributed random variables, and let M be a positive integer-valued random variable which is independent of the X_i's. Let $\{Y_j,\ j = 1, 2, \dots\}$ be another sequence of independent and identically distributed random variables, and let N be a positive integer-valued random variable which is independent of the Y_i's. Also, let $\{N_j, j = 1, 2, \dots\}$ be a sequence of independent random variables that are distributed as N. If for some positive integer K we have*

$$\sum_{i=1}^{K} X_i \le_{\mathrm{Lt}} Y_1 \quad \text{and} \quad M \le_{\mathrm{Lt}} \sum_{i=1}^{K} N_i,$$

or if we have

$$KX_1 \le_{\mathrm{Lt}} Y_1 \quad \text{and} \quad M \le_{\mathrm{Lt}} KN,$$

or if we have

$$KX_1 \le_{\mathrm{Lt}} Y_1 \quad \text{and} \quad M \le_{\mathrm{Lt}} \sum_{i=1}^{K} N_i,$$

then

$$\sum_{j=1}^{M} X_j \le_{\mathrm{Lt}} \sum_{j=1}^{N} Y_j.$$

Theorem 5.A.12. *Let $\{X_j,\ j = 1, 2, \dots\}$ be a sequence of nonnegative independent and identically distributed random variables, and let M be a positive integer-valued random variable which is independent of the X_i's. Let $\{Y_j,\ j = 1, 2, \dots\}$ be another sequence of independent and identically distributed random variables, and let N be a positive integer-valued random variable which is independent of the Y_i's. If for some positive integers K_1 and K_2, such that $K_1 \le K_2$, we have*

$$\sum_{i=1}^{K_1} X_i \le_{\mathrm{Lt}} \frac{K_1}{K_2} Y_1 \quad \text{and} \quad M \le_{\mathrm{Lt}} K_2 N,$$

then

$$\sum_{j=1}^{M} X_j \le_{\mathrm{Lt}} \sum_{j=1}^{N} Y_j.$$

Recall from page 2 the definition of the majorization order $\boldsymbol{a} \prec \boldsymbol{b}$ among n-dimensional vectors.

Theorem 5.A.13. *Let $X_1, X_2, \ldots, X_m$ be independent nonnegative random variables. Let $a_1 \geq a_2 \geq \cdots \geq a_m \geq 0$ and $b_1 \geq b_2 \geq \cdots \geq b_m \geq 0$ be constants such that $\boldsymbol{a} \prec \boldsymbol{b}$. If*

$$X_1 \leq_{\mathrm{rh}} X_2 \leq_{\mathrm{rh}} \cdots \leq_{\mathrm{rh}} X_m,$$

then

$$\sum_{i=1}^{m} a_i X_i \leq_{\mathrm{Lt}} \sum_{i=1}^{m} a_{m-i+1} X_i \quad \text{and} \quad \sum_{i=1}^{m} b_i X_i \leq_{\mathrm{Lt}} \sum_{i=1}^{m} a_i X_i.$$

Proof. By Theorem 5.A.7(d) the order $\leq_{\mathrm{Lt}}$ is closed under convolutions. Thus, it suffices to prove the stated results for $m = 2$.

Select an $s \geq 0$. In Theorem 1.B.50(b), take $\alpha(x) = \mathrm{e}^{-a_1 s x}$ and $\beta(x) = \mathrm{e}^{-a_2 s x}$ to obtain

$$E[\exp\{-s(a_1 X_1 + a_2 X_2)\}] \geq E[\exp\{-s(a_1 X_2 + a_2 X_1)\}], \quad s \geq 0;$$

that is, $a_1 X_1 + a_2 X_2 \leq_{\mathrm{Lt}} a_1 X_2 + a_2 X_1$.

In order to prove the second statement, take $\alpha(x) = \mathrm{e}^{-a_2 s x}$ and $\beta(x) = \mathrm{e}^{-b_2 s x}$ in Theorem 1.B.50(b) to obtain

$$\frac{E[\exp\{-b_2 X_2 s\}]}{E[\exp\{-a_2 X_2 s\}]} \geq \frac{E[\exp\{-b_2 X_1 s\}]}{E[\exp\{-a_2 X_1 s\}]}, \quad s \geq 0. \tag{5.A.7}$$

Also, by Theorem 3.A.35 we have $a_1 X_1 + a_2 X_1^* \leq_{\mathrm{cx}} b_1 X_1 + b_2 X_1^*$, where X_1^* is an independent copy of X_1. Therefore, $a_1 X_1 + a_2 X_1^* \geq_{\mathrm{Lt}} b_1 X_1 + b_2 X_1^*$, and hence,

$$\frac{E[\exp\{-b_2 X_1 s\}]}{E[\exp\{-a_2 X_1 s\}]} = \frac{E[\exp\{-b_2 X_1^* s\}]}{E[\exp\{-a_2 X_1^* s\}]} \geq \frac{E[\exp\{-a_1 X_1 s\}]}{E[\exp\{-b_1 X_1 s\}]}, \quad s \geq 0. \tag{5.A.8}$$

Combining (5.A.7) and (5.A.8) we obtain $b_1 X_1 + b_2 X_2 \leq_{\mathrm{Lt}} a_1 X_1 + a_2 X_2$. □

The Laplace transform order is closed under linear convex combinations as the following theorem shows. This result is an analog of Theorem 3.A.36, and its proof is similar to the proof of that theorem; therefore the proof is omitted. Similar results are Theorems 5.C.8 and 5.C.18.

Theorem 5.A.14. *Let $X_1, X_2, \ldots, X_n$ and Y be $n+1$ random variables. If $X_i \geq_{\mathrm{Lt}} Y$, $i = 1, 2, \ldots, n$, then*

$$\sum_{i=1}^{n} a_i X_i \geq_{\mathrm{Lt}} Y,$$

whenever $a_i \geq 0$, $i = 1, 2, \ldots, n$ and $\sum_{i=1}^{n} a_i = 1$.

A result that is similar to Theorems 1.A.8, 3.A.43, 3.A.60, 4.A.69, 6.B.19, 6.G.12, 6.G.13, and 7.A.14–7.A.16, is the following.

Theorem 5.A.15. *Let X and Y be two nonnegative random variables. Suppose that $X \leq_{\rm Lt} Y$ and that $E[X^\alpha] = E[Y^\alpha]$ for some $\alpha < 0$ or for some $\alpha \in (0,1)$, provided the expectations exist. Then $X =_{\rm st} Y$.*

The function ϕ defined by $\phi(x) = \exp\{-sx\}$ is decreasing and convex for each $s > 0$. Therefore $-\phi$ is increasing and concave. We thus obtain the next result.

Theorem 5.A.16. *Let X and Y be two nonnegative random variables. If $X \leq_{\rm icv} Y$, then $X \leq_{\rm Lt} Y$. In particular, if $X \leq_{\rm st} Y$, then $X \leq_{\rm Lt} Y$.*

In fact, from (4.A.41) it follows that if $X \leq_{m\text{-icv}} Y$, for any m, then $X \leq_{\rm Lt} Y$. For random variables with finite supports we have the following characterization of the Laplace transform order by means of the orders $\leq_{m\text{-icv}}$ that were studied in Section 4.A.7.

Theorem 5.A.17. *Let X and Y be two random variables with finite supports. Then $X \leq_{\rm Lt} Y$ if, and only if, $X \leq_{\infty\text{-icv}} Y$ (where $\leq_{\infty\text{-icv}}$ is defined in (4.A.42)).*

Another strengthening of Theorem 5.A.16 is stated and proven next. Recall from Section 4.A.7 the definition of the order $\leq_{p-}$.

Theorem 5.A.18. *Let X and Y be two nonnegative random variables. If $X \leq_{p-} Y$ for some $p \leq 1$, then $X \leq_{\rm Lt} Y$.*

Proof. Recall from (4.A.45) that if $X \leq_{p-} Y$, then $X^p \leq_{\rm icv} Y^p$. Select an $s > 0$. Define $\phi(x) \equiv \mathrm{e}^{-sx}$ and let

$$h(x) \equiv \phi\big(x^{1/p}\big) = \mathrm{e}^{-sx^{1/p}}.$$

It is easy to verify that the function h is decreasing and convex, and therefore $-h$ is increasing and concave. From the fact that $X^p \leq_{\rm icv} Y^p$ it follows that

$$-E[h(X^p)] \leq -E[h(Y^p)],$$

or, equivalently, that

$$E\big[\mathrm{e}^{-sX}\big] \geq E\big[\mathrm{e}^{-sY}\big].$$

Since the latter inequality holds for all $s > 0$ it follows that $X \leq_{\rm Lt} Y$. □

Closure properties of an order under the operation of taking minima are of importance in reliability theory. The next result gives conditions under which the order $\leq_{\rm Lt}$ is closed under this operation. We do not give the proof here.

Theorem 5.A.19. *Let the independent nonnegative random variables* $X_1, X_2, \dots, X_m, Y_1, Y_2, \dots, Y_m$ *have the survival functions* $\overline{F}_1, \overline{F}_2, \dots, \overline{F}_m, \overline{G}_1, \overline{G}_2, \dots, \overline{G}_m$, *respectively. If* $X_i \le_{\rm Lt} Y_i$, $i = 1, 2, \dots, m$, *and* $\overline{F}_i$ *and* $\overline{G}_i$, $i = 1, 2, \dots, m$, *are completely monotone, then*

$$\min\{X_1, X_2, \dots, X_m\} \le_{\rm Lt} \min\{Y_1, Y_2, \dots, Y_m\}.$$

Remark 5.A.20. Let $\{X, X_1, X_2, \dots\}$ be a set of nonnegative independent and identically distributed random variables, and let $\{Y, Y_1, Y_2, \dots\}$ be another set of nonnegative independent and identically distributed random variables. Denote by $X_{(i:n)}$ the ith order statistic in a sample of size n from $\{X_1, X_2, \dots\}$, and denote by $Y_{(i:n)}$ the ith order statistic in a sample of size n from $\{Y_1, Y_2, \dots\}$. If $X \le_{\rm disp} Y$, then, by Theorem 3.B.31, for $2 \le i \le n$ we have $X_{(i:n)} - X_{(i-1:n)} \le_{\rm st} Y_{(i:n)} - Y_{(i-1:n)}$, and therefore, by Theorem 5.A.16, we have

$$X_{(i:n)} - X_{(i-1:n)} \le_{\rm Lt} Y_{(i:n)} - Y_{(i-1:n)}, \quad 2 \le i \le n. \tag{5.A.9}$$

Bartoszewicz [46] proved a similar result. He showed that if $X \le_{\rm disp} Y$, and if the X_i's and the Y_i's are independent, then

$$X_{(i:n)} + Y_{(i-1:n)} \le_{\rm Lt} X_{(i-1:n)} + Y_{(i:n)}, \quad 2 \le i \le n. \tag{5.A.10}$$

This is different from (5.A.9) because $X_{(i-1:n)}$ and $X_{(i:n)}$ (and $Y_{(i-1:n)}$ and $Y_{(i:n)}$) in (5.A.9) have a particular joint distribution, whereas (5.A.10) involves only the marginal distributions of $X_{(i-1:n)}$ and $X_{(i:n)}$ (and of $Y_{(i-1:n)}$ and $Y_{(i:n)}$). Bartoszewicz [46] also proved that if $X \le_{\rm disp} Y$, and if the X_i's and the Y_i's are independent, then

$$X_{(n+1-i:n+1)} + Y_{(n-i:n)} \le_{\rm Lt} X_{(n-i:n)} + Y_{(n+1-i:n+1)}, \quad 0 \le i \le n-1,$$

and

$$X_{(i:n)} + Y_{(i:n+1)} \le_{\rm Lt} X_{(i:n+1)} + Y_{(i:n)}, \quad 1 \le i \le n.$$

In reliability theory, motivated by (3.A.62) and Theorem 5.A.16, one may consider the class of nonnegative random variables X which satisfy

$$X \ge_{\rm Lt} [\le_{\rm Lt}] \ {\rm Exp}(\mu) \tag{5.A.11}$$

or, equivalently,

$$\int_0^\infty {\rm e}^{-su} P\{X > u\} {\rm d}u \ge [\le] \ \frac{\mu}{1 + s\mu} \quad \text{for } s \ge 0,$$

where μ is the mean of X. Such random variables have interesting aging properties. From Theorems 3.A.55 and 5.A.16 it is seen that if X is NBUE [NWUE], then X satisfies (5.A.11).

Some researchers studied random variables X which satisfy

$$X \geq_{\rm Lt} {\rm Gamma}(\alpha, \beta),$$

where Gamma(α, β) denotes a Gamma random with shape parameter α and scale parameter β, which has the same mean as X. See Klar [300], Hu and Lin [228], and references therein.

Let X be a nonnegative random variable with a finite mean. Recall the definition of the asymptotic equilibrium age A_X whose distribution function is given in (1.A.20). Let Y be another nonnegative random variable with the corresponding asymptotic equilibrium age A_Y. From (5.A.3) it is seen at once that if $EX = EY$, then

$$X \leq_{\rm Lt} Y \Longleftrightarrow A_X \geq_{\rm Lt} A_Y. \tag{5.A.12}$$

The next result indicates the "minimal" and the "maximal" random variables, with respect to the order $\leq_{\rm Lt}$, when the mean and the variance are given. It is worthwhile to contrast it with Theorem 3.A.24.

Theorem 5.A.21. *Let Y be a nonnegative random variable with mean μ and variance σ^2. Let X be a random variable such that $P\{X = 0\} = 1 - P\{X = (\mu^2 + \sigma^2)/\mu\} = \sigma^2/(\mu^2 + \sigma^2)$ (so that $EX = \mu$ and $\mathrm{Var}(X) = \sigma^2$) and let Z be a random variable degenerate at μ. Then*

$$X \leq_{\rm Lt} Y \leq_{\rm Lt} Z. \tag{5.A.13}$$

Proof. The right-side inequality in (5.A.13) follows at once from Jensen's Inequality. Let $\overline{F}$ and $\overline{G}$ be, respectively, the survival functions of X and Y. In order to obtain the left-side inequality in (5.A.13) we will show that

$$\int_0^\infty {\rm e}^{-sx}\overline{F}(x){\rm d}x \leq \int_0^\infty {\rm e}^{-sx}\overline{G}(x){\rm d}x \quad \text{for all } s \geq 0. \tag{5.A.14}$$

The result will then follow from Theorem 5.A.1.

Define the functions α and β on $(0, \infty)$ by $\alpha(x) = \overline{F}(x)/\mu$ and $\beta(x) = \overline{G}(x)/\mu$. It is easy to see that both α and β are density functions with a common mean $(\mu^2 + \sigma^2)/2\mu$. In fact, α is the density function of the uniform distribution over the interval $[0, (\mu^2 + \sigma^2)/\mu)$, whereas β is a density which is decreasing on $[0, \infty)$. From Theorem 3.A.46 it now follows that

$$\int_0^\infty \phi(x)\frac{\overline{F}(x)}{\mu}{\rm d}x \leq \int_0^\infty \phi(x)\frac{\overline{G}(x)}{\mu}{\rm d}x$$

for all convex functions ϕ, and in particular (5.A.14) holds. □

A characterization of the hazard rate order, by means of the Laplace transform order, is described in the following theorem. Recall from Section 1.A.3 that for any random variable Z and an event A we denote by $[Z|A]$ any random variable that has as its distribution the conditional distribution of Z given A.

Theorem 5.A.22. *Let X and Y be two continuous random variables with right support endpoints u_X and u_Y, respectively. Then $X \le_{\rm hr} Y$ if, and only if,*

$$[X - t \big| X > t] \le_{\rm Lt} [Y - t \big| Y > t] \quad \textit{for all } t < \min\{u_X, u_Y\}. \tag{5.A.15}$$

Proof. The fact that $X \le_{\rm hr} Y$ implies (5.A.15) follows from (1.B.6) and Theorem 5.A.16. In order to prove the converse, let us assume, for simplicity, that $u_X = u_Y = \infty$. Denote by $\overline{F}$ and $\overline{G}$ the survival functions of X and Y, respectively. Now note that

$$\begin{aligned}
&[X - t \big| X > t] \le_{\rm Lt} [Y - t \big| Y > t] \quad \text{for all } t \\
\Longleftrightarrow\ & \int_0^\infty \mathrm{e}^{-su} \frac{\overline{F}(u+t)}{\overline{F}(t)} \mathrm{d}u \le \int_0^\infty \mathrm{e}^{-su} \frac{\overline{G}(u+t)}{\overline{G}(t)} \mathrm{d}u \quad \text{for all } t \text{ and } s > 0 \\
\Longleftrightarrow\ & \frac{\int_t^\infty \mathrm{e}^{-su}\overline{G}(u)\mathrm{d}u}{\int_t^\infty \mathrm{e}^{-su}\overline{F}(u)\mathrm{d}u} \ge \frac{\overline{G}(t)}{\overline{F}(t)} \quad \text{for all } t \text{ and } s > 0 \\
\Longleftrightarrow\ & \frac{\int_t^\infty \mathrm{e}^{-su}\overline{G}(u)\mathrm{d}u}{\int_t^\infty \mathrm{e}^{-su}\overline{F}(u)\mathrm{d}u} \quad \text{is increasing in } t \text{ for all } s > 0 \qquad (5.A.16) \\
\Longleftrightarrow\ & \frac{\frac{1}{s}\mathrm{e}^{-st}\left[\overline{G}(t) - \mathrm{e}^{st}\int_t^\infty \mathrm{e}^{-su}\overline{G}(u)\mathrm{d}u\right]}{\frac{1}{s}\mathrm{e}^{-st}\left[\overline{F}(t) - \mathrm{e}^{st}\int_t^\infty \mathrm{e}^{-su}\overline{F}(u)\mathrm{d}u\right]} \quad \text{is increasing in } t \text{ for all } s > 0 \\
\Longleftrightarrow\ & \frac{\left[\overline{G}(t) - \mathrm{e}^{st}\int_t^\infty \mathrm{e}^{-su}\overline{G}(u)\mathrm{d}u\right]}{\left[\overline{F}(t) - \mathrm{e}^{st}\int_t^\infty \mathrm{e}^{-su}\overline{F}(u)\mathrm{d}u\right]} \quad \text{is increasing in } t \text{ for all } s > 0, \quad (5.A.17)
\end{aligned}$$

where the second from last equivalence follows by integration by parts. Using the Dominated Convergence Theorem, it is not hard to see that $\lim_{s\to 0} \mathrm{e}^{st}\int_t^\infty \mathrm{e}^{-su}\overline{F}(u)\mathrm{d}u = \lim_{s\to 0} \mathrm{e}^{st}\int_t^\infty \mathrm{e}^{-su}\overline{G}(u)\mathrm{d}u = 0$. Therefore, letting $s \to 0$ in (5.A.17) we obtain that $\overline{G}(t)/\overline{F}(t)$ is increasing in t; that is, $X \le_{\rm hr} Y$. □

Remark 5.A.23. The equivalence of $X \le_{\rm hr} Y$ and (5.A.16), together with (2.A.3), yield a proof of Theorem 2.A.6.

In the following example it is shown, under a proper condition which is stated by means of the Laplace transform order, that random minima and maxima are ordered in the usual stochastic order sense; see related results in Examples 1.C.46, 3.B.39, 4.B.16, and 5.B.13.

Example 5.A.24. Let $X_1, X_2, \ldots$ be a sequence of nonnegative independent and identically distributed random variables with a common distribution function F_{X_1} and a common survival function $\overline{F}_{X_1}$. Let N_1 and N_2 be two positive integer-valued random variables, which are independent of the X_i's, and which have the Laplace transforms L_{N_1} and L_{N_2}. Denote $X_{(1:N_j)} \equiv \min\{X_1, X_2, \ldots, X_{N_j}\}$ and $X_{(N_j:N_j)} \equiv \max\{X_1, X_2, \ldots, X_{N_j}\}$, $j = 1, 2$. Then the survival function of $X_{(1:N_j)}$ is given by

$$\overline{F}_{X_{(1:N_j)}}(x) = L_{N_j}(-\log \overline{F}_{X_1}(x)), \quad j = 1, 2.$$

It is thus seen that if $N_1 \le_{\rm Lt} N_2$, then $X_{(1:N_1)} \ge_{\rm st} X_{(1:N_2)}$. In a similar manner it can be shown that if $N_1 \le_{\rm Lt} N_2$, then also $X_{(N_1:N_1)} \le_{\rm st} X_{(N_2:N_2)}$.

An example with a similar spirit is the following.

Example 5.A.25. Consider a compound Poisson process with rate λ and distribution ϕ. Suppose that this process is the (random) hazard rate function of a random variable X. Then the survival function $\overline{F}$ of X is given by

$$\overline{F}(t) = \exp\Big\{-\int_0^t \lambda[1 - L_\phi(s)]{\rm d}s\Big\}, \quad t \ge 0, \tag{5.A.18}$$

where L_ϕ is the Laplace transform of ϕ (see Kebir [280, page 873]). Similarly let Y have the survival function $\overline{G}$ given by

$$\overline{G}(t) = \exp\Big\{-\int_0^t \lambda[1 - L_\varphi(s)]{\rm d}s\Big\}, \quad t \ge 0, \tag{5.A.19}$$

where φ is a distribution function, and where L_φ is the Laplace transform of φ. It is now seen that if the random variable associated with ϕ is larger, in the Laplace transform order, than the random variable associated with φ, then $\overline{G}(t)/\overline{F}(t)$ is increasing in $t \ge 0$; that is (see (1.B.3)), $X \le_{\rm hr} Y$.

A variation of this result is given in Example 5.B.14.

When X is a nonnegative integer-valued random variable, then it is customary and convenient to analyze it using its probability generating function $E[t^X]$, $t \in (0,1)$, rather than its Laplace transform $E[{\rm e}^{-sX}]$, $s \ge 0$. This fact suggests the following definition.

Let X and Y be two nonnegative integer-valued random variables such that

$$E[t^X] \ge E[t^Y] \quad \text{for all } t \in (0,1). \tag{5.A.20}$$

Then X is said to be *smaller than Y in the probability generating function order* (denoted as $X \le_{\rm pgf} Y$).

It is not hard to verify the following relation which holds for any nonnegative integer-valued random variable X:

$$\sum_{j=1}^{\infty} t^j P\{X \ge j\} = t \sum_{j=0}^{\infty} t^j P\{X > j\} = \frac{t\,(1 - E[t^X])}{1 - t} \quad \text{for all } t \in (0,1).$$

We thus obtain the following analog of Theorem 5.A.1.

Theorem 5.A.26. *Let X and Y be two nonnegative integer-valued random variables. Then $X \le_{\rm pgf} Y$ if, and only if,*

$$\sum_{j=1}^{\infty} t^j P\{X \ge j\} \le \sum_{j=1}^{\infty} t^j P\{Y \ge j\} \quad \textit{for all } t \in (0,1).$$

It is easy to see that (5.A.20) holds if, and only if, (5.A.1) holds. That is,

$$X \le_{\rm pgf} Y \Longleftrightarrow X \le_{\rm Lt} Y.$$

5.B Orders Based on Ratios of Laplace Transforms

5.B.1 Definitions and equivalent conditions

In this section, for a nonnegative random variable X with distribution function F and survival function $\overline{F} \equiv 1 - F$, let us denote by

$$L_X(s) = \int_0^\infty \mathrm{e}^{-sx} \mathrm{d}F(x) \quad \text{and} \quad L_X^*(s) = \int_0^\infty \mathrm{e}^{-sx} \overline{F}(x) \mathrm{d}x$$

the Laplace-Stieltjes transform of F (or the Laplace transform of X) and the Laplace transform of $\overline{F}$, respectively. If Y is another nonnegative random variable, we similarly define L_Y and L_Y^*. If

$$\frac{L_Y(s)}{L_X(s)} \quad \text{is decreasing in } s > 0, \tag{5.B.1}$$

then X is said to be *smaller than Y in the Laplace transform ratio order* (denoted by $X \leq_{\text{Lt-r}} Y$). If

$$\frac{1 - L_Y(s)}{1 - L_X(s)} \quad \text{is decreasing in } s > 0, \tag{5.B.2}$$

then X is said to be *smaller than Y in the reverse Laplace transform ratio order* (denoted by $X \leq_{\text{r-Lt-r}} Y$).

Since $L_X^*(s) = s^{-1}(1 - L_X(s))$ and $L_Y^*(s) = s^{-1}(1 - L_Y(s))$ for all $s > 0$, it follows that

$$X \leq_{\text{Lt-r}} Y \Longleftrightarrow \frac{1 - sL_Y^*(s)}{1 - sL_X^*(s)} \quad \text{is decreasing in } s > 0,$$

and that

$$X \leq_{\text{r-Lt-r}} Y \Longleftrightarrow \frac{L_Y^*(s)}{L_X^*(s)} \quad \text{is decreasing in } s > 0.$$

Using (5.A.2) it is easy to verify the statements that are given in the following remark.

Remark 5.B.1. Let X and Y be two positive random variables, and let E_1 be a mean 1 exponential random variable which is independent of both X and Y. Define $\widetilde{X} = E_1/X$ and $\widetilde{Y} = E_1/Y$; that is, the distributions of both $\widetilde{X}$ and $\widetilde{Y}$ are scale mixtures of exponential distributions. Then

$$X \leq_{\text{Lt-r}} Y \Longleftrightarrow \widetilde{Y} \leq_{\text{hr}} \widetilde{X} \qquad \text{and}$$
$$X \leq_{\text{r-Lt-r}} Y \Longleftrightarrow \widetilde{Y} \leq_{\text{rh}} \widetilde{X}.$$

See similar results in Example 4.B.7 and in Remark 5.A.2.

The next theorem characterizes the orders $\le_{\text{Lt-r}}$ and $\le_{\text{r-Lt-r}}$ by means of functions of the respective moments. The characterization is an analog of the characterization of the Laplace transform order given in Theorem 5.A.5.

Theorem 5.B.2. *Let X and Y be nonnegative random variables that possess moments μ_i and ν_i, respectively, $i = 1, 2, \dots$, $(\mu_0 = \nu_0 = 1)$. Then*

(a) $X \le_{\text{Lt-r}} Y$ *if, and only if,*

$$\frac{\sum_{n=0}^{\infty} \frac{(-s)^i}{i!}\nu_i}{\sum_{n=0}^{\infty} \frac{(-s)^i}{i!}\mu_i} \quad \text{is decreasing in } s > 0.$$

(b) $X \le_{\text{r-Lt-r}} Y$ *if, and only if,*

$$\frac{\sum_{n=1}^{\infty} \frac{(-s)^i}{i!}\nu_i}{\sum_{n=1}^{\infty} \frac{(-s)^i}{i!}\mu_i} \quad \text{is decreasing in } s > 0.$$

Proof. By writing $\mathrm{e}^{-st} = \sum_{i=0}^{\infty} \frac{(-s)^i}{i!} t^i$, the result follows easily from the definitions. □

5.B.2 Closure properties

We list below some preservation properties of the orders $\le_{\text{Lt-r}}$ and $\le_{\text{r-Lt-r}}$. Below, for any nonnegative random variable Z, we will denote by L_Z the Laplace transform of Z.

Theorem 5.B.3. *Let $X_1, X_2, \dots$ be independent, identically distributed nonnegative random variables, and let N_1 and N_2 be positive integer-valued random variables which are independent of the X_i's. Then*

$$N_1 \le_{\text{Lt-r}} [\le_{\text{r-Lt-r}}]\ N_2 \Longrightarrow \sum_{i=1}^{N_1} X_i \le_{\text{Lt-r}} [\le_{\text{r-Lt-r}}] \sum_{i=1}^{N_2} X_i.$$

Proof. For $j = 1, 2$, we have

$$\begin{aligned} L_{X_1+X_2+\cdots+X_{N_j}}(s) &= \sum_{i=1}^{\infty} P\{N_j = i\} L_{X_1+X_2+\cdots+X_i}(s) \\ &= \sum_{i=1}^{\infty} P\{N_j = i\} L_{X_1}^i(s) \\ &= L_{N_j}(-\log L_{X_1}(s)). \end{aligned}$$

The stated results now follow from the assumptions. □

If the X_i's are not assumed to be identically distributed, then stronger assumptions on the relationship between N_1 and N_2 yield the same conclusion. This is shown in the next two theorems.

Theorem 5.B.4. *Let $X_1, X_2, \ldots$ be independent nonnegative random variables, and let N_1 and N_2 be positive integer-valued random variables which are independent of the X_i's. If $N_1 \le_{\rm rh} N_2$, then*

$$\sum_{i=1}^{N_1} X_i \le_{\text{Lt-r}} \sum_{i=1}^{N_2} X_i.$$

Proof. For $j = 1, 2$, we have

$$L_{X_1+X_2+\cdots+X_{N_j}}(s) = \sum_{i=1}^{\infty} P\{N_j = i\} \prod_{k=1}^{i} L_{X_k}(s).$$

For $0 < s_1 < s_2$ we need to show that

$$\left[\sum_{m=1}^{\infty} P\{N_1 = m\} \prod_{k=1}^{m} L_{X_k}(s_1)\right] \left[\sum_{n=1}^{\infty} P\{N_2 = n\} \prod_{k=1}^{n} L_{X_k}(s_2)\right] - \left[\sum_{m=1}^{\infty} P\{N_2 = m\} \prod_{k=1}^{m} L_{X_k}(s_1)\right] \left[\sum_{n=1}^{\infty} P\{N_1 = n\} \prod_{k=1}^{n} L_{X_k}(s_2)\right] \le 0.$$

This follows from the remark after Theorem 2.1 of Joag-Dev, Kochar, and Proschan [259] by noting that

$$\big(g_1(m), g_2(m)\big) \equiv \Big(\prod_{k=1}^{m} L_{X_k}(s_2), \prod_{k=1}^{m} L_{X_k}(s_1)\Big)$$

is a pair of what Joag-Dev, Kochar, and Proschan [259] call DP_2 functions of m, whenever $0 < s_1 < s_2$, and $g_1(m)$ is decreasing in m. □

Theorem 5.B.5. *Let $X_1, X_2, \ldots$ be independent nonnegative random variables, and let N_1 and N_2 be positive integer-valued random variables which are independent of the X_i's. If $N_1 \le_{\rm hr} N_2$, and if $X_i \le_{\text{r-Lt-r}} X_{i+1}$, then*

$$\sum_{i=1}^{N_1} X_i \le_{\text{r-Lt-r}} \sum_{i=1}^{N_2} X_i.$$

Proof. For $j = 1, 2$, we have

$$\begin{aligned} 1 - L_{X_1+X_2+\cdots+X_{N_j}}(s) &= \sum_{m=1}^{\infty} P\{N_j = m\} \left[1 - \prod_{k=1}^{m} L_{X_k}(s)\right] \\ &= \sum_{m=0}^{\infty} P\{N_j > m\} \prod_{k=1}^{m} L_{X_k}(s) \big[1 - L_{X_{m+1}}(s)\big], \end{aligned}$$

where $\prod_{k=1}^{0} L_{X_k}(s) \equiv 1$. So for $0 < s_1 < s_2$ we have that

$$\begin{aligned}
&\left[1 - L_{X_1+X_2+\cdots+X_{N_1}}(s_1)\right]\left[1 - L_{X_1+X_2+\cdots+X_{N_2}}(s_2)\right] \\
&\qquad - \left[1 - L_{X_1+X_2+\cdots+X_{N_2}}(s_1)\right]\left[1 - L_{X_1+X_2+\cdots+X_{N_1}}(s_2)\right] \\
&= \sum_{m=1}^{\infty}\sum_{n=0}^{m-1}\Big[P\{N_1 > m\}P\{N_2 > n\} - P\{N_2 > m\}P\{N_1 > n\}\Big] \\
&\qquad \times \prod_{k=1}^{n} L_{X_k}(s_1)\prod_{k=1}^{n} L_{X_k}(s_2) \\
&\qquad\qquad \times \Bigg[\prod_{k=n+1}^{m} L_{X_k}(s_1)(1 - L_{X_{m+1}}(s_1))(1 - L_{X_{n+1}}(s_2)) \\
&\qquad\qquad\qquad - \prod_{k=n+1}^{m} L_{X_k}(s_2)(1 - L_{X_{m+1}}(s_2))(1 - L_{X_{n+1}}(s_1))\Bigg] \\
&\le 0.
\end{aligned}$$

The last inequality follows since $N_1 \le_{\text{hr}} N_2$ implies that

$$P\{N_1 > m\}P\{N_2 > n\} - P\{N_2 > m\}P\{N_1 > n\} \le 0 \quad \text{for } m > n,$$

and $X_i \le_{\text{r-Lt-r}} X_{i+1}$ implies that

$$(1 - L_{X_{m+1}}(s_1))(1 - L_{X_{n+1}}(s_2)) - (1 - L_{X_{m+1}}(s_2))(1 - L_{X_{n+1}}(s_1)) \ge 0 \quad \text{for } m > n.$$

The stated result now follows. □

Some other preservation results are given in the following theorems.

Theorem 5.B.6. *Let $X_1, X_2, \ldots, X_n$ be a set of independent nonnegative random variables and let $Y_1, Y_2, \ldots, Y_n$ be another set of independent nonnegative random variables. If $X_j \le_{\text{Lt-r}} Y_j$, $j = 1, 2, \ldots, n$, then $X_1 + X_2 + \cdots + X_n \le_{\text{Lt-r}} Y_1 + Y_2 + \cdots + Y_n$.*

Proof. Since $L_{X_1+X_2+\cdots+X_n}(s) = \prod_{i=1}^{n} L_{X_i}(s)$, we see that if $\frac{L_{Y_j}(s)}{L_{X_j}(s)}$ is decreasing in s, $j = 1, 2, \ldots, n$, then $\frac{L_{Y_1+Y_2+\cdots+Y_n}(s)}{L_{X_1+X_2+\cdots+X_n}(s)}$ is also decreasing in s. □

As a special case of Theorem 5.B.6 we see that if X and Y are nonnegative independent random variables, then

$$X \le_{\text{Lt-r}} X + Y. \tag{5.B.3}$$

Theorem 5.B.7. *Let $\{X_j\}$ and $\{Y_j\}$ be two sequences of random variables such that $X_j \to_{\text{st}} X$ and $Y_j \to_{\text{st}} Y$ as $j \to \infty$. If $X_j \le_{\text{Lt-r}} [\le_{\text{r-Lt-r}}]\ Y_j$, $j = 1, 2, \ldots$, then $X \le_{\text{Lt-r}} [\le_{\text{r-Lt-r}}]\ Y$.*

Theorem 5.B.8. *Let X, Y, and Θ be random variables such that $[X\big|\Theta = \theta] \le_{\text{Lt-r}} [\le_{\text{r-Lt-r}}] [Y\big|\Theta = \theta']$ for all θ and θ' in the support of Θ. Then $X \le_{\text{Lt-r}} [\le_{\text{r-Lt-r}}] Y$.*

Proof. We only give the proof for the $\le_{\text{Lt-r}}$ order. The proof for the order $\le_{\text{r-Lt-r}}$ is similar. Note that

$$\frac{L_X(s)}{L_Y(s)} = \frac{E_\Theta\left[L_{[X|\Theta]}(s)\right]}{E_\Theta\left[L_{[Y|\Theta]}(s)\right]}.$$

It can be verified that $\frac{\mathrm{d}}{\mathrm{d}s}\left[\frac{L_X(s)}{L_Y(s)}\right] \ge 0$ if $\frac{\mathrm{d}}{\mathrm{d}s}\left[\frac{L_{[X|\theta]}(s)}{L_{[Y|\theta']}(s)}\right] \ge 0$ for all θ and θ' in the support of Θ. □

In the next result it is shown that a random variable, whose distribution is the mixture of two distributions of $\le_{\text{Lt-r}}$ $[\le_{\text{r-Lt-r}}]$ ordered random variables, is bounded from below and from above, in the $\le_{\text{Lt-r}}$ $[\le_{\text{r-Lt-r}}]$ order sense, by these two random variables.

Theorem 5.B.9. *Let X and Y be two nonnegative random variables with distribution functions F and G, respectively. Let W be a random variable with the distribution function $pF + (1-p)G$ for some $p \in (0, 1)$. If $X \le_{\text{Lt-r}} [\le_{\text{r-Lt-r}}] Y$, then $X \le_{\text{Lt-r}} [\le_{\text{r-Lt-r}}] W \le_{\text{Lt-r}} [\le_{\text{r-Lt-r}}] Y$.*

The proof of Theorem 5.B.9 is similar to the proof of Theorem 1.B.22, but it uses (5.B.1) [(5.B.2)] instead of (1.B.3). We omit the details.

5.B.3 Relationship to other stochastic orders

In this subsection we describe some relationships between the Laplace ratio orders and some other stochastic orders. We also mention some known counterimplications.

Theorem 5.B.10. *Let X and Y be positive random variables. Then*

$$X \le_{\text{Lt-r}} Y \Longrightarrow X \le_{\text{Lt}} Y$$

and

$$X \le_{\text{r-Lt-r}} Y \Longrightarrow X \le_{\text{Lt}} Y.$$

Proof. Denote $L_X(\infty) = \lim_{s\to\infty} L_X(s)$ and $L_Y(\infty) = \lim_{s\to\infty} L_Y(s)$. Since $L_X(0) = L_Y(0) = 1$ and $L_X(\infty) = L_Y(\infty) = 0$, we see that if $X \le_{\text{Lt-r}} Y$, then

$$\frac{L_Y(s)}{L_X(s)} \le \frac{L_Y(0)}{L_X(0)} = 1,$$

and if $X \le_{\text{r-Lt-r}} Y$, then

$$\frac{1 - L_Y(s)}{1 - L_X(s)} \ge \frac{1 - L_Y(\infty)}{1 - L_X(\infty)} = 1.$$

This proves the stated results. □

As a corollary of Theorem 5.B.10 and (5.A.5) we see that

$$X \leq_{\text{Lt-r}} Y \Longrightarrow EX \leq EY,$$

and that

$$X \leq_{\text{r-Lt-r}} Y \Longrightarrow EX \leq EY$$

provided the expectations exist.

The proof of the next theorem will not be given here.

Theorem 5.B.11. *Let X and Y be nonnegative absolutely continuous or integer-valued random variables. Then*

$$X \leq_{\text{rh}} Y \Longrightarrow X \leq_{\text{Lt-r}} Y$$

and

$$X \leq_{\text{hr}} Y \Longrightarrow X \leq_{\text{r-Lt-r}} Y.$$

The following result gives a relationship between the orders $\leq_{\text{mrl}}$ and $\leq_{\text{Lt-r}}$.

Theorem 5.B.12. *Let X and Y be two nonnegative absolutely continuous random variables that possess all moments and with bounded support $[0, b]$. If $X \leq_{\text{mrl}} Y$, then $b - Y \leq_{\text{Lt-r}} b - X$.*

Proof. Denote $g(1, n) = E[X^n]$ and $g(2, n) = E[Y^n]$. Since $X \leq_{\text{mrl}} Y$ it follows from (2.A.10) that $g(i, n)$ is totally positive of order 2 in $i = 1, 2$, and in $n \geq 0$. Therefore, by the Basic Composition Formula (Karlin [275]) we have that

$$h(i, s) \equiv \sum_{n=0}^{\infty} \frac{s^n}{n!} g(i, n)$$

is totally positive of order 2 in $i = 1, 2$, and in $s \geq 0$. That is,

$$\frac{h(2, s)}{h(1, s)} = \frac{Ee^{sY}}{Ee^{sX}} \quad \text{is increasing in } s \geq 0. \tag{5.B.4}$$

It is easy to verify that (5.B.4) implies $b - Y \leq_{\text{Lt-r}} b - X$. □

Counterexamples in the literature show that for nonnegative integer-valued random variables X and Y we have

$$X \leq_{\text{hr}} Y \not\Longrightarrow X \leq_{\text{Lt-r}} Y \not\Longrightarrow X \leq_{\text{icv}} Y$$

and

$$X \leq_{\text{rh}} Y \not\Longrightarrow X \leq_{\text{r-Lt-r}} Y \not\Longrightarrow X \leq_{\text{icv}} Y.$$

It is of interest to compare the above counterimplications, and the implications given in Theorems 5.B.10 and 5.B.11, with the implication $X \leq_{\text{icv}} Y \Longrightarrow X \leq_{\text{Lt}} Y$ given in Theorem 5.A.16.

From the above counterimplications it follows that for nonnegative integer-valued random variables X and Y we have

$$X \le_{\text{Lt-r}} Y \not\Rightarrow X \le_{\text{r-Lt-r}} Y$$

and

$$X \le_{\text{r-Lt-r}} Y \not\Rightarrow X \le_{\text{Lt-r}} Y.$$

Counterexamples in the literature also show that for nonnegative integer-valued random variables X and Y we have

$$X \le_{\text{Lt-r}} Y \not\Rightarrow X \le_{\text{icx}} Y$$

and

$$X \le_{\text{r-Lt-r}} Y \not\Rightarrow X \le_{\text{icx}} Y.$$

From (1.D.2) it follows that for nonnegative random variables,

$$X \le_{\text{conv}} Y \Longrightarrow X \le_{\text{Lt-r}} Y. \tag{5.B.5}$$

Example 5.B.13. The Laplace ratio orders are useful for the purpose of stochastically comparing random minima and maxima. Let $X_1, X_2, \ldots$ be a sequence of nonnegative independent and identically distributed random variables. Let N_1 and N_2 be two positive integer-valued random variables which are independent of the X_i's. Denote $X_{(1:N_j)} \equiv \min\{X_1, X_2, \ldots, X_{N_j}\}$ and $X_{(N_j:N_j)} \equiv \max\{X_1, X_2, \ldots, X_{N_j}\}$, $j = 1, 2$. Let the common distribution function, and the common survival function, of the X_i's be denoted, respectively, by F_{X_1} and $\overline{F}_{X_1}$, and let $F_{X_{(N_j:N_j)}}$ and $\overline{F}_{X_{(1:N_j)}}$ denote, respectively, the distribution function of $X_{(N_j:N_j)}$ and the survival function of $X_{(1:N_j)}$, $j = 1, 2$. Note that

$$F_{X_{(N_j:N_j)}}(x) = \sum_{n=1}^{\infty} F_{X_1}^n(x) p_{N_j}(n) = L_{N_j}(-\log F_{X_1}(x)), \quad j = 1, 2,$$

and that

$$\overline{F}_{X_{(1:N_j)}}(x) = \sum_{n=1}^{\infty} \overline{F}_{X_1}^n(x) p_{N_j}(n) = L_{N_j}(-\log \overline{F}_{X_1}(x)), \quad j = 1, 2.$$

Thus,

$$\frac{\overline{F}_{X_{(1:N_2)}}(x)}{\overline{F}_{X_{(1:N_1)}}(x)} = \frac{L_{N_2}(-\log \overline{F}_{X_1}(x))}{L_{N_1}(-\log \overline{F}_{X_1}(x))}.$$

Therefore

$$N_1 \le_{\text{Lt-r}} N_2 \Longrightarrow X_{(1:N_1)} \ge_{\text{hr}} X_{(1:N_2)}.$$

In a similar manner it can be shown that

$$N_1 \le_{\text{Lt-r}} N_2 \Longrightarrow X_{(N_1:N_1)} \le_{\text{rh}} X_{(N_2:N_2)},$$

$$N_1 \le_{\text{r-Lt-r}} N_2 \Longrightarrow X_{(1:N_1)} \ge_{\text{rh}} X_{(1:N_2)},$$

and that

$$N_1 \le_{\text{r-Lt-r}} N_2 \Longrightarrow X_{(N_1:N_1)} \le_{\text{hr}} X_{(N_2:N_2)}.$$

From Theorem 5.B.11 and the above implications it follows that

$$N_1 \le_{\text{rh}} N_2 \Longrightarrow X_{(1:N_1)} \ge_{\text{hr}} X_{(1:N_2)}, \tag{5.B.6}$$

$$N_1 \le_{\text{rh}} N_2 \Longrightarrow X_{(N_1:N_1)} \le_{\text{rh}} X_{(N_2:N_2)},$$

$$N_1 \le_{\text{hr}} N_2 \Longrightarrow X_{(1:N_1)} \ge_{\text{rh}} X_{(1:N_2)}, \tag{5.B.7}$$

and that

$$N_1 \le_{\text{hr}} N_2 \Longrightarrow X_{(N_1:N_1)} \le_{\text{hr}} X_{(N_2:N_2)}.$$

See related results in Examples 1.C.46, 3.B.39, 4.B.16, and 5.A.24.

The following example is a variation of Example 5.A.25 — under a stronger condition (the order $\le_{\text{Lt-r}}$ is stronger than the order $\le_{\text{Lt}}$) we obtain a stronger conclusion.

Example 5.B.14. As in Example 5.A.25, let X have a compound Poisson process, with rate λ and distribution ϕ, as its (random) hazard rate function. The survival function of X is given in (5.A.18), and it follows that its density function f is given by

$$f(t) = \lambda[1 - L_\phi(t)] \exp\Big\{ - \int_0^t \lambda[1 - L_\phi(s)]\mathrm{d}s \Big\}, \quad t \ge 0.$$

Similarly let Y have a compound Poisson process, with rate λ and distribution φ, as its (random) hazard rate function. Its survival function is given in (5.A.19), and its density function g is given by

$$g(t) = \lambda[1 - L_\varphi(t)] \exp\Big\{ - \int_0^t \lambda[1 - L_\varphi(s)]\mathrm{d}s \Big\}, \quad t \ge 0.$$

It is now seen that if the random variable associated with ϕ is larger, in the reverse Laplace transform order (and hence, by Theorem 5.B.10, also larger in the Laplace transform order), than the random variable associated with φ, then $g(t)/f(t)$ is increasing in $t \ge 0$; that is, $X \le_{\text{lr}} Y$.

5.C Some Related Orders

5.C.1 The factorial moments order

The factorial moments of a random variable X are $\mu_i = E[X(X-1)\cdots(X-i+1)]$, $i = 1, 2, \ldots$. They are particularly useful when X is a nonnegative integer-valued random variable, since they can be easily obtained from the probability generating function of X by repeated differentiation. Throughout this subsection we consider only nonnegative integer-valued random variables. The ith factorial moment of such a random variable X can also be written as $\mu_i = i!E\binom{X}{i}$, where $\binom{x}{i}$ is defined as 0 when $i > x$.

Let X and Y be two nonnegative integer-valued random variables such that

$$E\binom{X}{i} \le E\binom{Y}{i} \quad \text{for all } i \in \mathbb{N}_{++}. \tag{5.C.1}$$

Then X is said to be *smaller than* Y *in the factorial moments order* (denoted by $X \le_{\text{fm}} Y$).

For a real function ϕ defined on $\mathbb{N}_+$, define $\Delta^0\phi(x) = \phi(x)$, and $\Delta^j\phi(x) = \Delta^{j-1}\phi(x+1) - \Delta^{j-1}\phi(x)$, $x \in \mathbb{N}_+$, $j = 1, 2, \ldots$. It can be shown that for every $\phi : \mathbb{N}_+ \to [0, \infty)$, one has

$$\phi(x) = \sum_{j=0}^{\infty} \Delta^j\phi(0)\binom{x}{j}, \quad x \in \mathbb{N}_+. \tag{5.C.2}$$

The following characterization of the order $\le_{\text{fm}}$ is a direct consequence of (5.C.2).

Theorem 5.C.1. *Let X and Y be two nonnegative integer-valued random variables. Then $X \le_{\text{fm}} Y$ if, and only if,*

$$E[\phi(X)] \le E[\phi(Y)] \quad \textit{for all } \phi \textit{ such that } \Delta^j\phi(0) \ge 0, \ j \in \mathbb{N}_+.$$

It is easy to see that

$$X \le_{\text{fm}} Y \Longrightarrow EX \le EY.$$

Some closure properties of the order $\le_{\text{fm}}$ are given in the next theorem.

Theorem 5.C.2. (a) *Let X and Y be two nonnegative integer-valued random variables. If $X \le_{\text{fm}} Y$, then $X + k \le_{\text{fm}} Y + k$ for every $k \in \mathbb{N}_+$.*

(b) *Let X and Y be two nonnegative integer-valued random variables. If $X \le_{\text{fm}} Y$, then $kX \le_{\text{fm}} kY$ for every $k \in \mathbb{N}_+$.*

(c) *Let $X_1, X_2, \ldots, X_m$ be a set of independent nonnegative integer-valued random variables. Let $Y_1, Y_2, \ldots, Y_m$ be another set of independent nonnegative integer-valued random variables. If $X_i \le_{\text{fm}} Y_i$, $i = 1, 2, \ldots, m$, then*

$$\sum_{i=1}^{m} X_i \le_{\text{fm}} \sum_{i=1}^{m} Y_i.$$

Proof. It is enough to prove part (a) for $k = 1$; the proof can then be completed by induction. But for $k = 1$ the desired result follows directly from the identity

$$\binom{x+1}{i+1} = \binom{x}{i+1} + \binom{x}{i}, \quad i, x \in \mathbb{N}_+.$$

A lengthy straightforward calculation yields

$$\Delta^j \binom{kx}{i}\bigg|_{x=0} \geq 0, \quad i, j, k \in \mathbb{N}_+.$$

Part (b) then follows.

Finally, in order to prove part (c) it is enough to consider the case $m = 2$. This case follows immediately from the identity

$$\binom{x_1 + x_2}{i} = \sum_{j=0}^{i} \binom{x_1}{j} \binom{x_2}{i-j}, \quad x_1, x_2, i \in \mathbb{N}_+. \qquad \square$$

The next result shows that under some conditions the order $\leq_{\text{fm}}$ is closed under formation of random sums. We do not give the proof here.

Theorem 5.C.3. *Let $X_1, X_2, \ldots$ and $Y_1, Y_2, \ldots$ each be a sequence of nonnegative independent integer-valued random variables such that $X_i \leq_{\text{fm}} Y_i$, $i = 1, 2, \ldots$. Let M and N be integer-valued nonnegative random variables that are independent of the $\{X_i\}$ and the $\{Y_i\}$ sequences, respectively, such that $M \leq_{\text{icx}} N$. If the X_i's or the Y_i's are identically distributed, then*

$$\sum_{j=1}^{M} X_j \leq_{\text{fm}} \sum_{j=1}^{N} Y_j.$$

Select a positive integer i and consider the real function ϕ defined on $\{i+1, i+2, \ldots\}$ by $\phi(x) = \binom{x}{i}$. A straightforward computation yields that $\phi(x) + \phi(x+2) \geq 2\phi(x+1)$ for $x \in \{i+1, i+2, \ldots\}$. That is, the function ϕ is convex on $\{i+1, i+2, \ldots\}$. Thus we have proven the following result.

Theorem 5.C.4. *Let X and Y be two nonnegative integer-valued random variables. If $X \leq_{\text{icx}} Y$, then $X \leq_{\text{fm}} Y$. In particular, if $X \leq_{\text{st}} Y$, then $X \leq_{\text{fm}} Y$.*

A relationship between the orders $\leq_{\text{fm}}$ and $\leq_{\text{pgf}}$ is given in the next result.

Theorem 5.C.5. *Let X and Y be two nonnegative integer-valued random variables with bounded support $\{0, 1, 2, \ldots, b\}$. If $X \leq_{\text{fm}} Y$, then $b - Y \leq_{\text{pgf}} b - X$.*

Proof. For $a \geq 1$ define $M_X(a) = E[a^X]$ and $M_Y(a) = E[a^Y]$. Note that the ith derivative of M_X $[M_Y]$ at 1 is $M_X^{(i)}(1) = E[X(X-1)\cdots(X-i+1)]$ $[M_Y^{(i)}(1) = E[Y(Y-1)\cdots(Y-i+1)]]$. Expanding M_X and M_Y about 1, using the finiteness of the support for convergence, it is seen that

$$\begin{aligned} M_X(a) &= \sum_{i=0}^{\infty} \frac{M_X^{(i)}(1)}{i!}(a-1)^i \\ &= \sum_{i=0}^{\infty} \frac{E[X(X-1)\cdots(X-i+1)]}{i!}(a-1)^i \end{aligned}$$

$$\leq \sum_{i=0}^{\infty} \frac{E[Y(Y-1)\cdots(Y-i+1)]}{i!}(a-1)^i$$
$$= M_Y(a),$$

where the inequality follows from the assumption that $X \leq_{\rm fm} Y$ and from the fact that $a \geq 1$. Thus,

$$E[a^X] \leq E[a^Y] \quad \text{for all } a \geq 1. \tag{5.C.3}$$

Now it is easy to verify that (5.C.3) implies that $b - Y \leq_{\rm pgf} b - X$. □

5.C.2 The moments order

Consider now two general (that is, not necessarily integer-valued) nonnegative random variables X and Y such that

$$E[X^i] \leq E[Y^i] \quad \text{for all } i \in \mathbb{N}_{++}.$$

Then X is said to be *smaller than* Y *in the moments order* (denoted by $X \leq_{\rm mom} Y$). Thus $X \leq_{\rm mom} Y$ if, and only if,

$$E[\phi(X)] \leq E[\phi(Y)] \tag{5.C.4}$$

for all polynomials ϕ with nonnegative coefficients. In fact, $X \leq_{\rm mom} Y$ if, and only if, (5.C.4) holds for all absolutely monotone functions ϕ of the form $\phi(x) = \sum_{k=0}^{\infty} a_k x^k$, where the a_k's are nonnegative, provided the expectations exist.

Clearly,

$$X \leq_{\rm mom} Y \Longrightarrow EX \leq EY.$$

Some closure properties of the order $\leq_{\rm mom}$ are given in the next theorem. Its proof is similar to the proof of Theorem 5.C.2 (except that it is simpler) and is thus omitted.

Theorem 5.C.6. (a) *Let X and Y be two nonnegative random variables. If $X \leq_{\rm mom} Y$, then $X + k \leq_{\rm mom} Y + k$ for every $k \geq 0$.*

(b) *Let X and Y be two nonnegative random variables. If $X \leq_{\rm mom} Y$, then $kX \leq_{\rm mom} kY$ for every $k \geq 0$.*

(c) *Let $X_1, X_2, \ldots, X_m$ be a set of independent nonnegative random variables. Let $Y_1, Y_2, \ldots, Y_m$ be another set of independent nonnegative random variables. If $X_i \leq_{\rm mom} Y_i$, $i = 1, 2, \ldots, m$, then*

$$\sum_{i=1}^{m} X_i \leq_{\rm mom} \sum_{i=1}^{m} Y_i.$$

The next result shows that under some conditions the order $\leq_{\rm mom}$ is closed under formation of random sums. We do not give the proof here.

Theorem 5.C.7. *Let $X_1, X_2, \ldots$ and $Y_1, Y_2, \ldots$ each be a sequence of nonnegative independent random variables such that $X_i \leq_{\text{mom}} Y_i$, $i = 1, 2, \ldots$. Let M and N be integer-valued nonnegative random variables that are independent of the $\{X_i\}$ and the $\{Y_i\}$ sequences, respectively, such that $M \leq_{\text{icx}} N$. If the X_i's or the Y_i's are identically distributed, then*

$$\sum_{j=1}^{M} X_j \leq_{\text{mom}} \sum_{j=1}^{N} Y_j.$$

The moments order is closed under linear convex combinations as the following theorem shows. This result is an analog of Theorems 3.A.36 and 5.A.14. Its proof is similar to the proof of Theorem 3.A.36 and is therefore omitted. A similar result is Theorem 5.C.18.

Theorem 5.C.8. *Let $X_1, X_2, \ldots, X_n$ and Y be $n+1$ random variables. If $X_i \leq_{\text{mom}} Y$, $i = 1, 2, \ldots, n$, then*

$$\sum_{i=1}^{n} a_i X_i \leq_{\text{mom}} Y,$$

whenever $a_i \geq 0$, $i = 1, 2, \ldots, n$ and $\sum_{i=1}^{n} a_i = 1$.

The following result gives a relationship between the orders $\leq_{\text{fm}}$ and $\leq_{\text{mom}}$.

Theorem 5.C.9. *Let X and Y be two nonnegative integer-valued random variables. If $X \leq_{\text{fm}} Y$, then $X \leq_{\text{mom}} Y$. In particular, if $X \leq_{\text{icx}} Y$ (or if $X \leq_{\text{st}} Y$), then $X \leq_{\text{mom}} Y$.*

Proof. Denote $x_{[i]} = x(x-1)\cdots(x-i+1)$. The result will follow once we have shown that

$$x^i = \sum_{k=1}^{i} \alpha_k^{(i)} x_{[k]}, \quad i = 1, 2, \ldots, \tag{5.C.5}$$

where the $\alpha_k^{(i)}$'s are some nonnegative constants. The expression (5.C.5) can be found on page 4 of Johnson and Kotz [263]. The $\alpha_k^{(i)}$'s in (5.C.5) are the Stirling numbers of the second kind, which are known to be positive. □

In order to obtain a Laplace transform characterization of the order $\leq_{\text{mom}}$ we first prove the following result.

Theorem 5.C.10. *Let X_1 and X_2 be two nonnegative random variables, and let $N_\lambda(X_1)$ and $N_\lambda(X_2)$ be as described in Theorem 1.A.13. Then*

$$X_1 \leq_{\text{mom}} X_2 \Longrightarrow N_\lambda(X_1) \leq_{\text{fm}} N_\lambda(X_2) \quad \textit{for all } \lambda > 0.$$

Proof. For $k = 1, 2$, let F_k denote the distribution function of X_k. By (2.A.15) we have

$$\begin{aligned}
E[N_\lambda(X_k)(N_\lambda(X_k)-1)(N_\lambda(X_k)-2)\cdots(N_\lambda(X_k)-i+1)] \\
&= \sum_{n=0}^{\infty} n(n-1)(n-2)\cdots(n-i+1)\int_0^\infty \mathrm{e}^{-\lambda x}\frac{(\lambda x)^n}{n!}\mathrm{d}F_k(x) \\
&= \int_0^\infty \left[\sum_{n=0}^{\infty} n(n-1)(n-2)\cdots(n-i+1)\mathrm{e}^{-\lambda x}\frac{(\lambda x)^n}{n!}\right]\mathrm{d}F_k(x).
\end{aligned}$$

It is not difficult to verify that the ith factorial moment of a Poisson random variable with mean λx is given by

$$\sum_{n=0}^{\infty} n(n-1)(n-2)\cdots(n-i+1)\mathrm{e}^{-\lambda x}\frac{(\lambda x)^n}{n!} = (\lambda x)^i.$$

Therefore

$$\begin{aligned}
E[N_\lambda(X_1)(N_\lambda(X_1)-1)(N_\lambda(X_1)-2)\cdots(N_\lambda(X_1)-i+1)] \\
&= \lambda^i \int_0^\infty x^i \mathrm{d}F_1(x) \le \lambda^i \int_0^\infty x^i \mathrm{d}F_2(x) \\
&= E[N_\lambda(X_2)(N_\lambda(X_2)-1)(N_\lambda(X_2)-2)\cdots(N_\lambda(X_2)-i+1)],
\end{aligned}$$

where the inequality follows from $X_1 \le_{\text{mom}} X_2$. Thus $N_\lambda(X_1) \le_{\text{fm}} N_\lambda(X_2)$. □

A Laplace transform characterization of the order $\le_{\text{mom}}$ is given next. It may be compared to Theorems 1.A.13, 1.B.18, 1.B.53, 1.C.25, 2.A.16, 2.B.14, 4.A.21, and 5.A.6.

Theorem 5.C.11. *Let X_1 and X_2 be two nonnegative random variables, and let $N_\lambda(X_1)$ and $N_\lambda(X_2)$ be as described in Theorem* 1.A.13. *Then*

$$X_1 \le_{\text{mom}} X_2 \Longleftrightarrow N_\lambda(X_1) \le_{\text{mom}} N_\lambda(X_2) \quad \textit{for all } \lambda > 0.$$

Proof. If $X_1 \le_{\text{mom}} X_2$, then from Theorem 5.C.10 we get that $N_\lambda(X_1) \le_{\text{fm}} N_\lambda(X_2)$, and from Theorem 5.C.9 we get that $N_\lambda(X_1) \le_{\text{mom}} N_\lambda(X_2)$.

Now suppose that $N_\lambda(X_1) \le_{\text{mom}} N_\lambda(X_2)$ for all $\lambda > 0$. Then $E(N_\lambda(X_1))^i \le E(N_\lambda(X_2))^i$, $i = 1, 2, \ldots$. In particular, $E[N_\lambda(X_1)] \le E[N_\lambda(X_2)]$, therefore, by (2.A.16),

$$E[X_1] = E[N_\lambda(X_1)]/\lambda \le E[N_\lambda(X_2)]/\lambda = E[X_2].$$

Let the induction hypothesis be

$$E[X_1^i] \le E[X_2^i], \quad i = 1, 2, \ldots, m.$$

Now observe the following. From (2.A.15) it is seen that

$$E\big[(N_\lambda(X_k))^{m+1}\big] = \sum_{n=0}^{\infty} n^{m+1} \int_0^\infty \mathrm{e}^{-\lambda x} \frac{(\lambda x)^n}{n!} \mathrm{d}F_k(x)$$
$$= \int_0^\infty \left[\sum_{n=0}^{\infty} n^{m+1} \mathrm{e}^{-\lambda x} \frac{(\lambda x)^n}{n!} \right] \mathrm{d}F_k(x), \quad k = 1, 2.$$

The quantity $\big[\sum_{n=0}^{\infty} n^{m+1} \mathrm{e}^{-\lambda x} \frac{(\lambda x)^n}{n!}\big]$ is the $(m+1)$st moment of a Poisson random variable with mean λx. It is not difficult to verify that

$$\sum_{n=0}^{\infty} n^{m+1} \mathrm{e}^{-\lambda x} \frac{(\lambda x)^n}{n!} = a_{m+1}(\lambda x)^{m+1} + a_m(\lambda x)^m + \cdots + a_1(\lambda x) + a_0,$$

where $a_j > 0$, $j = 0, 1, 2, \ldots, m+1$. Therefore

$$E\big[(N_\lambda(X_k))^{m+1}\big] = \sum_{j=0}^{m+1} a_j \lambda^j \int_0^\infty x^j \mathrm{d}F_k(x), \quad k = 1, 2.$$

We know that $E\big[(N_\lambda(X_1))^{m+1}\big] \le E\big[(N_\lambda(X_2))^{m+1}\big]$ and therefore,

$$\sum_{j=0}^{m+1} a_j \lambda^j \int_0^\infty x^j \mathrm{d}F_1(x) \le \sum_{j=0}^{m+1} a_j \lambda^j \int_0^\infty x^j \mathrm{d}F_2(x)$$

for some $a_0, a_1, \ldots, a_{m+1} > 0$ and all $\lambda > 0$. Rewrite the inequality as

$$a_{m+1}\lambda^{m+1} \left\{E\left[X_1^{m+1}\right] - E\left[X_2^{m+1}\right]\right\} \le \sum_{j=1}^{m} a_j \lambda^j \left\{E\big[X_1^j\big] - E\big[X_2^j\big]\right\}.$$

The right-hand side is nonnegative by the induction hypothesis. If

$$E\left[X_1^{m+1}\right] - E\left[X_2^{m+1}\right] > 0,$$

then, by choosing sufficiently large λ, the left-hand side would be greater than the right-hand side, a contradiction. Thus we must have

$$E\left[X_1^{m+1}\right] - E\left[X_2^{m+1}\right] \le 0.$$

The result now follows by induction. □

The next result describes a relationship between the orders $\le_{\text{mom}}$ and $\le_{\text{r-Lt-r}}$; we omit its proof.

Theorem 5.C.12. *Let X and Y be two nonnegative random variables. Then*

$$X \le_{\text{r-Lt-r}} Y \Longrightarrow \frac{1}{Y} \le_{\text{mom}} \frac{1}{X}.$$

Finally we mention a related order. Let X and Y be two nonnegative random variables such that

$$\frac{E[Y^n]}{E[X^n]} \quad \text{is increasing in } n \in \mathbb{N}_+, \tag{5.C.6}$$

where, by convention, $E[X^0] = E[Y^0] = 1$. Then X is said to be *smaller than Y in the moments ratio order* (denoted as $X \leq_{\text{mom-r}} Y$).

From (5.C.6) it is seen that $\frac{E[Y^n]}{E[X^n]} \geq \frac{E[Y^0]}{E[X^0]} = 1$. Thus we see that

$$X \leq_{\text{mom-r}} Y \Longrightarrow X \leq_{\text{mom}} Y. \tag{5.C.7}$$

From (2.A.10) it is seen that

$$X \leq_{\text{mrl}} Y \Longrightarrow X \leq_{\text{mom-r}} Y.$$

Therefore, by Theorem 2.A.1, we also have that

$$X \leq_{\text{hr}} Y \Longrightarrow X \leq_{\text{mom-r}} Y. \tag{5.C.8}$$

In the proof of Theorem 5.B.12 it is essentially shown that for nonnegative random variables X and Y with bounded support $[0, b]$ we have

$$X \leq_{\text{mom-r}} Y \Longrightarrow b - Y \leq_{\text{Lt-r}} b - X.$$

This may be contrasted with (5.C.9) below (recall that $X \leq_{\text{Lt-r}} Y \Longrightarrow X \leq_{\text{Lt}} Y$; see Theorem 5.B.10).

The following result is obvious.

Theorem 5.C.13. *Let X and Y be two nonnegative random variables. If $X \leq_{\text{mom-r}} Y$, then $kX \leq_{\text{mom-r}} kY$ for every $k \geq 0$.*

The next result describes a relationship between the orders $\leq_{\text{mom-r}}$ and $\leq_{\text{Lt-r}}$; we omit its proof.

Theorem 5.C.14. *Let X and Y be two nonnegative random variables. Then*

$$X \leq_{\text{Lt-r}} Y \Longrightarrow \frac{1}{Y} \leq_{\text{mom-r}} \frac{1}{X}.$$

From (5.C.7) and Theorem 5.C.14 it is seen that if X and Y are nonnegative random variables, then

$$X \leq_{\text{Lt-r}} Y \Longrightarrow \frac{1}{Y} \leq_{\text{mom}} \frac{1}{X}.$$

5.C.3 The moment generating function order

Let X and Y be two nonnegative random variables such that $Ee^{s_0 Y} < \infty$ for some $s_0 > 0$, and

$$Ee^{sX} \leq Ee^{sY}, \quad \text{for all } s > 0.$$

Then X is said to be *smaller than Y in the moment generating function order* (denoted by $X \leq_{\text{mgf}} Y$).

A simple integration by parts shows that $X \leq_{\text{mgf}} Y$ if, and only if,

$$\int_0^\infty e^{sx}\overline{F}(x)\mathrm{d}x \leq \int_0^\infty e^{sx}\overline{G}(x)\mathrm{d}x \quad \text{for all } s > 0,$$

where $\overline{F}$ and $\overline{G}$ are the survival functions of X and of Y, respectively.

The following theorem is an analog of Theorem 5.A.5; its proof is similar to the proof of that result.

Theorem 5.C.15. *Let X and Y be two nonnegative random variables. Then $X \leq_{\text{mgf}} Y$ if, and only if,*

$$\sum_{i=0}^{\infty} \frac{s^i}{(i+1)!} EX^{i+1} \leq \sum_{i=0}^{\infty} \frac{s^i}{(i+1)!} EY^{i+1} \quad \textit{for all } s > 0.$$

It follows from Theorem 5.C.15 that

$$X \leq_{\text{mom}} Y \Longrightarrow X \leq_{\text{mgf}} Y.$$

Some closure properties of the order $\leq_{\text{mgf}}$ are given below (recall from Section 1.A.3 that for any random variable Z and any event A we denote by $[Z|A]$ any random variable whose distribution is the conditional distribution of Z given A.)

Theorem 5.C.16. *Let X and Y be two nonnegative random variables.*

(a) *If $X \leq_{\text{mgf}} Y$, then $X + k \leq_{\text{mgf}} Y + k$ for every $k > 0$.*
(b) *If $X \leq_{\text{mgf}} Y$, then $kX \leq_{\text{mgf}} kY$ for every $k > 0$.*
(c) *Let X, Y, and Θ be random variables such that $[X|\Theta = \theta] \leq_{\text{mgf}} [Y|\Theta = \theta]$ for all θ in the support of Θ. Then $X \leq_{\text{mgf}} Y$. That is, the moment generating function order is closed under mixtures.*
(d) *Let $X_1, X_2, \ldots, X_m$ be a set of independent random variables and let $Y_1, Y_2, \ldots, Y_m$ be another set of independent random variables. If $X_i \leq_{\text{mgf}} Y_i$ for $i = 1, 2, \ldots, m$, then*

$$\sum_{i=1}^{m} X_i \leq_{\text{mgf}} \sum_{i=1}^{m} Y_i;$$

that is, the moment generating function order is closed under convolutions.

The next result is an analog of Theorems 5.A.9 and 5.C.7.

Theorem 5.C.17. *Let $X_1, X_2, \ldots$ and $Y_1, Y_2, \ldots$ each be a sequence of nonnegative independent and identically distributed random variables such that $X_i \leq_{\text{mgf}} Y_i$, $i = 1, 2, \ldots$. Let M and N be integer-valued nonnegative random variables that are independent of the $\{X_i\}$ and the $\{Y_i\}$ sequences, respectively, such that $M \leq_{\text{mgf}} N$. Then*

$$\sum_{j=1}^{M} X_j \leq_{\text{mgf}} \sum_{j=1}^{N} Y_j.$$

The following result is an analog of Theorem 3.A.36; similar results are Theorems 5.A.14 and 5.C.8.

Theorem 5.C.18. *Let $X_1, X_2, \ldots, X_n$ and Y be $n+1$ random variables. If $X_i \leq_{\text{mgf}} Y$, $i = 1, 2, \ldots, n$, then*

$$\sum_{i=1}^{n} a_i X_i \leq_{\text{mgf}} Y,$$

whenever $a_i \geq 0$, $i = 1, 2, \ldots, n$ and $\sum_{i=1}^{n} a_i = 1$.

The next result is an analog of Theorem 5.C.5; it describes a relationship between the orders $\leq_{\text{mgf}}$ and $\leq_{\text{Lt}}$.

Theorem 5.C.19. *Let X and Y be two nonnegative random variables with bounded support $[0, b]$. Then $X \leq_{\text{mgf}} Y$ if, and only if, $b - Y \leq_{\text{Lt}} b - X$.*

In particular, for random variables as in Theorem 5.C.19,

$$X \leq_{\text{mom}} Y \Longrightarrow b - Y \leq_{\text{Lt}} b - X. \tag{5.C.9}$$

5.D Complements

Section 5.A: We used three main sources in order to collect the results regarding the Laplace transform order. These are Stoyan [540, Section 1.8], Kim and Proschan [294], and Alzaid, Kim, and Proschan [11]. The characterization (5.A.4) is taken from Denuit [141]. The characterization of the order $\leq_{\text{Lt}}$ in terms of exponential mixtures, given in Remark 5.A.2, is taken from Bartoszewicz [50]. The characterization described in Theorem 5.A.4 can be found in Bhattacharjee [84]. The Laplace transform characterization of the order $\leq_{\text{Lt}}$ given in Theorem 5.A.6 is essentially taken from Alzaid, Kim, and Proschan [11]. Some further characterizations of the Laplace transform order by means of infinitely divisible distributions are given in Bartoszewicz [48]. The closure property of the order $\leq_{\text{Lt}}$ under random sums (Theorem 5.A.8) is taken from Bhattacharjee [86]. The extensions of the closure property of the order $\leq_{\rm Lt}$ under random sums (Theorems 5.A.10–5.A.12) can be found in Pellerey [450]. The majorization result (Theorem 5.A.13) is taken from Ma [375]. The result which gives the closure of the Laplace transform order under linear convex combinations (Theorem 5.A.14) can be found in Pellerey [452]. The condition which implies stochastic equality (in Theorem 5.A.15) is a combination of results in Cai and Wu [116] and in Bhattacharjee [84], where some generalizations of this condition can also be found. The characterization of the Laplace transform order by means of the order $\leq_{\infty\text{-icv}}$ (Theorem 5.A.17) is taken from Thistle [548]; see also Fishburn and Lavalle [204] and further references in that paper. The implication of the order $\leq_{\rm Lt}$ from the order $\leq_{p-}$ (Theorem 5.A.18) is essentially taken from Bhattacharjee [83]; see also Cai and Wu [116]. The closure property of the order $\leq_{\rm Lt}$ under the operation of taking minima (Theorem 5.A.19) is taken from Alzaid, Kim, and Proschan [11]. Alzaid, Kim, and Proschan [11] also have a version of Theorem 5.A.19 which gives conditions under which the order $\leq_{\rm Lt}$ is closed under the operation of taking maxima, however their condition must be wrong, since it postulates that the F_i's and the G_i's (of Theorem 5.A.19) are completely monotone — but these functions are increasing, whereas all completely monotone functions must be decreasing. Looking over their proof it is seen that a sufficient condition, for the closure of the order $\leq_{\rm Lt}$ under the operation of taking maxima, is that $\mathrm{e}^{-tx}F_i(x)$ and $\mathrm{e}^{-tx}G_i(x)$ be completely monotone in x for each $t \geq 0$, $i = 1, 2, \dots, m$. We are not aware of any study of the latter condition. The class of random lifetimes, defined by (5.A.11), is studied in Klefsjö [302]. The equivalence of the Laplace transform ordering of nonnegative random variables with equal means, and their corresponding asymptotic equilibrium ages, given in (5.A.12), is taken from Denuit [141]. The lower bound in (5.A.13), in the sense of $\leq_{\rm Lt}$, when the mean and the variance are given, can be found in Stoyan [540, page 23], who credited it to Rolski. The characterization of the order $\leq_{\rm hr}$ by means of the order $\leq_{\rm Lt}$ (Theorem 5.A.22) is given in Belzunce, Gao, Hu, and Pellerey [67]. The results about the stochastic comparisons of random minima and maxima (Example 5.A.24) are taken from Shaked and Wong [526]. The hazard rate order comparison of two nonnegative random variables with random hazard rate functions (Example 5.A.25) is a special case of Theorem 3 of Di Crescenzo and Pellerey [166].

Section 5.B: Most of the results in this section can be found in Shaked and Wong [525]. The characterizations of the orders $\leq_{\rm Lt\text{-}r}$ and $\leq_{\rm r\text{-}Lt\text{-}r}$ in terms of exponential mixtures, given in Remark 5.B.1, are taken from Bartoszewicz [50]. Di Crescenzo and Shaked [167] used (5.B.3) in order to obtain Laplace transform ratio order comparisons of many pairs of random variables. The relationship between the orders $\leq_{\rm mrl}$ and $\leq_{\rm Lt\text{-}r}$ (Theorem 5.B.12) is essentially proven in Fagiuoli and Pellerey [187]. The relationship between the orders $\leq_{\rm Lt\text{-}r}$ and $\leq_{\rm conv}$, given in (5.B.5), was noted in Shaked and Suarez-Llorens [520]. Extensions of the implications (5.B.6) and (5.B.7) to order statistics other than the minimum can be found in Nanda, Misra, Paul, and Singh [427]. The likelihood ratio order comparison of two nonnegative random variables with random hazard rate functions (Example 5.B.14) is essentially Remark 3 of Di Crescenzo and Pellerey [166].

Section 5.C: Many of the results in this section are taken from Lefèvre and Picard [338]. A discussion about other related orders can also be found in Lefèvre and Picard [338]. The closure properties of the order $\leq_{\text{fm}}$ (Theorem 5.C.2), as well as the simple proof of Theorem 5.C.9, have been communicated to us by Lefèvre [335]. The results that give the closure under random convolutions property of the factorial moments order (Theorem 5.C.3) and of the moments order (Theorem 5.C.7) are taken from Jean-Marie and Liu [254]. Lefèvre and Utev [339] have noticed that for finite random variables with support $\{0, 1, \ldots, b\}$ the discrete versions of the orders $\leq_{m\text{-icx}}$, $m = 2, 3, \ldots, b$ (see Section 4.A.7), together with some conditions on the factorial moments, imply the order $\leq_{\text{fm}}$; thus they generalized Theorem 5.C.5. The result which gives the closure of the moments order under linear convex combinations (Theorem 5.C.8) can be found in Pellerey [452]. The Laplace transform characterizations of the order $\leq_{\text{mom}}$ (Theorems 5.C.10 and 5.C.11) are taken from Shaked and Wong [524]. The relationship between the orders $\leq_{\text{mom}}$ and $\leq_{\text{r-Lt-r}}$ (Theorem 5.C.12) can be found in Bartoszewicz [47]. The moments ratio order has been introduced by Whitt [565] who has also obtained the implications (5.C.7) and (5.C.8). The relationship between the orders $\leq_{\text{mom-r}}$ and $\leq_{\text{Lt-r}}$ (Theorem 5.C.14) can be found in Bartoszewicz [47]. The moment generating function order is called the *exponential order* in Kaas, Heerwaarden, and Goovaerts [269]. Most of the results in Section 5.C.3 can be found in Klar and Müller [301]. The result that gives the closure of the order $\leq_{\text{mgf}}$ under linear convex combinations (Theorem 5.C.18) is taken from Li [352].

6

Multivariate Stochastic Orders

In this chapter we describe various extensions, of the univariate stochastic orders in Chapters 1 and 2, to the multivariate case. The most important common orders that are studied in this chapter are the multivariate stochastic orders $\leq_{\rm st}$ and $\leq_{\rm lr}$. Multivariate extensions of the orders $\leq_{\rm hr}$ and $\leq_{\rm mrl}$ are also studied in this chapter. Also, we review here further analogs of the univariate order $\leq_{\rm st}$, such as the upper and lower orthants orders. In addition, some other related orders are investigated in this chapter as well.

6.A Notations and Preliminaries

In this chapter we will be concerned with random vectors that take on values in $\mathbb{R}^n \equiv (-\infty, \infty)^n$. When we say that the random vectors are nonnegative we mean that they take on values in $\mathbb{R}_+^n = [0, \infty)^n$. Elements in $\mathbb{R}^n$ will be denoted by $\boldsymbol{x}$, $\boldsymbol{y}$, and so forth, or, more explicitly, as $\boldsymbol{x} = (x_1, x_2, \ldots, x_n)$, $\boldsymbol{y} = (y_1, y_2, \ldots, y_n)$, and so on.

The space $\mathbb{R}^n$ is endowed with the usual componentwise partial order, which is defined as follows. Let $\boldsymbol{x} = (x_1, x_2, \ldots, x_n)$ and $\boldsymbol{y} = (y_1, y_2, \ldots, y_n)$ be two vectors in $\mathbb{R}^n$; then we denote $\boldsymbol{x} \leq \boldsymbol{y}$ if $x_i \leq y_i$ for $i = 1, 2, \ldots, n$. Let $\boldsymbol{x}$ be a vector in $\mathbb{R}^n$ and let $I = \{i_1, i_2, \ldots, i_k\} \subseteq \{1, 2, \ldots, n\}$; then we denote

$$\boldsymbol{x}_I = (x_{i_1}, x_{i_2}, \ldots, x_{i_k}). \tag{6.A.1}$$

For a random vector $\boldsymbol{X}$ that takes on values in $\mathbb{R}^n$, the interpretation of $\boldsymbol{X}_I$ is similar. The complement of I in $\{1, 2, \ldots, n\}$ is denoted by $\overline{I} \equiv \{1, 2, \ldots, n\} - I$. The vector of ones will be denoted by $\boldsymbol{e}$, that is, $\boldsymbol{e} = (1, 1, \ldots, 1)$. The dimension of $\boldsymbol{e}$ may vary from one formula to another, but it is always possible to determine it from the expression in which it appears. For example, if we write $\boldsymbol{x}_I \geq t\boldsymbol{e}$, then it is obvious that the dimension of $\boldsymbol{e}$ is $|I|$, that is, the cardinality of I.

Let ϕ be a univariate or a multivariate function with domain in $\mathbb{R}^n$. If $\phi(\boldsymbol{x}) \le [\ge]\ \phi(\boldsymbol{y})$ whenever $\boldsymbol{x} \le \boldsymbol{y}$, then we say that the function ϕ is increasing [decreasing]. A set $U \subseteq \mathbb{R}^n$ is called increasing or upper [decreasing or lower] if $\boldsymbol{y} \in U$ whenever $\boldsymbol{y} \ge [\le]\ \boldsymbol{x}$ and $\boldsymbol{x} \in U$. If U is Borel measurable, then it is increasing [decreasing] if, and only if, its indicator function I_U is increasing [decreasing]. In this chapter, and later in the book, when we consider increasing and decreasing sets, they are implicitly assumed to be Borel measurable.

6.B The Usual Multivariate Stochastic Order

6.B.1 Definition and equivalent conditions

Let $\boldsymbol{X}$ and $\boldsymbol{Y}$ be two random vectors such that

$$P\{\boldsymbol{X} \in U\} \le P\{\boldsymbol{Y} \in U\} \quad \text{for all upper sets } U \subseteq \mathbb{R}^n. \tag{6.B.1}$$

Then $\boldsymbol{X}$ is said to be *smaller than* $\boldsymbol{Y}$ *in the usual stochastic order* (denoted by $\boldsymbol{X} \le_{\rm st} \boldsymbol{Y}$). Roughly speaking, (6.B.1) says that $\boldsymbol{X}$ is less likely than $\boldsymbol{Y}$ to take on large values, where "large" means any value in an increasing set U for any increasing set U.

Another way of rewriting (6.B.1) is the following:

$$E[I_U(\boldsymbol{X})] \le E[I_U(\boldsymbol{Y})] \quad \text{for all upper sets } U \subseteq \mathbb{R}^n, \tag{6.B.2}$$

where I_U denotes the indicator function of U. From (6.B.2) it follows that if $\boldsymbol{X} \le_{\rm st} \boldsymbol{Y}$, then

$$E\Big[\sum_{i=1}^m a_i I_{U_i}(\boldsymbol{X})\Big] - b \le E\Big[\sum_{i=1}^m a_i I_{U_i}(\boldsymbol{Y})\Big] - b \tag{6.B.3}$$

for all $a_i \ge 0$, $i = 1, 2, \ldots, m$, $b \in \mathbb{R}^n$, and $m \ge 0$. Given an increasing real function ϕ on $\mathbb{R}^n$, it is possible, for each m, to define a sequence of U_i's, and a sequence of a_i's, and a b (all of which may depend on m), such that as $m \to \infty$, then (6.B.3) converges to

$$E[\phi(\boldsymbol{X})] \le E[\phi(\boldsymbol{Y})], \tag{6.B.4}$$

provided the expectations exist. It follows that $\boldsymbol{X} \le_{\rm st} \boldsymbol{Y}$ *if, and only if,* (6.B.4) *holds for all increasing functions* ϕ *for which the expectations exist.*

6.B.2 A characterization by construction on the same probability space

As in the univariate case, the usual multivariate stochastic order can be characterized as follows:

Theorem 6.B.1. *The random vectors* $\boldsymbol{X}$ *and* $\boldsymbol{Y}$ *satisfy* $\boldsymbol{X} \leq_{\rm st} \boldsymbol{Y}$ *if, and only if, there exist two random vectors* $\hat{\boldsymbol{X}}$ *and* $\hat{\boldsymbol{Y}}$*, defined on the same probability space, such that*

$$\hat{\boldsymbol{X}} =_{\rm st} \boldsymbol{X}, \tag{6.B.5}$$

$$\hat{\boldsymbol{Y}} =_{\rm st} \boldsymbol{Y}, \tag{6.B.6}$$

and

$$P\{\hat{\boldsymbol{X}} \leq \hat{\boldsymbol{Y}}\} = 1. \tag{6.B.7}$$

Obviously, if (6.B.5)–(6.B.7) hold, then $\boldsymbol{X} \leq_{\rm st} \boldsymbol{Y}$. We will not give the proof of Theorem 6.B.1 here; however, in the next subsection we point out an important special case in which the construction of $\hat{\boldsymbol{X}}$ and of $\hat{\boldsymbol{Y}}$ can be described explicitly.

As in the univariate case (see Theorem 1.A.2) Theorem 6.B.1 can be restated as follows.

Theorem 6.B.2. *The* n*-dimensional random vectors* $\boldsymbol{X}$ *and* $\boldsymbol{Y}$ *satisfy* $\boldsymbol{X} \leq_{\rm st} \boldsymbol{Y}$ *if, and only if, there exist a random variable* Z *and* $\mathbb{R}^n$*-valued functions* $\boldsymbol{\psi}_1$ *and* $\boldsymbol{\psi}_2$ *such that* $\boldsymbol{\psi}_1(z) \leq \boldsymbol{\psi}_2(z)$ *for all* $z \in \mathbb{R}$*, and* $\boldsymbol{X} =_{\rm st} \boldsymbol{\psi}_1(Z)$ *and* $\boldsymbol{Y} =_{\rm st} \boldsymbol{\psi}_2(Z)$.

In light of Theorem 6.B.1, the following question arises. Let $\{\boldsymbol{X}(\boldsymbol{\theta}), \boldsymbol{\theta} \in \Theta\}$ be a collection of n-dimensional random vectors indexed by $\boldsymbol{\theta}$, where Θ is a subset of $\mathbb{R}^m$ for some m (see the beginning of Chapter 8 for a discussion about the meaning of this notation). Suppose that $\boldsymbol{X}(\boldsymbol{\theta}) \leq_{\rm st} \boldsymbol{X}(\boldsymbol{\theta}')$ whenever $\boldsymbol{\theta} \leq \boldsymbol{\theta}'$; that is, that $\boldsymbol{X}(\boldsymbol{\theta})$ is stochastically increasing in $\boldsymbol{\theta}$. Is it possible then to construct, on some probability space, a family $\{\hat{\boldsymbol{X}}(\boldsymbol{\theta}), \boldsymbol{\theta} \in \Theta\}$ such that $\hat{\boldsymbol{X}}(\boldsymbol{\theta}) =_{\rm st} \boldsymbol{X}(\boldsymbol{\theta})$ for all $\boldsymbol{\theta} \in \Theta$, and such that $P\{\hat{\boldsymbol{X}}(\boldsymbol{\theta}) \leq \hat{\boldsymbol{X}}(\boldsymbol{\theta}')\} = 1$ whenever $\boldsymbol{\theta} \leq \boldsymbol{\theta}'$? It turns out that if $\Theta \in \mathbb{R}$ (that is, $m = 1$) the answer is in the affirmative. However, when $m \geq 2$ this need not be the case; see Fill and Machida [200] for a counterexample and a further discussion.

6.B.3 Conditions that lead to the multivariate usual stochastic order

The first basic result described in this subsection gives sufficient conditions for the usual multivariate stochastic order by means of the usual univariate stochastic order. The proof is based on the well-known *standard construction*: Suppose that we are given a distribution of a random vector $\boldsymbol{X} = (X_1, X_2, \ldots, X_n)$ and we want to construct a random vector $\hat{\boldsymbol{X}} = (\hat{X}_1, \hat{X}_2, \ldots, \hat{X}_n)$ such that $\hat{\boldsymbol{X}} =_{\rm st} \boldsymbol{X}$. The interest in such constructions is in simulation theory as well as in other areas of applications. In order to do it let $U_1, U_2, \ldots, U_n$ be independent uniform $[0, 1]$ random variables and define

$$\hat{X}_1 = \inf\{x_1 : P\{X_1 \leq x_1\} \geq U_1\},$$

$$\hat{X}_k = \inf\{x_k : P\{X_k \le x_k \big| X_1 = \hat{X}_1, \ldots, X_{k-1} = \hat{X}_{k-1}\} \ge U_k\},$$
$$k = 2, 3, \ldots, n.$$

Then $\hat{\boldsymbol{X}} =_{\rm st} \boldsymbol{X}$.

The conditions given in the next result are natural for a construction of $\hat{\boldsymbol{X}}$ and $\hat{\boldsymbol{Y}}$, as needed in Theorem 6.B.1, using the standard construction. The result then follows from Theorem 6.B.1. Recall from Section 1.A.3 that for any random vector $\boldsymbol{Z}$ and an event A we denote by $[\boldsymbol{Z}\big|A]$ any random vector that has as its distribution the conditional distribution of $\boldsymbol{Z}$ given A.

Theorem 6.B.3. *Let* $\boldsymbol{X} = (X_1, X_2, \ldots, X_n)$ *and* $\boldsymbol{Y} = (Y_1, Y_2, \ldots, Y_n)$ *be two* n*-dimensional random vectors. If*

$$X_1 \le_{\rm st} Y_1, \tag{6.B.8}$$

$$[X_2\big|X_1 = x_1] \le_{\rm st} [Y_2\big|Y_1 = y_1] \quad \textit{whenever } x_1 \le y_1, \tag{6.B.9}$$

and in general, for $i = 2, 3, \ldots, n$,

$$[X_i\big|X_1 = x_1, \ldots, X_{i-1} = x_{i-1}] \le_{\rm st} [Y_i\big|Y_1 = y_1, \ldots, Y_{i-1} = y_{i-1}]$$
$$\textit{whenever } x_j \le y_j,\ j = 1, 2, \ldots, i-1, \tag{6.B.10}$$

then $\boldsymbol{X} \le_{\rm st} \boldsymbol{Y}$.

Proof. First we construct $\hat{X}_1$ and $\hat{Y}_1$ on some probability space as described, for example, in Section 1.A.2. This is possible by (6.B.8). Any possible realization (x_1, y_1) of $(\hat{X}_1, \hat{Y}_1)$ must satisfy $x_1 \le y_1$. Conditioned on every such possible realization (x_1, y_1) we next construct $\hat{X}_2$ and $\hat{Y}_2$ on the same probability space again as described, for example, in Section 1.A.2. This, again, is possible by (6.B.9). We thus have constructed so far $(\hat{X}_1, \hat{X}_2)$ and $(\hat{Y}_1, \hat{Y}_2)$. Any possible realization $((x_1, x_2), (y_1, y_2))$ of $((\hat{X}_1, \hat{X}_2), (\hat{Y}_1, \hat{Y}_2))$ must satisfy $x_j \le y_j$, $j = 1, 2$. Therefore, conditioned on every such possible realization $((x_1, x_2), (y_1, y_2))$ we next can construct $\hat{X}_3$ and $\hat{Y}_3$ on the same probability space and so on. Continuing this procedure we finally arrive at random vectors $\hat{\boldsymbol{X}}$ and $\hat{\boldsymbol{Y}}$, which satisfy (6.B.7). By the standard construction they also satisfy (6.B.5) and (6.B.6). Therefore $\boldsymbol{X} \le_{\rm st} \boldsymbol{Y}$ by Theorem 6.B.1. □

Conditions (6.B.8)–(6.B.10) can be used to define a new stochastic order. More explicitly, if $\boldsymbol{X}$ and $\boldsymbol{Y}$ satisfy (6.B.8)–(6.B.10), then $\boldsymbol{X}$ is said to be *smaller than* $\boldsymbol{Y}$ *in the strong stochastic order* (denoted by $\boldsymbol{X} \le_{\rm sst} \boldsymbol{Y}$). Theorem 6.B.3 simply says that

$$\boldsymbol{X} \le_{\rm sst} \boldsymbol{Y} \Longrightarrow \boldsymbol{X} \le_{\rm st} \boldsymbol{Y}.$$

The order $\le_{\rm sst}$ is not an order in the usual sense; see Remark 6.B.5 below.

Suppose that $\boldsymbol{X} = (X_1, X_2, \ldots, X_n)$ satisfies, for $i = 2, 3, \ldots, n$, that

$$[X_i|X_1 = x_1, \dots, X_{i-1} = x_{i-1}] \leq_{\rm st} [X_i|X_1 = x'_1, \dots, X_{i-1} = x'_{i-1}]$$
$$\text{whenever } x_j \leq x'_j,\ j = 1, 2, \dots, i-1. \quad (6.B.11)$$

Then $\boldsymbol{X}$ is said to be *conditionally increasing in sequence* (CIS). It is easy to see that if $\boldsymbol{X}$ is CIS and if

$$[X_i|X_1 = x_1, \dots, X_{i-1} = x_{i-1}] \leq_{\rm st} [Y_i|Y_1 = x_1, \dots, Y_{i-1} = x_{i-1}]$$
$$\text{for all } x_j,\ j = 1, 2, \dots, i-1, \quad (6.B.12)$$

then (6.B.10) holds. Similarly, if $\boldsymbol{Y}$ is CIS and (6.B.12) holds, then (6.B.10) holds. We thus have proved the following result.

Theorem 6.B.4. *Let $\boldsymbol{X} = (X_1, X_2, \dots, X_n)$ and $\boldsymbol{Y} = (Y_1, Y_2, \dots, Y_n)$ be two n-dimensional random vectors. If either $\boldsymbol{X}$ or $\boldsymbol{Y}$ is* CIS *and* (6.B.8) *and* (6.B.12) *hold, then $\boldsymbol{X} \leq_{\rm st} \boldsymbol{Y}$.*

Remark 6.B.5. The order $\leq_{\rm sst}$ is not an order in the usual sense. In fact, it is obvious that

$$\boldsymbol{X} \leq_{\rm sst} \boldsymbol{X} \Longleftrightarrow \boldsymbol{X} \text{ is CIS.}$$

Remark 6.B.6. Let (U_1, U_2) be a bivariate random vector with uniform$[0,1]$ margins, and with an absolutely continuous distribution function F. Then, as can easily be verified, (U_1, U_2) is CIS if, and only if, $F(u_1, u_2)$ is a concave function of $u_1 \in [0,1]$ for any $u_2 \in [0,1]$.

A random vector $\boldsymbol{X} = (X_1, X_2, \dots, X_n)$ is said to be *weak conditionally increasing in sequence* (WCIS) if, for $i = 2, 3, \dots, n$, we have

$$[(X_i, \dots, X_n)|X_1 = x_1, \dots, X_{i-2} = x_{i-2}, X_{i-1} = x_{i-1}]$$
$$\leq_{\rm st} [(X_i, \dots, X_n)|X_1 = x_1, \dots, X_{i-2} = x_{i-2}, X_{i-1} = x'_{i-1}]$$
$$\text{for all } x_j,\ j = 1, 2, \dots, i-2, \text{ and } x_{i-1} \leq x'_{i-1}.$$

It can be shown that if a random vector is CIS, then it is WCIS. The next result thus strengthens Theorem 6.B.4. We do not give its proof here.

Theorem 6.B.7. *Let $\boldsymbol{X} = (X_1, X_2, \dots, X_n)$ and $\boldsymbol{Y} = (Y_1, Y_2, \dots, Y_n)$ be two n-dimensional random vectors. If either $\boldsymbol{X}$ or $\boldsymbol{Y}$ is* WCIS *and* (6.B.8) *and* (6.B.12) *hold, then $\boldsymbol{X} \leq_{\rm st} \boldsymbol{Y}$.*

The second basic result of this subsection is a multivariate analog of the univariate implication $X \leq_{\rm lr} Y \Longrightarrow X \leq_{\rm st} Y$ (the latter follows from Theorems 1.C.1 and 1.B.1). (Another multivariate analog is given in Theorem 6.E.8.) Recall the definition of association given in (3.A.53). Association, along with the notions of CIS and WCIS, are concepts that indicate positive dependence among the random variables $X_1, X_2, \dots, X_n$.

Theorem 6.B.8. *Let $\boldsymbol{X} = (X_1, X_2, \dots, X_n)$ and $\boldsymbol{Y} = (Y_1, Y_2, \dots, Y_n)$ be two n-dimensional random vectors with density functions f and g, respectively. If $\boldsymbol{X}$ is associated, and if $g(\boldsymbol{x})/f(\boldsymbol{x})$ is increasing in $\boldsymbol{x}$, then $\boldsymbol{X} \le_{\rm st} \boldsymbol{Y}$.*

Proof. Let ϕ be an increasing function for which $E[\phi(\boldsymbol{Y})]$ exists. Then

$$\begin{aligned}
E[\phi(\boldsymbol{Y})] &= \int \phi(\boldsymbol{y})g(\boldsymbol{y})\mathrm{d}\boldsymbol{y} \\
&= \int \phi(\boldsymbol{y})\frac{g(\boldsymbol{y})}{f(\boldsymbol{y})}f(\boldsymbol{y})\mathrm{d}\boldsymbol{y} \\
&\ge \int \phi(\boldsymbol{y})f(\boldsymbol{y})\mathrm{d}\boldsymbol{y} \int \frac{g(\boldsymbol{y})}{f(\boldsymbol{y})}f(\boldsymbol{y})\mathrm{d}\boldsymbol{y} \\
&= E[\phi(\boldsymbol{X})],
\end{aligned}$$

where the inequality follows from (3.A.53) and from the monotonicity of $\phi(\boldsymbol{x})$ and of $g(\boldsymbol{x})/f(\boldsymbol{x})$ in $\boldsymbol{x}$. The stated result now follows from (6.B.4). □

In order to motivate the third basic result of this subsection, consider m independent random variables $X_1, X_2, \dots, X_m$ and an increasing m-dimensional function ϕ. It seems reasonable to expect that $[(X_1, X_2, \dots, X_m) | \phi(X_1, X_2, \dots, X_m) = s]$ is stochastically increasing in s. This is not always true, but the next result indicates an important instance in which this is the case. We omit the proof.

Theorem 6.B.9. *Let $X_1, X_2, \dots, X_m$ be independent random variables, each with a logconcave density (that is, Polya frequency of order* 2 (PF$_2$)*; see Theorem* 1.C.52)*. Then*

$$\left[(X_1, X_2, \dots, X_m)\Big|\sum_{i=1}^m X_i = s\right] \le_{\rm st} \left[(X_1, X_2, \dots, X_m)\Big|\sum_{i=1}^m X_i = s'\right]$$

whenever $s \le s'$.

A variation of Theorem 6.B.9 is stated next. In stating the conditions of Theorem 6.B.10 below we use the discrete analog of the univariate down shifted likelihood ratio order (see Section 1.C.4). Explicitly, let X and Y be univariate discrete random variables, each with support $\mathbb{N}_+$. Then we denote $X \le_{\rm lr\downarrow} Y$ if

$$\frac{P\{Y = m + l\}}{P\{X = m\}} \quad \text{is increasing in } m \ge 0 \text{ for all } l \ge 0. \tag{6.B.13}$$

Note that (6.B.13) is a discrete analog of (1.C.21).

Theorem 6.B.10. *Let $X_1, X_2, \dots, X_m$ be independent random variables, each with support $\mathbb{N}_+$. Denote $S_i = \sum_{j=1}^i X_j$, $i = 1, 2, \dots, m$. If*

$$X_i \le_{\rm lr\downarrow} S_i, \quad i = 2, 3, \dots, m,$$

and if

$$S_i \leq_{\text{lr}\downarrow} S_{i+1}, \quad i = 1, 2, \ldots, m-1,$$

then

$$\Big[(X_1, X_2, \ldots, X_m)\Big|\sum_{i=1}^{m} X_i = s\Big] \leq_{\text{st}} \Big[(X_1, X_2, \ldots, X_m)\Big|\sum_{i=1}^{m} X_i = s'\Big]$$

whenever $s \leq s' \in \mathbb{N}_+$.

In Theorem 6.B.9, the function ϕ which is mentioned just before that theorem, is $\phi(x_1, x_2, \ldots, x_m) = \sum_{i=1}^{m} x_i$. Another case of interest is when $\phi(x_1, x_2, \ldots, x_m) = x_{(i)}$, for some $i \in \{1, 2, \ldots, m\}$, where $x_{(i)}$ is the ith smallest x_j. In fact we have the following result, whose proof we do not give. Note that it is not necessary to assume logconcavity in the next theorem.

Theorem 6.B.11. *Let $X_1, X_2, \ldots, X_m$ be independent and identically distributed random variables with a continuous distribution function. Let $X_{(1)} \leq X_{(2)} \leq \cdots \leq X_{(m)}$ denote the corresponding order statistics. Let $1 \leq r \leq m$. Then*

(a) *for $1 \leq k_1 < k_2 < \cdots < k_r \leq m$, one has that $[(X_1, X_2, \ldots, X_m)|X_{(k_1)} = s_1, X_{(k_2)} = s_2, \ldots, X_{(k_r)} = s_r]$ is stochastically increasing in $s_1 \leq s_2 \leq \cdots \leq s_r$;*
(b) *for $s_1 \leq s_2 \leq \cdots \leq s_r$, one has that $[(X_1, X_2, \ldots, X_m)|X_{(k_1)} = s_1, X_{(k_2)} = s_2, \ldots, X_{(k_r)} = s_r]$ is stochastically decreasing in $1 \leq k_1 < k_2 < \cdots < k_r \leq m$.*

A related result is given in the following theorem.

Theorem 6.B.12. *Let $X_1, X_2, \ldots, X_m$ be independent and identically distributed random variables with a continuous distribution function. Let $X_{(1)} \leq X_{(2)} \leq \cdots \leq X_{(m)}$ denote the corresponding order statistics. Let $1 \leq r \leq m$. Then for $1 \leq k \leq m$, and $s \in \mathbb{R}$, one has that $\big[(X_1, X_2, \ldots, X_m)\big|X_{(k-1)} < s < X_{(k)}\big]$ is stochastically increasing in s, and is stochastically decreasing in k.*

Another result that is related to Theorem 6.B.11 is the following.

Theorem 6.B.13. *Let $X_1, X_2, \ldots, X_m$ be independent exponential random variables with possibly different parameters. Let $X_{(1)} \leq X_{(2)} \leq \cdots \leq X_{(m)}$ denote the corresponding order statistics. Then $[(X_{(1)}, X_{(2)}, \ldots, X_{(m)})|X_{(1)} = s_1]$ is stochastically increasing in s_1.*

The proof of Theorem 6.B.13 uses ideas involving the total hazard construction which is described in Section 6.C.2. Therefore we defer the proof of this theorem to Remark 6.C.2.

For the next result we need the definition of a copula. Let F be an n-dimensional distribution function with univariate marginal distribution functions $F_1, F_2, \ldots, F_n$. Then there exists an n-dimensional distribution function C, with uniform$[0, 1]$ marginal distributions, such that

$$F(x_1, x_2, \dots, x_n) = C(F_1(x_1), F_2(x_2), \dots, F_n(x_n)), \quad (x_1, x_2, \dots, x_n) \in \mathbb{R}^n. \tag{6.B.14}$$

The function C is a *copula* associated with F. If F is continuous, then C is unique and can be obtained by

$$C(u_1, u_2, \dots, u_n) = F(F_1^{-1}(u_1), F_2^{-1}(u_2), \dots, F_n^{-1}(u_n)), \quad (u_1, u_2, \dots, u_n) \in [0, 1]^n; \tag{6.B.15}$$

see, for example, Nelsen [431]. Note that if $(U_1, U_2, \dots, U_n)$ has the distribution function C, then from (6.B.15) it follows that

$$(F_1^{-1}(U_1), F_2^{-1}(U_2), \dots, F_n^{-1}(U_n)) =_{\text{st}} (X_1, X_2, \dots, X_n). \tag{6.B.16}$$

Theorem 6.B.14. *Let the random vectors* $\boldsymbol{X} = (X_1, X_2, \dots, X_n)$ *and* $\boldsymbol{Y} = (Y_1, Y_2, \dots, Y_n)$ *have a common copula. If* $X_i \leq_{\text{st}} Y_i$, $i = 1, 2, \dots, n$, *then* $\boldsymbol{X} \leq_{\text{st}} \boldsymbol{Y}$.

Proof. We only give the proof for the continuous case. Let C be the common copula, and let $(U_1, U_2, \dots, U_n)$ be distributed according to C. Furthermore, let F_i and G_i denote the univariate distribution functions of X_i and Y_i, respectively, $i = 1, 2, \dots, n$. From $X_i \leq_{\text{st}} Y_i$ and (1.A.12) we get $F_i^{-1}(u_i) \leq G_i^{-1}(u_i)$ for all $u_i \in [0, 1]$, $i = 1, 2, \dots, n$. Hence

$$(F_1^{-1}(U_1), F_2^{-1}(U_2), \dots, F_n^{-1}(U_n)) \leq_{\text{a.s.}} (G_1^{-1}(U_1), G_2^{-1}(U_2), \dots, G_n^{-1}(U_n)).$$

The stated result now follows from (6.B.16). □

Theorem 6.B.14 may be compared with Theorem 7.A.38.

An interesting result, which gives conditions under which one can stochastically compare vectors of partial sums of independent random variables, is stated next.

Theorem 6.B.15. *Let* $\{Z_i\}_{i=1}^n$ *be a sequence of independent random variables. If*

$$Z_1 \leq_{\text{lr}} Z_2 \leq_{\text{lr}} \cdots \leq_{\text{lr}} Z_n$$

then

$$\Big(Z_1, Z_1 + Z_2, \dots, \sum_{i=1}^n Z_i\Big) \leq_{\text{st}} \Big(Z_{\pi_1}, Z_{\pi_1} + Z_{\pi_2}, \dots, \sum_{i=1}^n Z_{\pi_i}\Big) \leq_{\text{st}} \Big(Z_n, Z_n + Z_{n-1}, \dots, \sum_{i=1}^n Z_i\Big),$$

for every permutation $(\pi_1, \pi_2, \dots, \pi_n)$ *of* $(1, 2, \dots, n)$.

In particular it follows from Theorem 6.B.15 that if the random variables X and Y are such that $X \leq_{\rm lr} Y$, then

$$(X, X+Y) \leq_{\rm st} (Y, X+Y). \tag{6.B.17}$$

Conclusion (6.B.17) does not necessarily follow from merely assuming that $X \leq_{\rm st} Y$. This can be shown by a counterexample.

The proof of (6.B.17) can be obtained from Theorem 1.C.20 as follows. Let ψ be a bivariate increasing function. Then the function ϕ, defined by $\phi(x,y) = \psi(x, x+y)$, belongs to $\mathcal{G}_{\rm lr}$. Therefore, from (1.C.11) one sees that $\psi(X, X+Y) \leq_{\rm st} \psi(Y, X+Y)$ and this gives (6.B.17). The proof of Theorem 6.B.15 uses the same idea together with a conditioning argument.

6.B.4 Closure properties

Using (6.B.1) through (6.B.7) it is easy to prove each of the following closure results (note that parts (a) and (c) are special cases of part (b) in the next theorem).

Theorem 6.B.16. (a) *Let $\boldsymbol{X}$ and $\boldsymbol{Y}$ be two n-dimensional random vectors. If $\boldsymbol{X} \leq_{\rm st} \boldsymbol{Y}$ and $\boldsymbol{g} : \mathbb{R}^n \to \mathbb{R}^k$ is any k-dimensional increasing [decreasing] function, for any positive integer k, then the k-dimensional vectors $\boldsymbol{g}(\boldsymbol{X})$ and $\boldsymbol{g}(\boldsymbol{Y})$ satisfy $\boldsymbol{g}(\boldsymbol{X}) \leq_{\rm st} [\geq_{\rm st}]\ \boldsymbol{g}(\boldsymbol{Y})$.*

(b) *Let $\boldsymbol{X}_1, \boldsymbol{X}_2, \dots, \boldsymbol{X}_m$ be a set of independent random vectors where the dimension of $\boldsymbol{X}_i$ is k_i, $i = 1, 2, \dots, m$. Let $\boldsymbol{Y}_1, \boldsymbol{Y}_2, \dots, \boldsymbol{Y}_m$ be another set of independent random vectors where the dimension of $\boldsymbol{Y}_i$ is k_i, $i = 1, 2, \dots, m$. Denote $k = k_1 + k_2 + \cdots + k_m$. If $\boldsymbol{X}_i \leq_{\rm st} \boldsymbol{Y}_i$ for $i = 1, 2, \dots, m$, then, for any increasing function $\psi : \mathbb{R}^k \to \mathbb{R}$, one has*

$$\psi(\boldsymbol{X}_1, \boldsymbol{X}_2, \dots, \boldsymbol{X}_m) \leq_{\rm st} \psi(\boldsymbol{Y}_1, \boldsymbol{Y}_2, \dots, \boldsymbol{Y}_m).$$

That is, the usual multivariate stochastic order is closed under conjunctions. In particular, the usual multivariate stochastic order is closed under convolutions.

(c) *Let $\boldsymbol{X} = (X_1, X_2, \dots, X_n)$ and $\boldsymbol{Y} = (Y_1, Y_2, \dots, Y_n)$ be two n-dimensional random vectors. If $\boldsymbol{X} \leq_{\rm st} \boldsymbol{Y}$, then $\boldsymbol{X}_I \leq_{\rm st} \boldsymbol{Y}_I$ for each $I \subseteq \{1, 2, \dots, n\}$. That is, the usual multivariate stochastic order is closed under marginalization.*

(d) *Let $\{\boldsymbol{X}_j, j = 1, 2, \dots\}$ and $\{\boldsymbol{Y}_j, j = 1, 2, \dots\}$ be two sequences of random vectors such that $\boldsymbol{X}_j \to_{\rm st} \boldsymbol{X}$ and $\boldsymbol{Y}_j \to_{\rm st} \boldsymbol{Y}$ as $j \to \infty$, where $\to_{\rm st}$ denotes convergence in distribution. If $\boldsymbol{X}_j \leq_{\rm st} \boldsymbol{Y}_j$, $j = 1, 2, \dots$, then $\boldsymbol{X} \leq_{\rm st} \boldsymbol{Y}$.*

(e) *Let $\boldsymbol{X}$, $\boldsymbol{Y}$, and $\boldsymbol{\Theta}$ be random vectors such that $[\boldsymbol{X} | \boldsymbol{\Theta} = \boldsymbol{\theta}] \leq_{\rm st} [\boldsymbol{Y} | \boldsymbol{\Theta} = \boldsymbol{\theta}]$ for all $\boldsymbol{\theta}$ in the support of $\boldsymbol{\Theta}$. Then $\boldsymbol{X} \leq_{\rm st} \boldsymbol{Y}$. That is, the usual stochastic order is closed under mixtures.*

In (6.B.1) the random vectors $\boldsymbol{X}$ and $\boldsymbol{Y}$ can be taken to be of countable infinite dimension; that is, each of $\boldsymbol{X}$ and $\boldsymbol{Y}$ may correspond to an infinite sequence of random variables. In such a case, if (6.B.1) holds for all upper sets in $\mathbb{R}^\infty$, then we still say that $\boldsymbol{X}$ is smaller than $\boldsymbol{Y}$ in the usual stochastic order (denoted as $\boldsymbol{X} \le_{\rm st} \boldsymbol{Y}$). A generalization of this idea is described in Section 6.B.7. The inequality (6.B.4), as well as Theorem 6.B.1, are still valid when $\boldsymbol{X}$ and $\boldsymbol{Y}$ have countable infinite dimension. We thus get the following result which involves multivariate random sums. Below, an empty sum is understood to be 0.

Theorem 6.B.17. *Let $\boldsymbol{X}_1, \boldsymbol{X}_2, \dots, \boldsymbol{X}_m$ be m countably infinite vectors of nonnegative random variables, and let $\boldsymbol{Y}_1, \boldsymbol{Y}_2, \dots, \boldsymbol{Y}_m$ be other m such vectors. Let $\boldsymbol{M} = (M_1, M_2, \dots, M_m)$ and $\boldsymbol{N} = (N_1, N_2, \dots, N_m)$ be two vectors of nonnegative integers such that $\boldsymbol{M}$ is independent of $\boldsymbol{X}_1, \boldsymbol{X}_2, \dots, \boldsymbol{X}_m$, and $\boldsymbol{N}$ is independent of $\boldsymbol{Y}_1, \boldsymbol{Y}_2, \dots, \boldsymbol{Y}_m$. Denote by $X_{j,i}$ [$Y_{j,i}$] the ith element of $\boldsymbol{X}_j$ [$\boldsymbol{Y}_j$]. If $(\boldsymbol{X}_1, \boldsymbol{X}_2, \dots, \boldsymbol{X}_m) \le_{\rm st} (\boldsymbol{Y}_1, \boldsymbol{Y}_2, \dots, \boldsymbol{Y}_m)$, and if $\boldsymbol{M} \le_{\rm st} \boldsymbol{N}$, then*

$$\Big(\sum_{i=1}^{M_1} X_{1,i}, \sum_{i=1}^{M_2} X_{2,i}, \dots, \sum_{i=1}^{M_m} X_{m,i}\Big) \le_{\rm st} \Big(\sum_{i=1}^{N_1} Y_{1,i}, \sum_{i=1}^{N_2} Y_{2,i}, \dots, \sum_{i=1}^{N_m} Y_{m,i}\Big).$$

Consider now n families of univariate distribution functions $\{G_\theta^{(i)}, \theta \in \mathcal{X}_i\}$ where $\mathcal{X}_i$ is a subset of the real line $\mathbb{R}$, $i = 1, 2, \dots, n$. Let $X_i(\theta)$ denote a random variable with distribution function $G_\theta^{(i)}$, $i = 1, 2, \dots, n$. Let $\boldsymbol{\Theta} = (\Theta_1, \Theta_2, \dots, \Theta_n)$ be a random vector with support in $\prod_{i=1}^n \mathcal{X}_i$, and with distribution function F. Consider the n-dimensional distribution function H given by

$$H(y_1, y_2, \dots, y_n) = \int_{\mathcal{X}_1} \int_{\mathcal{X}_2} \cdots \int_{\mathcal{X}_n} \prod_{i=1}^n G_{\theta_i}^{(i)}(y_i) \mathrm{d}F(\theta_1, \theta_2, \dots, \theta_n),$$
$$(y_1, y_2, \dots, y_n) \in \mathbb{R}^n. \quad (6.B.18)$$

The following result is a generalization of Theorem 6.B.16(e), and is a multivariate extension of Theorem 1.A.6; see Theorems 6.G.8, 7.A.37, 9.A.7, and 9.A.15 for related results.

Theorem 6.B.18. *Let $\{G_\theta^{(i)}, \theta \in \mathcal{X}_i\}$, $i = 1, 2, \dots, n$, be n families of univariate distribution functions as above. Let $\boldsymbol{\Theta}_1$ and $\boldsymbol{\Theta}_2$ be two random vectors with supports in $\prod_{i=1}^n \mathcal{X}_i$ and distribution functions F_1 and F_2, respectively. Let $\boldsymbol{Y}_1$ and $\boldsymbol{Y}_2$ be two random vectors with distribution functions H_1 and H_2 given by*

$$H_j(y_1, y_2, \dots, y_n) = \int_{\mathcal{X}_1} \int_{\mathcal{X}_2} \cdots \int_{\mathcal{X}_n} \prod_{i=1}^n G_{\theta_i}^{(i)}(y_i) \mathrm{d}F_j(\theta_1, \theta_2, \dots, \theta_n),$$
$$(y_1, y_2, \dots, y_n) \in \mathbb{R}^n, \ j = 1, 2.$$

If

$$X_i(\theta) \le_{\text{st}} X_i(\theta') \quad \textit{whenever } \theta \le \theta', \; i = 1, 2, \ldots, n,$$

and if

$$\boldsymbol{\Theta}_1 \le_{\text{st}} \boldsymbol{\Theta}_2,$$

then

$$\boldsymbol{Y}_1 \le_{\text{st}} \boldsymbol{Y}_2.$$

6.B.5 Further properties

Clearly if $\boldsymbol{X} \le_{\text{st}} \boldsymbol{Y}$, then $E\boldsymbol{X} \le E\boldsymbol{Y}$. However, similar to the univariate case, if two random vectors are ordered in the usual multivariate stochastic order and have the same expected values, then they must have the same distribution. This is shown in the following result, which is a multivariate generalization of Theorem 1.A.8. Similar results are given in Theorems 3.A.43, 3.A.60, 4.A.69, 5.A.15, 6.G.12, and 7.A.14–7.A.16.

Theorem 6.B.19. *Let* $\boldsymbol{X} = (X_1, X_2, \ldots, X_m)$ *and* $\boldsymbol{Y} = (Y_1, Y_2, \ldots, Y_m)$ *be two random vectors. If* $\boldsymbol{X} \le_{\text{st}} \boldsymbol{Y}$ *and if* $E[h_i(X_i)] = E[h_i(Y_i)]$ *for some strictly increasing function* h_i, $i = 1, 2, \ldots, m$, *then* $\boldsymbol{X} =_{\text{st}} \boldsymbol{Y}$.

We will not give the complete proof of Theorem 6.B.19 here, but we will show a simple argument that proves it when $\boldsymbol{X}$ and $\boldsymbol{Y}$ are nonnegative random vectors. From the assumption $\boldsymbol{X} \le_{\text{st}} \boldsymbol{Y}$ and from Theorem 6.B.16(c) it follows that $X_i \le_{\text{st}} Y_i$. Since $E[h_i(X_i)] = E[h_i(Y_i)]$ it follows from Theorem 1.A.8 that $X_i =_{\text{st}} Y_i$, and thus, in particular, $EX_i = EY_i$ for $i = 1, 2, \ldots, m$. Therefore

$$E\Big[\sum_{i=1}^m \alpha_i X_i\Big] = \sum_{i=1}^m \alpha_i E[X_i] = \sum_{i=1}^m \alpha_i E[Y_i] = E\Big[\sum_{i=1}^m \alpha_i Y_i\Big]$$

whenever $\alpha_i \ge 0$, $i = 1, 2, \ldots, m$. Also, from $\boldsymbol{X} \le_{\text{st}} \boldsymbol{Y}$ it follows that

$$\sum_{i=1}^m \alpha_i X_i \le_{\text{st}} \sum_{i=1}^m \alpha_i Y_i \quad \text{whenever } \alpha_i \ge 0, \; i = 1, 2, \ldots, m.$$

Therefore, again by Theorem 1.A.8, we have that

$$\sum_{i=1}^m \alpha_i X_i =_{\text{st}} \sum_{i=1}^m \alpha_i Y_i \quad \text{whenever } \alpha_i \ge 0, \; i = 1, 2, \ldots, m.$$

Thus

$$E\Big[\exp\Big\{-\sum_{i=1}^m \alpha_i X_i\Big\}\Big] = E\Big[\exp\Big\{-\sum_{i=1}^m \alpha_i Y_i\Big\}\Big]$$

whenever $\alpha_i \ge 0$, $i = 1, 2, \ldots, m$. From the unicity property of the Laplace transform we obtain $\boldsymbol{X} =_{\text{st}} \boldsymbol{Y}$.

A straightforward analog of Theorem 1.A.15 is in general not true in the multivariate case. That is, if $\boldsymbol{X}$ is any random vector and if U_1 and U_2 are any increasing sets such that $U_1 \supseteq U_2$, then it is not necessarily true that $[\boldsymbol{X}|U_1] \le_{\text{st}} [\boldsymbol{X}|U_2]$; some property of positive dependence is needed to be imposed on $\boldsymbol{X}$ in order for this result to hold. We do not give the details here.

Recall from (6.B.4) that $\boldsymbol{X} = (X_1, X_2, \ldots, X_m) \le_{\text{st}} \boldsymbol{Y} = (Y_1, Y_2, \ldots, Y_m)$ if, and only if, $E[\phi(\boldsymbol{X})] \le E[\phi(\boldsymbol{Y})]$ for all increasing functions ϕ, and that (6.B.2) says that $\boldsymbol{X} \le_{\text{st}} \boldsymbol{Y}$ if, and only if, $E[\phi(\boldsymbol{X})] \le E[\phi(\boldsymbol{Y})]$ for all increasing indicator functions ϕ. When $m = 2$ we have a further similar characterization of the multivariate order $\le_{\text{st}}$, as is stated next. The proof is omitted.

Theorem 6.B.20. *Let (X_1, X_2) and (Y_1, Y_2) be two random vectors. Then $(X_1, X_2) \le_{\text{st}} (Y_1, Y_2)$ if, and only if,*

$$\phi_1(X_1) + \phi_2(X_2) \le_{\text{st}} \phi_1(Y_1) + \phi_2(Y_2)$$

for all increasing functions ϕ_1 and ϕ_2.

A random vector $(X_1, X_2, \ldots, X_m)$ or its distribution is said to be *permutation symmetric* or *exchangeable* if

$$(X_1, X_2, \ldots, X_m) =_{\text{st}} (X_{\pi_1}, X_{\pi_2}, \ldots, X_{\pi_m})$$

for every permutation $\boldsymbol{\pi}$ of $(1, 2, \ldots, m)$. A set U is said to be *symmetric* if

$$(x_1, x_2, \ldots, x_m) \in U \Longrightarrow (x_{\pi_1}, x_{\pi_2}, \ldots, x_{\pi_m}) \in U$$

for every permutation $\boldsymbol{\pi}$ of $(1, 2, \ldots, m)$. For permutation symmetric random vectors the result in the following theorem holds. The proof uses symmetry arguments and is omitted.

Theorem 6.B.21. *Let $\boldsymbol{X} = (X_1, X_2, \ldots, X_m)$ and $\boldsymbol{Y} = (Y_1, Y_2, \ldots, Y_m)$ be two permutation symmetric random vectors. Then $\boldsymbol{X} \le_{\text{st}} \boldsymbol{Y}$ if, and only if, $P\{\boldsymbol{X} \in U\} \le P\{\boldsymbol{Y} \in U\}$ for all symmetric upper sets $U \subseteq \mathbb{R}^m$.*

In the next result we obtain a comparison of order statistics with respect to $\le_{\text{st}}$, but first we need a lemma. Let $z_1, z_2, \ldots$ be a sequence of constants or of random variables. Denote by $z_{(i:m)}$ the ith smallest value among the first m z_i's.

Lemma 6.B.22. *For any sequence of constants $z_1, z_2, \ldots$ the following inequalities hold:*

$$z_{(i:m)} \le z_{(i+1:m)}, \quad 1 \le i \le m-1. \tag{6.B.19}$$

$$z_{(i:m+1)} \le z_{(i:m)}, \quad 1 \le i \le m. \tag{6.B.20}$$

$$z_{(i:m)} \le z_{(i+1:m+1)}, \quad 1 \le i \le m. \tag{6.B.21}$$

Proof. The proof of (6.B.19) is obvious from the definition of the $z_{(i:m)}$'s. The proof of (6.B.20) is also quite simple—just note that if $z_{m+1} \le z_{(i:m)}$, then $z_{(i:m+1)} \le z_{(i:m)}$, whereas if $z_{m+1} > z_{(i:m)}$, then $z_{(i:m+1)} = z_{(i:m)}$. Finally, in order to prove (6.B.21), note that if $z_{m+1} \le z_{(i:m)}$, then $z_{(i+1:m+1)} = z_{(i:m)}$, whereas if $z_{m+1} > z_{(i:m)}$, then $z_{(i:m)} \le z_{(i+1:m+1)}$. □

Theorem 6.B.23. *Let $\{X_1, X_2, \dots\}$ and $\{Y_1, Y_2, \dots\}$ be two sequences of random variables such that*

$$(X_1, X_2, \dots, X_k) \le_{\text{st}} (Y_1, Y_2, \dots, Y_k), \quad k \ge 1. \tag{6.B.22}$$

Then

$$X_{(i:m)} \le_{\text{st}} Y_{(j:n)} \quad \textit{whenever } i \le j \textit{ and } m - i \ge n - j. \tag{6.B.23}$$

Proof. First note that from (6.B.22) it follows that

$$X_{(i:m)} \le_{\text{st}} Y_{(i:m)}, \quad 1 \le i \le m. \tag{6.B.24}$$

Now, if $m \ge n$, then

$$\begin{aligned} X_{(i:m)} &\le_{\text{a.s.}} X_{(i:n)} && \text{(by (6.B.20) and } m \ge n) \\ &\le_{\text{st}} Y_{(i:n)} && \text{(by (6.B.24))} \\ &\le_{\text{a.s.}} Y_{(j:n)} && \text{(by (6.B.19) and } i \le j). \end{aligned}$$

And if $m < n$, then

$$\begin{aligned} X_{(i:m)} &\le_{\text{st}} Y_{(i:m)} && \text{(by (6.B.24))} \\ &\le_{\text{a.s.}} Y_{(i+n-m:n)} && \text{(by (6.B.21) and } m < n) \\ &\le_{\text{a.s.}} Y_{(j:n)} && \text{(by (6.B.19) and } j \ge i + n - m). \end{aligned}$$

Since the almost sure relation $\le_{\text{a.s.}}$ implies the relation $\le_{\text{st}}$, we obtain (6.B.23) from the above inequalities. □

If in Theorem 6.B.23 we take $Y_i = X_i$, $i = 1, 2, \dots$, then obviously (6.B.22) holds. Thus we obtain the following corollary.

Corollary 6.B.24. *Let $\{X_1, X_2, \dots\}$ be a sequence of (not necessarily independent) random variables. Then*

$$X_{(i:m)} \le_{\text{st}} X_{(j:n)} \quad \textit{whenever } i \le j \textit{ and } m - i \ge n - j.$$

The next example shows that if two random variables are ordered in the dispersive order, then the corresponding vectors of spacings are ordered in the usual stochastic order. Related results can be found in Theorems 1.C.45 and 4.B.17, and in Example 6.E.15.

Example 6.B.25. Let X and Y be two random variables. Let $X_{(1)} \le X_{(2)} \le \cdots \le X_{(n)}$ denote the order statistics from a sample $X_1, X_2, \ldots, X_n$ of independent and identically distributed random variables that have the same distribution as X. Similarly, let $Y_{(1)} \le Y_{(2)} \le \cdots \le Y_{(n)}$ denote the order statistics from another sample $Y_1, Y_2, \ldots, Y_n$ of independent and identically distributed random variables that have the same distribution as Y. The corresponding spacings are defined by $U_{(i)} \equiv X_{(i)} - X_{(i-1)}$ and $V_{(i)} \equiv Y_{(i)} - Y_{(i-1)}$, $i = 2, 3, \ldots, n$. Denote $\boldsymbol{U} = (U_{(2)}, U_{(3)}, \ldots, U_{(n)})$ and $\boldsymbol{V} = (V_{(2)}, V_{(3)}, \ldots, V_{(n)})$. We will now show that if $X \le_{\text{disp}} Y$, then $\boldsymbol{U} \le_{\text{st}} \boldsymbol{V}$. Let F and G denote the distribution functions of X and Y, respectively. Define $\hat{Y}_{(i)} = G^{-1}(F(X_{(i)}))$, $i = 1, 2, \ldots, n$, and $\hat{V}_{(i)} = \hat{Y}_{(i)} - \hat{Y}_{(i-1)}$, $i = 2, 3, \ldots, n$. Clearly, $(V_{(2)}, V_{(3)}, \ldots, V_{(n)}) =_{\text{st}} (\hat{V}_{(2)}, \hat{V}_{(3)}, \ldots, \hat{V}_{(n)})$. Furthermore, from (3.B.10) we have that

$$\hat{V}_{(i)} = G^{-1}(F(X_{(i)})) - G^{-1}(F(X_{(i-1)})) \ge X_{(i)} - X_{(i-1)} = U_{(i)} \text{ a.s.}, \quad i = 2, 3, \ldots, n.$$

Thus, it follows from Theorem 6.B.1 that $\boldsymbol{U} \le_{\text{st}} \boldsymbol{V}$. In particular, from Theorem 6.B.16(c) we get that $U_{(i)} \le_{\text{st}} V_{(i)}$ for $i = 2, 3, \ldots, n$, and this proves Theorem 3.B.31.

For the next two examples recall from page 2 the definition of the majorization order $\boldsymbol{a} \prec \boldsymbol{b}$ among n-dimensional vectors.

Example 6.B.26. Let $X_1, X_2, \ldots, X_n, Y_1, Y_2, \ldots, Y_n$ be independent Gamma random variables where X_i has the density function f_i defined by

$$f_i(x) = \frac{\lambda_i^\alpha}{\Gamma(\alpha)} x^{\alpha-1} e^{-\lambda_i x}, \quad x \ge 0,$$

where $\alpha > 0$ and $\lambda_i > 0$, $i = 1, 2, \ldots, n$, and Y_i has the density function g_i defined by

$$g_i(x) = \frac{\mu_i^\alpha}{\Gamma(\alpha)} x^{\alpha-1} e^{-\mu_i x}, \quad x \ge 0,$$

where $\alpha > 0$ is as above, and $\mu_i > 0$, $i = 1, 2, \ldots, n$. Denote the corresponding order statistics by $X_{(1)} \le X_{(2)} \le \cdots \le X_{(n)}$ and $Y_{(1)} \le Y_{(2)} \le \cdots \le Y_{(n)}$. Suppose that $(\lambda_1, \lambda_2, \ldots, \lambda_n) \prec (\mu_1, \mu_2, \ldots, \mu_n)$. If $\alpha \le 1$, then

$$(X_{(1)}, X_{(2)}, \ldots, X_{(n)}) \le_{\text{st}} (Y_{(1)}, Y_{(2)}, \ldots, Y_{(n)}),$$

and if $\alpha \ge 1$, then

$$X_{(1)} \ge_{\text{st}} Y_{(1)} \quad \text{and} \quad X_{(n)} \le_{\text{st}} Y_{(n)}.$$

In particular, by taking $\alpha = 1$, it is seen that the above inequalities hold for heterogeneous exponential random variables.

Example 6.B.27. Let $X_1, X_2, \ldots, X_n, Y_1, Y_2, \ldots, Y_n$ be independent Weibull random variables where X_i has the survival function $\overline{F}_i$ defined by

$$\overline{F}_i(x) = \mathrm{e}^{-(\lambda_i x)^\alpha}, \quad x \geq 0,$$

where $\alpha > 0$ and $\lambda_i > 0$, $i = 1, 2, \ldots, n$, and Y_i has the survival function $\overline{G}_i$ defined by

$$\overline{G}_i(x) = \mathrm{e}^{-(\mu_i x)^\alpha}, \quad x \geq 0,$$

where $\alpha > 0$ is as above, and $\mu_i > 0$, $i = 1, 2, \ldots, n$. Denote the corresponding order statistics by $X_{(1)} \leq X_{(2)} \leq \cdots \leq X_{(n)}$ and $Y_{(1)} \leq Y_{(2)} \leq \cdots \leq Y_{(n)}$. Suppose that $(\lambda_1, \lambda_2, \ldots, \lambda_n) \prec (\mu_1, \mu_2, \ldots, \mu_n)$. If $\alpha \leq 1$, then

$$(X_{(1)}, X_{(2)}, \ldots, X_{(n)}) \leq_{\mathrm{st}} (Y_{(1)}, Y_{(2)}, \ldots, Y_{(n)}).$$

Again, by taking $\alpha = 1$, it is seen that the above inequalities hold for heterogeneous exponential random variables.

Example 6.B.28. Let $\boldsymbol{X} = (X_1, X_2, \ldots, X_m)$ and $\boldsymbol{Y} = (Y_1, Y_2, \ldots, Y_m)$ be infinitely divisible random vectors with Lèvy measures $\nu_{\boldsymbol{X}}$ and $\nu_{\boldsymbol{Y}}$, respectively; that is, $\nu_{\boldsymbol{X}}$ and $\nu_{\boldsymbol{Y}}$ satisfy $\int_{\mathbb{R}^m} (1 \wedge |\boldsymbol{x}|) \nu_{\boldsymbol{X}}(\mathrm{d}\boldsymbol{x}) < \infty$ and $\int_{\mathbb{R}^m} (1 \wedge |\boldsymbol{y}|) \nu_{\boldsymbol{Y}}(\mathrm{d}\boldsymbol{y}) < \infty$, and the characteristic functions of $\boldsymbol{X}$ and of $\boldsymbol{Y}$ can be written in the form

$$\varphi_{\boldsymbol{X}}(\boldsymbol{t}) = \exp\left\{ \int_{\mathbb{R}^m \setminus \{\boldsymbol{0}\}} (\mathrm{e}^{i(\boldsymbol{t} \cdot \boldsymbol{x})} - 1) \nu_{\boldsymbol{X}}(\mathrm{d}\boldsymbol{x}) + i(\boldsymbol{t} \cdot \boldsymbol{b}_{\boldsymbol{X}}) \right\}$$

and

$$\varphi_{\boldsymbol{Y}}(\boldsymbol{t}) = \exp\left\{ \int_{\mathbb{R}^m \setminus \{\boldsymbol{0}\}} (\mathrm{e}^{i(\boldsymbol{t} \cdot \boldsymbol{y})} - 1) \nu_{\boldsymbol{Y}}(\mathrm{d}\boldsymbol{y}) + i(\boldsymbol{t} \cdot \boldsymbol{b}_{\boldsymbol{Y}}) \right\},$$

respectively, for some $\boldsymbol{b}_{\boldsymbol{X}}, \boldsymbol{b}_{\boldsymbol{Y}} \in \mathbb{R}^m$. Assume that $\nu_{\boldsymbol{X}}$ and $\nu_{\boldsymbol{Y}}$ are concentrated on $[0, \infty)^m$. If $\nu_{\boldsymbol{X}}(U) \leq \nu_{\boldsymbol{Y}}(U)$ for all Borel measurable upper sets in $\mathbb{R}^m$, and if $\boldsymbol{b}_{\boldsymbol{X}} \leq \boldsymbol{b}_{\boldsymbol{Y}}$, then $\boldsymbol{X} \leq_{\mathrm{st}} \boldsymbol{Y}$.

The following example gives necessary and sufficient conditions for the comparison of multivariate normal random vectors. See Examples 6.G.11, 7.A.13, 7.A.26, 7.A.39, 7.B.5, and 9.A.20 for related results.

Example 6.B.29. Let $\boldsymbol{X}$ be a multivariate normal random vector with mean vector $\boldsymbol{\mu}_{\boldsymbol{X}}$ and variance-covariance matrix $\boldsymbol{\Sigma}_{\boldsymbol{X}}$, and let $\boldsymbol{Y}$ be a multivariate normal random vector with mean vector $\boldsymbol{\mu}_{\boldsymbol{Y}}$ and variance-covariance matrix $\boldsymbol{\Sigma}_{\boldsymbol{Y}}$. Then $\boldsymbol{X} \leq_{\mathrm{st}} \boldsymbol{Y}$ if, and only if, $\boldsymbol{\mu}_{\boldsymbol{X}} \leq \boldsymbol{\mu}_{\boldsymbol{Y}}$ and $\boldsymbol{\Sigma}_{\boldsymbol{X}} = \boldsymbol{\Sigma}_{\boldsymbol{Y}}$.

6.B.6 A property in reliability theory

In this subsection we show how the multivariate order $\leq_{\mathrm{st}}$ can be used as a tool for the purpose of defining aging properties for components whose lifetimes are not necessarily independent. The notions and notations introduced in this subsection will also be used in the rest of this chapter.

Let $\boldsymbol{T} = (T_1, T_2, \dots, T_m)$ be a nonnegative random vector with an absolutely continuous distribution function. In this subsection it is helpful to think about $T_1, T_2, \dots, T_m$ as the lifetimes of m components $1, 2, \dots, m$ that make up some system. Suppose that an observer observes the system continuously in time and records the failure times and the identities of the components that fail as time passes. Thus, a typical "history" that the observer has observed by time $t \geq 0$ is of the form

$$h_t = \{\boldsymbol{T}_I = \boldsymbol{t}_I, \boldsymbol{T}_{\overline{I}} > t\boldsymbol{e}\}, \quad 0\boldsymbol{e} \leq \boldsymbol{t}_I \leq t\boldsymbol{e},\ I \subseteq \{1, 2, \dots, m\}. \tag{6.B.25}$$

In (6.B.25) I is the set of components that have already failed by time t (with failure times $\boldsymbol{t}_I$) and $\overline{I}$ is the set of components that are still alive at time t.

Let

$$h'_s = \{\boldsymbol{T}_J = \boldsymbol{s}_J, \boldsymbol{T}_{\overline{J}} > s\boldsymbol{e}\}, \quad 0\boldsymbol{e} \leq \boldsymbol{s}_J \leq s\boldsymbol{e},\ J \subseteq \{1, 2, \dots, m\}, \tag{6.B.26}$$

be another history. If $t \leq s$ and the histories h_t and h'_s are such that each component that failed in h_t also failed in h'_s, and, for components that failed in both histories, the failures in h'_s are earlier than the failures in h_t, then we say that the history h_t is *less severe* or "more pleasant" than the history h'_s and we denote it by $h_t \leq h'_s$. Note that if h_t and h'_s are as in (6.B.25) and (6.B.26), then $h_t \leq h'_s$ if, and only if, $I \subseteq J$ and $\boldsymbol{s}_I \leq \boldsymbol{t}_I$.

For every vector $\boldsymbol{a} = (a_1, a_2, \dots, a_m)$ denote by $\boldsymbol{a}_+$ the vector

$$\boldsymbol{a}_+ = ((a_1)_+, (a_2)_+, \dots, (a_m)_+).$$

Recalling Theorem 1.A.30 we can define a nonnegative random vector $\boldsymbol{T}$ as multivariate IFR if for $t \leq s$ we have

$$[(\boldsymbol{T} - t\boldsymbol{e})_+ \big| h_t] \geq_{\text{st}} [(\boldsymbol{T} - s\boldsymbol{e})_+ \big| h'_s] \quad \text{whenever } h_t \leq h'_s. \tag{6.B.27}$$

Another possibility is to call the nonnegative random vector $\boldsymbol{T}$ multivariate IFR if for $t \leq s$ we have

$$[(\boldsymbol{T} - t\boldsymbol{e})_+ \big| h_t] \geq_{\text{st}} [(\boldsymbol{T} - s\boldsymbol{e})_+ \big| h'_s] \quad \text{whenever } h_t \text{ and } h'_s \text{ coincide on } [0, t). \tag{6.B.28}$$

These two *different* definitions of multivariate IFR have some desirable properties. For example, a vector consisting of independent IFR random variables is multivariate IFR according to either one of these two definitions. However, perhaps the most important feature of these kinds of definitions is their intuitive interpretation. In the univariate case these two definitions coincide with the usual univariate definition of IFR.

Further notions of multivariate IFR are studied in Section 6.D.3.

6.B.7 Stochastic ordering of stochastic processes

In Section 1.A.1 we saw how to define the usual stochastic order between two univariate random variables. In Section 6.B.1 we saw how this comparison can be defined for two multivariate random vectors. The next level of generalization, then, is the stochastic comparison of two stochastic processes. In fact, several levels of generalization can be studied. The stochastic processes can be univariate (if their common state space $\mathcal{S}$ is a subset of $\mathbb{R}$). Or they can be multivariate (if their common state space $\mathcal{S}$ is a subset of $\mathbb{R}^m$ for some m). Or, more generally, the common state space $\mathcal{S}$ can be any general space, according to the requirements of the particular application in which the order is to be used. In this subsection we consider only the case in which the random processes are univariate. Section 6.H contains some references for the more general results.

Let $\{X(t), t \in \mathcal{T}\}$ and $\{Y(t), t \in \mathcal{T}\}$ be two stochastic processes with state space $\mathcal{S} \subseteq \mathbb{R}$ and time parameter space $\mathcal{T}$ (usually $\mathcal{T} = [0, \infty)$ or $\mathcal{T} = \mathbb{N}_+$). Suppose that, for all choices of an integer m and $t_1 < t_2 < \cdots < t_m$ in $\mathcal{T}$, it holds that

$$(X(t_1), X(t_2), \ldots, X(t_m)) \leq_{\rm st} (Y(t_1), Y(t_2), \ldots, Y(t_m)),$$

where here $\leq_{\rm st}$ is in the sense of Section 6.B.1. Then $\{X(t),\ t \in \mathcal{T}\}$ is said to be *smaller than* $\{Y(t),\ t \in \mathcal{T}\}$ *in the usual stochastic order* (denoted by $\{X(t),\ t \in \mathcal{T}\} \leq_{\rm st} \{Y(t),\ t \in \mathcal{T}\}$).

It can be shown that $\{X(t),\ t \in \mathcal{T}\} \leq_{\rm st} \{Y(t),\ t \in \mathcal{T}\}$ if, and only if,

$$E\{g(\{X(t),\ t \in \mathcal{T}\})\} \leq E\{g(\{Y(t),\ t \in \mathcal{T}\})\}, \tag{6.B.29}$$

for every increasing functional g for which the expectations in (6.B.29) exist (a functional g is called increasing if $g(\{x(t),\ t \in \mathcal{T}\}) \leq g(\{y(t),\ t \in \mathcal{T}\})$ whenever $x(t) \leq y(t)$, $t \in \mathcal{T}$).

An analog of (6.B.1) can also be stated and proved, but it is not included here. However, we do state the following important property of the order $\leq_{\rm st}$, which is a generalization of Theorem 6.B.1.

Theorem 6.B.30. *The random processes* $\{X(t),\ t \in \mathcal{T}\}$ *and* $\{Y(t),\ t \in \mathcal{T}\}$ *satisfy* $\{X(t), t \in \mathcal{T}\} \leq_{\rm st} \{Y(t), t \in \mathcal{T}\}$ *if, and only if, there exist two random processes* $\{\hat{X}(t),\ t \in \mathcal{T}\}$ *and* $\{\hat{Y}(t),\ t \in \mathcal{T}\}$, *defined on the same probability space, such that*

$$\{\hat{X}(t),\ t \in \mathcal{T}\} =_{\rm st} \{X(t),\ t \in \mathcal{T}\},$$
$$\{\hat{Y}(t),\ t \in \mathcal{T}\} =_{\rm st} \{Y(t),\ t \in \mathcal{T}\},$$

and

$$P\{\hat{X}(t) \leq \hat{Y}(t),\ t \in \mathcal{T}\} = 1.$$

For discrete-time processes ($\mathcal{T} = \mathbb{N}_+$), an analog of Theorem 6.B.3 is given in Theorem 6.B.31. The proof of it is the same as the proof of Theorem 6.B.3, except that Theorem 6.B.30 is applied at the end of the proof rather than Theorem 6.B.1.

Theorem 6.B.31. *Let* $\{X(n),\ n \in \mathbb{N}_+\} = \{X(0), X(1), X(2), \dots\}$ *and* $\{Y(n),\ n \in \mathbb{N}_+\} = \{Y(0), Y(1), Y(2), \dots\}$ *be two discrete-time stochastic processes. If*

$$X(0) \le_{\rm st} Y(0),$$

and if

$$\begin{aligned} &[X(i) \big| X(1) = x_1, \dots, X(i-1) = x_{i-1}] \\ &\qquad \le_{\rm st} [Y(i) \big| Y(1) = y_1, \dots, Y(i-1) = y_{i-1}] \\ &\qquad\quad \text{whenever } x_j \le y_j,\ j = 1, 2, \dots, i-1,\ i = 1, 2, 3, \dots, \end{aligned}$$

then $\{X(n),\ n \in \mathbb{N}_+\} \le_{\rm st} \{Y(n),\ n \in \mathbb{N}_+\}$.

Theorems 6.B.2 and 6.B.4 also have straightforward analogs that we do not state here.

The order $\le_{\rm st}$ for stochastic processes is closed under operations similar to those described in Theorem 6.B.16. In particular, $\{X(t), t \in \mathcal{T}\} \le_{\rm st} \{Y(t), t \in \mathcal{T}\} \Longrightarrow \{g(\{X(t), t \in \mathcal{T}\})\} \le_{\rm st} \{g(\{Y(t), t \in \mathcal{T}\})\}$ for all increasing functionals g. The order is also closed under mixtures.

To see an important application of these ideas, consider two discrete-time homogeneous Markov processes $\{X_1(n),\ n \in \mathbb{N}_+\}$ and $\{X_2(n),\ n \in \mathbb{N}_+\}$ with a common state space $\mathcal{S} \subseteq \mathbb{R}$. Denote $Y_{X_1}(x) =_{\rm st} [X_1(n+1) \big| X_1(n) = x]$ and $Y_{X_2}(x) =_{\rm st} [X_2(n+1) \big| X_2(n) = x]$, $x \in \mathcal{S}$. The proof of the next result follows directly from Theorem 6.B.31.

Theorem 6.B.32. *Let* $\{X_1(n), n \in \mathbb{N}_+\}$ *and* $\{X_2(n), n \in \mathbb{N}_+\}$ *be two Markov processes as described above. Suppose that* $X_1(0) \le_{\rm st} X_2(0)$ *and that*

$$Y_{X_1}(x) \le_{\rm st} Y_{X_2}(x') \quad \text{whenever } x \le x'.$$

Then $\{X_1(n),\ n \in \mathbb{N}_+\} \le_{\rm st} \{X_2(n),\ n \in \mathbb{N}_+\}$.

A variation of Theorem 6.B.32 for Markov chains (that is, discrete-time homogeneous Markov process with state space in $\mathbb{N}$) is given next. Recall that a Markov chain is called skip-free positive if it does not have positive jumps of magnitude more than one. For a Markov chain $\{X(n),\ n \in \mathbb{N}_+\}$ with state space $\mathcal{S} \subseteq \mathbb{N}$ we denote $Y_X(i) =_{\rm st} [X(n+1) \big| X(n) = i]$, $i \in \mathcal{S}$. The proof of the following result is obtained by a straightforward construction of the two underlying Markov chains on the same probability space, and then using Theorem 6.B.30.

Theorem 6.B.33. *Let* $\{X_1(n), n \in \mathbb{N}_+\}$ *and* $\{X_2(n), n \in \mathbb{N}_+\}$ *be two Markov chains. Suppose that* $X_1(0) \le_{\rm st} X_2(0)$, *that*

$$Y_{X_1}(i) \le_{\rm st} Y_{X_2}(i) \quad \text{for all } i,$$

and

$$Y_{X_2}(i) \ge i \quad \text{for all } i, \tag{6.B.30}$$

and that $\{X_1(n),\ n \in \mathbb{N}_+\}$ *is skip-free positive. Then* $\{X_1(n),\ n \in \mathbb{N}_+\} \le_{\rm st} \{X_2(n),\ n \in \mathbb{N}_+\}$.

The discrete-time homogeneous Markov process $\{X(n),\, n \in \mathbb{N}_+\}$ is said to be *stochastically monotone* if $Y_X(x) =_{\text{st}} [X(n+1) | X(n) = x]$ is stochastically increasing in $x \in \mathcal{S}$. Note that stochastic monotonicity is a different condition than the almost sure monotonicity condition (6.B.30) — none of these implies the other. Denote by $\{X^{(x)}(n),\, n \in \mathbb{N}_+\}$ the process $\{X(n),\, n \in \mathbb{N}_+\}$ under the condition that $X(0) = x$. The following result is a direct consequence of Theorem 6.B.32.

Theorem 6.B.34. *Let $\{X(n),\, n \in \mathbb{N}_+\}$ be a discrete-time homogeneous Markov process that is stochastically monotone. Then*

$$\{X^{(x)}(n),\, n \in \mathbb{N}_+\} \le_{\text{st}} \{X^{(x')}(n),\, n \in \mathbb{N}_+\} \tag{6.B.31}$$

whenever $x \le x'$.

For example, a discrete-time birth and death chain (with state space $\mathbb{N}$) with birth probabilities $P\{X(n+1) = i+1 | X(n) = i\} = p_i$ and death probabilities $P\{X(n+1) = i-1 | X(n) = i\} = 1 - p_i$, $i \in \mathbb{N}$, is stochastically monotone if p_i increases in $i \in \mathbb{N}$. Hence it satisfies (6.B.31).

If two processes $\{X(t),\, t \in \mathcal{T}\}$ and $\{Y(t),\, t \in \mathcal{T}\}$ satisfy $\{X(t),\, t \in \mathcal{T}\} \le_{\text{st}} \{Y(t),\, t \in \mathcal{T}\}$, then, by Theorem 6.B.30, the first passage times $T_X(a) \equiv \inf\{t : X(t) > a\}$ and $T_Y(a) \equiv \inf\{t : Y(t) > a\}$ (where $\inf \varnothing = \infty$) satisfy $T_X(a) \ge_{\text{st}} T_Y(a)$ for all a. The reverse implication need not be true. By removing (6.B.30) from Theorem 6.B.33 we obtain the following result. Its proof consists of a proper construction of the two underlying Markov chains on the same probability space, and then using Theorem 6.B.30.

Theorem 6.B.35. *Let $\{X_1(n), n \in \mathbb{N}_+\}$ and $\{X_2(n), n \in \mathbb{N}_+\}$ be two Markov chains. Suppose that $X_1(0) \le_{\text{st}} X_2(0)$, that*

$$Y_{X_1}(i) \le_{\text{st}} Y_{X_2}(i) \quad \textit{for all } i,$$

and that $\{X_1(n),\, n \in \mathbb{N}_+\}$ is skip-free positive. Then $T_X(a) \ge_{\text{st}} T_Y(a)$ for all a.

Suppose now that the two processes that we want to compare are point processes that, for distinction, we denote by $\{K(t),\, t \ge 0\}$ and $\{N(t),\, t \ge 0\}$. That is, for each $t \ge 0$, $K(t)$ and $N(t)$ are the numbers of jumps that the corresponding processes have experienced over the time interval $(0, t]$. In addition to the possible relationship $\{K(t),\, t \ge 0\} \le_{\text{st}} \{N(t),\, t \ge 0\}$ between these processes, we will consider also two other stronger possible relationships.

For any positive integer m, let $B_1, B_2, \ldots, B_m$ be bounded Borel sets of $[0, \infty)$. Let $K(B_i)$ and $N(B_i)$ denote the number of jumps of the corresponding processes over the set B_i, $i = 1, 2, \ldots, m$. Suppose that, for all choices of an integer m and bounded Borel sets $B_1, B_2, \ldots, B_m$, it holds that

$$(K(B_1), K(B_2), \ldots, K(B_m)) \le_{\text{st}} (N(B_1), N(B_2), \ldots, N(B_m)).$$

Then $\{K(t),\ t \geq 0\}$ is said to be *smaller than* $\{N(t),\ t \geq 0\}$ *in the usual stochastic order over* $\mathcal{N}$ (denoted by $\{K(t),\ t \geq 0\} \leq_{\text{st-}\mathcal{N}} \{N(t),\ t \geq 0\}$). (Here $\mathcal{N}$ denotes the space of integer-valued Radon measures.) The usual stochastic order over $\mathcal{N}$ gives a "global" comparison of the point processes $\{K(t),\ t \geq 0\}$ and $\{N(t),\ t \geq 0\}$.

Let $X_1 < X_2 < \cdots$ be the sequence of interpoint distances of the process $\{K(t),\ t \geq 0\}$, and let $Y_1 < Y_2 < \cdots$ be the sequence of interpoint distances of the process $\{N(t),\ t \geq 0\}$. We assume that the X_i's and that the Y_i's are almost surely positive. Also we assume that the processes are nonexplosive in the sense that $\lim_{n\to\infty} \sum_{i=1}^{n} X_i = \infty$ and $\lim_{n\to\infty} \sum_{i=1}^{n} Y_i = \infty$ almost surely. Suppose that, for all choices of an integer m and indices $i_1, i_2, \ldots, i_m$, it holds that

$$(X_{i_1}, X_{i_2}, \ldots, X_{i_m}) \geq_{\text{st}} (Y_{i_1}, Y_{i_2}, \ldots, Y_{i_m}).$$

Then $\{K(t),\ t \geq 0\}$ is said to be *smaller than* $\{N(t),\ t \geq 0\}$ *in the usual stochastic order over* $\mathbb{R}^\infty$ (denoted by $\{K(t),\ t \geq 0\} \leq_{\text{st-}\infty} \{N(t),\ t \geq 0\}$). The usual stochastic order over $\mathbb{R}^\infty$ gives a "local" comparison of the point processes $\{K(t),\ t \geq 0\}$ and $\{N(t),\ t \geq 0\}$.

Analogs of (6.B.29) can be stated and proven for the orders $\leq_{\text{st-}\mathcal{N}}$ and $\leq_{\text{st-}\infty}$. Also, "almost sure" constructions, that are analogs of Theorem 6.B.30, can be shown for these orders. We do not give the technical details here. We note, however, that in such constructions the counterparts $\hat{K} = \{\hat{K}(t), t \geq 0\}$ and $\hat{N} = \{\hat{N}(t), t \geq 0\}$ of $\{K(t), t \geq 0\}$ and $\{N(t), t \geq 0\}$, respectively, satisfy the following properties: The relationship $\{K(t),\ t \geq 0\} \leq_{\text{st-}\mathcal{N}} \{N(t),\ t \geq 0\}$ means that $\hat{K}$ is a thinning of $\hat{N}$. The relationship $\{K(t), t \geq 0\} \leq_{\text{st}} \{N(t), t \geq 0\}$ means that $\hat{N}$ has a.s. earlier and more numerous points than $\hat{K}$ before each time instant t. The relationship $\{K(t), t \geq 0\} \leq_{\text{st-}\infty} \{N(t), t \geq 0\}$ means that the corresponding interpoint distances are shorter for $\hat{N}$ than for $\hat{K}$ a.s. From this it is immediate that

$$\{K(t),\ t \geq 0\} \leq_{\text{st-}\mathcal{N}} \{N(t),\ t \geq 0\} \Longrightarrow \{K(t),\ t \geq 0\} \leq_{\text{st}} \{N(t),\ t \geq 0\},$$

and that

$$\{K(t),\ t \geq 0\} \leq_{\text{st-}\infty} \{N(t),\ t \geq 0\} \Longrightarrow \{K(t),\ t \geq 0\} \leq_{\text{st}} \{N(t),\ t \geq 0\}. \tag{6.B.32}$$

It can be shown that, in general, $\{K(t),\ t \geq 0\} \leq_{\text{st-}\mathcal{N}} \{N(t),\ t \geq 0\} \not\Longrightarrow \{K(t),\ t \geq 0\} \leq_{\text{st-}\infty} \{N(t),\ t \geq 0\}$ and also that $\{K(t),\ t \geq 0\} \leq_{\text{st-}\infty} \{N(t),\ t \geq 0\} \not\Longrightarrow \{K(t),\ t \geq 0\} \leq_{\text{st-}\mathcal{N}} \{N(t),\ t \geq 0\}$.

For renewal processes we have the following results.

Theorem 6.B.36. *Consider two nondelayed renewal processes* $\{K(t),\ t \geq 0\}$ *and* $\{N(t),\ t \geq 0\}$ *with generic interpoint distances* X *and* Y*, respectively. The following three statements are equivalent.*

(i) $Y \leq_{\text{st}} X$,
(ii) $\{K(t),\ t \geq 0\} \leq_{\text{st}} \{N(t),\ t \geq 0\}$,

(iii) $\{K(t),\, t \ge 0\} \le_{\text{st-}\infty} \{N(t),\, t \ge 0\}$.

Proof. Note that from the independence of the interpoint distances it follows that (i)$\Longleftrightarrow$(iii). From (6.B.32) it follows that (iii)$\Longrightarrow$(ii). The implication (ii)$\Longrightarrow$(i) is obvious. □

Theorem 6.B.37. *Consider two nondelayed renewal processes* $\{K(t),\, t \ge 0\}$ *and* $\{N(t),\, t \ge 0\}$ *with generic interpoint distances* X *and* Y*, respectively. Let* r_X *and* r_Y *denote the hazard rate functions corresponding to* X *and* Y*, respectively. If*

$$r_X(t) \le r_Y(s) \quad \textit{for all } 0 \le s \le t, \tag{6.B.33}$$

then

$$\{K(t),\, t \ge 0\} \le_{\text{st-}\mathcal{N}} \{N(t),\, t \ge 0\}.$$

Theorem 6.B.37 can be easily proven using the fact, mentioned above, that $\hat{K}$ is a thinning of $\hat{N}$. We do not give a detailed proof of it here.

Note that (6.B.33) holds if $Y \le_{\text{hr}} X$ and if X is DFR or if Y is DFR.

The proofs of the next two theorems are similar to the proofs of Theorems 6.B.36 and 6.B.37, respectively.

Theorem 6.B.38. *Consider two delayed renewal processes* $\{K^d(t),\, t \ge 0\}$ *and* $\{N^d(t),\, t \ge 0\}$*, with the corresponding delays* X^d *and* Y^d *and with the* same *interrenewal distribution after the delay. The following statements are equivalent.*

(i) $Y^d <_{\text{st}} X^d$,
(ii) $\{K^d(t),\, t \ge 0\} \le_{\text{st}} \{N^d(t),\, t \ge 0\}$,
(iii) $\{K^d(t),\, t \ge 0\} \le_{\text{st-}\infty} \{N^d(t),\, t \ge 0\}$.

Theorem 6.B.39. *Consider two delayed renewal processes* $\{K^d(t),\, t \ge 0\}$ *and* $\{N^d(t),\, t \ge 0\}$*, with the corresponding delays* X^d *and* Y^d *and with the* same *interrenewal distribution after the delay. Let* r_{X^d} *denote the hazard rate function corresponding to* X^d*. If* $Y^d \le_{\text{hr}} X^d$ *and if*

$$r_{X^d}(t) \le r(s) \quad \textit{for all } 0 \le s \le t, \tag{6.B.34}$$

where r *is the hazard rate function associated with the common interrenewal distribution function, then*

$$\{K^d(t),\, t \ge 0\} \le_{\text{st-}\mathcal{N}} \{N^d(t),\, t \ge 0\}.$$

Note that (6.B.34) holds, for example, if $X \le_{\text{hr}} X^d$, and if X is DFR or if X^d is DFR.

Finally we give conditions for two nonhomogeneous Poisson processes to be ordered according to the above orders.

Theorem 6.B.40. *Let* $\{K(t),\, t \ge 0\}$ *and* $\{N(t),\, t \ge 0\}$ *be two nonhomogeneous Poisson processes with mean functions* M_K *and* M_N*, respectively, and with intensity functions* λ_K *and* λ_N*, respectively.*

(i) *If* $M_K(t) \le M_N(t)$, $t \ge 0$, *then* $\{K(t),\, t \ge 0\} \le_{\text{st}} \{N(t),\, t \ge 0\}$.
(ii) *If* $\lambda_K(t) \le \lambda_N(t)$, $t \ge 0$, *then* $\{K(t),\, t \ge 0\} \le_{\text{st-}\mathcal{N}} \{N(t),\, t \ge 0\}$.
(iii) *If* $M_K^{-1}(M_N(t)) - t$ *is increasing in* $t \ge 0$, *then* $\{K(t),\, t \ge 0\} \le_{\text{st-}\infty} \{N(t),\, t \ge 0\}$.

In the following example, parts (i) and (iii) of Theorem 6.B.40 are restated in the terminology of Examples 1.B.24, 1.C.48, 2.A.22, 3.B.38, 4.B.14, 6.D.8, 6.E.13, and 7.B.13.

Example 6.B.41. Let X and Y be two absolutely continuous nonnegative random variables with survival functions $\overline{F}$ and $\overline{G}$, respectively. Denote $\Lambda_1 = -\log \overline{F}$ and $\Lambda_2 = -\log \overline{G}$, $i = 1, 2$. Consider two nonhomogeneous Poisson processes $N_1 = \{N_1(t),\, t \ge 0\}$ and $N_2 = \{N_2(t),\, t \ge 0\}$ with mean functions Λ_1 and Λ_2 (see Example 1.B.13), respectively. Let $T_{i,1}, T_{i,2}, \ldots$ be the successive epoch times of process N_i, $i = 1, 2$. Note that $X =_{\text{st}} T_{1,1}$ and $Y =_{\text{st}} T_{2,1}$.

It turns out that the usual stochastic ordering of the first two epoch times implies the multivariate usual stochastic ordering of all the corresponding later epoch times. Explicitly, part (i) of Theorem 6.B.40 says that if $X \le_{\text{st}} Y$, then $(T_{1,1}, T_{1,2}, \ldots, T_{1,n}) \le_{\text{st}} (T_{2,1}, T_{2,2}, \ldots, T_{2,n})$, $n \ge 1$.

Now let $X_{i,n} \equiv T_{i,n} - T_{i,n-1}$, $n \ge 1$ (where $T_{i,0} \equiv 0$), be the inter-epoch times of the process N_i, $i = 1, 2$. Part (iii) of Theorem 6.B.40 says that if $X \le_{\text{disp}} Y$, then $(X_{1,1}, X_{1,2}, \ldots, X_{1,n}) \le_{\text{st}} (X_{2,1}, X_{2,2}, \ldots, X_{2,n})$, $n \ge 1$.

6.C The Cumulative Hazard Order

6.C.1 Definition

Let $\boldsymbol{T} = (T_1, T_2, \ldots, T_m)$ be a nonnegative random vector with an absolutely continuous distribution function. In this section, as in Section 6.B.6, it is helpful to think about $T_1, T_2, \ldots, T_m$ as the lifetimes of m components $1, 2, \ldots, m$ that make up some system. Consider a typical "history" of $\boldsymbol{T}$ at time $t \ge 0$, which is of the form (see (6.B.25))

$$h_t = \{\boldsymbol{T}_I = \boldsymbol{t}_I, \boldsymbol{T}_{\overline{I}} > t\boldsymbol{e}\}, \quad 0\boldsymbol{e} \le \boldsymbol{t}_I \le t\boldsymbol{e},\ I \subseteq \{1, 2, \ldots, m\}. \tag{6.C.1}$$

Given the history h_t in (6.C.1), let $i \in \overline{I}$ be a component that is still alive at time t. Its multivariate conditional hazard rate, at time t, is defined as follows:

$$\lambda_{i|I}(t|\boldsymbol{t}_I) = \lim_{\Delta t \downarrow 0} \frac{1}{\Delta t} P\{t < T_i \le t + \Delta t | \boldsymbol{T}_I = \boldsymbol{t}_I, \boldsymbol{T}_{\overline{I}} > t\boldsymbol{e}\}, \tag{6.C.2}$$

where, of course, $0\boldsymbol{e} \le \boldsymbol{t}_I \le t\boldsymbol{e}$, and $I \subseteq \{1, 2, \ldots, m\}$. As long as the item is alive it accumulates hazard at the rate of $\lambda_{i|I}(t|\boldsymbol{t}_I)$ at time t. If $I = \{i_1, i_2, \ldots, i_k\}$ and

$$t_{i_1} \leq t_{i_2} \leq \cdots \leq t_{i_k},$$

then the cumulative hazard of component $i \in \overline{I}$ at time t is

$$\begin{aligned}\Psi_{i|i_1,i_2,\ldots,i_k}(t|t_{i_1},t_{i_2},\ldots,t_{i_k}) \\ = \int_0^{t_{i_1}} \lambda_{i|\varnothing}(u|\boldsymbol{t}_\varnothing)\mathrm{d}u + \sum_{j=2}^{k} \int_{t_{i_{j-1}}}^{t_{i_j}} \lambda_{i|i_1,i_2,\ldots,i_{j-1}}(u|t_{i_1},t_{i_2},\ldots,t_{i_{j-1}})\mathrm{d}u \\ + \int_{t_{i_k}}^{t} \lambda_{i|i_1,i_2,\ldots,i_k}(u|t_{i_1},t_{i_2},\ldots,t_{i_k})\mathrm{d}u. \quad (6.C.3)\end{aligned}$$

Let $\boldsymbol{S} = (S_1, S_2, \ldots, S_m)$ be another nonnegative random vector with an absolutely continuous distribution function and with cumulative hazard functions $\Phi_{\cdot|\cdot}(\cdot|\cdot)$, which are defined analogously to the Ψ's in (6.C.3). Select two integers j and l such that $j \leq l \leq m$. Let $t_1, t_2, \ldots, t_j$ and $s_1, \ldots, s_j, \ldots, s_l$ be such that $0 \leq t_1 \leq t_2 \leq \cdots \leq t_j$, and $0 \leq s_i \leq t_i$, $i = 1, 2, \ldots, j$, and $s_i \geq 0$, $i = j+1, \ldots, l$. Let

$$s_{k_1} \leq s_{k_2} \leq \cdots \leq s_{k_l}$$

be the ordered s_i's. If for any integer $\alpha > l$ we have

$$\Phi_{\alpha|k_1,k_2,\ldots,k_l}(u|s_{k_1},s_{k_2},\ldots,s_{k_l}) \geq \Psi_{\alpha|1,2,\ldots,j}(u|t_1,t_2,\ldots,t_j) \qquad (6.C.4)$$

whenever $u \geq \max\{t_j, s_{j+1}, s_{j+2}, \ldots, s_l\}$, and if the same holds with $1, 2, \ldots, l$ replaced by $\pi_1, \pi_2, \ldots, \pi_l$ for every permutation $\boldsymbol{\pi}$ of $(1, 2, \ldots, m)$, then $\boldsymbol{S}$ is said to be *smaller than* $\boldsymbol{T}$ *in the cumulative hazard order* (denoted as $\boldsymbol{S} \leq_{\mathrm{ch}} \boldsymbol{T}$).

The order $\leq_{\mathrm{ch}}$ is not an order in the usual sense; a comment, similar to the comment in Remark 6.B.5, applies to this order too. Explicitly, $\boldsymbol{X} \leq_{\mathrm{ch}} \boldsymbol{X}$ means that $\boldsymbol{X}$ has the positive dependence property of "supporting lifetimes" discussed in Norros [437] and in Shaked and Shanthikumar [511].

Condition (6.C.4) simply states that at any time t the cumulative hazard of S_α is larger than the cumulative hazard of T_α whenever the history of the components corresponding to $\boldsymbol{S}$ is more "severe" than the history of the components corresponding to $\boldsymbol{T}$. Thus (6.C.4) can be written as (see Section 6.B.6 for the definition of histories and for the definition of their comparison)

$$\Phi_\alpha(h'_u) \geq \Psi_\alpha(h_u) \quad \text{whenever } h'_u \geq h_u,$$

where α denotes a component that has not failed by time u in the history h'_u.

In the univariate case (that is, $m = 1$) condition (6.C.4) simply says that $-\log P\{S_1 > u\} \geq -\log P\{T_1 > u\}$. Therefore, in the univariate case

$$S_1 \leq_{\mathrm{ch}} T_1 \Longleftrightarrow S_1 \leq_{\mathrm{st}} T_1.$$

Thus, if the components of $\boldsymbol{S}$ are independent, and if the components of $\boldsymbol{T}$ are independent, then $\boldsymbol{S} \leq_{\mathrm{ch}} \boldsymbol{T} \Longleftrightarrow \boldsymbol{S} \leq_{\mathrm{st}} \boldsymbol{T}$. In the general multivariate case the two orders are not equivalent, but it will be shown below that if $\boldsymbol{S} \leq_{\mathrm{ch}} \boldsymbol{T}$, then $\boldsymbol{S} \leq_{\mathrm{st}} \boldsymbol{T}$.

6.C.2 The relationship between the cumulative hazard order and the usual multivariate stochastic order

The *total hazard accumulated by the failure time* T_i, given that T_i was the time of the kth failure and that the previous failure times were $T_{j_1}, T_{j_2}, \ldots, T_{j_{k-1}}$, is $\Psi_{i|j_1,j_2,\ldots,j_{k-1}}(T_i | T_{j_1}, T_{j_2}, \ldots, T_{j_{k-1}})$. It can be shown that the total hazards accumulated by the failure times T_i's are independent standard (that is, mean one) exponential random variables. This fact motivates the following *total hazard construction*, which is of independent interest but we will use it here in order to show that if $\boldsymbol{S} \leq_{\text{ch}} \boldsymbol{T}$, then $\boldsymbol{S} \leq_{\text{st}} \boldsymbol{T}$.

The idea of the construction is as follows. The components accumulate hazard as long as they are alive with the rates given in (6.C.2). Each one of them dies when its accumulated hazard crosses a random threshold. The random thresholds are independent standard exponential random variables. Thus, by continuously comparing the accumulated hazards to the independent exponential random thresholds it is possible to determine the times in which the accumulated hazards cross the respective thresholds, and these times have the desired distribution. From this heuristic description it is seen that the multivariate conditional cumulative hazard functions, given in (6.C.3), determine the distribution of the generated random variables. This, indeed, is well known.

Let $\boldsymbol{T} = (T_1, T_2, \ldots, T_m)$ be a nonnegative random vector with an absolutely continuous distribution function. Given the functions $\Psi_{\cdot|\cdot}(\cdot|\cdot)$ that are associated with $\boldsymbol{T}$, as described in (6.C.3), we will describe now how to generate a random vector $\hat{\boldsymbol{T}} = (\hat{T}_1, \hat{T}_2, \ldots, \hat{T}_m)$ such that $\hat{\boldsymbol{T}} =_{\text{st}} \boldsymbol{T}$. Let $X_1, X_2, \ldots, X_m$ be independent standard exponential random variables. The total hazard construction will be described in m steps.

STEP 1. In this step we determine the identity i_1 of the component that fails first and its time of failure $\hat{T}_{i_1}$. This is determined by

$$\hat{T}_{i_1} = \min\{\tilde{T}_1, \tilde{T}_2, \ldots, \tilde{T}_m\},$$

where

$$\tilde{T}_j = \min\{t \geq 0 : \Psi_{j|\varnothing}(t|\varnothing) \geq X_j\}, \quad j = 1, 2, \ldots, m,$$

and i_1 is the index of the smallest $\tilde{T}_j$.

STEP k. ($k = 2, 3, \ldots, m$). Suppose that Steps $1, 2, \ldots, k-1$ have already yielded

$$\hat{T}_{i_1}, \hat{T}_{i_2}, \ldots, \hat{T}_{i_{k-1}}.$$

Let $I = \{i_1, i_2, \ldots, i_{k-1}\}$ and denote $\overline{I} = \{j_1, j_2, \ldots, j_{m-k+1}\}$. In this step we determine the identity i_k of the component that is the kth one to fail and its failure time $\hat{T}_{i_k}$. This is determined by

$$\hat{T}_{i_k} = \min\{\tilde{T}_{j_1}, \tilde{T}_{j_2}, \ldots, \tilde{T}_{j_{m-k+1}}\},$$

where here, for $j \in \overline{I}$,

$$\tilde{T}_j = \min\{t \geq \hat{T}_{i_{k-1}} : \Psi_{j|i_1,i_2,\ldots,i_{k-1}}(t|\hat{T}_{i_1}, \hat{T}_{i_2}, \ldots, \hat{T}_{i_{k-1}}) \geq X_j\},$$

and i_k is the index of the smallest $\tilde{T}_j$, $j \in \overline{I}$.

It can be shown that indeed $\hat{\boldsymbol{T}} =_{\text{st}} \boldsymbol{T}$.

Let $\boldsymbol{S} = (S_1, S_2, \ldots, S_m)$ be another nonnegative random vector with an absolutely continuous distribution function and multivariate conditional cumulative hazard functions $\Phi_{\cdot|\cdot}(\cdot|\cdot)$. Using the same independent standard exponential random variables $X_1, X_2, \ldots, X_m$, construct $\hat{\boldsymbol{S}} = (\hat{S}_1, \hat{S}_2, \ldots, \hat{S}_m)$ using the total hazard construction described above. Thus $\hat{\boldsymbol{S}}$ and $\hat{\boldsymbol{T}}$ are constructed on the same probability space and they satisfy $\hat{\boldsymbol{T}} =_{\text{st}} \boldsymbol{T}$ and $\hat{\boldsymbol{S}} =_{\text{st}} \boldsymbol{S}$. Also, if (6.C.4) holds, that is, if $\boldsymbol{S} \leq_{\text{ch}} \boldsymbol{T}$, then it is clear that $P\{\hat{\boldsymbol{S}} \leq \hat{\boldsymbol{T}}\} = 1$. Thus, from Theorem 6.B.1, we see that we have proved the following theorem.

Theorem 6.C.1. *Let $\boldsymbol{S}$ and $\boldsymbol{T}$ be two nonnegative random vectors with absolutely continuous distribution functions. If $\boldsymbol{S} \leq_{\text{ch}} \boldsymbol{T}$, then $\boldsymbol{S} \leq_{\text{st}} \boldsymbol{T}$.*

It is worth mentioning that the total hazard construction is theoretically and practically different from the standard construction discussed in Section 6.B.3. In the standard construction the uniform random variables $U_1, U_2, \ldots, U_n$, which are used to generate the desired $\hat{T}_1, \hat{T}_2, \ldots, \hat{T}_n$, can be used sequentially, that is, U_i can be used to generate $\hat{T}_i$, once $\hat{T}_1, \hat{T}_2, \ldots, \hat{T}_{i-1}$ have already been generated, $i = 1, 2, \ldots, n$. On the other hand, in the total hazard construction, the exponential random variables $X_1, X_2, \ldots, X_m$ are all used simultaneously in the generation of each $\hat{T}_i$.

Remark 6.C.2. Looking at Step 1 of the total hazard construction it is seen that it can be split into two substeps. First the value of first order statistic, $\hat{T}_{(1)}$ say, of the $\hat{T}_j$'s is determined, and then the identity (index) of $\hat{T}_{(1)}$ is selected. Similarly Step k can be split into two substeps. Suppose now that $\boldsymbol{T} = (T_1, T_2, \ldots, T_m)$ is a vector of exponential random variables with possibly different parameters. Then also $\hat{\boldsymbol{T}} = (\hat{T}_1, \hat{T}_2, \ldots, \hat{T}_m)$ is such a vector. Furthermore, $\hat{T}_{(1)}$ is also an exponential random variable. If it is known that $\hat{T}_{(1)} = s_1$ say, and if the identity of the smallest $\hat{T}_j$ is also known, then, conditionally, the residual lives of the remaining $m-1$ components are independent exponential random variables, and they do not depend on s_1. If the identity of the smallest $\hat{T}_j$ is not known known, then the conditional distribution of the residual lives of the remaining $m-1$ components is a mixture of distributions of independent exponential random variables, and it still does not depend on s_1 (notice that the probabilities of the mixture do not depend on s_1). Therefore the conditional distribution of $(T_{(2)} - s_1, T_{(3)} - s_1, \ldots, T_{(m)} - s_1)$, given $\hat{T}_{(1)} = s_1$, does not depend on s_1. It follows that $[(\hat{T}_{(1)}, \hat{T}_{(2)}, \ldots, \hat{T}_{(m)})|\hat{T}_{(1)} = s_1]$ is stochastically increasing in s_1. Since $\boldsymbol{T} =_{\text{st}} \hat{\boldsymbol{T}}$ we obtain a proof of Theorem 6.B.13.

6.D Multivariate Hazard Rate Orders

6.D.1 Definitions and basic properties

The following notation will be used below. For any two real numbers x and y we denote $x \vee y = \max\{x, y\}$ and $x \wedge y = \min\{x, y\}$. If $\boldsymbol{x} = (x_1, x_2, \ldots, x_n)$ and $\boldsymbol{y} = (y_1, y_2, \ldots, y_n)$ are two vectors in $\mathbb{R}^n$, then we denote $\boldsymbol{x} \vee \boldsymbol{y} = (x_1 \vee y_1, x_2 \vee y_2, \ldots, x_n \vee y_n)$ and $\boldsymbol{x} \wedge \boldsymbol{y} = (x_1 \wedge y_1, x_2 \wedge y_2, \ldots, x_n \wedge y_n)$.

Let $\boldsymbol{X} = (X_1, X_2, \ldots, X_n)$ and $\boldsymbol{Y} = (Y_1, Y_2, \ldots, Y_n)$ be two random vectors with respective survival functions $\overline{F}$ and $\overline{G}$ defined by $\overline{F}(\boldsymbol{x}) = P\{\boldsymbol{X} > \boldsymbol{x}\}$ and $\overline{G}(\boldsymbol{x}) = P\{\boldsymbol{Y} > \boldsymbol{x}\}$, $\boldsymbol{x} \in \mathbb{R}^n$. We say that $\boldsymbol{X}$ is smaller than $\boldsymbol{Y}$ in the *multivariate hazard rate order* (denoted by $\boldsymbol{X} \le_{\text{hr}} \boldsymbol{Y}$) if

$$\overline{F}(\boldsymbol{x})\overline{G}(\boldsymbol{y}) \le \overline{F}(\boldsymbol{x} \wedge \boldsymbol{y})\overline{G}(\boldsymbol{x} \vee \boldsymbol{y}) \quad \text{for every } \boldsymbol{x} \text{ and } \boldsymbol{y} \text{ in } \mathbb{R}^n. \tag{6.D.1}$$

We say that $\boldsymbol{X}$ is smaller than $\boldsymbol{Y}$ in the *weak multivariate hazard rate order* (denoted by $\boldsymbol{X} \le_{\text{whr}} \boldsymbol{Y}$) if

$$\frac{\overline{G}(\boldsymbol{x})}{\overline{F}(\boldsymbol{x})} \text{ is increasing in } \boldsymbol{x} \in \{\boldsymbol{x} : \overline{G}(\boldsymbol{x}) > \boldsymbol{0}\}, \tag{6.D.2}$$

where in (6.D.2) we use the convention $a/0 \equiv \infty$ whenever $a > 0$. Note that (6.D.2) can be written equivalently as

$$\overline{F}(\boldsymbol{y})\overline{G}(\boldsymbol{x}) \le \overline{F}(\boldsymbol{x})\overline{G}(\boldsymbol{y}) \quad \text{whenever } \boldsymbol{x} \le \boldsymbol{y}. \tag{6.D.3}$$

Thus, from (6.D.1) and (6.D.3) it follows that

$$\boldsymbol{X} \le_{\text{hr}} \boldsymbol{Y} \Longrightarrow \boldsymbol{X} \le_{\text{whr}} \boldsymbol{Y}. \tag{6.D.4}$$

Note that from (6.D.3) it follows that if $\boldsymbol{y} \in \{\boldsymbol{x} : \overline{G}(\boldsymbol{x}) = 0\}$, then $\boldsymbol{y} \in \{\boldsymbol{x} : \overline{F}(\boldsymbol{x}) = 0\}$. That is, if $\boldsymbol{X} \le_{\text{whr}} \boldsymbol{Y}$, then

$$\{\boldsymbol{x} : \overline{F}(\boldsymbol{x}) > 0\} \subseteq \{\boldsymbol{x} : \overline{G}(\boldsymbol{x}) > 0\}.$$

It can be shown that the implication (6.D.4) is strict. However, when at least one of the survival functions of $\boldsymbol{X}$ and of $\boldsymbol{Y}$ is MTP$_2$ (recall from Karlin and Rinott [278] that a function $K : \mathbb{R}^n \to \mathbb{R}_+$ is said to be multivariate totally positive of order 2 (MTP$_2$) if $K(\boldsymbol{x})K(\boldsymbol{y}) \le K(\boldsymbol{x} \wedge \boldsymbol{y})K(\boldsymbol{x} \vee \boldsymbol{y})$ for all $\boldsymbol{x}, \boldsymbol{y} \in \mathbb{R}^n$), then, under some regularity conditions, the orders $\le_{\text{hr}}$ and $\le_{\text{whr}}$ are equivalent. This is shown next. Recall that a set $S \subseteq \mathbb{R}^n$ is called a lattice if for all $\boldsymbol{x}, \boldsymbol{y}$ in S we have that $\boldsymbol{x} \wedge \boldsymbol{y}$ and $\boldsymbol{x} \vee \boldsymbol{y}$ are in S.

Theorem 6.D.1. *Let $\boldsymbol{X}$ and $\boldsymbol{Y}$ be two random vectors with respective survival functions $\overline{F}$ and $\overline{G}$, and with a common support S which is a lattice. If $\overline{F}$ and/or $\overline{G}$ are/is* MTP$_2$, *then*

$$\boldsymbol{X} \le_{\text{whr}} \boldsymbol{Y} \Longrightarrow \boldsymbol{X} \le_{\text{hr}} \boldsymbol{Y}. \tag{6.D.5}$$

Proof. Note that the left hand side of the implication (6.D.5) implies

$$\overline{F}(\boldsymbol{x} \vee \boldsymbol{y})\overline{G}(\boldsymbol{y}) \le \overline{F}(\boldsymbol{y})\overline{G}(\boldsymbol{x} \vee \boldsymbol{y}), \quad \boldsymbol{x}, \boldsymbol{y} \in \mathbb{R}^n,$$

and that the MTP$_2$-ness of $\overline{F}$ implies

$$\overline{F}(\boldsymbol{x})\overline{F}(\boldsymbol{y}) \le \overline{F}(\boldsymbol{x} \wedge \boldsymbol{y})\overline{F}(\boldsymbol{x} \vee \boldsymbol{y}), \quad \boldsymbol{x}, \boldsymbol{y} \in \mathbb{R}^n.$$

Multiplication of these two inequalities yields

$$\overline{F}(\boldsymbol{x} \vee \boldsymbol{y})\overline{G}(\boldsymbol{y})\overline{F}(\boldsymbol{x})\overline{F}(\boldsymbol{y}) \le \overline{F}(\boldsymbol{y})\overline{G}(\boldsymbol{x} \vee \boldsymbol{y})\overline{F}(\boldsymbol{x} \wedge \boldsymbol{y})\overline{F}(\boldsymbol{x} \vee \boldsymbol{y}).$$

Now, from the assumption that S is a lattice it follows that if $\overline{F}(\boldsymbol{x})\overline{G}(\boldsymbol{y}) > 0$, then $\overline{F}(\boldsymbol{y})$ and $\overline{F}(\boldsymbol{x} \vee \boldsymbol{y})$ are positive. Canceling these we obtain that (6.D.1) holds in this case. If $\overline{F}(\boldsymbol{x})\overline{G}(\boldsymbol{y}) = 0$, then (6.D.1) obviously holds too. Therefore $\boldsymbol{X} \le_{\rm hr} \boldsymbol{Y}$. In a similar manner the implication (6.D.5) can be shown when $\overline{G}$ is MTP$_2$. □

The order $\le_{\rm hr}$ is not an order in the usual sense (that is, it is not reflexive) because from (6.D.1) it follows that

$$\boldsymbol{X} \le_{\rm hr} \boldsymbol{X} \Longleftrightarrow P\{\boldsymbol{X} > \boldsymbol{x}\} \text{ is MTP}_2.$$

Consider now a random vector $\boldsymbol{X} = (X_1, X_2, \ldots, X_n)$ with a partially differentiable survival function $\overline{F}$. Let $\boldsymbol{r_X} = (r_{\boldsymbol{X}}^{(1)}, r_{\boldsymbol{X}}^{(2)}, \ldots, r_{\boldsymbol{X}}^{(n)})$ be its hazard gradient as defined in (1.B.28). Let $\boldsymbol{Y}$ be another n-dimensional random vector with hazard gradient $\boldsymbol{r_Y} = (r_{\boldsymbol{Y}}^{(1)}, r_{\boldsymbol{Y}}^{(2)}, \ldots, r_{\boldsymbol{Y}}^{(n)})$. The following result, which can be obtained by differentiation of (6.D.2), justifies the terminology "hazard rate order" for the orders that were introduced in (6.D.1) and (6.D.2).

Theorem 6.D.2. *Let $\boldsymbol{X}$ and $\boldsymbol{Y}$ be n-dimensional random vectors with hazard gradients $\boldsymbol{r_X}$ and $\boldsymbol{r_Y}$, respectively. Then $\boldsymbol{X} \le_{\rm whr} \boldsymbol{Y}$ if, and only if,*

$$r_{\boldsymbol{X}}^{(i)}(\boldsymbol{x}) \ge r_{\boldsymbol{Y}}^{(i)}(\boldsymbol{x}), \quad i = 1, 2, \ldots, n, \ \boldsymbol{x} \in \mathbb{R}^n.$$

A useful inequality is described next; we omit its proof.

Theorem 6.D.3. *Let $\boldsymbol{X} = (X_1, X_2, \ldots, X_n)$ be a random vector, and let $\boldsymbol{X}^I = (Y_1, Y_2, \ldots, Y_n)$ be a vector of independent random variables such that $X_i =_{\rm st} Y_i$, $i = 1, 2, \ldots, n$. If the survival function of $\boldsymbol{X}$ is* MTP$_2$, *then*

$$\boldsymbol{X}^I \le_{\rm hr} \boldsymbol{X}.$$

The relation $\boldsymbol{X} \le_{\rm hr} \boldsymbol{Y}$ does not necessarily imply $\boldsymbol{X} \le_{\rm st} \boldsymbol{Y}$, where $\le_{\rm st}$ denotes the usual multivariate stochastic order discussed in Section 6.B. However, a generalization of the univariate Theorem 1.B.1 is given in (6.G.10) in Section 6.G.1. Theorem 6.G.9 is a multivariate generalization of (1.B.7).

6.D.2 Preservation properties

The orders $\le_{\rm hr}$ and $\le_{\rm whr}$ are closed under some common operations.

Theorem 6.D.4. (a) *Let $(X_1, X_2, \dots, X_n)$ and $(Y_1, Y_2, \dots, Y_n)$ be two n-dimensional random vectors. If $(X_1, X_2, \dots, X_n) \le_{\rm hr} [\le_{\rm whr}] (Y_1, Y_2, \dots, Y_n)$, then*

$$(g_1(X_1), g_2(X_2), \dots, g_n(X_n)) \le_{\rm hr} [\le_{\rm whr}] (g_1(Y_1), g_2(Y_2), \dots, g_n(Y_n))$$

whenever $g_i : \mathbb{R} \to \mathbb{R}$ is an increasing function, $i = 1, 2, \dots, n$.

(b) *Let $\boldsymbol{X}_1, \boldsymbol{X}_2, \dots, \boldsymbol{X}_m$ be a set of independent random vectors where the dimension of $\boldsymbol{X}_i$ is k_i, $i = 1, 2, \dots, m$. Let $\boldsymbol{Y}_1, \boldsymbol{Y}_2, \dots, \boldsymbol{Y}_m$ be another set of independent random vectors where the dimension of $\boldsymbol{Y}_i$ is k_i, $i = 1, 2, \dots, m$. If $\boldsymbol{X}_i \le_{\rm hr} [\le_{\rm whr}] \boldsymbol{Y}_i$ for $i = 1, 2, \dots, m$, then*

$$(\boldsymbol{X}_1, \boldsymbol{X}_2, \dots, \boldsymbol{X}_m) \le_{\rm hr} [\le_{\rm whr}] (\boldsymbol{Y}_1, \boldsymbol{Y}_2, \dots, \boldsymbol{Y}_m).$$

That is, the multivariate hazard rate orders are closed under conjunctions.

(c) *Let $\boldsymbol{X} = (X_1, X_2, \dots, X_n)$ and $\boldsymbol{Y} = (Y_1, Y_2, \dots, Y_n)$ be two n-dimensional random vectors. If $\boldsymbol{X} \le_{\rm hr} [\le_{\rm whr}] \boldsymbol{Y}$, then $\boldsymbol{X}_I \le_{\rm hr} [\le_{\rm whr}] \boldsymbol{Y}_I$ for each $I \subseteq \{1, 2, \dots, n\}$. That is, the multivariate hazard rate orders are closed under marginalization.*

(d) *Let $\{\boldsymbol{X}_j, j = 1, 2, \dots\}$ and $\{\boldsymbol{Y}_j, j = 1, 2, \dots\}$ be two sequences of random vectors such that $\boldsymbol{X}_j \to_{\rm st} \boldsymbol{X}$ and $\boldsymbol{Y}_j \to_{\rm st} \boldsymbol{Y}$ as $j \to \infty$, where $\to_{\rm st}$ denotes convergence in distribution. If $\boldsymbol{X}_j \le_{\rm hr} [\le_{\rm whr}] \boldsymbol{Y}_j$, $j = 1, 2, \dots$, then $\boldsymbol{X} \le_{\rm hr} [\le_{\rm whr}] \boldsymbol{Y}$.*

We will now describe some preservation properties of the multivariate hazard rate orders under random compositions.

Let $\{\overline{F}_\theta,\ \theta \in \mathcal{X}\}$ be a family of n-dimensional survival functions, where $\mathcal{X}$ is a subset of the real line. Let $\boldsymbol{X}(\theta)$ denote a random vector with survival function $\overline{F}_\theta$. For any random variable Θ with support in $\mathcal{X}$, and with distribution function H, let us denote by $\boldsymbol{X}(\Theta)$ a random vector with survival function $\overline{G}$ given by

$$\overline{G}(\boldsymbol{x}) = \int_{\mathcal{X}} \overline{F}_\theta(\boldsymbol{x}) \mathrm{d}H(\theta), \quad \boldsymbol{x} \in \mathbb{R}^n.$$

Theorem 6.D.5. *Let $\{\overline{F}_\theta,\ \theta \in \mathcal{X}\}$ be a family of n-dimensional survival functions as above. Let Θ_1 and Θ_2 be two random variables with supports in $\mathcal{X}$ and distribution functions H_1 and H_2, respectively. Let $\boldsymbol{Y}_1$ and $\boldsymbol{Y}_2$ be two random vectors such that $\boldsymbol{Y}_i =_{\rm st} \boldsymbol{X}(\Theta_i)$, $i = 1, 2$; that is, suppose that the survival function of $\boldsymbol{Y}_i$ is given by*

$$\overline{G}_i(\boldsymbol{x}) = \int_{\mathcal{X}} \overline{F}_\theta(\boldsymbol{x}) \mathrm{d}H_i(\theta), \quad \boldsymbol{x} \in \mathbb{R}^n, \ i = 1, 2.$$

If

$$\boldsymbol{X}(\theta) \leq_{\text{whr}} \boldsymbol{X}(\theta') \quad \textit{whenever } \theta \leq \theta', \tag{6.D.6}$$

and if Θ_1 *and* Θ_2 *are ordered in the univariate hazard rate order; that is, if*

$$\Theta_1 \leq_{\text{hr}} \Theta_2, \tag{6.D.7}$$

then

$$\boldsymbol{Y}_1 \leq_{\text{whr}} \boldsymbol{Y}_2. \tag{6.D.8}$$

Proof. Assumption (6.D.6) means that for each $j \in \{1, 2, \ldots, n\}$, the function $\overline{F}_\theta(x_1, x_2, \ldots, x_n)$ is TP_2 (totally positive of order 2; that is, bivariate MTP_2) as a function of $\theta \in \mathcal{X}$ and of $x_j \in \mathbb{R}$. Assumption (6.D.7) means that $\overline{H}_i(\theta)$ is TP_2 as a function of $i \in \{1, 2\}$ and of $\theta \in \mathcal{X}$. Therefore, by Theorem 2.1 of Joag-Dev, Kochar, and Proschan [259], $\overline{G}_i(x_1, x_2, \ldots, x_n)$ is TP_2 in $i \in \{1, 2\}$ and in $x_j \in \mathbb{R}$, $j = 1, 2, \ldots, n$. That is,

$$\frac{\overline{G}_2(x_1, x_2, \ldots, x_n)}{\overline{G}_1(x_1, x_2, \ldots, x_n)} \text{ is increasing in } x_j, \quad j = 1, 2, \ldots, n.$$

By (6.D.2), this yields the stated result. □

In the case where $\boldsymbol{Y}_1$ and $\boldsymbol{Y}_2$ in Theorem 6.D.5 are vectors of conditionally independent random variables, the conclusion (6.D.8) can be strengthened. For this purpose, consider n families of univariate survival functions $\{\overline{F}_{j,\theta},\ \theta \in \mathcal{X}\}$, $j = 1, 2, \ldots, n$, where $\mathcal{X}$ is a subset of the real line. Let $X_j(\theta)$ denote a univariate random variable with survival function $\overline{F}_{j,\theta}$. For any random variable Θ with support in $\mathcal{X}$, and with distribution function H, let $X_j(\Theta)$ denote a univariate random variable with survival function given by $\int_{\mathcal{X}} \overline{F}_{j,\theta}(x) \mathrm{d}H(\theta)$, $x \in \mathbb{R}$, $j = 1, 2, \ldots, n$.

Theorem 6.D.6. *Let* $\{\overline{F}_{j,\theta},\ \theta \in \mathcal{X}\}$ *be* n *families of univariate survival functions as above,* $j = 1, 2, \ldots, n$. *Assume that for each* $j = 1, 2, \ldots, n$, *the univariate supports corresponding to all the* $\overline{F}_{j,\theta}$*'s are identical,* $\mathcal{Y}_j$, *say. Let* Θ_1 *and* Θ_2 *be two random variables with supports in* $\mathcal{X}$ *and distribution functions* H_1 *and* H_2, *respectively. Let* $\boldsymbol{Y}_1 = (Y_{11}, Y_{12}, \ldots, Y_{1n})$ *and* $\boldsymbol{Y}_2 = (Y_{21}, Y_{22}, \ldots, Y_{2n})$ *be two vectors of conditionally independent random variables such that* $Y_{ij} =_{\text{st}} X_j(\Theta_i)$, $i = 1, 2$, $j = 1, 2, \ldots, n$; *that is, suppose that the survival function of* $\boldsymbol{Y}_i$ *is given by*

$$\overline{G}_i(x_1, x_2, \ldots, x_n) = \int_{\mathcal{X}} \prod_{j=1}^{n} \overline{F}_{j,\theta}(x_j) \mathrm{d}H_i(\theta),$$

$$(x_1, x_2, \ldots, x_n) \in \mathbb{R}^n, \ i = 1, 2. \tag{6.D.9}$$

If

$$X_j(\theta) \leq_{\text{hr}} X_j(\theta') \quad \textit{whenever } \theta \leq \theta',\ j = 1, 2, \ldots, n, \tag{6.D.10}$$

and if

$$\Theta_1 \leq_{\rm hr} \Theta_2,$$

then

$$\boldsymbol{Y}_1 \leq_{\rm hr} \boldsymbol{Y}_2.$$

Proof. Let $\theta \leq \theta'$. From assumption (6.D.10), from the conditional independence of the $X_j(\theta)$'s, and from the conditional independence of the $X_j(\theta')$'s, it follows by Theorem 6.D.4(b) that

$$(X_1(\theta), X_2(\theta), \ldots, X_n(\theta)) \leq_{\rm hr} (X_1(\theta'), X_2(\theta'), \ldots, X_n(\theta')) \quad \text{whenever } \theta \leq \theta'.$$

Therefore, by Theorem 6.D.5 we get

$$\boldsymbol{Y}_1 \leq_{\rm whr} \boldsymbol{Y}_2. \tag{6.D.11}$$

Next, it is easy to verify that $\overline{G}_i$ in (6.D.9) is TP_2 in each pair of its variables when the other variables are held fixed, $i = 1, 2$. Therefore $\overline{G}_i$ is MTP_2, $i = 1, 2$. Furthermore, from the assumption that for $j = 1, 2, \ldots, n$, all the $\overline{F}_{j,\theta}$'s have a corresponding univariate common support $\mathcal{Y}_j$, it follows that $\boldsymbol{Y}_1$ and $\boldsymbol{Y}_2$ have a common support which is a lattice. The stated result now follows from (6.D.11) and Theorem 6.D.1. □

An interesting property of the order $\leq_{\rm whr}$, for nonnegative random vectors, is given next; see Theorem 6.G.15 for a related result.

Theorem 6.D.7. *Let $\boldsymbol{X} = (X_1, X_2, \ldots, X_n)$ and $\boldsymbol{Y} = (Y_1, Y_2, \ldots, Y_n)$ be two nonnegative random vectors. If $\boldsymbol{X} \leq_{\rm whr} \boldsymbol{Y}$, then*

$$\min\{a_1X_1, \ldots, a_nX_n\} \leq_{\rm hr} \min\{a_1Y_1, \ldots, a_nY_n\}$$
$$\textit{whenever } a_i > 0, \; i = 1, 2, \ldots, n. \tag{6.D.12}$$

6.D.3 The dynamic multivariate hazard rate order

Let $\boldsymbol{T} = (T_1, T_2, \ldots, T_m)$ be a nonnegative random vector with an absolutely continuous distribution function. Denote the multivariate conditional hazard rate functions of $\boldsymbol{T}$ by $\lambda_{\cdot|\cdot}(\cdot|\cdot)$ as defined in (6.C.2). Clearly, the higher the multivariate conditional hazard rate functions are, the smaller $\boldsymbol{T}$ should be stochastically. This is the motivation for the order discussed in this subsection.

Let $\boldsymbol{S} = (S_1, S_2, \ldots, S_m)$ be another nonnegative random vector with an absolutely continuous distribution function. Denote its multivariate conditional hazard rate functions by $\eta_{\cdot|\cdot}(\cdot|\cdot)$, where the η's are defined analogously to the λ's in (6.C.2). Suppose that

$$\eta_{i|I\cup J}(u|\boldsymbol{s}_I, \boldsymbol{s}_J) \geq \lambda_{i|I}(u|\boldsymbol{t}_I)$$
$$\text{whenever } J \cap I = \varnothing, \; \boldsymbol{s}_I \leq \boldsymbol{t}_I \leq u\boldsymbol{e}, \text{ and } \boldsymbol{s}_J \leq u\boldsymbol{e}, \tag{6.D.13}$$

where $i \in \overline{I \cup J}$. Then $\boldsymbol{S}$ is said to be *smaller than* $\boldsymbol{T}$ *in the dynamic multivariate hazard rate order* (denoted as $\boldsymbol{S} \le_{\text{dyn-hr}} \boldsymbol{T}$).

The order $\le_{\text{dyn-hr}}$ is not an order in the usual sense; a comment, similar to the comment in Remark 6.B.5, applies to this order too. Explicitly, $\boldsymbol{X} \le_{\text{dyn-hr}} \boldsymbol{X}$ means that $\boldsymbol{X}$ has the positive dependence property of "hazard rate increasing upon failures" discussed in Shaked and Shanthikumar [511].

Note that (6.D.13) can be written as (see Section 6.B.6 for the definition of histories and for the definition of their comparison)

$$\eta_i(h'_u) \ge \lambda_i(h_u) \quad \text{whenever } h'_u \ge h_u,$$

where i denotes a component that has not failed by time u in the history h'_u.

The following example illustrates how the dynamic multivariate hazard rate order can be verified. This example may be compared with Examples 1.B.24, 1.C.48, 2.A.22, 3.B.38, 6.B.41, 6.E.13, and 7.B.13.

Example 6.D.8. Let X and Y be two absolutely continuous nonnegative random variables with survival functions $\overline{F}$ and $\overline{G}$, respectively. Denote $\Lambda_1 = -\log \overline{F}$, $\Lambda_2 = -\log \overline{G}$, and $\lambda_i = \Lambda'_i$, $i = 1, 2$. Consider two nonhomogeneous Poisson processes $N_1 = \{N_1(t),\ t \ge 0\}$ and $N_2 = \{N_2(t),\ t \ge 0\}$ with mean functions Λ_1 and Λ_2 (see Example 1.B.13), respectively. Let $T_{i,1}, T_{i,2}, \dots$ be the successive epoch times of process N_i, $i = 1, 2$. Note that $X =_{\text{st}} T_{1,1}$ and $Y =_{\text{st}} T_{2,1}$.

It turns out that the univariate hazard rate ordering of the first two epoch times implies the dynamic multivariate hazard rate ordering of the corresponding vectors of the later epoch times. Explicitly, it will be shown below that if $X \le_{\text{hr}} Y$, then $(T_{1,1}, T_{1,2}, \dots, T_{1,n}) \le_{\text{dyn-hr}} (T_{2,1}, T_{2,2}, \dots, T_{2,n})$ for each $n \ge 1$.

Fix an $n \ge 1$. Let $\eta_{\cdot|\cdot}(\cdot|\cdot)$ be the multivariate conditional hazard rate functions associated with $(T_{1,1}, T_{1,2}, \dots, T_{1,n})$ and let $\zeta_{\cdot|\cdot}(\cdot|\cdot)$ be the multivariate conditional hazard rate functions associated with $(T_{2,1}, T_{2,2}, \dots, T_{2,n})$.

First let us obtain an explicit expression for $\zeta_{i|I}(u|\boldsymbol{t}_I)$ under the restrictions on $\boldsymbol{t}$ and u in (6.D.13). Since $T_{2,1} \le T_{2,2} \le \cdots \le T_{2,n}$ a.s., it follows that $\boldsymbol{t}_I$ in (6.D.13) can be a realization ("history") of observations up to time u only if I is of the form $I = \{1, 2, \dots, m\}$ for some $m \ge 1$, or $I = \varnothing$ (that is, $m = 0$). Then we have

$$\zeta_{i|I}(u|\boldsymbol{t}_I) = \begin{cases} \lambda_2(u), & \text{if } i = m+1; \\ 0, & \text{if } i > m+1; \end{cases} \qquad \text{where } I = \{1, 2, \dots, m\}.$$

Next, let us obtain an explicit expression for $\eta_{i|I\cup J}(u|\boldsymbol{s}_{I\cup J})$ under the restrictions on $\boldsymbol{s}$, $\boldsymbol{t}$, and u in (6.D.13). Since $T_{1,1} \le T_{1,2} \le \cdots \le T_{1,n}$ a.s., we see that when $I = \{1, 2, \dots, m\}$, then $\boldsymbol{s}_{I\cup J}$ in (6.D.13) can be a realization of observations up to time u only if J is of the form $J = \{m+1, m+2, \dots, k\}$ for some $k \ge m+1$, or $J = \varnothing$ (that is, $k = m$). Then we have

$$\eta_{i|I\cup J}(u|\boldsymbol{s}_{I\cup J}) = \begin{cases} \lambda_1(u), & \text{if } i = k+1; \\ 0, & \text{if } i > k+1; \end{cases}$$

where $I = \{1, 2, \ldots, m\}$ and $J = \{m+1, m+2, \ldots, k\}$.

Suppose that $X \leq_{\text{hr}} Y$. Since i in (6.D.13) must satisfy $i \in \overline{I \cup J}$ (that is, $i > k$), we see that if $k > m$, then

$$\begin{aligned} \eta_{i|I\cup J}(u|\boldsymbol{s}_{I\cup J}) &= \lambda_1(u) \geq 0 = \zeta_{i|I}(u|\boldsymbol{t}_I) & \text{if } i = k+1; \\ \eta_{i|I\cup J}(u|\boldsymbol{s}_{I\cup J}) &= 0 = \zeta_{i|I}(u|\boldsymbol{t}_I) & \text{if } i > k+1; \end{aligned}$$

so (6.D.13) holds with $\zeta_{\cdot|\cdot}(\cdot|\cdot)$ replacing $\lambda_{\cdot|\cdot}(\cdot|\cdot)$. If $k = m$ (that is, $J = \emptyset$), then, using $X \leq_{\text{hr}} Y$, we get

$$\begin{aligned} \eta_{i|I\cup J}(u|\boldsymbol{s}_{I\cup J}) &= \lambda_1(u) \geq \lambda_2(u) = \zeta_{i|I}(u|\boldsymbol{t}_I) & \text{if } i = k+1; \\ \eta_{i|I\cup J}(u|\boldsymbol{s}_{I\cup J}) &= 0 = \zeta_{i|I}(u|\boldsymbol{t}_I) & \text{if } i > k+1; \end{aligned}$$

so (6.D.13), with $\zeta_{\cdot|\cdot}(\cdot|\cdot)$ replacing $\lambda_{\cdot|\cdot}(\cdot|\cdot)$, holds in this case too. Thus $(T_{1,1}, T_{1,2}, \ldots, T_{1,n}) \leq_{\text{dyn-hr}} (T_{2,1}, T_{2,2}, \ldots, T_{2,n})$.

It should be noted that in Example 1.B.24 it was shown that if $X \leq_{\text{hr}} Y$, then we have the univariate stochastic inequality $T_{1,n} \leq_{\text{hr}} T_{2,n}$ for each $n \geq 1$. This stochastic inequality does not follow from the above result because the dynamic multivariate hazard rate order is not closed under marginalization.

In the univariate case ($m = 1$) condition (6.D.13) reduces to (1.B.2) [with a different notation]. We have already seen that in the univariate case

$$S_1 \leq_{\text{hr}} T_1 \Longrightarrow S_1 \leq_{\text{st}} T_1.$$

This is also true in the general dynamic multivariate case. In order to see it, note that if (6.D.13) holds, then (6.C.4) holds, where in (6.C.4) the functions Ψ's are defined by means of the functions λ's as in (6.C.3) and the functions Φ's are analogously defined by means of the functions η's. We thus have proven the following result.

Theorem 6.D.9. *If $\boldsymbol{S}$ and $\boldsymbol{T}$ are two nonnegative random vectors such that $\boldsymbol{S} \leq_{\text{dyn-hr}} \boldsymbol{T}$, then $\boldsymbol{S} \leq_{\text{ch}} \boldsymbol{T}$.*

Let $X_{(1)} \leq X_{(2)} \leq \cdots \leq X_{(n)}$ be the order statistics corresponding to a sample of independent and identically distributed nonnegative random variables $X_1, X_2, \ldots, X_n$. Similarly, let $Y_{(1)} \leq Y_{(2)} \leq \cdots \leq Y_{(n)}$ be the order statistics corresponding to a sample of independent and identically distributed nonnegative random variables $Y_1, Y_2, \ldots, Y_n$. In the next result, the vectors of order statistics are compared in the order $\leq_{\text{dyn-hr}}$; it may be compared with Theorems 6.E.12, 7.B.4, and 7.B.12. The proof of the next result is similar to the proof of the main result in Example 6.D.8.

Theorem 6.D.10. *Let $X_{(1)}, X_{(2)}, \ldots, X_{(n)}$ and $Y_{(1)}, Y_{(2)}, \ldots, Y_{(n)}$ be order statistics as described above. If $X_1 \leq_{\rm hr} Y_1$, then*

$$(X_{(1)}, X_{(2)}, \ldots, X_{(n)}) \leq_{\text{dyn-hr}} (Y_{(1)}, Y_{(2)}, \ldots, Y_{(n)}).$$

We will now see a property of the order $\leq_{\text{dyn-hr}}$ in reliability theory. Recall from Section 1.B.5 that a nonnegative random variable T is IFR if, and only if, either one of the following equivalent conditions holds:

$$[T - t \big| T > t] \geq_{\rm hr} [T - t' \big| T > t'] \qquad \text{whenever } t \leq t', \tag{6.D.14}$$

$$T \geq_{\rm hr} [T - t \big| T > t] \qquad \text{for all } t \geq 0. \tag{6.D.15}$$

With the dynamic multivariate analog of the order $\geq_{\rm hr}$, one can generalize (6.D.14) and (6.D.15) to the multivariate case, thus introducing notions of multivariate IFR distributions. This can be done in several ways. Below we show that various generalizations of (6.D.14) and (6.D.15) actually yield the same notion of multivariate IFR.

Let $\boldsymbol{T}$ be a nonnegative random vector. Recall from Section 6.B.6 the definition, the notation h_t, and the comparison of histories associated with $\boldsymbol{T}$. One possible multivariate analog of (6.D.14) is to require $\boldsymbol{T}$ to satisfy, for $t \leq s$ and histories h_t and h'_s,

$$[(\boldsymbol{T} - t\boldsymbol{e})_+ \big| h_t] \geq_{\text{dyn-hr}} [(\boldsymbol{T} - s\boldsymbol{e})_+ \big| h'_s] \quad \text{whenever } h_t \leq h'_s. \tag{6.D.16}$$

Still another possible multivariate analog of (6.D.14) is to require $\boldsymbol{T}$ to satisfy, for $t \leq s$,

$$[(\boldsymbol{T} - t\boldsymbol{e})_+ \big| h_t] \geq_{\text{dyn-hr}} [(\boldsymbol{T} - t\boldsymbol{e})_+ \big| h'_s] \quad \text{whenever } h_t \text{ and } h'_s \text{ coincide on } [0, t). \tag{6.D.17}$$

An analog of (6.D.15) is to require $\boldsymbol{T}$ to satisfy (6.D.16) or (6.D.17) with $t = 0$; that is,

$$\boldsymbol{T} \geq_{\text{dyn-hr}} [(\boldsymbol{T} - s\boldsymbol{e})_+ \big| h'_s] \quad \text{for any history } h'_s,\ s \geq 0. \tag{6.D.18}$$

It turns out that these three conditions are equivalent. If we say that the nonnegative random $\boldsymbol{T}$ is *multivariate* IFR if it satisfies (6.D.16), then we have the following result, the proof of which can be found elsewhere.

Theorem 6.D.11. *Let $\boldsymbol{T}$ be a nonnegative random vector. The following three statements are equivalent.*

(i) $\boldsymbol{T}$ *is multivariate* IFR.
(ii) $\boldsymbol{T}$ *satisfies* (6.D.17).
(iii) $\boldsymbol{T}$ *satisfies* (6.D.18).

Note that if $\boldsymbol{T}$ is multivariate IFR in the sense of Theorem 6.D.11, it is also multivariate IFR in the sense of both (6.B.27) and (6.B.28).

6.E The Multivariate Likelihood Ratio Order

6.E.1 Definition

A multivariate analog of the univariate order $\le_{\rm lr}$ from Section 1.C will be introduced in this subsection. This order is sometimes also called the TP_2 order.

Let $\boldsymbol{X} = (X_1, X_2, \dots, X_n)$ and $\boldsymbol{Y} = (Y_1, Y_2, \dots, Y_n)$ be two n-dimensional random vectors with absolutely continuous [or discrete] distribution functions and let f and g denote their [continuous or discrete] density functions, respectively. Suppose that

$$f(\boldsymbol{x})g(\boldsymbol{y}) \le f(\boldsymbol{x} \wedge \boldsymbol{y})g(\boldsymbol{x} \vee \boldsymbol{y}) \quad \text{for every } \boldsymbol{x} \text{ and } \boldsymbol{y} \text{ in } \mathbb{R}^n. \tag{6.E.1}$$

Then $\boldsymbol{X}$ is said to be *smaller than* $\boldsymbol{Y}$ *in the multivariate likelihood ratio order* (denoted as $\boldsymbol{X} \le_{\rm lr} \boldsymbol{Y}$). Indeed, in the univariate case ($n = 1$), (6.E.1) reduces to (1.C.2).

The order $\le_{\rm lr}$ is not an order in the usual sense; a comment, similar to the comment in Remark 6.B.5, applies to this order too. Explicitly, $\boldsymbol{X} \le_{\rm lr} \boldsymbol{X}$ means that $\boldsymbol{X}$ has the positive dependence property of "multivariate TP_2" discussed in Karlin and Rinott [278] and in Whitt [563]; see its definition in Example 6.E.16 below.

In the slightly more general case, when $\boldsymbol{X}$ and $\boldsymbol{Y}$ are nonnegative, some of the X_i's may be identically zero and the joint distribution of the rest is absolutely continuous or discrete. Suppose that $X_1, X_2, \dots, X_m$ are those that are identically zero for some $0 < m < n$. Let f now denote the joint density of $(X_{m+1}, X_{m+2}, \dots, X_n)$. In that case we denote $\boldsymbol{X} \le_{\rm lr} \boldsymbol{Y}$ if

$$\begin{aligned} f(\boldsymbol{x})g(\boldsymbol{y}) \le {} & f(\boldsymbol{x} \wedge (y_{m+1}, y_{m+2}, \dots, y_n)) \\ & \times g((y_1, y_2, \dots, y_m), \boldsymbol{x} \vee (y_{m+1}, y_{m+2}, \dots, y_n)) \end{aligned} \tag{6.E.2}$$

for every $\boldsymbol{x} = (x_{m+1}, x_{m+2}, \dots, x_n)$ and $\boldsymbol{y} = (y_1, y_2, \dots, y_n)$.

At a first glance (6.E.1) and (6.E.2) seem to be unintuitive technical conditions. However, it turns out that in many situations they are very easy to verify and this is one of the major reasons for the usefulness and importance of the order $\le_{\rm lr}$.

Another possible analog of (1.C.2) is to require that $f(\boldsymbol{y})g(\boldsymbol{x}) \le f(\boldsymbol{x})g(\boldsymbol{y})$ whenever $\boldsymbol{x} \le \boldsymbol{y}$. However, this does not yield an intuitive notion; see Remark 6.E.10.

6.E.2 Some properties

The multivariate likelihood ratio order is preserved under conditioning on any rectangular set A (that is, A of the form $A = A_1 \times A_2 \times \cdots \times A_n$ where $A_i \subseteq \mathbb{R}$, $i = 1, 2, \dots, n$). This is shown in the next result. The proof is quite trivial and is omitted.

Theorem 6.E.1. *If $\boldsymbol{X}$ and $\boldsymbol{Y}$ are two n-dimensional random vectors such that $\boldsymbol{X} \le_{\rm lr} \boldsymbol{Y}$, then, for any measurable rectangular set $A \subseteq \mathbb{R}^n$, we have that $[\boldsymbol{X}|\boldsymbol{X} \in A] \le_{\rm lr} [\boldsymbol{Y}|\boldsymbol{Y} \in A]$.*

The above theorem can be generalized as follows. For $A, B \subseteq \mathbb{R}^n$ we denote $A \vee B = \{\boldsymbol{x} \vee \boldsymbol{y} : \boldsymbol{x} \in A, \boldsymbol{y} \in B\}$ and $A \wedge B = \{\boldsymbol{x} \wedge \boldsymbol{y} : \boldsymbol{x} \in A, \boldsymbol{y} \in B\}$.

Theorem 6.E.2. *Let $A, B \subseteq \mathbb{R}^n$ satisfy $A \vee B \subseteq B$ and $A \wedge B \subseteq A$. If $\boldsymbol{X}$ and $\boldsymbol{Y}$ are two n-dimensional random vectors such that $\boldsymbol{X} \le_{\rm lr} \boldsymbol{Y}$, then*

$$[\boldsymbol{X}|\boldsymbol{X} \in A] \le_{\rm lr} [\boldsymbol{Y}|\boldsymbol{Y} \in B].$$

Proof. Let f and g denote the density functions of $\boldsymbol{X}$ and $\boldsymbol{Y}$, respectively. For any set C, let I_C denote its indicator function. The assumptions imply

$$I_A(\boldsymbol{x})I_B(\boldsymbol{y}) \le I_A(\boldsymbol{x} \wedge \boldsymbol{y})I_B(\boldsymbol{x} \vee \boldsymbol{y}) \quad \text{and} \quad f(\boldsymbol{x})g(\boldsymbol{y}) \le f(\boldsymbol{x} \wedge \boldsymbol{y})g(\boldsymbol{x} \vee \boldsymbol{y}).$$

Therefore

$$\frac{f(\boldsymbol{x})I_A(\boldsymbol{x})}{P\{\boldsymbol{X} \in A\}} \cdot \frac{g(\boldsymbol{y})I_B(\boldsymbol{y})}{P\{\boldsymbol{Y} \in B\}} \le \frac{f(\boldsymbol{x} \wedge \boldsymbol{y})I_A(\boldsymbol{x} \wedge \boldsymbol{y})}{P\{\boldsymbol{X} \in A\}} \cdot \frac{g(\boldsymbol{x} \vee \boldsymbol{y})I_B(\boldsymbol{x} \vee \boldsymbol{y})}{P\{\boldsymbol{Y} \in B\}}. \qquad \square$$

The following result shows that the order $\le_{\rm lr}$ is preserved under strictly monotone transformations of each individual coordinate of the underlying random vectors. The proof follows the lines of the proof of Theorem 1.C.8 and is omitted.

Theorem 6.E.3. *Let ψ_i be any increasing function, $i = 1, 2, \dots, n$. Let $\boldsymbol{X} = (X_1, X_2, \dots, X_n)$ and $\boldsymbol{Y} = (Y_1, Y_2, \dots, Y_n)$ be two n-dimensional random vectors. If $\boldsymbol{X} \le_{\rm lr} \boldsymbol{Y}$, then*

$$(\psi_1(X_1), \psi_2(X_2), \dots, \psi_n(X_n)) \le_{\rm lr} (\psi_1(Y_1), \psi_2(Y_2), \dots, \psi_n(Y_n)).$$

The order $\le_{\rm lr}$ is closed under marginalization and under conjunctions as the following result shows. The first part of the theorem can easily be proven from the definitions. The proof of the second part uses ideas from the theory of total positivity and is not given here.

Theorem 6.E.4. (a) *Let $\boldsymbol{X}_1, \boldsymbol{X}_2, \dots, \boldsymbol{X}_m$ be a set of independent random vectors where the dimension of $\boldsymbol{X}_i$ is k_i, $i = 1, 2, \dots, m$. Let $\boldsymbol{Y}_1, \boldsymbol{Y}_2, \dots, \boldsymbol{Y}_m$ be another set of independent random vectors where the dimension of $\boldsymbol{Y}_i$ is k_i, $i = 1, 2, \dots, m$. If $\boldsymbol{X}_i \le_{\rm lr} \boldsymbol{Y}_i$ for $i = 1, 2, \dots, m$, then*

$$(\boldsymbol{X}_1, \boldsymbol{X}_2, \dots, \boldsymbol{X}_m) \le_{\rm lr} (\boldsymbol{Y}_1, \boldsymbol{Y}_2, \dots, \boldsymbol{Y}_m).$$

That is, the multivariate likelihood ratio order is closed under conjunctions.

(b) *Let $\boldsymbol{X} = (X_1, X_2, \dots, X_n)$ and $\boldsymbol{Y} = (Y_1, Y_2, \dots, Y_n)$ be two n-dimensional random vectors. If $\boldsymbol{X} \le_{\rm lr} \boldsymbol{Y}$, then $\boldsymbol{X}_I \le_{\rm lr} \boldsymbol{Y}_I$ for each $I \subseteq \{1, 2, \dots, n\}$. That is, the multivariate likelihood ratio order is closed under marginalization.*

A result which shows the preservation of the order $\leq_{\rm lr}$ under random summations is stated next. The proof is based on standard arguments from the theory of total positivity, and is omitted.

Theorem 6.E.5. *Let $\boldsymbol{X}_1, \boldsymbol{X}_2, \dots, \boldsymbol{X}_m$ be m countably infinite vectors of independent nonnegative random variables. Assume that $\boldsymbol{X}_1, \boldsymbol{X}_2, \dots, \boldsymbol{X}_m$ are independent. Let $\boldsymbol{M} = (M_1, M_2, \dots, M_m)$ and $\boldsymbol{N} = (N_1, N_2, \dots, N_m)$ be two vectors of nonnegative integers which are independent of $\boldsymbol{X}_1, \boldsymbol{X}_2, \dots, \boldsymbol{X}_m$. Denote by $X_{j,i}$ the ith element of $\boldsymbol{X}_j$. If $X_{j,i}$ has a logconcave density function for all $j = 1, 2, \dots, m$ and $i \geq 1$, and if $\boldsymbol{M} \leq_{\rm lr} \boldsymbol{N}$, then*

$$\Big(\sum_{i=1}^{M_1} X_{1,i}, \sum_{i=1}^{M_2} X_{2,i}, \dots, \sum_{i=1}^{M_m} X_{m,i}\Big) \leq_{\rm lr} \Big(\sum_{i=1}^{N_1} X_{1,i}, \sum_{i=1}^{N_2} X_{2,i}, \dots, \sum_{i=1}^{N_m} X_{m,i}\Big).$$

In the univariate case the likelihood ratio order implies the hazard rate order. It turns out that this is also the case in the multivariate case as the following two results show.

Theorem 6.E.6. *If $\boldsymbol{X}$ and $\boldsymbol{Y}$ are two n-dimensional random vectors such that $\boldsymbol{X} \leq_{\rm lr} \boldsymbol{Y}$, then $\boldsymbol{X} \leq_{\rm hr} \boldsymbol{Y}$.*

Proof. This result follows from Theorem 2.4 in Karlin and Rinott [278] with the MTP$_2$ kernel K defined by $K(\boldsymbol{x}, \boldsymbol{u}) = \prod_{i=1}^n 1_{(x_i,\infty)}(u_i)$. □

Theorem 6.E.7. *If $\boldsymbol{X}$ and $\boldsymbol{Y}$ are two nonnegative n-dimensional random vectors such that $\boldsymbol{X} \leq_{\rm lr} \boldsymbol{Y}$, then $\boldsymbol{X} \leq_{\text{dyn-hr}} \boldsymbol{Y}$.*

Proof. First suppose that $\boldsymbol{X} > 0\boldsymbol{e}$ a.s. Split $\{1, 2, \dots, n\}$ into three mutually exclusive sets, I, J, and L (so that $L = \overline{I \cup J}$). Select $\boldsymbol{x}_I$, $\boldsymbol{x}_J$, $\boldsymbol{y}_I$, and t such that $\boldsymbol{x}_I \leq \boldsymbol{y}_I \leq t\boldsymbol{e}$ and $\boldsymbol{x}_J \leq t\boldsymbol{e}$. Denote the densities of $(\boldsymbol{X}_I, \boldsymbol{X}_J, \boldsymbol{X}_L)$ and of $(\boldsymbol{Y}_I, \boldsymbol{Y}_J, \boldsymbol{Y}_L)$ by $\tilde{f}$ and $\tilde{g}$, respectively. The density of $[\boldsymbol{X}_L \big| \boldsymbol{X}_I = \boldsymbol{x}_I, \boldsymbol{X}_J = \boldsymbol{x}_J]$, with argument $\boldsymbol{x}_L$, is then $\tilde{f}(\boldsymbol{x}_I, \boldsymbol{x}_J, \boldsymbol{x}_L)/\tilde{f}_{I,J}(\boldsymbol{x}_I, \boldsymbol{x}_J)$ where $\tilde{f}_{I,J}$ is the marginal density of $(\boldsymbol{X}_I, \boldsymbol{X}_J)$. The density of $[\boldsymbol{Y}_L \big| \boldsymbol{Y}_I = \boldsymbol{y}_I, \boldsymbol{Y}_J > t\boldsymbol{e}]$, with argument $\boldsymbol{y}_L$, is then

$$\frac{\int_{\boldsymbol{y}_J > t\boldsymbol{e}} \tilde{g}(\boldsymbol{y}_I, \boldsymbol{y}_J, \boldsymbol{y}_L) \mathrm{d}\boldsymbol{y}_J}{\int_{\boldsymbol{y}_J > t\boldsymbol{e}} \tilde{g}_{I,J}(\boldsymbol{y}_I, \boldsymbol{y}_J) \mathrm{d}\boldsymbol{y}_J},$$

where $\tilde{g}_{I,J}$ is the marginal density of $(\boldsymbol{Y}_I, \boldsymbol{Y}_J)$.

Now select a $\boldsymbol{y}_J > t\boldsymbol{e}$. Since $\boldsymbol{y}_J > t\boldsymbol{e}$ and $\boldsymbol{x}_J \leq t\boldsymbol{e}$ it follows that $\boldsymbol{x}_J \leq \boldsymbol{y}_J$. Also $\boldsymbol{x}_I \leq \boldsymbol{y}_I$. Therefore, from the assumption that $\boldsymbol{X} \leq_{\rm lr} \boldsymbol{Y}$ it follows that

$$\tilde{f}(\boldsymbol{x}_I, \boldsymbol{x}_J, \boldsymbol{x}_L)\tilde{g}(\boldsymbol{y}_I, \boldsymbol{y}_J, \boldsymbol{y}_L) \leq \tilde{f}(\boldsymbol{x}_I, \boldsymbol{x}_J, \boldsymbol{x}_L \wedge \boldsymbol{y}_L)\tilde{g}(\boldsymbol{y}_I, \boldsymbol{y}_J, \boldsymbol{x}_L \vee \boldsymbol{y}_L). \quad (6.E.3)$$

Integration of (6.E.3) over the region $\{\boldsymbol{y}_J : \boldsymbol{y}_J > t\boldsymbol{e}\}$ yields

$$\int_{\boldsymbol{y}_J > t\boldsymbol{e}} \tilde{f}(\boldsymbol{x}_I, \boldsymbol{x}_J, \boldsymbol{x}_L)\tilde{g}(\boldsymbol{y}_I, \boldsymbol{y}_J, \boldsymbol{y}_L) \mathrm{d}\boldsymbol{y}_J$$
$$\leq \int_{\boldsymbol{y}_J > t\boldsymbol{e}} \tilde{f}(\boldsymbol{x}_I, \boldsymbol{x}_J, \boldsymbol{x}_L \wedge \boldsymbol{y}_L)\tilde{g}(\boldsymbol{y}_I, \boldsymbol{y}_J, \boldsymbol{x}_L \vee \boldsymbol{y}_L) \mathrm{d}\boldsymbol{y}_J$$

which, in turn, yields

$$\frac{\tilde{f}(\boldsymbol{x}_I, \boldsymbol{x}_J, \boldsymbol{x}_L)}{\tilde{f}_{I,J}(\boldsymbol{x}_I, \boldsymbol{x}_J)} \times \frac{\int_{\boldsymbol{y}_J > t\boldsymbol{e}} \tilde{g}(\boldsymbol{y}_I, \boldsymbol{y}_J, \boldsymbol{y}_L) \mathrm{d}\boldsymbol{y}_J}{\int_{\boldsymbol{y}_J > t\boldsymbol{e}} \tilde{g}_{I,J}(\boldsymbol{y}_I, \boldsymbol{y}_J)\, d\boldsymbol{y}_J}$$
$$\leq \frac{\tilde{f}(\boldsymbol{x}_I, \boldsymbol{x}_J, \boldsymbol{x}_L \wedge \boldsymbol{y}_L)}{\tilde{f}_{I,J}(\boldsymbol{x}_I, \boldsymbol{x}_J)} \times \frac{\int_{\boldsymbol{y}_J > t\boldsymbol{e}} \tilde{g}(\boldsymbol{y}_I, \boldsymbol{y}_J, \boldsymbol{x}_L \vee \boldsymbol{y}_L) \mathrm{d}\boldsymbol{y}_J}{\int_{\boldsymbol{y}_J > t\boldsymbol{e}} \tilde{g}_{I,J}(\boldsymbol{y}_I, \boldsymbol{y}_J) \mathrm{d}\boldsymbol{y}_J}.$$

That is, we have shown so far that

$$[\boldsymbol{X}_L \big| \boldsymbol{X}_I = \boldsymbol{x}_I, \boldsymbol{X}_J = \boldsymbol{x}_J] \leq_{\mathrm{lr}} [\boldsymbol{Y}_L \big| \boldsymbol{Y}_I = \boldsymbol{y}_I, \boldsymbol{Y}_J > t\boldsymbol{e}]. \tag{6.E.4}$$

From Theorems 6.E.1 and 6.E.3 it now follows that

$$[\boldsymbol{X}_L - t\boldsymbol{e} \big| \boldsymbol{X}_I = \boldsymbol{x}_I, \boldsymbol{X}_J = \boldsymbol{x}_J, \boldsymbol{X}_L > t\boldsymbol{e}]$$
$$\leq_{\mathrm{lr}} [\boldsymbol{Y}_L - t\boldsymbol{e} \big| \boldsymbol{Y}_I = \boldsymbol{y}_I, \boldsymbol{Y}_J > t\boldsymbol{e}, \boldsymbol{Y}_L > t\boldsymbol{e}],$$

and from Theorem 6.E.4(b) it follows that, for $k \in L$, we have

$$[X_k - t \big| \boldsymbol{X}_I = \boldsymbol{x}_I, \boldsymbol{X}_J = \boldsymbol{x}_J, \boldsymbol{X}_L > t\boldsymbol{e}]$$
$$\leq_{\mathrm{lr}} [Y_k - t \big| \boldsymbol{Y}_I = \boldsymbol{y}_I, \boldsymbol{Y}_J > t\boldsymbol{e}, \boldsymbol{Y}_L > t\boldsymbol{e}], \tag{6.E.5}$$

where here $\leq_{\mathrm{lr}}$ denotes the univariate likelihood ratio order discussed in Section 1.C.

From (6.E.5) it follows that the density of $[X_k - t \big| \boldsymbol{X}_I = \boldsymbol{x}_I, \boldsymbol{X}_J = \boldsymbol{x}_J, \boldsymbol{X}_L > t\boldsymbol{e}]$ at zero is larger than the density of $[Y_k - t \big| \boldsymbol{Y}_I = \boldsymbol{y}_I, \boldsymbol{Y}_J > t\boldsymbol{e}, \boldsymbol{Y}_L > t\boldsymbol{e}]$ at zero. But the density of $[X_k - t \big| \boldsymbol{X}_I = \boldsymbol{x}_I, \boldsymbol{X}_J = \boldsymbol{x}_J, \boldsymbol{X}_L > t\boldsymbol{e}]$ at zero is $\eta_{k|I\cup J}(t \big| \boldsymbol{x}_I, \boldsymbol{x}_J)$ and the density of $[Y_k - t \big| \boldsymbol{Y}_I = \boldsymbol{y}_I, \boldsymbol{Y}_J > t\boldsymbol{e}, \boldsymbol{Y}_L > t\boldsymbol{e}]$ at zero is $\lambda_{k|I}(t \big| \boldsymbol{y}_I)$, where $\lambda_{\cdot|\cdot}(\cdot|\cdot)$ and $\eta_{\cdot|\cdot}(\cdot|\cdot)$ denote the multivariate conditional hazard rate functions of $\boldsymbol{X}$ and $\boldsymbol{Y}$, respectively. We thus have shown that $\boldsymbol{X}$ and $\boldsymbol{Y}$ satisfy (6.D.13) and this completes the proof of the theorem when $\boldsymbol{X} > 0\boldsymbol{e}$ a.s.

If $\boldsymbol{X}$ has some components that are identically zero a.s., then the above arguments still apply after some simple modifications. □

A combination of Theorems 6.C.1, 6.D.9, and 6.E.7 shows that for nonnegative random vectors $\boldsymbol{X}$ and $\boldsymbol{Y}$ one has $\boldsymbol{X} \leq_{\mathrm{lr}} \boldsymbol{Y} \Longrightarrow \boldsymbol{X} \leq_{\mathrm{st}} \boldsymbol{Y}$. But this is true in general as is stated in the next result, the proof of which we omit.

Theorem 6.E.8. *If $\boldsymbol{X}$ and $\boldsymbol{Y}$ are two n-dimensional random vectors such that $\boldsymbol{X} \leq_{\mathrm{lr}} \boldsymbol{Y}$, then $\boldsymbol{X} \leq_{\mathrm{st}} \boldsymbol{Y}$.*

Remark 6.E.9. A combination of Theorems 6.E.1 and 6.E.8 shows that

$$\boldsymbol{X} \leq_{\mathrm{lr}} \boldsymbol{Y} \Longrightarrow [\boldsymbol{X} \big| A] \leq_{\mathrm{st}} [\boldsymbol{Y} \big| A] \text{ for all measurable rectangular sets } A \subseteq \mathbb{R}^n. \tag{6.E.6}$$

The conclusion in (6.E.6) is a generalization of (1.C.6). However, the characterization of the order $\le_{\rm lr}$ in the univariate case, given in (1.C.6), does not generalize to the multivariate case. That is, $\boldsymbol{X} \le_{\rm lr} \boldsymbol{Y}$ does not necessarily imply that $[\boldsymbol{X}|A] \le_{\rm st} [\boldsymbol{Y}|A]$ for all measurable sets $A \in \mathbb{R}^n$.

Remark 6.E.10. Let $\boldsymbol{X}$ and $\boldsymbol{Y}$ be two n-dimensional random vectors with (continuous or discrete) density functions f and g, respectively. If it is only assumed that $f(\boldsymbol{y})g(\boldsymbol{x}) \le f(\boldsymbol{x})g(\boldsymbol{y})$ whenever $\boldsymbol{x} \le \boldsymbol{y}$ (rather than (6.E.1)), then it is not necessarily true that $\boldsymbol{X} \le_{\rm st} \boldsymbol{Y}$; counterexamples can be found in the literature. Note, however, that, under some additional conditions, the monotonicity of $g(\boldsymbol{x})/f(\boldsymbol{x})$ in $\boldsymbol{x}$ implies that $\boldsymbol{X} \le_{\rm st} \boldsymbol{Y}$; see, for example, Theorem 6.B.8.

A result that may be viewed as a generalization of Theorems 1.C.9 and 1.C.52 is stated next.

Theorem 6.E.11. *Let $\boldsymbol{X}$ be an n-dimensional random vector.*

(a) *$\boldsymbol{X} \le_{\rm lr} \boldsymbol{X}+\boldsymbol{a}$ for all $\boldsymbol{a} \ge \boldsymbol{0}$ if, and only if, $\boldsymbol{X}$ has independent components with logconcave density functions.*
(b) *If $\boldsymbol{X}$ has independent components with logconcave density functions, then $\boldsymbol{X} \le_{\rm lr} \boldsymbol{X}+\boldsymbol{Y}$ for any random vector $\boldsymbol{Y} \ge \boldsymbol{0}$ independent of $\boldsymbol{X}$.*

In the next result, vectors of order statistics are compared in the multivariate order $\le_{\rm lr}$. The result may be compared with Theorems 6.D.10, 7.B.4, and 7.B.12.

Theorem 6.E.12. *Let $X_{(1)}, X_{(2)}, \ldots, X_{(n)}$ and $Y_{(1)}, Y_{(2)}, \ldots, Y_{(n)}$ be order statistics as in Theorem* 6.D.10. *If $X_1 \le_{\rm lr} Y_1$, then*

$$(X_{(1)}, X_{(2)}, \ldots, X_{(n)}) \le_{\rm lr} (Y_{(1)}, Y_{(2)}, \ldots, Y_{(n)}).$$

The following example may be compared with Examples 1.B.24, 1.C.48, 2.A.22, 3.B.38, 6.B.41, 6.D.8, and 7.B.13.

Example 6.E.13. Let X and Y be two absolutely continuous nonnegative random variables with survival functions $\overline{F}$ and $\overline{G}$, and density functions f and g, respectively. Denote $\Lambda_1 = -\log \overline{F}$, $\Lambda_2 = -\log \overline{G}$, and $\lambda_i = \Lambda_i'$, $i = 1, 2$. Consider two nonhomogeneous Poisson processes $N_1 = \{N_1(t),\ t \ge 0\}$ and $N_2 = \{N_2(t),\ t \ge 0\}$ with mean functions Λ_1 and Λ_2 (see Example 1.B.13), respectively. Let $T_{i,1}, T_{i,2}, \ldots$ be the successive epoch times of process N_i, $i = 1, 2$. Note that $X =_{\rm st} T_{1,1}$ and $Y =_{\rm st} T_{2,1}$.

It turns out that, under some conditions, the univariate likelihood ratio ordering of the first two epoch times implies the multivariate likelihood ratio ordering of the corresponding vectors of the later epoch times. Explicitly, it will be shown below that if $X \le_{\rm hr} Y$, and if (1.B.25) holds, then $(T_{1,1}, T_{1,2}, \ldots, T_{1,n}) \le_{\rm lr} (T_{2,1}, T_{2,2}, \ldots, T_{2,n})$ for each $n \ge 1$. (Note that the condition $X \le_{\rm hr} Y$, together with (1.B.25), is stronger than merely assuming $X \le_{\rm lr} Y$; see Theorem 1.C.4.)

As is mentioned above, the stated result is true for $n = 1$. So let $n \geq 2$. The density functions of $(T_{i,1}, T_{i,2}, \ldots, T_{i,n})$, $i = 1, 2$, are given by

$$h_{1,n}(x_1, x_2, \ldots, x_n) = \lambda_1(x_1)\lambda_1(x_2)\cdots\lambda_1(x_{n-1})f(x_n) \qquad \text{for } x_1 \leq x_2 \leq \cdots \leq x_n,$$

and

$$h_{2,n}(x_1, x_2, \ldots, x_n) = \lambda_2(x_1)\lambda_2(x_2)\cdots\lambda_2(x_{n-1})g(x_n) \qquad \text{for } x_1 \leq x_2 \leq \cdots \leq x_n.$$

Consider now $(x_1, x_2, \ldots, x_n)$ and $(y_1, y_2, \ldots, y_n)$ such that $x_1 \leq x_2 \leq \cdots \leq x_n$ and $y_1 \leq y_2 \leq \cdots \leq y_n$. We want to prove that

$$\begin{aligned}\lambda_1(x_1 \wedge y_1)\lambda_1(x_2 \wedge y_2)\cdots\lambda_1(x_{n-1} \wedge y_{n-1})f(x_n \wedge y_n) \\ \times \lambda_2(x_1 \vee y_1)\lambda_2(x_2 \vee y_2)\cdots\lambda_2(x_{n-1} \vee y_{n-1})g(x_n \vee y_n) \\ \geq \lambda_1(x_1)\lambda_1(x_2)\cdots\lambda_1(x_{n-1})f(x_n) \\ \times \lambda_2(y_1)\lambda_2(y_2)\cdots\lambda_2(y_{n-1})g(y_n). \qquad (6.E.7)\end{aligned}$$

Let $E = \{i \leq n-1 : x_i \geq y_i\}$. Then (6.E.7) reduces to

$$\Big(\prod_{i\in E}\lambda_1(y_i)\lambda_2(x_i)\Big)f(x_n \wedge y_n)g(x_n \vee y_n) \geq \Big(\prod_{i\in E}\lambda_1(x_i)\lambda_2(y_i)\Big)f(x_n)g(y_n),$$

and this follows from (1.B.25) and $X \leq_{\mathrm{lr}} Y$.

From the above result, and the closure of the likelihood ratio order under marginalization (Theorem 6.E.4(b)), it follows that if $X \leq_{\mathrm{hr}} Y$, and if (1.B.25) holds, then $T_{1,n} \leq_{\mathrm{lr}} T_{2,n}$, $n \geq 1$. However, a stronger result is given in Example 1.C.48—this is so because the conditions $X \leq_{\mathrm{hr}} Y$ and (1.B.25), together, imply the conditions $X \leq_{\mathrm{lr}} Y$ and (1.C.15).

Now let $X_{i,n} \equiv T_{i,n} - T_{i,n-1}$, $n \geq 1$ (where $T_{i,0} \equiv 0$), be the inter-epoch times of the process N_i, $i = 1, 2$. Again, note that $X =_{\mathrm{st}} X_{1,1}$ and $Y =_{\mathrm{st}} X_{2,1}$. It turns out that, under some conditions, the univariate likelihood ratio ordering of the first two inter-epoch times implies the multivariate likelihood ratio ordering of the corresponding vectors of the later inter-epoch times. Explicitly, if $X \leq_{\mathrm{hr}} Y$, and if f and/or g are logconvex, and if λ_1 and/or λ_2 are logconvex, and if (1.B.25) holds, then $(X_{1,1}, X_{1,2}, \ldots, X_{1,n}) \leq_{\mathrm{lr}} (X_{2,1}, X_{2,2}, \ldots, X_{2,n})$ for each $n \geq 1$. The proof of this statement will not be detailed here.

From the above result, and the closure of the likelihood ratio order under marginalization (Theorem 6.E.4(b)), it follows that if $X \leq_{\mathrm{hr}} Y$, and if f and/or g are logconvex, and if λ_1 and/or λ_2 are logconvex, and if (1.B.25) holds, then $X_{1,n} \leq_{\mathrm{lr}} X_{2,n}$, $n \geq 1$. This is a different set of conditions for the last stochastic inequality than the set of conditions in Example 1.C.48.

Example 6.E.14. Recall that the spacings that correspond to the nonnegative random variables $X_1, X_2, \ldots, X_n$ are denoted by $U_{(i)} = X_{(i)} - X_{(i-1)}$, $i = 1, 2, \ldots, n$, where the $X_{(i)}$'s are the corresponding order statistics (here we take $X_{(0)} \equiv 0$). The normalized spacings are defined by $D_{(i)} = (n - i - 1)U_{(i)}$, $i = 1, 2, \ldots, n$. Now, let $D_{(1)}, D_{(2)}, \ldots, D_{(n)}$ be the normalized spacings associated with exponential random variables $X_1, X_2, \ldots, X_n$, where X_i has the hazard rate λ_i, $i = 1, 2, \ldots, n$. Let $D^*_{(1)}, D^*_{(2)}, \ldots, D^*_{(n)}$ be the normalized spacings associated with a sample of n independent and identically distributed exponential random variables that have the hazard rate $(1/n)\sum_{i=1}^{n} \lambda_i$. Then

$$(D^*_{(1)}, D^*_{(2)}, \ldots, D^*_{(n)}) \leq_{\text{lr}} (D_{(1)}, D_{(2)}, \ldots, D_{(n)}).$$

The following example is similar to Example 6.B.25 except that under a different assumption we obtain a stronger conclusion. Other results which give related comparisons can be found in Theorems 1.C.45 and 4.B.17.

Example 6.E.15. Let X and Y be two random variables. Let $X_{(1)} \leq X_{(2)} \leq \cdots \leq X_{(n)}$ denote the order statistics from a sample $X_1, X_2, \ldots, X_n$ of independent and identically distributed random variables that have the same distribution as X. Similarly, let $Y_{(1)} \leq Y_{(2)} \leq \cdots \leq Y_{(n)}$ denote the order statistics from another sample $Y_1, Y_2, \ldots, Y_n$ of independent and identically distributed random variables that have the same distribution as Y. The corresponding spacings are defined by $U_{(i)} \equiv X_{(i)} - X_{(i-1)}$ and $V_{(i)} \equiv Y_{(i)} - Y_{(i-1)}$, $i = 2, 3, \ldots, n$. Denote $\boldsymbol{U} = (U_{(2)}, U_{(3)}, \ldots, U_{(n)})$ and $\boldsymbol{V} = (V_{(2)}, V_{(3)}, \ldots, V_{(n)})$. Kochar [311] has shown that if $X \leq_{\text{lr}} Y$, and if either X or Y have logconvex densities, then

$$\boldsymbol{U} \leq_{\text{lr}} \boldsymbol{V}.$$

The next example extends Example 1.C.57 to the multivariate likelihood ratio order.

Example 6.E.16. Let $\boldsymbol{X}$ be an n-dimensional random vector whose distribution function depends on the m-dimensional parameter $\boldsymbol{\Theta}$. Denote the prior density function of $\boldsymbol{\Theta}$ by $\pi(\cdot)$, and denote the conditional density of $\boldsymbol{X}$, given $\boldsymbol{\Theta} = \boldsymbol{\theta}$, by $f(\cdot|\boldsymbol{\theta})$. Suppose that the m-dimensional density function of $\boldsymbol{\Theta}$ is MTP$_2$ (multivariate totally positive of order 2), that is, suppose that $\boldsymbol{\Theta} \leq_{\text{lr}} \boldsymbol{\Theta}$, or, equivalently (see (6.E.1)), that $\pi(\boldsymbol{\theta})\pi(\boldsymbol{\theta}') \leq \pi(\boldsymbol{\theta} \wedge \boldsymbol{\theta}')\pi(\boldsymbol{\theta} \vee \boldsymbol{\theta}')$ for every $\boldsymbol{\theta}$ and $\boldsymbol{\theta}'$ in $\mathbb{R}^m$. Then, if $f(\boldsymbol{x}|\boldsymbol{\theta})$ is ($(m+n)$-dimensional) MTP$_2$, then $\boldsymbol{\Theta}$ is increasing in $\boldsymbol{X}$ in the likelihood ratio sense (that is, $[\boldsymbol{\Theta}|\boldsymbol{X} = \boldsymbol{x}] \leq_{\text{lr}} [\boldsymbol{\Theta}|\boldsymbol{X} = \boldsymbol{x}']$ whenever $\boldsymbol{x} \leq \boldsymbol{x}'$). The proof of this statement is similar to the proof of the statement in Example 1.C.57 and is omitted.

6.E.3 A property in reliability theory

In Theorem 1.C.52 it was shown that a nonnegative random variable T has a logconcave density if, and only if, either one of the following equivalent conditions holds:

$$[T - t | T > t] \geq_{\text{lr}} [T - t' | T > t'] \quad \text{whenever} \quad t \leq t', \tag{6.E.8}$$

$$T \geq_{\text{lr}} [T - t | T > t] \quad \text{for all} \quad t \geq 0. \tag{6.E.9}$$

We commented there that logconcavity can thus be interpreted as an aging notion in reliability theory. Having a multivariate analog of the order $\geq_{\text{lr}}$ one can generalize (6.E.8) and (6.E.9) to the multivariate case, thus introducing notions which can be considered as multivariate analogs of distributions with logconcave densities. This can be done in several ways. In this subsection we show that various generalizations of (6.E.8) and (6.E.9) actually yield the same notion of multivariate PF_2 distributions.

Let $\boldsymbol{T}$ be a nonnegative random vector. Recall from Section 6.B.6 the definition, the notation h_t, and the comparison of histories associated with $\boldsymbol{T}$. One possible multivariate analog of (6.E.8) is to require $\boldsymbol{T}$ to satisfy, for $t \leq s$ and histories h_t and h'_s,

$$[(\boldsymbol{T} - t\boldsymbol{e})_+ | h_t] \geq_{\text{lr}} [(\boldsymbol{T} - s\boldsymbol{e})_+ | h'_s] \quad \text{whenever } h_t \leq h'_s. \tag{6.E.10}$$

Still another possible multivariate analog of (6.E.8) is to require $\boldsymbol{T}$ to satisfy, for $t \leq s$,

$$[(\boldsymbol{T} - t\boldsymbol{e})_+ | h_t] \geq_{\text{lr}} [(\boldsymbol{T} - s\boldsymbol{e})_+ | h'_s] \quad \text{whenever } h_t \text{ and } h'_s \text{ coincide on } [0, t). \tag{6.E.11}$$

An analog of (6.E.9) is to require $\boldsymbol{T}$ to satisfy (6.E.10) or (6.E.11) with $t = 0$, that is,

$$\boldsymbol{T} \geq_{\text{lr}} [(\boldsymbol{T} - s\boldsymbol{e})_+ | h'_s] \quad \text{for any history } h'_s, \ s \geq 0. \tag{6.E.12}$$

It turns out that these three conditions are equivalent. If we say that the nonnegative random vector $\boldsymbol{T}$ is *multivariate* PF_2 if it satisfies (6.E.10), then we have the following result, the proof of which is similar to the proof of Theorem 6.D.11.

Theorem 6.E.17. *Let $\boldsymbol{T}$ be a nonnegative random vector. The following three statements are equivalent.*

(i) *$\boldsymbol{T}$ is multivariate* PF_2.
(ii) *$\boldsymbol{T}$ satisfies* (6.E.11).
(iii) *$\boldsymbol{T}$ satisfies* (6.E.12).

6.F The Multivariate Mean Residual Life Order

6.F.1 Definition

Let $\boldsymbol{T} = (T_1, T_2, \ldots, T_m)$ be a nonnegative random vector with a finite mean vector. Consider a typical history of $\boldsymbol{T}$ at time $t \geq 0$, which is of the form (see (6.B.25))

$$h_t = \{\boldsymbol{T}_I = \boldsymbol{t}_I, \boldsymbol{T}_{\overline{I}} > t\boldsymbol{e}\}, \quad 0\boldsymbol{e} \le \boldsymbol{t}_I \le t\boldsymbol{e},\ I \subseteq \{1,2,\ldots,m\}. \tag{6.F.1}$$

Given the history h_t as in (6.F.1), let $i \in \overline{I}$ be a component that is still alive at time t. Its multivariate mean residual life, at time t, is defined as follows:

$$m_{i|I}(t|\boldsymbol{t}_I) = E[T_i - t|\boldsymbol{T}_I = \boldsymbol{t}_I,\ \boldsymbol{T}_{\overline{I}} > t\boldsymbol{e}], \tag{6.F.2}$$

where, of course, $0\boldsymbol{e} \le \boldsymbol{t}_I \le t\boldsymbol{e}$ and $I \subseteq \{1,2,\ldots,m\}$.

Clearly, the smaller the mrl function is, the smaller $\boldsymbol{T}$ should be in some stochastic sense. This is the motivation for the order discussed in this section.

Let $\boldsymbol{S}$ be another nonnegative random vector with a finite mean vector. Denote its multivariate mean residual life functions by $l_{\cdot|\cdot}(\cdot|\cdot)$, where the l's are defined analogously as the m's in (6.F.2). Suppose that

$$l_{i|I\cup J}(u|\boldsymbol{s}_I, \boldsymbol{s}_J) \le m_{i|I}(u|\boldsymbol{t}_I)$$
$$\text{whenever } J \cap I = \varnothing,\ \boldsymbol{s}_I \le \boldsymbol{t}_I \le u\boldsymbol{e}, \text{ and } \boldsymbol{s}_J \le u\boldsymbol{e}, \tag{6.F.3}$$

where $i \in \overline{I \cup J}$. Then $\boldsymbol{S}$ is said to be *smaller than* $\boldsymbol{T}$ *in the multivariate mean residual life order* (denoted as $\boldsymbol{S} \le_{\text{mrl}} \boldsymbol{T}$).

The order $\le_{\text{mrl}}$ is not an order in the usual sense; a comment, similar to the comment in Remark 6.B.5, applies to this order too. Explicitly, $\boldsymbol{X} \le_{\text{mrl}} \boldsymbol{X}$ means that $\boldsymbol{X}$ has the positive dependence property of "mrl decreasing upon failure" discussed in Shaked and Shanthikumar [513].

Note that (6.F.3) can be written as

$$l_i(h'_u) \le m_i(h_u) \quad \text{whenever } h'_u \ge h_u,$$

where i denotes a component that has not failed by time u in the history h'_u.

In the univariate case ($m = 1$) condition (6.F.3) reduces to (2.A.2) [with a different notation]. We have already seen that in the univariate case

$$S_1 \le_{\text{hr}} T_1 \Longrightarrow S_1 \le_{\text{mrl}} T_1.$$

This is also true in the general multivariate case as will be shown in the next subsection.

6.F.2 The relation between the multivariate mean residual life and the dynamic multivariate hazard rate orders

Theorem 6.F.1. *If* $\boldsymbol{S}$ *and* $\boldsymbol{T}$ *are two nonnegative random vectors with finite mean vectors such that* $\boldsymbol{S} \le_{\text{dyn-hr}} \boldsymbol{T}$*, then* $\boldsymbol{S} \le_{\text{mrl}} \boldsymbol{T}$.

Proof. Select a $t > 0$ and two histories h_t and h'_t such that $h_t \le h'_t$. It is not hard to verify that if $\boldsymbol{S} \le_{\text{dyn-hr}} \boldsymbol{T}$, then $[(\boldsymbol{S} - t\boldsymbol{e})_+|h'_t] \le_{\text{dyn-hr}} [(\boldsymbol{T} - t\boldsymbol{e})_+|h_t]$. From Theorems 6.D.9 and 6.C.1 it is seen that if $[(\boldsymbol{S} - t\boldsymbol{e})_+|h'_t] \le_{\text{dyn-hr}} [(\boldsymbol{T} - t\boldsymbol{e})_+|h_t]$, then $[(\boldsymbol{S} - t\boldsymbol{e})_+|h'_t] \le_{\text{st}} [(\boldsymbol{T} - t\boldsymbol{e})_+|h_t]$. Therefore, for a component i, which is still alive at time t in history h'_t, we have $l_i(h'_t) = E[S_i - t|h'_t] \le E[T_i - t|h_t] = m_i(h_t)$, that is, $\boldsymbol{S} \le_{\text{mrl}} \boldsymbol{T}$. □

6.F.3 A property in reliability theory

Recall from Section 2.A.4 that a nonnegative random variable T with a finite mean is DMRL if, and only if, either one of the following equivalent conditions holds:

$$[T-t|T>t] \geq_{\rm mrl} [T-t'|T>t'] \quad \text{whenever } t \leq t', \tag{6.F.4}$$

$$T \geq_{\rm mrl} [T-t|T>t] \quad \text{for all } t \geq 0. \tag{6.F.5}$$

With the multivariate analog of the order $\geq_{\rm mrl}$ one can generalize (6.F.4) and (6.F.5) to the multivariate case, thus introducing notions of multivariate DMRL distributions. This can be done in several ways. In this subsection we show that various generalizations of (6.F.4) and (6.F.5) actually yield the same notion of multivariate DMRL.

Let $\boldsymbol{T}$ be a nonnegative random vector with a finite mean vector. A possible multivariate analog of (6.F.4) is to require, for $t \leq s$ and histories h_t and h'_s, that $\boldsymbol{T}$ satisfies

$$[(\boldsymbol{T}-t\boldsymbol{e})_+|h_t] \geq_{\rm mrl} [(\boldsymbol{T}-s\boldsymbol{e})_+|h'_s] \quad \text{whenever } h_t \leq h'_s. \tag{6.F.6}$$

Still another possible multivariate analog of (6.F.4) is to require, for $t \leq s$, that $\boldsymbol{T}$ satisfies

$$[(\boldsymbol{T}-t\boldsymbol{e})_+|h_t] \geq_{\rm mrl} [(\boldsymbol{T}-s\boldsymbol{e})_+|h'_s] \quad \text{whenever } h_t \text{ and } h'_s \text{ coincide on } [0,t). \tag{6.F.7}$$

An analog of (6.F.5) is to require that $\boldsymbol{T}$ satisfies (6.F.6) or (6.F.7) with $t=0$, that is,

$$\boldsymbol{T} \geq_{\rm mrl} [(\boldsymbol{T}-s\boldsymbol{e})_+|h'_s] \quad \text{for any history } h'_s,\ s \geq 0. \tag{6.F.8}$$

It turns out that these three conditions are equivalent. If we say that the nonnegative random vector $\boldsymbol{T}$ is *multivariate* DMRL if it satisfies (6.F.6), then we have the following result, the proof of which is similar to the proof of Theorem 6.D.11 and is omitted.

Theorem 6.F.2. *Let $\boldsymbol{T}$ be a nonnegative random vector with a finite mean vector. The following three statements are equivalent.*

(i) $\boldsymbol{T}$ *is multivariate* DMRL.
(ii) $\boldsymbol{T}$ *satisfies* (6.F.7).
(iii) $\boldsymbol{T}$ *satisfies* (6.F.8).

6.G Other Multivariate Stochastic Orders

6.G.1 The orthant orders

The usual multivariate stochastic order, discussed in Section 6.B, is a possible multivariate generalization of (1.A.4) or (1.A.7). In this section we discuss a few other possible generalizations of the univariate order $\leq_{\rm st}$ which are straightforward analogs of (1.A.1) and of (1.A.2). These generalizations yield orders that are strictly weaker than the usual multivariate stochastic order.

For a random vector $\boldsymbol{X} = (X_1, X_2, \ldots, X_n)$ with distribution function F, let $\overline{F}$ be the multivariate survival function of $\boldsymbol{X}$, that is,

$$\overline{F}(x_1, x_2, \ldots, x_n) \equiv P\{X_1 > x_1, X_2 > x_2, \ldots, X_n > x_n\} \quad \text{for all } \boldsymbol{x}.$$

Let $\boldsymbol{Y}$ be another n-dimensional random vector with distribution function G and survival function $\overline{G}$. If

$$\overline{F}(x_1, x_2, \ldots, x_n) \leq \overline{G}(x_1, x_2, \ldots, x_n) \quad \text{for all } \boldsymbol{x}, \tag{6.G.1}$$

then we say that $\boldsymbol{X}$ is *smaller than* $\boldsymbol{Y}$ *in the upper orthant order* (denoted by $\boldsymbol{X} \leq_{\rm uo} \boldsymbol{Y}$). If

$$F(x_1, x_2, \ldots, x_n) \geq G(x_1, x_2, \ldots, x_n) \quad \text{for all } \boldsymbol{x}, \tag{6.G.2}$$

then we say that $\boldsymbol{X}$ is *smaller than* $\boldsymbol{Y}$ *in the lower orthant order* (denoted by $\boldsymbol{X} \leq_{\rm lo} \boldsymbol{Y}$). The reason for this terminology is that sets of the form $\{\boldsymbol{x} : x_1 > a_1, x_2 > a_2, \ldots, x_n > a_n\}$, for some fixed $\boldsymbol{a}$, are called *upper orthants*, and sets of the form $\{\boldsymbol{x} : x_1 \leq a_1, x_2 \leq a_2, \ldots, x_n \leq a_n\}$, for some fixed $\boldsymbol{a}$, are called *lower orthants*.

Note that (6.G.1) can be written as

$$E[I_U(\boldsymbol{X})] \leq E[I_U(\boldsymbol{Y})] \quad \text{for all upper orthants } U. \tag{6.G.3}$$

Similarly, (6.G.2) can be written as

$$E[I_L(\boldsymbol{X})] \geq E[I_L(\boldsymbol{Y})] \quad \text{for all lower orthants } L. \tag{6.G.4}$$

Let ψ be an n-variate function of the form

$$\psi(x_1, x_2, \ldots, x_n) = \prod_{i=1}^{n} g_i(x_i), \quad (x_1, x_2, \ldots, x_n) \in \mathbb{R}^n,$$

where the g_i's are univariate nonnegative increasing functions. Every such function can be approximated by positive linear combinations of indicator functions of upper orthants. Therefore, using (6.G.3), we obtain the first part of the next theorem. The other part can be obtained similarly using (6.G.4).

Theorem 6.G.1. *Let* $\boldsymbol{X}$ *and* $\boldsymbol{Y}$ *be two n-dimensional random vectors. Then*

(a) $\boldsymbol{X} \leq_{\rm uo} \boldsymbol{Y}$ *if, and only if,*

$$E\left[\prod_{i=1}^{n} g_i(X_i)\right] \leq E\left[\prod_{i=1}^{n} g_i(Y_i)\right] \tag{6.G.5}$$

for every collection $\{g_1, g_2, \ldots, g_n\}$ *of univariate nonnegative increasing functions.*

(b) $\boldsymbol{X} \le_{\text{lo}} \boldsymbol{Y}$ *if, and only if,*

$$E\left[\prod_{i=1}^{n} h_i(X_i)\right] \ge E\left[\prod_{i=1}^{n} h_i(Y_i)\right] \tag{6.G.6}$$

for every collection $\{h_1, h_2, \dots, h_n\}$ *of univariate nonnegative decreasing functions.*

For a real n-variate function g, the multivariate difference operator Δ is defined by

$$\Delta_{\boldsymbol{x}}^{\boldsymbol{y}} g = \sum_{(\epsilon_1, \epsilon_2, \dots, \epsilon_n) \in \{0,1\}^n} (-1)^{\sum_{i=1}^{n} \epsilon_i} g(\epsilon_1 x_1 + (1-\epsilon_1) y_1, \dots, \epsilon_n x_n + (1-\epsilon_n) y_n),$$

where $\boldsymbol{x}$ and $\boldsymbol{y}$ are elements of $\mathbb{R}^n$. The function g is called Δ-monotone if

$$\Delta_{\boldsymbol{x}}^{\boldsymbol{y}} g \ge 0 \quad \text{whenever } \boldsymbol{x} \le \boldsymbol{y}.$$

Let M be the set of all n-variate functions that are Δ-monotone in any of their k coordinates when the other $n-k$ coordinates are held fixed, $1 \le k \le n$. It can be shown that if $\psi \in M$ and $\boldsymbol{X} \le_{\text{uo}} \boldsymbol{Y}$, then $E[\psi(\boldsymbol{X})] \le E[\psi(\boldsymbol{Y})]$. Every distribution function is a member of M. Thus we have proven the first part of the following theorem. The other part can be shown similarly.

Theorem 6.G.2. *Let* $\boldsymbol{X}$ *and* $\boldsymbol{Y}$ *be two n-dimensional random vectors. Then*

(a) $\boldsymbol{X} \le_{\text{uo}} \boldsymbol{Y}$ *if, and only if,*

$$E[\psi(\boldsymbol{X})] \le E[\psi(\boldsymbol{Y})] \quad \text{for every distribution function } \psi. \tag{6.G.7}$$

(b) $\boldsymbol{X} \le_{\text{lo}} \boldsymbol{X}$ *if, and only if,*

$$E[\psi(\boldsymbol{X})] \ge E[\psi(\boldsymbol{Y})] \quad \text{for every survival function } \psi. \tag{6.G.8}$$

It is clear, for example from Theorem 6.G.2, that

$$\boldsymbol{X} \le_{\text{st}} \boldsymbol{Y} \Longrightarrow \big(\boldsymbol{X} \le_{\text{uo}} \boldsymbol{Y} \text{ and } \boldsymbol{X} \le_{\text{lo}} \boldsymbol{Y}\big). \tag{6.G.9}$$

Let $\boldsymbol{X} = (X_1, X_2, \dots, X_n)$ and $\boldsymbol{Y} = (Y_1, Y_2, \dots, Y_n)$ be two random vectors. Note that if $\boldsymbol{X} \le_{\text{uo}} \boldsymbol{Y}$, or if $\boldsymbol{X} \le_{\text{lo}} \boldsymbol{Y}$, then $X_i \le_{\text{st}} Y_i$, $i = 1, 2, \dots, n$. It follows that

$$\boldsymbol{X} \le_{\text{uo}} \boldsymbol{Y} \Longrightarrow E\boldsymbol{X} \le E\boldsymbol{Y}, \quad \text{and}$$
$$\boldsymbol{X} \le_{\text{lo}} \boldsymbol{Y} \Longrightarrow E\boldsymbol{X} \le E\boldsymbol{Y}.$$

The following closure properties of the orthant orders can be easily verified using (6.G.1)–(6.G.4).

Theorem 6.G.3. (a) *Let* $(X_1, X_2, \ldots, X_n)$ *and* $(Y_1, Y_2, \ldots, Y_n)$ *be two* n*-dimensional random vectors. If* $(X_1, X_2, \ldots, X_n) \le_{\text{uo}} [\le_{\text{lo}}] (Y_1, Y_2, \ldots, Y_n)$*, then*

$$(g_1(X_1), g_2(X_2), \ldots, g_n(X_n)) \le_{\text{uo}} [\le_{\text{lo}}] (g_1(Y_1), g_2(Y_2), \ldots, g_n(Y_n))$$

whenever $g_i : \mathbb{R} \to \mathbb{R}$ *is an increasing function,* $i = 1, 2, \ldots, n$.

(b) *Let* $\boldsymbol{X}_1, \boldsymbol{X}_2, \ldots, \boldsymbol{X}_m$ *be a set of independent random vectors where the dimension of* $\boldsymbol{X}_i$ *is* k_i, $i = 1, 2, \ldots, m$*. Let* $\boldsymbol{Y}_1, \boldsymbol{Y}_2, \ldots, \boldsymbol{Y}_m$ *be another set of independent random vectors where the dimension of* $\boldsymbol{Y}_i$ *is* k_i, $i = 1, 2, \ldots, m$*. If* $\boldsymbol{X}_i \le_{\text{uo}} [\le_{\text{lo}}] \boldsymbol{Y}_i$ *for* $i = 1, 2, \ldots, m$*, then*

$$(\boldsymbol{X}_1, \boldsymbol{X}_2, \ldots, \boldsymbol{X}_m) \le_{\text{uo}} [\le_{\text{lo}}] (\boldsymbol{Y}_1, \boldsymbol{Y}_2, \ldots, \boldsymbol{Y}_m).$$

That is, the orthant orders are closed under conjunctions.

(c) *Let* $\boldsymbol{X} = (X_1, X_2, \ldots, X_n)$ *and* $\boldsymbol{Y} = (Y_1, Y_2, \ldots, Y_n)$ *be two* n*-dimensional random vectors. If* $\boldsymbol{X} \le_{\text{uo}} [\le_{\text{lo}}] \boldsymbol{Y}$*, then* $\boldsymbol{X}_I \le_{\text{uo}} [\le_{\text{lo}}] \boldsymbol{Y}_I$ *for each* $I \subseteq \{1, 2, \ldots, n\}$*. That is, the orthant orders are closed under marginalization.*

(d) *Let* $\{\boldsymbol{X}_j, j = 1, 2, \ldots\}$ *and* $\{\boldsymbol{Y}_j, j = 1, 2, \ldots\}$ *be two sequences of random vectors such that* $\boldsymbol{X}_j \to_{\text{st}} \boldsymbol{X}$ *and* $\boldsymbol{Y}_j \to_{\text{st}} \boldsymbol{Y}$ *as* $j \to \infty$*, where* $\to_{\text{st}}$ *denotes convergence in distribution. If* $\boldsymbol{X}_j \le_{\text{uo}} [\le_{\text{lo}}] \boldsymbol{Y}_j$, $j = 1, 2, \ldots$*, then* $\boldsymbol{X} \le_{\text{uo}} [\le_{\text{lo}}] \boldsymbol{Y}$.

(e) *Let* $\boldsymbol{X}$, $\boldsymbol{Y}$*, and* $\boldsymbol{\Theta}$ *be random vectors such that* $[\boldsymbol{X} | \boldsymbol{\Theta} = \boldsymbol{\theta}] \le_{\text{uo}} [\le_{\text{lo}}] [\boldsymbol{Y} | \boldsymbol{\Theta} = \boldsymbol{\theta}]$ *for all* $\boldsymbol{\theta}$ *in the support of* $\boldsymbol{\Theta}$*. Then* $\boldsymbol{X} \le_{\text{uo}} [\le_{\text{lo}}] \boldsymbol{Y}$*. That is, the orthant orders are closed under mixtures.*

From parts (a) and (e) of Theorem 6.G.3 we obtain the following corollary.

Corollary 6.G.4. *Let* $\boldsymbol{X} = (X_1, X_2, \ldots, X_n)$ *and* $\boldsymbol{Y} = (Y_1, Y_2, \ldots, Y_n)$ *be two random vectors such that* $\boldsymbol{X} \le_{\text{uo}} [\le_{\text{lo}}] \boldsymbol{Y}$*, and let* $\boldsymbol{Z}$ *be an* m*-dimensional random vector which is independent of* $\boldsymbol{X}$ *and* $\boldsymbol{Y}$*. Then*

$$\begin{aligned}(h_1(X_1, \boldsymbol{Z}), h_2(X_2, \boldsymbol{Z}), \ldots, h_n(X_n, \boldsymbol{Z}))& \\ \le_{\text{uo}} [\le_{\text{lo}}] (h_1(Y_1, \boldsymbol{Z}), h_2(Y_2, \boldsymbol{Z}), \ldots, h_n(Y_n, \boldsymbol{Z})),&\end{aligned}$$

whenever $h_i(x, \boldsymbol{z})$, $i = 1, 2, \ldots, n$*, are increasing in* x *for every* $\boldsymbol{z}$.

By applying Corollary 6.G.4 twice (letting $\boldsymbol{Z}$ there be an n-dimensional random vector, and letting each h_i depend only on its first argument and on the ith component of the second argument, $i = 1, 2, \ldots, n$), we get the following result. A strengthening of the following result is Theorem 6.G.18 below.

Theorem 6.G.5. *Let* $\boldsymbol{X}$, $\boldsymbol{Y}$, $\boldsymbol{Z}$*, and* $\boldsymbol{W}$ *be* n*-dimensional random vectors such that* $\boldsymbol{X}$ *and* $\boldsymbol{Z}$ *are independent and* $\boldsymbol{Y}$ *and* $\boldsymbol{W}$ *are independent. Let* $c_i : [0, \infty)^2 \to [0, \infty)$ *be a continuous increasing function,* $i = 1, 2, \ldots, n$*. If* $\boldsymbol{X} \le_{\text{uo}} [\le_{\text{lo}}] \boldsymbol{Y}$ *and* $\boldsymbol{Z} \le_{\text{uo}} [\le_{\text{lo}}] \boldsymbol{W}$*, then*

$$\begin{aligned}(c_1(X_1, Z_1), c_2(X_2, Z_2), \ldots, c_n(X_n, Z_n))& \\ \le_{\text{uo}} [\le_{\text{lo}}] (c_1(Y_1, W_1), c_2(Y_2, W_2), \ldots, c_n(Y_n, W_n)).&\end{aligned}$$

Example 6.G.6. Consider an n-dimensional Markov chain $\{\boldsymbol{X}_k = (X_{k,1},\dots, X_{k,n}),\ k \geq 0\}$ defined by $\boldsymbol{X}_0 = (0,\dots,0)$ and

$$\boldsymbol{X}_{k+1} = (g_1(X_{k,1}, U^1_{k,1},\dots,U^m_{k,1}),\dots, g_n(X_{k,n}, U^1_{k,n},\dots,U^m_{k,n}), \quad n \geq 1,$$

where, for each $1 \leq l \leq m$, the random vectors $\boldsymbol{U}^l_k = (U^l_{k,1},\dots,U^l_{k,n})$, $k = 1,2,\dots$, are independent and identically distributed, and the g_i's are some deterministic $(m+1)$-dimensional functions. Consider another n-dimensional Markov chain $\{\boldsymbol{Y}_k = (Y_{k,1},\dots,Y_{k,n}),\ k \geq 0\}$ similarly defined by $\boldsymbol{Y}_0 = (0,\dots,0)$ and

$$\boldsymbol{Y}_{k+1} = (g_1(Y_{k,1}, V^1_{k,1},\dots,V^m_{k,1}),\dots, g_n(Y_{k,n}, V^1_{k,n},\dots,V^m_{k,n}), \quad n \geq 1,$$

where, for each $1 \leq l \leq m$, the random vectors $\boldsymbol{V}^l_k = (V^l_{k,1},\dots,V^l_{k,n})$, $k = 1,2,\dots$, are independent and identically distributed. If the g_i's are increasing in their $m+1$ arguments, if $\boldsymbol{U}^l = \{\boldsymbol{U}^l_k,\ k \geq 0\}$, $l = 1,\dots,m$, are independent, if $\boldsymbol{V}^l = \{\boldsymbol{V}^l_k,\ k \geq 0\}$, $l = 1,\dots,m$, are independent, and if $\boldsymbol{U}^l_k \leq_{\rm uo} [\leq_{\rm lo}]\ \boldsymbol{V}^l_k$, $l = 1,\dots,m$, $k \geq 0$, then, for each $k \geq 0$ we have

$$(\boldsymbol{X}_0,\dots,\boldsymbol{X}_k) \leq_{\rm uo} [\leq_{\rm lo}]\ (\boldsymbol{Y}_0,\dots,\boldsymbol{Y}_k).$$

The proof uses Theorem 6.G.5, Corollary 6.G.4, and Theorem 6.G.3(b). We omit the details.

Another preservation property of the orthant orders is described in the next theorem. In the following theorem we define $\sum_{j=1}^{0} x_j \equiv 0$ for any sequence $\{x_j,\ j = 1,2,\dots\}$. Similar results are Theorems 9.A.6 and 9.A.14.

Theorem 6.G.7. *Let $\boldsymbol{X}_j = (X_{j,1}, X_{j,2},\dots,X_{j,m})$, $j = 1,2,\dots$, be a sequence of nonnegative random vectors, and let $\boldsymbol{M} = (M_1, M_2,\dots,M_m)$ and $\boldsymbol{N} = (N_1,N_2,\dots,N_m)$ be two vectors of nonnegative integer-valued random variables. Assume that both $\boldsymbol{M}$ and $\boldsymbol{N}$ are independent of the $\boldsymbol{X}_j$'s. If $\boldsymbol{M} \leq_{\rm uo} [\leq_{\rm lo}]\ \boldsymbol{N}$, then*

$$\Big(\sum_{j=1}^{M_1} X_{j,1}, \sum_{j=1}^{M_2} X_{j,2},\dots,\sum_{j=1}^{M_m} X_{j,m}\Big) \leq_{\rm uo} [\leq_{\rm lo}] \Big(\sum_{j=1}^{N_1} X_{j,1}, \sum_{j=1}^{N_2} X_{j,2},\dots,\sum_{j=1}^{N_m} X_{j,m}\Big).$$

Proof. We only give the proof for the upper orthant order; the proof for the lower orthant order is similar. For $\boldsymbol{t} = (t_1, t_2,\dots,t_m)$ we have

$$\begin{aligned} &P\Big\{\bigcap_{i=1}^{m}\Big\{\sum_{j=1}^{M_i} X_{j,i} > t_i\Big\}\Big\} \\ &\quad = \sum_{n_1=0}^{\infty}\sum_{n_2=0}^{\infty}\cdots\sum_{n_m=0}^{\infty} P\Big\{\bigcap_{i=1}^{m}\Big\{\sum_{j=1}^{n_i} X_{j,i} \leq t_i < \sum_{j=1}^{n_i+1} X_{j,i}\Big\}\Big\} \\ &\qquad\qquad \times P\{\boldsymbol{M} > (n_1,n_2,\dots,n_m)\}\end{aligned}$$

$$\leq \sum_{n_1=0}^{\infty} \sum_{n_2=0}^{\infty} \cdots \sum_{n_m=0}^{\infty} P\left\{ \bigcap_{i=1}^{m} \left\{ \sum_{j=1}^{n_i} X_{j,i} \leq t_i < \sum_{j=1}^{n_i+1} X_{j,i} \right\} \right\}$$

$$\times P\{\boldsymbol{N} > (n_1, n_2, \ldots, n_m)\}$$

$$= P\left\{ \bigcap_{i=1}^{m} \left\{ \sum_{j=1}^{N_i} X_{j,i} > t_i \right\} \right\}. \qquad \square$$

Consider now, as in Section 6.B.4, n families of univariate distribution functions $\{G_\theta^{(i)},\ \theta \in \mathcal{X}_i\}$ where $\mathcal{X}_i$ is a subset of the real line $\mathbb{R}$, $i = 1, 2, \ldots, n$. Let $X_i(\theta)$ denote a random variable with distribution function $G_\theta^{(i)}$, $i = 1, 2, \ldots, n$. Below we give a result which provides comparisons of two random vectors, with distribution functions of the form (6.B.18), in the upper and lower orthant orders. The following result is a generalization of Theorem 6.G.3(e), and is a multivariate extension of Theorem 1.A.6 (an extension that is different than Theorem 6.B.18); see Theorems 7.A.37, 9.A.7, and 9.A.15 for related results.

Theorem 6.G.8. *Let $\{G_\theta^{(i)},\ \theta \in \mathcal{X}_i\}$, $i = 1, 2, \ldots, n$, be n families of univariate distribution functions as above. Let $\boldsymbol{\Theta}_1$ and $\boldsymbol{\Theta}_2$ be two random vectors with supports in $\prod_{i=1}^{n} \mathcal{X}_i$ and distribution functions F_1 and F_2, respectively. Let $\boldsymbol{Y}_1$ and $\boldsymbol{Y}_2$ be two random vectors with distribution functions H_1 and H_2 given by*

$$H_j(y_1, y_2, \ldots, y_n) = \int_{\mathcal{X}_1} \int_{\mathcal{X}_2} \cdots \int_{\mathcal{X}_n} \prod_{i=1}^{n} G_{\theta_i}^{(i)}(y_i) \mathrm{d}F_j(\theta_1, \theta_2, \ldots, \theta_n),$$

$$(y_1, y_2, \ldots, y_n) \in \mathbb{R}^n, \ j = 1, 2.$$

If

$$X_i(\theta) \leq_{\mathrm{st}} X_i(\theta') \text{ whenever } \theta \leq \theta', \quad i = 1, 2, \ldots, n,$$

and if

$$\boldsymbol{\Theta}_1 \leq_{\mathrm{uo}} [\leq_{\mathrm{lo}}]\ \boldsymbol{\Theta}_2,$$

then

$$\boldsymbol{Y}_1 \leq_{\mathrm{uo}} [\leq_{\mathrm{lo}}]\ \boldsymbol{Y}_2.$$

An interesting relationship between the orders $\leq_{\mathrm{uo}}$ and $\leq_{\mathrm{whr}}$ is stated next; its proof follows easily from (6.D.3). Note that the following result is a multivariate generalization of (1.B.7).

Theorem 6.G.9. *Let $\boldsymbol{X}$ and $\boldsymbol{Y}$ be two n-dimensional random vectors. Then $\boldsymbol{X} \leq_{\mathrm{whr}} \boldsymbol{Y}$ if, and only if,*

$$[\boldsymbol{X} \mid \boldsymbol{X} > \boldsymbol{x}] \leq_{\mathrm{uo}} [\boldsymbol{Y} \mid \boldsymbol{Y} > \boldsymbol{x}] \quad \text{for all } \boldsymbol{x} \in \mathbb{R}^n,$$

for which these conditional random vectors are well defined.

It follows from Theorem 6.G.9 that

$$\boldsymbol{X} \leq_{\rm whr} \boldsymbol{Y} \Longrightarrow \boldsymbol{X} \leq_{\rm uo} \boldsymbol{Y}; \tag{6.G.10}$$

this is a multivariate generalization of Theorem 1.B.1.

An interesting relationship between the order $\leq_{\rm uo}$ and the orders $\leq_{m\text{-cx}}^{\mathcal{S}}$ and $\leq_{m\text{-icx}}$ (defined in Sections 3.A.5 and 4.A.7, respectively) is given in the next theorem.

Theorem 6.G.10. *Let $\boldsymbol{X} = (X_1, X_2, \ldots, X_m)$ and $\boldsymbol{Y} = (Y_1, Y_2, \ldots, Y_m)$ be random vectors such that the $(m-1)$st moment exists for each X_i and Y_i, $i = 1, 2, \ldots, m$.*

(a) *If $\boldsymbol{X} \leq_{\rm uo} \boldsymbol{Y}$, and if $E\big(\sum_{i=1}^m X_i\big)^k = E\big(\sum_{i=1}^m Y_i\big)^k$, $k = 1, 2, \ldots, m-1$, then $\sum_{i=1}^m X_i \leq_{m\text{-cx}}^{\mathcal{S}} \sum_{i=1}^m Y_i$, where $\mathcal{S}$ is the assumed common support of $\sum_{i=1}^m X_i$ and of $\sum_{i=1}^m Y_i$, and $\mathcal{S}$ is also assumed to be an interval.*
(b) *If $\boldsymbol{X} \leq_{\rm uo} \boldsymbol{Y}$, and if $\boldsymbol{X}$ and $\boldsymbol{Y}$ are nonnegative, then $\sum_{i=1}^m X_i \leq_{m\text{-icx}} \sum_{i=1}^m Y_i$.*

It is of interest to compare Theorem 6.G.10 with Theorem 7.A.30 and with implication (9.A.19).

The following example gives sufficient conditions for the comparison of multivariate normal random vectors. See Examples 6.B.29, 7.A.13, 7.A.26, 7.A.39, 7.B.5, and 9.A.20 for related results.

Example 6.G.11. Let $\boldsymbol{X}$ be a multivariate normal random vector with mean vector $\boldsymbol{\mu_X}$ and variance-covariance matrix $\boldsymbol{\Sigma}$, and let $\boldsymbol{Y}$ be a multivariate normal random vector with mean vector $\boldsymbol{\mu_Y}$ and variance-covariance matrix $\boldsymbol{\Sigma} + \boldsymbol{D}$, where $\boldsymbol{D}$ is a matrix with zero diagonal elements such that $\boldsymbol{\Sigma} + \boldsymbol{D}$ is nonnegative definite. If $\boldsymbol{\mu_x} \leq \boldsymbol{\mu_Y}$ and $\boldsymbol{D} \geq \boldsymbol{0}$, then $\boldsymbol{X} \leq_{\rm uo} \boldsymbol{Y}$.

The following results give conditions that ensure stochastic equality; see Theorems 1.A.8, 3.A.43, 3.A.60, 4.A.69, 5.A.15, 6.B.19, and 7.A.14–7.A.16 for similar results.

First, in the bivariate case ($n = 2$) we have the following result; its proof is not given here since it is a special case of Theorem 6.G.13.

Theorem 6.G.12. *Let $\boldsymbol{X} = (X_1, X_2)$ and $\boldsymbol{Y} = (Y_1, Y_2)$ be two bivariate random vectors. If $X_1 =_{\rm st} Y_1$, $X_2 =_{\rm st} Y_2$, $\boldsymbol{X} \leq_{\rm uo} \boldsymbol{Y}$, and $\boldsymbol{X} \leq_{\rm lo} \boldsymbol{Y}$, then $\boldsymbol{X} =_{\rm st} \boldsymbol{Y}$.*

Note that when $n = 2$, Theorem 6.B.19 is a special case of Theorem 6.G.12, as can be seen from (6.G.9).

If $n \geq 3$, then the conclusion of Theorem 6.G.12 need not hold. The following theorem gives conditions under which the conclusion $\boldsymbol{X} =_{\rm st} \boldsymbol{Y}$ holds.

Theorem 6.G.13. *Let $\boldsymbol{X} = (X_1, X_2, \ldots, X_n)$ and $\boldsymbol{Y} = (Y_1, Y_2, \ldots, Y_n)$ be two random vectors with distributions and survival functions F, $\overline{F}$, G, and $\overline{G}$,* *respectively. If the m-dimensional marginals of* $\boldsymbol{X}$ *and* $\boldsymbol{Y}$ *are equal* ($m \le n-1$) *and if* $\boldsymbol{X} \le_{\text{uo}} \boldsymbol{Y}$, *that is,*

$$\overline{F}(\boldsymbol{x}) \le \overline{G}(\boldsymbol{x}) \quad \textit{for all } \boldsymbol{x} \in \mathbb{R}^n, \tag{6.G.11}$$

and if

$$(-1)^n F(\boldsymbol{x}) \ge (-1)^n G(\boldsymbol{x}) \quad \textit{for all } \boldsymbol{x} \in \mathbb{R}^n, \tag{6.G.12}$$

then $\boldsymbol{X} =_{\text{st}} \boldsymbol{Y}$.

Proof. Write

$$\begin{aligned}
\overline{F}(\boldsymbol{x}) &= 1 - \sum_i P\{X_i \le x_i\} + \sum_{i \ne j} P\{X_i \le x_i, X_j \le x_j\} - \cdots + (-1)^n F(\boldsymbol{x}) \\
&\ge 1 - \sum_i P\{Y_i \le x_i\} + \sum_{i \ne j} P\{Y_i \le x_i, Y_j \le x_j\} - \cdots + (-1)^n G(\boldsymbol{x}) \\
&= \overline{G}(\boldsymbol{x}), \quad \boldsymbol{x} \in \mathbb{R}^n,
\end{aligned}$$

where the equality of the m-dimensional marginals and also assumption (6.G.12) were used. Thus we get that for each $\boldsymbol{x} \in \mathbb{R}^n$, $\overline{F}(\boldsymbol{x}) \ge \overline{G}(\boldsymbol{x})$. This, together with (6.G.11), yields the stated result. □

An interesting relationship between the orders $\le_{\text{lo}}$ and $\le_{\text{Lt}}$ (see Section 5.A) is revealed in the following theorem.

Theorem 6.G.14. *Let* $\boldsymbol{X}$ *and* $\boldsymbol{Y}$ *be two nonnegative random vectors. If* $(X_1, X_2, \ldots, X_n) \le_{\text{lo}} (Y_1, Y_2, \ldots, Y_n)$, *then*

$$\sum_{i=1}^n a_i X_i \le_{\text{Lt}} \sum_{i=1}^n a_i Y_i \quad \textit{whenever } a_i \ge 0,\ i = 1, 2, \ldots, n.$$

Proof. Select an $s \ge 0$ and $a_i \ge 0$, $i = 1, 2, \ldots, n$. The function g_i defined by $g_i(x) = \exp\{-a_i s x\}$ is decreasing and nonnegative. Therefore, from (6.G.6), we obtain that

$$E\Big[\exp\Big(-s\sum_{i=1}^n a_i X_i\Big)\Big] \ge E\Big[\exp\Big(-s\sum_{i=1}^n a_i Y_i\Big)\Big] \quad \text{for all } s \ge 0,$$

and this yields the stated result. □

6.G.2 The scaled order statistics orders

Consider now nonnegative random vectors $\boldsymbol{X} = (X_1, X_2, \ldots, X_n)$ and $\boldsymbol{Y} = (Y_1, Y_2, \ldots, Y_n)$. For any $\boldsymbol{z} = (z_1, z_2, \ldots, z_n)$ denote by $\boldsymbol{z}_{(k)} = (z_1, z_2, \ldots, z_n)_{(k)}$ the kth smallest z_i in $\{z_1, z_2, \ldots, z_n\}$. Thus, for a random vector $\boldsymbol{Z} = (Z_1, Z_2, \ldots, Z_n)$, the kth order statistic of $Z_1, Z_2, \ldots, Z_n$ is $\boldsymbol{Z}_{(k)} = (Z_1, Z_2, \ldots, Z_n)_{(k)}$. In particular, $\boldsymbol{Z}_{(1)} = \min\{Z_1, Z_2, \ldots, Z_n\}$ and $\boldsymbol{Z}_{(n)} = \max\{Z_1, Z_2, \ldots, Z_n\}$. The next result describes the orders $\le_{\text{uo}}$ and $\le_{\text{lo}}$ in a new fashion when the underlying random vectors are nonnegative (see Theorem 6.D.7 for a related result).

Theorem 6.G.15. *Let $\boldsymbol{X} = (X_1, X_2, \ldots, X_n)$ and $\boldsymbol{Y} = (Y_1, Y_2, \ldots, Y_n)$ be two nonnegative random vectors. Then*

(a) $\boldsymbol{X} \le_{\rm uo} \boldsymbol{Y}$ *if, and only if,*

$$\min\{a_1X_1, \ldots, a_nX_n\} \le_{\rm st} \min\{a_1Y_1, \ldots, a_nY_n\} \tag{6.G.13}$$

whenever $a_i > 0$, $i = 1, 2, \ldots, n$.

(b) $\boldsymbol{X} \le_{\rm lo} \boldsymbol{Y}$ *if, and only if,*

$$\max\{a_1X_1, \ldots, a_nX_n\} \le_{\rm st} \max\{a_1Y_1, \ldots, a_nY_n\} \tag{6.G.14}$$

whenever $a_i > 0$, $i = 1, 2, \ldots, n$.

Proof. Condition (6.G.13) is the same as

$$\overline{F}(\frac{t}{a_1}, \frac{t}{a_2}, \ldots, \frac{t}{a_n}) \le \overline{G}(\frac{t}{a_1}, \frac{t}{a_2}, \ldots, \frac{t}{a_n})$$

whenever $t \ge 0$, $a_i > 0$, $i = 1, 2, \ldots, n$, which is the same as

$$\overline{F}(t_1, t_2, \ldots, t_n) \le \overline{G}(t_1, t_2, \ldots, t_n) \tag{6.G.15}$$

whenever $t_i > 0$, $i = 1, 2, \ldots, n$. Using standard limiting arguments it is seen that (6.G.15) is the same as $\boldsymbol{X} \le_{\rm uo} \boldsymbol{Y}$. This proves (a). The proof of (b) is similar. $\square$

Theorem 6.G.15 suggests the following class of orders which contains the orders $\le_{\rm uo}$ and $\le_{\rm lo}$ as special cases. Let $\boldsymbol{X} = (X_1, X_2, \ldots, X_n)$ and $\boldsymbol{Y} = (Y_1, Y_2, \ldots, Y_n)$ be two nonnegative random vectors. Suppose that

$$(a_1X_1, a_2X_2, \ldots, a_nX_n)_{(k)} \le_{\rm st} (a_1Y_1, a_2Y_2, \ldots, a_nY_n)_{(k)} \tag{6.G.16}$$

whenever $a_i > 0$, $i = 1, 2, \ldots, n$. Then we say that $\boldsymbol{X}$ is *smaller than* $\boldsymbol{Y}$ *in the kth scaled order statistic order* (denoted by $\boldsymbol{X} \le_{(k)} \boldsymbol{Y}$), $k = 1, 2, \ldots, n$. So $\boldsymbol{X} \le_{\rm uo} \boldsymbol{Y} \Longleftrightarrow \boldsymbol{X} \le_{(1)} \boldsymbol{Y}$ and $\boldsymbol{X} \le_{\rm lo} \boldsymbol{Y} \Longleftrightarrow \boldsymbol{X} \le_{(n)} \boldsymbol{Y}$.

The next theorem identifies a rich class of functions ψ such that $E[\psi(\boldsymbol{X})] \le E[\psi(\boldsymbol{Y})]$ whenever $\boldsymbol{X} \le_{(k)} \boldsymbol{Y}$. First we need to introduce some notation. For $m \in \{1, 2, \ldots, n\}$ let A_m be the set of all subsets of $\{1, 2, \ldots, n\}$ of size m. As in Section 6.A, for $I = \{i_1, i_2, \ldots, i_m\} \in A_m$ and a vector $\boldsymbol{x} = (x_1, x_2, \ldots, x_n)$, we denote

$$\boldsymbol{x}_I = (x_{i_1}, x_{i_2}, \ldots, x_{i_m}).$$

Let $M_{1,n}$ denote the class of all distribution functions corresponding to nonnegative finite measures on $\mathbb{R}_+^n$. For $\boldsymbol{x} \in \mathbb{R}_+^n$, $I \in A_m$, and $\psi \in M_{1,n}$, we denote

$$\tilde{\psi}(\boldsymbol{x}_I, \infty \boldsymbol{e}) = \lim_{\boldsymbol{x}_{\overline{I}} \to \infty \boldsymbol{e}} \psi(x_1, x_2, \ldots, x_n).$$

For $k \in \{1, 2, \ldots, n\}$ let $M_{k,n}$ be the class of functions $\phi : \mathbb{R}_+^n \to \mathbb{R}$ of the form

$$\phi(x_1, x_2, \ldots, x_n) = \sum_{m=n-k+1}^{n} (-1)^{m-n+k-1} \binom{m-1}{n-k} \sum_{I \in A_m} \tilde{\psi}(\boldsymbol{x}_I, \infty \boldsymbol{e}),$$

for some $\psi \in M_{1,n}$, where $\sum_{I \in A_m}$ denotes the sum over all the $\binom{n}{m}$ elements of A_m. Note that for $k = 1$ the two definitions of $M_{1,n}$ coincide. The proof of the next result is not given here; it can be found elsewhere.

Theorem 6.G.16. *Let* $\boldsymbol{X} = (X_1, X_2, \ldots, X_n)$ *and* $\boldsymbol{Y} = (Y_1, Y_2, \ldots, Y_n)$ *be two nonnegative random vectors. Then* $\boldsymbol{X} \le_{(k)} \boldsymbol{Y}$ *if, and only if,*

$$E[\phi(\boldsymbol{X})] \le E[\phi(\boldsymbol{Y})]$$

for every $\phi \in M_{k,n}$ *for which the expectations exist.*

Note that both parts of Theorem 6.G.2 are special cases of Theorem 6.G.16.

The orders $\le_{(k)}$ are closed under general monotone increasing transformations as the following theorem shows. The proof is easy and is omitted.

Theorem 6.G.17. *Let* $\boldsymbol{X} = (X_1, X_2, \ldots, X_n)$ *and* $\boldsymbol{Y} = (Y_1, Y_2, \ldots, Y_n)$ *be two nonnegative random vectors. Let* $b_i : \mathbb{R}_+ \to \mathbb{R}_+$ *be a right continuous increasing function,* $i = 1, 2, \ldots, n$. *If* $\boldsymbol{X} \le_{(k)} \boldsymbol{Y}$, *then*

$$\big(b_1(X_1), b_2(X_2), \ldots, b_n(X_n)\big) \le_{(k)} \big(b_1(Y_1), b_2(Y_2), \ldots, b_n(Y_n)\big).$$

The orders $\le_{(k)}$ also satisfy the following general closure property, the proof of which can be found elsewhere and is omitted.

Theorem 6.G.18. *Let* $\boldsymbol{X}$, $\boldsymbol{Y}$, $\boldsymbol{Z}$, *and* $\boldsymbol{W}$ *be* n*-dimensional nonnegative random vectors such that* $\boldsymbol{X}$ *and* $\boldsymbol{Z}$ *are independent, and* $\boldsymbol{Y}$ *and* $\boldsymbol{W}$ *are independent. Let* $c_i : \mathbb{R}_+^2 \to \mathbb{R}_+$ *be a right continuous increasing function,* $i = 1, 2, \ldots, n$. *If* $\boldsymbol{X} \le_{(k)} \boldsymbol{Y}$ *and* $\boldsymbol{Z} \le_{(k)} \boldsymbol{W}$, *then*

$$\begin{aligned}\big(c_1(X_1, Z_1), c_2(X_2, Z_2), \ldots, c_n(X_n, Z_n)\big)& \\ \le_{(k)} \big(c_1(Y_1, W_1), c_2(Y_2, W_2), \ldots, c_n(Y_n, W_n)\big)&.\end{aligned}$$

From Theorem 6.G.18 we obtain the following two results as corollaries.

Theorem 6.G.19. *Let* $\boldsymbol{X}$, $\boldsymbol{Y}$, $\boldsymbol{Z}$, *and* $\boldsymbol{W}$ *be* n*-dimensional nonnegative random vectors such that* $\boldsymbol{X}$ *and* $\boldsymbol{Z}$ *are independent and* $\boldsymbol{Y}$ *and* $\boldsymbol{W}$ *are independent. If* $\boldsymbol{X} \le_{(k)} \boldsymbol{Y}$ *and* $\boldsymbol{Z} \le_{(k)} \boldsymbol{W}$, *then*

$$\boldsymbol{X} + \boldsymbol{Z} \le_{(k)} \boldsymbol{Y} + \boldsymbol{W};$$

that is, the orders $\le_{(k)}$ *are closed under convolutions.*

Theorem 6.G.20. *Let $\boldsymbol{X}$, $\boldsymbol{Y}$, $\boldsymbol{Z}$, and $\boldsymbol{W}$ be n-dimensional nonnegative random vectors such that $\boldsymbol{X}$ and $\boldsymbol{Z}$ are independent and $\boldsymbol{Y}$ and $\boldsymbol{W}$ are independent. If $\boldsymbol{X} \leq_{(k)} \boldsymbol{Y}$ and $\boldsymbol{Z} \leq_{(k)} \boldsymbol{W}$, then*

$$(\min(X_1, Z_1), \min(X_2, Z_2), \dots, \min(X_n, Z_n)) \\ \leq_{(k)} (\min(Y_1, W_1), \min(Y_2, W_2), \dots, \min(Y_n, W_n))$$

and

$$(\max(X_1, Z_1), \max(X_2, Z_2), \dots, \max(X_n, Z_n)) \\ \leq_{(k)} (\max(Y_1, W_1), \max(Y_2, W_2), \dots, \max(Y_n, W_n)).$$

The next result states a closure under marginalization property. In its statement $\boldsymbol{X}^{(i)}$ denotes $(X_1, \dots, X_{i-1}, X_{i+1}, \dots, X_n)$ and $\boldsymbol{Y}^{(i)}$ denotes $(Y_1, \dots, Y_{i-1}, Y_{i+1}, \dots, Y_n)$, $i = 1, 2, \dots, n$.

Theorem 6.G.21. *Let $\boldsymbol{X} = (X_1, X_2, \dots, X_n)$ and $\boldsymbol{Y} = (Y_1, Y_2, \dots, Y_n)$ be two nonnegative random vectors. Suppose that $\boldsymbol{X} \leq_{(k)} \boldsymbol{Y}$.*

(a) *If $1 < k \leq n$, then*

$$\boldsymbol{X}^{(i)} \leq_{(k-1)} \boldsymbol{Y}^{(i)}.$$

(b) *If $\boldsymbol{X}$ and $\boldsymbol{Y}$ are positive with probability one and if $1 \leq k \leq n-1$, then*

$$\boldsymbol{X}^{(i)} \leq_{(k)} \boldsymbol{Y}^{(i)}.$$

It is clear from (6.G.16) that

$$\boldsymbol{X} \leq_{\text{st}} \boldsymbol{Y} \Longrightarrow \boldsymbol{X} \leq_{(k)} \boldsymbol{Y}.$$

Let $\boldsymbol{X} = (X_1, X_2, \dots, X_n)$ and $\boldsymbol{Y} = (Y_1, Y_2, \dots, Y_n)$ be two random nonnegative vectors. By letting $k-1$ of the a_i's in (6.G.16) go to 0, and by letting $n-k$ of the other a_i's be ∞, it is seen that if $\boldsymbol{X} \leq_{(k)} \boldsymbol{Y}$, then $X_i \leq_{\text{st}} Y_i$, $i = 1, 2, \dots, n$ (this fact can also be obtained from Theorem 6.G.21). It follows that

$$\boldsymbol{X} \leq_{(k)} \boldsymbol{Y} \Longrightarrow E\boldsymbol{X} \leq E\boldsymbol{Y}.$$

6.H Complements

Section 6.B: Many of the results described in Section 6.B can be found, or are alluded to, in Marshall and Olkin [383]. For example, the result given in Theorem 6.B.2 can be found there. Some studies of so called *integral stochastic orders*, which have as their starting point relations such as (6.B.4) or (6.G.7), can be found in Marshall [382], in Mosler and Scarsini [400], in Müller [408], and in Dubra, Maccheroni, and Ok [172]. A proof of fact that the usual stochastic order is equivalent to an almost sure construction (Theorem 6.B.1) can be found in Kamae, Krengel, and O'Brien [272], where this result is obtained for spaces that are more general than $\mathbb{R}^n$. Theorem 6.B.3 was obtained originally in Veinott [556], but various versions of it appear elsewhere and it is often rediscovered; Shanthikumar [527] has identified a condition that is weaker than (6.B.8)–(6.B.10) and which still implies $\boldsymbol{X} \leq_{\rm st} \boldsymbol{Y}$. A standard reference for notions of positive dependence such as association and CIS is Barlow and Proschan [36]. The condition under which CIS random vectors are stochastically ordered (Theorem 6.B.4) can be found, for example, in Langberg [332]. An extension of Theorem 6.B.4 can be found in Shanthikumar [527]. The notation $\leq_{\rm sst}$ and the result in Remark 6.B.5 are taken from Li, Scarsini, and Shaked [348]. The characterization of the CIS notion for bivariate distribution functions with uniform$[0,1]$ margins (Remark 6.B.6) is taken from Nelsen [431, Corollary 5.2.11], where this result is derived in the context of copulas. The notion of positive dependence WCIS is introduced in Cohen and Sackrowitz [131], from which Theorem 6.B.7 is taken. The fact that association, together with the monotonicity of the ratio of the densities, implies the multivariate usual stochastic order (Theorem 6.B.8), is essentially proved in Proposition 2.6 of Perlman and Olkin [457]. The stochastic monotonicity of a random vector conditioned on the sum of its elements (Theorem 6.B.9) is taken from Efron [181], who credited it to Karlin; extensions of it can be found in Shanthikumar [527] as well as in Efron [181]. This theorem is put into the context of queuing theory in Daduna and Szekli [137]. The result which gives conditions, by means of the univariate down shifted likelihood ratio order, under which a random vector is stochastically increasing in its given sum (Theorem 6.B.10) can be found in Liggett [360]. The results that involve the stochastic monotonicity of a random vector conditioned on some of its order statistics (Theorems 6.B.11 and 6.B.12) are taken from Block, Bueno, Savits, and Shaked [91] and from Shanthikumar [527]; related results can be found in Bueno [113] and in Joag-Dev [257]. The stochastic monotonicity of the order statistics, of heterogeneous exponential random variables, in the first order statistic (Theorem 6.B.13), is a strengthening of a result of Kochar and Korwar [314]; its conclusion also holds if it is merely assumed that $X_1, X_2, \ldots, X_m$ have proportional hazard functions (rather than having exponential distributions). The stochastic comparison of random vectors with a common copula (Theorem 6.B.14) can be found in Scarsini [491]; an extension of it is given in Li, Scarsini, and Shaked [348]. The result on the comparison of the vector of partial sums (Theorem 6.B.15) is taken from Boland, Proschan, and Tong [100], where the counterexample, mentioned after the theorem, can also be found. Some extensions of this result are given in Shaked, Shanthikumar, and Tong [519]. The result which compares random sums (Theorem 6.B.17) is taken from Pellerey [451], whereas the comparison of mixtures result (Theorem 6.B.18) is taken from Denuit and Müller [157]. The conditions for stochastic equality (Theorem 6.B.19) can be found in Baccelli and Makowski [27]. The proof of Theorem 6.B.19 that is given in Section 6.B.5 follows the ideas of Scarsini and Shaked [494]. Lemma 2.1 of Costantini and Pasqualucci [135] is an interesting variation of Theorem 6.B.19. The characterizations of the usual stochastic order given in Theorems 6.B.20 and 6.B.21 are taken from Scarsini and Shaked [495]. The comparisons of order statistics, given in Theorem 6.B.23 and Corollary 6.B.24, can be found in Mi and Shaked [395]. These comparisons extend some results of Nanda and Shaked [428] and of Belzunce, Franco, Ruiz, and Ruiz [66, Corollary 3.2]; see a related result in Belzunce, Mercader, and Ruiz [70]. The result that is given in Example 6.B.25 is stated in Bartoszewicz [39], but without a detailed proof; an extension of it is given in Belzunce, Mercader, and Ruiz [70]. The usual stochastic order of vectors of order statistics of Gamma and Weibull random variables with different scale parameters (Examples 6.B.26 and 6.B.27) are taken from Hu [229] and from Sun and Zhang [543]. Several other examples of this kind can be found in Hu [229], and a general method for identifying such examples can be found in Hu [230]. The conditions under which infinitely divisible random vectors are comparable in the usual multivariate stochastic order (Example 6.B.28) can be found in Samorodnitsky and Taqqu [487]; see also Braverman [108] who has mistakenly confused the usual stochastic order with the upper orthant order. The necessary and sufficient conditions for the comparison of multivariate normal random vectors (Example 6.B.29) can be found in Müller [413]; extensions of this result to Kotz-type distributions are given in Ding and Zhang [168]. The multivariate IFR notions described in Section 6.B.6 are taken from Shaked and Shanthikumar [512]; however, the notion corresponding to (6.B.28) is equivalent to a multivariate IFR notion of Arjas [18]. General results concerning the usual stochastic comparison of stochastic processes (that is, results that are more general than Theorems 6.B.30 and 6.B.31) can be found in Kamae, Krengel, and O'Brien [272]; see also Block, Langberg, and Savits [93] and Rolski and Szekli [474]. Versions of the results regarding the usual stochastic comparison of Markov chains (Theorems 6.B.32 and 6.B.34) can be found in Stoyan [540, Chapter 4]. The comparison of Markov chains, one of which is skip-free positive (Theorem 6.B.35), is taken from Ferreira and Pacheco [199]; they obtained stronger results than Theorem 6.B.33 although they use a different terminology than the one used in this theorem. The discussion about the stochastic orders of point processes is based on Shaked and Szekli [521] and Szekli [544], although the definition of the orders $\leq_{\rm st}$ and $\leq_{{\rm st}\text{-}\mathcal{N}}$ for point processes can be found already in Ebrahimi [176]; see related results in Schöttl [497]. Kulik and Szekli [325] extended these orders to k-variate point processes. The statements about the stochastic comparisons of the epoch and inter-epoch times of two nonhomogeneous Poisson processes (Example 6.B.41) are taken from Belzunce, Lillo, Ruiz, and Shaked [69].

Section 6.C: The development in this section follows the works of Norros [436, 437] and of Shaked and Shanthikumar [504]. A result that is similar to Theorem 6.C.1, but that gives conditions under which two point processes are stochastically ordered, can be found in Kwieciński and Szekli [328]. The fact (which is mentioned in Section 6.C.2) that the cumulative hazards of the components, by the time that they fail, are independent standard exponential random variables, follows from more general results of Aalen and Hoem [1, Section 4.5], Kurtz [326, Theorem 6.19(b)], and Jacobsen [252, Proposition 2.2.11)].

Section 6.D: The development in Sections 6.D.1 and 6.D.2 follows the work of Hu, Khaledi, and Shaked [235], although the definition of the order $\leq_{\text{whr}}$ (with a different name), and its characterization by means of the hazard gradients (Theorem 6.D.2) can be found in Jain and Nanda [253]. In Hu, Khaledi, and Shaked [235] it is claimed that (6.D.12) in Theorem 6.D.7 is equivalent to $\boldsymbol{X} \leq_{\text{whr}} \boldsymbol{Y}$, but this is erroneous, as was communicated to us by Antonio Colangelo. An order that is stronger than the order $\leq_{\text{whr}}$ is mentioned in Collet, López, and Martínez [134]. The development in Section 6.D.3 follows the work of Shaked and Shanthikumar [505]. The dynamic multivariate hazard rate order comparison of the epoch times of two nonhomogeneous Poisson processes (Example 6.D.8) is taken from Belzunce, Lillo, Ruiz, and Shaked [69]. The comparison, in the dynamic hazard rate order, of vectors of order statistics (Theorem 6.D.10), can be found in Belzunce, Ruiz, and Ruiz [75]; an extension of it is given in Belzunce, Mercader, and Ruiz [70]. The multivariate IFR notions described in Section 6.D.3 are taken from Shaked and Shanthikumar [512]; some related notion and results can be found in Bassan and Spizzichino [56].

Section 6.E: The multivariate likelihood ratio order (though using a different terminology) is studied in Karlin and Rinott [278] and in Whitt [563]. The preservation under conditioning result (Theorem 6.E.2) can be found in Rinott and Scarsini [468]. The result which shows a preservation property of the order $\leq_{\text{lr}}$ under random summations (Theorem 6.E.5) is taken from Pellerey [451]. The result about the relationship between the multivariate likelihood ratio and the multivariate hazard rate order (Theorem 6.E.6) is taken from Hu, Khaledi, and Shaked [235]. The relationship between the multivariate likelihood ratio and the dynamic multivariate hazard rate order (Theorem 6.E.7) can be found in Shaked and Shanthikumar [511], whereas the notion of multivariate PF_2 distributions is taken from Shaked and Shanthikumar [512]. Theorem 6.E.8 has been proved in the literature in various generalities; see, for example, Holley [226] or Preston [460]. For a proof of the present statement of Theorem 6.E.8 see Karlin and Rinott [278]. The implication (6.E.6) can be found in Whitt [563]. Shanthikumar and Koo [528] studied an order which is defined as in (6.E.6), except that rather than requiring A there to be a rectangular set, they require the right-hand side of (6.E.6) to hold for all planar regions

A. The statement in Remark 6.E.9 that (1.C.6) does not generalize to the multivariate case, follows from Rüschendorf [485, Theorem 8]. The order mentioned in Remark 6.E.10 is studied in Whitt [563], where other orders, related to the multivariate likelihood ratio order, are also studied. One of the counterexamples, mentioned in Remark 6.E.10, can be found in Whitt [563]. Other counterexamples can be found in Lehmann [341]; in that paper it is also claimed that Theorem 6.B.2 is wrong, but that claim is based on erroneous examples. The conditions for the monotonicity of the order $\leq_{\rm lr}$, given in Theorem 6.E.11, are taken from Rinott and Scarsini [468]. The comparison, in the multivariate likelihood ratio order, of vectors of order statistics (Theorem 6.E.12), can be found in Belzunce, Ruiz, and Ruiz [75]; an extension of it is given in Belzunce, Mercader, and Ruiz [70]. The multivariate likelihood ratio comparisons of epoch and inter-epoch times of nonhomogeneous Poisson processes (Example 6.E.13) can be found in Belzunce, Lillo, Ruiz, and Shaked [69]; in that paper these results are also extended to nonhomogeneous pure birth processes. The likelihood ratio order comparison of the vectors of the normalized spacings associated with exponential random variables (Example 6.E.14) is taken from Kochar and Rojo [318]. The result about the likelihood ratio ordering of the posterior distributions (Example 6.E.16) can be found in Fahmy, Pereira, Proschan, and Shaked [189]; see also Purcaru and Denuit [462, Proposition 5.1]. A modification of Example 6.E.16 is Theorem 3.61 of Spizzichino [539]. The proof of the equivalence of the various notions of multivariate PF_2 notions (Theorem 6.E.17) is given in Shaked and Shanthikumar [512].

Section 6.F: The development in this section follows the work of Shaked and Shanthikumar [513]. A notion that is related to the multivariate DMRL concept in Section 6.F.3 can be found in Bassan, Kochar, and Spizzichino [53].

Section 6.G: The orthant orders, which are already mentioned in Marshall and Olkin [383], have been studied further by several authors. Some of the results in Section 6.G.1 can be found in Tchen [547], Rüschendorf [481], and Mosler [401]. Several extensions of these orders can be found in Bergmann [82]. The closure results of the orthant orders given in Theorem 6.G.5, and the application to Markov chains given in Example 6.G.6, are taken from Li and Xu [350]. The result about the preservation of the orthant orders under random sums (Theorem 6.G.7) is taken from Wong [568]; this result also appeared in Denuit, Genest, and Marceau [145], and in Pellerey [451] there is an equivalent result with an alternative proof. The comparison of mixtures result (Theorem 6.G.8) can be found in Denuit and Müller [157]. The relationship between the orders $\leq_{\rm uo}$ and $\leq_{\rm whr}$, given in (6.G.10), can be found in Hu, Khaledi, and Shaked [235]. The relationship between the order $\leq_{\rm uo}$ and the orders $\leq^{\mathcal{S}}_{m\text{-cx}}$ and $\leq_{m\text{-icx}}$ (Theorem 6.G.10) is taken from Boutsikas and Vaggelatou [107]. The sufficient conditions for the comparison of multivariate normal random vectors (Example 6.G.11) can be found in Müller [413]. Theorem 6.G.13 is taken from Scarsini and Shaked [494], whereas Theorem 6.G.14 is adopted from Baccelli and Makowski [27]. Dyckerhoff and Mosler [173] introduced some relatively easy conditions for verifying $\boldsymbol{X} \le_{\rm uo} \boldsymbol{Y}$ or $\boldsymbol{X} \le_{\rm lo} \boldsymbol{Y}$ when $\boldsymbol{X}$ and $\boldsymbol{Y}$ have finite discrete supports. The development in Section 6.G.2 follows the work of Scarsini and Shaked [493]. Hennessy [220] considered the order which is defined by taking all the a_i's in (6.G.16) to be equal to 1; he obtained for this order a result which is analogous to Theorem 6.G.16.

A generalization of the order $\le_{\rm uo}$ is mentioned and studied in Daduna and Szekli [138].

7

Multivariate Variability and Related Orders

In this chapter we describe various extensions, of the univariate variability orders in Chapters 3 and 4, to the multivariate case. The most important common orders that are studied in this chapter are the increasing and the directional convex and concave orders. Multivariate extensions of the order $\leq_{\text{disp}}$ are also studied in this chapter. Some multivariate extensions of the transform orders, and of the Laplace transform order, are investigated in this chapter as well.

7.A The Monotone Convex and Monotone Concave Orders

7.A.1 Definitions

The multivariate orders $\leq_{\text{icx}}$ and $\leq_{\text{icv}}$ are defined in a similar fashion to their univariate counterparts discussed in Section 4.A. Let $\boldsymbol{X}$ and $\boldsymbol{Y}$ be two n-dimensional random vectors such that

$$E[\phi(\boldsymbol{X})] \leq E[\phi(\boldsymbol{Y})] \quad \text{for all increasing convex [concave] functions } \phi : \mathbb{R}^n \rightarrow \mathbb{R}, \tag{7.A.1}$$

provided the expectations exist. Then $\boldsymbol{X}$ is said to be *smaller than* $\boldsymbol{Y}$ *in the increasing convex* [*concave*] *order* (denoted by $\boldsymbol{X} \leq_{\text{icx}} \boldsymbol{Y}$ [$\boldsymbol{X} \leq_{\text{icv}} \boldsymbol{Y}$]).

One can also define a decreasing convex [concave] order by requiring (7.A.1) to hold for all decreasing convex [concave] functions ϕ. But the terms "decreasing convex" and "decreasing concave" orders are counterintuitive because if $\boldsymbol{X}$ is smaller than $\boldsymbol{Y}$ in the sense of either of these two orders, then $\boldsymbol{X}$ is "larger" than $\boldsymbol{Y}$ in some stochastic sense. These orders can easily be characterized using the orders $\leq_{\text{icx}}$ and $\leq_{\text{icv}}$. It is therefore not necessary to have a separate discussion about these orders.

For any i, $i = 1, 2, \dots, n$, the function ϕ_i, defined by $\phi_i(\boldsymbol{x}) = \phi_i(x_1, x_2, \dots, x_n) = x_i$, is increasing and is both convex and concave. Therefore, from (7.A.1) it easily follows that

$$\boldsymbol{X} \le_{\text{icx}} \boldsymbol{Y} \Longrightarrow E[\boldsymbol{X}] \le E[\boldsymbol{Y}] \tag{7.A.2}$$

and that

$$\boldsymbol{X} \le_{\text{icv}} \boldsymbol{Y} \Longrightarrow E[\boldsymbol{X}] \le E[\boldsymbol{Y}], \tag{7.A.3}$$

provided the expectations exist.

If the two n-dimensional random vectors $\boldsymbol{X}$ and $\boldsymbol{Y}$ are such that

$$E[\phi(\boldsymbol{X})] \le E[\phi(\boldsymbol{Y})] \quad \text{for all convex functions } \phi : \mathbb{R}^n \to \mathbb{R}, \tag{7.A.4}$$

provided the expectations exist, then $\boldsymbol{X}$ is said to be *smaller than* $\boldsymbol{Y}$ *in the convex order* (denoted by $\boldsymbol{X} \le_{\text{cx}} \boldsymbol{Y}$). For any i, $i = 1, 2, \dots, n$, the function ϕ_i, defined as above, and the function ψ_i, defined by $\psi_i(\boldsymbol{x}) = \psi_i(x_1, x_2, \dots, x_n) = -x_i$, are both convex. Therefore, from (7.A.4) it follows that

$$\boldsymbol{X} \le_{\text{cx}} \boldsymbol{Y} \Longrightarrow E[\boldsymbol{X}] = E[\boldsymbol{Y}], \tag{7.A.5}$$

provided the expectations exist.

The multivariate convex order can be characterized by construction on the same probability space as the univariate convex order (see Theorem 3.A.4). This is stated next.

Theorem 7.A.1. *The random vectors* $\boldsymbol{X}$ *and* $\boldsymbol{Y}$ *satisfy* $\boldsymbol{X} \le_{\text{cx}} \boldsymbol{Y}$ *if, and only if, there exist two random vectors* $\hat{\boldsymbol{X}}$ *and* $\hat{\boldsymbol{Y}}$*, defined on the same probability space, such that*

$$\hat{\boldsymbol{X}} =_{\text{st}} \boldsymbol{X}, \tag{7.A.6}$$

$$\hat{\boldsymbol{Y}} =_{\text{st}} \boldsymbol{Y}, \tag{7.A.7}$$

and $\{\hat{\boldsymbol{X}}, \hat{\boldsymbol{Y}}\}$ *is a martingale, that is,*

$$E[\hat{\boldsymbol{Y}} | \hat{\boldsymbol{X}}] = \hat{\boldsymbol{X}} \quad a.s. \tag{7.A.8}$$

Similarly, the multivariate extension of Theorem 4.A.5 is the following.

Theorem 7.A.2. *Two random vectors* $\boldsymbol{X}$ *and* $\boldsymbol{Y}$ *satisfy* $\boldsymbol{X} \le_{\text{icx}} \boldsymbol{Y}$ $[\boldsymbol{X} \le_{\text{icv}} \boldsymbol{Y}]$ *if, and only if, there exist two random vectors* $\hat{\boldsymbol{X}}$ *and* $\hat{\boldsymbol{Y}}$*, defined on the same probability space, such that*

$$\hat{\boldsymbol{X}} =_{\text{st}} \boldsymbol{X},$$

$$\hat{\boldsymbol{Y}} =_{\text{st}} \boldsymbol{Y},$$

and $\{\hat{\boldsymbol{X}}, \hat{\boldsymbol{Y}}\}$ *is a submartingale* $[\{\hat{\boldsymbol{Y}}, \hat{\boldsymbol{X}}\}$ *is a supermartingale*]*, that is,*

$$E[\hat{\boldsymbol{Y}} | \hat{\boldsymbol{X}}] \ge \hat{\boldsymbol{X}} \quad [E[\hat{\boldsymbol{X}} | \hat{\boldsymbol{Y}}] \le \hat{\boldsymbol{Y}}] \quad a.s.$$

The next theorem is a multivariate analog of Theorem 4.A.6. The proof of the next theorem is similar to the proof of Theorem 4.A.6, and is therefore omitted.

Theorem 7.A.3. (a) *Two random vectors* $\boldsymbol{X}$ *and* $\boldsymbol{Y}$ *satisfy* $\boldsymbol{X} \le_{\text{icx}} \boldsymbol{Y}$ *if, and only if, there exists a random vector* $\boldsymbol{Z}$ *such that*

$$\boldsymbol{X} \le_{\text{st}} \boldsymbol{Z} \le_{\text{cx}} \boldsymbol{Y}.$$

(b) *Two random vectors* $\boldsymbol{X}$ *and* $\boldsymbol{Y}$ *satisfy* $\boldsymbol{X} \le_{\text{icx}} \boldsymbol{Y}$ *if, and only if, there exists a random vector* $\boldsymbol{Z}$ *such that*

$$\boldsymbol{X} \le_{\text{cx}} \boldsymbol{Z} \le_{\text{st}} \boldsymbol{Y}.$$

The next result is similar to a result of Veinott that can be found in Section 6.B.3. Veinott's result deals with the multivariate usual stochastic order (rather than the convex order) and does not assume independence of either the X_j's or the Y_j's. However, the convex order is harder to work with as compared to the usual stochastic order. Thus we have the following result.

Theorem 7.A.4. *Let* $\boldsymbol{X} = (X_1, X_2, \ldots, X_n)$ *and* $\boldsymbol{Y} = (Y_1, Y_2, \ldots, Y_n)$ *be two* n*-dimensional random vectors. If* $Y_1, Y_2, \ldots, Y_n$ *are independent, and if*

$$X_1 \le_{\text{cx}} Y_1, \tag{7.A.9}$$

$$[X_2 \mid X_1 = x_1] \le_{\text{cx}} Y_2 \quad \textit{for all } x_1, \tag{7.A.10}$$

and, in general, for $i = 2, 3, \ldots, n$,

$$[X_i \mid X_1 = x_1, \ldots, X_{i-1} = x_{i-1}] \le_{\text{cx}} Y_i \quad \textit{for all } x_j \ , j = 1, 2, \ldots, i-1, \tag{7.A.11}$$

then

$$\boldsymbol{X} \le_{\text{cx}} \boldsymbol{Y}. \tag{7.A.12}$$

The proof consists of constructing $\hat{\boldsymbol{X}}$ and $\hat{\boldsymbol{Y}}$ on the same probability space such that (7.A.6)–(7.A.8) hold. This can be done by first constructing independent $\hat{Y}_1, \hat{Y}_2, \ldots, \hat{Y}_n$ such that $\hat{\boldsymbol{Y}} =_{\text{st}} \boldsymbol{Y}$. To construct the $\hat{X}_i$'s, note that by Theorem 3.A.4 (using (7.A.9)) it is possible to construct an $\hat{X}_1$ on the same probability space such that $E[\hat{Y}_1 \mid \hat{X}_1] = \hat{X}_1$ a.s. Next, given $\hat{X}_1 = x_1$, it is possible to construct, again using Theorem 3.A.4 and (7.A.10), an $\hat{X}_2$ on the same probability space such that $E[\hat{Y}_2 \mid \hat{X}_1, \hat{X}_2] = \hat{X}_2$ a.s. Continuing this way, using Theorem 3.A.4 and (7.A.11), the vector $\hat{\boldsymbol{X}}$ is constructed. The vectors $\hat{\boldsymbol{X}}$ and $\hat{\boldsymbol{Y}}$ satisfy the conditions of Theorem 7.A.1, and thus (7.A.12) follows.

Note that under the conditions of Theorem 7.A.4 one has

$$\sum_{j=1}^{n} X_j \le_{\text{cx}} \sum_{j=1}^{n} Y_j. \tag{7.A.13}$$

This inequality gives a stronger result than Theorem 3.A.12(d).

7.A.2 Closure properties

The proofs of the following closure properties are similar to the univariate counterparts and are omitted.

Theorem 7.A.5. (a) *Let $\boldsymbol{X}$ and $\boldsymbol{Y}$ be n-dimensional random vectors. If $\boldsymbol{X} \le_{\rm icx} \boldsymbol{Y}$ [$\boldsymbol{X} \le_{\rm icv} \boldsymbol{Y}$] and $\boldsymbol{g} : \mathbb{R}^n \to \mathbb{R}^m$ is any increasing convex [concave] function, then $\boldsymbol{g}(\boldsymbol{X}) \le_{\rm icx} [\le_{\rm icv}]\ \boldsymbol{g}(\boldsymbol{Y})$.*

(b) *Let $\boldsymbol{X}$, $\boldsymbol{Y}$, and $\boldsymbol{\Theta}$ be random vectors such that $[\boldsymbol{X} \mid \boldsymbol{\Theta} = \boldsymbol{\theta}] \le_{\rm icx} [\le_{\rm icv}]$ $[\boldsymbol{Y} \mid \boldsymbol{\Theta} = \boldsymbol{\theta}]$ for all $\boldsymbol{\theta}$ in the support of $\boldsymbol{\Theta}$. Then $\boldsymbol{X} \le_{\rm icx} [\le_{\rm icv}]\ \boldsymbol{Y}$. That is, the increasing convex [concave] order is closed under mixtures.*

(c) *Let $\{\boldsymbol{X}_j, j = 1, 2, \dots\}$ and $\{\boldsymbol{Y}_j, j = 1, 2, \dots\}$ be two sequences of random vectors such that $\boldsymbol{X}_j \to_{\rm st} \boldsymbol{X}$ and $\boldsymbol{Y}_j \to_{\rm st} \boldsymbol{Y}$ as $j \to \infty$. Assume that $E\boldsymbol{X}_j \to E\boldsymbol{X}$ and that $E\boldsymbol{Y}_j \to E\boldsymbol{Y}$ as $j \to \infty$. If $\boldsymbol{X}_j \le_{\rm cx} [\le_{\rm icx}, \le_{\rm icv}]\ \boldsymbol{Y}_j$, $j = 1, 2, \dots$, then $\boldsymbol{X} \le_{\rm cx} [\le_{\rm icx}, \le_{\rm icv}]\ \boldsymbol{Y}$.*

(d) *Let $\boldsymbol{X}_1, \boldsymbol{X}_2, \dots, \boldsymbol{X}_m$ be a set of independent random vectors and let $\boldsymbol{Y}_1, \boldsymbol{Y}_2, \dots, \boldsymbol{Y}_m$ be another set of independent random vectors. If $\boldsymbol{X}_i \le_{\rm icx}$ $[\le_{\rm icv}]\ \boldsymbol{Y}_i$ for $i = 1, 2, \dots, m$, then*

$$\sum_{j=1}^{m} \boldsymbol{X}_j \le_{\rm icx} [\le_{\rm icv}] \sum_{j=1}^{m} \boldsymbol{Y}_j.$$

That is, the increasing convex [concave] order is closed under convolutions.

Parts (a) and (d) of Theorem 7.A.5 can be generalized as follows.

Theorem 7.A.6. *Let $\boldsymbol{X}_1, \boldsymbol{X}_2, \dots, \boldsymbol{X}_m$ be a set of independent random vectors, let $\boldsymbol{Y}_1, \boldsymbol{Y}_2, \dots, \boldsymbol{Y}_m$ be another set of independent random vectors, and assume that $\boldsymbol{X}_i$ and $\boldsymbol{Y}_i$ have the same dimension, $i = 1, 2, \dots, m$. If $\boldsymbol{X}_i \le_{\rm icx} \boldsymbol{Y}_i$ for $i = 1, 2, \dots, m$, then*

$$\boldsymbol{g}(\boldsymbol{X}_1, \boldsymbol{X}_2, \dots, \boldsymbol{X}_m) \le_{\rm icx} \boldsymbol{g}(\boldsymbol{Y}_1, \boldsymbol{Y}_2, \dots, \boldsymbol{Y}_m)$$

for every function $\boldsymbol{g}$ of a proper dimension that is increasing and convex in each argument.

A generalization of Theorem 7.A.5(d) is the following result which deals with vectors of random partial sums of random variables.

Theorem 7.A.7. *Let $\{X_i\}$ and $\{Y_i\}$ each be a sequence of independent random variables. Also, let $\{M_i\}$ and $\{N_i\}$ each be a sequence of independent positive integer-valued random variables, and suppose that the X_i's and the M_i's are independent and also that Y_i's and the N_i's are independent. Let*

$$\tilde{M}_j = \sum_{i=1}^{j} M_i, \quad \tilde{N}_j = \sum_{i=1}^{j} N_i, \quad U_j = \sum_{i=1}^{\tilde{M}_j} X_i, \quad V_j = \sum_{i=1}^{\tilde{N}_j} Y_i, \quad j = 1, 2, \dots, m.$$

If

$$Y_i \geq 0 \ a.s., \quad i = 1, 2, \ldots, \tag{7.A.14}$$
$$M_i \leq_{\rm st} N_i, \quad i = 1, 2, \ldots,$$

and

$$X_i \leq_{\rm icx} Y_i, \quad i = 1, 2, \ldots,$$

then

$$(U_1, U_2, \ldots, U_m) \leq_{\rm icx} (V_1, V_2, \ldots, V_m). \tag{7.A.15}$$

Proof. According to Theorems 1.A.1 and 4.A.5 there exist sequences of random variables $\{\hat{X}_i\}$, $\{\hat{Y}_i\}$, $\{\hat{M}_i\}$, and $\{\hat{N}_i\}$ such that

$$\hat{X}_i =_{\rm st} X_i, \quad \hat{Y}_i =_{\rm st} Y_i, \quad \hat{M}_i =_{\rm st} M_i, \quad \hat{N}_i =_{\rm st} N_i, \quad i = 1, 2, \ldots,$$

and

$$\hat{M}_i \leq \hat{N}_i \text{ a.s.}, \quad \hat{X}_i \leq E[\hat{Y}_i | \hat{X}_i] \text{ a.s.}, \quad i = 1, 2, \ldots.$$

Define

$$\tilde{\hat{M}}_j = \sum_{i=1}^{j} \hat{M}_i, \quad \tilde{\hat{N}}_j = \sum_{i=1}^{j} \hat{N}_i, \quad \hat{U}_j = \sum_{i=1}^{\tilde{\hat{M}}_j} \hat{X}_i, \quad \hat{V}_j = \sum_{i=1}^{\tilde{\hat{N}}_j} \hat{Y}_i, \quad j = 1, 2, \ldots, m.$$

From (7.A.14) it is seen that

$$\hat{U}_j = \sum_{i=1}^{\tilde{\hat{M}}_j} \hat{X}_i \leq E\Big[\sum_{i=1}^{\tilde{\hat{N}}_j} \hat{Y}_i | \{\hat{X}_k\}\Big] = E\big[\hat{V}_j | \{\hat{X}_k\}\big] \text{ a.s.}, \quad j = 1, 2, \ldots, m.$$

Let ϕ be an increasing convex real n-dimensional function. Then

$$\begin{aligned} E[\phi(\hat{U}_1, \hat{U}_2, \ldots, \hat{U}_m)] &\leq E[\phi(E[(\hat{V}_1, \hat{V}_2, \ldots, \hat{V}_m) | \{\hat{X}_k\}])] \\ &\leq E[E[\phi(\hat{V}_1, \hat{V}_2, \ldots, \hat{V}_m) | \{\hat{X}_k\}]] \\ &= E[\phi(\hat{V}_1, \hat{V}_2, \ldots, \hat{V}_m)], \end{aligned}$$

where the second inequality follows from Jensen's Inequality. Since $(\hat{U}_1, \hat{U}_2, \ldots, \hat{U}_m) =_{\rm st} (U_1, U_2, \ldots, U_m)$ and $(\hat{V}_1, \hat{V}_2, \ldots, \hat{V}_m) =_{\rm st} (V_1, V_2, \ldots, V_m)$ we obtain (7.A.15). □

Let $\boldsymbol{X}_1, \boldsymbol{X}_2, \ldots, \boldsymbol{X}_m$ be m countably infinite vectors of independent nonnegative random variables, and let $\boldsymbol{M} = (M_1, M_2, \ldots, M_m)$ and $\boldsymbol{N} = (N_1, N_2, \ldots, N_m)$ be two vectors of nonnegative integers which are independent of $\boldsymbol{X}_i$'s. Denote by $X_{j,i}$ the ith element of $\boldsymbol{X}_j$. From Theorems 2.3 and 2.4 of Pellerey [451] it seems that if $\boldsymbol{M} \leq_{\rm cx} [\leq_{\rm icx}] \boldsymbol{N}$, then $\big(\sum_{i=1}^{M_1} X_{1,i}, \sum_{i=1}^{M_2} X_{2,i}, \ldots, \sum_{i=1}^{M_m} X_{m,i}\big) \leq_{\rm cx} [\leq_{\rm icx}] \big(\sum_{i=1}^{N_1} X_{1,i}, \sum_{i=1}^{N_2} X_{2,i}, \ldots, \sum_{i=1}^{N_m} X_{m,i}\big)$. However, the proofs given in that paper yield somewhat different results; see Theorem 7.A.36 for the details.

The following two results can easily be proven using Theorem 7.A.1.

Theorem 7.A.8. *Let $X_1, X_2, \dots, X_m$ be a set of independent random variables and let $Y_1, Y_2, \dots, Y_m$ be another set of independent random variables. If $X_i \le_{\rm cx} Y_i$ for $i = 1, 2, \dots, m$, then*

$$(X_1, X_2, \dots, X_m) \le_{\rm cx} (Y_1, Y_2, \dots, Y_m).$$

A result that is slightly stronger than Theorem 7.A.8 is given in Theorem 7.A.24.

Theorem 7.A.9. *Let the random vector $\boldsymbol{X}$ and the nonnegative random variable U be independent. If $E[U] = 1$, then $\boldsymbol{X} \le_{\rm cx} U\boldsymbol{X}$.*

From Theorem 3.B.15 [Theorem 4.B.23] and Theorem 7.A.8 we obtain the following result.

Theorem 7.A.10. *Let $X_1, X_2, \dots, X_m$ be a set of nonnegative independent random variables, let $Y_1, Y_2, \dots, Y_m$ be another set of nonnegative independent random variables, and assume that $EX_i = EY_i$, $i = 1, 2, \dots, m$. If $X_i \le_{\rm disp}$ [$\le_{\rm nbue}$] Y_i for $i = 1, 2, \dots, m$, then*

$$(X_1, X_2, \dots, X_m) \le_{\rm cx} (Y_1, Y_2, \dots, Y_m).$$

An application of Theorem 7.A.1 is illustrated in the following example (which is, in fact, an extension of Example 3.A.29).

Example 7.A.11. Let $\boldsymbol{X}_1, \boldsymbol{X}_2, \dots$ be independent and identically distributed m-dimensional random variables. Denote by $\overline{\boldsymbol{X}}_n$ the sample mean of $\boldsymbol{X}_1, \boldsymbol{X}_2, \dots, \boldsymbol{X}_n$. That is, $\overline{\boldsymbol{X}}_n = (\boldsymbol{X}_1 + \boldsymbol{X}_2 + \cdots + \boldsymbol{X}_n)/n$. If the expectation of $\boldsymbol{X}_1$ exists, then for any choice of positive integers $n \le n'$ one has

$$\overline{\boldsymbol{X}}_{n'} \le_{\rm cx} \overline{\boldsymbol{X}}_n.$$

In order to see it note that by the symmetry of $\boldsymbol{X}_1, \boldsymbol{X}_2, \dots, \boldsymbol{X}_{n'}$ it follows that $E[\boldsymbol{X}_i | \overline{\boldsymbol{X}}_{n'}] = \overline{\boldsymbol{X}}_{n'}$ for all $i \le n'$. Therefore $E[\overline{\boldsymbol{X}}_n | \overline{\boldsymbol{X}}_{n'}] = \overline{\boldsymbol{X}}_{n'}$. That is, $\{\overline{\boldsymbol{X}}_{n'}, \overline{\boldsymbol{X}}_n\}$ is a martingale. The result now follows from Theorem 7.A.1.

7.A.3 Further properties

Let $\boldsymbol{X}$ and $\boldsymbol{Y}$ be random vectors. If $E[\phi(\boldsymbol{X})] \le E[\phi(\boldsymbol{Y})]$ for all increasing functions ϕ, then (7.A.1) obviously holds. Thus we obtain the following result.

Theorem 7.A.12. *Let $\boldsymbol{X}$ and $\boldsymbol{Y}$ be two random vectors. If $\boldsymbol{X} \le_{\rm st} \boldsymbol{Y}$, then $\boldsymbol{X} \le_{\rm icx} \boldsymbol{Y}$ and $\boldsymbol{X} \le_{\rm icv} \boldsymbol{Y}$.*

The following example gives necessary (and sufficient) conditions for the comparison of multivariate normal random vectors. See Examples 6.B.29, 6.G.11, 7.A.26, 7.A.39, 7.B.5, and 9.A.20 for related results.

Example 7.A.13. Let $\boldsymbol{X}$ be a multivariate normal random vector with mean vector $\boldsymbol{\mu}_{\boldsymbol{X}}$ and variance-covariance matrix $\boldsymbol{\Sigma}_{\boldsymbol{X}}$, and let $\boldsymbol{Y}$ be a multivariate normal random vector with mean vector $\boldsymbol{\mu}_{\boldsymbol{Y}}$ and variance-covariance matrix $\boldsymbol{\Sigma}_{\boldsymbol{Y}}$.

(a) If $\boldsymbol{\mu}_{\boldsymbol{X}} \le \boldsymbol{\mu}_{\boldsymbol{Y}}$ and if $\boldsymbol{\Sigma}_{\boldsymbol{Y}} - \boldsymbol{\Sigma}_{\boldsymbol{X}}$ is positive semidefinite, then $\boldsymbol{X} \le_{\text{icx}} \boldsymbol{Y}$.
(b) $\boldsymbol{X} \le_{\text{cx}} \boldsymbol{Y}$ if, and only if, $\boldsymbol{\mu}_{\boldsymbol{X}} = \boldsymbol{\mu}_{\boldsymbol{Y}}$ and $\boldsymbol{\Sigma}_{\boldsymbol{Y}} - \boldsymbol{\Sigma}_{\boldsymbol{X}}$ is positive semidefinite.

Using Theorem 4.A.48 we can obtain conditions under which two nonnegative random vectors, that are comparable in the $\le_{\text{icx}}$ or in the $\le_{\text{icv}}$ orders, have the same distribution; related results are Theorems 1.A.8, 3.A.43, 3.A.60, 4.A.69, 5.A.15, 6.B.19, 6.G.12, and 6.G.13.

Theorem 7.A.14. *Let* $\boldsymbol{X} = (X_1, X_2, \dots, X_n)$ *and* $\boldsymbol{Y} = (Y_1, Y_2, \dots, Y_n)$ *be two nonnegative random vectors.*

(a) *If* $\boldsymbol{X} \le_{\text{icx}} \boldsymbol{Y}$*, and if* $E[X_iX_j] = E[Y_iY_j]$ *for all* i *and* j*, then* $\boldsymbol{X} =_{\text{st}} \boldsymbol{Y}$.
(b) *If* $\boldsymbol{X} \le_{\text{icv}} \boldsymbol{Y}$*, and if* $E\boldsymbol{X} = E\boldsymbol{Y}$*, and if* $E[X_iX_j] = E[Y_iY_j]$ *for all* i *and* j*, then* $\boldsymbol{X} =_{\text{st}} \boldsymbol{Y}$.

Proof. First we prove (a). From the assumption that $\boldsymbol{X} \le_{\text{icx}} \boldsymbol{Y}$ it follows that $\sum_{i=1}^n a_iX_i \le_{\text{icx}} \sum_{i=1}^n a_iY_i$ for all $a_i \ge 0$, $i = 1, 2, \dots, n$. Also

$$E\Big(\sum_{i=1}^n a_iX_i\Big)^2 = \sum_{i=1}^n\sum_{j=1}^n a_ia_jE[X_iX_j] = \sum_{i=1}^n\sum_{j=1}^n a_ia_jE[Y_iY_j] = E\Big(\sum_{i=1}^n a_iY_i\Big)^2.$$

It then follows, from Theorem 4.A.48, that $\sum_{i=1}^n a_iX_i =_{\text{st}} \sum_{i=1}^n a_iY_i$ for all $a_i \ge 0$, $i = 1, 2, \dots, n$. Thus we have that $E[\exp\{-\sum_{i=1}^n a_iX_i\}] = E[\exp\{-\sum_{i=1}^n a_iY_i\}]$ for all $a_i \ge 0$, $i = 1, 2, \dots, n$. From the unicity property of the Laplace transform we obtain $\boldsymbol{X} =_{\text{st}} \boldsymbol{Y}$.

The proof of part (b) follows from part (a) and from the observation that if $\sum_{i=1}^n a_iX_i \le_{\text{icv}} \sum_{i=1}^n a_iY_i$ and if $E\big[\sum_{i=1}^n a_iX_i\big] = E\big[\sum_{i=1}^n a_iY_i\big]$, then $\sum_{i=1}^n a_iX_i \ge_{\text{icx}} \sum_{i=1}^n a_iY_i$. □

In a similar manner, using now Theorem 3.A.42 rather than Theorem 4.A.48, we can obtain conditions under which two (not necessarily nonnegative) random vectors, that are comparable in the $\le_{\text{cx}}$ order, have the same distribution.

Theorem 7.A.15. *Let* $\boldsymbol{X} = (X_1, X_2, \dots, X_n)$ *and* $\boldsymbol{Y} = (Y_1, Y_2, \dots, Y_n)$ *be two (not necessarily nonnegative) random vectors. If* $\boldsymbol{X} \le_{\text{cx}} \boldsymbol{Y}$*, and if* $\text{Var}(X_i) = \text{Var}(Y_i)$*,* $i = 1, 2, \dots, n$*, then* $\boldsymbol{X} =_{\text{st}} \boldsymbol{Y}$.

Proof. From the assumption that $\boldsymbol{X} \le_{\text{cx}} \boldsymbol{Y}$ it follows that for $i \ne j$ we have

$$\begin{aligned} a_i^2EX_i^2 + a_j^2EX_j^2 + a_ia_jE[X_iX_j] &= E(a_iX_i + a_jX_j)^2 \\ &\le E(a_iY_i + a_jY_j)^2 = a_i^2EY_i^2 + a_j^2EY_j^2 + a_ia_jE[Y_iY_j], \end{aligned}$$

where a_i and a_j are any constants. Since, by assumption, $EX_i^2 = EY_i^2$ and $EX_j^2 = EY_j^2$, we have that $a_i a_j E[X_i X_j] \le a_i a_j E[Y_i Y_j]$. Since a_i and a_j are arbitrary, we see that $E[X_i X_j] = E[Y_i Y_j]$.

Now, again from the assumption that $\boldsymbol{X} \le_{\rm cx} \boldsymbol{Y}$ it follows that $\sum_{i=1}^n a_i X_i \le_{\rm cx} \sum_{i=1}^n a_i Y_i$ for all a_i, $i = 1, 2, \ldots, n$. As in the proof of Theorem 7.A.14 we can show that $E\big(\sum_{i=1}^n a_i X_i\big)^2 = E\big(\sum_{i=1}^n a_i Y_i\big)^2$. It then follows, from Theorem 3.A.42, that $\sum_{i=1}^n a_i X_i =_{\rm st} \sum_{i=1}^n a_i Y_i$ for all a_i, $i = 1, 2, \ldots, n$. Therefore the characteristic functions of $\boldsymbol{X}$ and of $\boldsymbol{Y}$ are identical. This implies that $\boldsymbol{X} =_{\rm st} \boldsymbol{Y}$. □

An interesting application of the orthant order in the context of the increasing convex and concave orders is given in the following result. The proof, which can be found elsewhere (see Section 7.D), is not given here.

Theorem 7.A.16. *Let* $\boldsymbol{X} = (X_1, X_2, \ldots, X_n)$ *and* $\boldsymbol{Y} = (Y_1, Y_2, \ldots, Y_n)$ *be two random vectors. Suppose that* $\boldsymbol{X} \le_{\rm lo} \boldsymbol{Y}$ *[respectively,* $\boldsymbol{X} \le_{\rm uo} \boldsymbol{Y}$*] and that*

$$-\infty < E[\phi_i(X_i)] = E[\phi_i(Y_i)] < \infty, \quad i = 1, 2, \ldots, n,$$

for some nonnegative strictly increasing convex functions ϕ_i, $i = 1, 2, \ldots, n$. *If* $\boldsymbol{X}$ *and* $\boldsymbol{Y}$ *are comparable in the order* $\le_{\rm icx}$ *[respectively,* $\le_{\rm icv}$*], then* $\boldsymbol{X} =_{\rm st} \boldsymbol{Y}$.

Two orders related to the multivariate monotone convex order are discussed in Sections 7.A.6 and 7.A.7 below.

7.A.4 Convex and concave ordering of stochastic processes

In Section 6.B.7 we showed that some of the results regarding the usual stochastic ordering of random vectors can be extended to the usual stochastic ordering of stochastic processes. It turns out that some of the results regarding the monotone convex and concave orderings of random vectors can also be extended to the analogous orderings of stochastic processes. In this subsection we describe a basic result that formally states that two stochastic processes are comparable in the sense of any of these orders if, and only if, any finite dimensional marginals of them are comparable in the same sense.

Let $\{X(n),\ n \in \mathbb{N}_{++}\}$ and $\{Y(n),\ n \in \mathbb{N}_{++}\}$ be two discrete-time stochastic processes with state space $\mathbb{R}$. Suppose that, for all choices of an integer m, it holds that

$$(X(1), X(2), \ldots, X(m)) \le_{\rm cx} [\le_{\rm icx}, \le_{\rm icv}]\ (Y(1), Y(2), \ldots, Y(m)),$$

then $\{X(n),\ n \in \mathbb{N}_{++}\}$ is said to be *smaller than* $\{Y(n)\ , n \in \mathbb{N}_{++}\}$ *in the convex [increasing convex, increasing concave] order* (denoted by $\{X(n),\ n \in \mathbb{N}_{++}\} \le_{\rm cx} [\le_{\rm icx}, \le_{\rm icv}]\ \{Y(n),\ n \in \mathbb{N}_{++}\}$). Below, a functional g is called convex [concave] if $g(\{\alpha x(n) + (1-\alpha) y(n),\ n \in \mathbb{N}_{++}\}) \le [\ge]\ \alpha g(\{x(n),\ n \in \mathbb{N}_{++}\}) + (1-\alpha) g(\{y(n),\ n \in \mathbb{N}_{++}\})$ for all $\alpha \in [0, 1]$ and $\{x(n),\ n \in \mathbb{N}_{++}\}$ and $\{y(n),\ n \in \mathbb{N}_{++}\}$.

Theorem 7.A.17. *Let $\{X(n),\ n \in \mathbb{N}_{++}\}$ and $\{Y(n),\ n \in \mathbb{N}_{++}\}$ be two discrete-time stochastic processes with state space $\mathbb{R}$. Then $\{X(n),\ n \in \mathbb{N}_{++}\} \le_{\rm cx} [\le_{\rm icx}, \le_{\rm icv}]\ \{Y(n),\ n \in \mathbb{N}_{++}\}$ if, and only if,*

$$E\{g(\{X(n),\ n \in \mathbb{N}_{++}\})\} \le E\{g(\{Y(n),\ n \in \mathbb{N}_{++}\})\} \tag{7.A.16}$$

for every continuous (with respect to the product topology in $\mathbb{R}^\infty$) convex [increasing convex, increasing concave] functional g for which the expectations in (7.A.16) exist.

Notice that the assumption of continuity with respect to the product topology is quite restrictive, but, as far as we know, it is the best result available.

7.A.5 The (m_1, m_2)-icx orders

The multivariate $\le_{\rm icx}$ can be extended in a manner similar to the way in which the univariate order $\le_{m\text{-icx}}$ in Section 4.A.7 extends the univariate $\le_{\rm icx}$ order. Only the bivariate extension will be described here.

Let (X_1, X_2) and (Y_1, Y_2) be two random vectors with a common support $I \times J$, where I and J are finite, or half infinite, or infinite intervals in $\mathbb{R}$. If $E[\phi(X_1, X_2)] \le E[\phi(Y_1, Y_2)]$ for all $(m_1 + m_2)$-differentiable functions ϕ such that $\frac{\partial^{k_1+k_2}}{\partial x_1^{k_1} \partial x_2^{k_2}} \phi(x_1, x_2) \ge 0$ on $I \times J$ whenever $0 \le k_1 \le m_1$, $0 \le k_2 \le m_2$, and $k_1 + k_2 \ge 1$, then (X_1, X_2) is said to be *smaller than* (Y_1, Y_2) *in the* (m_1, m_2)*-icx order* (denoted by $(X_1, X_2) \le^{I \times J}_{(m_1,m_2)\text{-icx}} (Y_1, Y_2)$). If $E[\phi(X_1, X_2)] \le E[\phi(Y_1, Y_2)]$ for all $(m_1 + m_2)$-differentiable functions ϕ such that $(-1)^{k_1+k_2+1} \frac{\partial^{k_1+k_2}}{\partial x_1^{k_1} \partial x_2^{k_2}} \phi(x_1, x_2) \ge 0$ on $I \times J$ whenever $0 \le k_1 \le m_1$, $0 \le k_2 \le m_2$, and $k_1 + k_2 \ge 1$, then (X_1, X_2) is said to be *smaller than* (Y_1, Y_2) *in the* (m_1, m_2)*-icv order* (denoted by $(X_1, X_2) \le^{I \times J}_{(m_1,m_2)\text{-icv}} (Y_1, Y_2)$).

The (m_1, m_2)-icx and the (m_1, m_2)-icv orders are related as follows

$$(X_1, X_2) \le^{[a_1,b_1]\times[a_2,b_2]}_{(m_1,m_2)\text{-icv}} (Y_1, Y_2) \iff (b_1 - Y_1, b_2 - Y_2) \le^{[0,b_1-a_1]\times[0,b_2-a_2]}_{(m_1,m_2)\text{-icx}} (b_1 - X_1, b_2 - X_2),$$

and

$$(X_1, X_2) \le^{\mathbb{R}^2}_{(m_1,m_2)\text{-icv}} (Y_1, Y_2) \iff -(Y_1, Y_2) \le^{\mathbb{R}^2}_{(m_1,m_2)\text{-icx}} -(X_1, X_2).$$

Thus it suffices for most purposes to focus on the (m_1, m_2)-icx order only.

Note that the orders $\le^{\mathbb{R}^2}_{(1,1)\text{-icx}}$ and $\le^{\mathbb{R}^2}_{(1,1)\text{-icv}}$ are the orders $\le_{\rm uo}$ and $\le_{\rm lo}$ (see Section 6.G.1). The orders $\le^{\mathbb{R}^2}_{(2,2)\text{-icx}}$ and $\le^{\mathbb{R}^2}_{(2,2)\text{-icv}}$ are the orders $\le_{\text{uo-cx}}$ and $\le_{\text{uo-cx}}$ which are discussed in Section 7.A.9 below. Also, the order $\le^{\mathbb{R}^2}_{(m,m)\text{-icv}}$ is the order $\le^2_m$ which is discussed in Section 7.A.9.

Some closure properties of the (m_1, m_2)-icx order are given in the next theorem. Some of the results below are stated for simplicity only for the case in which $I = J = [0, \infty)$, but they can be rewritten for the general case.

Theorem 7.A.18. (a) *Let* (X_1, X_2) *and* (Y_1, Y_2) *be two random vectors with a common support* $I \times J$. *Let* K *and* L *be two intervals in* $\mathbb{R}$, *and let* $\phi_1 : I \to K$ *and* $\phi_2 : J \to L$ *be two univariate functions with nonnegative first* m_1 *and* m_2 *derivatives, respectively. If* $(X_1, X_2) \le^{I\times J}_{(m_1,m_2)\text{-icx}} (Y_1, Y_2)$, *then* $(\phi_1(X_1), \phi_2(X_2)) \le^{K\times L}_{(m_1,m_2)\text{-icx}} (\phi_1(Y_1), \phi_2(Y_2))$.

(b) *Let* (X_1, X_2), (Y_1, Y_2), *and* $\boldsymbol{\Theta}$ *be random vectors such that* $[(X_1, X_2)\big|\boldsymbol{\Theta} = \boldsymbol{\theta}] \le^{[0,\infty)^2}_{(m_1,m_2)\text{-icx}} [(Y_1, Y_2)\big|\boldsymbol{\Theta} = \boldsymbol{\theta}]$ *for all* $\boldsymbol{\theta}$ *in the support of* $\boldsymbol{\Theta}$. *Then* $(X_1, X_2) \le^{[0,\infty)^2}_{(m_1,m_2)\text{-icx}} (Y_1, Y_2)$. *That is, the* (m_1, m_2)*-icx order is closed under mixtures.*

(c) *Let* $\{(X_{11}, X_{12}), (X_{21}, X_{22}), \dots\}$ *be a sequence of independent random vectors and let* $\{(Y_{11}, Y_{12}), (Y_{21}, Y_{22}), \dots\}$ *be another set of independent random vectors. Furthermore, let* N *be a positive integer-valued random variable which is independent of the above random vectors. If* $(X_{j1}, X_{j2}) \le^{[0,\infty)^2}_{(m_1,m_2)\text{-icx}} (Y_{j1}, Y_{j2})$ *for* $j = 1, 2, \dots$, *then*

$$\sum_{j=1}^{N}(X_{j1}, X_{j2}) \le^{[0,\infty)^2}_{(m_1,m_2)\text{-icx}} \sum_{j=1}^{N}(Y_{j1}, Y_{j2}).$$

In particular, the (m_1, m_2)*-icx order is closed under convolutions.*

Part (c) of Theorem 7.A.18 can be used, for example, to prove (9.A.11) in Chapter 9.

The bivariate (m_1, m_2)-icx orders imply some interesting results on their univariate components.

Theorem 7.A.19. *Let* (X_1, X_2) *and* (Y_1, Y_2) *be two random vectors with a common support* $[0, \infty)^2$. *Let* ϕ *be a bivariate function which satisfies* $\frac{\partial^{k_1+k_2}}{\partial x_1^{k_1} \partial x_2^{k_2}} \phi(x_1, x_2) \ge 0$ *on* $[0, \infty)^2$ *whenever* $0 \le k_1 \le m_1$, $0 \le k_2 \le m_2$, *and* $k_1 + k_2 \ge 1$. *If* $(X_1, X_2) \le^{[0,\infty)^2}_{(m_1,m_2)\text{-icx}} (Y_1, Y_2)$, *then* $\phi(X_1, X_2) \le_{(m_1+m_2)\text{-icx}} \phi(Y_1, Y_2)$.

This result can be used, for example, to prove the second inequality in Theorem 9.A.18 in Chapter 9.

Theorem 7.A.20. *Let* (X_1, X_2) *and* (Y_1, Y_2) *be two random vectors, of independent components, with a common support* $[0, \infty)^2$. *Then*

$$(X_1, X_2) \le^{[0,\infty)^2}_{(m_1,m_2)\text{-icx}} (Y_1, Y_2) \Longleftrightarrow \Big(X_1 \le_{m_1\text{-icx}} Y_1 \text{ and } X_2 \le_{m_2\text{-icx}} Y_2\Big).$$

7.A.6 The symmetric convex order

Let $\boldsymbol{X} = (X_1, X_2, \dots, X_n)$ and $\boldsymbol{Y} = (Y_1, Y_2, \dots, Y_n)$ be two random vectors. When $\boldsymbol{X}$ and $\boldsymbol{Y}$ have exchangeable (that is, permutation symmetric) distribution functions, it is of interest to consider orders defined by the condition

$E\phi(\boldsymbol{X}) \le E\phi(\boldsymbol{Y})$ for all functions in a certain class of (permutation) symmetric functions. One such order is defined as follows.

Suppose that $\boldsymbol{X}$ and $\boldsymbol{Y}$ are such that

$$E\phi(\boldsymbol{X}) \le E\phi(\boldsymbol{Y}) \quad \text{for all symmetric convex functions } \phi : \mathbb{R}^n \to \mathbb{R},$$

provided the expectations exist. Then $\boldsymbol{X}$ is said to be *smaller than* $\boldsymbol{Y}$ *in the symmetric convex order* (denoted as $\boldsymbol{X} \le_{\text{symcx}} \boldsymbol{Y}$).

The following relationship between the orders $\le_{\text{cx}}$ and $\le_{\text{symcx}}$ is obvious.

Theorem 7.A.21. *Let* $\boldsymbol{X}$ *and* $\boldsymbol{Y}$ *be two random vectors. If* $\boldsymbol{X} \le_{\text{cx}} \boldsymbol{Y}$, *then* $\boldsymbol{X} \le_{\text{symcx}} \boldsymbol{Y}$.

A further discussion regarding the order $\le_{\text{symcx}}$ can be found in Chapter 7 by Tong in [515].

7.A.7 The componentwise convex order

Let $\boldsymbol{X} = (X_1, X_2, \ldots, X_n)$ and $\boldsymbol{Y} = (Y_1, Y_2, \ldots, Y_n)$ be two random vectors. Suppose that $\boldsymbol{X}$ and $\boldsymbol{Y}$ are such that

$$E\phi(\boldsymbol{X}) \le E\phi(\boldsymbol{Y}) \quad \begin{array}{l}\text{for all [increasing] functions } \phi : \mathbb{R}^n \to \mathbb{R} \\ \text{that are convex in each argument when} \\ \text{the other arguments are held fixed,}\end{array}$$

provided the expectations exist. Then $\boldsymbol{X}$ is said to be *smaller than* $\boldsymbol{Y}$ *in the [increasing] componentwise convex order* (denoted by $\boldsymbol{X}$ $[\le_{\text{iccx}}]$ $\le_{\text{ccx}}$ $\boldsymbol{Y}$).

The following relationship between the orders $\le_{\text{ccx}}$ $[\le_{\text{iccx}}]$ and $\le_{\text{cx}}$ $[\le_{\text{icx}}]$ is obvious.

Theorem 7.A.22. *Let* $\boldsymbol{X}$ *and* $\boldsymbol{Y}$ *be two random vectors. If* $\boldsymbol{X} \le_{\text{ccx}} [\le_{\text{iccx}}]\ \boldsymbol{Y}$, *then* $\boldsymbol{X} \le_{\text{cx}} [\le_{\text{icx}}]\ \boldsymbol{Y}$.

The functions $\phi_1(x_1, x_2, \ldots, x_n) = x_i x_j$ and $\phi_2(x_1, x_2, \ldots, x_n) = -x_i x_j$ are both componentwise convex, $1 \le i < j \le n$. The next result thus follows from Theorem 7.A.22 and (7.A.5).

Theorem 7.A.23. *Let* $\boldsymbol{X} = (X_1, X_2, \ldots, X_n)$ *and* $\boldsymbol{Y} = (Y_1, Y_2, \ldots, Y_n)$ *be two random vectors. If* $\boldsymbol{X} \le_{\text{ccx}} \boldsymbol{Y}$, *then* $\text{Cov}(X_i, X_j) = \text{Cov}(Y_i, Y_j)$, $1 \le i < j \le n$.

Theorem 7.A.24. *Let* $X_1, X_2, \ldots, X_m$ *be a set of independent random variables and let* $Y_1, Y_2, \ldots, Y_m$ *be another set of independent random variables. If* $X_i \le_{\text{cx}} [\le_{\text{icx}}]\ Y_i$ *for* $i = 1, 2, \ldots, m$, *then* $(X_1, X_2, \ldots, X_n) \le_{\text{ccx}} [\le_{\text{iccx}}]$ $(Y_1, Y_2, \ldots, Y_n)$.

Proof. The parenthetical statement follows at once from Theorem 4.A.15. The proof of the other statement is similar to the proof of that theorem. As in there, we can assume, without loss of generality, that all the $2m$ random variables are independent. The proof is by induction on m. For $m = 1$ the result is obvious. Assume that the stated result is true for vectors of size $m-1$. Let ϕ be a componentwise convex function. Then

$$\begin{aligned}E[\phi(X_1, X_2, \dots, X_m) \big| X_1 = x] &= E[\phi(x, X_2, \dots, X_m)] \\ &\le E[\phi(x, Y_2, \dots, Y_m)] \\ &= E[\phi(X_1, Y_2, \dots, Y_m) \big| X_1 = x],\end{aligned}$$

where the equalities above follow from the independence assumption and the inequality follows from the induction hypothesis. Taking expectations with respect to X_1, we obtain

$$E[\phi(X_1, X_2, \dots, X_m)] \le E[\phi(X_1, Y_2, \dots, Y_m)].$$

Repeating the argument, but now conditioning on $Y_2, \dots, Y_m$ and using $X_1 \le_{\text{cx}} Y_1$, we see that

$$E[\phi(X_1, Y_2, \dots, Y_m)] \le E[\phi(Y_1, Y_2, \dots, Y_m)],$$

and this proves the result. □

It is not hard to show that if $\boldsymbol{X} = (X_1, X_2, \dots, X_n)$ and $\boldsymbol{Y} = (Y_1, Y_2, \dots, Y_n)$ satisfy conditions (7.A.9)–(7.A.11) of Theorem 7.A.4, and if $Y_1, Y_2, \dots, Y_n$ are independent, then, in fact, $\boldsymbol{X} \le_{\text{ccx}} \boldsymbol{Y}$. This observation provides an alternative proof for the $\le_{\text{ccx}}$ case of Theorem 7.A.24

The following results may be compared with Theorem 6.B.17.

Theorem 7.A.25. *Let $\boldsymbol{X}_1, \boldsymbol{X}_2, \dots, \boldsymbol{X}_m$ be m countably infinite vectors of independent nonnegative random variables. Let $\boldsymbol{M} = (M_1, M_2, \dots, M_m)$ and $\boldsymbol{N} = (N_1, N_2, \dots, N_m)$ be two vectors of nonnegative integers which are independent of $\boldsymbol{X}_1, \boldsymbol{X}_2, \dots, \boldsymbol{X}_m$. Denote by $X_{j,i}$ the ith element of $\boldsymbol{X}_j$. If $X_{j,i} \le_{\text{cx}} [\le_{\text{icx}}] X_{j,i+1}$ for $j = 1, 2, \dots, m$, and $i \ge 1$, and if $\boldsymbol{M} \le_{\text{ccx}} [\le_{\text{iccx}}] \boldsymbol{N}$, then*

$$\Big(\sum_{i=1}^{M_1} X_{1,i}, \sum_{i=1}^{M_2} X_{2,i}, \dots, \sum_{i=1}^{M_m} X_{m,i}\Big) \\ \le_{\text{ccx}} [\le_{\text{iccx}}] \Big(\sum_{i=1}^{N_1} X_{1,i}, \sum_{i=1}^{N_2} X_{2,i}, \dots, \sum_{i=1}^{N_m} X_{m,i}\Big).$$

The following example gives sufficient conditions for the comparison of multivariate normal random vectors. See Examples 6.B.29, 6.G.11, 7.A.13, 7.A.39, 7.B.5, and 9.A.20 for related results.

Example 7.A.26. Let $\boldsymbol{X}$ be a multivariate normal random vector with mean vector $\mathbf{0}$ and variance-covariance matrix $\boldsymbol{\Sigma}$, and let $\boldsymbol{Y}$ be a multivariate normal random vector with mean vector $\mathbf{0}$ and variance-covariance matrix $\boldsymbol{\Sigma}+\boldsymbol{D}$, where $\boldsymbol{D}$ is a nonnegative diagonal matrix. Then $\boldsymbol{X} \leq_{\text{ccx}} \boldsymbol{Y}$.

7.A.8 The directional convex and concave orders

Let $\leq$ denote the coordinatewise ordering in $\mathbb{R}^n$. For $\boldsymbol{x}, \boldsymbol{y}, \boldsymbol{z} \in \mathbb{R}^n$ we use the notation $[\boldsymbol{x}, \boldsymbol{y}] \leq \boldsymbol{z}$ as a shorthand for $\boldsymbol{x} \leq \boldsymbol{z}$ and $\boldsymbol{y} \leq \boldsymbol{z}$. Also, the notation $\boldsymbol{z} \leq [\boldsymbol{x}, \boldsymbol{y}]$ stands for $\boldsymbol{z} \leq \boldsymbol{x}$ and $\boldsymbol{z} \leq \boldsymbol{y}$. A function $\phi : \mathbb{R}^n \to \mathbb{R}$ is said to be directionally convex [concave] if for any $\boldsymbol{x}_i \in \mathbb{R}^n$, $i = 1, 2, 3, 4$, such that $\boldsymbol{x}_1 \leq [\boldsymbol{x}_2, \boldsymbol{x}_3] \leq \boldsymbol{x}_4$ and $\boldsymbol{x}_1 + \boldsymbol{x}_4 = \boldsymbol{x}_2 + \boldsymbol{x}_3$, one has

$$\phi(\boldsymbol{x}_2) + \phi(\boldsymbol{x}_3) \leq [\geq]\ \phi(\boldsymbol{x}_1) + \phi(\boldsymbol{x}_4). \tag{7.A.17}$$

A function $\boldsymbol{\phi} : \mathbb{R}^n \to \mathbb{R}^m$ is called directionally convex [concave] if the coordinate functions ϕ_i, $i = 1, 2, \ldots, m$, defined by $\boldsymbol{\phi}(\boldsymbol{x}) = (\phi_1(\boldsymbol{x}), \phi_2(\boldsymbol{x}), \ldots, \phi_n(\boldsymbol{x}))$, are directionally convex [concave].

Directional convexity neither implies, nor is implied by, conventional convexity. However, a univariate function is directionally convex [concave] if, and only if, it is convex [concave].

A function $\phi : \mathbb{R}^n \to \mathbb{R}$ is said to be supermodular [submodular] if for any $\boldsymbol{x}, \boldsymbol{y} \in \mathbb{R}^n$ it satisfies

$$\phi(\boldsymbol{x}) + \phi(\boldsymbol{y}) \leq [\geq]\ \phi(\boldsymbol{x} \wedge \boldsymbol{y}) + \phi(\boldsymbol{x} \vee \boldsymbol{y}),$$

where the operators $\wedge$ and $\vee$ denote coordinatewise minimum and maximum, respectively. If $\phi : \mathbb{R}^n \to \mathbb{R}$ has second partial derivatives, then it is supermodular if, and only if, $\frac{\partial^2}{\partial x_i \partial x_j}\phi \geq 0$ for all $i \neq j$. Many examples of supermodular functions can be found in Marshall and Olkin [383, Chapter 6].

Proposition 7.A.27. *The following statements are equivalent:*

(a) *The function ϕ is directionally convex [concave].*
(b) *The function ϕ is supermodular [submodular] and coordinatewise convex [concave].*
(c) *For any $\boldsymbol{x}_1, \boldsymbol{x}_2, \boldsymbol{y} \in \mathbb{R}^n$, such that $\boldsymbol{x}_1 \leq \boldsymbol{x}_2$ and $\boldsymbol{y} \geq \mathbf{0}$, one has*

$$\phi(\boldsymbol{x}_1 + \boldsymbol{y}) - \phi(\boldsymbol{x}_1) \leq [\geq]\ \phi(\boldsymbol{x}_2 + \boldsymbol{y}) - \phi(\boldsymbol{x}_2).$$

If ϕ is twice differentiable, then it is directionally convex [concave] if, and only if, all its second derivatives are nonnegative [nonpositive]. Another useful property of directionally convex [concave] functions is stated next.

Proposition 7.A.28. (a) *If $\boldsymbol{\psi} : \mathbb{R}^m \to \mathbb{R}^k$ is increasing and directionally convex [concave] and $\boldsymbol{\phi} : \mathbb{R}^n \to \mathbb{R}^m$ is increasing and directionally convex [concave], then the composition $\boldsymbol{\psi}(\boldsymbol{\phi})$ is increasing and directionally* *convex [concave]. In particular, if* $\psi : \mathbb{R} \to \mathbb{R}$ *is increasing and convex [concave] and* $\phi : \mathbb{R}^n \to \mathbb{R}$ *is increasing and directionally convex [concave], then the composition* $\psi(\phi)$ *is increasing and directionally convex [concave].*

(b) *If* $\boldsymbol{\psi} : \mathbb{R}^m \to \mathbb{R}^k$ *is increasing and directionally convex [concave] and* $\boldsymbol{\phi} : \mathbb{R}^n \to \mathbb{R}^m$ *is decreasing and directionally convex [concave], then the composition* $\boldsymbol{\psi}(\boldsymbol{\phi})$ *is decreasing and directionally convex [concave]. In particular, if* $\psi : \mathbb{R} \to \mathbb{R}$ *is increasing and convex [concave] and* $\phi : \mathbb{R}^n \to \mathbb{R}$ *is decreasing and directionally convex [concave], then the composition* $\psi(\phi)$ *is decreasing and directionally convex [concave].*

Let $\boldsymbol{X} = (X_1, X_2, \ldots, X_n)$ and $\boldsymbol{Y} = (Y_1, Y_2, \ldots, Y_n)$ be two random vectors. Suppose that $\boldsymbol{X}$ and $\boldsymbol{Y}$ are such that

$$E\phi(\boldsymbol{X}) \le E\phi(\boldsymbol{Y}) \quad \text{for all [increasing] functions } \phi : \mathbb{R}^n \to \mathbb{R} \text{ that are directionally convex,}$$

provided the expectations exist. Then $\boldsymbol{X}$ is said to be *smaller than* $\boldsymbol{Y}$ *in the [increasing] directionally convex order* (denoted by $\boldsymbol{X}$ $[\le_{\text{idir-cx}}]$ $\le_{\text{dir-cx}}$ $\boldsymbol{Y}$). The orders $\le_{\text{dir-cv}}$ and $\le_{\text{idir-cv}}$ are defined similarly.

The following relationships among the orders $\le_{\text{dir-cx}}$ $[\le_{\text{idir-cx}}]$ and $\le_{\text{ccx}}$ $[\le_{\text{iccx}}]$ follow from Proposition 7.A.27. The last assertion in the next theorem follows from the observation that $-\phi$ is directionally concave if, and only if, ϕ is directionally convex.

Theorem 7.A.29. *Let* $\boldsymbol{X}$ *and* $\boldsymbol{Y}$ *be two random vectors. If* $\boldsymbol{X} \le_{\text{ccx}} [\le_{\text{iccx}}]\ \boldsymbol{Y}$, *then* $\boldsymbol{X} \le_{\text{dir-cx}} [\le_{\text{idir-cx}}]\ \boldsymbol{Y}$. *Also, if* $\boldsymbol{X} \le_{\text{dir-cx}} \boldsymbol{Y}$, *then* $\boldsymbol{X} \le_{\text{idir-cx}} \boldsymbol{Y}$ *and* $\boldsymbol{X} \ge_{\text{dir-cv}} \boldsymbol{Y}$.

From Proposition 7.A.28 we obtain the following result (which may be compared with Theorems 6.G.10 and 9.A.16).

Theorem 7.A.30. *Let* $\boldsymbol{X}$ *and* $\boldsymbol{Y}$ *be two n-dimensional random vectors. If* $\boldsymbol{X} \le_{\text{idir-cx}} \boldsymbol{Y}$, *then* $\boldsymbol{\phi}(\boldsymbol{X}) \le_{\text{idir-cx}} \boldsymbol{\phi}(\boldsymbol{Y})$ *for any increasing and directionally convex function* $\boldsymbol{\phi} : \mathbb{R}^n \to \mathbb{R}^m$. *In particular,* $\phi(\boldsymbol{X}) \le_{\text{icx}} \phi(\boldsymbol{Y})$ *for any increasing and directionally convex function* $\phi : \mathbb{R}^n \to \mathbb{R}$.

Theorem 7.A.31. *Let* $\{\boldsymbol{X}_j,\ j = 1, 2, \ldots\}$ *and* $\{\boldsymbol{Y}_j,\ j = 1, 2, \ldots\}$ *be two sequences of random vectors such that* $\boldsymbol{X}_j \to_{\text{st}} \boldsymbol{X}$ *and* $\boldsymbol{Y}_j \to_{\text{st}} \boldsymbol{Y}$ *as* $j \to \infty$. *Assume that* $E\boldsymbol{X}_j \to E\boldsymbol{X}$ *and that* $E\boldsymbol{Y}_j \to E\boldsymbol{Y}$ *as* $j \to \infty$. *If* $\boldsymbol{X}_j \le_{\text{dir-cx}} \boldsymbol{Y}_j$, $j = 1, 2, \ldots$, *then* $\boldsymbol{X} \le_{\text{dir-cx}} \boldsymbol{Y}$.

From Theorems 7.A.24 and 7.A.29 we immediately obtain the next result.

Theorem 7.A.32. *Let* $X_1, X_2, \ldots, X_m$ *be a set of independent random variables and let* $Y_1, Y_2, \ldots, Y_m$ *be another set of independent random variables. If* $X_i \le_{\text{cx}} [\le_{\text{icx}}]\ Y_i$ *for* $i = 1, 2, \ldots, m$, *then* $(X_1, X_2, \ldots, X_n) \le_{\text{dir-cx}} [\le_{\text{idir-cx}}]$ $(Y_1, Y_2, \ldots, Y_n)$.

A stronger result than the $\le_{\rm cx}$ and $\le_{\rm dir\text{-}cx}$ part of Theorem 7.A.32 is Theorem 7.A.38 below. Also, the $\le_{\rm icx}$ and $\le_{\rm idir\text{-}cx}$ part of Theorem 7.A.32 still holds if it is merely assumed that $(Y_1, Y_2, \dots, Y_m)$ is CIS (as defined in (6.B.11)) rather than assuming that it consists of independent components.

The following result (which is a generalization of Theorem 7.A.32) shows that the directionally convex orders are closed under conjunctions.

Theorem 7.A.33. *Let $\boldsymbol{X}_1, \boldsymbol{X}_2, \dots, \boldsymbol{X}_m$ be a set of independent random vectors where the dimension of $\boldsymbol{X}_i$ is k_i, $i = 1, 2, \dots, m$. Let $\boldsymbol{Y}_1, \boldsymbol{Y}_2, \dots, \boldsymbol{Y}_m$ be another set of independent random vectors where the dimension of $\boldsymbol{Y}_i$ is k_i, $i = 1, 2, \dots, m$. If $\boldsymbol{X}_i \le_{\rm dir\text{-}cx} [\le_{\rm idir\text{-}cx}]\ \boldsymbol{Y}_i$ for $i = 1, 2, \dots, m$, then*

$$(\boldsymbol{X}_1, \boldsymbol{X}_2, \dots, \boldsymbol{X}_m) \le_{\rm dir\text{-}cx} [\le_{\rm idir\text{-}cx}]\ (\boldsymbol{Y}_1, \boldsymbol{Y}_2, \dots, \boldsymbol{Y}_m).$$

Proof. It is enough to show that if $\boldsymbol{X}_1$ and $\boldsymbol{Y}_1$ are of the same dimension k_1, and if $\boldsymbol{Z}$ is another random vector, of dimension k, which is independent of $\boldsymbol{X}_1$ and $\boldsymbol{Y}_1$, and if $\boldsymbol{X}_1 \le_{\rm dir\text{-}cx} [\le_{\rm idir\text{-}cx}]\ \boldsymbol{Y}_1$, then $(\boldsymbol{X}_1, \boldsymbol{Z}) \le_{\rm dir\text{-}cx} [\le_{\rm idir\text{-}cx}]$ $(\boldsymbol{Y}_1, \boldsymbol{Z})$. The rest of the proof can then be obtained by induction and pairwise interchanges.

So let ϕ be a (k_1+k)-dimensional [increasing] directionally convex function. Note that $\phi(\boldsymbol{x}, \boldsymbol{z})$ is [increasing] directionally convex in $\boldsymbol{x}$ for any $\boldsymbol{z}$, where the dimensions of $\boldsymbol{x}$ and $\boldsymbol{z}$ are k_1 and k, respectively. Thus from $\boldsymbol{X}_1 \le_{\rm dir\text{-}cx}$ $[\le_{\rm idir\text{-}cx}]\ \boldsymbol{Y}_1$ and the independence assumption we obtain

$$E\phi(\boldsymbol{X}_1, \boldsymbol{Z}) = E\big[E\phi(\boldsymbol{X}_1, \boldsymbol{Z})\big|\boldsymbol{Z}\big] \le E\big[E\phi(\boldsymbol{Y}_1, \boldsymbol{Z})\big|\boldsymbol{Z}\big] = E\phi(\boldsymbol{Y}_1, \boldsymbol{Z}),$$

and the proof is complete. □

The next result shows that the directionally convex orders are closed under convolutions.

Theorem 7.A.34. *Let $\boldsymbol{X}_1, \boldsymbol{X}_2, \dots, \boldsymbol{X}_m$ be a set of independent random vectors and let $\boldsymbol{Y}_1, \boldsymbol{Y}_2, \dots, \boldsymbol{Y}_m$ be another set of independent random vectors, all of the same dimension k. If $\boldsymbol{X}_i \le_{\rm dir\text{-}cx} [\le_{\rm idir\text{-}cx}]\ \boldsymbol{Y}_i$ for $i = 1, 2, \dots, m$, then*

$$\sum_{i=1}^{m} \boldsymbol{X}_i \le_{\rm dir\text{-}cx} [\le_{\rm idir\text{-}cx}] \sum_{i=1}^{m} \boldsymbol{Y}_i.$$

Proof. Let $\phi : \mathbb{R}^k \to \mathbb{R}$ be any [increasing] directionally convex function. Then the function $\psi : \mathbb{R}^{km} \to \mathbb{R}$, defined by $\psi(\boldsymbol{x}_1, \boldsymbol{x}_2, \dots, \boldsymbol{x}_m) = \phi\big(\sum_{i=1}^m \boldsymbol{x}_i\big)$, is [increasing] directionally convex function. The stated result now follows from Theorem 7.A.33. (The idir-cx part also follows directly from Theorems 7.A.30 and 7.A.33.) □

A continuous analog of Theorem 7.A.34 (where the sums are replaced by integrals) is the following result.

Theorem 7.A.35. *Let $\{X(\boldsymbol{t})\}_{\boldsymbol{t}\in\mathbb{R}^d}$ and $\{Y(\boldsymbol{t})\}_{\boldsymbol{t}\in\mathbb{R}^d}$ be two $\mathbb{R}$-valued random fields which are a.s. Riemann-integrable. Suppose that $(X(\boldsymbol{t}_1), X(\boldsymbol{t}_2), \dots, X(\boldsymbol{t}_k)) \le_{\text{idir-cx}} (Y(\boldsymbol{t}_1), Y(\boldsymbol{t}_2), \dots, Y(\boldsymbol{t}_k))$ for all $\boldsymbol{t}_1, \boldsymbol{t}_2, \dots, \boldsymbol{t}_k \in \mathbb{R}^d$, $k = 1, 2, \dots$. Then*

$$\left(\int_{B_1} X(\boldsymbol{t})\mathrm{d}\boldsymbol{t}, \int_{B_2} X(\boldsymbol{t})\mathrm{d}\boldsymbol{t}, \dots, \int_{B_k} X(\boldsymbol{t})\mathrm{d}\boldsymbol{t}\right)$$
$$\le_{\text{idir-cx}} \left(\int_{B_1} Y(\boldsymbol{t})\mathrm{d}\boldsymbol{t}, \int_{B_2} Y(\boldsymbol{t})\mathrm{d}\boldsymbol{t}, \dots, \int_{B_k} Y(\boldsymbol{t})\mathrm{d}\boldsymbol{t}\right)$$

for any disjoint bounded Borel-measurable sets $B_1, B_2, \dots, B_k$ in $\mathbb{R}^d$, $k = 1, 2, \dots$.

The following result may be compared with Theorem 7.A.25.

Theorem 7.A.36. *Let $\boldsymbol{X}_1, \boldsymbol{X}_2, \dots, \boldsymbol{X}_m$ be m countably infinite vectors of independent nonnegative random variables. Let $\boldsymbol{M} = (M_1, M_2, \dots, M_m)$ and $\boldsymbol{N} = (N_1, N_2, \dots, N_m)$ be two vectors of nonnegative integers which are independent of $\boldsymbol{X}_1, \boldsymbol{X}_2, \dots, \boldsymbol{X}_m$. Denote by $X_{j,i}$ the ith element of $\boldsymbol{X}_j$. If $X_{j,i} \le_{\text{cx}}$ $[\le_{\text{icx}}]$ $X_{j,i+1}$ for $j = 1, 2, \dots, m$, and $i \ge 1$, and if $\boldsymbol{M} \le_{\text{dir-cx}} [\le_{\text{idir-cx}}] \boldsymbol{N}$, then*

$$\left(\sum_{i=1}^{M_1} X_{1,i}, \sum_{i=1}^{M_2} X_{2,i}, \dots, \sum_{i=1}^{M_m} X_{m,i}\right)$$
$$\le_{\text{dir-cx}} [\le_{\text{idir-cx}}] \left(\sum_{i=1}^{N_1} X_{1,i}, \sum_{i=1}^{N_2} X_{2,i}, \dots, \sum_{i=1}^{N_m} X_{m,i}\right).$$

Consider now, as in Section 6.B.4, n families of univariate distribution functions $\{G_\theta^{(i)}, \ \theta \in \mathcal{X}_i\}$ where $\mathcal{X}_i$ is a subset of the real line $\mathbb{R}$, $i = 1, 2, \dots, n$. Let $X_i(\theta)$ denote a random variable with distribution function $G_\theta^{(i)}$, $i = 1, 2, \dots, n$. Below we give a result which provides comparisons of two random vectors, with distribution functions of the form (6.B.18), in the [increasing] directionally convex order. The following result is a multivariate extension of Theorems 3.A.21 and 4.A.18; see Theorems 6.B.17, 6.G.8, 9.A.7, and 9.A.15 for related results.

Theorem 7.A.37. *Let $\{G_\theta^{(i)}, \ \theta \in \mathcal{X}_i\}$, $i = 1, 2, \dots, n$, be n families of univariate distribution functions as above. Let $\boldsymbol{\Theta}_1$ and $\boldsymbol{\Theta}_2$ be two random vectors with supports in $\prod_{i=1}^n \mathcal{X}_i$ and distribution functions F_1 and F_2, respectively. Let $\boldsymbol{Y}_1$ and $\boldsymbol{Y}_2$ be two random vectors with distribution functions H_1 and H_2 given by*

$$H_j(y_1, y_2, \dots, y_n) = \int_{\mathcal{X}_1}\int_{\mathcal{X}_2} \cdots \int_{\mathcal{X}_n} \prod_{i=1}^n G_{\theta_i}^{(i)}(y_i)\mathrm{d}F_j(\theta_1, \theta_2, \dots, \theta_n),$$
$$(y_1, y_2, \dots, y_n) \in \mathbb{R}^n, \ j = 1, 2.$$

If for every [increasing] convex function ϕ,

$$E[\phi(X_i(\theta))] \text{ is [increasing] convex in } \theta, \quad i = 1, 2, \ldots, n,$$

and if

$$\boldsymbol{\Theta}_1 \leq_{\text{dir-cx}} [\leq_{\text{idir-cx}}] \boldsymbol{\Theta}_2,$$

then

$$\boldsymbol{Y}_1 \leq_{\text{dir-cx}} [\leq_{\text{idir-cx}}] \boldsymbol{Y}_2.$$

The following result compares, with respect to $\leq_{\text{dir-cx}}$, two random vectors with the same dependence structure. Recall the definition of CIS given in (6.B.11). If every permutation of the coordinates of a random vector is CIS, then the vector is said to be *conditionally increasing* (CI). Recall also the definition of a copula, given in (6.B.14).

Theorem 7.A.38. *Let the random vectors $\boldsymbol{X} = (X_1, X_2, \ldots, X_n)$ and $\boldsymbol{Y} = (Y_1, Y_2, \ldots, Y_n)$ have a common copula that is* CI. *If $X_i \leq_{\text{cx}} Y_i$, $i = 1, 2, \ldots, n$, then $\boldsymbol{X} \leq_{\text{dir-cx}} \boldsymbol{Y}$.*

Theorem 7.A.38 may be compared with Theorems 6.B.14 and 7.A.32. A result that is stronger than Theorem 7.A.38 is Theorem 9.A.25 in Section 9.A.

The following example gives necessary and sufficient conditions for the comparison of multivariate normal random vectors. See Examples 6.B.29, 6.G.11, 7.A.13, 7.B.5, and 9.A.20 for related results.

Example 7.A.39. Let $\boldsymbol{X}$ be a multivariate normal random vector with mean vector $\boldsymbol{\mu}_{\boldsymbol{X}}$ and variance-covariance matrix $\boldsymbol{\Sigma}_{\boldsymbol{X}}$, and let $\boldsymbol{Y}$ be a multivariate normal random vector with mean vector $\boldsymbol{\mu}_{\boldsymbol{Y}}$ and variance-covariance matrix $\boldsymbol{\Sigma}_{\boldsymbol{Y}}$. Then $\boldsymbol{X} \leq_{\text{dir-cx}} \boldsymbol{Y}$ if, and only if, $\boldsymbol{\mu}_{\boldsymbol{X}} = \boldsymbol{\mu}_{\boldsymbol{Y}}$ and $\boldsymbol{\Sigma}_{\boldsymbol{X}} \leq \boldsymbol{\Sigma}_{\boldsymbol{Y}}$.

It is worth mentioning that the result in Example 7.A.26 implies the sufficiency part in Example 7.A.39.

In closing this subsection it is worthwhile to mention that a stochastic order, which is defined by requiring $E\phi(\boldsymbol{X}) \leq E\phi(\boldsymbol{Y})$ to hold for all supermodular [rather than supermodular and componentwise convex, that is, directionally convex] functions ϕ, is studied in Section 9.A.4.

7.A.9 The orthant convex and concave orders

Analogous to the orthant orders studied in Section 6.G.1, one can introduce and study orthant convex and concave orders. This is done in this subsection.

Let $\boldsymbol{X} = (X_1, X_2, \ldots, X_n)$ be a random vector with distribution function F and multivariate survival function $\overline{F}$ (see the exact definition of a multivariate survival function in Section 6.G.1). Let $\boldsymbol{Y}$ be another n-dimensional random vector with distribution function G and survival function $\overline{G}$. If

$$\int_{x_1}^{\infty}\int_{x_2}^{\infty}\cdots\int_{x_n}^{\infty}\overline{F}(u_1,u_2,\ldots,u_n)du_1du_2\cdots du_n$$
$$\leq\int_{x_1}^{\infty}\int_{x_2}^{\infty}\cdots\int_{x_n}^{\infty}\overline{G}(u_1,u_2,\ldots,u_n)du_1du_2\cdots du_n \quad \text{for all } \boldsymbol{x},$$

then we say that $\boldsymbol{X}$ is *smaller than* $\boldsymbol{Y}$ *in the upper orthant-convex order* (denoted by $\boldsymbol{X} \leq_{\text{uo-cx}} \boldsymbol{Y}$). If

$$\int_{-\infty}^{x_1}\int_{-\infty}^{x_2}\cdots\int_{-\infty}^{x_n}F(u_1,u_2,\ldots,u_n)du_1du_2\cdots du_n$$
$$\geq\int_{-\infty}^{x_1}\int_{-\infty}^{x_2}\cdots\int_{-\infty}^{x_n}G(u_1,u_2,\ldots,u_n)du_1du_2\cdots du_n \quad \text{for all } \boldsymbol{x},$$

then we say that $\boldsymbol{X}$ is *smaller than* $\boldsymbol{Y}$ *in the lower orthant-concave order* (denoted by $\boldsymbol{X} \leq_{\text{lo-cv}} \boldsymbol{Y}$).

In analogy with Theorem 6.G.1 it is not hard to obtain the following characterizations of the orders $\leq_{\text{uo-cx}}$ and $\leq_{\text{lo-cv}}$.

Theorem 7.A.40. *Let* $\boldsymbol{X}$ *and* $\boldsymbol{Y}$ *be two* n*-dimensional random vectors. Then*

(a) $\boldsymbol{X} \leq_{\text{uo-cx}} \boldsymbol{Y}$ *if, and only if,*

$$E\Big[\prod_{i=1}^{n} g_i(X_i)\Big] \leq E\Big[\prod_{i=1}^{n} g_i(Y_i)\Big]$$

for every collection $\{g_1, g_2, \ldots, g_n\}$ *of univariate nonnegative increasing convex functions.*

(b) $\boldsymbol{X} \leq_{\text{lo-cv}} \boldsymbol{Y}$ *if, and only if,*

$$E\Big[\prod_{i=1}^{n} h_i(X_i)\Big] \leq E\Big[\prod_{i=1}^{n} h_i(Y_i)\Big]$$

for every collection $\{h_1, h_2, \ldots, h_n\}$ *of univariate nonnegative increasing functions such that* h_i *is concave on the union of the supports of* X_i *and* Y_i, $i = 1, 2, \ldots, n$.

From Theorem 7.A.40 it is easy to obtain the following result which is an extension of the fact that if the random variables X and Y satisfy $X \leq_{\text{icx}} Y$, then $\phi(X) \leq_{\text{icx}} \phi(Y)$ for all real increasing convex functions ϕ on $\mathbb{R}$ (see Theorem 4.A.15).

Theorem 7.A.41. *Let* $\boldsymbol{X} = (X_1, X_2, \ldots, X_n)$ *and* $\boldsymbol{Y} = (Y_1, Y_2, \ldots, Y_n)$ *be two random vectors.*

(a) *If* $\boldsymbol{X} \leq_{\text{uo-cx}} \boldsymbol{Y}$, *then* $(\phi_1(X_1), \phi_2(X_2), \ldots, \phi_n(X_n)) \leq_{\text{uo-cx}} (\phi_1(Y_1), \phi_2(Y_2), \ldots, \phi_n(Y_n))$ *whenever* $\phi_1, \phi_2, \ldots, \phi_n$ *are increasing convex functions.*

(b) *If* $\boldsymbol{X} \leq_{\text{lo-cv}} \boldsymbol{Y}$, *then* $(\phi_1(X_1), \phi_2(X_2), \ldots, \phi_n(X_n)) \leq_{\text{lo-cv}} (\phi_1(Y_1), \phi_2(Y_2), \ldots, \phi_n(Y_n))$ *whenever* $\phi_1, \phi_2, \ldots, \phi_n$ *are increasing concave functions.*

From Theorems 7.A.40 and 6.G.1 it follows that

$$\boldsymbol{X} \leq_{\text{uo}} \boldsymbol{Y} \Longrightarrow \boldsymbol{X} \leq_{\text{uo-cx}} \boldsymbol{Y}$$

and that

$$\boldsymbol{X} \leq_{\text{lo}} \boldsymbol{Y} \Longrightarrow \boldsymbol{X} \leq_{\text{lo-cv}} \boldsymbol{Y}.$$

The following results may be compared with Theorems 7.A.25 and 7.A.36.

Theorem 7.A.42. *Let $\boldsymbol{X}_1, \boldsymbol{X}_2, \ldots, \boldsymbol{X}_m$ be m countably infinite vectors of independent nonnegative random variables. Let $\boldsymbol{M} = (M_1, M_2, \ldots, M_m)$ and $\boldsymbol{N} = (N_1, N_2, \ldots, N_m)$ be two vectors of nonnegative integers which are independent of $\boldsymbol{X}_1, \boldsymbol{X}_2, \ldots, \boldsymbol{X}_m$. Denote by $X_{j,i}$ the ith element of $\boldsymbol{X}_j$. If $X_{j,i} \leq_{\text{icx}} [\geq_{\text{icv}}]\ X_{j,i+1}$ for $j = 1, 2, \ldots, m$, and $i \geq 1$, and if $\boldsymbol{M} \leq_{\text{uo-cx}} [\leq_{\text{lo-cv}}]$ $\boldsymbol{N}$, then*

$$\Big(\sum_{i=1}^{M_1} X_{1,i}, \sum_{i=1}^{M_2} X_{2,i}, \ldots, \sum_{i=1}^{M_m} X_{m,i}\Big) \leq_{\text{uo-cx}} [\leq_{\text{lo-cv}}] \Big(\sum_{i=1}^{N_1} X_{1,i}, \sum_{i=1}^{N_2} X_{2,i}, \ldots, \sum_{i=1}^{N_m} X_{m,i}\Big).$$

Consider now the function $\phi : \mathbb{R}^n \to \mathbb{R}$ which is defined by $\phi(x_1, x_2, \ldots, x_n) = \prod_{i=1}^n g_i(x_i)$, where each $g_i : \mathbb{R} \to \mathbb{R}$ is increasing and convex [concave]. It is easy to verify that ϕ is increasing and directionally convex [concave]. Thus, from Theorem 7.A.40 we obtain that

$$\boldsymbol{X} \leq_{\text{idir-cx}} \boldsymbol{Y} \Longrightarrow \boldsymbol{X} \leq_{\text{uo-cx}} \boldsymbol{Y}$$

and

$$\boldsymbol{X} \leq_{\text{idir-cv}} \boldsymbol{Y} \Longrightarrow \boldsymbol{X} \leq_{\text{lo-cv}} \boldsymbol{Y}.$$

It is worth mentioning that the supermodular order, studied in Section 9.A.4, implies the orders $\leq_{\text{uo}}$, $\leq_{\text{lo}}$, and $\leq_{\text{idir-cx}}$, mentioned above.

We now describe a multivariate extension of the univariate order $\leq_{m\text{-icv}}$ (see Section 4.A.7). A special case of this extension is the order $\leq^{\mathbb{R}^2}_{(m,m)\text{-icv}}$ which is discussed in Section 7.A.5. A similar extension of the univariate order $\leq_{m\text{-icx}}$ can also be defined and studied.

For $\boldsymbol{x} \in \mathbb{R}^n$, let $L(\boldsymbol{x}) = \{\boldsymbol{y} : \boldsymbol{y} \leq \boldsymbol{x}\}$. For an n-dimensional distribution function F define

$$F_1(\boldsymbol{x}) = F(\boldsymbol{x})$$

and

$$F_m(\boldsymbol{x}) = \int_{L(\boldsymbol{x})} F_{m-1}(\boldsymbol{u}) \mathrm{d}\boldsymbol{u}.$$

For n-dimensional distribution functions F and G denote

$$F \leq^n_m G \Longleftrightarrow F_m(\boldsymbol{x}) \geq G_m(\boldsymbol{x}) \quad \text{for all } \boldsymbol{x} \in \mathbb{R}^n.$$

When $m = 1$ the above order is equivalent to the lower orthant order defined in (6.G.2). When $m = 2$ the above order is a multivariate (left-sided) analog of (4.A.5). If $\boldsymbol{X}$ and $\boldsymbol{Y}$ have the distribution functions F and G, respectively, then (as can be easily seen by taking $m = 2$) the relationship $F \leq_2^n G$ is the same as $\boldsymbol{X} \leq_{\text{lo-cv}} \boldsymbol{Y}$. Also, the order $\leq_m^2$ is the order $\leq_{(m,m)\text{-icv}}^{\mathbb{R}^2}$ which is discussed in Section 7.A.5.

For any n-dimensional distribution function F, its $(n-1)$-dimensional marginal distribution functions are defined by

$$F^{(i)}(x_1, \ldots, x_{i-1}, x_{i+1}, \ldots, x_n) = F(x_1, \ldots, x_{i-1}, \infty, x_{i+1}, \ldots, x_n),$$
$$i = 1, 2, \ldots, n.$$

The next result shows that the order $\leq_m^n$ is preserved under marginalization. Before stating the next result we need the following definition. The distribution function F is said to be *margin-regular* for $m > 1$ and $i \leq n$ if for each $\boldsymbol{x}^{(i)} = (x_1, \ldots, x_{i-1}, x_{i+1}, \ldots, x_n)$ for which $F^{(i)}(\boldsymbol{x}^{(i)}) < \infty$, there is an $x_i \in \mathbb{R}$ such that $F(x_1, x_2, \ldots, x_n) < \infty$.

Theorem 7.A.43. *For $n > 1$, $m > 1$, and $i \leq n$, let F and G be two n-dimensional distribution functions such that $F \leq_m^n G$ and F is margin-regular for m and i. Then $F^{(i)} \leq_m^{n-1} G^{(i)}$.*

7.B Multivariate Dispersion Orders

Different characterizations of the univariate order $\leq_{\text{disp}}$ give rise to different multivariate dispersive orders. In this section we describe some such orders.

7.B.1 A strong multivariate dispersion order

Recall from (3.B.13) that for univariate random variables we have that $X \leq_{\text{disp}} Y$ if, and only if, $Y =_{\text{st}} \phi(X)$ for some ϕ that satisfies $\phi(x') - \phi(x) \geq x' - x$ whenever $x \leq x'$. An extension of this definition of the univariate dispersion order gives the multivariate dispersion order that is discussed in this subsection.

A function $\boldsymbol{\phi} : \mathbb{R}^n \to \mathbb{R}^n$ is called an *expansion* if

$$\|\boldsymbol{\phi}(\boldsymbol{x}) - \boldsymbol{\phi}(\boldsymbol{x}')\| \geq \|\boldsymbol{x} - \boldsymbol{x}'\| \quad \text{for all } \boldsymbol{x} \text{ and } \boldsymbol{x}' \text{ in } \mathbb{R}^n.$$

Let $\boldsymbol{X}$ and $\boldsymbol{Y}$ be two n-dimensional random vectors. Suppose that

$$\boldsymbol{Y} =_{\text{st}} \boldsymbol{\phi}(\boldsymbol{X}) \quad \text{for some expansion } \boldsymbol{\phi}. \tag{7.B.1}$$

Then we say that $\boldsymbol{X}$ is *less than* $\boldsymbol{Y}$ *in the strong multivariate dispersive order* (denoted by $\boldsymbol{X} \leq_{\text{SD}} \boldsymbol{Y}$).

Let $\boldsymbol{J}_{\boldsymbol{\phi}}(\boldsymbol{x})$ denote the Jacobian matrix of $\boldsymbol{\phi}$ at $\boldsymbol{x}$, that is,

$$\boldsymbol{J}_{\boldsymbol{\phi}}(\boldsymbol{x}) = \left\{ \frac{\partial \phi_i}{\partial x_j} \right\}.$$

It is useful to note that $\boldsymbol{\phi}$ is an expansion if, and only if,

$$\boldsymbol{J}_{\boldsymbol{\phi}}^T(\boldsymbol{x})\boldsymbol{J}_{\boldsymbol{\phi}}(\boldsymbol{x}) - \boldsymbol{I} \text{ is nonnegative definite,}$$

where $\boldsymbol{I}$ is the identity matrix; see Giovagnoli and Wynn [211].

It is very easy to show that the strong multivariate dispersion order $\le_{\rm SD}$ is closed under conjunctions as the following result states.

Theorem 7.B.1. *Let $\boldsymbol{X}_1, \boldsymbol{X}_2, \ldots, \boldsymbol{X}_m$ be a set of independent random vectors where the dimension of $\boldsymbol{X}_i$ is k_i, $i = 1, 2, \ldots, m$. Let $\boldsymbol{Y}_1, \boldsymbol{Y}_2, \ldots, \boldsymbol{Y}_m$ be another set of independent random vectors where the dimension of $\boldsymbol{Y}_i$ is k_i, $i = 1, 2, \ldots, m$. If $\boldsymbol{X}_i \le_{\rm SD} \boldsymbol{Y}_i$ for $i = 1, 2, \ldots, m$, then*

$$(\boldsymbol{X}_1, \boldsymbol{X}_2, \ldots, \boldsymbol{X}_m) \le_{\rm SD} (\boldsymbol{Y}_1, \boldsymbol{Y}_2, \ldots, \boldsymbol{Y}_m).$$

The strong multivariate dispersion order $\le_{\rm SD}$ also satisfies the following closure property, the proof of which is omitted.

Theorem 7.B.2. *Let $\boldsymbol{X}$ and $\boldsymbol{Y}$ be two n-dimensional random vectors. Let $\boldsymbol{A}$ be an $n \times n$ matrix such that for any orthogonal matrix $\boldsymbol{\Gamma}$ there exists an orthogonal matrix $\tilde{\boldsymbol{\Gamma}}$ such that $\boldsymbol{\Gamma}\boldsymbol{A}\tilde{\boldsymbol{\Gamma}} = \boldsymbol{A}$. If $\boldsymbol{X} \le_{\rm SD} \boldsymbol{Y}$, then $\boldsymbol{A}\boldsymbol{X} \le_{\rm SD} \boldsymbol{A}\boldsymbol{Y}$.*

The following result compares, with respect to the order $\le_{\rm SD}$, two random vectors with the same dependence structure. Recall the definition of a copula, given in (6.B.14).

Theorem 7.B.3. *Let the random vectors $\boldsymbol{X} = (X_1, X_2, \ldots, X_n)$ and $\boldsymbol{Y} = (Y_1, Y_2, \ldots, Y_n)$ have a common copula. If $X_i \le_{\rm disp} Y_i$, $i = 1, 2, \ldots, n$, then $\boldsymbol{X} \le_{\rm SD} \boldsymbol{Y}$.*

An interesting application of Theorem 7.B.3 is the following result which may be compared with Theorems 6.D.10, 6.E.12, and 7.B.12.

Theorem 7.B.4. *Let $X_{(1)}, X_{(2)}, \ldots, X_{(n)}$ and $Y_{(1)}, Y_{(2)}, \ldots, Y_{(n)}$ be order statistics as in Theorem 6.D.10. If $X_1 \le_{\rm disp} Y_1$, then*

$$(X_{(1)}, X_{(2)}, \ldots, X_{(n)}) \le_{\rm SD} (Y_{(1)}, Y_{(2)}, \ldots, Y_{(n)}).$$

Proof. The vectors $(X_{(1)}, X_{(2)}, \ldots, X_{(n)})$ and $(Y_{(1)}, Y_{(2)}, \ldots, Y_{(n)})$ have the same copula. By Theorem 3.B.26, $X_1 \le_{\rm disp} Y_1$ implies that $X_{(i)} \le_{\rm disp} Y_{(i)}$, $i = 1, 2, \ldots, n$. The stated result now follows from Theorem 7.B.3. □

An interesting example in which the order $\le_{\rm SD}$ arises naturally is the following. See also Examples 6.B.29, 6.G.11, 7.A.13, 7.A.26, 7.A.39, and 9.A.20.

Example 7.B.5. Let $\boldsymbol{X} = (X_1, X_2, \dots, X_n)$ be a multivariate normal random vector with mean vector $\boldsymbol{\mu}_1$, and let $\boldsymbol{Y} = (Y_1, Y_2, \dots, Y_n)$ be a multivariate normal random vector with mean vector $\boldsymbol{\mu}_2$. If $\boldsymbol{X}$ and $\boldsymbol{Y}$ have the same correlation matrix, and if $\mathrm{Var}(X_i) \le \mathrm{Var}(Y_i)$, $i = 1, 2, \dots, n$, then $\boldsymbol{X} \le_{\mathrm{SD}} \boldsymbol{Y}$. This can be seen from Theorem 7.B.3 by noting that $\boldsymbol{X}$ and $\boldsymbol{Y}$ have the same copula, and that $\mathrm{Var}(X_i) \le \mathrm{Var}(Y_i)$ implies $X_i \le_{\mathrm{disp}} Y_i$, $i = 1, 2, \dots, n$.

Arias-Nicolás, Fernández-Ponce, Luque-Calvo, and Suárez-Llorens [17] and Fernández-Ponce and Rodríguez-Griñolo [196] compared, respectively, some multivariate t and Wishart random vectors with respect to the order $\le_{\mathrm{SD}}$.

According to Oja [441], an n-dimensional random vector $\boldsymbol{Y}$ is said to be *more scattered* than another n-dimensional random vector $\boldsymbol{X}$ (denoted as $\boldsymbol{X} \le_{\Delta} \boldsymbol{Y}$) if $\boldsymbol{Y} =_{\mathrm{st}} \boldsymbol{\phi}(\boldsymbol{X})$ for some function $\boldsymbol{\phi} : \mathbb{R}^n \to \mathbb{R}^n$ that has the property that

$$\Delta(\boldsymbol{\phi}(\boldsymbol{x}_1), \boldsymbol{\phi}(\boldsymbol{x}_2), \dots, \boldsymbol{\phi}(\boldsymbol{x}_{n+1})) \ge \Delta(\boldsymbol{x}_1, \boldsymbol{x}_2, \dots, \boldsymbol{x}_{n+1}) \tag{7.B.2}$$

for all $\{\boldsymbol{x}_1, \boldsymbol{x}_2, \dots, \boldsymbol{x}_{n+1}\} \subset \mathbb{R}^n$, where $\Delta(\boldsymbol{x}_1, \boldsymbol{x}_2, \dots, \boldsymbol{x}_{n+1})$ is the volume of the simplex with vertices $\boldsymbol{x}_1, \boldsymbol{x}_2, \dots, \boldsymbol{x}_{n+1}$. It is useful to note that a function $\boldsymbol{\phi}$ satisfies (7.B.2) for all $\{\boldsymbol{x}_1, \boldsymbol{x}_2, \dots, \boldsymbol{x}_{n+1}\} \subset \mathbb{R}^n$ if, and only if, the determinant of the Jacobian matrix of $\boldsymbol{\phi}$ satisfies

$$|\mathrm{Det}(\boldsymbol{J}_{\boldsymbol{\phi}}(\boldsymbol{x}))| \ge 1 \quad \text{for all } \boldsymbol{x} \in \mathbb{R}^n.$$

The order $\le_{\Delta}$, as the order $\le_{\mathrm{SD}}$, is a multivariate extension of the characterization (3.B.13) of the univariate order $\le_{\mathrm{disp}}$.

We have the following relationship between the orders $\le_{\Delta}$ and $\le_{\mathrm{SD}}$:

$$\boldsymbol{X} \le_{\mathrm{SD}} \boldsymbol{Y} \Longrightarrow \boldsymbol{X} \le_{\Delta} \boldsymbol{Y}.$$

Fernandez-Ponce and Suarez-Llorens [198] introduced a multivariate dispersion order that is even stronger than $\le_{\mathrm{SD}}$. They did it by essentially requiring (7.B.1) to hold for a particular expansion $\boldsymbol{\phi}$ which is a multivariate analog of the univariate function $\phi = G^{-1}F$ in (3.B.13) in Section 3.B.

7.B.2 A weak multivariate dispersion order

The property (3.B.34) of the univariate dispersive order has an obvious multivariate analog, which is used in this subsection in order to define a multivariate dispersion order.

Let $\boldsymbol{X}$ and $\boldsymbol{Y}$ be two n-dimensional random vectors. Let $\boldsymbol{X}'$ and $\boldsymbol{Y}'$ be such that $\boldsymbol{X} =_{\mathrm{st}} \boldsymbol{X}'$ and $\boldsymbol{Y} =_{\mathrm{st}} \boldsymbol{Y}'$ and such that $\boldsymbol{X}$ and $\boldsymbol{X}'$ are independent and $\boldsymbol{Y}$ and $\boldsymbol{Y}'$ are independent. Suppose that

$$\|\boldsymbol{X} - \boldsymbol{X}'\| \le_{\mathrm{st}} \|\boldsymbol{Y} - \boldsymbol{Y}'\|,$$

where $\|\cdot\|$ is the Euclidean norm and $\le_{\mathrm{st}}$ is the usual univariate stochastic order discussed in Section 1.A. Then we say that $\boldsymbol{X}$ is *smaller than* $\boldsymbol{Y}$ *in the multivariate dispersion order* (denoted by $\boldsymbol{X} \le_{\mathrm{D}} \boldsymbol{Y}$).

The multivariate dispersion order $\leq_{\rm D}$ has the desirable property that the traces of the corresponding covariance matrices are ordered as expected. This multivariate analog of (3.B.25) is shown in the next theorem.

Theorem 7.B.6. *Let $\boldsymbol{X}$ and $\boldsymbol{Y}$ be two n-dimensional random vectors. If $\boldsymbol{X} \leq_{\rm D} \boldsymbol{Y}$, then*

$$\mathrm{tr}(\mathrm{Cov}(\boldsymbol{X})) \leq \mathrm{tr}(\mathrm{Cov}(\boldsymbol{Y})). \tag{7.B.3}$$

Proof. Let $\boldsymbol{X}'$ and $\boldsymbol{Y}'$ be such that $\boldsymbol{X} =_{\rm st} \boldsymbol{X}'$ and $\boldsymbol{Y} =_{\rm st} \boldsymbol{Y}'$ and such that $\boldsymbol{X}$ and $\boldsymbol{X}'$ are independent and $\boldsymbol{Y}$ and $\boldsymbol{Y}'$ are independent. Then $\mathrm{Cov}(\boldsymbol{X}) = \frac{1}{2}E[(\boldsymbol{X}-\boldsymbol{X}')^T(\boldsymbol{X}-\boldsymbol{X}')]$, and $\mathrm{Cov}(\boldsymbol{Y}) = \frac{1}{2}E[(\boldsymbol{Y}-\boldsymbol{Y}')^T(\boldsymbol{Y}-\boldsymbol{Y}')]$. Therefore

$$\begin{aligned}
\mathrm{tr}(\mathrm{Cov}(\boldsymbol{X})) &= \frac{1}{2}E\big[\mathrm{tr}(\boldsymbol{X}-\boldsymbol{X}')(\boldsymbol{X}-\boldsymbol{X}')^T\big] \\
&= \frac{1}{2}E\big[\|\boldsymbol{X}-\boldsymbol{X}'\|^2\big] \\
&\leq \frac{1}{2}E\big[\|\boldsymbol{Y}-\boldsymbol{Y}'\|^2\big] \\
&= \mathrm{tr}(\mathrm{Cov}(\boldsymbol{Y}))
\end{aligned}$$

and (7.B.3) is obtained. □

The multivariate dispersion order $\leq_{\rm D}$ is location-free and rotation-free as the next result shows. The proof is simple and is omitted.

Theorem 7.B.7. *Let $\boldsymbol{X}$ and $\boldsymbol{Y}$ be two n-dimensional random vectors. If $\boldsymbol{X} \leq_{\rm D} \boldsymbol{Y}$, then*

$$\boldsymbol{\Gamma X} + \boldsymbol{a} \leq_{\rm D} \boldsymbol{\Lambda Y} + \boldsymbol{b},$$

for all orthogonal matrices $\boldsymbol{\Gamma}$ and $\boldsymbol{\Lambda}$ and for all vectors $\boldsymbol{a}$ and $\boldsymbol{b}$.

The multivariate dispersion order $\leq_{\rm D}$ is also closed under conjunctions as the following result states.

Theorem 7.B.8. *Let $\boldsymbol{X}_1, \boldsymbol{X}_2, \ldots, \boldsymbol{X}_m$ be a set of independent random vectors where the dimension of $\boldsymbol{X}_i$ is k_i, $i = 1, 2, \ldots, m$. Let $\boldsymbol{Y}_1, \boldsymbol{Y}_2, \ldots, \boldsymbol{Y}_m$ be another set of independent random vectors where the dimension of $\boldsymbol{Y}_i$ is k_i, $i = 1, 2, \ldots, m$. If $\boldsymbol{X}_i \leq_{\rm D} \boldsymbol{Y}_i$ for $i = 1, 2, \ldots, m$, then*

$$(\boldsymbol{X}_1, \boldsymbol{X}_2, \ldots, \boldsymbol{X}_m) \leq_{\rm D} (\boldsymbol{Y}_1, \boldsymbol{Y}_2, \ldots, \boldsymbol{Y}_m).$$

Proof. It is sufficient to prove the result when $m = 2$. Let $\boldsymbol{X}_1'$, $\boldsymbol{X}_2'$, $\boldsymbol{Y}_1'$, and $\boldsymbol{Y}_2'$ be such that

$$\boldsymbol{X}_1' =_{\rm st} \boldsymbol{X}_1, \ \boldsymbol{X}_2' =_{\rm st} \boldsymbol{X}_2, \ \boldsymbol{Y}_1' =_{\rm st} \boldsymbol{Y}_1, \quad \text{and} \quad \boldsymbol{Y}_2' =_{\rm st} \boldsymbol{Y}_2.$$

Let

$$\boldsymbol{X} = (\boldsymbol{X}_1, \boldsymbol{X}_2), \ \boldsymbol{X}' = (\boldsymbol{X}_1', \boldsymbol{X}_2'), \ \boldsymbol{Y} = (\boldsymbol{Y}_1, \boldsymbol{Y}_2), \quad \text{and} \quad \boldsymbol{Y}' = (\boldsymbol{Y}_1', \boldsymbol{Y}_2').$$

Then

$$\begin{aligned}\|\boldsymbol{X}-\boldsymbol{X}'\|^2 &= \|\boldsymbol{X}_1-\boldsymbol{X}_1'\|^2+\|\boldsymbol{X}_2-\boldsymbol{X}_2'\|^2\\ &\le_{\mathrm{st}} \|\boldsymbol{Y}_1-\boldsymbol{Y}_1'\|^2+\|\boldsymbol{Y}_2-\boldsymbol{Y}_2'\|^2\\ &= \|\boldsymbol{Y}-\boldsymbol{Y}'\|^2.\end{aligned}$$

That is, $\boldsymbol{X} \le_{\mathrm{D}} \boldsymbol{Y}$. □

By construction on the same probability space (see Section 6.B.2), it is easy to prove the following result.

Theorem 7.B.9. *Let* $\boldsymbol{X}$ *and* $\boldsymbol{Y}$ *be two n-dimensional random vectors. Then*

$$\boldsymbol{X} \le_{\mathrm{SD}} \boldsymbol{Y} \Longrightarrow \boldsymbol{X} \le_{\mathrm{D}} \boldsymbol{Y}.$$

7.B.3 Dispersive orders based on constructions

The *standard construction* of an n-dimensional random vector $\boldsymbol{X} = (X_1, X_2, \dots, X_n)$, from a vector $(U_1, U_2, \dots, U_n)$ of independent uniform$[0,1]$ random variables, was described in Section 6.B.3. Here we first describe explicitly the function that transforms $(U_1, U_2, \dots, U_n)$ into $(X_1, X_2, \dots, X_n)$. Let F be the distribution function of $\boldsymbol{X}$. Denote by $F_1(\cdot)$ the marginal distribution function of X_1, and denote by $F_{i+1|1,2,\dots,i}(\cdot|x_1, x_2, \dots, x_i)$ the conditional distribution function of X_{i+1} given that $X_1 = x_1, X_2 = x_2, \dots, X_i = x_i$, $i = 1, 2, \dots, n-1$. The inverse of F_1 will be denoted by $F_1^{-1}(\cdot)$ and the inverse of $F_{i+1|1,2,\dots,i}(\cdot|x_1, x_2, \dots, x_i)$ will be denoted by $F_{i+1|1,2,\dots,i}^{-1}(\cdot|x_1, x_2, \dots, x_i)$ for every $(x_1, x_2, \dots, x_i)$ in the support of $(X_1, X_2, \dots, X_i)$, $i = 1, 2, \dots, n-1$. For $(u_1, u_2, \dots, u_n) \in (0,1)^n$ denote

$$x_1 = F_1^{-1}(u_1), \tag{7.B.4}$$

and, by induction,

$$x_i = F_{i|1,2,\dots,i-1}^{-1}(u_i|x_1, x_2, \dots, x_{i-1}), \quad i = 2, 3, \dots, n. \tag{7.B.5}$$

Denote the transformation $(u_1, u_2, \dots, u_n) \to (x_1, x_2, \dots, x_n)$ described in (7.B.4) and (7.B.5) by $\boldsymbol{\Psi}_F^* : (0,1)^n \to \mathbb{R}^n$. It is well known that

$$\boldsymbol{\Psi}_F^*(U_1, U_2, \dots, U_n) =_{\mathrm{st}} (X_1, X_2, \dots, X_n).$$

Let $\boldsymbol{Y} = (Y_1, Y_2, \dots, Y_n)$ be another random vector with distribution function G, and denote the corresponding transformation by $\boldsymbol{\Psi}_G^*$. Note that $\boldsymbol{\Psi}_F^*$ and $\boldsymbol{\Psi}_G^*$ can be thought of as "inverses" of F and of G, respectively. The following order is a multivariate extension of the characterization (3.B.7) of the univariate order $\le_{\mathrm{disp}}$. Suppose that

$$\boldsymbol{\Psi}_G^*(\boldsymbol{u}) - \boldsymbol{\Psi}_F^*(\boldsymbol{u}) \text{ is increasing in } \boldsymbol{u} \in (0,1)^n.$$

Then $\boldsymbol{X}$ is said to be *smaller than* $\boldsymbol{Y}$ *in the multivariate dispersion order* (denoted by $\boldsymbol{X} \leq_{\text{disp}} \boldsymbol{Y}$).

It is easy to prove that the order $\leq_{\text{disp}}$ is closed under conjunctions as the following result states.

Theorem 7.B.10. *Let $\boldsymbol{X}_1, \boldsymbol{X}_2, \ldots, \boldsymbol{X}_m$ be a set of independent random vectors where the dimension of $\boldsymbol{X}_i$ is k_i, $i = 1, 2, \ldots, m$. Let $\boldsymbol{Y}_1, \boldsymbol{Y}_2, \ldots, \boldsymbol{Y}_m$ be another set of independent random vectors where the dimension of $\boldsymbol{Y}_i$ is k_i, $i = 1, 2, \ldots, m$. If $\boldsymbol{X}_i \leq_{\text{disp}} \boldsymbol{Y}_i$ for $i = 1, 2, \ldots, m$, then*

$$(\boldsymbol{X}_1, \boldsymbol{X}_2, \ldots, \boldsymbol{X}_m) \leq_{\text{disp}} (\boldsymbol{Y}_1, \boldsymbol{Y}_2, \ldots, \boldsymbol{Y}_m).$$

In particular, if the random variables $X_1, X_2, \ldots, X_n$ and $Y_1, Y_2, \ldots, Y_n$ are independent and satisfy $X_i \leq_{\text{disp}} Y_i$, $i = 1, 2, \ldots, n$, then $(X_1, X_2, \ldots, X_n)$ $\leq_{\text{disp}} (Y_1, Y_2, \ldots, Y_n)$.

A useful property of the multivariate order $\leq_{\text{disp}}$ is given next. Recall from Section 6.B.3 the definition of a CIS random vector, and recall from Section 7.A.8 the definition of directionally convex functions. The proof of the following result is not given here.

Theorem 7.B.11. *Let $\boldsymbol{X}$ and $\boldsymbol{Y}$ be two nonnegative* CIS *random vectors. If $\boldsymbol{X} \leq_{\text{disp}} \boldsymbol{Y}$, then*

$$\text{Var}[\phi(\boldsymbol{X})] \leq \text{Var}[\phi(\boldsymbol{Y})] \quad \textit{for all increasing directionally convex functions } \phi.$$

In particular, if $(X_1, X_2, \ldots, X_n) \leq_{\text{disp}} (Y_1, Y_2, \ldots, Y_n)$, then

$$\text{Var}[X_1 + X_2 + \cdots + X_n] \leq \text{Var}[Y_1 + Y_2 + \cdots + Y_n].$$

The following result may be compared with Theorems 6.D.10, 6.E.12, and 7.B.4.

Theorem 7.B.12. *Let $X_{(1)}, X_{(2)}, \ldots, X_{(n)}$ and $Y_{(1)}, Y_{(2)}, \ldots, Y_{(n)}$ be order statistics as in Theorem* 6.D.10. *If $X_1 \leq_{\text{disp}} Y_1$, then*

$$(X_{(1)}, X_{(2)}, \ldots, X_{(n)}) \leq_{\text{disp}} (Y_{(1)}, Y_{(2)}, \ldots, Y_{(n)}).$$

The following example may be compared with Examples 1.B.24, 1.C.48, 2.A.22, 3.B.38, 6.B.41, 6.D.8, and 6.E.13.

Example 7.B.13. Let X and Y be two absolutely continuous nonnegative random variables with survival functions $\overline{F}$ and $\overline{G}$, respectively. Denote $\Lambda_1 = -\log \overline{F}$ and $\Lambda_2 = -\log \overline{G}$. Consider two nonhomogeneous Poisson processes $N_1 = \{N_1(t),\, t \geq 0\}$ and $N_2 = \{N_2(t),\, t \geq 0\}$ with mean functions Λ_1 and Λ_2 (see Example 1.B.13), respectively. Let $T_{i,1}, T_{i,2}, \ldots$ be the successive epoch times of process N_i, $i = 1, 2$. Note that $X =_{\text{st}} T_{1,1}$ and $Y =_{\text{st}} T_{2,1}$. If $X \leq_{\text{disp}} Y$, then $(T_{1,1}, T_{1,2}, \ldots, T_{1,n}) \leq_{\text{disp}} (T_{2,1}, T_{2,2}, \ldots, T_{2,n})$ for each $n \geq 1$.

The *total hazard construction* of a nonnegative n-dimensional random vector $\boldsymbol{T} = (T_1, T_2, \ldots, T_n)$ with distribution function F, from a vector $(X_1, X_2, \ldots, X_n)$ of independent standard exponential random variables, was described in Section 6.C.2. The construction defines a transformation of $(X_1, X_2, \ldots, X_n)$ to $\hat{\boldsymbol{T}} = (\hat{T}_1, \hat{T}_2, \ldots, \hat{T}_n)$ such that $\boldsymbol{T} =_{\rm st} \hat{\boldsymbol{T}}$. Denote this transformation from $[0,\infty)^n$ to $[0,\infty)^n$ by $\boldsymbol{R}^*_F$. Thus

$$\boldsymbol{R}^*_F(X_1, X_2, \ldots, X_n) =_{\rm st} (T_1, T_2, \ldots, T_n).$$

Let $\boldsymbol{S} = (S_1, S_2, \ldots, S_n)$ be another nonnegative random vector with distribution function G, and denote the corresponding transformation by $\boldsymbol{R}^*_G$. Note that $\boldsymbol{R}^*_F$ and $\boldsymbol{R}^*_G$ can be thought of as "inverses" of the "total hazards" $-\log \overline{F}$ and $-\log \overline{G}$, respectively. The following order is a multivariate extension of the characterization (3.B.9) of the univariate order $\le_{\rm disp}$. Suppose that

$$\boldsymbol{R}^*_G(\boldsymbol{x}) - \boldsymbol{R}^*_F(\boldsymbol{x}) \text{ is increasing in } \boldsymbol{x} \in [0,\infty)^n.$$

Then $\boldsymbol{T}$ is said to be *smaller than* $\boldsymbol{S}$ *in the dynamic multivariate dispersion order* (denoted by $\boldsymbol{T} \le_{\text{dyn-disp}} \boldsymbol{S}$).

The order $\le_{\text{dyn-disp}}$ is closed under conjunctions as the following, easy to prove, result states.

Theorem 7.B.14. *Let $\boldsymbol{T}_1, \boldsymbol{T}_2, \ldots, \boldsymbol{T}_m$ be a set of independent random vectors where the dimension of $\boldsymbol{T}_i$ is k_i, $i = 1, 2, \ldots, m$. Let $\boldsymbol{S}_1, \boldsymbol{S}_2, \ldots, \boldsymbol{S}_m$ be another set of independent random vectors where the dimension of $\boldsymbol{S}_i$ is k_i, $i = 1, 2, \ldots, m$. If $\boldsymbol{T}_i \le_{\text{dyn-disp}} \boldsymbol{S}_i$ for $i = 1, 2, \ldots, m$, then*

$$(\boldsymbol{T}_1, \boldsymbol{T}_2, \ldots, \boldsymbol{T}_m) \le_{\text{dyn-disp}} (\boldsymbol{S}_1, \boldsymbol{S}_2, \ldots, \boldsymbol{S}_m).$$

A version of Theorem 7.B.11 holds for the order $\le_{\text{dyn-disp}}$, and is given next. Recall from Section 6.C.1 that a nonnegative random vector $\boldsymbol{T}$ has the positive dependence property of "supporting lifetimes" if $\boldsymbol{T} \le_{\rm ch} \boldsymbol{T}$. The proof of the following result is not given here.

Theorem 7.B.15. *Let $\boldsymbol{T}$ and $\boldsymbol{S}$ be two nonnegative random vectors with the supporting lifetimes property. If $\boldsymbol{T} \le_{\text{dyn-disp}} \boldsymbol{S}$, then*

$$\mathrm{Var}[\phi(\boldsymbol{T})] \le \mathrm{Var}[\phi(\boldsymbol{S})] \quad \textit{for all increasing directionally convex functions } \phi.$$

7.C Multivariate Transform Orders: Convex, Star, and Superadditive Orders

In this section we review some extensions of the univariate orders $\le_{\rm c}$, $\le_*$, and $\le_{\rm su}$, which were studied in Section 4.B.

Let $\boldsymbol{X} = (X_1, X_2, \ldots, X_n)$ and $\boldsymbol{Y} = (Y_1, Y_2, \ldots, Y_n)$ be two nonnegative random vectors with survival functions $\overline{F}$ and $\overline{G}$, respectively. Denote

$$\overline{F}_i(\boldsymbol{x}) = \frac{\overline{F}(x_1, \ldots, x_{i-1}, x_i, x_{i+1}, \ldots, x_n)}{\overline{F}(x_1, \ldots, x_{i-1}, 0, x_{i+1}, \ldots, x_n)}, \qquad \boldsymbol{x} \geq \boldsymbol{0},$$

and

$$\overline{G}_i(\boldsymbol{x}) = \frac{\overline{G}(x_1, \ldots, x_{i-1}, x_i, x_{i+1}, \ldots, x_n)}{\overline{G}(x_1, \ldots, x_{i-1}, 0, x_{i+1}, \ldots, x_n)}, \qquad \boldsymbol{x} \geq \boldsymbol{0}.$$

For any $(x_1, x_2, \ldots, x_n) \geq \boldsymbol{0}$ and for any $i = 1, 2, \ldots, n$, let u_i be the solution of

$$\overline{G}_i(x_1, \ldots, x_{i-1}, u_i, x_{i+1}, \ldots, x_n) = \overline{F}_i(x_1, \ldots, x_{i-1}, x_i, x_{i+1}, \ldots, x_n).$$

If, for every $i = 1, 2, \ldots, n$ and every $(x_1, \ldots, x_{i-1}, x_{i+1}, \ldots, x_n)$, we have that u_i is convex in x_i, then $\boldsymbol{X}$ is said to be *smaller than* $\boldsymbol{Y}$ *in the multivariate convex transform order* (denoted as $\boldsymbol{X} \leq_{\text{mc}} \boldsymbol{Y}$). If, for every $i = 1, 2, \ldots, n$ and every $(x_1, \ldots, x_{i-1}, x_{i+1}, \ldots, x_n)$, we have that u_i is starshaped in x_i, then $\boldsymbol{X}$ is said to be *smaller than* $\boldsymbol{Y}$ *in the multivariate star order* (denoted as $\boldsymbol{X} \leq_{\text{m*}} \boldsymbol{Y}$). Finally, if, for every $i = 1, 2, \ldots, n$ and every $(x_1, \ldots, x_{i-1}, x_{i+1}, \ldots, x_n)$, we have that u_i is superadditive in x_i, then $\boldsymbol{X}$ is said to be *smaller than* $\boldsymbol{Y}$ *in the multivariate superadditive order* (denoted as $\boldsymbol{X} \leq_{\text{msu}} \boldsymbol{Y}$).

Obviously,

$$\boldsymbol{X} \leq_{\text{mc}} \boldsymbol{Y} \Longrightarrow \boldsymbol{X} \leq_{\text{m*}} \boldsymbol{Y} \Longrightarrow \boldsymbol{X} \leq_{\text{msu}} \boldsymbol{Y}.$$

The above three orders are partial orders in the sense that each of them is transitive and reflexive. They are also closed under marginalization:

Theorem 7.C.1. *Let* $\boldsymbol{X} = (X_1, X_2, \ldots, X_n)$ *and* $\boldsymbol{Y} = (Y_1, Y_2, \ldots, Y_n)$ *be two nonnegative* n*-dimensional random vectors. If* $\boldsymbol{X} \leq_{\text{mc}} [\leq_{\text{m*}}, \leq_{\text{msu}}] \boldsymbol{Y}$, *then* $\boldsymbol{X}_I \leq_{\text{mc}} [\leq_{\text{m*}}, \leq_{\text{msu}}] \boldsymbol{Y}_I$ *for each* $I \subseteq \{1, 2, \ldots, n\}$.

In analogy with Theorem 4.B.11, the above three orders can be used to define multivariate notions of the IFR, IFRA, and NBU aging notions.

7.D The Multivariate Laplace Transform and Related Orders

The orders we studied in Section 5.A have multivariate extensions, which we will briefly review in this section.

7.D.1 The multivariate Laplace transform order

Extending (5.A.1), we have the following definition of the multivariate Laplace transform order. Let $\boldsymbol{X} = (X_1, X_2, \ldots, X_n)$ and $\boldsymbol{Y} = (Y_1, Y_2, \ldots, Y_n)$ be two nonnegative n-dimensional random vectors such that

$$E\Big[\exp\Big\{-\sum_{i=1}^{n} s_i X_i\Big\}\Big] \geq E\Big[\exp\Big\{-\sum_{i=1}^{n} s_i Y_i\Big\}\Big] \quad \text{for all } \boldsymbol{s} > \boldsymbol{0}. \tag{7.D.1}$$

Then $\boldsymbol{X}$ is said to be *smaller than* $\boldsymbol{Y}$ *in the Laplace transform order* (denoted as $\boldsymbol{X} \leq_{\text{Lt}} \boldsymbol{Y}$). Throughout this section we consider only nonnegative random vectors.

As in the univariate case (see Theorem 5.A.7), the multivariate order $\leq_{\text{Lt}}$ is closed under mixtures, limits in distribution, and convolutions. We do not formally state and prove these closure properties here.

The following property of the multivariate Laplace transform order can be verified easily. Recall the notation $\boldsymbol{X}_I$ and $\boldsymbol{Y}_I$ from (6.A.1).

Theorem 7.D.1. *Let* $\boldsymbol{X} = (X_1, X_2, \ldots, X_n)$ *and* $\boldsymbol{Y} = (Y_1, Y_2, \ldots, Y_n)$ *be two nonnegative* n*-dimensional random vectors. If* $\boldsymbol{X} \leq_{\text{Lt}} \boldsymbol{Y}$*, then* $\boldsymbol{X}_I \leq_{\text{Lt}} \boldsymbol{Y}_I$ *for each* $I \subseteq \{1, 2, \ldots, n\}$*. That is, the multivariate Laplace transform order is closed under marginalization.*

From Theorem 7.D.1 and (5.A.5) we see that

$$\boldsymbol{X} \leq_{\text{Lt}} \boldsymbol{Y} \Longrightarrow E[X_i] \leq E[Y_i], \quad i = 1, 2, \ldots, n, \tag{7.D.2}$$

provided the expectations exist.

The following property is also easy to verify.

Theorem 7.D.2. *Let* $\boldsymbol{X}_1, \boldsymbol{X}_2, \ldots, \boldsymbol{X}_m$ *be a set of independent random vectors where the dimension of* $\boldsymbol{X}_i$ *is* k_i*,* $i = 1, 2, \ldots, m$*. Let* $\boldsymbol{Y}_1, \boldsymbol{Y}_2, \ldots, \boldsymbol{Y}_m$ *be another set of independent random vectors where the dimension of* $\boldsymbol{Y}_i$ *is* k_i*,* $i = 1, 2, \ldots, m$*. If* $\boldsymbol{X}_i \leq_{\text{Lt}} \boldsymbol{Y}_i$ *for* $i = 1, 2, \ldots, m$*, then*

$$(\boldsymbol{X}_1, \boldsymbol{X}_2, \ldots, \boldsymbol{X}_m) \leq_{\text{Lt}} (\boldsymbol{Y}_1, \boldsymbol{Y}_2, \ldots, \boldsymbol{Y}_m).$$

That is, the multivariate Laplace transform order is closed under conjunctions.

Another closure property of the multivariate Laplace transform order is given in Theorem 7.D.7.

Theorem 7.D.3. *Let* $\boldsymbol{X}$ *and* $\boldsymbol{Y}$ *be two nonnegative random vectors. If* $\boldsymbol{X} \leq_{\text{lo}} \boldsymbol{Y}$ *or* $\boldsymbol{X} \leq_{\text{icv}} \boldsymbol{Y}$ *or* $\boldsymbol{X} \geq_{\text{dir-cx}} \boldsymbol{Y}$*, then* $\boldsymbol{X} \leq_{\text{Lt}} \boldsymbol{Y}$*. In particular, if* $\boldsymbol{X} \leq_{\text{st}} \boldsymbol{Y}$*, then* $\boldsymbol{X} \leq_{\text{Lt}} \boldsymbol{Y}$*.*

Proof. The function h_i, defined by $h_i(x) = \exp\{-s_i x\}$, is nonnegative and decreasing for each $s_i > 0$, $i = 1, 2, \ldots, n$. Therefore $\boldsymbol{X} \leq_{\text{lo}} \boldsymbol{Y} \Longrightarrow \boldsymbol{X} \leq_{\text{Lt}} \boldsymbol{Y}$ by (6.G.6) and (7.D.1). The implication $\boldsymbol{X} \leq_{\text{icv}} \boldsymbol{Y} \Longrightarrow \boldsymbol{X} \leq_{\text{Lt}} \boldsymbol{Y}$ follows from the fact that the function ϕ, defined by $\phi(\boldsymbol{x}) = \exp\{-\sum_{i=1}^{n} s_i x_i\}$, is decreasing and convex for each $\boldsymbol{s} > \boldsymbol{0}$ (and therefore $-\phi$ is increasing and concave). Finally, the implication $\boldsymbol{X} \geq_{\text{dir-cx}} \boldsymbol{Y} \Longrightarrow \boldsymbol{X} \leq_{\text{Lt}} \boldsymbol{Y}$ follows from the fact that the function ϕ above is directionally convex for each $\boldsymbol{s} > \boldsymbol{0}$. □

The following result is a multivariate analog of the right side of (5.A.13). It can be obtained from Jensen's Inequality.

Theorem 7.D.4. *Let $\boldsymbol{Y}$ be a nonnegative random vector with mean vector $(\mu_1, \mu_2, \ldots, \mu_n)$. Let $\boldsymbol{Z}$ be a random vector degenerate at $(\mu_1, \mu_2, \ldots, \mu_n)$. Then*

$$\boldsymbol{X} \le_{\text{Lt}} \boldsymbol{Z}.$$

The next result follows easily from (7.D.1).

Theorem 7.D.5. *Let $\boldsymbol{X} = (X_1, X_2, \ldots, X_n)$ and $\boldsymbol{Y} = (Y_1, Y_2, \ldots, Y_n)$ be two nonnegative n-dimensional random vectors. If $\boldsymbol{X} \le_{\text{Lt}} \boldsymbol{Y}$, then*

$$\sum_{i=1}^n a_i X_i \le_{\text{Lt}} \sum_{i=1}^n a_i Y_i, \quad \textit{whenever } a_i \ge 0,\ i = 1, 2, \ldots, n.$$

A multivariate analog of Theorem 5.A.3 is the following result. Its proof is similar to the proof of Theorem 5.A.3 and is therefore omitted.

Theorem 7.D.6. *Let $\boldsymbol{X} = (X_1, X_2, \ldots, X_n)$ and $\boldsymbol{Y} = (Y_1, Y_2, \ldots, Y_n)$ be two nonnegative n-dimensional random vectors. Then $\boldsymbol{X} \le_{\text{Lt}} \boldsymbol{Y}$ if, and only if,*

$$E\Big[\prod_{i=1}^n \phi_i(X_i)\Big] \ge E\Big[\prod_{i=1}^n \phi_i(Y_i)\Big]$$

for all completely monotone functions ϕ_i, $i = 1, 2, \ldots, n$, provided the expectations exist.

When $\boldsymbol{X}$ and $\boldsymbol{Y}$ are vectors of nonnegative integer-valued random variables, it is customary and convenient to work with their probability generating functions, rather than with their Laplace transforms. This suggests the following definition. Let $\boldsymbol{X} = (X_1, X_2, \ldots, X_n)$ and $\boldsymbol{Y} = (Y_1, Y_2, \ldots, Y_n)$ be two vectors, of nonnegative integer-valued random variables, such that

$$E\Big[\prod_{i=1}^n t_i^{X_i}\Big] \ge E\Big[\prod_{i=1}^n t_i^{Y_i}\Big] \quad \text{for all } \boldsymbol{t} \in (0,1)^n. \tag{7.D.3}$$

Then $\boldsymbol{X}$ is said to be *smaller than $\boldsymbol{Y}$ in the multivariate probability generating function order* (denoted by $\boldsymbol{X} \le_{\text{pgf}} \boldsymbol{Y}$).

It is easy to see that (7.D.3) holds if, and only if, (7.D.1) holds. That is,

$$\boldsymbol{X} \le_{\text{pgf}} \boldsymbol{Y} \Longleftrightarrow \boldsymbol{X} \le_{\text{Lt}} \boldsymbol{Y}.$$

A preservation property of the Laplace transform order is described in the next theorem. It is a multivariate extension of Theorem 5.A.9.

Theorem 7.D.7. *For $i = 1, 2, \ldots, m$, let $\{X_{j,i},\ j = 1, 2, \ldots\}$ be a sequence of nonnegative identically distributed random vectors, and assume that all the $X_{j,i}$'s are mutually independent. Let $\boldsymbol{M} = (M_1, M_2, \ldots, M_m)$ and $\boldsymbol{N} = (N_1, N_2, \ldots, N_m)$ be two vectors of nonnegative integer-valued random variables. Assume that both $\boldsymbol{N}$ and $\boldsymbol{N}$ are independent of the $X_{j,i}$'s. If $\boldsymbol{M} \le_{\text{pgf}} \boldsymbol{N}$, then*

$$\Big(\sum_{j=1}^{M_1} X_{j,1}, \sum_{j=1}^{M_2} X_{j,2}, \ldots, \sum_{j=1}^{M_m} X_{j,m}\Big) \le_{\text{Lt}} \Big(\sum_{j=1}^{N_1} X_{j,1}, \sum_{j=1}^{N_2} X_{j,2}, \ldots, \sum_{j=1}^{N_m} X_{j,m}\Big).$$

Proof. For fixed $(n_1, n_2, \ldots, n_m)$ and fixed $b_i > 0$, $i = 1, 2, \ldots, m$, we compute

$$E\big[\mathrm{e}^{-\sum_{i=1}^m b_i \sum_{j=1}^{n_i} X_{j,i}}\big] = \prod_{i=1}^m E\big[\mathrm{e}^{-b_i \sum_{j=1}^{n_i} X_{j,i}}\big] = \prod_{i=1}^m \big(L_{X_{1,i}}(b_i)\big)^{n_i},$$

where $L_{X_{1,i}}$ denotes the Laplace transform of $X_{1,i}$, $i = 1, 2, \ldots, m$. Therefore

$$\begin{aligned} E\big[\mathrm{e}^{-\sum_{i=1}^m b_i \sum_{j=1}^{M_i} X_{j,i}}\big] &= E\Big[\prod_{i=1}^m \big(L_{X_{1,i}}(b_i)\big)^{M_i}\Big] \\ &\ge E\Big[\prod_{i=1}^m \big(L_{X_{1,i}}(b_i)\big)^{N_i}\Big] \\ &= E\big[\mathrm{e}^{-\sum_{i=1}^m b_i \sum_{j=1}^{N_i} X_{j,i}}\big]. \end{aligned}$$

□

7.D.2 The multivariate factorial moments order

Let $\boldsymbol{X}$ and $\boldsymbol{Y}$ be two vectors of nonnegative integer-valued random variables such that

$$E\Big[\prod_{i=1}^n \binom{X_i}{j_i}\Big] \le E\Big[\prod_{i=1}^n \binom{Y_i}{j_i}\Big] \quad \text{for all } j_i \in \mathbb{N}_+,\ i = 1, 2, \ldots, n. \tag{7.D.4}$$

Then $\boldsymbol{X}$ is said to be *smaller than $\boldsymbol{Y}$ in the factorial moments order* (denoted by $\boldsymbol{X} \le_{\text{fm}} \boldsymbol{Y}$).

It is easy to see that

$$\boldsymbol{X} \le_{\text{fm}} \boldsymbol{Y} \Longrightarrow E\boldsymbol{X} \le E\boldsymbol{Y}.$$

The proofs of the following three results are similar to the proofs of Theorems 5.C.2, 5.C.4, and 5.C.5, respectively. We omit the straightforward details.

Theorem 7.D.8. (a) *Let $\boldsymbol{X} = (X_1, X_2, \ldots, X_n)$ and $\boldsymbol{Y} = (Y_1, Y_2, \ldots, Y_n)$ be two vectors of nonnegative integer-valued random variables. If $\boldsymbol{X} \le_{\text{fm}} \boldsymbol{Y}$, then $\boldsymbol{X} + \boldsymbol{k} \le_{\text{fm}} \boldsymbol{Y} + \boldsymbol{k}$ for every $\boldsymbol{k} \in \mathbb{N}_+^n$.*

(b) *Let* $(X_1, X_2, \ldots, X_n)$ *and* $(Y_1, Y_2, \ldots, Y_n)$ *be two vectors of nonnegative integer-valued random variables. If* $(X_1, X_2, \ldots, X_n) \le_{\rm fm} (Y_1, Y_2, \ldots, Y_n)$, *then* $(k_1X_1, k_2X_2, \ldots, k_nX_n) \le_{\rm fm} (k_1Y_1, k_2Y_2, \ldots, k_nY_n)$ *for every* $(k_1, k_2, \ldots, k_n) \in \mathbb{N}_+^n$.

(c) *Let* $\boldsymbol{X}_1, \boldsymbol{X}_2, \ldots, \boldsymbol{X}_m$ *be a set of independent* n*-dimensional vectors of nonnegative integer-valued random variables. Let* $\boldsymbol{Y}_1, \boldsymbol{Y}_2, \ldots, \boldsymbol{Y}_m$ *be another set of independent* n*-dimensional vectors of nonnegative integer-valued random variables. If* $\boldsymbol{X}_i \le_{\rm fm} \boldsymbol{Y}_i$, $i = 1, 2, \ldots, m$, *then*

$$\sum_{i=1}^{m} \boldsymbol{X}_i \le_{\rm fm} \sum_{i=1}^{m} \boldsymbol{Y}_i.$$

Theorem 7.D.9. *Let* $\boldsymbol{X}$ *and* $\boldsymbol{Y}$ *be two vectors of nonnegative integer-valued random variables. If* $\boldsymbol{X} \le_{\rm icx} \boldsymbol{Y}$, *then* $\boldsymbol{X} \le_{\rm fm} \boldsymbol{Y}$. *In particular, if* $\boldsymbol{X} \le_{\rm st} \boldsymbol{Y}$, *then* $\boldsymbol{X} \le_{\rm fm} \boldsymbol{Y}$.

Theorem 7.D.10. *Let* $\boldsymbol{X}$ *and* $\boldsymbol{Y}$ *be two vectors of nonnegative integer-valued random variables with bounded support* $\prod_{i=1}^{n}\{0, 1, 2, \ldots, b_i\}$. *If* $\boldsymbol{X} \le_{\rm fm} \boldsymbol{Y}$, *then* $\boldsymbol{b} - \boldsymbol{Y} \le_{\rm pgf} \boldsymbol{b} - \boldsymbol{X}$.

7.D.3 The multivariate moments order

Consider now two vectors, of general (that is, not necessarily integer-valued) nonnegative random variables, $\boldsymbol{X}$ and $\boldsymbol{Y}$ such that

$$E\left[\prod_{i=1}^{n} X_i^{j_i}\right] \le E\left[\prod_{i=1}^{n} Y_i^{j_i}\right] \quad \text{for all } j_i \in \mathbb{N}_+,\ i = 1, 2, \ldots, n.$$

Then $\boldsymbol{X}$ is said to be *smaller than* $\boldsymbol{Y}$ *in the moments order* (denoted as $\boldsymbol{X} \le_{\rm mom} \boldsymbol{Y}$).

Clearly,

$$\boldsymbol{X} \le_{\rm mom} \boldsymbol{Y} \Longrightarrow E\boldsymbol{X} \le E\boldsymbol{Y}.$$

The following three results are analogs of Theorems 5.C.6, 5.C.9, and 5.C.19. We omit the straightforward proofs.

Theorem 7.D.11. (a) *Let* $\boldsymbol{X}$ *and* $\boldsymbol{Y}$ *be two vectors of nonnegative random variables. If* $\boldsymbol{X} \le_{\rm mom} \boldsymbol{Y}$, *then* $\boldsymbol{X} + \boldsymbol{k} \le_{\rm mom} \boldsymbol{Y} + \boldsymbol{k}$ *for every* $\boldsymbol{k} \ge \boldsymbol{0}$.

(b) *Let* $(X_1, X_2, \ldots, X_n)$ *and* $(Y_1, Y_2, \ldots, Y_n)$ *be two vectors of nonnegative random variables. If* $(X_1, X_2, \ldots, X_n) \le_{\rm mom} (Y_1, Y_2, \ldots, Y_n)$, *then* $(k_1X_1, k_2X_2, \ldots, k_nX_n) \le_{\rm mom} (k_1Y_1, k_2Y_2, \ldots, k_nY_n)$ *for every* $(k_1, k_2, \ldots, k_n) \ge \boldsymbol{0}$.

(c) *Let* $\boldsymbol{X}_1, \boldsymbol{X}_2, \ldots, \boldsymbol{X}_m$ *be a set of independent* n*-dimensional vectors of nonnegative random variables. Let* $\boldsymbol{Y}_1, \boldsymbol{Y}_2, \ldots, \boldsymbol{Y}_m$ *be another set of independent* n*-dimensional vectors of nonnegative random variables. If* $\boldsymbol{X}_i \le_{\rm mom} \boldsymbol{Y}_i$, $i = 1, 2, \ldots, m$, *then*

$$\sum_{i=1}^{m} \boldsymbol{X}_i \leq_{\text{mom}} \sum_{i=1}^{m} \boldsymbol{Y}_i.$$

Theorem 7.D.12. *Let $\boldsymbol{X}$ and $\boldsymbol{Y}$ be two vectors of nonnegative integer-valued random variables. If $\boldsymbol{X} \leq_{\text{fm}} \boldsymbol{Y}$, then $\boldsymbol{X} \leq_{\text{mom}} \boldsymbol{Y}$. In particular, if $\boldsymbol{X} \leq_{\text{icx}} \boldsymbol{Y}$ (or if $\boldsymbol{X} \leq_{\text{st}} \boldsymbol{Y}$), then $\boldsymbol{X} \leq_{\text{mom}} \boldsymbol{Y}$.*

Theorem 7.D.13. *Let $\boldsymbol{X}$ and $\boldsymbol{Y}$ be two vectors of nonnegative random variables with bounded support $\prod_{i=1}^{n}[0, b_i]$. If $\boldsymbol{X} \leq_{\text{mom}} \boldsymbol{Y}$, then $\boldsymbol{b}-\boldsymbol{Y} \leq_{\text{Lt}} \boldsymbol{b}-\boldsymbol{X}$.*

The $\leq_{\text{uo-cx}}$ order implies the multivariate moments order as it is described in the following result. This result follows at once from Theorem 7.A.40.

Theorem 7.D.14. *Let $\boldsymbol{X}$ and $\boldsymbol{Y}$ be two vectors of nonnegative random variables. If $\boldsymbol{X} \leq_{\text{uo-cx}} \boldsymbol{Y}$, then $\boldsymbol{X} \leq_{\text{mom}} \boldsymbol{Y}$.*

7.E Complements

Section 7.A: The proofs of Theorems 7.A.1 and 7.A.2 can be derived from results of Strassen [541]; see, for instance, Rüschendorf [482]. Elton and Hill [183] derived a constructive proof of Theorem 7.A.1. Further references regarding these theorems and several variations of them can be found in Elton and Hill [182]. Most of the other results in this section are easy to derive. The first characterization of the order $\leq_{\text{icx}}$, given in Theorem 7.A.3, can be found in Müller and Stoyan [419]. The result about the convex order comparison of two sums (7.A.13) is taken from Berger [79]. The comparisons of vectors of random partial sums of random variables (Theorem 7.A.7) is taken from Jean-Marie and Liu [254]. Theorems 7.A.8 and 7.A.9 can be found in Arnold [19]. Results similar to the conclusions of Theorem 7.A.10 can be found in Alzaid and Proschan [14]. The convex order comparison of multivariate means (Example 7.A.11) is a variation of Lemma 1 of Bäuerle [59]. The necessary (and sufficient) conditions for the comparison of multivariate normal random vectors (Example 7.A.13) can be found in Müller [413]; some variations of the results in this example are given in Ding and Zhang [168]. The conditions which yield the stochastic equality of $\boldsymbol{X}$ and $\boldsymbol{Y}$ (Theorems 7.A.14 and 7.A.15) are taken from Li and Zhu [351] and from Scarsini [492], whereas Theorem 7.A.16 is taken from Baccelli and Makowski [27]. Some orders that are weaker than the multivariate convex order are studied in Mosler [399, Chapter 8]; for example, he studies the order defined by $E\phi(a_1X_1 + a_2X_2 + \cdots + a_nX_n) \leq E\phi(a_1Y_1 + a_2Y_2 + \cdots + a_nY_n)$ for all univariate convex functions ϕ and constants $a_1, a_2, \ldots, a_n$ for which the expectations exist. Fernández and Molchanov [194] studied related orders. The material in Section 7.A.5 follows Denuit, Lefèvre, and Mesfioui [148]; a version of the (m_1, m_2)-icx order for discrete random vectors is studied in Denuit, Lefèvre, and Mesfioui [150]. A more general version of Theorem 7.A.17 can be found in Bassan and Scarsini [54]. The order $\leq_{\text{symcx}}$ is defined and studied in Marshall and Olkin [383, page 282]. The fact that random vectors, that are comparable in the order $\leq_{\text{ccx}}$, must have the same covariance matrix (Theorem 7.A.23), can be found in Müller and Stoyan [419]. The "preservation property" of the convex order under independence (Theorem 7.A.24) can be found in Müller and Scarsini [417]. The results which compare random sums (Theorem 7.A.25) are taken from Pellerey [451]. The result about the ordering of multivariate normal random vectors according to the $\leq_{\text{ccx}}$ order (Example 7.A.26) is taken from Block and Sampson [94, Section 3]. The notion of directionally convex functions is studied in Shaked and Shanthikumar [509], though Fan and Lorentz [190], Marshall and Olkin [383, page 157], and Rüschendorf [483] mentioned such functions earlier. Most of the results about the directionally convex order (Section 7.A.8) are taken from Chang, Chao, Pinedo, and Shanthikumar [125] and from Meester and Shanthikumar [387]. The closure under limits property of the directionally convex order (Theorem 7.A.31) can be found in Müller and Stoyan [419]. The comparison of integrals result (Theorem 7.A.35) is taken from Miyoshi [397]. The results which compare random sums (Theorem 7.A.36) are corrected versions of Theorem 2.3 and a part of Theorem 2.4 of Pellerey [451]. The comparison of mixtures result (Theorem 7.A.37) can be found in Denuit and Müller [157], whereas the comparison of vectors with the same dependence structure (Theorem 7.A.38) can be found in Müller and Scarsini [417]. The necessary and sufficient conditions for the comparison of multivariate normal random vectors (Example 7.A.39) are taken from Müller [413]; an extension of this result to Kotz-type distributions is given in Ding and Zhang [168]. A discussion about the order $\leq_{\text{uo-cx}}$ can be found in Bergmann [82], where other orders, related to several unimodality notions, are also studied; the characterization given in Theorem 7.A.40(a) is taken from that paper. The preservation property of the order $\leq_{\text{uo-cx}}$ given in Theorem 7.A.41(a) can be found in Bergmann [80]. The results which compare random sums (Theorem 7.A.42) are taken from Pellerey [451]. Dyckerhoff and Mosler [173] introduced some relatively easy conditions for verifying $\boldsymbol{X} \leq_{\text{uo-cx}} \boldsymbol{Y}$ or $\boldsymbol{X} \leq_{\text{lo-cv}} \boldsymbol{Y}$ when $\boldsymbol{X}$ and $\boldsymbol{Y}$ have finite discrete supports. The material about the orders $\leq_m^n$ is taken from O'Brien and Scarsini [438]. Scarsini [490] has studied the order $\leq_m^2$ in some detail; in particular, he has identified a class $\mathcal{U}$ of functions such that $(X_1, X_2) \leq_m^2 (Y_1, Y_2)$ if, and only if, $E[\phi(X_1, X_2)] \leq E[\phi(Y_1, Y_2)]$ for all $\phi \in \mathcal{U}$.

Müller [412] studied stochastic orders that are defined by requiring (7.A.1) to hold for all quasiconcave or increasing quasiconcave functions.

Arnold [20], building on previous ideas, introduced a multivariate Lorenz order that is based on the characterization of the univariate Lorenz order given in Theorem 3.A.11.

Section 7.B: The development in Sections 7.B.1 and 7.B.2 follows the work of Giovagnoli and Wynn [211]. The comparison of vectors with the same dependence structure (Theorem 7.B.3) can be found in Arias-Nicolás, Fernández-Ponce, Luque-Calvo, and Suárez-Llorens [17]. The conditions under which normal random vectors can be compared with respect to the order $\leq_{\rm SD}$ (Example 7.B.5) are taken from Arias-Nicolás, Fernández-Ponce, Luque-Calvo, and Suárez-Llorens [17]. The comparison, in the order $\leq_{\rm SD}$, of vectors of order statistics (Theorem 7.B.4), has been communicated to us by Suárez-Llorens [542]. The orders that are studied in Section 7.B.3 were introduced in Shaked and Shanthikumar [518]; the properties of these orders, given in Theorems 7.B.11 and 7.B.15, can be found in that paper. The comparison, in the multivariate dispersive order, of vectors of order statistics (Theorem 7.B.12), can be found in Belzunce, Ruiz, and Ruiz [75]; an extension of it is given in Belzunce, Mercader, and Ruiz [70]. The result that compares vectors of epoch times of nonhomogeneous Poisson processes (Example 7.B.13) is taken from Belzunce and Ruiz [73]; an extension of it is given in Belzunce, Mercader, and Ruiz [70].

Khaledi and Kochar [290] and Belzunce, Ruiz, and Suárez-Llorens [76] introduced and studied multivariate dispersive orders that are generalizations, respectively, of characterizations (3.B.12) and (3.B.13) of the univariate order $\leq_{\rm disp}$.

Section 7.C: The multivariate transform orders in this section were introduced and studied in Roy [480].

Section 7.D: A basic paper on the multivariate Laplace transform order is Denuit [141], where many of the results in Section 7.D.1 can be found. The result about the preservation of the multivariate Laplace transform order under random sums (Theorem 7.D.7) is taken from Wong [568]; see also Pellerey [451]. The multivariate factorial moment order is studied in Lefèvre and Picard [337], where Theorems 7.D.9 and 7.D.10 can be found. That paper also mentions and studies the multivariate moments order. The closure properties of the multivariate order $\leq_{\rm fm}$ (Theorem 7.D.8) have been communicated to us by Lefèvre [335].

8

Stochastic Convexity and Concavity

In this chapter we study stochastic monotonicities of parametric families of distributions with respect to various stochastic orders. We have already encountered stochastic monotonicities earlier in this book. For example, condition (1.A.13) in Theorem 1.A.6, condition (3.A.47) in Theorem 3.A.21, and condition (4.A.17) in Theorem 4.A.18 describe such monotonicities. In this chapter a systematic study of such stochastic monotonicities is given. Various notions of stochastic convexity and concavity are reviewed. A multivariate extension of the notion of stochastic convexity, namely, stochastic directional convexity, is investigated in this chapter as well.

Let $\{P_\theta,\ \theta \in \Theta\}$ be a family of univariate distributions. Throughout this chapter Θ is a convex set (that is, an interval) of the real line $\mathbb{R}$ or of the set $\mathbb{N}_+$. Let $X(\theta)$ denote a random variable with distribution P_θ. It is convenient and intuitive to replace the notation $\{P_\theta,\ \theta \in \Theta\}$ by $\{X(\theta),\ \theta \in \Theta\}$, which we do throughout this chapter. Note that when we write $\{X(\theta),\ \theta \in \Theta\}$ we do not assume (and often we are not concerned with) any dependence (or independence) properties among the $X(\theta)$'s. We are only interested in the "marginal distributions" $\{P_\theta,\ \theta \in \Theta\}$ of $\{X(\theta),\ \theta \in \Theta\}$ even when in some circumstances $\{X(\theta),\ \theta \in \Theta\}$ is a well-defined stochastic process. Note also that $X(\theta)$ does not mean that X is a function of θ; it only indicates that the distribution of $X(\theta)$ is P_θ.

8.A Regular Stochastic Convexity

We start our discussion with the weakest notion of stochastic convexity and concavity and show its usefulness by a list of examples. Then, in the following sections, we introduce stronger notions which provide a systematic way of verifying the weak notion of this section.

8.A.1 Definitions

In the following definitions SI, SCX, SCV, SICX, SIL, SD, SDCV, and so forth, stand, respectively, for stochastically increasing, stochastically convex, stochastically concave, stochastically increasing and convex, stochastically increasing and linear, stochastically decreasing, stochastically decreasing and concave, and so forth.

Let $\{X(\theta),\, \theta \in \Theta\}$ be a set of random variables. Denote

(a) $\{X(\theta),\, \theta \in \Theta\} \in \mathrm{SI}$ [or SD] if $E\phi(X(\theta))$ is increasing [or decreasing] for all increasing functions ϕ,
(b) $\{X(\theta),\, \theta \in \Theta\} \in \mathrm{SCX}$ [or SCV] if $E\phi(X(\theta))$ is convex [or concave] for all convex [or concave] functions ϕ,
(c) $\{X(\theta),\, \theta \in \Theta\} \in \mathrm{SICX}$ [or SICV] if $\{X(\theta),\, \theta \in \Theta\} \in \mathrm{SI}$ and $E\phi(X(\theta))$ is increasing convex [or concave] in θ for all increasing convex [or concave] functions ϕ,
(d) $\{X(\theta),\, \theta \in \Theta\} \in \mathrm{SDCX}$ [or SDCV] if $\{X(\theta),\, \theta \in \Theta\} \in \mathrm{SD}$ and $E\phi(X(\theta))$ is decreasing convex [or concave] in θ for all increasing convex [or concave] functions ϕ,
(e) $\{X(\theta),\, \theta \in \Theta\} \in \mathrm{SIL}$ if $\{X(\theta),\, \theta \in \Theta\} \in \mathrm{SI}$ and $E\phi(X(\theta))$ is increasing convex in θ for all increasing convex functions ϕ, and is increasing concave in θ for all increasing concave functions ϕ,
(f) $\{X(\theta),\, \theta \in \Theta\} \in \mathrm{SDL}$ if $\{X(\theta),\, \theta \in \Theta\} \in \mathrm{SD}$ and $E\phi(X(\theta))$ is decreasing convex in θ for all increasing convex functions ϕ, and is decreasing concave in θ for all increasing concave functions ϕ.

Note that

$$\{X(\theta),\, \theta \in \Theta\} \in \mathrm{SIL} \Longleftrightarrow \{X(\theta),\, \theta \in \Theta\} \in \mathrm{SICX} \cap \mathrm{SICV}$$

and

$$\{X(\theta),\, \theta \in \Theta\} \in \mathrm{SDL} \Longleftrightarrow \{X(\theta),\, \theta \in \Theta\} \in \mathrm{SDCX} \cap \mathrm{SDCV}.$$

Also, since a function is convex if, and only if, its negative is concave, we see that

$$\{X(\theta),\, \theta \in \Theta\} \in \mathrm{SCX} \Longleftrightarrow \{X(\theta),\, \theta \in \Theta\} \in \mathrm{SCV}.$$

Example 8.A.1. Let $X(\mu, \sigma)$ be a normal random variable with mean μ and standard deviation σ. Then, for each $\sigma > 0$, one has $\{X(\mu, \sigma),\, \mu \in \mathbb{R}\} \in \mathrm{SIL}$. This follows from Example 8.D.4 and Theorem 8.D.11 below.

Example 8.A.2. Let $X(\lambda)$ be a Poisson random variable with mean λ. Then $\{X(\lambda),\, \lambda \in [0, \infty)\} \in \mathrm{SIL}$. This follows from Example 8.A.7 below.

Equivalently, Example 8.A.2 shows that a homogeneous Poisson process $\{K(t),\, t \geq 0\}$ is SIL.

Lynch [370] has found conditions under which a stationary renewal process $\{K(t), t \geq 0\}$ is SCX. Explicitly, let $X_2, X_3, \ldots$ be independent and identically distributed interrenewal times with a distribution function F. Let the time until the first renewal, X_1, have the equilibrium distribution function G given by $G(x) = \frac{\int_0^x \overline{F}(u)du}{EX_2}$, $x \geq 0$. Lynch [370] has shown that if X_2 has a logconcave density function, then $\{K(t),\ t \in [0, \infty)\} \in \text{SCX}$.

Example 8.A.3. Let $X(n,p)$ be a binomial random variable with mean np and variance $np(1-p)$. Then, for each $p \in (0,1)$, one has $\{X(n,p), n \in \mathbb{N}_{++}\} \in \text{SIL}$ and, for each $n \in \mathbb{N}_{++}$, one has $\{X(n,p),\ p \in (0,1)\} \in \text{SIL}$. These follow from Example 8.B.3 and Theorem 8.B.9 below.

Example 8.A.4. Let $Y(n)$, $n = 1, 2, \ldots$, be a sequence of nonnegative independent and identically distributed random variables with mean 1. For $\mu > 0$ define $X(\mu, n) = \mu \sum_{k=1}^{n} Y(k)$, $n \in \mathbb{N}_{++}$. Then, for each $n \in \mathbb{N}_{++}$, one has $\{X(\mu,n),\ \mu \in [0,\infty)\} \in \text{SIL}$ and, for each $\mu > 0$, one has $\{X(\mu,n),\ n \in \mathbb{N}_{++}\} \in \text{SIL}$. The first result follows from Example 8.D.5 and Theorem 8.D.11 below. The second result follows from Example 8.B.4 and Theorem 8.B.9 below.

Specifically, when $Y(n)$ in Example 8.A.4 is an exponential random variable we have the following example.

Example 8.A.5. Let $X(\mu, n)$ be an Erlang-n random variable with mean $n\mu$ and variance $n\mu^2$. Then, for each $n \in \mathbb{N}_{++}$, one has $\{X(\mu,n),\ \mu \in [0,\infty)\} \in$ SIL and, for each $\mu > 0$, one has $\{X(\mu,n),\ n \in \mathbb{N}_{++}\} \in \text{SIL}$.

By taking $n = 1$ in Example 8.A.4 we obtain the following result.

Example 8.A.6. Let Y be a nonnegative random variable. For $\mu > 0$ define $X(\mu) = \mu Y$. Then $\{X(\mu),\ \mu \in [0,\infty)\} \in \text{SIL}$.

Example 8.A.7. Suppose that Θ is $[0,\infty)$ or $\mathbb{N}_{++}$. The family of nonnegative random variables $\{X(\theta),\ \theta \in \Theta\}$ is said to have the semigroup property if, for all θ_1 and θ_2 in Θ, one has

$$X(\theta_1 + \theta_2) =_{\text{st}} X(\theta_1) + X(\theta_2), \tag{8.A.1}$$

where $X(\theta_1)$ and $X(\theta_2)$ in (8.A.1) are independent. Note that $\{X(\lambda),\ \lambda \in [0,\infty)\}$ of Example 8.A.2 has the semigroup property. Also, for each $\mu > 0$, it is seen that $\{X(\mu,n), n \in \mathbb{N}_{++}\}$ of Example 8.A.4 has the semigroup property. If $\{X(\theta), \theta \in \Theta\}$ has the semigroup property, then $\{X(\theta), \theta \in \Theta\} \in \text{SIL}$. This result follows from Example 8.B.7 and Theorem 8.B.9 below.

Example 8.A.8. The Beta distribution with parameters $\alpha > 0$ and $\beta > 0$ is the one that has the density function defined as

$$f_{\alpha,\beta}(x) = \frac{1}{B(\alpha,\beta)} x^{\alpha-1}(1-x)^{\beta-1}, \quad 0 < x < 1,$$

where $B(\alpha,\beta) \equiv \int_0^1 x^{\alpha-1}(1-x)^{\beta-1}\mathrm{d}x$. The beta distribution of the second kind with parameters $\alpha > 0$ and $\beta > 0$ is the one that has the density function defined as

$$g_{\alpha,\beta}(x) = \frac{1}{B(\alpha,\beta)} \frac{x^{\alpha-1}}{(1-x)^{\alpha+\beta}}, \qquad x > 0.$$

Fix a $t > 0$. Adell, Badía, and de la Cal [2] proved the following results:

(a) If $X(\theta)$ has the density function $f_{t\theta,t(1-\theta)}$, $\theta \in (0,1)$, then $\{X(\theta),\ \theta \in (0,1)\} \in \text{SICX}$.
(b) If $Y(\theta)$ has the density function $f_{t\theta+1,t(1-\theta)+1}$, $\theta \in (0,1)$, then $\{Y(\theta),\ \theta \in (0,1)\} \in \text{SICX}$.
(c) If $Z(\theta)$ has the density function $g_{t\theta,t}$, $\theta > 0$, then $\{Z(\theta),\ \theta > 0\} \in \text{SICX}$.

For a random variable Y, let F_Y and $\overline{F}_Y$ denote its distribution and survival functions, respectively. Similarly, for a random variable $X(\theta)$, let $F_X(\cdot,\theta)$ and $\overline{F}_X(\cdot,\theta)$ denote the corresponding distribution and survival functions. Since the class of functions $f_a(x) = \max\{x-a,0\}$ [$\min\{x-a,0\}$] for all $a \in \mathbb{R}$ generates all the increasing and convex [concave] functions, and since $E(\max\{X-a,0\}) = \int_a^\infty \overline{F}_X(x)\mathrm{d}x$ [$E(\min\{X-a,0\}) = -\int_{-\infty}^a F_X(x)\mathrm{d}x$] (see Section 4.A.1), we have the following equivalences.

Theorem 8.A.9. (a) $\{X(\theta),\ \theta \in \Theta\} \in \text{SICX}$ [SICV] *if, and only if,* $\{X(\theta),\ \theta \in \Theta\} \in \text{SI}$ *and* $\int_x^\infty \overline{F}_X(y,\theta)\mathrm{d}y$ [$\int_{-\infty}^x F_X(y,\theta)\mathrm{d}y$] *is increasing* [*decreasing*] *convex in* θ *for all* x, *and*
(b) $\{X(\theta),\ \theta \in \Theta\} \in \text{SDCX}$ [SDCV] *if, and only if,* $\{X(\theta),\ \theta \in \Theta\} \in \text{SD}$ *and* $\int_x^\infty \overline{F}_X(y,\theta)\mathrm{d}y$ [$\int_{-\infty}^x F_X(y,\theta)\mathrm{d}y$] *is decreasing* [*increasing*] *convex in* θ *for all* x.

For discrete random variables we have the following analog of Theorem 8.A.9.

Theorem 8.A.10. *Suppose that for each* $\theta \in \Theta$, *the support of* $X(\theta)$ *is in* $\mathbb{N}$. *Then*

(a) $\{X(\theta),\ \theta \in \Theta\} \in \text{SICX}$ [SICV] *if, and only if,* $\{X(\theta),\ \theta \in \Theta\} \in \text{SI}$ *and* $\sum_{l=k}^\infty P\{X(\theta) \geq l\}$ [$\sum_{l=-\infty}^k P\{X(\theta) \leq l\}$] *is increasing* [*decreasing*] *convex in* θ *for all* $k \in \mathbb{N}$, *and*
(b) $\{X(\theta),\ \theta \in \Theta\} \in \text{SDCX}$ [SDCV] *if, and only if,* $\{X(\theta),\ \theta \in \Theta\} \in \text{SD}$ *and* $\sum_{l=k}^\infty P\{X(\theta) \geq l\}$ [$\sum_{l=-\infty}^k P\{X(\theta) \leq l\}$] *is decreasing* [*increasing*] *convex in* θ *for all* $k \in \mathbb{N}$.

Recall the following identity which holds for any random variable Z with mean EZ:

$$EZ = -\int_{-\infty}^0 F(u)\mathrm{d}u + \int_0^\infty \overline{F}(u)\mathrm{d}u, \tag{8.A.2}$$

where F and $\overline{F}$ are the distribution function and the survival function of Z, respectively. From Theorem 8.A.9 and (8.A.2) we thus obtain the next result.

Theorem 8.A.11. *Suppose that $EX(\theta)$ is a linear function of θ.*

(a) *If $\{X(\theta),\ \theta \in \Theta\} \in$ SICX [SICV], then $\{X(\theta),\ \theta \in \Theta\} \in$ SICV [SICX], and therefore $\{X(\theta),\ \theta \in \Theta\} \in$ SIL.*
(b) *If $\{X(\theta), \theta \in \Theta\} \in$ SDCX [SDCV], then $\{X(\theta), \theta \in \Theta\} \in$ SDCV [SDCX], and therefore $\{X(\theta),\ \theta \in \Theta\} \in$ SDL.*

From Example 8.A.6 it follows that if $X(\theta)$ is uniformly distributed on $[0, \theta]$, then $\{X(\theta),\ \theta \in [0, \infty)\} \in$ SIL. However, in order to obtain the discrete analog of this result we need to proceed in a different route as in the next example.

Example 8.A.12. Let $X(n)$ be uniformly distributed on $\{0, 1, \ldots, n-1\}$. Then $\{X(n),\ n \in \mathbb{N}_+\} \in$ SIL. In order to see it first note that $EX(n)$ is a linear function of n. Thus, by Theorem 8.A.11 it is sufficient to show that $\{X(n), n \in \mathbb{N}_+\} \in$ SICV. Clearly, $\{X(n),\ n \in \mathbb{N}_+\} \in$ SI. Now we compute

$$\sum_{l=0}^{k} P\{X(n) \leq l\} = \frac{1}{n} \cdot \frac{(k+1)k}{2}.$$

This is a decreasing convex function of n. Thus the stated result follows from Theorem 8.A.10(a).

We will now present an application of these notions in establishing a stochastic inequality.

Theorem 8.A.13. *Let $\{Y_k, k \in \mathbb{N}_{++}\}$ be a sequence of independent and identically distributed nonnegative random variables independent of the two nonnegative discrete random variables M and N. Then*

(a) $M \leq_{\text{icx}} [\leq_{\text{icv}}]\ N \Longrightarrow \sum_{k=1}^{M} Y_k \leq_{\text{icx}} [\leq_{\text{icv}}]\ \sum_{k=1}^{N} Y_k$, *and*
(b) $M \leq_{\text{cx}} N \Longrightarrow \sum_{k=1}^{M} Y_k \leq_{\text{cx}} \sum_{k=1}^{N} Y_k$.

Proof. Let ϕ be an increasing and convex [concave] function and define $\psi(n) = E\phi\big(\sum_{k=1}^{n} Y_k\big)$. Then ψ is an increasing and convex [concave] function (see Example 8.A.4). Therefore $M \leq_{\text{icx}} [\leq_{\text{icv}}]\ N$ implies that $E\psi(M) = E\phi\big(\sum_{k=1}^{M} Y_k\big) \leq E\phi\big(\sum_{k=1}^{N} Y_k\big) = E\psi(N)$. This establishes part (a). When $M \leq_{\text{cx}} N$ one has $E\big(\sum_{k=1}^{M} Y_k\big) = E\big(\sum_{k=1}^{N} Y_k\big)$ (see Theorem 4.A.35). This observation combined with part (a) completes the proof for part (b). □

A stronger result than Theorem 8.A.13(b) is stated as Theorem 3.A.13 in Chapter 3. A stronger result than Theorem 8.A.13(a) is stated as Theorem 4.A.9 in Chapter 4. Theorem 8.A.13 can also be obtained from Theorem 4.A.18. In fact, we next restate Theorems 3.A.21 and 4.A.18 in terms of the terminology of this section (the assumption in Theorem 8.A.14(a) below is slightly stronger than the assumption in Theorem 4.A.18; see a comment after Theorem 4.A.18).

Theorem 8.A.14. *Let $\{X(\theta),\ \theta \in \mathcal{X}\}$ be a collection of random variables, and let Θ_1 and Θ_2 be two $\mathcal{X}$-valued random variables that are independent of $\{X(\theta),\ \theta \in \mathcal{X}\}$.*

(a) *If $\{X(\theta), \theta \in \mathcal{X}\} \in$ SICX [SICV] and if $\Theta_1 \leq_{\text{icx}} [\leq_{\text{icv}}]\Theta_2$, then $X(\Theta_1) \leq_{\text{icx}} [\leq_{\text{icv}}]\, X(\Theta_2)$.*
(b) *If $\{X(\theta),\ \theta \in \mathcal{X}\} \in$ SCX and if $\Theta_1 \leq_{\text{cx}} \Theta_2$, then $X(\Theta_1) \leq_{\text{cx}} X(\Theta_2)$.*

8.A.2 Closure properties

Closure properties of the notions that were introduced in Section 8.A.1 serve as the basis for studying the convexity and concavity properties of the performance measures of stochastic systems. In this subsection we describe some of these closure properties.

Theorem 8.A.15. *Suppose that $\{X(\theta),\ \theta \in \Theta\}$ and $\{Y(\theta),\ \theta \in \Theta\}$ are two collections of random variables such that $X(\theta)$ and $Y(\theta)$ are independent for each θ. If $\{X(\theta),\ \theta \in \Theta\} \in$ SICX [or SICV] and $\{Y(\theta),\ \theta \in \Theta\} \in$ SICX [or SICV], then $\{X(\theta) + Y(\theta),\ \theta \in \Theta\} \in$ SICX [or SICV].*

Proof. We prove the convex case only. The concave case can be similarly proven. Let $\theta_i \in \Theta$, $i = 1,2,3,4$, be such that $\theta_1 \leq \theta_2 = \theta_3 \leq \theta_4$ and $\theta_1 + \theta_4 = \theta_2 + \theta_3$. The stochastic monotonicity of $X(\theta)$ and $Y(\theta)$ can be used to construct four random variables $\hat{X}_1$, $\hat{X}_4$, $\hat{Y}_1$, and $\hat{Y}_4$ such that $\hat{X}_i =_{\text{st}} X(\theta_i)$, $\hat{Y}_i =_{\text{st}} Y(\theta_i)$, $i = 1,4$, $\hat{X}_1 \leq \hat{X}_4$ a.s., and $\hat{Y}_1 \leq \hat{Y}_4$ a.s. (see Theorem 1.A.1). Furthermore $(\hat{X}_1, \hat{X}_4)$ and $(\hat{Y}_1, \hat{Y}_4)$ can be constructed so that they are independent. Let I_1 and I_2 be independent random variables, independent of $\hat{X}_1$, $\hat{X}_4$, $\hat{Y}_1$, and $\hat{Y}_4$, such that $P\{I_1 = 0\} = P\{I_1 = 1\} = P\{I_2 = 0\} = P\{I_2 = 1\} = \frac{1}{2}$. Define $\hat{X}_2 = (1 - I_1)\hat{X}_1 + I_1\hat{X}_4$, $\hat{X}_3 = I_1\hat{X}_1 + (1 - I_1)\hat{X}_4$, $\hat{Y}_2 = (1 - I_2)\hat{Y}_1 + I_2\hat{Y}_4$, and $\hat{Y}_3 = I_2\hat{Y}_1 + (1 - I_2)\hat{Y}_4$. It is then not hard to see that $\hat{X}_2 =_{\text{st}} \hat{X}_3$, $\hat{Y}_2 =_{\text{st}} \hat{Y}_3$,

$$(\hat{X}_1, \hat{Y}_1) \leq \big[(\hat{X}_2, \hat{Y}_2), (\hat{X}_3, \hat{Y}_3)\big] \leq (\hat{X}_4, \hat{Y}_4) \quad \text{a.s.}$$

(where, for any four numbers a, b, c, and d, the notation $a \leq [b, c] \leq d$ means $a \leq \min\{b, c\}$ and $\max\{b, c\} \leq d$), and

$$(\hat{X}_1 + \hat{Y}_1) + (\hat{X}_4 + \hat{Y}_4) = (\hat{X}_2 + \hat{Y}_2) + (\hat{X}_3 + \hat{Y}_3) \quad \text{a.s.}$$

Then, for any increasing convex function ϕ, one has

$$E\phi(\hat{X}_1 + \hat{Y}_1) + E\phi(\hat{X}_4 + \hat{Y}_4) \geq E\phi(\hat{X}_2 + \hat{Y}_2) + E\phi(\hat{X}_3 + \hat{Y}_3).$$

Observe that $\hat{X}_2 \geq_{\text{icx}} X(\theta_2)$ and $\hat{Y}_2 \geq_{\text{icx}} Y(\theta_2)$. So by the preservation of the order $\geq_{\text{icx}}$ under convolution (see Theorem 4.A.8) it follows that $\hat{X}_2 + \hat{Y}_2 \geq_{\text{icx}} X(\theta_2) + Y(\theta_2)$. That is, for any increasing convex function ϕ, one has

$$E\phi(\hat{X}_2 + \hat{Y}_2) \geq E\phi(X(\theta_2) + Y(\theta_2)).$$

Similarly,

$$E\phi(\hat{X}_3 + \hat{Y}_3) \geq E\phi(X(\theta_3) + Y(\theta_3)).$$

Therefore,

$$E\phi(X(\theta_1) + Y(\theta_1)) + E\phi(X(\theta_4) + Y(\theta_4)) \geq E\phi(X(\theta_2) + Y(\theta_2)) + E\phi(X(\theta_3) + Y(\theta_3)).$$

Combining this with the preservation of stochastic monotonicity under convolution (see Theorem 1.A.3), one has $\{X(\theta) + Y(\theta),\ \theta \in \Theta\} \in \mathrm{SICX}$. □

A combination of Example 8.A.4 and Theorem 8.A.15 yields the following generalization of Example 8.A.4 which will be used later.

Example 8.A.16. Let $Y(n)$, $n = 1, 2, \ldots$, be a sequence of nonnegative independent and identically distributed random variables with mean 1, and let Z be a random variable which is independent of the $Y(n)$'s. For $\mu > 0$ define $X(\mu, n) = Z + \mu \sum_{k=1}^{n} Y(k)$, $n \in \mathbb{N}_{++}$. Then, for each $n \in \mathbb{N}_{++}$, one has $\{X(\mu, n),\ \mu \in \mathbb{R}_+\} \in \mathrm{SIL}$ and, for each $\mu > 0$, one has $\{X(\mu, n),\ n \in \mathbb{N}_{++}\} \in \mathrm{SIL}$.

Theorem 8.A.17. *Let $\{X(\theta),\ \theta \in \Theta\}$ be a family of Λ-valued random variables, where $\Lambda \subseteq \mathbb{R}$ is a convex set, and let $\{Y(\lambda),\ \lambda \in \Lambda\}$ be another family of random variables. Suppose that $X(\theta)$ and $Y(\lambda)$ are independent for any choice of $\theta \in \Theta$ and $\lambda \in \Lambda$.*

(a) *If $\{X(\theta),\ \theta \in \Theta\} \in$ SICX [SICV, SIL] and $\{Y(\lambda),\ \lambda \in \Lambda\} \in$ SICX [SICV, SIL], then $\{Y(X(\theta)),\ \theta \in \Theta\} \in$ SICX [SICV, SIL].*
(b) *If $\{X(\theta),\ \theta \in \Theta\} \in$ SDCX [SDCV, SDL] and $\{Y(\lambda),\ \lambda \in \Lambda\} \in$ SICX [SICV, SIL], then $\{Y(X(\theta)),\ \theta \in \Theta\} \in$ SDCX [SDCV, SDL].*

Proof. We will prove the increasing convex case only. The other cases can be proven similarly. Using the construction in the proof of Theorem 1.A.1 for the usual stochastic order, it is easily verified that $\{Y(X(\theta)),\ \theta \in \Theta\} \in \mathrm{SI}$. Let ϕ be an increasing and convex function. Consider

$$E\phi(Y(X(\theta))) = E\psi(X(\theta)), \tag{8.A.3}$$

where $\psi(\lambda) = E\phi(Y(\lambda))$. Since $\{Y(\lambda),\ \lambda \in \Lambda\} \in \mathrm{SICX}$, we see that ψ is an increasing and convex function. Therefore, since $\{X(\theta),\ \theta \in \Theta\} \in \mathrm{SICX}$, one sees from (8.A.3) that $E\phi(Y(X(\theta)))$ is increasing and convex in θ. Therefore $\{Y(X(\theta)),\ \theta \in \Theta\} \in \mathrm{SICX}$. □

Example 8.A.18. Let $Y(n)$, $n = 1, 2, \ldots$, be a sequence of nonnegative independent and identically distributed random variables as in Example 8.A.4, but here, since we are interested only in convexity properties with respect to n, we let the common mean of the $Y(n)$'s be a fixed $\mu > 0$. Denote

$X(n) = \sum_{k=1}^{n} Y(k)$, $n \in \mathbb{N}_{++}$, and let $\tilde{X}(n)$ be the forward recurrence time associated with $X(n)$, that is, let $\tilde{X}(n)$ have the survival function given by

$$P\{\tilde{X}(n) > x\} = \frac{\int_x^\infty P\{X(n) > u\}du}{n\mu}, \quad x \geq 0, \ n \in \mathbb{N}_{++}.$$

Then $\{\tilde{X}(n),\ n \in \mathbb{N}_{++}\} \in \text{SIL}$. This follows, by Examples 8.A.12 and 8.A.16, and by Theorem 8.A.17, from the relation (proven below)

$$\tilde{X}(n) =_{\text{st}} \tilde{Y} + \sum_{k=1}^{U(n)} Y(k), \tag{8.A.4}$$

where $U(n)$ is a random variable which is uniformly distributed on $\{0, 1, \ldots, n-1\}$, and $\tilde{Y}$ is the forward recurrence time associated with $Y(1)$, that is,

$$P\{\tilde{Y} > x\} = \frac{\int_x^\infty P\{Y(1) > u\}du}{\mu}, \quad x \geq 0.$$

The relation (8.A.4) can be proven as follows: Consider n independent renewal processes $\{N_i(t),\ t \geq 0\}$, $i = 1, 2, \ldots, n$, all with interrenewal times that are distributed as $Y(1)$, and consider the renewal process $\{N(t),\ t \geq 0\}$ with interrenewal intervals which are the sums of the corresponding interrenewal intervals of the n independent renewal processes $\{N_i(t), t \geq 0\}$, $i = 1, 2, \ldots, n$. That is, the interrenewal times that are associated with $\{N(t), t \geq 0\}$ are distributed as $X(n)$. Select a $t > 0$ and consider the associated forward recurrence time in the process $\{N(t),\ t \geq 0\}$. Clearly the value t falls in an interrenewal interval which is the sum of the n interrenewal intervals corresponding to $\{N_1(t),\ t \geq 0\}, \{N_2(t),\ t \geq 0\}, \ldots, \{N_n(t),\ t \geq 0\}$. With probability $1/n$, t falls in the interrenewal interval corresponding to the process $\{N_i(t),\ t \geq 0\}$, $i = 1, 2, \ldots, n$. Let $U(n)+1$ be the index of the process in whose interrenewal interval t falls. Then $U(n)$ is uniformly distributed on $\{0, 1, \ldots, n-1\}$. If t falls in an interval corresponding to $\{N_i(t),\ t \geq 0\}$ (that is, when $U(n) = i - 1$), then its forward recurrence time is $\tilde{Y} + \sum_{k=i+1}^{n} Y(k) =_{\text{st}} \tilde{Y} + \sum_{k=1}^{n-i} Y(k)$. Unconditioning with respect to the value i of $U(n) + 1$ we obtain

$$\tilde{X}(n) =_{\text{st}} \tilde{Y} + \sum_{k=1}^{n-U(n)-1} Y(k) =_{\text{st}} \tilde{Y} + \sum_{k=1}^{U(n)} Y(k),$$

and the proof of (8.A.4) is complete. In Example 8.B.12 of Section 8.B the reader may find a related result.

Let $\{X(n),\ n \in \mathbb{N}_+\}$ be a Markov chain with state space $\mathcal{S}$ ($\mathcal{S} = [0, \infty)$ or $\mathbb{N}_+$). Let $Y(x)$ and $Z(x)$ denote generic random variables representing $[X(n+1)|X(n) = x]$ and $[X(n+1) - x|X(n) = x]$, respectively (recall that, for a random variable U and an event A, we denote by $[U|A]$ any random variable whose distribution is the conditional distribution of U given A). Note that $Y(x) =_{\text{st}} x + Z(x)$, $x \in \mathcal{S}$.

Theorem 8.A.19. *Suppose that* $X(0) = 0$ *a.s. If* $\{Z(x),\ x \in \mathcal{S}\} \in \mathrm{SD}$ *and* $Z(x) \geq 0$ *a.s. for each* $x \in \mathcal{S}$, *then* $\{X(n),\ n \in \mathbb{N}_+\} \in \mathrm{SICV}$.

Proof. Since $Z(x) \geq 0$ a.s. we have $Y(x) \geq x$ a.s., and therefore $X(n)$ is a.s. increasing in n. For any increasing and concave function ϕ we have that $\phi(x+y) - \phi(y)$ increasing in x and decreasing in y. Therefore, since $\{Z(y),\ y \in \mathcal{S}\} \in \mathrm{SD}$, we see that $E\phi(Z(y) + y) - \phi(y)$ is decreasing in y. Since $X(n)$ is a.s. increasing in n, we have

$$
\begin{aligned}
E\phi(Z(X(n+1)) + X(n+1)) - E\phi(X(n+1))& \\
\leq E\phi(Z(X(n)) + X(n)) - E\phi(X(n)).&
\end{aligned}
$$

Noting that $X(n+1) =_{\mathrm{st}} Z(X(n)) + X(n)$, from the above equation one obtains

$$
E\phi(X(n+2)) + E\phi(X(n)) \leq E\phi(X(n+1)) + E\phi(X(n+1)).
$$

That is, $\{X(n),\ n \in \mathbb{N}_+\} \in \mathrm{SICV}$. □

Let $X(n)$ be the historical record value of a sequence of independent and identically distributed random variables $\{D_n,\ n \in \mathbb{N}_{++}\}$. That is, $X(n) = \max\{X(n-1), D_n\} = \max\{X(0), D_1, D_2, \ldots, D_n\}$, $n \in \mathbb{N}_{++}$.

Theorem 8.A.20. *If* $X(0) = 0$ *a.s., then* $\{X(n),\ n \in \mathbb{N}_+\} \in \mathrm{SICV}$.

Proof. We apply Theorem 8.A.19. Here $Y(x) =_{\mathrm{st}} \max\{D_n, x\}$ and $Z(x) =_{\mathrm{st}} \max\{D_n - x, 0\}$. Clearly, $\{Z(x),\ x \geq 0\}$ satisfies the conditions of Theorem 8.A.19. □

8.A.3 Stochastic m-convexity

Let $\mathcal{S}$ be a subinterval of the real line. Recall from Section 3.A.5 the class $\mathcal{M}^{\mathcal{S}}_{m\text{-cx}}$ of all functions $\phi : \mathcal{S} \to \mathbb{R}$ whose mth derivative $\phi^{(m)}$ exists and satisfies $\phi^{(m)}(x) \geq 0$, for all $x \in \mathcal{S}$, or which are limits of sequences of functions whose mth derivative is continuous and nonnegative on $\mathcal{S}$, $m = 1, 2, \ldots$. A function $\phi : \mathcal{S} \to \mathbb{R}$ is said to be m-increasing convex if $\phi \in \bigcap_{k=1}^{m} \mathcal{M}^{\mathcal{S}}_{k\text{-cx}}$. A set of random variables $\{X(\theta),\ \theta \in \Theta\}$ (Θ is a subinterval of the real line) is said to be *stochastically m-increasing convex* if $E\phi(X(\theta))$ is m-increasing convex in θ whenever ϕ is m-increasing convex. If Θ is a subinterval of $\mathbb{N}_{++}$, then the definition of stochastic m-increasing convexity is similar; we do not give the details here — they can be found in Denuit, Lefèvre, and Utev [155].

The proofs of most of the following examples, as well as many other examples, can be found in Denuit, Lefèvre, and Utev [155].

Example 8.A.21. Let $X(\lambda)$ be a Poisson random variable with mean λ. Then $\{X(\lambda),\ \lambda \in [0, \infty)\}$ is stochastically m-increasing convex for each $m \in \mathbb{N}_{++}$.

Example 8.A.22. Let $X(n,p)$ be a binomial random variable with mean np and variance $np(1-p)$. Then, for each $p \in (0,1)$, one has that $\{X(n,p),\, n \in \mathbb{N}_{++}\}$ is stochastically m-increasing convex, and for each $n \in \mathbb{N}_{++}$, one has that $\{X(n,p),\, p \in (0,1)\}$ is stochastically m-increasing convex, for each $m \in \mathbb{N}_{++}$.

Example 8.A.23. Let $Y(n)$, $n = 1,2,\ldots$, be a sequence of nonnegative independent and identically distributed random variables with mean 1. For $\mu > 0$ define $X(\mu,n) = \mu\sum_{k=1}^{n} Y(k)$, $n \in \mathbb{N}_{++}$. Then, for each $n \in \mathbb{N}_{++}$, one has that $\{X(\mu,n),\, \mu \in [0,\infty)\}$ is stochastically m-increasing convex, and for each $\mu > 0$, one has that $\{X(\mu,n),\, n \in \mathbb{N}_{++}\}$ is stochastically m-increasing convex, for each $m \in \mathbb{N}_{++}$.

Specifically, when $Y(n)$ in Example 8.A.23 is an exponential random variable we have the following example.

Example 8.A.24. Let $X(\mu,n)$ be an Erlang-n random variable with mean $n\mu$ and variance $n\mu^2$. Then, for each $n \in \mathbb{N}_{++}$, one has that $\{X(\mu,n),\, \mu \in [0,\infty)\}$ is stochastically m-increasing convex, and for each $\mu > 0$, one has $\{X(\mu,n),\, n \in \mathbb{N}_{++}\}$ is stochastically m-increasing convex, for each $m \in \mathbb{N}_{++}$.

By taking $n = 1$ in Example 8.A.23 we obtain the following result.

Example 8.A.25. Let Y be a nonnegative random variable. For $\mu > 0$ define $X(\mu) = \mu Y$. Then $\{X(\mu),\, \mu \in [0,\infty)\}$ is stochastically m-increasing convex for each $m \in \mathbb{N}_{++}$.

When the set of random variables is parametrized by a location parameter then we have:

Example 8.A.26. Let Y be a real random variable. For $\mu > 0$ define $X(\mu) = Y + \mu$. Then $\{X(\mu),\, \mu \in [0,\infty)\}$ is stochastically m-increasing convex for each $m \in \mathbb{N}_{++}$.

Another example of interest is the following.

Example 8.A.27. Let $X(n)$ be uniformly distributed on $\{0,1,\ldots,n-1\}$. Then $\{X(n),\, n \in \mathbb{N}_{+}\}$ is stochastically m-increasing convex for each $m \in \mathbb{N}_{++}$.

Since the composition of two m-increasing functions is m-increasing, we obtain the following closure properties of stochastic m-convexity.

Theorem 8.A.28. (a) *Let $\varphi : \mathcal{S} \to \mathbb{R}$ be an m-increasing convex function. If $\{X(\theta),\, \theta \in \Theta\}$ is stochastically m-increasing convex, then $\{\varphi(X(\theta)),\, \theta \in \Theta\}$ is also stochastically m-increasing convex.*

(b) *Let $\vartheta : \Theta \to \Theta$ be an m-increasing convex function. If $\{X(\theta),\, \theta \in \Theta\}$ is stochastically m-increasing convex, then $\{X(\vartheta(\theta)),\, \theta \in \Theta\}$ is also stochastically m-increasing convex.*

From Theorem 8.A.28(a) and Example 8.A.23 we obtain the following result.

Theorem 8.A.29. *Let $\{Y_n, n \geq 1\}$ be a sequence of nonnegative, independent and identically distributed random variables. Let $\{N(\theta),\ \theta \in \Theta\}$ be a set of nonnegative integer-valued random variables, independent of the Y_n's. Define $X(\theta) = \sum_{n=1}^{N(\theta)} Y_n$. If $\{N(\theta),\ \theta \in \Theta\}$ is stochastically m-increasing convex, then $\{X(\theta),\ \theta \in \Theta\}$ is stochastically m-increasing convex.*

8.B Sample Path Convexity

Sample path convexity is one powerful tool that can be used for the purpose of obtaining the regular convexity notions presented in Section 8.A. Two other related tools will be described in Sections 8.C and 8.D.

8.B.1 Definitions

Consider a family $\{X(\theta), \theta \in \Theta\}$ of random variables. Let $\theta_i \in \Theta$, $i = 1, 2, 3, 4$, be any four values such that $\theta_1 \leq \theta_2 \leq \theta_3 \leq \theta_4$ and $\theta_1 + \theta_4 = \theta_2 + \theta_3$.

If there exist four random variables $\hat{X}_i$, $i = 1, 2, 3, 4$, defined on a common probability space, such that $\hat{X}_i =_{\text{st}} X(\theta_i)$, $i = 1, 2, 3, 4$, and

(a) (i) $\max[\hat{X}_2, \hat{X}_3] \leq \hat{X}_4$ a.s. and (ii) $\hat{X}_2 + \hat{X}_3 \leq \hat{X}_1 + \hat{X}_4$ a.s., then $\{X(\theta), \theta \in \Theta\}$ is said to be *stochastically increasing and convex in the sample path sense* (denoted by $\{X(\theta),\ \theta \in \Theta\} \in \text{SICX(sp)}$);
(b) (i) $\hat{X}_1 \leq \min[\hat{X}_2, \hat{X}_3]$ a.s. and (ii) $\hat{X}_1 + \hat{X}_4 \leq \hat{X}_2 + \hat{X}_3$ a.s., then $\{X(\theta), \theta \in \Theta\}$ is said to be *stochastically increasing and concave in the sample path sense* (denoted by $\{X(\theta),\ \theta \in \Theta\} \in \text{SICV(sp)}$);
(c) (i) $\hat{X}_1 \geq \max[\hat{X}_2, \hat{X}_3]$ a.s. and (ii) $\hat{X}_1 + \hat{X}_4 \geq \hat{X}_2 + \hat{X}_3$ a.s., then $\{X(\theta), \theta \in \Theta\}$ is said to be *stochastically decreasing and convex in the sample path sense* (denoted by $\{X(\theta),\ \theta \in \Theta\} \in \text{SDCX(sp)}$);
(d) (i) $\hat{X}_4 \leq \min[\hat{X}_2, \hat{X}_3]$ a.s. and (ii) $\hat{X}_1 + \hat{X}_4 \leq \hat{X}_2 + \hat{X}_3$ a.s., then $\{X(\theta), \theta \in \Theta\}$ is said to be *stochastically decreasing and concave in the sample path sense* (denoted by $\{X(\theta),\ \theta \in \Theta\} \in \text{SDCV(sp)}$);
(e) (i) $\max[\hat{X}_2, \hat{X}_3] \leq \hat{X}_4$ a.s. and (ii) $\hat{X}_1 + \hat{X}_4 = \hat{X}_2 + \hat{X}_3$ a.s., then $\{X(\theta), \theta \in \Theta\}$ is said to be *stochastically increasing and linear in the sample path sense* (denoted by $\{X(\theta),\ \theta \in \Theta\} \in \text{SIL(sp)}$);
(f) (i) $\hat{X}_1 \geq \max[\hat{X}_2, \hat{X}_3]$ a.s. and (ii) $\hat{X}_1 + \hat{X}_4 = \hat{X}_2 + \hat{X}_3$ a.s., then $\{X(\theta), \theta \in \Theta\}$ is said to be *stochastically decreasing and linear in the sample path sense* (denoted by $\{X(\theta),\ \theta \in \Theta\} \in \text{SDL(sp)}$).

Although Condition (i) in these definitions requires stochastic monotonicity in X_i, $i = 1, 2, 3, 4$, we do not require the construction of $\hat{X}_i$, $i = 2, 3$, to satisfy any a.s. monotonicity property (that is, we do not require that either $\hat{X}_2 \geq \hat{X}_3$ a.s. or $\hat{X}_2 \leq \hat{X}_3$ a.s. be satisfied).

Example 8.B.1. Let $X(\mu, \sigma)$ be a normal random variable with mean μ and standard deviation σ. Then, for each $\sigma > 0$, one has $\{X(\mu, \sigma),\ \mu \in \mathbb{R}\} \in \text{SIL(sp)}$. This follows from Example 8.D.4 and Theorem 8.D.11 below.

Example 8.B.2. Let $X(\lambda)$ be a Poisson random variable with mean λ. Then $\{X(\lambda),\ \lambda \in \mathbb{R}_+\} \in \mathrm{SIL(sp)}$. This follows from Example 8.B.7 below.

Example 8.B.3. Let $X(n,p)$ be a binomial random variable with mean np and variance $np(1-p)$. Then, for each $p \in (0,1)$, one has $\{X(n,p),\ n \in \mathbb{N}_{++}\} \in \mathrm{SIL(sp)}$ and, for each $n \in \mathbb{N}_{++}$, one has $\{X(n,p),\ p \in (0,1)\} \in \mathrm{SIL(sp)}$. The first result follows from Example 8.B.4 below. In order to prove the second result, first note that $X(n,p) =_{\mathrm{st}} X_1(p) + X_2(p) + \cdots + X_n(p)$, where $X_j(p)$, $j = 1, 2, \ldots, n$, are independent and identically distributed Bernoulli random variables with $P\{X_j(p)\} = p$. We will show that

$$\{X_1(p),\ p \in (0,1)\} \in \mathrm{SIL(sp)}. \tag{8.B.1}$$

The second result above then follows from Theorem 8.B.10 below. To prove (8.B.1) let p_i, $i = 1, 2, 3, 4$, be such that $0 < p_1 \le p_2 \le p_3 \le p_4 < 1$ and $p_1 + p_4 = p_2 + p_3$. Let U be a uniform $(0,1)$ random variable. Let I_A denote the indicator function of A. Define

$$\hat{X}_1 = I_{\{U \le p_1\}}, \qquad \hat{X}_2 = I_{\{U \le p_2\}},$$
$$\hat{X}_3 = I_{\{U \le p_1\}} + I_{\{p_2 \le U \le p_4\}}, \qquad \hat{X}_4 = I_{\{U \le p_4\}}.$$

Then $\hat{X}_i =_{\mathrm{st}} X_1(p)$, $i = 1, 2, 3, 4$, and $\hat{X}_i$, $i = 1, 2, 3, 4$, satisfy the conditions given in the definitions of SICX(sp) and SICV(sp). This proves (8.B.1).

Example 8.B.4. Let $Y(n)$, $n = 1, 2, \ldots$, be a sequence of nonnegative independent and identically distributed random variables with mean 1. For $\mu > 0$ define $X(\mu, n) = \mu \sum_{k=1}^{n} Y(k)$, $n \in \mathbb{N}_{++}$. Then, for each $n \in \mathbb{N}_{++}$, one has $\{X(\mu,n),\ \mu \in \mathbb{R}_+\} \in \mathrm{SIL(sp)}$ and, for each $\mu > 0$, one has $\{X(\mu,n),\ n \in \mathbb{N}_{++}\} \in \mathrm{SIL(sp)}$. The first result follows from Example 8.D.5 and Theorem 8.D.11 below. The second result follows from Example 8.B.7 below.

Specifically, when $Y(n)$ in Example 8.B.4 is an exponential random variable we have the following example.

Example 8.B.5. Let $X(\mu,n)$ be an Erlang-n random variable with mean $n\mu$ and variance $n\mu^2$. Then, for each $n \in \mathbb{N}_{++}$, one has $\{X(\mu,n),\ \mu \in \mathbb{R}_+\} \in \mathrm{SIL(sp)}$ and, for each $\mu > 0$, one has $\{X(\mu,n),\ n \in \mathbb{N}_{++}\} \in \mathrm{SIL(sp)}$.

By taking $n = 1$ in Example 8.B.4 we obtain the following result.

Example 8.B.6. Let Y be a nonnegative random variable. For $\mu > 0$ define $X(\mu) = \mu Y$. Then $\{X(\mu),\ \mu \in \mathbb{R}_+\} \in \mathrm{SIL(sp)}$.

Example 8.B.7. If $\{X(\theta),\ \theta \in \Theta\}$ has the semigroup property (see Example 8.A.7), then $\{X(\theta), \theta \in \Theta\} \in \mathrm{SIL(sp)}$. In order to see it let $\theta_i \in \Theta, = 1, 2, 3, 4$, be such that $\theta_1 \le \theta_2 \le \theta_3 \le \theta_4$ and $\theta_1 + \theta_4 = \theta_2 + \theta_3$. Let Z_i, $i = 1, 2, 3, 4$, be independent random variables such that

$$Z_1 =_{\text{st}} X(\theta_1),$$
$$Z_2 =_{\text{st}} X(\theta_2 - \theta_1),$$
$$Z_3 =_{\text{st}} X(\theta_3 - \theta_2),$$

and

$$Z_4 =_{\text{st}} X(\theta_4 - \theta_3),$$

where, by convention, $X(0) \equiv 0$. Define

$$\hat{X}_1 = Z_1,$$
$$\hat{X}_2 = Z_1 + Z_2,$$
$$\hat{X}_3 = Z_1 + Z_3 + Z_4,$$

and

$$\hat{X}_4 = Z_1 + Z_2 + Z_3 + Z_4.$$

Then $\hat{X}_i =_{\text{st}} X(\theta_i)$, $i = 1, 2, 3, 4$, and $\hat{X}_i$, $i = 1, 2, 3, 4$, satisfy the conditions given in the definitions of SICX(sp) and SICV(sp). This proves the result stated above.

The following theorem is obvious. A more general result is proven in Theorem 8.B.13 (see Corollary 8.B.14).

Theorem 8.B.8. (a) *If* $\{X(\theta),\ \theta \in \Theta\} \in$ SICX(sp) [*or* SICV(sp)] *and if* ϕ *is an increasing convex* [*or concave*] *function, then* $\{\phi(X(\theta)),\ \theta \in \Theta\} \in$ SICX(sp) [*or* SICV(sp)].

(b) *If* $\{X(\theta),\ \theta \in \Theta\} \in$ SDCX(sp) [*or* SDCV(sp)] *and if* ϕ *is an increasing convex* [*or concave*] *function, then* $\{\phi(X(\theta)),\ \theta \in \Theta\} \in$ SDCX(sp) [*or* SDCV(sp)].

Theorem 8.B.8 shows that the sample path notions imply the regular notions of stochastic convexity/concavity. Counterexamples can be constructed to show that the reverse need not be true. We have the following results.

Theorem 8.B.9.

$$\text{SICX(sp)} \Longrightarrow \text{SICX},$$
$$\text{SICV(sp)} \Longrightarrow \text{SICV},$$
$$\text{SDCX(sp)} \Longrightarrow \text{SDCX},$$
$$\text{SDCV(sp)} \Longrightarrow \text{SDCV}.$$

8.B.2 Closure properties

In this section we present some closure properties of the sample path convexity notions.

Theorem 8.B.10. *Let $\{X(\theta),\ \theta \in \Theta\}$ and $\{Y(\theta),\ \theta \in \Theta\}$ be two families of random variables such that for each $\theta \in \Theta$, $X(\theta)$ and $Y(\theta)$ are independent. Then*

(a) $\{X(\theta),\ \theta \in \Theta\} \in \text{SICX(sp)}$ *and* $\{Y(\theta),\ \theta \in \Theta\} \in \text{SICX(sp)} \Longrightarrow \{X(\theta) + Y(\theta),\ \theta \in \Theta\} \in \text{SICX(sp)}$,
(b) $\{X(\theta),\ \theta \in \Theta\} \in \text{SICV(sp)}$ *and* $\{Y(\theta),\ \theta \in \Theta\} \in \text{SICV(sp)} \Longrightarrow \{X(\theta) + Y(\theta),\ \theta \in \Theta\} \in \text{SICV(sp)}$,
(c) $\{X(\theta), \theta \in \Theta\} \in \text{SDCX(sp)}$ *and* $\{X(\theta), \theta \in \Theta\} \in \text{SDCX(sp)} \Longrightarrow \{X(\theta)+ Y(\theta),\ \theta \in \Theta\} \in \text{SDCX(sp)}$, *and*
(d) $\{X(\theta),\ \theta \in \Theta\} \in \text{SDCV(sp)}$ *and* $\{Y(\theta), \theta \in \Theta\} \in \text{SDCV(sp)} \Longrightarrow \{X(\theta)+ Y(\theta),\ \theta \in \Theta\} \in \text{SDCV(sp)}$.

Proof. We will prove part (a) only, since the other parts can be similarly proven. Let $\theta_i \in \Theta$, $i = 1, 2, 3, 4$, be any four values such that $\theta_1 \le \theta_2 \le \theta_3 \le \theta_4$ and $\theta_1 + \theta_4 = \theta_2 + \theta_3$. From the definition of SICX(sp) one sees that there exist eight random variables $\hat{X}_i, \hat{Y}_i$, $i = 1, 2, 3, 4$, defined on a common probability space, such that $\hat{X}_i =_{\text{st}} X(\theta_i)$, $\hat{Y}_i =_{\text{st}} Y(\theta_i)$, $i = 1, 2, 3, 4$, and

$$\max[\hat{X}_2, \hat{X}_3] \le \hat{X}_4 \text{ a.s.,} \qquad \max[\hat{Y}_2, \hat{Y}_3] \le \hat{Y}_4 \text{ a.s.,}$$
$$\hat{X}_2 + \hat{X}_3 \le \hat{X}_1 + \hat{X}_4 \text{ a.s.,} \qquad \hat{Y}_2 + \hat{Y}_3 \le \hat{Y}_1 + \hat{Y}_4 \text{ a.s.,}$$

and $\hat{X}_i$ and $\hat{Y}_i$ are independent, $i = 1, 2, 3, 4$. Let $\hat{Z}_i = \hat{X}_i + \hat{Y}_i$, $i = 1, 2, 3, 4$. Then $Z_i =_{\text{st}} X(\theta_i) + Y(\theta_i)$, $i = 1, 2, 3, 4$, and

$$\max[\hat{Z}_2, \hat{Z}_3] \le \hat{Z}_4 \text{ a.s.} \quad \text{and} \quad \hat{Z}_2 + \hat{Z}_3 \le \hat{Z}_1 + \hat{Z}_4 \text{ a.s.}$$

Therefore, $\{X(\theta) + Y(\theta),\ \theta \in \Theta\} \in \text{SICX(sp)}$. □

A combination of Example 8.B.4 and Theorem 8.B.10 yields the following generalization of Example 8.B.4 which will be used later.

Example 8.B.11. Let $Y(n)$, $n = 1, 2, \ldots$, be a sequence of nonnegative independent and identically distributed random variables with mean 1, and let Z be a random variable which is independent of the $Y(n)$'s. For $\mu > 0$ define $X(\mu, n) = Z + \mu \sum_{k=1}^{n} Y(k)$, $n \in \mathbb{N}_{++}$. Then, for each $n \in \mathbb{N}_{++}$, one has $\{X(\mu, n),\ \mu \in \mathbb{R}_{+}\} \in \text{SIL(sp)}$, and, for each $\mu > 0$, one has $\{X(\mu, n),\ n \in \mathbb{N}_{++}\} \in \text{SIL(sp)}$.

Example 8.B.12. Let $Y(n)$, $n = 1, 2, \ldots$, be a sequence of nonnegative independent and identically distributed random variables with a common mean $\mu > 0$, as in Example 8.A.18. Let Y^* be the spread of the renewal process generated by the $Y(n)$'s; that is, if f is the density function of $Y(1)$, then the density function of Y^* is $(1/\mu)xf(x)$. Denote $X(n) = \sum_{k=1}^{n} Y(k)$, $n \in \mathbb{N}_{++}$, and let $X^*(n)$ be the spread corresponding to $X(n)$. Then $\{X^*(n),\ n \in \mathbb{N}_{++}\} \in \text{SIL(sp)}$. This follows, by Example 8.B.11, from the relation

$$X^*(n) =_{\text{st}} Y^* + \sum_{k=1}^{n-1} Y(k). \tag{8.B.2}$$

The relation (8.B.2) can be proven as follows: Consider n independent renewal processes $\{N_i(t),\ t \geq 0\}$, $i = 1, 2, \ldots, n$, all with interrenewal times that are distributed as $Y(1)$, and consider the renewal process $\{N(t),\ t \geq 0\}$ with interrenewal intervals which are the sums of the corresponding interrenewal intervals of the n independent renewal processes $\{N_i(t), t \geq 0\}$, $i = 1, 2, \ldots, n$. That is, the interrenewal times that are associated with $\{N(t),\ t \geq 0\}$ are distributed as $X(n)$. Select a $t > 0$ and consider the spread corresponding to the process $\{N(t),\ t \geq 0\}$. Clearly the value t falls in an interrenewal interval which is the sum of the n interrenewal intervals corresponding to $\{N_1(t),\ t \geq 0\}, \{N_2(t),\ t \geq 0\}, \ldots, \{N_n(t),\ t \geq 0\}$. With probability $1/n$, t falls in the interrenewal interval corresponding to the process $\{N_i(t),\ t \geq 0\}$, $i = 1, 2, \ldots, n$. Let $U(n)$ be the index of the process in whose interrenewal interval t falls. Then $U(n)$ is uniformly distributed on $\{1, 2, \ldots, n\}$. If t falls in an interval corresponding to $\{N_i(t),\ t \geq 0\}$ (that is, when $U(n) = i$), then its spread is $Y^* + \sum_{k \neq i} Y(k) =_{\text{st}} Y^* + \sum_{k=1}^{n-1} Y(k)$. Note that the distribution of the spread is independent of i. Therefore, by unconditioning with respect to the value i of $U(n)$ we obtain (8.B.2).

Theorem 8.B.13. *Let $\{X(\theta),\ \theta \in \Theta\}$ be a family of Λ-valued random variables, where $\Lambda \subset \mathbb{R}$ is a convex set. Also, let $\{Y(\lambda),\ \lambda \in \Lambda\}$ be another family of random variables. Suppose that $X(\theta)$ and $Y(\lambda)$ are independent for any choice of $\theta \in \Theta$ and $\lambda \in \Lambda$.*

(a) *If $\{X(\theta),\ \theta \in \Theta\} \in$ SICX(sp) [SICV(sp)] and $\{Y(\lambda),\ \lambda \in \Lambda\} \in$ SICX(sp) [SICV(sp)], then $\{Y(X(\theta)),\ \theta \in \Theta\} \in$ SICX(sp) [SICV(sp)].*
(b) *If $\{X(\theta), \theta \in \Theta\} \in$ SDCX(sp) [SDCV(sp)] and $\{Y(\lambda), \lambda \in \Lambda\} \in$ SICX(sp) [SICV(sp)], then $\{Y(X(\theta)),\ \theta \in \Theta\} \in$ SDCX(sp) [SDCV(sp)].*

Proof. We will prove the convex case of part (a) only, as the proofs of the other cases are similar. Let $\theta_i \in \Theta$, $i = 1, 2, 3, 4$, be any four values such that $\theta_1 \leq \theta_2 \leq \theta_3 \leq \theta_4$ and $\theta_1 + \theta_4 = \theta_2 + \theta_3$. Since $\{X(\theta),\ \theta \in \Theta\} \in$ SICX(sp), there exist four random variables $\hat{X}_i$, $i = 1, 2, 3, 4$, defined on a common probability space, such that $\hat{X}_i =_{\text{st}} X(\theta_i)$, $i = 1, 2, 3, 4$, and

$$[\hat{X}_2, \hat{X}_3] \leq \hat{X}_4 \text{ a.s.} \quad \text{and} \quad \hat{X}_2 + \hat{X}_3 \leq \hat{X}_1 + \hat{X}_4 \text{ a.s.}$$

Let

$$X_2^* = \min[\hat{X}_4, \hat{X}_1 + \hat{X}_4 - \hat{X}_3], \tag{8.B.3}$$

and

$$X_1^* = X_2^* + \hat{X}_3 - \hat{X}_4.$$

Clearly, X_1^* and $X_2^* \in \Lambda$ a.s., and

$$X_1^* \le \hat{X}_1 \quad \text{and} \quad X_2^* \ge \hat{X}_2. \tag{8.B.4}$$

Also,

$$[X_2^*, \hat{X}_3] \le \hat{X}_4 \quad \text{and} \quad X_1^* + \hat{X}_4 = X_2^* + \hat{X}_3.$$

Therefore, since $\{Y(\lambda),\ \lambda \in \Lambda\} \in \text{SICX(sp)}$, there exist four random variables Z_1^*, Z_2^*, $\hat{Z}_3$, and $\hat{Z}_4$, defined on a common probability space, such that $Z_1^* =_{\text{st}} Y(X_1^*)$, $Z_2^* =_{\text{st}} Y(X_2^*)$, $\hat{Z}_3 =_{\text{st}} Y(\hat{X}_3)$, $\hat{Z}_4 =_{\text{st}} Y(\hat{X}_4)$, and

$$[Z_2^*, \hat{Z}_3] \le \hat{Z}_4 \text{ a.s.} \quad \text{and} \quad Z_2^* + \hat{Z}_3 \le Z_1^* + \hat{Z}_4 \text{ a.s.} \tag{8.B.5}$$

Since $Y(\lambda)$ is stochastically increasing in λ, from (8.B.4) it is seen that there exist random variables $\hat{Z}_i$, $i = 1, 2$, such that $\hat{Z}_i =_{\text{st}} Y(\hat{X}_i)$, $i = 1, 2$, and

$$Z_1^* \le \hat{Z}_1 \quad \text{and} \quad Z_2^* \ge \hat{Z}_2.$$

Then from (8.B.5) one sees that

$$[\hat{Z}_2, \hat{Z}_3] \le \hat{Z}_4 \text{ a.s.} \quad \text{and} \quad \hat{Z}_2 + \hat{Z}_3 \le \hat{Z}_1 + \hat{Z}_4 \text{ a.s.}$$

The proof is completed by observing that $Y(X(\theta_i)) =_{\text{st}} \hat{Z}_i$, $i = 1, 2, 3, 4$. □

By letting $\{Y(\lambda),\ \lambda \in \Lambda\}$ of Theorem 8.B.13 be deterministic (we denote it then as a real function $\phi : \Lambda \to \mathbb{R}$) we obtain the following corollary.

Corollary 8.B.14. *Let $\{X(\theta),\ \theta \in \Theta\}$ be a family of Λ-valued random variables, where $\Lambda \subset \mathbb{R}$ is a convex set, and let ϕ be a real function on Λ.*

(a) *If $\{X(\theta),\ \theta \in \Theta\} \in$ SICX(sp) [or SICV(sp)] and ϕ is increasing and convex [or concave], then $\{\phi(X(\theta)),\ \theta \in \Theta\} \in$ SICX(sp) [or SICV(sp)].*
(b) *If $\{X(\theta),\ \theta \in \Theta\} \in$ SDCX(sp) [or SDCV(sp)] and ϕ is increasing and convex [or concave], then $\{\phi(X(\theta)),\ \theta \in \Theta\} \in$ SDCX(sp) [or SDCV(sp)].*

By letting $\{X(\theta),\ \theta \in \Theta\}$ of Theorem 8.B.13 be deterministic (we denote it then as a real function $\phi : \Theta \to \Lambda$) we obtain the following corollary.

Corollary 8.B.15. *Let $\{Y(\lambda),\ \lambda \in \Lambda\}$ be a family of real-valued random variables, where $\Lambda \subset \mathbb{R}$ is a convex set, and let ϕ be a Λ-valued function on Θ, where $\Theta \subset \mathbb{R}$ is a convex set.*

(a) *If $\{Y(\lambda),\ \lambda \in \Lambda\} \in$ SICX(sp) [or SICV(sp)] and ϕ is increasing and convex [or concave], then $\{Y(\phi(\theta)),\ \theta \in \Theta\} \in$ SICX(sp) [or SICV(sp)].*
(b) *If $\{Y(\lambda),\ \lambda \in \Lambda\} \in$ SDCX(sp) [or SDCV(sp)] and ϕ is increasing and convex [or concave], then $\{Y(\phi(\theta)),\ \theta \in \Theta\} \in$ SDCX(sp) [or SDCV(sp)].*

Let $\{X(n),\ n \in \mathbb{N}_+\}$ be a Markov chain with state space $\mathcal{S}$ ($\mathcal{S} = \mathbb{R}_+$ or $\mathbb{N}_+$). Let $Y(x) =_{\text{st}} [X(n+1) \big| X(n) = x]$ and $Z(x) = Y(x) - x$, $x \in \mathcal{S}$.

Theorem 8.B.16. *Suppose* $X(0) = x_0$ *a.s. If* $Z(x) \geq 0$ *a.s. for each* $x \in \mathcal{S}$ *and* $\{Z(x),\ x \in \mathcal{S}\} \in \text{SI}$, *then* $\{X(n),\ n \in \mathbb{N}_+\} \in \text{SICX(sp)}$.

Proof. Since $Z(x) \geq 0$ a.s., for $n_1 \leq n_2$ we have $X(n_1) \leq X(n_2)$ a.s. Let n_3 and n_4 be such that $n_1 \leq n_2 \leq n_3 \leq n_4$ and $n_1 + n_4 = n_2 + n_3$. Define $m = n_4 - n_2 = n_3 - n_1$ and $Z^{(m)}(x) =_{\text{st}} [X(m) - x \big| X(0) = x]$. Since $Z(x)$ is stochastically increasing in x, using sample path construction (as in the proof of Theorem 6.B.3 when it applies to Theorem 6.B.34 through Theorems 6.B.32 and 6.B.31), it can be established that $Z^{(m)}(x)$ is also stochastically increasing in x. Then there exist two random vectors $(\hat{X}_1, \hat{Z}_1)$ and $(\hat{X}_2, \hat{Z}_2)$ defined on a common probability space such that $(\hat{X}_i, \hat{Z}_i) =_{\text{st}} (X(n_i), Z^{(m)}(X(n_i)))$, $i = 1, 2$, and

$$(\hat{X}_1, \hat{Z}_1) \leq (\hat{X}_2, \hat{Z}_2) \text{ a.s.} \tag{8.B.6}$$

Set

$$\hat{X}_3 = \hat{X}_1 + \hat{Z}_1 \quad \text{and} \quad \hat{X}_4 = \hat{X}_2 + \hat{Z}_2.$$

Since $Z^{(m)}(x) \geq 0$ a.s., from (8.B.6) it follows that

$$\max[\hat{X}_2, \hat{X}_3] \leq \hat{X}_4 \quad \text{and} \quad \hat{X}_1 + \hat{X}_4 \geq \hat{X}_2 + \hat{X}_3.$$

The proof is now completed by noting that $X(n_i) =_{\text{st}} \hat{X}_i$, $i = 1, 2, 3, 4$. □

Next consider a Galton-Watson branching process $\{X(n),\ n \in \mathbb{N}_+\}$ in discrete time. Let D_i, $i = 1, 2, \ldots$, be independent and identically distributed random variables such that D_i has the same distribution as the number of offsprings of an ancestor. Then, for this process, $Y(x) =_{\text{st}} \sum_{i=1}^{x} D_i$, $x \in \mathbb{N}_+$.

Theorem 8.B.17. *Suppose* $D_i \geq 1$ *a.s. and* $P\{D_i > 1\} > 0$. *If* $X(0) \geq 1$ *a.s., then* $\{X(n),\ n \in \mathbb{N}_+\} \in \text{SICX(sp)}$.

Proof. First, condition on $X(0) = x_0$. Since $Z(x) = Y(x) - x =_{\text{st}} \sum_{i=1}^{x}(D_i - 1)$ and $D_i \geq 1$ a.s., one sees that $Z(x) \geq 0$ a.s. Also, it is easily seen that $\{Z(x),\ x \in \mathbb{N}_+\} \in \text{SI}$. Then, conditioned on $X(0) = x_0$, the result of Theorem 8.B.17 follows immediately from Theorem 8.B.16. From the definition of sample path convexity, it is clear that by unconditioning with respect to $X(0)$, the sample path convexity of $\{X(n),\ n \in \mathbb{N}_+\}$ is preserved. □

Now consider a nonhomogeneous Poisson process $\{N(t),\ t \geq 0\}$ with mean value function $M(t) = EN(t)$. To avoid trivialities we assume that M is strictly increasing. Denote by R_n the nth epoch time of this process.

Theorem 8.B.18. *If* M *is concave* [*convex*], *then* $\{R_n, n \in \mathbb{N}_{++}\} \in \text{SICX(sp)}$ [SICV(sp)].

Proof. Let $\{K(t),\ t \geq 0\}$ be a Poisson process with rate 1, and let T_n denote the nth epoch time of this process. By Example 8.B.4 we have $\{T_n,\ n \in \mathbb{N}_{++}\} \in \text{SIL(sp)}$. Now,

$$\{R_n,\, n \in \mathbb{N}_{++}\} =_{\text{st}} \{M^{-1}(T_n),\, n \in \mathbb{N}_{++}\}.$$

Since M is increasing and concave [convex] it follows that M^{-1} is increasing and convex [concave]. The result now follows from Corollary 8.B.14. □

Theorem 8.B.19. *If M is convex [concave], then $\{N(t),\, t \in [0,\infty)\} \in$ SICX(sp) [SICV(sp)].*

Proof. Let $\{K(t),\, t \geq 0\}$ be a Poisson process with rate 1. By Example 8.B.2 we have $\{K(t),\, t \in [0,\infty)\} \in$ SIL(sp). Now,

$$\{N(t),\, t \in [0,\infty)\} =_{\text{st}} \{K(M(t)),\, t \in [0,\infty)\}.$$

The result now follows from Corollary 8.B.15. □

8.C Convexity in the Usual Stochastic Order

In some applications it is hard to find the construction needed to establish the sample path convexity of Section 8.B. Then the stochastic convexity notions of this section may be useful.

8.C.1 Definitions

Let $\{X(\theta),\, \theta \in \Theta\}$ be a family of random variables with survival functions $\overline{F}_\theta(x) = P\{X(\theta) > x\}$, $\theta \in \Theta$. The family $\{X(\theta),\, \theta \in \Theta\}$ is said to be *stochastically increasing [decreasing] and convex [concave, linear] in the sense of the usual stochastic ordering* if $E\phi(X(\theta))$ is increasing [decreasing] and convex [concave, linear] for all increasing functions ϕ. We denote this by $\{X(\theta),\, \theta \in \Theta\} \in$ SICX(st) [SICV(st), SIL(st), SDCX(st), SDCV(st), SDL(st)].

It is easy to see the following characterization.

Theorem 8.C.1. *The family $\{X(\theta),\, \theta \in \Theta\}$ satisfies $\{X(\theta),\, \theta \in \Theta\} \in$ SICX(st) [SICV(st), SDCX(st), SDCV(st)] if, and only if, $\overline{F}(x,\theta)$ is increasing and convex [increasing and concave, decreasing and convex, decreasing and concave] in θ for each fixed x.*

The following are other characterizations of these notions.

Theorem 8.C.2. *The family $\{X(\theta),\, \theta \in \Theta\}$ satisfies $\{X(\theta),\, \theta \in \Theta\} \in$ SICX(st) [SICV(st), SDCX(st), SDCV(st)] if, and only if, for any $\theta_i \in \Theta$, $i = 1,2,3,4$, such that $\theta_1 \leq \theta_2 \leq \theta_3 \leq \theta_4$ and $\theta_1 + \theta_4 = \theta_2 + \theta_3$, there exist four random variables $\hat{X}_i$, $i = 1,2,3,4$, defined on a common probability space, such that $\hat{X}_i =_{\text{st}} X(\theta_i)$, $i = 1,2,3,4$, and $\hat{X}_1 \leq [\leq, \geq, \geq]\ \hat{X}_4$ a.s., $\min\{\hat{X}_1, \hat{X}_4\} \geq [\leq, \geq, \leq]\ \min\{\hat{X}_2, \hat{X}_3\}$ a.s., $\max\{\hat{X}_1, \hat{X}_4\} \geq [\leq, \geq, \leq]\ \max\{\hat{X}_2, \hat{X}_3\}$ a.s., and hence $\hat{X}_1 + \hat{X}_4 \geq [\leq, \geq, \leq]\ \hat{X}_2 + \hat{X}_3$ a.s.*

Proof. We prove the increasing convex case only since the other cases can be proven similarly. Since $X(\theta)$ is stochastically increasing in θ there exist, on a common probability space, random variables $\hat{X}_1$ and $\hat{X}_4$ such that $\hat{X}_i =_{\rm st} X(\theta_i)$, $i = 1, 4$ and $\hat{X}_1 \le \hat{X}_4$ a.s. Let U be a uniform random variable on $(0, 1)$ and define

$$X_2^* = I\Big\{U \le \frac{\theta_2 - \theta_1}{\theta_4 - \theta_1}\Big\}\hat{X}_4 + \Big[1 - I\Big\{U \le \frac{\theta_2 - \theta_1}{\theta_4 - \theta_1}\Big\}\Big]\hat{X}_1,$$

and

$$X_3^* = I\Big\{U \le \frac{\theta_2 - \theta_1}{\theta_4 - \theta_1}\Big\}\hat{X}_1 + \Big[1 - I\Big\{U \le \frac{\theta_2 - \theta_1}{\theta_4 - \theta_1}\Big\}\Big]\hat{X}_4.$$

Then

$$\min[X_2^*, X_3^*] = \min[\hat{X}_1, \hat{X}_4] \tag{8.C.1}$$

and

$$\max[X_2^*, X_3^*] = \max[\hat{X}_1, \hat{X}_4]. \tag{8.C.2}$$

Also note that $P\{X_2^* > x\} = \frac{\theta_2-\theta_1}{\theta_4-\theta_1}\overline{F}(x, \theta_4) + \frac{\theta_4-\theta_2}{\theta_4-\theta_1}\overline{F}(x, \theta_1)$, and $P\{X_3^* > x\} = \frac{\theta_2-\theta_1}{\theta_4-\theta_1}\overline{F}(x, \theta_1) + \frac{\theta_4-\theta_2}{\theta_4-\theta_1}\overline{F}(x, \theta_4)$. Since $\overline{F}(x, \theta)$ is increasing and convex in θ, it is then obvious that

$$X_2^* \ge_{\rm st} X(\theta_2) \quad \text{and} \quad X_3^* \ge_{\rm st} X(\theta_3).$$

Therefore, there exist $\hat{X}_2$ and $\hat{X}_3$ such that $\hat{X}_i =_{\rm st} X(\theta_i)$, $i = 2, 3$, and $X_i^* \ge \hat{X}_i$, $i = 2, 3$. Then, from (8.C.1) and (8.C.2), one sees that

$$\min[\hat{X}_2, \hat{X}_3] \le \min[\hat{X}_1, \hat{X}_4] \quad \text{and} \quad \max[\hat{X}_2, \hat{X}_3] \le \max[\hat{X}_1, \hat{X}_4].$$

The proof is now completed by observing that $X(\theta_i) =_{\rm st} \hat{X}_i$, $i = 1, 2, 3, 4$. □

Example 8.C.3. Let $X(p)$ be a geometric random variable with mean $1/(1-p)$. Then $\{X(p),\ p \in (0, 1)\} \in$ SICX(st).

Example 8.C.4. Let $X(\lambda)$ be an exponential random variable with mean $1/\lambda$. Then $\{X(\lambda),\ \lambda \in (0, \infty)\} \in$ SDCX(st).

It is evident from Theorems 8.C.2 and 8.B.9 that one has the following results.

Theorem 8.C.5.

$$\begin{aligned}
\text{SICX(st)} &\Longrightarrow \text{SICX(sp)} \Longrightarrow \text{SICX},\\
\text{SICV(st)} &\Longrightarrow \text{SICV(sp)} \Longrightarrow \text{SICV},\\
\text{SDCX(st)} &\Longrightarrow \text{SDCX(sp)} \Longrightarrow \text{SDCX},\\
\text{SDCV(st)} &\Longrightarrow \text{SDCV(sp)} \Longrightarrow \text{SDCV}.
\end{aligned}$$

Observing that $\{\psi(\theta),\ \theta \in \Theta\} \in$ SICX(sp) for any increasing convex function ψ, and that it is not SICX(st), it is clear that the implications in Theorem 8.C.5 are strict.

8.C.2 Closure properties

Unlike the two previous notions, stochastic convexity in the usual stochastic ordering does not have many closure properties. For example, there are no counterparts to Theorems 8.A.15 and 8.A.17 or Theorems 8.B.10 and 8.B.13 for this stochastic convexity notion. Instead, we present some specialized closure properties under random summation.

Theorem 8.C.6. *Let $\{N(\theta),\, \theta \in \Theta\}$ be a family of discrete random variables on $\mathbb{N}_+$, let $\{X(n),\, n = 1, 2, \dots\}$ be a sequence of independent and identically distributed nonnegative random variables, and let $X(0) = 0$. Suppose that $\{N(\theta),\, \theta \in \Theta\}$ and $\{X(n),\, n \in \mathbb{N}_+\}$ are mutually independent. Set $Y(\theta) = \sum_{n=0}^{N(\theta)} X(n)$, $\theta \in \Theta$. If $\{N(\theta),\, \theta \in \Theta\} \in \text{SICX(st)}$ [SICV(st), SDCX(st), SDCV(st)], then $\{Y(\theta), \theta \in \Theta\} \in \text{SICX(st)}$ [SICV(st), SDCX(st), SDCV(st)].*

Proof. Consider the case $\{N(\theta),\, \theta \in \Theta\} \in \text{SICX(st)}$. The other three cases can be similarly proven. From Theorem 8.C.2 one knows that for any $\theta_i \in \Theta$, $i = 1, 2, 3, 4$, such that $\theta_1 \le \theta_2 \le \theta_3 \le \theta_4$, and $\theta_1 + \theta_4 = \theta_2 + \theta_3$, there exist four random variables $\hat{N}_i$, $i = 1, 2, 3, 4$, defined on a common probability space, such that $\hat{N}_i =_{\text{st}} N(\theta_i)$, $i = 1, 2, 3, 4$, and

$$\hat{N}_4 \ge \hat{N}_1 \quad \text{a.s.,} \tag{8.C.3}$$

$$\min\{\hat{N}_1, \hat{N}_4\} \ge \min\{\hat{N}_2, \hat{N}_3\} \quad \text{a.s.,} \tag{8.C.4}$$

$$\max\{\hat{N}_1, \hat{N}_4\} \ge \max\{\hat{N}_2, \hat{N}_3\} \quad \text{a.s., and hence} \tag{8.C.5}$$

$$\hat{N}_1 + \hat{N}_4 \ge \hat{N}_2 + \hat{N}_3 \quad \text{a.s.} \tag{8.C.6}$$

Define $\hat{Y}_i = \sum_{n=0}^{\hat{N}_i} X(n)$, $i = 1, 2, 3, 4$. Then, clearly, $\hat{Y}_i =_{\text{st}} Y(\theta_i)$, $i = 1, 2, 3, 4$. Furthermore, from (8.C.3)–(8.C.6), one sees that

$$\hat{Y}_4 \ge \hat{Y}_1 \quad \text{a.s.,} \tag{8.C.7}$$

$$\min\{\hat{Y}_1, \hat{Y}_4\} \ge \min\{\hat{Y}_2, \hat{Y}_3\} \quad \text{a.s.,} \tag{8.C.8}$$

$$\max\{\hat{Y}_1, \hat{Y}_4\} \ge \max\{\hat{Y}_2, \hat{Y}_3\} \quad \text{a.s., and hence} \tag{8.C.9}$$

$$\hat{Y}_1 + \hat{Y}_4 \ge \hat{Y}_2 + \hat{Y}_3 \quad \text{a.s.} \tag{8.C.10}$$

Theorem 8.C.6 then follows from Theorem 8.C.2. □

Theorem 8.C.7. *Consider $\{X(\theta),\, \theta \in \Theta\}$ and $\{Y(\theta),\, \theta \in \Theta\}$ and suppose that, for each θ, $X(\theta)$ and $Y(\theta)$ are independent. Define*

$$V(\theta) = \max\{X(\theta), Y(\theta)\}$$

and

$$W(\theta) = \min\{X(\theta), Y(\theta)\}.$$

(i) *If $\{X(\theta),\, \theta \in \Theta\} \in \text{SICX(st)}$ [SDCX(st)] and $\{Y(\theta),\, \theta \in \Theta\} \in \text{SICX(st)}$ [SDCX(st)], then $\{W(\theta),\, \theta \in \Theta\} \in \text{SICX(st)}$ [SDCX(st)].*

(ii) *If* $\{X(\theta),\ \theta \in \Theta\} \in \text{SICV(st)}$ [SDCV(st)] *and* $\{Y(\theta),\ \theta \in \Theta\} \in \text{SICV(st)}$ [SDCV(st)], *then* $\{V(\theta),\ \theta \in \Theta\} \in \text{SICV(st)}$ [SDCV(st)].

Proof. The stated results follow immediately from the observations that (i) the survival function of $W(\theta)$ at x is equal to $P\{X(\theta) > x\}P\{Y(\theta) > x\}$, (ii) the survival function of $V(\theta)$ at x is equal to $1-(1-P\{X(\theta) > x\})(1-P\{Y(\theta) > x\})$, and from Theorem 8.C.1. □

Consider the imperfect repair model. A new item with an absolutely continuous survival function $\overline{F}$ undergoes an imperfect repair each time it fails before it is scrapped. With probability p the repair is unsuccessful and the item is scrapped. With probability $1-p$ the repair is successful and minimal, that is, after a successful repair at time t the item is as good as a working item at age t. It is well known that if $X(p)$ denotes the time to scrap, then the survival function of $X(p)$ is $\overline{F}^p$. Thus, the following result is apparent.

Theorem 8.C.8. *Let $\overline{F}$ be an absolutely continuous survival function such that $\overline{F}(0) = 1$. Then $\{X(p),\ p \in (0,1)\} \in \text{SDCX(st)}$.*

8.D Strong Stochastic Convexity

Another notion which is sometimes useful in verifying the sample path convexity of Section 8.B is described in this section.

8.D.1 Definitions

Let $\{X(\theta), \theta \in \Theta\}$ be a family of random variables. The family $\{X(\theta), \theta \in \Theta\}$ is said to be *stochastically* [*increasing, decreasing*] *and convex* [*concave, linear*] *almost everywhere* if there exist $\{\hat{X}(\theta),\ \theta \in \Theta\}$ such that $\hat{X}(\theta) =_{\text{st}} X(\theta)$ for each $\theta \in \Theta$ and $\hat{X}(\theta)$ is [increasing, decreasing] and convex [concave, linear] in θ. We denote this by $\{X(\theta),\ \theta \in \Theta\} \in \text{SCX(ae)}$ [SCV(ae), SL(ae), SICX(ae), SICV(ae), SIL(ae), SDCX(ae), SDCV(ae), SDL(ae)].

Although it appears that the definition of strong stochastic convexity/concavity is restrictive, several families of random variables do satisfy the conditions of this class of convexity/concavity. This is shown in the next theorem and in the corollaries and examples which follow it.

Theorem 8.D.1. *Suppose that $X(\theta) = \phi(\theta, \boldsymbol{Z})$, where ϕ is a real-valued deterministic function, and $\boldsymbol{Z}$ is a random vector. If ϕ is convex* [*concave, linear, increasing convex, increasing concave, increasing linear, decreasing convex, decreasing concave, decreasing linear*] *in $\theta \in \Theta$, then $\{X(\theta),\ \theta \in \Theta\} \in \text{SCX(ae)}$* [SCV(ae), SL(ae), SICX(ae), SICV(ae), SIL(ae), SDCX(ae), SDCV(ae), SDL(ae)].

Corollary 8.D.2. *Suppose that $X(\theta) = Z + \psi(\theta)$, where ψ is a real-valued deterministic function, and Z is a random variable. If ψ is convex [concave, linear, increasing convex, increasing concave, increasing linear, decreasing convex, decreasing concave, decreasing linear] in $\theta \in \Theta$, then $\{X(\theta),\ \theta \in \Theta\} \in$ SCX(ae) [SCV(ae), SL(ae), SICX(ae), SICV(ae), SIL(ae), SDCX(ae), SDCV(ae), SDL(ae)].*

Corollary 8.D.3. *Suppose that $X(\theta) = Z \cdot \psi(\theta)$, where ψ is a real-valued deterministic function, and Z is a nonnegative random variable. If ψ is convex [concave, linear, increasing convex, increasing concave, increasing linear, decreasing convex, decreasing concave, decreasing linear] in $\theta \in \Theta$, then $\{X(\theta),\ \theta \in \Theta\} \in$ SCX(ae) [SCV(ae), SL(ae), SICX(ae), SICV(ae), SIL(ae), SDCX(ae), SDCV(ae), SDL(ae)].*

Example 8.D.4. Let $X(\mu,\sigma)$ be a normal random variable with mean μ and standard deviation σ. Since for a unit normal random variable $N(0,1)$, we have $\hat{X}(\mu,\sigma) = \mu + \sigma N(0,1) =_{\text{st}} X(\mu,\sigma)$, $\mu \in \mathbb{R}$, $\sigma \in \mathbb{R}_+$, we see that, for each $\sigma > 0$,

$$\{X(\mu,\sigma),\ \mu \in \mathbb{R}\} \in \text{SIL(ae)},$$

and, for each $\mu \in \mathbb{R}$,

$$\{X(\mu,\sigma),\ \sigma \in \mathbb{R}_+\} \in \text{SL(ae)}.$$

Similarly one can prove the result in the next example.

Example 8.D.5. Let $Y(n)$, $n = 1, 2, \ldots$, be a sequence of nonnegative independent and identically distributed random variables with mean 1. For $\mu > 0$ define $X(\mu) = \mu \sum_{k=1}^{n} Y(k)$, $n \in \mathbb{N}_+$. Then, for each $n \in \mathbb{N}_{++}$, one has $\{X(\mu),\ \mu \in \mathbb{R}_+\} \in$ SIL(ae).

Specifically, when $Y(n)$ in Example 8.D.5 is an exponential random variable we have the following example.

Example 8.D.6. Let $X(\mu, n)$ be an Erlang-n random variable with mean $n\mu$ and variance $n\mu^2$. Then, for each $n \in \mathbb{N}_{++}$, one has $\{X(\mu,n),\ \mu \in \mathbb{R}_+\} \in$ SIL(ae).

By taking $n = 1$ in Example 8.D.5 we obtain the following result.

Example 8.D.7. Let Y be a nonnegative random variable. For $\mu > 0$ define $X(\mu) = \mu Y$. Then $\{X(\mu),\ \mu \in \mathbb{R}_+\} \in$ SIL(ae).

The following generalization of Example 8.D.5 is easily observed.

Example 8.D.8. Let $Y(n)$, $n = 1, 2, \ldots$, be a sequence of nonnegative independent and identically distributed random variables with mean 1, and let Z be a random variable which is independent of the $Y(n)$'s. For $\mu > 0$ define $X(\mu) = Z + \mu \sum_{k=1}^{n} Y(k)$, $n \in \mathbb{N}_{++}$. Then, for each $n \in \mathbb{N}_{++}$, one has $\{X(\mu),\ \mu \in \mathbb{R}_+\} \in$ SIL(ae).

Another sufficient condition (in addition to Theorem 8.D.1 and Corollaries 8.D.2 and 8.D.3) for strong convexity and concavity is described next. Let $\{X(\theta),\ \theta \in \Theta\}$ be a family of random variables, and let F_θ denote the distribution function of $X(\theta)$. If U is a uniform$[0,1]$ random variable, then $F_\theta^{-1}(U) =_{\rm st} X(\theta)$. The following result follows at once from this observation.

Theorem 8.D.9. *Suppose that $F_\theta^{-1}(u)$ is convex [concave, linear, increasing convex, increasing concave, increasing linear, decreasing convex, decreasing concave, decreasing linear] in $\theta \in \Theta$, for all $u \in (0,1)$, then $\{X(\theta),\ \theta \in \Theta\} \in$ SCX(ae) [SCV(ae), SL(ae), SICX(ae), SICV(ae), SIL(ae), SDCX(ae), SDCV(ae), SDL(ae)].*

A sufficient condition for strong convexity and concavity, which is stated on F_θ (rather than on F_θ^{-1} as in Theorem 8.D.9), is described next. Recall the definition of supermodular and submodular functions given in Section 7.A.8.

Theorem 8.D.10. *Let $\{X(\theta),\ \theta \in \Theta\}$ be a family of random variables, and suppose that all the partial second derivatives of $F_\theta(x)$ exist.*

(a) *If $F_\theta(x)$ is concave and strictly increasing in x, and is decreasing and concave in θ, and if $F_\theta(x)$ is submodular in (x,θ), then $\{X(\theta),\ \theta \in \Theta\} \in$ SICX(ae).*
(b) *If $F_\theta(x)$ is convex and strictly increasing in x, and is decreasing and convex in θ, and if $F_\theta(x)$ is supermodular in (x,θ), then $\{X(\theta),\ \theta \in \Theta\} \in$ SICV(ae).*
(c) *If $F_\theta(x)$ is concave and strictly increasing in x, and is increasing and concave in θ, and if $F_\theta(x)$ is supermodular in (x,θ), then $\{X(\theta),\ \theta \in \Theta\} \in$ SDCX(ae).*
(d) *If $F_\theta(x)$ is convex and strictly increasing in x, and is increasing and convex in θ, and if $F_\theta(x)$ is submodular in (x,θ), then $\{X(\theta),\ \theta \in \Theta\} \in$ SDCV(ae).*

Proof. Only the proof of part (a) is given; the proofs of the other parts are similar. Let U be a uniform$[0,1]$ random variable and define $\hat{X}(\theta)$ by

$$F_\theta(\hat{X}(\theta)) = U. \tag{8.D.1}$$

Differentiating (8.D.1) for a fixed value of U, we obtain

$$\frac{\partial}{\partial x}F \cdot \frac{\partial}{\partial \theta}\hat{X} + \frac{\partial}{\partial \theta}F = 0, \tag{8.D.2}$$

and

$$\frac{\partial}{\partial x}F \cdot \frac{\partial^2}{\partial \theta^2}\hat{X} + \left(\frac{\partial^2}{\partial x^2}F\right)\left(\frac{\partial}{\partial \theta}\hat{X}\right)^2 + 2\frac{\partial^2}{\partial x \partial \theta}F \cdot \frac{\partial}{\partial \theta}\hat{X} + \frac{\partial^2}{\partial \theta^2}F = 0. \tag{8.D.3}$$

The conditions stated in part (a) can be written as

$$\frac{\partial}{\partial x}F > 0, \quad \frac{\partial}{\partial \theta}F \le 0, \quad \frac{\partial^2}{\partial x^2}F \le 0, \quad \frac{\partial^2}{\partial \theta^2}F \le 0, \quad \text{and} \quad \frac{\partial^2}{\partial x \partial \theta}F \le 0. \tag{8.D.4}$$

From (8.D.2), (8.D.3), and (8.D.4) it is seen that $\frac{\partial}{\partial\theta}\hat{X} \ge 0$ and $\frac{\partial^2}{\partial\theta^2}\hat{X} \ge 0$, that is, $\{X(\theta),\ \theta \in \Theta\} \in \text{SICX(ae)}$. □

The following theorem is easily verified.

Theorem 8.D.11.

$$\begin{aligned}
\text{SICX(ae)} &\Longrightarrow \text{SICX(sp)} \Longrightarrow \text{SICX},\\
\text{SICV(ae)} &\Longrightarrow \text{SICV(sp)} \Longrightarrow \text{SICV},\\
\text{SDCX(ae)} &\Longrightarrow \text{SDCX(sp)} \Longrightarrow \text{SDCX},\\
\text{SDCV(ae)} &\Longrightarrow \text{SDCV(sp)} \Longrightarrow \text{SDCV}.
\end{aligned}$$

These are strict implications. It can be verified that the stochastic convexity in the usual stochastic order neither implies nor is implied by the strong stochastic convexity.

8.D.2 Closure properties

In this subsection we present some closure properties of the strong convexity notions. These results trivially follow from the closure properties of deterministic functions. Thus we will not give the proofs here.

Theorem 8.D.12. *Let $\{X(\theta),\ \theta \in \Theta\}$ and $\{Y(\theta),\ \theta \in \Theta\}$ be two families of random variables such that for each $\theta \in \Theta$, $X(\theta)$ and $Y(\theta)$ are independent.*

(a) $\{X(\theta),\ \theta \in \Theta\} \in \text{SICX(ae)}$ *and* $\{Y(\theta),\ \theta \in \Theta\} \in \text{SICX(ae)}$ *imply that* $\{f(X(\theta), Y(\theta)),\ \theta \in \Theta\} \in \text{SICX(ae)}$ *for any increasing and convex function f.*
(b) $\{X(\theta),\ \theta \in \Theta\} \in \text{SICV(ae)}$ *and* $\{Y(\theta),\ \theta \in \Theta\} \in \text{SICV(ae)}$ *imply that* $\{f(X(\theta), Y(\theta)),\ \theta \in \Theta\} \in \text{SICV(ae)}$ *for any increasing and concave function f.*
(c) $\{X(\theta),\ \theta \in \Theta\} \in \text{SDCX(ae)}$ *and* $\{X(\theta),\ \theta \in \Theta\} \in \text{SDCX(ae)}$ *imply that* $\{f(X(\theta), Y(\theta)),\ \theta \in \Theta\} \in \text{SDCX(ae)}$ *for any increasing and convex function f.*
(d) $\{X(\theta),\ \theta \in \Theta\} \in \text{SDCV(ae)}$ *and* $\{Y(\theta),\ \theta \in \Theta\} \in \text{SDCV(ae)}$ *imply that* $\{f(X(\theta), Y(\theta)),\ \theta \in \Theta\} \in \text{SDCV(ae)}$ *for any increasing and concave function f.*

Theorem 8.D.13. *Let $\{X(\theta),\ \theta \in \Theta\}$ be a family of Λ-valued random variables, where $\Lambda \subset \mathbb{R}$ is a convex set. Also, let $\{Y(\lambda),\ \lambda \in \Lambda\}$ be another family of random variables. Suppose that $X(\theta)$ and $Y(\lambda)$ are independent for any choice of $\theta \in \Theta$ and $\lambda \in \Lambda$.*

(a) *If* $\{X(\theta),\ \theta \in \Theta\} \in \text{SICX(ae)}$ [SICV(ae)] *and* $\{Y(\lambda),\ \lambda \in \Lambda\} \in \text{SICX(ae)}$ [SICV(ae)], *then* $\{Y(X(\theta)),\ \theta \in \Theta\} \in \text{SICX(ae)}$ [SICV(ae)].
(b) *If* $\{X(\theta), \theta \in \Theta\} \in \text{SDCX(ae)}$ [SDCV(ae)] *and* $\{Y(\lambda), \lambda \in \Lambda\} \in \text{SICX(ae)}$ [SICV(ae)], *then* $\{Y(X(\theta)),\ \theta \in \Theta\} \in \text{SDCX(ae)}$ [SDCV(ae)].

8.E Stochastic Directional Convexity

8.E.1 Definitions

In Sections 8.A–8.D of this chapter, the parameter space Θ, of the families of random variables $\{X(\theta),\ \theta \in \Theta\}$ that we studied, was a subset of the real line $\mathbb{R}$. However, in some applications the parameter space is multidimensional, that is, Θ is a subset of $\mathbb{R}^m$ for some positive integer $m \geq 2$. In this section we study such families of random variables or vectors. In such cases one is interested in convexity [concavity] properties with respect to the vector $\boldsymbol{\theta} = (\theta_1, \theta_2, \ldots, \theta_m)$. Rather than studying convexity [concavity] properties of $\{X(\boldsymbol{\theta}),\ \boldsymbol{\theta} \in \boldsymbol{\Theta}\}$, we will study here directional convexity [concavity] properties of such families of random variables or vectors. The reader may recall the definition of directional convexity [concavity] given in (7.A.17) of Section 7.A.8. Below $\boldsymbol{\Theta}$ will always be a sublattice of $\mathbb{R}^m$.

Let $\{\boldsymbol{X}(\boldsymbol{\theta}),\ \boldsymbol{\theta} \in \boldsymbol{\Theta}\}$ be a family of random vectors. The family $\{\boldsymbol{X}(\boldsymbol{\theta}),\ \boldsymbol{\theta} \in \boldsymbol{\Theta}\}$ is said to be

(a) *stochastically increasing and directionally convex* [*concave*] if $\{\boldsymbol{X}(\boldsymbol{\theta}),\ \boldsymbol{\theta} \in \boldsymbol{\Theta}\} \in$ SI and if $E\phi(\boldsymbol{X}(\boldsymbol{\theta}))$ is directionally convex [concave] in $\boldsymbol{\theta}$ for any increasing directionally convex [concave] function $\boldsymbol{\phi}$. We denote it by $\{\boldsymbol{X}(\boldsymbol{\theta}),\ \boldsymbol{\theta} \in \boldsymbol{\Theta}\} \in$ SI-DIR-CX [SI-DIR-CV];
(b) *stochastically increasing and directionally linear* if $\{\boldsymbol{X}(\boldsymbol{\theta}),\ \boldsymbol{\theta} \in \boldsymbol{\Theta}\} \in$ SI-DIR-CX $\cap$ SI-DIR-CV. We denote it by $\{\boldsymbol{X}(\boldsymbol{\theta}),\ \boldsymbol{\theta} \in \boldsymbol{\Theta}\} \in$ SI-DIR-L;
(c) *stochastically decreasing and directionally convex* [*concave*] if $\{\boldsymbol{X}(\boldsymbol{\theta}),\ \boldsymbol{\theta} \in \boldsymbol{\Theta}\} \in$ SD and if $E\phi(\boldsymbol{X}(\boldsymbol{\theta}))$ is directionally convex [concave] in $\boldsymbol{\theta}$ for any increasing directionally convex [concave] function $\boldsymbol{\phi}$. We denote it by $\{\boldsymbol{X}(\boldsymbol{\theta}),\ \boldsymbol{\theta} \in \boldsymbol{\Theta}\} \in$ SD-DIR-CX [SD-DIR-CV];
(d) *stochastically decreasing and directionally linear* if $\{\boldsymbol{X}(\boldsymbol{\theta}),\ \boldsymbol{\theta} \in \boldsymbol{\Theta}\} \in$ SD-DIR-CX$\cap$SD-DIR-CV. We denote it by $\{\boldsymbol{X}(\boldsymbol{\theta}),\ \boldsymbol{\theta} \in \boldsymbol{\Theta}\} \in$ SD-DIR-L.

In particular, if $X(\boldsymbol{\theta})$ is a univariate random variable for all $\boldsymbol{\theta} \in \boldsymbol{\Theta}$, then $\{X(\boldsymbol{\theta}), \boldsymbol{\theta} \in \boldsymbol{\Theta}\} \in$ SI-DIR-CX [SI-DIR-CV] if, and only if, $\{X(\boldsymbol{\theta}), \boldsymbol{\theta} \in \boldsymbol{\Theta}\} \in$ SI and $E\phi(X(\boldsymbol{\theta}))$ is directionally convex [concave] in $\boldsymbol{\theta}$ for any increasing convex [concave] function ϕ. Similarly, $\{X(\boldsymbol{\theta}),\ \boldsymbol{\theta} \in \boldsymbol{\Theta}\} \in$ SD-DIR-CX [SD-DIR-CV] if, and only if, $\{X(\boldsymbol{\theta}),\ \boldsymbol{\theta} \in \boldsymbol{\Theta}\} \in$ SD and $E\phi(X(\boldsymbol{\theta}))$ is directionally convex [concave] in $\boldsymbol{\theta}$ for any increasing convex [concave] function ϕ. If both the parameter and the random variables are univariate, then the notions of SI-DIR-CX, SI-DIR-CV, SI-DIR-L, SD-DIR-CX, SD-DIR-CV, and SD-DIR-L, reduce to the notions of SICX, SICV, SIL, SDCX, SDCV, and SDL, respectively.

In order to define stochastic directional convexity [concavity] in the sample path sense let $\{\boldsymbol{X}(\boldsymbol{\theta}),\ \boldsymbol{\theta} \in \boldsymbol{\Theta}\}$ be a family of random vectors as above. Let $\boldsymbol{\theta}_i \in \boldsymbol{\Theta}$, $i = 1, 2, 3, 4$, be any four vectors such that $\boldsymbol{\theta}_1 \leq [\boldsymbol{\theta}_2, \boldsymbol{\theta}_3] \leq \boldsymbol{\theta}_4$ and $\boldsymbol{\theta}_1 + \boldsymbol{\theta}_4 = \boldsymbol{\theta}_2 + \boldsymbol{\theta}_3$.

If there exist four random variables $\hat{\boldsymbol{X}}_i$, $i = 1, 2, 3, 4$, defined on a common probability space, such that $\hat{\boldsymbol{X}}_i =_{\rm st} \boldsymbol{X}(\boldsymbol{\theta}_i)$, $i = 1, 2, 3, 4$, and

(a) (i) $[\hat{\boldsymbol{X}}_2, \hat{\boldsymbol{X}}_3] \le \hat{\boldsymbol{X}}_4$ a.s. and (ii) $\hat{\boldsymbol{X}}_2 + \hat{\boldsymbol{X}}_3 \le \hat{\boldsymbol{X}}_1 + \hat{\boldsymbol{X}}_4$ a.s., then $\{\boldsymbol{X}(\boldsymbol{\theta}), \boldsymbol{\theta} \in \boldsymbol{\Theta}\}$ is said to be *stochastically increasing and directionally convex in the sample path sense* (denoted by $\{\boldsymbol{X}(\boldsymbol{\theta}),\ \boldsymbol{\theta} \in \boldsymbol{\Theta}\} \in$ SI-DIR-CX(sp));
(b) (i) $\hat{\boldsymbol{X}}_1 \le [\hat{\boldsymbol{X}}_2, \hat{\boldsymbol{X}}_3]$ a.s. and (ii) $\hat{\boldsymbol{X}}_1 + \hat{\boldsymbol{X}}_4 \le \hat{\boldsymbol{X}}_2 + \hat{\boldsymbol{X}}_3$ a.s., then $\{\boldsymbol{X}(\boldsymbol{\theta}), \boldsymbol{\theta} \in \boldsymbol{\Theta}\}$ is said to be *stochastically increasing and directionally concave in the sample path sense* (denoted by $\{\boldsymbol{X}(\boldsymbol{\theta}),\ \boldsymbol{\theta} \in \boldsymbol{\Theta}\} \in$ SI-DIR-CV(sp));
(c) (i) $\hat{\boldsymbol{X}}_1 \ge [\hat{\boldsymbol{X}}_2, \hat{\boldsymbol{X}}_3]$ a.s. and (ii) $\hat{\boldsymbol{X}}_1 + \hat{\boldsymbol{X}}_4 \ge \hat{\boldsymbol{X}}_2 + \hat{\boldsymbol{X}}_3$ a.s., then $\{\boldsymbol{X}(\boldsymbol{\theta}), \boldsymbol{\theta} \in \boldsymbol{\Theta}\}$ is said to be *stochastically decreasing and directionally convex in the sample path sense* (denoted by $\{\boldsymbol{X}(\boldsymbol{\theta}),\ \boldsymbol{\theta} \in \boldsymbol{\Theta}\} \in$ SD-DIR-CX(sp));
(d) (i) $\hat{\boldsymbol{X}}_4 \le [\hat{\boldsymbol{X}}_2, \hat{\boldsymbol{X}}_3]$ a.s. and (ii) $\hat{\boldsymbol{X}}_1 + \hat{\boldsymbol{X}}_4 \le \hat{\boldsymbol{X}}_2 + \hat{\boldsymbol{X}}_3$ a.s., then $\{\boldsymbol{X}(\boldsymbol{\theta}), \boldsymbol{\theta} \in \boldsymbol{\Theta}\}$ is said to be *stochastically decreasing and directionally concave in the sample path sense* (denoted by $\{\boldsymbol{X}(\boldsymbol{\theta}),\ \boldsymbol{\theta} \in \boldsymbol{\Theta}\} \in$ SD-DIR-CV(sp));
(e) (i) $[\hat{\boldsymbol{X}}_2, \hat{\boldsymbol{X}}_3] \le \hat{\boldsymbol{X}}_4$ a.s. and (ii) $\hat{\boldsymbol{X}}_2 + \hat{\boldsymbol{X}}_3 = \hat{\boldsymbol{X}}_1 + \hat{\boldsymbol{X}}_4$ a.s., then $\{\boldsymbol{X}(\boldsymbol{\theta}), \boldsymbol{\theta} \in \boldsymbol{\Theta}\}$ is said to be *stochastically increasing and directionally linear in the sample path sense* (denoted by $\{\boldsymbol{X}(\boldsymbol{\theta}),\ \boldsymbol{\theta} \in \boldsymbol{\Theta}\} \in$ SI-DIR-L(sp));
(f) (i) $\hat{\boldsymbol{X}}_1 \ge [\hat{\boldsymbol{X}}_2, \hat{\boldsymbol{X}}_3]$ a.s. and (ii) $\hat{\boldsymbol{X}}_1 + \hat{\boldsymbol{X}}_4 = \hat{\boldsymbol{X}}_2 + \hat{\boldsymbol{X}}_3$ a.s., then $\{\boldsymbol{X}(\boldsymbol{\theta}), \boldsymbol{\theta} \in \boldsymbol{\Theta}\}$ is said to be *stochastically decreasing and directionally linear in the sample path sense* (denoted by $\{\boldsymbol{X}(\boldsymbol{\theta}),\ \boldsymbol{\theta} \in \boldsymbol{\Theta}\} \in$ SD-DIR-L(sp)).

If both the parameter and the random variables are univariate, then the notions of SI-DIR-CX(sp), SI-DIR-CV(sp), SI-DIR-L(sp), SD-DIR-CX(sp), SD-DIR-CV(sp), and SD-DIR-L(sp), reduce to the notions of SICX(sp), SICV(sp), SIL(sp), SDCX(sp), SDCV(sp), and SDL(sp), respectively.

8.E.2 Closure properties

The following two results are extensions of Theorems 8.A.17 and 8.B.13 to the stochastic directional convexity setting. The proof of Theorem 8.E.1 is similar to the proof of Theorem 8.A.17, using Proposition 7.A.28. The proof of Theorem 8.E.2 is similar to the proof of Theorem 8.B.13, where the minimum in (8.B.3) is performed coordinatewise.

Theorem 8.E.1. *Let $\{\boldsymbol{X}(\boldsymbol{\theta}), \boldsymbol{\theta} \in \boldsymbol{\Theta}\}$ be a family of $\boldsymbol{\Lambda}$-valued random vectors, and let $\{\boldsymbol{Y}(\boldsymbol{\lambda}),\ \boldsymbol{\lambda} \in \boldsymbol{\Lambda}\}$ be another family of random vectors. Suppose that $\boldsymbol{X}(\boldsymbol{\theta})$ and $\boldsymbol{Y}(\boldsymbol{\lambda})$ are independent for any choice of $\boldsymbol{\theta} \in \boldsymbol{\Theta}$ and $\boldsymbol{\lambda} \in \boldsymbol{\Lambda}$.*

(a) *If* $\{\boldsymbol{X}(\boldsymbol{\theta}),\ \boldsymbol{\theta} \in \boldsymbol{\Theta}\} \in$ SI-DIR-CX [SI-DIR-CV, SI-DIR-L] *and* $\{\boldsymbol{Y}(\boldsymbol{\lambda}), \boldsymbol{\lambda} \in \boldsymbol{\Lambda}\} \in$ SI-DIR-CX [SI-DIR-CV, SI-DIR-L], *then* $\{\boldsymbol{Y}(\boldsymbol{X}(\boldsymbol{\theta})),\ \boldsymbol{\theta} \in \boldsymbol{\Theta}\} \in$ SI-DIR-CX [SI-DIR-CV, SI-DIR-L].
(b) *If* $\{\boldsymbol{X}(\boldsymbol{\theta}),\ \boldsymbol{\theta} \in \boldsymbol{\Theta}\} \in$ SD-DIR-CX [SD-DIR-CV, SD-DIR-L] *and* $\{\boldsymbol{Y}(\boldsymbol{\lambda}), \boldsymbol{\lambda} \in \boldsymbol{\Lambda}\} \in$ SI-DIR-CX [SI-DIR-CV, SI-DIR-L], *then* $\{\boldsymbol{Y}(\boldsymbol{X}(\boldsymbol{\theta})),\ \boldsymbol{\theta} \in \boldsymbol{\Theta}\} \in$ SD-DIR-CX [SD-DIR-CV, SD-DIR-L].

Theorem 8.E.2. *Let $\{\boldsymbol{X}(\boldsymbol{\theta}), \boldsymbol{\theta} \in \boldsymbol{\Theta}\}$ be a family of $\boldsymbol{\Lambda}$-valued random vectors, and let $\{\boldsymbol{Y}(\boldsymbol{\lambda}),\ \boldsymbol{\lambda} \in \boldsymbol{\Lambda}\}$ be another family of random vectors. Suppose that $\boldsymbol{X}(\boldsymbol{\theta})$ and $\boldsymbol{Y}(\boldsymbol{\lambda})$ are independent for any choice of $\boldsymbol{\theta} \in \boldsymbol{\Theta}$ and $\boldsymbol{\lambda} \in \boldsymbol{\Lambda}$.*

(a) *If* $\{\boldsymbol{X}(\boldsymbol{\theta}),\ \boldsymbol{\theta} \in \boldsymbol{\Theta}\} \in$ SI-DIR-CX(sp) [SI-DIR-CV(sp), SI-DIR-L(sp)] *and* $\{\boldsymbol{Y}(\boldsymbol{\lambda}),\ \boldsymbol{\lambda} \in \boldsymbol{\Lambda}\} \in$ SI-DIR-CX(sp) [SI-DIR-CV(sp), SI-DIR-L(sp)], *then* $\{\boldsymbol{Y}(\boldsymbol{X}(\boldsymbol{\theta})),\ \boldsymbol{\theta} \in \boldsymbol{\Theta}\} \in$ SI-DIR-CX(sp) [SI-DIR-CV(sp), SI-DIR-L(sp)].

(b) *If* $\{\boldsymbol{X}(\boldsymbol{\theta}),\ \boldsymbol{\theta} \in \boldsymbol{\Theta}\} \in$ SD-DIR-CX(sp) [SD-DIR-CV(sp), SD-DIR-L(sp)] *and* $\{\boldsymbol{Y}(\boldsymbol{\lambda}),\ \boldsymbol{\lambda} \in \boldsymbol{\Lambda}\} \in$ SI-DIR-CX(sp) [SI-DIR-CV(sp), SI-DIR-L(sp)], *then* $\{\boldsymbol{Y}(\boldsymbol{X}(\boldsymbol{\theta})),\ \boldsymbol{\theta} \in \boldsymbol{\Theta}\} \in$ SD-DIR-CX(sp) [SD-DIR-CV(sp), SD-DIR-L(sp)].

From Theorem 8.E.2 it is easy to verify the following results.

Theorem 8.E.3.

$$\begin{aligned}
\text{SI-DIR-CX(sp)} &\Longrightarrow \text{SI-DIR-CX},\\
\text{SI-DIR-CV(sp)} &\Longrightarrow \text{SI-DIR-CV},\\
\text{SD-DIR-CX(sp)} &\Longrightarrow \text{SD-DIR-CX},\\
\text{SD-DIR-CV(sp)} &\Longrightarrow \text{SD-DIR-CV}.
\end{aligned}$$

The next results will be stated only for the increasing convex cases, however, they have versions that apply to the decreasing convex, the increasing concave, and the decreasing concave cases.

By combining independent SI-DIR-CX [SI-DIR-CX(sp)] families of random vectors, one obtains a new SI-DIR-CX [SI-DIR-CX(sp)] family of random vectors.

Theorem 8.E.4. *Let* $\{\boldsymbol{X}_i(\boldsymbol{\theta}_i),\ \boldsymbol{\theta}_i \in \boldsymbol{\Theta}_i\} \in$ SD-DIR-CX [SI-DIR-CX(sp)], $i = 1, 2, \ldots, m$, *be mutually independent collections of random vectors. Define* $\boldsymbol{X}(\boldsymbol{\theta}) = (\boldsymbol{X}_1(\boldsymbol{\theta}_1), \boldsymbol{X}_2(\boldsymbol{\theta}_2), \ldots, \boldsymbol{X}_m(\boldsymbol{\theta}_m))$. *Then* $\{\boldsymbol{X}(\boldsymbol{\theta}),\ \boldsymbol{\theta} \in \times_{i=1}^m \boldsymbol{\Theta}_i\} \in$ SD-DIR-CX [SI-DIR-CX(sp)].

The (sp) part of Theorem 8.E.4 can be proven by observing that, by independence, the constructions required by the definition of the SI-DIR-CX(sp) notion can be done coordinatewise. The other part of Theorem 8.E.4 can be verified by noticing that an m-variate directionally convex function is also directionally convex in any subset of the m coordinates, and again using the independence assumption.

As a special case of Theorem 8.E.4 it is seen that if the families of random variables $\{X_i(\theta_i),\ \theta_i \in \Theta_i\} \in$ SICX [SICX(sp)], $i = 1, 2, \ldots, m$, then $\{(X_1(\theta_1), X_2(\theta_2), \ldots, X_m(\theta_m)),\ (\theta_1, \theta_2, \ldots, \theta_m) \in \times_{i=1}^m \Theta_i\} \in$ SD-DIR-CX [SI-DIR-CX(sp)].

A version of Theorem 8.E.4, in which some or all of the parameters are the same, can also be stated and proven. For example, if the families of random variables $\{X_i(\theta),\ \theta \in \Theta\} \in$ SICX [SICX(sp)], $i = 1, 2, \ldots, m$, then $\{(X_1(\theta), X_2(\theta), \ldots, X_m(\theta)),\ \theta \in \Theta\} \in$ SD-DIR-CX [SI-DIR-CX(sp)] (here all the parameters are the same).

Example 8.E.5. Recall from Example 8.B.4 that if $Y(n)$, $n = 1, 2, \ldots$, are nonnegative independent and identically distributed random variables, then

$\{\sum_{k=1}^{n} Y(k),\ n \in \mathbb{N}_{++}\} \in \text{SIL(sp)}$. Now, let $\{Y_i(n),\ n = 1,2,\dots\}$, $i = 1,2,\dots,m$, be independent sequences of nonnegative independent and identically distributed random variables. Then $\{\big(\sum_{k=1}^{n_1} Y_1(k), \sum_{k=1}^{n_2} Y_2(k), \dots, \sum_{k=1}^{n_m} Y_m(k)\big),\ (n_1, n_2, \dots, n_m) \in \mathbb{N}_{++}^m\} \in \text{SI-DIR-CX(sp)}$.

Similar examples can be constructed from the other examples in Section 8.B.

The following result illustrates the use of Theorems 8.E.1 and 8.E.2. For each $\theta \in \Theta$ (where Θ is a convex subset of $\mathbb{R}$ or $\mathbb{N}$) let $\{X(n,\theta),\ n \in \mathbb{N}_+\}$ be a Markov chain with state space $\mathcal{S}$ ($\mathcal{S} = [0,\infty)$ or $\mathbb{N}_+$). Let $Y(x,\theta) =_{\text{st}} [X(n+1,\theta) \big| X(n,\theta) = x]$, $x \in \mathcal{S}$.

Theorem 8.E.6. *Suppose that* $\{Y(x,\theta),\ (x,\theta) \in \mathcal{S} \times \Theta\} \in$ SI-DIR-CX [SI-DIR-CV, SI-DIR-CX(sp), SI-DIR-CV(sp)]. *If* $\{X(0,\theta), \theta \in \Theta\} \in$ SICX [SICV, SICX(sp), SICV(sp)], *then* $\{X(n,\theta),\ \theta \in \Theta\} \in$ SICX [SICV, SICX(sp), SICV(sp)] *for each* $n \in \mathbb{N}_+$.

Proof. As an induction hypothesis assume that for some n we have

$$\{X(n,\theta),\ \theta \in \Theta\} \in \text{SICX [SICV, SICX(sp), SICV(sp)]}. \tag{8.E.1}$$

Note that

$$X(n+1,\theta) =_{\text{st}} Y(X(n,\theta),\theta). \tag{8.E.2}$$

Now, from (8.E.1), (8.E.2), and from a straightforward extension of Theorem 8.E.2(a) (for the (sp) cases) [or of Theorem 8.E.1(a) (for the other cases)], one obtains that

$$\{X(n+1,\theta),\ \theta \in \Theta\} \in \text{SICX [SICV, SICX(sp), SICV(sp)]}. \qquad \square$$

8.F Complements

Section 8.A: The notion of (regular) stochastic convexity/concavity is introduced in Shaked and Shanthikumar [508]. However, the condition $\{X(\theta),\ \theta \in \Theta\} \in$ SCX was encountered earlier by Schweder [499] who described it by saying that $\{X(\theta),\ \theta \in \Theta\}$ is "convexly parametrized." The basic closure properties (Theorems 8.A.15 and 8.A.17) are established in Shaked and Shanthikumar [508]. As an example of the use of these results, we note that Theorem 1(b) of Lefèvre and Malice [336] can be obtained from a combination of Example 8.A.3 with Theorems 8.A.15 and 8.A.17. A slightly weaker version of the example regarding the forward recurrence times (Example 8.A.18) can be found in Makowski and Philips [380]. Temporal convexity of Markov processes (Theorems 8.A.19 and 8.A.20) are studied in Shaked and Shanthikumar [507, 509], Shanthikumar and Yao [534], and Li and Shaked [349]. Extensions of these notions to random vectors can be found in Chang, Chao, Pinedo, and Shanthikumar [125], and to arbitrary random variables can be found in Meester [385] and in Meester and Shanthikumar [388]. A study of regular stochastic convexity by means of operators is developed in Adell and Perez-Palomares [5]. The results about stochastic m-convexity (Section 8.A.3) are mostly taken from Denuit, Lefèvre and Utev [155]. The stochastic m-increasing convexity of a family with a location parameter (Example 8.A.26) can be found in Denuit and Lefèvre [147].

"Derivatives" of stochastically convex and m-convex processes are introduced and studied in Adell and Lekuona [4].

Section 8.B: The notion of (sample path) stochastic convexity/concavity is introduced in Shaked and Shanthikumar [508]. A generalization of the notion of the semigroup property can be found in Shaked, Shanthikumar, and Tong [519]; Example 8.B.7 is a special case of a result due to them. The closure properties (Theorems 8.B.10 and 8.B.13) are established in Shaked and Shanthikumar [508]. The relation (8.B.2) between the spreads can be found in Goldstein and Rinott [212]. Temporal sample path convexity of Markov processes (Theorems 8.B.16 and 8.B.17) is studied in Shaked and Shanthikumar [508, 509]. Extensions of these notions to random vectors can be found in Chang, Chao, Pinedo, and Shanthikumar [125], and to arbitrary random variables can be found in Meester [385] and in Meester and Shanthikumar [388]. Theorem 8.B.18 is essentially proved in Kirmani and Gupta [299].

Section 8.C: Stochastic convexity/concavity in the usual stochastic ordering is introduced in Shaked and Shanthikumar [510]. Theorem 8.C.6 is established in Shaked and Shanthikumar [514], and Theorem 8.C.7 is established in Shaked and Shanthikumar [510].

Some variations of the stochastic convexity notions in Section 8.C, and also of the notions in Section 8.A, can be found in Atakan [24].

Section 8.D: The notion of strong stochastic convexity (in a different form) is introduced in Shanthikumar and Yao [531, 533]. The definition presented here is given in Meester and Shanthikumar [386].

Section 8.E: The notion of multivariate stochastic directional convexity is introduced in Meester and Shanthikumar [387]. Most of the results of this section are taken from that paper. The results yielding the parametric stochastic convexity and concavity of Markov processes (Theorem 8.E.6) can be found in Shaked and Shanthikumar [509].

Yao [573] introduced notions of stochastic supermodularity and submodularity that are weaker than the notions of stochastic directional convexity and concavity, respectively.

9

Positive Dependence Orders

Notions of positive dependence of two random variables X_1 and X_2 have been introduced in the literature in an effort to mathematically describe the property that "large (respectively, small) values of X_1 tend to go together with large (respectively, small) values of X_2." Many of the notions of positive dependence are defined by means of some comparison of the joint distribution of X_1 and X_2 with their distribution under the theoretical assumption that X_1 and X_2 are independent. Often such a comparison can be extended to general pairs of bivariate distributions with given marginals. This fact led researchers to introduce various notions of positive dependence orders. These orders are designed to compare the strength of the positive dependence of the two underlying bivariate distributions. In this chapter we describe some such notions.

In many sections of this chapter we first describe a positive dependence order which compares two *bivariate* random vectors (or distributions). When the order can be extended to general *n-dimensional* ($n > 2$) random vectors, we will describe the extension in a later part of that section.

Most of the orders that we describe in this chapter are defined on the Fréchet class $\mathcal{M}(F_1, F_2)$ of bivariate distributions with fixed marginals F_1 and F_2. The upper bound of this class is the distribution defined by $\min\{F_1(x_1), F_2(x_2)\}$ (whose probability mass is concentrated on the set $\{(x_1, x_2) : F_1(x_1) = F_2(x_2)\}$). The lower bound of this class is the distribution defined by $\max\{F_1(x_1) + F_2(x_2) - 1, 0\}$ (whose probability mass is concentrated on the set $\{(x_1, x_2) : F_1(x_1) + F_2(x_2) = 1\}$).

9.A The PQD and the Supermodular Orders

9.A.1 Definition and basic properties: The bivariate case

Let the random vector (X_1, X_2) have the distribution function F, and let F_1 and F_2 denote, respectively, the marginal distributions of X_1 and X_2.

Lehmann [343] defined (X_1, X_2) (or F) to be positive quadrant dependent (PQD) if

$$F(x_1, x_2) \geq F_1(x_1)F_2(x_2) \quad \text{for all } x_1 \text{ and } x_2. \tag{9.A.1}$$

Note that (9.A.1) can be rewritten as

$$F(x_1, x_2) \geq F^I(x_1, x_2) \quad \text{for all } x_1 \text{ and } x_2, \tag{9.A.2}$$

where $F^I(x_1, x_2) \equiv F_1(x_1)F_2(x_2)$ for all x_1 and x_2. This characterization of the PQD notion leads naturally to the definition of the PQD order that is described next.

For a random vector (X_1, X_2) with distribution function F, let $\overline{F}$ be the bivariate survival function of (X_1, X_2), that is, $\overline{F}(x_1, x_2) \equiv P\{X_1 > x_1, X_2 > x_2\}$ for all x_1 and x_2. Let (Y_1, Y_2) be another bivariate random vector with distribution function G and survival function $\overline{G}$. Suppose that F and G have the same univariate marginals; that is, suppose that both belong to $\mathcal{M}(F_1, F_2)$ for some univariate distribution functions F_1 and F_2. If

$$F(x_1, x_2) \leq G(x_1, x_2) \quad \text{for all } x_1 \text{ and } x_2, \tag{9.A.3}$$

then we say that (X_1, X_2) is *smaller than* (Y_1, Y_2) *in the* PQD *order* (denoted by $(X_1, X_2) \leq_{\text{PQD}} (Y_1, Y_2)$). Sometimes it will be useful to write this as $F \leq_{\text{PQD}} G$. Using the assumption that F and G have the same univariate marginals, it is easy to see that (9.A.3) is equivalent to

$$\overline{F}(x_1, x_2) \leq \overline{G}(x_1, x_2) \quad \text{for all } x_1 \text{ and } x_2.$$

Note that for random vectors (X_1, X_2) and (Y_1, Y_2), with distribution functions in $\mathcal{M}(F_1, F_2)$, we have

$$(X_1, X_2) \leq_{\text{PQD}} (Y_1, Y_2) \Longleftrightarrow (X_1, X_2) \leq_{\text{uo}} (Y_1, Y_2)$$

and

$$(X_1, X_2) \leq_{\text{PQD}} (Y_1, Y_2) \Longleftrightarrow (X_1, X_2) \geq_{\text{lo}} (Y_1, Y_2);$$

see (6.G.1) and (6.G.2) in Section 6.G.1. The reader should notice, however, that in (6.G.1) and (6.G.2) it is not required that (X_1, X_2) and (Y_1, Y_2) have the same marginals. Therefore, whereas the upper and lower orthant orders measure the size (or the location) of the underlying random vectors, the PQD order measures the amount of positive dependence of the underlying random vectors.

From (9.A.2) it is seen that F is PQD if, and only if,

$$F^I \leq_{\text{PQD}} F.$$

By Hoeffding's Lemma (see Lehmann [343, page 1139]) we see that if (X_1, X_2) and (Y_1, Y_2) have distributions F and G in $\mathcal{M}(F_1, F_2)$, then

$$\mathrm{Cov}(X_1, X_2) = \int_{-\infty}^{\infty}\int_{-\infty}^{\infty}[F(x_1, x_2) - F_1(x_1)F_2(x_2)]\mathrm{d}x_1\mathrm{d}x_2$$

and

$$\mathrm{Cov}(Y_1, Y_2) = \int_{-\infty}^{\infty}\int_{-\infty}^{\infty}[G(x_1, x_2) - F_1(x_1)F_2(x_2)]\mathrm{d}x_1\mathrm{d}x_2,$$

provided the covariances are well defined. It thus follows from (9.A.3) that if $(X_1, X_2) \leq_{\mathrm{PQD}} (Y_1, Y_2)$, then

$$\mathrm{Cov}(X_1, X_2) \leq \mathrm{Cov}(Y_1, Y_2), \tag{9.A.4}$$

and therefore, since $\mathrm{Var}(X_i) = \mathrm{Var}(Y_i)$, $i = 1, 2$, we have that

$$\rho_{X_1,X_2} \leq \rho_{Y_1,Y_2},$$

where ρ_{X_1,X_2} and ρ_{Y_1,Y_2} denote the correlation coefficients associated with (X_1, X_2) and (Y_1, Y_2), respectively, provided the underlying variances are well defined. Yanagimoto and Okamoto [570] have shown that some other correlation measures, such as Kendall's τ, Spearman's ρ, and Blomquist's q, are preserved under the PQD order. The inequality (9.A.4), and the monotonicity of other correlation measures under the PQD order, can also be obtained as corollaries from (9.A.17) below.

Let (X_1, X_2) and (Y_1, Y_2) be random vectors with distribution functions F and G. If $(X_1, X_2) \leq_{\mathrm{PQD}} (Y_1, Y_2)$, then

$$\overline{F}(x_1, x_2) \leq \overline{G}(x_1, x_2) \quad \text{for all } x_1 \text{ and } x_2,$$

and

$$P\{X_1 > x_1, X_2 \leq x_2\} \geq P\{Y_1 > x_1, Y_2 \leq x_2\} \quad \text{for all } x_1 \text{ and } x_2.$$

Therefore

$$P\{X_2 > x_2 | X_1 > x_1\} \leq P\{Y_2 > x_2 | Y_1 > x_1\} \quad \text{for all } x_1 \text{ and } x_2,$$

and

$$P\{X_2 \leq x_2 | X_1 > x_1\} \geq P\{Y_2 \leq x_2 | Y_1 > x_1\} \quad \text{for all } x_1 \text{ and } x_2.$$

Thus, for all x_1 we have

$$\begin{aligned}
E[X_2 | X_1 > x_1] &= -\int_{-\infty}^{0} P\{X_2 \leq x_2 | X_1 > x_1\}\mathrm{d}x_2 \\
&\qquad\qquad + \int_{0}^{\infty} P\{X_2 > x_2 | X_1 > x_1\}\mathrm{d}x_2 \\
&\leq -\int_{-\infty}^{0} P\{Y_2 \leq x_2 | Y_1 > x_1\}\mathrm{d}x_2 \\
&\qquad\qquad + \int_{0}^{\infty} P\{Y_2 > x_2 | Y_1 > x_1\}\mathrm{d}x_2 \\
&= E[Y_2 | Y_1 > x_1].
\end{aligned}$$

For random vectors (X_1, X_2) and (Y_1, Y_2) with distribution functions in $\mathcal{M}(F_1, F_2)$, the condition

$$E[X_2|X_1 > x_1] \le E[Y_2|Y_1 > x_1] \quad \text{for all } x_1 \tag{9.A.5}$$

can be used to define a positive dependence stochastic order. Such an order is discussed in Muliere and Petrone [405]. We see that if $(X_1, X_2) \le_{\text{PQD}} (Y_1, Y_2)$, then (9.A.5) holds.

Let F_L and F_U denote the Fréchet lower and upper bounds in the class $\mathcal{M}(F_1, F_2)$. Then, for every distribution $F \in \mathcal{M}(F_1, F_2)$ we have

$$F_L \le_{\text{PQD}} F \le_{\text{PQD}} F_U. \tag{9.A.6}$$

9.A.2 Closure properties

A powerful closure property of the PQD order is given in the next theorem.

Theorem 9.A.1. *Suppose that the four random vectors (X_1, X_2), (Y_1, Y_2), (U_1, U_2), and (V_1, V_2) satisfy*

$$(X_1, X_2) \le_{\text{PQD}} (Y_1, Y_2) \quad \textit{and} \quad (U_1, U_2) \le_{\text{PQD}} (V_1, V_2), \tag{9.A.7}$$

and suppose that (X_1, X_2) and (U_1, U_2) are independent, and also that (Y_1, Y_2) and (V_1, V_2) are independent. Then

$$(\phi(X_1, U_1), \psi(X_2, U_2)) \le_{\text{PQD}} (\phi(Y_1, V_1), \psi(Y_2, V_2)),$$
$$\textit{for all increasing functions } \phi \textit{ and } \psi. \tag{9.A.8}$$

Proof. From the monotonicity of ϕ and ψ it follows that the set $\{(u_1, u_2) : \phi(x_1, u_1) \le a_1, \psi(x_2, u_2) \le a_2\}$ is a lower quadrant for all x_1, x_2, a_1, and a_2. Therefore, for all a_1 and a_2 we have

$$\begin{aligned} P\{\phi(X_1, U_1) \le a_1, &\psi(X_2, U_2) \le a_2\} \\ &= \iint P\{\phi(X_1, u_1) \le a_1, \psi(X_2, u_2) \le a_2\} \mathrm{d}H(u_1, u_2) \\ &\le \iint P\{\phi(Y_1, u_1) \le a_1, \psi(Y_2, u_2) \le a_2\} \mathrm{d}H(u_1, u_2) \\ &= P\{\phi(Y_1, U_1) \le a_1, \psi(Y_2, U_2) \le a_2\}, \end{aligned}$$

where H is the distribution function of (U_1, U_2). Thus,

$$(\phi(X_1, U_1), \psi(X_2, U_2)) \le_{\text{PQD}} (\phi(Y_1, U_1), \psi(Y_2, U_2)),$$
$$\text{for all increasing functions } \phi \text{ and } \psi. \tag{9.A.9}$$

In a similar manner one can show that

$$(\phi(Y_1, U_1), \psi(Y_2, U_2)) \le_{\text{PQD}} (\phi(Y_1, V_1), \psi(Y_2, V_2)),$$
$$\text{for all increasing functions } \phi \text{ and } \psi. \tag{9.A.10}$$

From (9.A.9) and (9.A.10) one obtains (9.A.8). □

In particular, if (9.A.7) holds, then

$$(X_1 + U_1, X_2 + U_2) \leq_{\mathrm{PQD}} (Y_1 + V_1, Y_2 + V_2), \tag{9.A.11}$$

that is, the PQD order is closed under convolutions. From Theorem 9.A.1 it also follows that

$$(X_1, X_2) \leq_{\mathrm{PQD}} (Y_1, Y_2) \Longrightarrow (\phi(X_1), \psi(X_2)) \leq_{\mathrm{PQD}} (\phi(Y_1), \psi(Y_2)),$$

for all increasing functions ϕ and ψ.

The closure properties that are stated in the next theorem are easy to verify.

Theorem 9.A.2. (a) *Let* $\{(X_1^{(j)}, X_2^{(j)}),\ j = 1, 2, \ldots\}$ *and* $\{(Y_1^{(j)}, Y_2^{(j)}),\ j = 1, 2, \ldots\}$ *be two sequences of random vectors such that* $(X_1^{(j)}, X_2^{(j)}) \to_{\mathrm{st}} (X_1, X_2)$ *and* $(Y_1^{(j)}, Y_2^{(j)}) \to_{\mathrm{st}} (Y_1, Y_2)$ *as* $j \to \infty$, *where* $\to_{\mathrm{st}}$ *denotes convergence in distribution. If* $(X_1^{(j)}, X_2^{(j)}) \leq_{\mathrm{PQD}} (Y_1^{(j)}, Y_2^{(j)})$, $j = 1, 2, \ldots$, *then* $(X_1, X_2) \leq_{\mathrm{PQD}} (Y_1, Y_2)$.

(b) *Let* (X_1, X_2), (Y_1, Y_2), *and* $\boldsymbol{\Theta}$ *be random vectors such that* $[(X_1, X_2) | \boldsymbol{\Theta} = \boldsymbol{\theta}] \leq_{\mathrm{PQD}} [(Y_1, Y_2) | \boldsymbol{\Theta} = \boldsymbol{\theta}]$ *for all* $\boldsymbol{\theta}$ *in the support of* $\boldsymbol{\Theta}$. *Then* $(X_1, X_2) \leq_{\mathrm{PQD}} (Y_1, Y_2)$. *That is, the* PQD *order is closed under mixtures.*

Fang, Hu, and Joe [191] applied the idea of the PQD order to stationary Markov chains and showed that, if the process is stochastically increasing, then dependence (in the sense of the PQD order) is decreasing with the lag, namely, if $\{X_1, X_2, \ldots\}$ is a Markov chain and X_i is distributed according to F and if (X_1, X_n) is distributed according to F_{1n}, $n = 2, 3, \ldots$, then

$$F_{12} \geq_{\mathrm{PQD}} F_{13} \geq_{\mathrm{PQD}} \cdots \geq_{\mathrm{PQD}} F_{1n} \geq_{\mathrm{PQD}} \cdots \geq_{\mathrm{PQD}} F^{(2)}, \tag{9.A.12}$$

where $F^{(2)}(x, y) = F(x)F(y)$. See also Remark 9.A.29 below.

Another example is the following.

Example 9.A.3. Let ϕ and ψ be two Laplace transforms of positive random variables. Then F and G, defined by

$$F(x_1, x_2) = \phi(\phi^{-1}(x_1) + \phi^{-1}(x_2)), \quad (x_1, x_2) \in [0, 1]^2,$$

and

$$G(y_1, y_2) = \psi(\psi^{-1}(y_1) + \psi^{-1}(y_2)), \quad (y_1, y_2) \in [0, 1]^2,$$

are bivariate distribution functions with uniform$[0, 1]$ marginals (such F and G are called *Archimedean copulas*). Let (X_1, X_2) and (Y_1, Y_2) be distributed according to F and G, respectively. Then $(X_1, X_2) \leq_{\mathrm{PQD}} (Y_1, Y_2)$ if, and only if, $\psi^{-1}\phi$ is superadditive (that is, $\psi^{-1}\phi(x + y) \geq \psi^{-1}\phi(x) + \psi^{-1}\phi(y)$ for all $x, y \geq 0$). Also, if $\phi^{-1}\psi$ has a completely monotone derivative, then $(X_1, X_2) \leq_{\mathrm{PQD}} (Y_1, Y_2)$.

9.A.3 The multivariate case

Let $\boldsymbol{X} = (X_1, X_2, \ldots, X_n)$ be a random vector with distribution function F and survival function $\overline{F}$. Let $\boldsymbol{Y} = (Y_1, Y_2, \ldots, Y_n)$ be another random vector with distribution function G and survival function $\overline{G}$. If

$$F(\boldsymbol{x}) \le G(\boldsymbol{x}) \quad \text{for all } \boldsymbol{x}, \tag{9.A.13}$$

and

$$\overline{F}(\boldsymbol{x}) \le \overline{G}(\boldsymbol{x}) \quad \text{for all } \boldsymbol{x}, \tag{9.A.14}$$

then we say that $\boldsymbol{X}$ is *smaller than* $\boldsymbol{Y}$ *in the* PQD *order* (denoted by $\boldsymbol{X} \le_{\text{PQD}} \boldsymbol{Y}$). From (9.A.13) and (9.A.14) it follows that only random vectors with the same univariate marginals can be compared in the PQD order.

From (9.A.13) and (9.A.14) it follows that

$$\boldsymbol{X} \le_{\text{PQD}} \boldsymbol{Y} \Longleftrightarrow \big\{\boldsymbol{X} \le_{\text{uo}} \boldsymbol{Y} \text{ and } \boldsymbol{X} \ge_{\text{lo}} \boldsymbol{Y}\big\}. \tag{9.A.15}$$

An extension of Theorem 9.A.1 to the general multivariate case is the following. The proof of Theorem 9.A.4 is a straightforward extension of the proof of Theorem 9.A.1, and therefore it is omitted.

Theorem 9.A.4. *Suppose that the four random vectors* $\boldsymbol{X} = (X_1, X_2, \ldots, X_n)$, $\boldsymbol{Y} = (Y_1, Y_2, \ldots, Y_n)$, $\boldsymbol{U} = (U_1, U_2, \ldots, U_n)$, *and* $\boldsymbol{V} = (V_1, V_2, \ldots, V_n)$ *satisfy*

$$\boldsymbol{X} \le_{\text{PQD}} \boldsymbol{Y} \quad \textit{and} \quad \boldsymbol{U} \le_{\text{PQD}} \boldsymbol{V}, \tag{9.A.16}$$

and suppose that $\boldsymbol{X}$ *and* $\boldsymbol{U}$ *are independent, and also that* $\boldsymbol{Y}$ *and* $\boldsymbol{V}$ *are independent. Then*

$$\begin{aligned}
&(\phi_1(X_1, U_1), \phi_2(X_2, U_2), \ldots, \phi_n(X_n, U_n)) \\
&\qquad \le_{\text{PQD}} (\phi_1(Y_1, V_1), \phi_2(Y_2, V_2), \ldots, \phi_n(Y_n, V_n)), \\
&\qquad\qquad \textit{for all increasing functions } \phi_i,\ i = 1, 2, \ldots, n.
\end{aligned}$$

In particular, if (9.A.16) holds, then

$$\boldsymbol{X} + \boldsymbol{U} \le_{\text{PQD}} \boldsymbol{Y} + \boldsymbol{V},$$

that is, the PQD order is closed under convolutions. Also, from Theorem 9.A.4 it follows that

$$\begin{aligned}
&(X_1, X_2, \ldots, X_n) \le_{\text{PQD}} (Y_1, Y_2, \ldots, Y_n) \\
&\quad \Longrightarrow (\phi_1(X_1), \phi_2(X_2), \ldots, \phi_n(X_n)) \le_{\text{PQD}} (\phi_1(Y_1), \phi_2(Y_2), \ldots, \phi_n(Y_n)),
\end{aligned}$$

for all increasing functions ϕ_i, $i = 1, 2, \ldots, n$.

The closure properties that are stated in the next theorem are easy to verify.

Theorem 9.A.5. (a) *Let $\boldsymbol{X}_1, \boldsymbol{X}_2, \ldots, \boldsymbol{X}_m$ be a set of independent random vectors where the dimension of $\boldsymbol{X}_i$ is k_i, $i = 1, 2, \ldots, m$. Let $\boldsymbol{Y}_1, \boldsymbol{Y}_2, \ldots, \boldsymbol{Y}_m$ be another set of independent random vectors where the dimension of $\boldsymbol{Y}_i$ is k_i, $i = 1, 2, \ldots, m$. If $\boldsymbol{X}_i \le_{\text{PQD}} \boldsymbol{Y}_i$ for $i = 1, 2, \ldots, m$, then*

$$(\boldsymbol{X}_1, \boldsymbol{X}_2, \ldots, \boldsymbol{X}_m) \le_{\text{PQD}} (\boldsymbol{Y}_1, \boldsymbol{Y}_2, \ldots, \boldsymbol{Y}_m).$$

That is, the PQD *order is closed under conjunctions.*

(b) *Let $\boldsymbol{X} = (X_1, X_2, \ldots, X_n)$ and $\boldsymbol{Y} = (Y_1, Y_2, \ldots, Y_n)$ be two n-dimensional random vectors. If $\boldsymbol{X} \le_{\text{PQD}} \boldsymbol{Y}$, then $\boldsymbol{X}_I \le_{\text{PQD}} \boldsymbol{Y}_I$ for each $I \subseteq \{1, 2, \ldots, n\}$. That is, the* PQD *order is closed under marginalization.*

(c) *Let $\{\boldsymbol{X}_j, j = 1, 2, \ldots\}$ and $\{\boldsymbol{Y}_j, j = 1, 2, \ldots\}$ be two sequences of random vectors such that $\boldsymbol{X}_j \to_{\text{st}} \boldsymbol{X}$ and $\boldsymbol{Y}_j \to_{\text{st}} \boldsymbol{Y}$ as $j \to \infty$, where $\to_{\text{st}}$ denotes convergence in distribution. If $\boldsymbol{X}_j \le_{\text{PQD}} \boldsymbol{Y}_j$, $j = 1, 2, \ldots$, then $\boldsymbol{X} \le_{\text{PQD}} \boldsymbol{Y}$.*

(d) *Let $\boldsymbol{X}$, $\boldsymbol{Y}$, and $\boldsymbol{\Theta}$ be random vectors such that $[\boldsymbol{X} \big| \boldsymbol{\Theta} = \boldsymbol{\theta}] \le_{\text{PQD}} [\boldsymbol{Y} \big| \boldsymbol{\Theta} = \boldsymbol{\theta}]$ for all $\boldsymbol{\theta}$ in the support of $\boldsymbol{\Theta}$. Then $\boldsymbol{X} \le_{\text{PQD}} \boldsymbol{Y}$. That is, the* PQD *order is closed under mixtures.*

From Theorem 9.A.5(b) and (9.A.4) it follows that if $(X_1, X_2, \ldots, X_n) \le_{\text{PQD}} (Y_1, Y_2, \ldots, Y_n)$, then, for all $i_1 \ne i_2$, we have that

$$\text{Cov}(X_{i_1}, X_{i_2}) \le \text{Cov}(Y_{i_1}, Y_{i_2}).$$

Since the univariate marginals of $\boldsymbol{X}$ and $\boldsymbol{Y}$ are equal, it also follows that

$$\rho_{X_{i_1}, X_{i_2}} \le \rho_{Y_{i_1}, Y_{i_2}},$$

where $\rho_{X_{i_1}, X_{i_2}}$ and $\rho_{Y_{i_1}, Y_{i_2}}$ denote the correlation coefficients associated with (X_{i_1}, X_{i_2}) and (Y_{i_1}, Y_{i_2}), respectively, provided the underlying variances are well defined. Joe [260] has shown that some multivariate versions of the correlation measures Kendall's τ, Spearman's ρ, and Blomquist's q, are monotone with respect to the PQD order.

Another preservation property of the PQD order is described in the next theorem. In the following theorem we define $\sum_{j=1}^{0} x_j \equiv 0$ for any sequence $\{x_j, j = 1, 2, \ldots\}$. Similar results are Theorems 6.G.7 and 9.A.15.

Theorem 9.A.6. *Let $\boldsymbol{X}_j = (X_{j,1}, X_{j,2}, \ldots, X_{j,m})$, $j = 1, 2, \ldots$, be a sequence of nonnegative random vectors, and let $\boldsymbol{M} = (M_1, M_2, \ldots, M_m)$ and $\boldsymbol{N} = (N_1, N_2, \ldots, N_m)$ be two vectors of nonnegative integer-valued random variables. Assume that both $\boldsymbol{M}$ and $\boldsymbol{N}$ are independent of the $\boldsymbol{X}_j$'s. If $\boldsymbol{M} \le_{\text{PQD}} \boldsymbol{N}$, then*

$$\Big(\sum_{j=1}^{M_1} X_{j,1}, \sum_{j=1}^{M_2} X_{j,2}, \ldots, \sum_{j=1}^{M_m} X_{j,m}\Big) \le_{\text{PQD}} \Big(\sum_{j=1}^{N_1} X_{j,1}, \sum_{j=1}^{N_2} X_{j,2}, \ldots, \sum_{j=1}^{N_m} X_{j,m}\Big).$$

Consider now, as in Section 6.B.4, n families of univariate distribution functions $\{G_\theta^{(i)},\ \theta \in \mathcal{X}_i\}$ where $\mathcal{X}_i$ is a subset of the real line $\mathbb{R}$, $i = 1, 2, \ldots, n$. Let $X_i(\theta)$ denote a random variable with distribution function $G_\theta^{(i)}$, $i = 1, 2, \ldots, n$. Below we give a result which provides comparisons of two random vectors, with distribution functions of the form (6.B.18), in the PQD order. The following result is obtained easily from Theorem 6.G.8; see Theorems 6.B.17, 7.A.37, and 9.A.15 for related results.

Theorem 9.A.7. *Let $\{G_\theta^{(i)},\ \theta \in \mathcal{X}_i\}$, $i = 1, 2, \ldots, n$, be n families of univariate distribution functions as above. Let $\boldsymbol{\Theta}_1$ and $\boldsymbol{\Theta}_2$ be two random vectors with supports in $\prod_{i=1}^n \mathcal{X}_i$ and distribution functions F_1 and F_2, respectively. Let $\boldsymbol{Y}_1$ and $\boldsymbol{Y}_2$ be two random vectors with distribution functions H_1 and H_2 given by*

$$H_j(y_1, y_2, \ldots, y_n) = \int_{\mathcal{X}_1} \int_{\mathcal{X}_2} \cdots \int_{\mathcal{X}_n} \prod_{i=1}^n G_{\theta_i}^{(i)}(y_i) \mathrm{d}F_j(\theta_1, \theta_2, \ldots, \theta_n),$$
$$(y_1, y_2, \ldots, y_n) \in \mathbb{R}^n, \ j = 1, 2.$$

If

$$X_i(\theta) \le_{\mathrm{st}} X_i(\theta') \text{ whenever } \theta \le \theta', \quad i = 1, 2, \ldots, n,$$

and if

$$\boldsymbol{\Theta}_1 \le_{\mathrm{PQD}} \boldsymbol{\Theta}_2,$$

then

$$\boldsymbol{Y}_1 \le_{\mathrm{PQD}} \boldsymbol{Y}_2.$$

Example 9.A.8. Let $\boldsymbol{X}$ be an n-dimensional random vector with a density function f of the form

$$f(\boldsymbol{x}) = |\boldsymbol{\Sigma}|^{-1/2} g(\boldsymbol{x}\boldsymbol{\Sigma}^{-1}\boldsymbol{x}),$$

where $\boldsymbol{\Sigma} = (\sigma_{ij})$ is a positive definite $n \times n$ matrix, and g satisfies $\int_0^\infty r^{n-1} g(r^2)\mathrm{d}r < \infty$. Such density functions are called elliptically contoured. Let $\boldsymbol{Y}$ be an n-dimensional random vector with a density function h of the form

$$h(\boldsymbol{x}) = |\boldsymbol{\Lambda}|^{-1/2} g(\boldsymbol{x}\boldsymbol{\Lambda}^{-1}\boldsymbol{x}),$$

where $\boldsymbol{\Lambda} = (\lambda_{ij})$ is a positive definite $n \times n$ matrix. If $\sigma_{ii} = \lambda_{ii}$, $i = 1, 2, \ldots, n$, and $\sigma_{ij} \le \lambda_{ij}$, $1 \le i < j \le n$, then

$$\boldsymbol{X} \le_{\mathrm{PQD}} \boldsymbol{Y}.$$

In particular, multivariate normal random vectors with mean $\mathbf{0}$ and the same variances are ordered in the PQD order if their covariances are pointwise ordered.

9.A.4 The supermodular order

The supermodular order, which is described in this subsection, is a sufficient condition that implies the PQD order, but it is also of independent interest.

Recall from Section 7.A.8 that a function $\phi : \mathbb{R}^n \to \mathbb{R}$ is said to be supermodular if for any $\boldsymbol{x}, \boldsymbol{y} \in \mathbb{R}^n$ it satisfies

$$\phi(\boldsymbol{x}) + \phi(\boldsymbol{y}) \leq \phi(\boldsymbol{x} \wedge \boldsymbol{y}) + \phi(\boldsymbol{x} \vee \boldsymbol{y}),$$

where the operators $\wedge$ and $\vee$ denote coordinatewise minimum and maximum, respectively. Note that if $\phi : \mathbb{R}^n \to \mathbb{R}$ is supermodular, then the function ψ, defined by $\psi(x_1, x_2, \ldots, x_n) = \phi(g_1(x_1), g_2(x_2), \ldots, g_n(x_n))$, is also supermodular, whenever $g_i : \mathbb{R} \to \mathbb{R}$, $i = 1, 2, \ldots, n$, are all increasing or are all decreasing.

Let $\boldsymbol{X}$ and $\boldsymbol{Y}$ be two n-dimensional random vectors such that

$$E[\phi(\boldsymbol{X})] \leq E[\phi(\boldsymbol{Y})] \quad \text{for all supermodular functions } \phi : \mathbb{R}^n \to \mathbb{R},$$

provided the expectations exist. Then $\boldsymbol{X}$ is said to be *smaller than* $\boldsymbol{Y}$ *in the supermodular order* (denoted by $\boldsymbol{X} \leq_{\text{sm}} \boldsymbol{Y}$).

Since the functions $\phi_{\boldsymbol{x}} = I_{\{\boldsymbol{y}:\boldsymbol{y}>\boldsymbol{x}\}}$ and $\psi_{\boldsymbol{x}} = I_{\{\boldsymbol{y}:\boldsymbol{y}\leq\boldsymbol{x}\}}$ are supermodular for each fixed $\boldsymbol{x}$, it is immediate that

$$\boldsymbol{X} \leq_{\text{sm}} \boldsymbol{Y} \Longrightarrow \boldsymbol{X} \leq_{\text{PQD}} \boldsymbol{Y}. \tag{9.A.17}$$

These implications also follow from Theorem 6.G.2 and (9.A.15) since every n-dimensional ($n \geq 2$) distribution function, and any n-dimensional survival function, are supermodular functions. In fact, when $n = 2$ we have that

$$(X_1, X_2) \leq_{\text{sm}} (Y_1, Y_2) \Longleftrightarrow (X_1, X_2) \leq_{\text{PQD}} (Y_1, Y_2); \tag{9.A.18}$$

see, for example, Tchen [547]. From (9.A.17) it is seen that if $\boldsymbol{X} \leq_{\text{sm}} \boldsymbol{Y}$, then $\boldsymbol{X}$ and $\boldsymbol{Y}$ must have the same univariate marginals.

Some closure properties of the supermodular order are described in the next theorem.

Theorem 9.A.9. (a) *Let* $(X_1, X_2, \ldots, X_n)$ *and* $(Y_1, Y_2, \ldots, Y_n)$ *be two* n*-dimensional random vectors. If* $(X_1, X_2, \ldots, X_n) \leq_{\text{sm}} (Y_1, Y_2, \ldots, Y_n)$, *then*

$$(g_1(X_1), g_2(X_2), \ldots, g_n(X_n)) \leq_{\text{sm}} (g_1(Y_1), g_2(Y_2), \ldots, g_n(Y_n))$$

whenever $g_i : \mathbb{R} \to \mathbb{R}$, $i = 1, 2, \ldots, n$, *are all increasing or are all decreasing.*

(b) *Let* $\boldsymbol{X}_1, \boldsymbol{X}_2, \ldots, \boldsymbol{X}_m$ *be a set of independent random vectors where the dimension of* $\boldsymbol{X}_i$ *is* k_i, $i = 1, 2, \ldots, m$. *Let* $\boldsymbol{Y}_1, \boldsymbol{Y}_2, \ldots, \boldsymbol{Y}_m$ *be another set of independent random vectors where the dimension of* $\boldsymbol{Y}_i$ *is* k_i, $i = 1, 2, \ldots, m$. *If* $\boldsymbol{X}_i \leq_{\text{sm}} \boldsymbol{Y}_i$ *for* $i = 1, 2, \ldots, m$, *then*

$$(\boldsymbol{X}_1, \boldsymbol{X}_2, \ldots, \boldsymbol{X}_m) \leq_{\text{sm}} (\boldsymbol{Y}_1, \boldsymbol{Y}_2, \ldots, \boldsymbol{Y}_m).$$

That is, the supermodular order is closed under conjunctions.

(c) *Let* $\boldsymbol{X} = (X_1, X_2, \dots, X_n)$ *and* $\boldsymbol{Y} = (Y_1, Y_2, \dots, Y_n)$ *be two* n*-dimensional random vectors. If* $\boldsymbol{X} \le_{\rm sm} \boldsymbol{Y}$*, then* $\boldsymbol{X}_I \le_{\rm sm} \boldsymbol{Y}_I$ *for each* $I \subseteq \{1, 2, \dots, n\}$*. That is, the supermodular order is closed under marginalization.*

(d) *Let* $\boldsymbol{X}$*,* $\boldsymbol{Y}$*, and* $\boldsymbol{\Theta}$ *be random vectors such that* $[\boldsymbol{X}|\boldsymbol{\Theta} = \boldsymbol{\theta}] \le_{\rm sm} [\boldsymbol{Y}|\boldsymbol{\Theta} = \boldsymbol{\theta}]$ *for all* $\boldsymbol{\theta}$ *in the support of* $\boldsymbol{\Theta}$*. Then* $\boldsymbol{X} \le_{\rm sm} \boldsymbol{Y}$*. That is, the supermodular order is closed under mixtures.*

(e) *Let* $\{\boldsymbol{X}_j, j = 1, 2, \dots\}$ *and* $\{\boldsymbol{Y}_j, j = 1, 2, \dots\}$ *be two sequences of random vectors such that* $\boldsymbol{X}_j \to_{\rm st} \boldsymbol{X}$ *and* $\boldsymbol{Y}_j \to_{\rm st} \boldsymbol{Y}$ *as* $j \to \infty$*, where* $\to_{\rm st}$ *denotes convergence in distribution. If* $\boldsymbol{X}_j \le_{\rm sm} \boldsymbol{Y}_j$*,* $j = 1, 2, \dots$*, then* $\boldsymbol{X} \le_{\rm sm} \boldsymbol{Y}$*.*

Proof. Part (a) follows from the fact that a composition of a supermodular function with coordinatewise functions, that are all increasing or are all decreasing, is a supermodular function.

In order to see part (b) let $\boldsymbol{X}_1$ and $\boldsymbol{X}_2$ be two independent random vectors, and let $\boldsymbol{Y}_1$ and $\boldsymbol{Y}_2$ be two other independent random vectors. Suppose that $\boldsymbol{X}_1 \le_{\rm sm} \boldsymbol{Y}_1$ and that $\boldsymbol{X}_2 \le_{\rm sm} \boldsymbol{Y}_2$. Then, for any supermodular function ϕ (of the proper dimension) we have that

$$\begin{aligned} E\phi(\boldsymbol{X}_1, \boldsymbol{X}_2) &= E\big[E\phi(\boldsymbol{X}_1, \boldsymbol{X}_2)\big|\boldsymbol{X}_2\big] \\ &\le E\big[E\phi(\boldsymbol{Y}_1, \boldsymbol{X}_2)\big|\boldsymbol{X}_2\big] \\ &= E\phi(\boldsymbol{Y}_1, \boldsymbol{X}_2) \\ &\le E\phi(\boldsymbol{Y}_1, \boldsymbol{Y}_2), \end{aligned}$$

where the first inequality follows from the fact that $\phi(\boldsymbol{x}_1, \boldsymbol{x}_2)$ is supermodular in $\boldsymbol{x}_1$ when $\boldsymbol{x}_2$ is fixed, and the second inequality follows in a similar manner. Part (b) of Theorem 9.A.9 follows from the above by induction.

Parts (c) and (d) are easy to prove. A proof of part (e) can be found in Müller and Scarsini [416]. □

From parts (a) and (d) of Theorem 9.A.9 we obtain the following corollary.

Corollary 9.A.10. *Let* $\boldsymbol{X} = (X_1, X_2, \dots, X_n)$ *and* $\boldsymbol{Y} = (Y_1, Y_2, \dots, Y_n)$ *be two random vectors such that* $\boldsymbol{X} \le_{\rm sm} \boldsymbol{Y}$*, and let* $\boldsymbol{Z}$ *be an* m*-dimensional random vector which is independent of* $\boldsymbol{X}$ *and* $\boldsymbol{Y}$*. Then*

$$\begin{aligned} &(h_1(X_1, \boldsymbol{Z}), h_2(X_2, \boldsymbol{Z}), \dots, h_n(X_n, \boldsymbol{Z})) \\ &\qquad \le_{\rm sm} (h_1(Y_1, \boldsymbol{Z}), h_2(Y_2, \boldsymbol{Z}), \dots, h_n(Y_n, \boldsymbol{Z})), \end{aligned}$$

whenever $h_i(x, \boldsymbol{z})$*,* $i = 1, 2, \dots, n$*, are all increasing or are all decreasing in* x *for every* $\boldsymbol{z}$*.*

Example 9.A.11. Let $\boldsymbol{X}$ and $\boldsymbol{Y}$ be two n-dimensional random vectors such that $\boldsymbol{X} \le_{\rm sm} \boldsymbol{Y}$, and let $\boldsymbol{Z}$ be an n-dimensional random vector which is independent of $\boldsymbol{X}$ and $\boldsymbol{Y}$. Then from Corollary 9.A.10 it follows that

$$\boldsymbol{X} \wedge \boldsymbol{Z} \le_{\rm sm} \boldsymbol{Y} \wedge \boldsymbol{Z},$$

and that

$$\boldsymbol{X} + \boldsymbol{Z} \le_{\rm sm} \boldsymbol{Y} + \boldsymbol{Z}.$$

By applying Corollary 9.A.10 twice (letting $\boldsymbol{Z}$ there be an n-dimensional random vector, and letting each h_i depend only on its first argument and on the ith component of the second argument, $i = 1, 2, \ldots, n$), we get the following result.

Theorem 9.A.12. *Let $\boldsymbol{X}$, $\boldsymbol{Y}$, $\boldsymbol{Z}$, and $\boldsymbol{W}$ be n-dimensional random vectors such that $\boldsymbol{X}$ and $\boldsymbol{Z}$ are independent and $\boldsymbol{Y}$ and $\boldsymbol{W}$ are independent. Let $c_i : [0,\infty)^2 \to [0,\infty)$ be a continuous increasing function, $i = 1, 2, \ldots, n$. If $\boldsymbol{X} \le_{\rm sm} \boldsymbol{Y}$ and $\boldsymbol{Z} \le_{\rm sm} \boldsymbol{W}$, then*

$$(c_1(X_1, Z_1), c_2(X_2, Z_2), \ldots, c_n(X_n, Z_n)) \\ \le_{\rm sm} (c_1(Y_1, W_1), c_2(Y_2, W_2), \ldots, c_n(Y_n, W_n)).$$

Example 9.A.13. Let $\{\boldsymbol{X}_k = (X_{k,1}, \ldots, X_{k,n}),\ k \ge 0\}$ and $\{\boldsymbol{Y}_k = (Y_{k,1}, \ldots, Y_{k,n}),\ k \ge 0\}$ be two Markov chains as described in Example 6.G.6. If the g_i's are increasing in their $m+1$ arguments, if $\boldsymbol{U}^l = \{\boldsymbol{U}^l_k,\ k \ge 0\}$, $l = 1, \ldots, m$, are independent, if $\boldsymbol{V}^l = \{\boldsymbol{V}^l_k,\ k \ge 0\}$, $l = 1, \ldots, m$, are independent, and if $\boldsymbol{U}^l_k \le_{\rm sm} \boldsymbol{V}^l_k$, $l = 1, \ldots, m$, $k \ge 0$, then, for each $k \ge 0$ we have

$$(\boldsymbol{X}_0, \ldots, \boldsymbol{X}_k) \le_{\rm sm} (\boldsymbol{Y}_0, \ldots, \boldsymbol{Y}_k).$$

The proof uses Theorem 9.A.12, Corollary 9.A.10, and Theorem 9.A.9(b). We omit the details.

Another preservation property of the supermodular order is described in the next theorem. In the following theorem we define $\sum_{j=1}^{0} x_j \equiv 0$ for any sequence $\{x_j,\ j = 1, 2, \ldots\}$. Similar results are Theorems 6.G.7 and 9.A.6.

Theorem 9.A.14. *Let $\boldsymbol{X}_j = (X_{j,1}, X_{j,2}, \ldots, X_{j,m})$, $j = 1, 2, \ldots$, be a sequence of nonnegative random vectors, and let $\boldsymbol{M} = (M_1, M_2, \ldots, M_m)$ and $\boldsymbol{N} = (N_1, N_2, \ldots, N_m)$ be two vectors of nonnegative integer-valued random variables. Assume that both $\boldsymbol{M}$ and $\boldsymbol{N}$ are independent of the $\boldsymbol{X}_j$'s. If $\boldsymbol{M} \le_{\rm sm} \boldsymbol{N}$, then*

$$\Big(\sum_{j=1}^{M_1} X_{j,1}, \sum_{j=1}^{M_2} X_{j,2}, \ldots, \sum_{j=1}^{M_m} X_{j,m}\Big) \le_{\rm sm} \Big(\sum_{j=1}^{N_1} X_{j,1}, \sum_{j=1}^{N_2} X_{j,2}, \ldots, \sum_{j=1}^{N_m} X_{j,m}\Big).$$

Proof. Let ϕ be a supermodular function. Conditioning on the possible realizations of $(\boldsymbol{X}_1, \boldsymbol{X}_2, \ldots)$ we can write

$$E\Big[\phi\Big(\sum_{j=1}^{M_1} X_{j,1}, \sum_{j=1}^{M_2} X_{j,2}, \ldots, \sum_{j=1}^{M_m} X_{j,m}\Big)\Big] \\ = E\Big\{E\Big[\phi\Big(\sum_{j=1}^{M_1} X_{j,1}, \sum_{j=1}^{M_2} X_{j,2}, \ldots, \sum_{j=1}^{M_m} X_{j,m}\Big)\Big|(\boldsymbol{X}_1, \boldsymbol{X}_2, \ldots)\Big]\Big\}.$$

Now, it is easy to see that for any realization $(\boldsymbol{x}_1, \boldsymbol{x}_2, \dots)$ of $(\boldsymbol{X}_1, \boldsymbol{X}_2, \dots)$, the function ψ, defined by $\psi(n_1, n_2, \dots, n_m) = \phi\big(\sum_{j=1}^{n_1} x_{j,1}, \sum_{j=1}^{n_2} x_{j,2}, \dots, \sum_{j=1}^{n_m} x_{j,m}\big)$, is supermodular. Therefore, since $\boldsymbol{M} \le_{\text{sm}} \boldsymbol{N}$, we have that

$$E\Big[\phi\Big(\sum_{j=1}^{M_1} X_{j,1}, \sum_{j=1}^{M_2} X_{j,2}, \dots, \sum_{j=1}^{M_m} X_{j,m}\Big)\Big|(\boldsymbol{X}_1, \boldsymbol{X}_2, \dots) = (\boldsymbol{x}_1, \boldsymbol{x}_2, \dots)\Big]$$
$$\le E\Big[\phi\Big(\sum_{j=1}^{N_1} X_{j,1}, \sum_{j=1}^{N_2} X_{j,2}, \dots, \sum_{j=1}^{N_m} X_{j,m}\Big)\Big|(\boldsymbol{X}_1, \boldsymbol{X}_2, \dots) = (\boldsymbol{x}_1, \boldsymbol{x}_2, \dots)\Big],$$

and thus

$$E\Big[\phi\Big(\sum_{j=1}^{M_1} X_{j,1}, \sum_{j=1}^{M_2} X_{j,2}, \dots, \sum_{j=1}^{M_m} X_{j,m}\Big)\Big]$$
$$\le E\Big\{E\Big[\phi\Big(\sum_{j=1}^{N_1} X_{j,1}, \sum_{j=1}^{N_2} X_{j,2}, \dots, \sum_{j=1}^{N_m} X_{j,m}\Big)\Big|(\boldsymbol{X}_1, \boldsymbol{X}_2, \dots)\Big]\Big\}$$
$$= E\Big[\phi\Big(\sum_{j=1}^{N_1} X_{j,1}, \sum_{j=1}^{N_2} X_{j,2}, \dots, \sum_{j=1}^{N_m} X_{j,m}\Big)\Big]. \qquad \square$$

Consider now, as in Section 6.B.4, n families of univariate distribution functions $\{G_\theta^{(i)},\ \theta \in \mathcal{X}_i\}$ where $\mathcal{X}_i$ is a subset of the real line $\mathbb{R}$, $i = 1, 2, \dots, n$. Let $X_i(\theta)$ denote a random variable with distribution function $G_\theta^{(i)}$, $i = 1, 2, \dots, n$. Below we give a result which provides comparisons of two random vectors, with distribution functions of the form (6.B.18), in the supermodular order. The following result is a generalization of Theorem 9.A.9(d); see Theorems 6.B.17, 6.G.8, 7.A.37, and 9.A.7 for related results.

Theorem 9.A.15. *Let $\{G_\theta^{(i)},\ \theta \in \mathcal{X}_i\}$, $i = 1, 2, \dots, n$, be n families of univariate distribution functions as above. Let $\boldsymbol{\Theta}_1$ and $\boldsymbol{\Theta}_2$ be two random vectors with supports in $\prod_{i=1}^n \mathcal{X}_i$ and distribution functions F_1 and F_2, respectively. Let $\boldsymbol{Y}_1$ and $\boldsymbol{Y}_2$ be two random vectors with distribution functions H_1 and H_2 given by*

$$H_j(y_1, y_2, \dots, y_n) = \int_{\mathcal{X}_1}\int_{\mathcal{X}_2}\cdots\int_{\mathcal{X}_n} \prod_{i=1}^n G_{\theta_i}^{(i)}(y_i)\mathrm{d}F_j(\theta_1, \theta_2, \dots, \theta_n),$$
$$(y_1, y_2, \dots, y_n) \in \mathbb{R}^n,\ j = 1, 2.$$

If

$$X_i(\theta) \le_{\text{st}} X_i(\theta') \text{ whenever } \theta \le \theta', \quad i = 1, 2, \dots, n,$$

and if

$$\boldsymbol{\Theta}_1 \le_{\text{sm}} \boldsymbol{\Theta}_2,$$

then

$$\boldsymbol{Y}_1 \le_{\text{sm}} \boldsymbol{Y}_2.$$

Before stating the next result, it is worthwhile to mention that from Proposition 7.A.27 it follows that

$$\boldsymbol{X} \le_{\text{sm}} \boldsymbol{Y} \Longrightarrow \boldsymbol{X} \le_{\text{dir-cx}} \boldsymbol{Y}.$$

The following result may be compared with Theorems 6.G.10 and 7.A.30.

Theorem 9.A.16. *Let $\boldsymbol{X}$ and $\boldsymbol{Y}$ be two random vectors. If $\boldsymbol{X} \le_{\text{sm}} \boldsymbol{Y}$, then $\phi(\boldsymbol{X}) \le_{\text{icx}} \phi(\boldsymbol{Y})$ for any increasing supermodular function $\phi : \mathbb{R}^n \to \mathbb{R}$.*

A consequence of Theorem 9.A.16, that is useful in queuing theory, is described in the following example.

Example 9.A.17. Let $\{A_i\}_{i=0}^{\infty}$ be a sequence of random variables, and let c be some constant. Define inductively

$$Q_0 = q; \quad Q_{i+1} = [Q_i + A_i - c]_+, \ i = 1, 2, \ldots,$$

for some fixed q. Similarly, let $\{A_i'\}_{i=0}^{\infty}$ be another sequence of random variables, and define inductively

$$Q_0' = q; \quad Q_{i+1}' = [Q_i' + A_i' - c]_+, \ i = 1, 2, \ldots.$$

If $(A_0, A_1, \ldots, A_i) \le_{\text{sm}} (A_0', A_1', \ldots, A_i')$ for all $i = 1, 2, \ldots$, then $Q_i \le_{\text{icx}} Q_i'$ for all $i = 1, 2, \ldots$. In fact, the above result holds even if Q_0 and Q_0' are random variables satisfying $Q_0 \le_{\text{icx}} Q_0'$.

As a particular case of Theorem 9.A.16 we have that

$$(X_1, X_2, \ldots, X_n) \le_{\text{sm}} (Y_1, Y_2, \ldots, Y_n) \Longrightarrow \sum_{i=1}^{n} X_i \le_{\text{cx}} \sum_{i=1}^{n} Y_i \qquad (9.A.19)$$

(since $\boldsymbol{X} \le_{\text{sm}} \boldsymbol{Y} \Longrightarrow E\boldsymbol{X} = E\boldsymbol{Y}$).

A related result is the following. It shows that the larger in the supermodular order a random vector is, the "closer" are its coordinates in the proper stochastic sense.

Theorem 9.A.18. *Let (X_1, X_2) and (Y_1, Y_2) be two random vectors. If $(X_1, X_2) \le_{\text{sm}} (Y_1, Y_2)$ (that is, $(X_1, X_2) \le_{\text{PQD}} (Y_1, Y_2)$; see (9.A.18)), then*

$$Y_1 - Y_2 \le_{\text{cx}} X_1 - X_2.$$

Proof. Let ϕ be a univariate convex function. Then the function ψ, defined by

$$\psi(x_1, x_2) = -\phi(x_1 - x_2),$$

is easily seen to be supermodular. Thus $E\phi(Y_1 - Y_2) \le E\phi(X_1 - X_2)$. This proves the inequality. □

A consequence of Theorem 9.A.14 and (9.A.19) is described in the following example.

Example 9.A.19. Let $X_1, X_2, \ldots$ and $Y_1, Y_2, \ldots$ be two sequences of random variables. Let N_1 and N_2 be two independent and identically distributed positive integer-valued random variables independent of the X_i's and of the Y_i's. Then

$$\sum_{i=1}^{N_1} X_i + \sum_{i=1}^{N_2} Y_i \le_{\text{cx}} \sum_{i=1}^{N_1} (X_i + Y_i).$$

In order to see it, note that $(N_1, N_2) \le_{\text{sm}} (N_1, N_1)$, and use Theorem 9.A.14 and (9.A.19). This proof was communicated to us by Taizhong Hu.

An interesting example in which the supermodular order arises naturally is the following. See also Examples 6.B.29, 6.G.11, 7.A.13, 7.A.26, 7.A.39, and 7.B.5.

Example 9.A.20. Let $\boldsymbol{X}$ be a multivariate normal random vector with mean vector $\mathbf{0}$ and variance-covariance matrix $\boldsymbol{\Sigma}$, and let $\boldsymbol{Y}$ be a multivariate normal random vector with mean vector $\mathbf{0}$ and variance-covariance matrix $\boldsymbol{\Sigma}+\boldsymbol{D}$, where $\boldsymbol{D}$ is a matrix with zero diagonal elements such that $\boldsymbol{\Sigma} + \boldsymbol{D}$ is nonnegative definite. Then $\boldsymbol{X} \le_{\text{sm}} \boldsymbol{Y}$ if, and only if, all the entries of $\boldsymbol{D}$ are nonnegative.

The supermodular order can be used to bound some quite general random vectors. This is shown in the next three theorems. The proofs of the these theorems are omitted. Theorem 9.A.21 can be considered to be an extension of the right-hand side of (9.A.6).

Theorem 9.A.21. *Let $\boldsymbol{X} = (X_1, X_2, \ldots, X_n)$ be a random vector and let F_{X_i} be the marginal distribution of X_i, $i = 1, 2, \ldots, n$. Then, for a uniform$[0,1]$ random variable U we have that*

$$\boldsymbol{X} \le_{\text{sm}} (F_{X_1}^{-1}(U), F_{X_2}^{-1}(U), \ldots, F_{X_n}^{-1}(U)),$$

and therefore

$$\boldsymbol{X} \le_{\text{PQD}} (F_{X_1}^{-1}(U), F_{X_2}^{-1}(U), \ldots, F_{X_n}^{-1}(U)).$$

In particular, if the X_i's in Theorem 9.A.21, marginally, have the same (univariate) distribution function, then

$$\boldsymbol{X} \le_{\text{sm}} (X_1, X_1, \ldots, X_1),$$

and therefore

$$\boldsymbol{X} \le_{\text{PQD}} (X_1, X_1, \ldots, X_1).$$

Combining (9.A.19) and Theorem 9.A.21 it is seen, using the notation of Theorem 9.A.21, that

$$X_1 + X_2 + \cdots + X_n \leq_{\rm cx} F_{X_1}^{-1}(U) + F_{X_2}^{-1}(U) + \cdots + F_{X_n}^{-1}(U). \qquad (9.\text{A}.20)$$

A more detailed result is described next. Let $X_1, X_2, \ldots, X_n$, Z, and U be random variables, where U has the uniform$[0,1]$ distribution. Let $F_{X_i|Z}^{-1}(U)$ denote the random variable $g_i(U, Z)$, where g_i is defined by $g_i(u,z) = F_{X_i|Z=z}^{-1}(u)$, $i = 1, 2, \ldots, n$.

Proposition 9.A.22. *Let $\boldsymbol{X} = (X_1, X_2, \ldots, X_n)$ be a random vector, and let F_{X_i} be the marginal distribution of X_i, $i = 1, 2, \ldots, n$. Let Z and U be two other random variables, such that U has a uniform$[0,1]$ distribution, and is independent of Z. Then*

$$\begin{aligned} X_1 + X_2 + \cdots + X_n &\leq_{\rm cx} F_{X_1|Z}^{-1}(U) + F_{X_2|Z}^{-1}(U) + \cdots + F_{X_n|Z}^{-1}(U) \\ &\leq_{\rm cx} F_{X_1}^{-1}(U) + F_{X_2}^{-1}(U) + \cdots + F_{X_n}^{-1}(U). \qquad (9.\text{A}.21) \end{aligned}$$

Proof. From (9.A.20) it is seen that for any convex function ϕ we have (below F_Z denotes the distribution function of Z)

$$\begin{aligned} &E[\phi(X_1 + \cdots + X_n)] \\ &\quad = \int_{-\infty}^{\infty} E[\phi(X_1 + \cdots + X_n) \big| Z = z] \mathrm{d}F_Z(z) \\ &\quad \leq \int_{-\infty}^{\infty} E[\phi(F_{X_1|Z=z}^{-1}(U) + \cdots + F_{X_n|Z=z}^{-1}(U)) \big| Z = z] \mathrm{d}F_Z(z) \\ &\quad = E[\phi(F_{X_1|Z}^{-1}(U) + \cdots + F_{X_n|Z}^{-1}(U))], \end{aligned}$$

and the first inequality in (9.A.21) follows.

Next, note that the random vector $(F_{X_1|Z}^{-1}(U), F_{X_2|Z}^{-1}(U), \ldots, F_{X_n|Z}^{-1}(U))$ has the same marginals as $(X_1, X_2, \ldots, X_n)$ because

$$\begin{aligned} P(X_i \leq x) &= \int_{-\infty}^{\infty} P(X_i \leq x \big| Z = z) \mathrm{d}F_Z(z) \\ &= \int_{-\infty}^{\infty} P(F_{X_i|Z=z}^{-1}(U) \leq x) \mathrm{d}F_Z(z) \\ &= P(F_{X_i|Z}^{-1}(U) \leq x), \quad -\infty \leq x \leq \infty, \ i = 1, 2, \ldots, n, \end{aligned}$$

and the second inequality in (9.A.21) therefore follows from (9.A.20). □

The next result has been motivated by the desire to generalize and unify Theorems 3.A.34 and 4.A.17. Recall the definition of negative association in (3.A.54). If the inequality (3.A.54) is reversed, that is, if the random variables $X_1, X_2, \ldots, X_n$ satisfy

$$\mathrm{Cov}(h_1(X_{i_1}, X_{i_2}, \ldots, X_{i_k}), h_2(X_{j_1}, X_{j_2}, \ldots, X_{j_{n-k}})) \geq 0 \qquad (9.\text{A}.22)$$

for all choices of disjoint subsets $\{i_1, i_2, \ldots, i_k\}$ and $\{j_1, j_2, \ldots, j_{n-k}\}$ of $\{1, 2, \ldots, n\}$, and for all increasing functions h_1 and h_2 for which the above covariance is defined, then $X_1, X_2, \ldots, X_n$ are said to be *weakly positively associated.*

Theorem 9.A.23. *Let $\boldsymbol{X} = (X_1, X_2, \ldots, X_n)$ be a random vector, and let $\boldsymbol{Y} = (Y_1, Y_2, \ldots, Y_n)$ be a vector of independent random variables such that, marginally, $X_i =_{\rm st} Y_i$, $i = 1, 2, \ldots, n$.*

(a) *If $X_1, X_2, \ldots, X_n$ are weakly positively associated, then $\boldsymbol{X} \ge_{\rm sm} \boldsymbol{Y}$.*
(b) *If $X_1, X_2, \ldots, X_n$ are negatively associated, then $\boldsymbol{X} \le_{\rm sm} \boldsymbol{Y}$.*

A result that is stronger than Theorem 9.A.23 is given in Section 9.E below; see details in Remark 9.E.9.

Combining Theorem 9.A.23 with Theorem 9.A.16 (and using the fact that positive association implies weak positive association) one obtains Theorems 3.A.34 and 4.A.17 (for the latter, note that the function $\phi(\boldsymbol{x}) = \max_{1 \le k \le n} \sum_{i=1}^{k} x_i$ is increasing and supermodular).

Theorem 9.A.24. *Let $\boldsymbol{X} = (X_1, X_2, \ldots, X_n)$ be a vector of nonnegative random variables, and let F_i denote the marginal distribution of X_i, $i = 1, 2, \ldots, n$. Suppose that*

$$\sum_{1}^{n} \overline{F}_i(0) \le 1.$$

Then there exists a unique random vector $\boldsymbol{Y} = (Y_1, Y_2, \ldots, Y_n)$ with marginal distributions F_i, $i = 1, 2, \ldots, n$, such that

$$P\{Y_i > 0, Y_j > 0\} = 0 \quad \text{for all } i \ne j, \tag{9.A.23}$$

and this $\boldsymbol{Y}$ satisfies

$$\boldsymbol{Y} \le_{\rm sm} \boldsymbol{X}.$$

The following result strengthens Theorem 7.A.38; the terminology that is used there is also used in the theorem below.

Theorem 9.A.25. *Let the random vectors $\boldsymbol{X} = (X_1, X_2, \ldots, X_n)$ and $\boldsymbol{Y} = (Y_1, Y_2, \ldots, Y_n)$ have the respective copulas $C_{\boldsymbol{X}}$ and $C_{\boldsymbol{Y}}$. Let $\boldsymbol{U}_{\boldsymbol{X}}$ and $\boldsymbol{U}_{\boldsymbol{Y}}$ be distributed according to $C_{\boldsymbol{X}}$ and $C_{\boldsymbol{Y}}$. If $X_i \le_{\rm cx} Y_i$, $i = 1, 2, \ldots, n$, if $\boldsymbol{U}_{\boldsymbol{X}} \le_{\rm sm} \boldsymbol{U}_{\boldsymbol{Y}}$, and if $\boldsymbol{U}_{\boldsymbol{Y}}$ is* CI, *then $\boldsymbol{X} \le_{\text{dir-cx}} \boldsymbol{Y}$.*

Example 9.A.26. Let $Z_1, Z_2, \ldots, Z_n$ be a collection of independent and identically distributed random variables, let $U_1, U_2, \ldots, U_n$ be another collection of independent and identically distributed random variables, and let V be still another random variable that is independent of the U_i's. Consider the random vectors $\boldsymbol{Y}$ and $\boldsymbol{X}$ defined as

$$(Y_1, Y_2, \ldots, Y_n) = (g_1(Z_1), g_2(Z_2), \ldots, g_n(Z_n))$$
$$(X_1, X_2, \ldots, X_n) = (\tilde{g}_1(U_1, V), \tilde{g}_2(U_2, V), \ldots, \tilde{g}_n(U_n, V)),$$

where $g_i : \mathbb{R} \to \mathbb{R}$ and $\tilde{g}_i : \mathbb{R}^2 \to \mathbb{R}$ are measurable functions that satisfy

$$g_i(Z_i) =_{\text{st}} \tilde{g}_i(U_i, V), \qquad i = 1, 2, \ldots, n.$$

If $\tilde{g}_i$ is increasing in its second variable, $i = 1, 2, \ldots, n$, then it is known that for fixed values $u_1, u_2, \ldots, u_n$ of $U_1, U_2, \ldots, U_n$ we have that $\tilde{g}_1(u_1, V), \tilde{g}_2(u_2, V), \ldots, \tilde{g}_n(u_n, V)$ are weakly positively associated. Thus, for a supermodular function $\phi : \mathbb{R}^n \to \mathbb{R}$ we have (here $V_1, V_2, \ldots, V_n$ are independent copies of V)

$$\begin{aligned} E\phi(X_1, X_2, \ldots, X_n) & \\ &= E\big[E\big[\phi(\tilde{g}_1(U_1, V), \tilde{g}_2(U_2, V), \ldots, \tilde{g}_n(U_n, V))\big|U_1, U_2, \ldots, U_n\big]\big] \\ &\geq E\big[E\big[\phi(\tilde{g}_1(U_1, V_1), \tilde{g}_2(U_2, V_2), \ldots, \tilde{g}_n(U_n, V_n))\big|U_1, U_2, \ldots, U_n\big]\big] \\ &= E\phi(Y_1, Y_2, \ldots, Y_n), \end{aligned}$$

where the inequality follows from Theorem 9.A.23. Thus $\boldsymbol{Y} \leq_{\text{sm}} \boldsymbol{X}$.

Example 9.A.27. Let $\Omega = \{a_1, a_2, \ldots, a_N\}$ be a finite population. Let $X_1, X_2, \ldots, X_n$ be a sample without replacement of size $n \leq N$ from Ω; that is,

$$P\{(X_1, X_2, \ldots, X_n) = (x_1, x_2, \ldots, x_n)\} = \frac{1}{N(N-1)\cdots(N-n+1)}, \qquad (x_1, x_2, \ldots, x_n) \in \Omega^n,$$

provided all the x_i's comprise different elements of Ω. Let $Y_1, Y_2, \ldots, Y_n$ be a sample with replacement of size n from Ω; that is,

$$P\{(Y_1, Y_2, \ldots, Y_n) = (x_1, x_2, \ldots, x_n)\} = \frac{1}{N^n}, \quad (x_1, x_2, \ldots, x_n) \in \Omega^n.$$

Then $(X_1, X_2, \ldots, X_n) \leq_{\text{sm}} (Y_1, Y_2, \ldots, Y_n)$.

Example 9.A.28. Let ϕ and ψ be two Laplace transforms of positive random variables. Then F and G, defined by

$$F(x_1, x_2, \ldots, x_n) = \phi(\phi^{-1}(x_1) + \phi^{-1}(x_2) + \cdots + \phi^{-1}(x_n)), \qquad (x_1, x_2, \ldots, x_n) \in [0, 1]^n,$$

and

$$G(y_1, y_2, \ldots, y_n) = \psi(\psi^{-1}(y_1) + \psi^{-1}(y_2) + \cdots + \psi^{-1}(y_n)), \qquad (y_1, y_2, \ldots, y_n) \in [0, 1]^n,$$

are multivariate distribution functions with uniform$[0, 1]$ marginals (see Example 9.A.3). Let $\boldsymbol{X} = (X_1, X_2, \ldots, X_n)$ and $\boldsymbol{Y} = (Y_1, Y_2, \ldots, Y_n)$ be distributed according to F and G, respectively. If $\phi^{-1}\psi$ has a completely monotone derivative, then $\boldsymbol{X} \leq_{\text{sm}} \boldsymbol{Y}$.

Hu, Xie, and Ruan [241] described various sets of conditions under which two multivariate Bernoulli random vectors are ordered with respect to the supermodular order.

Remark 9.A.29. Hu and Pan [239] elegantly extended (9.A.12) to the supermodular order. They also identified conditions under which any n corresponding values of two stationary Markov chains are comparable in the order $\le_{\rm sm}$. See also Miyoshi and Rolski [398].

9.B The Orthant Ratio Orders

Some multivariate stochastic orders, that compare the "location" or "magnitude" of two random vectors, may be thought of as stochastic orders of positive dependence if the compared random vectors have the same univariate marginal distributions. For example, in the bivariate case, when this is the situation, the orthant orders $\le_{\rm uo}$ and $\le_{\rm lo}$ (see Section 6.G.1) become the order $\le_{\rm PQD}$, or, equivalently (see (9.A.18)), the order $\le_{\rm sm}$. On the other hand, some multivariate location orders do not give anything meaningful once the marginals are held fixed. For instance, the usual multivariate stochastic order $\le_{\rm st}$ can order two random vectors, with marginals that are stochastically equal, only if they have the same distributions (see Theorem 6.B.19).

In this section we study, among other things, some stochastic orders of positive dependence that arise when the underlying random vectors are ordered with respect to some multivariate hazard rate stochastic orders that were discussed in Section 6.D, and have the same univariate marginal distributions.

9.B.1 The (weak) orthant ratio orders

Let $\boldsymbol{X} = (X_1, X_2, \dots, X_n)$ and $\boldsymbol{Y} = (Y_1, Y_2, \dots, Y_n)$ be two random vectors with respective distribution functions F and G, and with survival functions $\overline{F}$ and $\overline{G}$. We suppose that F and G belong to the same Fréchet class; that is, have the same univariate marginals.

We say that $\boldsymbol{X}$ is smaller than $\boldsymbol{Y}$ in the *lower orthant decreasing ratio order* (denoted by $\boldsymbol{X} \le_{\rm lodr} \boldsymbol{Y}$ or $F \le_{\rm lodr} G$) if

$$F(\boldsymbol{y})G(\boldsymbol{x}) \ge F(\boldsymbol{x})G(\boldsymbol{y}) \quad \text{whenever } \boldsymbol{x} \le \boldsymbol{y}. \tag{9.B.1}$$

This is equivalent to

$$\frac{G(\boldsymbol{x})}{F(\boldsymbol{x})} \quad \text{is decreasing in } \boldsymbol{x} \in \{\boldsymbol{x} : G(\boldsymbol{x}) > 0\}, \tag{9.B.2}$$

where in (9.B.2) we use the convention $a/0 \equiv \infty$ whenever $a > 0$. Note that (9.B.2) can be written equivalently as

$$\frac{F(\boldsymbol{x}-\boldsymbol{u})}{F(\boldsymbol{x})} \le \frac{G(\boldsymbol{x}-\boldsymbol{u})}{G(\boldsymbol{x})}, \quad \boldsymbol{u} \ge \boldsymbol{0},\ \boldsymbol{x} \in \{\boldsymbol{x} : F(\boldsymbol{x}) > \boldsymbol{0}\} \cap \{\boldsymbol{x} : G(\boldsymbol{x}) > \boldsymbol{0}\}, \tag{9.B.3}$$

and it is also equivalent to

$$[\boldsymbol{X}-\boldsymbol{x}\big|\boldsymbol{X}\le\boldsymbol{x}]\ge_{\mathrm{lo}}[\boldsymbol{Y}-\boldsymbol{x}\big|\boldsymbol{Y}\le\boldsymbol{x}],\quad \boldsymbol{x}\in\{\boldsymbol{x}:F(\boldsymbol{x})>0\}\cap\{\boldsymbol{x}:G(\boldsymbol{x})>0\}. \tag{9.B.4}$$

Note that from (9.B.2) it follows that $\{\boldsymbol{x}:F(\boldsymbol{x})>0\}\subseteq\{\boldsymbol{x}:G(\boldsymbol{x})>0\}$. Thus, in (9.B.3) and (9.B.4) we can formally replace the expression $\{\boldsymbol{x}:F(\boldsymbol{x})>0\}\cap\{\boldsymbol{x}:G(\boldsymbol{x})>0\}$ by the simpler expression $\{\boldsymbol{x}:F(\boldsymbol{x})>0\}$.

We say that $\boldsymbol{X}$ is smaller than $\boldsymbol{Y}$ in the *upper orthant increasing ratio order* (denoted by $\boldsymbol{X}\le_{\mathrm{uoir}}\boldsymbol{Y}$ or $F\le_{\mathrm{uoir}}G$) if

$$\overline{F}(\boldsymbol{y})\overline{G}(\boldsymbol{x})\le\overline{F}(\boldsymbol{x})\overline{G}(\boldsymbol{y})\quad\text{whenever } \boldsymbol{x}\le\boldsymbol{y}. \tag{9.B.5}$$

This is equivalent to

$$\frac{\overline{G}(\boldsymbol{x})}{\overline{F}(\boldsymbol{x})}\quad\text{is increasing in } \boldsymbol{x}\in\{\boldsymbol{x}:\overline{G}(\boldsymbol{x})>0\},$$

where here, again, we use the convention $a/0\equiv\infty$ whenever $a>0$. Note that the above can be written equivalently as

$$\frac{\overline{F}(\boldsymbol{x}+\boldsymbol{u})}{\overline{F}(\boldsymbol{x})}\le\frac{\overline{G}(\boldsymbol{x}+\boldsymbol{u})}{\overline{G}(\boldsymbol{x})},\quad \boldsymbol{u}\ge\boldsymbol{0},\ \boldsymbol{x}\in\{\boldsymbol{x}:\overline{F}(\boldsymbol{x})>0\}\cap\{\boldsymbol{x}:\overline{G}(\boldsymbol{x})>0\}, \tag{9.B.6}$$

and it is also equivalent to

$$[\boldsymbol{X}-\boldsymbol{x}\big|\boldsymbol{X}>\boldsymbol{x}]\le_{\mathrm{uo}}[\boldsymbol{Y}-\boldsymbol{x}\big|\boldsymbol{Y}>\boldsymbol{x}],\ \boldsymbol{x}\in\{\boldsymbol{x}:\overline{F}(\boldsymbol{x})>0\}\cap\{\boldsymbol{x}:\overline{G}(\boldsymbol{x})>0\}. \tag{9.B.7}$$

Formally the expression $\{\boldsymbol{x}:\overline{F}(\boldsymbol{x})>0\}\cap\{\boldsymbol{x}:\overline{G}(\boldsymbol{x})>0\}$ in (9.B.6) and (9.B.7) can be replaced by the simpler expression $\{\boldsymbol{x}:\overline{F}(\boldsymbol{x})>0\}$.

We note that if $\boldsymbol{X}$ and $\boldsymbol{Y}$ have the same marginals, then $\boldsymbol{X}\le_{\mathrm{uoir}}\boldsymbol{Y}$ if, and only if, $\boldsymbol{X}\le_{\mathrm{whr}}\boldsymbol{Y}$; see (6.D.2).

The two orders $\le_{\mathrm{lodr}}$ and $\le_{\mathrm{uoir}}$ are closely related, as is indicated in the next result.

Theorem 9.B.1. *Let $\boldsymbol{X}=(X_1,X_2,\dots,X_n)$ and $\boldsymbol{Y}=(Y_1,Y_2,\dots,Y_n)$ be two random vectors in the same Fréchet class.*

(a) *If $\boldsymbol{X}\le_{\mathrm{lodr}}\boldsymbol{Y}$, then $(\phi_1(X_1),\phi_2(X_2),\dots,\phi_n(X_n))\le_{\mathrm{uoir}}(\phi_1(Y_1),\phi_2(Y_2),\dots,\phi_n(Y_n))$ for any decreasing functions $\phi_1,\phi_2,\dots,\phi_n$. Conversely, if $(\phi_1(X_1),\phi_2(X_2),\dots,\phi_n(X_n))\le_{\mathrm{uoir}}(\phi_1(Y_1),\phi_2(Y_2),\dots,\phi_n(Y_n))$ for some strictly decreasing functions $\phi_1,\phi_2,\dots,\phi_n$, then $\boldsymbol{X}\le_{\mathrm{lodr}}\boldsymbol{Y}$.*

(b) *If $\boldsymbol{X}\le_{\mathrm{uoir}}\boldsymbol{Y}$, then $(\phi_1(X_1),\phi_2(X_2),\dots,\phi_n(X_n))\le_{\mathrm{lodr}}(\phi_1(Y_1),\phi_2(Y_2),\dots,\phi_n(Y_n))$ for any decreasing functions $\phi_1,\phi_2,\dots,\phi_n$. Conversely, if $(\phi_1(X_1),\phi_2(X_2),\dots,\phi_n(X_n))\le_{\mathrm{lodr}}(\phi_1(Y_1),\phi_2(Y_2),\dots,\phi_n(Y_n))$ for some strictly decreasing functions $\phi_1,\phi_2,\dots,\phi_n$, then $\boldsymbol{X}\le_{\mathrm{uoir}}\boldsymbol{Y}$.*

The next result is similar to Theorem 9.B.1, but it involves increasing, rather than decreasing, functions. It shows that the orders $\le_{\mathrm{lodr}}$ and $\le_{\mathrm{uoir}}$ are closed under componentwise increasing transformations.

Theorem 9.B.2. *Let* $\boldsymbol{X} = (X_1, X_2, \ldots, X_n)$ *and* $\boldsymbol{Y} = (Y_1, Y_2, \ldots, Y_n)$ *be two random vectors in the same Fréchet class.*

(a) *If* $\boldsymbol{X} \le_{\rm lodr} \boldsymbol{Y}$*, then* $(\phi_1(X_1), \phi_2(X_2), \ldots, \phi_n(X_n)) \le_{\rm lodr} (\phi_1(Y_1), \phi_2(Y_2), \ldots, \phi_n(Y_n))$ *for any increasing functions* $\phi_1, \phi_2, \ldots, \phi_n$*. Conversely, if* $(\phi_1(X_1), \phi_2(X_2), \ldots, \phi_n(X_n)) \le_{\rm lodr} (\phi_1(Y_1), \phi_2(Y_2), \ldots, \phi_n(Y_n))$ *for some strictly increasing functions* $\phi_1, \phi_2, \ldots, \phi_n$*, then* $\boldsymbol{X} \le_{\rm lodr} \boldsymbol{Y}$*.*
(b) *If* $\boldsymbol{X} \le_{\rm uoir} \boldsymbol{Y}$*, then* $(\phi_1(X_1), \phi_2(X_2), \ldots, \phi_n(X_n)) \le_{\rm uoir} (\phi_1(Y_1), \phi_2(Y_2), \ldots, \phi_n(Y_n))$ *for any increasing functions* $\phi_1, \phi_2, \ldots, \phi_n$*. Conversely, if* $(\phi_1(X_1), \phi_2(X_2), \ldots, \phi_n(X_n)) \le_{\rm uoir} (\phi_1(Y_1), \phi_2(Y_2), \ldots, \phi_n(Y_n))$ *for some strictly increasing functions* $\phi_1, \phi_2, \ldots, \phi_n$*, then* $\boldsymbol{X} \le_{\rm uoir} \boldsymbol{Y}$*.*

Since the order $\le_{\rm uoir}$ is equivalent to the order $\le_{\rm whr}$ when the compared random vectors have the same marginals, it follows from Theorem 6.D.4 that the order $\le_{\rm uoir}$ is closed under conjunctions, marginalization, and convergence in distribution. Using Theorem 9.B.1 it is seen that also the order $\le_{\rm lodr}$ is closed under these operations.

If $\boldsymbol{X} \le_{\rm lodr} \boldsymbol{Y}$, then from (9.B.4) it follows that $[\boldsymbol{X} | \boldsymbol{X} \le \boldsymbol{x}] \ge_{\rm lo} [\boldsymbol{Y} | \boldsymbol{Y} \le \boldsymbol{x}]$ for all relevant $\boldsymbol{x}$. Letting $\boldsymbol{x} \to -\infty$ it is seen that (9.A.13) holds (with F and G being the distributions functions of $\boldsymbol{X}$ and $\boldsymbol{Y}$, respectively). Similarly, if $\boldsymbol{X} \le_{\rm uoir} \boldsymbol{Y}$, then (9.A.14) holds. Thus we have that

$$\Big(\boldsymbol{X} \le_{\rm lodr} \boldsymbol{Y} \text{ and } \boldsymbol{X} \le_{\rm uoir} \boldsymbol{Y}\Big) \Longrightarrow \boldsymbol{X} \le_{\rm PQD} \boldsymbol{Y}.$$

Example 9.B.3. Recall from page 387 the definition of the Fréchet class $\mathcal{M}(F_1, F_2)$ and the Fréchet lower bound in that class which we denote here by F^-. Suppose that (X_1, X_2) has a distribution function in $\mathcal{M}(F_1, F_2)$. Then $F^- \le_{\rm lodr} F$ and $F^- \le_{\rm uoir} F$.

Example 9.B.4. Let $\boldsymbol{X}$ and $\boldsymbol{Y}$ be two n-dimensional random vectors with Marshall-Olkin exponential distributions F and G with the survival functions given, for $\boldsymbol{x} \ge \boldsymbol{0}$, by

$$\overline{F}(\boldsymbol{x}) = \exp\Big\{ -\sum_{i=1}^n \lambda_i x_i - \sum_{1 \le i_1 \le i_2 \le n} \lambda_{i_1 i_2}(x_{i_1} \vee x_{i_2}) \\ - \cdots - \lambda_{12\cdots n}(x_1 \vee x_2 \vee \cdots \vee x_n)\Big\},$$

and

$$\overline{G}(\boldsymbol{x}) = \exp\Big\{ -\sum_{i=1}^n \theta_i x_i - \sum_{1 \le i_1 \le i_2 \le n} \theta_{i_1 i_2}(x_{i_1} \vee x_{i_2}) \\ - \cdots - \theta_{12\cdots n}(x_1 \vee x_2 \vee \cdots \vee x_n)\Big\},$$

where the λ's and the θ's are positive constants. Denote $\nu_A = \lambda_A - \theta_A$, $A \subseteq \{1, 2, \ldots, n\}$. Then $\boldsymbol{X} \le_{\rm uoir} \boldsymbol{Y}$ if, and only if,

$$\nu_i \geq 0, \quad i \in \{1, 2, \ldots, n\},$$
$$\nu_{i_1} + \nu_{i_1 i_2} \geq 0, \quad \{i_1, i_2\} \in \{1, 2, \ldots, n\},$$
$$\nu_{i_1} + \nu_{i_1 i_2} + \nu_{i_1 i_3} + \nu_{i_1 i_2 i_3} \geq 0, \quad \{i_1, i_2, i_3\} \in \{1, 2, \ldots, n\},$$
$$\vdots$$

and

$$\sum_{\substack{A \ni i \\ A \subseteq \{1,2,\ldots,n\}}} \nu_A = 0.$$

9.B.2 The strong orthant ratio orders

Let $\boldsymbol{X} = (X_1, X_2, \ldots, X_n)$ and $\boldsymbol{Y} = (Y_1, Y_2, \ldots, Y_n)$ be two random vectors with respective distribution functions F and G, and with survival functions $\overline{F}$ and $\overline{G}$. As in Section 9.B.1, we suppose that F and G belong to the same Fréchet class; that is, have the same univariate marginals.

We say that $\boldsymbol{X}$ is smaller than $\boldsymbol{Y}$ in the *strong lower orthant decreasing ratio order* (denoted by $\boldsymbol{X} \leq_{\text{slodr}} \boldsymbol{Y}$ or $F \leq_{\text{slodr}} G$) if

$$F(\boldsymbol{x})G(\boldsymbol{y}) \leq F(\boldsymbol{x} \vee \boldsymbol{y})G(\boldsymbol{y} \wedge \boldsymbol{x}), \quad \boldsymbol{x}, \boldsymbol{y} \in \mathbb{R}^n. \tag{9.B.8}$$

We say that $\boldsymbol{X}$ is smaller than $\boldsymbol{Y}$ in the *strong upper orthant increasing ratio order* (denoted by $\boldsymbol{X} \leq_{\text{suoir}} \boldsymbol{Y}$ or $F \leq_{\text{suoir}} G$) if

$$\overline{F}(\boldsymbol{x})\overline{G}(\boldsymbol{y}) \leq \overline{F}(\boldsymbol{x} \wedge \boldsymbol{y})\overline{G}(\boldsymbol{y} \vee \boldsymbol{x}), \quad \boldsymbol{x}, \boldsymbol{y} \in \mathbb{R}^n. \tag{9.B.9}$$

We note that if $\boldsymbol{X}$ and $\boldsymbol{Y}$ have the same marginals, then $\boldsymbol{X} \leq_{\text{suoir}} \boldsymbol{Y}$ if, and only if, $\boldsymbol{X} \leq_{\text{hr}} \boldsymbol{Y}$; see (6.D.1).

By choosing $\boldsymbol{x} \leq \boldsymbol{y}$ in (9.B.8) we get (9.B.1), and by choosing $\boldsymbol{x} \geq \boldsymbol{y}$ in (9.B.9) we get (9.B.5), that is,

$$\boldsymbol{X} \leq_{\text{slodr}} \boldsymbol{Y} \Longrightarrow \boldsymbol{X} \leq_{\text{lodr}} \quad \text{and} \quad \boldsymbol{X} \leq_{\text{suoir}} \boldsymbol{Y} \Longrightarrow \boldsymbol{X} \leq_{\text{uoir}} . \tag{9.B.10}$$

Thus the orders $\leq_{\text{slodr}}$ and $\leq_{\text{suoir}}$ are often useful as a tool to identify random vectors that are ordered with respect to the orders $\leq_{\text{lodr}}$ and $\leq_{\text{uoir}}$.

The two orders $\leq_{\text{slodr}}$ and $\leq_{\text{suoir}}$ are closely related, and are preserved under componentwise increasing transformations, as is indicated in the next analog of Theorems 9.B.1 and 9.B.2.

Theorem 9.B.5. *Let $\boldsymbol{X} = (X_1, X_2, \ldots, X_n)$ and $\boldsymbol{Y} = (Y_1, Y_2, \ldots, Y_n)$ be two random vectors in the same Fréchet class.*

(a) *If $\boldsymbol{X} \leq_{\text{slodr}} \boldsymbol{Y}$, then $(\phi_1(X_1), \phi_2(X_2), \ldots, \phi_n(X_n)) \leq_{\text{suoir}} (\phi_1(Y_1), \phi_2(Y_2), \ldots, \phi_n(Y_n))$ for any decreasing functions $\phi_1, \phi_2, \ldots, \phi_n$. On the other hand, if $(\phi_1(X_1), \phi_2(X_2), \ldots, \phi_n(X_n)) \leq_{\text{suoir}} (\phi_1(Y_1), \phi_2(Y_2), \ldots, \phi_n(Y_n))$ for some strictly decreasing functions $\phi_1, \phi_2, \ldots, \phi_n$, then $\boldsymbol{X} \leq_{\text{slodr}} \boldsymbol{Y}$.*

(b) *If* $\boldsymbol{X} \le_{\text{suoir}} \boldsymbol{Y}$, *then* $(\phi_1(X_1), \phi_2(X_2), \dots, \phi_n(X_n)) \le_{\text{slodr}} (\phi_1(Y_1), \phi_2(Y_2), \dots, \phi_n(Y_n))$ *for any decreasing functions* $\phi_1, \phi_2, \dots, \phi_n$. *On the other hand, if* $(\phi_1(X_1), \phi_2(X_2), \dots, \phi_n(X_n)) \le_{\text{slodr}} (\phi_1(Y_1), \phi_2(Y_2), \dots, \phi_n(Y_n))$ *for some strictly decreasing functions* $\phi_1, \phi_2, \dots, \phi_n$, *then* $\boldsymbol{X} \le_{\text{suoir}} \boldsymbol{Y}$.
(c) *If* $\boldsymbol{X} \le_{\text{slodr}} \boldsymbol{Y}$, *then* $(\phi_1(X_1), \phi_2(X_2), \dots, \phi_n(X_n)) \le_{\text{slodr}} (\phi_1(Y_1), \phi_2(Y_2), \dots, \phi_n(Y_n))$ *for any increasing functions* $\phi_1, \phi_2, \dots, \phi_n$. *On the other hand, if* $(\phi_1(X_1), \phi_2(X_2), \dots, \phi_n(X_n)) \le_{\text{slodr}} (\phi_1(Y_1), \phi_2(Y_2), \dots, \phi_n(Y_n))$ *for some strictly increasing functions* $\phi_1, \phi_2, \dots, \phi_n$, *then* $\boldsymbol{X} \le_{\text{slodr}} \boldsymbol{Y}$.
(d) *If* $\boldsymbol{X} \le_{\text{suoir}} \boldsymbol{Y}$, *then* $(\phi_1(X_1), \phi_2(X_2), \dots, \phi_n(X_n)) \le_{\text{suoir}} (\phi_1(Y_1), \phi_2(Y_2), \dots, \phi_n(Y_n))$ *for any increasing functions* $\phi_1, \phi_2, \dots, \phi_n$. *On the other hand, if* $(\phi_1(X_1), \phi_2(X_2), \dots, \phi_n(X_n)) \le_{\text{suoir}} (\phi_1(Y_1), \phi_2(Y_2), \dots, \phi_n(Y_n))$ *for some strictly increasing functions* $\phi_1, \phi_2, \dots, \phi_n$, *then* $\boldsymbol{X} \le_{\text{suoir}} \boldsymbol{Y}$.

Since the order $\le_{\text{suoir}}$ is equivalent to the order $\le_{\text{hr}}$ when the compared random vectors have the same marginals, it follows from Theorem 6.D.4 that the order $\le_{\text{uoir}}$ is closed under conjunctions, marginalization, and convergence in distribution. Using Theorem 9.B.5 it is seen that also the order $\le_{\text{slodr}}$ is closed under these operations.

The converses of the implications in (9.B.10) are not true in general. However, under an additional assumption they are valid; these are given in the following theorem.

Theorem 9.B.6. *Let* $\boldsymbol{X}$ *and* $\boldsymbol{Y}$ *be two random vectors in the same Fréchet class with respective distribution functions* F *and* G, *and respective survival functions* $\overline{F}$ *and* $\overline{G}$.

(a) *If* F *and/or* G *are/is* MTP$_2$, *then* $\boldsymbol{X} \le_{\text{lodr}} \boldsymbol{Y} \Longrightarrow \boldsymbol{X} \le_{\text{slodr}} \boldsymbol{Y}$.
(b) *If* $\overline{F}$ *and/or* $\overline{G}$ *are/is* MTP$_2$, *then* $\boldsymbol{X} \le_{\text{uoir}} \boldsymbol{Y} \Longrightarrow \boldsymbol{X} \le_{\text{suoir}} \boldsymbol{Y}$.

Part (b) of the above theorem is similar to Theorem 6.D.1. However, it turns out that since the compared random vectors are in the same Fréchet class, it is not needed, in Theorem 9.B.6(b), that they have a common support which is a lattice.

9.C The LTD, RTI, and PRD Orders

For any random vector (X_1, X_2) with distribution function $F \in \mathcal{M}(F_1, F_2)$ (see page 387 for the definition of $\mathcal{M}(F_1, F_2)$) we define the conditional distribution function F_x^L by

$$F_{x_1}^L(x_2) = P\{X_2 \le x_2 \big| X_1 \le x_1\} \tag{9.C.1}$$

for all x_1 for which this conditional distribution is well defined. Barlow and Proschan [36] defined F (or X_1 and X_2) to be left tail decreasing (LTD) if

$$F_{x_1}^L(x_2) \ge F_{x_1'}^L(x_2) \quad \text{for all } x_1 \le x_1' \text{ and } x_2,$$

or, equivalently, if

$$(F_{x_1}^L)^{-1}(u) \le (F_{x_1'}^L)^{-1}(u) \quad \text{for all } x_1 \le x_1' \text{ and } u \in [0,1]. \tag{9.C.2}$$

Note that when $(F_{x_1}^L)^{-1}(u)$ is continuous in u for all x_1, then (9.C.2) can be equivalently written as

$$F_{x_1'}^L\left[(F_{x_1}^L)^{-1}(u)\right] \le u \quad \text{for all } x_1 \le x_1' \text{ and } u \in [0,1]. \tag{9.C.3}$$

This notion leads to the following definition.

Let (X_1, X_2) be a bivariate random vector with distribution function $F \in \mathcal{M}(F_1, F_2)$, and let (Y_1, Y_2) be another bivariate random vector with distribution function $G \in \mathcal{M}(F_1, F_2)$. Suppose that for any $x_1 \le x_1'$ we have

$$(F_{x_1}^L)^{-1}(u) \le (F_{x_1'}^L)^{-1}(v) \Longrightarrow (G_{x_1}^L)^{-1}(u) \le (G_{x_1'}^L)^{-1}(v) \quad \text{for all } u, v \in [0,1]. \tag{9.C.4}$$

Then we say that (X_1, X_2) is *smaller than* (Y_1, Y_2) *in the* LTD *order* (denoted by $(X_1, X_2) \le_{\text{LTD}} (Y_1, Y_2)$ or $F \le_{\text{LTD}} G$).

Note that (9.C.4) can be equivalently written as

$$G_{x_1'}^L\left[(G_{x_1}^L)^{-1}(u)\right] \le F_{x_1'}^L\left[(F_{x_1}^L)^{-1}(u)\right] \quad \text{for all } x_1 \le x_1' \text{ and } u \in [0,1]. \tag{9.C.5}$$

It can be shown that if $F_{x_1}^L(x_2)$ and $G_{x_1}^L(x_2)$ are continuous in x_2 for all x_1, then $(X_1, X_2) \le_{\text{LTD}} (Y_1, Y_2)$ if, and only if, for any $x_1 \le x_1'$,

$$F_{x_1}^L(x_2) \ge G_{x_1}^L(x_2') \Longrightarrow F_{x_1'}^L(x_2) \ge G_{x_1'}^L(x_2') \quad \text{for any } x_2 \text{ and } x_2'. \tag{9.C.6}$$

Note that (9.C.6) can be equivalently written as

$$(G_{x_1}^L)^{-1}\left[F_{x_1}^L(x_2)\right] \le (G_{x_1'}^L)^{-1}\left[F_{x_1'}^L(x_2)\right] \quad \text{for all } x_1 \le x_1' \text{ and } x_2,$$

that is, $(G_{x_1}^L)^{-1}\left[F_{x_1}^L(x_2)\right]$ is increasing in x_1 for all x_2.

In the continuous case, it is immediate from (9.C.3) and (9.C.5) that F is LTD if, and only if,

$$F^I \le_{\text{LTD}} F,$$

where F^I is defined in Section 9.A, but this is true also when F is not continuous.

Theorem 9.C.1. *Let (X_1, X_2) and (Y_1, Y_2) be two random vectors with distribution functions $F, G \in \mathcal{M}(F_1, F_2)$, such that $F_{x_1}^L(x_2)$ and $G_{x_1}^L(x_2)$ are continuous in x_2 for all x_1. Then*

$$(X_1, X_2) \le_{\text{LTD}} (Y_1, Y_2) \Longrightarrow (X_1, X_2) \le_{\text{PQD}} (Y_1, Y_2).$$

Proof. Since F and G have the same marginals, we see from (9.C.6) that $(X_1, X_2) \le_{\text{LTD}} (Y_1, Y_2)$ if, and only if,

$$F(x_1, x_2) \geq G(x_1, x_2') \Longrightarrow F(x_1', x_2) \geq G(x_1', x_2')$$
$$\text{for any } x_2,\ x_2',\ \text{and } x_1 \leq x_1'. \quad (9.C.7)$$

If $(X_1, X_2) \leq_{\text{PQD}} (Y_1, Y_2)$ did not hold, then there would have existed a point (x_0, y_0) such that $F(x_0, y_0) > G(x_0, y_0)$. Let $y < y_0$ be such that $F(x_0, y_0) > F(x_0, y) > G(x_0, y_0)$. Since $F_2(y) < F_2(y_0)$, one can then find an x such that $x > x_0$ and $F(x, y) < G(x, y_0)$. But then $F(x_0, y) > G(x_0, y_0)$ and $F(x, y) < G(x, y_0)$ contradict (9.C.7). □

The LTD order is not symmetric in the sense that $(X_1, X_2) \leq_{\text{LTD}} (Y_1, Y_2)$ does not necessarily imply that $(X_2, X_1) \leq_{\text{LTD}} (Y_2, Y_1)$. However, it satisfies the following closure under monotone transformations property.

Theorem 9.C.2. *Let (X_1, X_2) and (Y_1, Y_2) be two random vectors with distribution functions in the same Fréchet class. If $(X_1, X_2) \leq_{\text{LTD}} (Y_1, Y_2)$, then $(\phi(X_1), \psi(X_2)) \leq_{\text{LTD}} (\phi(Y_1), \psi(Y_2))$ for all increasing functions ϕ and ψ.*

Example 9.C.3. Let $\phi_\theta(t) \equiv (1 - t^\theta)^{1/\theta}$, $t \in [0, 1]$, $\theta \in (0, 1)$. Then the function C_θ, defined as

$$C_{\phi_\theta}(x, y) = \phi_\theta^{-1}\{\phi_\theta(x) + \phi_\theta(y)\}, \quad x, y \in [0, 1],$$

is a bivariate distribution function with uniform$[0, 1]$ marginals (it is a particular Archimedean copula). If $\theta_1 \leq \theta_2$, then $C_{\phi_{\theta_2}} \leq_{\text{LTD}} C_{\phi_{\theta_1}}$.

An order that is similar to the LTD order, but which is based on conditioning on right tails, rather than on left tails, is described next.

For any random vector (X_1, X_2) with distribution function $F \in \mathcal{M}(F_1, F_2)$ we define the conditional distribution function F_x^R by

$$F_{x_1}^R(x_2) = P\{X_2 \leq x_2 \big| X_1 > x_1\} \tag{9.C.8}$$

for all x_1 for which this conditional distribution is well defined. Barlow and Proschan [36] defined F (or X_1 and X_2) to be right tail increasing (RTI) if

$$F_{x_1}^R(x_2) \geq F_{x_1'}^R(x_2) \quad \text{for all } x_1 \leq x_1' \text{ and } x_2,$$

or, equivalently, if

$$(F_{x_1}^R)^{-1}(u) \leq (F_{x_1'}^R)^{-1}(u) \quad \text{for all } x_1 \leq x_1' \text{ and } u \in [0, 1]. \tag{9.C.9}$$

When $(F_{x_1}^R)^{-1}(u)$ is continuous in u for all x_1 then (9.C.9) can be written as

$$F_{x_1'}^R\left[(F_{x_1}^R)^{-1}(u)\right] \leq u \quad \text{for all } x_1 \leq x_1' \text{ and } u \in [0, 1]. \tag{9.C.10}$$

This notion leads to the following definition.

Let (X_1, X_2) be a bivariate random vector with distribution function $F \in \mathcal{M}(F_1, F_2)$, and let (Y_1, Y_2) be another bivariate random vector with distribution function $G \in \mathcal{M}(F_1, F_2)$. Suppose that for any $x_1 \leq x_1'$ we have

$$(F_{x_1}^R)^{-1}(u) \le (F_{x_1'}^R)^{-1}(v) \Longrightarrow (G_{x_1}^R)^{-1}(u) \le (G_{x_1'}^R)^{-1}(v) \quad \text{for all } u, v \in [0,1]. \tag{9.C.11}$$

Then we say that (X_1, X_2) is *smaller than* (Y_1, Y_2) *in the* RTI *order* (denoted by $(X_1, X_2) \le_{\text{RTI}} (Y_1, Y_2)$ or $F \le_{\text{RTI}} G$).

In analogy to (9.C.5) we note that (9.C.11) can be written as

$$G_{x_1'}^R\left[(G_{x_1}^R)^{-1}(u)\right] \le F_{x_1'}^R\left[(F_{x_1}^R)^{-1}(u)\right] \quad \text{for all } x_1 \le x_1' \text{ and } u \in [0,1]. \tag{9.C.12}$$

It can be shown that if $F_{x_1}^R(x_2)$ and $G_{x_1}^R(x_2)$ are continuous in x_2 for all x_1, then $(X_1, X_2) \le_{\text{RTI}} (Y_1, Y_2)$ if, and only if, for any $x_1 \le x_1'$,

$$F_{x_1}^R(x_2) \ge G_{x_1}^R(x_2') \Longrightarrow F_{x_1'}^R(x_2) \ge G_{x_1'}^R(x_2') \quad \text{for any } x_2 \text{ and } x_2'. \tag{9.C.13}$$

Note that (9.C.13) can be written as

$$(G_{x_1}^R)^{-1}\left[F_{x_1}^R(x_2)\right] \le (G_{x_1'}^R)^{-1}\left[F_{x_1'}^R(x_2)\right] \quad \text{for all } x_1 \le x_1' \text{ and } x_2,$$

that is, $(G_{x_1}^R)^{-1}\left[F_{x_1}^R(x_2)\right]$ is increasing in x_1 for all x_2.

In the continuous case, it is immediate from (9.C.10) and (9.C.12) that F is RTI if, and only if,

$$F^I \le_{\text{RTI}} F,$$

where F^I is defined in Section 9.A, but this is true also when F is not continuous.

The following result is an analog of Theorem 9.C.1; its proof is similar to the proof of that theorem, and is therefore omitted.

Theorem 9.C.4. *Let (X_1, X_2) and (Y_1, Y_2) be two random vectors with distribution functions $F, G \in \mathcal{M}(F_1, F_2)$, such that $F_{x_1}^R(x_2)$ and $G_{x_1}^R(x_2)$ are continuous in x_2 for all x_1. Then*

$$(X_1, X_2) \le_{\text{RTI}} (Y_1, Y_2) \Longrightarrow (X_1, X_2) \le_{\text{PQD}} (Y_1, Y_2).$$

The RTI order is not symmetric in the sense that $(X_1, X_2) \le_{\text{RTI}} (Y_1, Y_2)$ does not necessarily imply that $(X_2, X_1) \le_{\text{RTI}} (Y_2, Y_1)$. However, it satisfies the following closure under monotone transformations property.

Theorem 9.C.5. *Let (X_1, X_2) and (Y_1, Y_2) be two random vectors with distribution functions in the same Fréchet class. If $(X_1, X_2) \le_{\text{RTI}} (Y_1, Y_2)$, then $(\phi(X_1), \psi(X_2)) \le_{\text{RTI}} (\phi(Y_1), \psi(Y_2))$ for all increasing functions ϕ and ψ.*

The LTD and RTI orders are related to each other as follows.

Theorem 9.C.6. *Let (X_1, X_2) and (Y_1, Y_2) be two random vectors in the same Fréchet class.*

(a) *If $(X_1, X_2) \le_{\text{LTD}} (Y_1, Y_2)$, then $(\phi_1(X_1), \phi_2(X_2))) \le_{\text{RTI}} (\phi_1(Y_1), \phi_2(Y_2))$ for any decreasing functions ϕ_1 and ϕ_2. Conversely, if $(\phi_1(X_1), \phi_2(X_2))$ $\le_{\text{RTI}} (\phi_1(Y_1), \phi_2(Y_2))$ for some strictly decreasing functions ϕ_1 and ϕ_2, then $(X_1, X_2) \le_{\text{LTD}} (Y_1, Y_2)$.*

(b) *If* $(X_1, X_2) \leq_{\rm RTI} (Y_1, Y_2)$, *then* $(\phi_1(X_1), \phi_2(X_2)) \leq_{\rm LTD} (\phi_1(Y_1), \phi_2(Y_2))$ *for any decreasing functions* ϕ_1 *and* ϕ_2. *Conversely, if* $(\phi_1(X_1), \phi_2(X_2)) \leq_{\rm LTD} (\phi_1(Y_1), \phi_2(Y_2))$ *for some strictly decreasing functions* ϕ_1 *and* ϕ_2, *then* $(X_1, X_2) \leq_{\rm RTI} (Y_1, Y_2)$.

The orders $\leq_{\rm slodr}$ and $\leq_{\rm suoir}$ imply the LTD and RTI orders under some regularity conditions. This is shown in the next result.

Theorem 9.C.7. *Let* F *and* G *be in the Fréchet class* $\mathcal{M}(F_1, F_2)$. *Assume that, for every* x, *the conditional distributions* F_x^L *and* F_x^R *(see (9.C.1) and (9.C.8)) are strictly increasing and continuous on their supports. Then*

$$F \leq_{\rm slodr} G \Longrightarrow F \leq_{\rm LTD} G \quad \textit{and} \quad F \leq_{\rm suoir} G \Longrightarrow F \leq_{\rm RTI} G.$$

Proof. It is enough to prove the first implication; the other implication then follows from Theorems 9.B.5 and 9.C.6.

By (9.C.7), we need to show that for $x \leq x'$, and for any y, y', it holds that

$$F(x, y) \geq G(x, y') \Longrightarrow F(x', y) \geq G(x', y'). \tag{9.C.14}$$

Now assume that $F \leq_{\rm slodr} G$. So, for $x \leq x'$ and $y' \leq y$ we have

$$F(x, y)G(x', y') \leq F(x', y)G(x, y'). \tag{9.C.15}$$

In the bivariate case, $F \leq_{\rm slodr} G$ implies that $F \leq_{\rm PQD} G$. So the left-hand side inequality in (9.C.14) can hold only for $y' \leq y$. If it does hold, then (9.C.15) implies the inequality on the right-hand side of (9.C.14). □

In light of Theorem 9.C.7 it is of interest to note that the (weak) orthant ratio orders $\leq_{\rm lodr}$ and $\leq_{\rm uoir}$ do not imply the orders $\leq_{\rm LTD}$ and $\leq_{\rm RTI}$, respectively. Counterexamples can be found in the literature.

An order that is of the same type as the LTD and RTI orders is the one that we study next. For any random vector (X_1, X_2), with distribution function $F \in \mathcal{M}(F_1, F_2)$, let F_{x_1} denote the conditional distribution of X_2 given that $X_1 = x_1$. Lehmann [343] defined F (or X_1 and X_2) to be positive regression dependent (PRD) if X_2 is stochastically increasing in X_1, that is, if

$$F_{x_1}(x_2) \geq F_{x_1'}(x_2) \quad \text{for all } x_1 \leq x_1' \text{ and } x_2,$$

or, equivalently, if

$$F_{x_1}^{-1}(u) \leq F_{x_1'}^{-1}(u) \quad \text{for all } x_1 \leq x_1' \text{ and } u \in [0,1]. \tag{9.C.16}$$

Note that when $F_{x_1}^{-1}(u)$ is continuous in u for all x_1, then (9.C.16) can be written as

$$F_{x_1'}\left(F_{x_1}^{-1}(u)\right) \leq u \quad \text{for all } x_1 \leq x_1' \text{ and } u \in [0,1]. \tag{9.C.17}$$

This notion leads to the following definition.

Let (X_1, X_2) be a bivariate random vector with distribution function $F \in \mathcal{M}(F_1, F_2)$, and let (Y_1, Y_2) be another bivariate random vector with distribution function $G \in \mathcal{M}(F_1, F_2)$. Suppose that for any $x_1 \le x_1'$ we have

$$F_{x_1}^{-1}(u) \le F_{x_1'}^{-1}(v) \Longrightarrow G_{x_1}^{-1}(u) \le G_{x_1'}^{-1}(v) \quad \text{for all } u, v \in [0,1]. \qquad (9.C.18)$$

Then we say that (X_1, X_2) is *smaller than* (Y_1, Y_2) *in the* PRD *order* (denoted by $(X_1, X_2) \le_{\text{PRD}} (Y_1, Y_2)$ or $F \le_{\text{PRD}} G$).

Note that (9.C.18) can be written as

$$G_{x_1'}\big(G_{x_1}^{-1}(u)\big) \le F_{x_1'}\big(F_{x_1}^{-1}(u)\big) \quad \text{for all } x_1 \le x_1' \text{ and } u \in [0,1]. \qquad (9.C.19)$$

It can be shown that if $F_{x_1}(x_2)$ and $G_{x_1}(x_2)$ are continuous in x_2 for all x_1, then $(X_1, X_2) \le_{\text{PRD}} (Y_1, Y_2)$ if, and only if, for any $x_1 \le x_1'$,

$$F_{x_1}(x_2) \ge G_{x_1}(x_2') \Longrightarrow F_{x_1'}(x_2) \ge G_{x_1'}(x_2') \quad \text{for any } x_2 \text{ and } x_2'. \qquad (9.C.20)$$

Note that (9.C.20) can be written as

$$G_{x_1}^{-1}\big(F_{x_1}(x_2)\big) \le G_{x_1'}^{-1}\big(F_{x_1'}(x_2)\big) \quad \text{for all } x_1 \le x_1' \text{ and } x_2, \qquad (9.C.21)$$

that is, $G_{x_1}^{-1}\big(F_{x_1}(x_2)\big)$ is increasing in x_1 for all x_2.

In the continuous case, it is immediate from (9.C.17) and (9.C.19) that F is PRD if, and only if,

$$F^I \le_{\text{PRD}} F,$$

where F^I is defined in Section 9.A, but this is true also when F is not continuous.

The next result shows the relationship between the PRD, LTD, and RTI orders. We do not give the proof of it here.

Theorem 9.C.8. *Let (X_1, X_2) and (Y_1, Y_2) be two random vectors with absolutely continuous distribution functions $F, G \in \mathcal{M}(F_1, F_2)$. Then*

$$(X_1, X_2) \le_{\text{PRD}} (Y_1, Y_2) \Longrightarrow (X_1, X_2) \le_{\text{LTD}} (Y_1, Y_2) \quad \textit{and}$$
$$(X_1, X_2) \le_{\text{PRD}} (Y_1, Y_2) \Longrightarrow (X_1, X_2) \le_{\text{RTI}} (Y_1, Y_2).$$

The PRD order is not symmetric in the sense that $(X_1, X_2) \le_{\text{PRD}} (Y_1, Y_2)$ does not necessarily imply that $(X_2, X_1) \le_{\text{PRD}} (Y_2, Y_1)$. However, it satisfies the following closure under monotone transformations property.

Theorem 9.C.9. *Let (X_1, X_2) and (Y_1, Y_2) be two random vectors. If $(X_1, X_2) \le_{\text{PRD}} (Y_1, Y_2)$, then $(\phi(X_1), \psi(X_2)) \le_{\text{PRD}} (\phi(Y_1), \psi(Y_2))$ for all increasing functions ϕ and ψ.*

Example 9.C.10. Let U and V be any independent random variables, each having a continuous distribution. Define

$$X = U, \quad Y_\rho = \rho U + (1-\rho^2)^{1/2} V, \quad \text{for } -1 \le \rho \le 1.$$

Then $(X, Y_{\rho_1}) \le_{\text{PRD}} (X, Y_{\rho_2})$ whenever $\rho_1 \le \rho_2$. A bivariate normal distribution is a particular case of this example when U and V are normally distributed.

Example 9.C.11. Let U and V be any independent random variables, each having a continuous distribution. Define

$$X = U, \quad Y_\alpha = \alpha U + V, \quad \text{for } -\infty \le \alpha \le \infty.$$

Then $(X, Y_{\alpha_1}) \le_{\text{PRD}} (X, Y_{\alpha_2})$ whenever $\alpha_1 \le \alpha_2$.

Example 9.C.12. Let U and V be any independent random variables, each having a continuous distribution, such that U is distributed on $(0, 1)$, while V is nonnegative. Define

$$X = U, \quad Y_\alpha = (1 + \alpha U)V, \quad \text{for } \alpha \ge -1.$$

Then $(X, Y_{\alpha_1}) \le_{\text{PRD}} (X, Y_{\alpha_2})$ whenever $\alpha_1 \le \alpha_2$.

9.D The PLRD Order

Let the random variables X_1 and X_2 have the joint distribution F. For any two intervals I_1 and I_2 of the real line, let us denote $I_1 \le I_2$ if $x_1 \in I_1$ and $x_2 \in I_2$ imply that $x_1 \le x_2$. For any two intervals I and J of the real line denote $F(I, J) \equiv P\{X_1 \in I, X_2 \in J\}$. Block, Savits, and Shaked [95] essentially defined F (or X_1 and X_2) to be positive likelihood ratio dependent if

$$F(I_1, J_1)F(I_2, J_2) \ge F(I_1, J_2)F(I_2, J_1), \quad \text{whenever } I_1 \le I_2 \text{ and } J_1 \le J_2. \tag{9.D.1}$$

In fact, Block, Savits and Shaked [95] called F totally positive of order 2 (TP_2) if (9.D.1) holds. When F has a (continuous or discrete) density f, then (9.D.1) is equivalent to the condition that f is TP_2, that is,

$$f(x_1, y_1)f(x_2, y_2) \ge f(x_1, y_2)f(x_2, y_1), \quad \text{whenever } x_1 \le x_2 \text{ and } y_1 \le y_2.$$

Then (9.D.1) is the same as the condition for the positive dependence notion that Lehmann [343] called positive likelihood ratio dependence (PLRD). This notion leads naturally to the order that is described below.

Let (X_1, X_2) be a bivariate random vector with distribution function $F \in \mathcal{M}(F_1, F_2)$, and let (Y_1, Y_2) be another bivariate random vector with distribution function $G \in \mathcal{M}(F_1, F_2)$. Suppose that

$$\begin{aligned} F(I_1, J_1)F(I_2, J_2)&G(I_1, J_2)G(I_2, J_1) \\ &\le F(I_1, J_2)F(I_2, J_1)G(I_1, J_1)G(I_2, J_2), \\ &\qquad \text{whenever } I_1 \le I_2 \text{ and } J_1 \le J_2. \end{aligned} \tag{9.D.2}$$

where the generic notation $G(I, J)$ is obvious. Then we say that (X_1, X_2) is *smaller than* (Y_1, Y_2) *in the* PLRD *order* (denoted by $(X_1, X_2) \le_{\text{PLRD}}$

(Y_1, Y_2) or $F \leq_{\text{PLRD}} G$). Since only random vectors with the same univariate marginals can be compared in the PLRD order, we will implicitly assume this fact throughout this section.

When F and G have (continuous or discrete) densities f and g, then (9.D.2) is equivalent to

$$f(x_1, y_1)f(x_2, y_2)g(x_1, y_2)g(x_2, y_1) \leq f(x_1, y_2)f(x_2, y_1)g(x_1, y_1)g(x_2, y_2),$$
$$\text{whenever } x_1 \leq x_2 \text{ and } y_1 \leq y_2.$$

If $\frac{\partial^2}{\partial x \partial y} f$ and $\frac{\partial^2}{\partial x \partial y} g$ exist, then (9.D.2) is equivalent to

$$f^2 \Delta_g - g^2 \Delta_f \geq 0,$$

where

$$\Delta_f \equiv f \frac{\partial^2 f}{\partial x \partial y} - \frac{\partial f}{\partial x} \cdot \frac{\partial f}{\partial y} \quad \text{and} \quad \Delta_g \equiv g \frac{\partial^2 g}{\partial x \partial y} - \frac{\partial g}{\partial x} \cdot \frac{\partial g}{\partial y}.$$

Obviously F is PLRD if, and only if,

$$F^I \leq_{\text{PLRD}} F,$$

where F^I is defined in Section 9.A.

Theorem 9.D.1. *Let* (X_1, X_2) *and* (Y_1, Y_2) *be two random vectors with distribution functions* $F, G \in \mathcal{M}(F_1, F_2)$*. Then*

$$(X_1, X_2) \leq_{\text{PLRD}} (Y_1, Y_2) \Longrightarrow (X_1, X_2) \leq_{\text{PQD}} (Y_1, Y_2).$$

Proof. Assume $(X_1, X_2) \leq_{\text{PLRD}} (Y_1, Y_2)$ and suppose that $(X_1, X_2) \not\leq_{\text{PQD}} (Y_1, Y_2)$. Then

$$F(x, y) > G(x, y) \tag{9.D.3}$$

for some (x, y). Let $I_1 = (-\infty, x]$, $I_2 = (x, \infty)$, $J_1 = (-\infty, y]$ and $J_2 = (y, \infty)$. Then from (9.D.3), and from the fact that F and G have the same marginals, it follows that

$$F(I_1, J_1) > G(I_1, J_1),$$
$$F(I_2, J_2) > G(I_2, J_2),$$
$$G(I_1, J_2) > F(I_1, J_2)$$

and

$$G(I_2, J_1) > F(I_2, J_1).$$

Multiplying these four inequalities we obtain a contradiction to (9.D.2). □

We do not know whether $(X_1, X_2) \le_{\rm PLRD} (Y_1, Y_2) \Longrightarrow (X_1, X_2) \le_{\rm PRD} (Y_1, Y_2)$.

The following closure properties of the PLRD order are easy to prove.

Theorem 9.D.2. (a) *Let* (X_1, X_2) *and* (Y_1, Y_2) *be two random vectors such that* $(X_1, X_2) \le_{\rm PLRD} (Y_1, Y_2)$. *Then* $(\phi(X_1), \psi(X_2)) \le_{\rm PLRD} (\phi(Y_1), \psi(Y_2))$ *for all increasing functions* ϕ *and* ψ.

(b) *Let* $\{(X_1^{(j)}, X_2^{(j)}),\ j = 1, 2, \ldots\}$ *and* $\{(Y_1^{(j)}, Y_2^{(j)}),\ j = 1, 2, \ldots\}$ *be two sequences of random vectors such that* $(X_1^{(j)}, X_2^{(j)}) \to_{\rm st} (X_1, X_2)$ *and* $(Y_1^{(j)}, Y_2^{(j)}) \to_{\rm st} (Y_1, Y_2)$ *as* $j \to \infty$, *where* $\to_{\rm st}$ *denotes convergence in distribution. If* $(X_1^{(j)}, X_2^{(j)}) \le_{\rm PLRD} (Y_1^{(j)}, Y_2^{(j)})$, $j = 1, 2, \ldots$, *then* $(X_1, X_2) \le_{\rm PLRD} (Y_1, Y_2)$.

Let F_L and F_U denote the Fréchet lower and upper bounds in the class $\mathcal{M}(F_1, F_2)$. Since F_L assigns all its mass to some decreasing curve in $\mathbb{R}^2$, and F_U assigns all its mass to some increasing curve in $\mathbb{R}^2$, it follows that for every distribution $F \in \mathcal{M}(F_1, F_2)$ we have

$$F_L \le_{\rm PLRD} F \le_{\rm PLRD} F_U.$$

By Theorem 9.D.1, this is a stronger result than (9.A.6).

The proof of the next result is similar to the proof of Theorem 9.D.1 and is therefore omitted.

Theorem 9.D.3. *Let* (X_1, X_2) *and* (Y_1, Y_2) *be two random vectors such that* $(X_1, X_2) \le_{\rm PLRD} (Y_1, Y_2)$ *and* $(X_1, X_2) \ge_{\rm PLRD} (Y_1, Y_2)$. *Then* $(X_1, X_2) =_{\rm st} (Y_1, Y_2)$.

Example 9.D.4. Let H and K be two continuous univariate distribution functions. For $-1 \le \alpha \le 1$, define the following distribution function

$$F_\alpha(x, y) = H(x)K(y)\{1 + \alpha[1 - H(x)][1 - K(y)]\}, \quad \text{for all } x \text{ and } y.$$

Then $F_{\alpha_1} \le_{\rm PLRD} F_{\alpha_2}$ whenever $\alpha_1 \le \alpha_2$.

Example 9.D.5. Let ϕ and ψ be two Laplace transforms of positive random variables and let the random vectors (X_1, X_2) and (Y_1, Y_2) be distributed according to F and G as in Example 9.A.3. If $\phi^{-1}\psi$ has a completely monotone derivative, then $(X_1, X_2) \le_{\rm PLRD} (Y_1, Y_2)$.

Example 9.D.6. Let (X_1, X_2) and (Y_1, Y_2) be bivariate normal random vectors with the same marginals, and with correlation coefficients ρ_X and ρ_Y, respectively. If $\rho_X \le \rho_Y$, then $(X_1, X_2) \le_{\rm PLRD} (Y_1, Y_2)$.

9.E Association Orders

The random variables X_1 and X_2 are said to be associated if

$$\mathrm{Cov}(K(X_1, X_2), L(X_1, X_2)) \geq 0$$

for all increasing functions K and L for which the covariance is well defined (see (3.A.53)). This notion leads to the order that is described below.

Let (X_1, X_2) be a bivariate random vector with distribution function $F \in \mathcal{M}(F_1, F_2)$, and let (Y_1, Y_2) be another bivariate random vector with distribution function $G \in \mathcal{M}(F_1, F_2)$. Suppose that

$$(Y_1, Y_2) =_{\text{st}} (K(X_1, X_2), L(X_1, X_2)), \tag{9.E.1}$$

for some increasing functions K and L which satisfy

$$K(x_1, y_1) < K(x_2, y_2),\ L(x_1, y_1) > L(x_2, y_2) \Longrightarrow x_1 < x_2,\ y_1 > y_2. \tag{9.E.2}$$

Then we say that (X_1, X_2) is *smaller than* (Y_1, Y_2) *in the association order* (denoted by $(X_1, X_2) \leq_{\text{assoc}} (Y_1, Y_2)$ or $F \leq_{\text{assoc}} G$). Since only random vectors with the same univariate marginals are compared in the association order, we will implicitly assume this fact throughout this section.

The restriction (9.E.2) on the functions K and L is for the purpose of making the association order applicable in situations which are not symmetric in the X_1 and X_2 variables. [In case (9.E.2) is dropped, $(X_1, X_2) \geq_{\text{assoc}} (X_2, X_1) \geq_{\text{assoc}} (X_1, X_2)$.] If K and L are partially differentiable increasing functions, then (9.E.2) is equivalent to

$$\frac{\partial}{\partial x}K(x,y) \cdot \frac{\partial}{\partial y}L(x,y) \geq \frac{\partial}{\partial y}K(x,y) \cdot \frac{\partial}{\partial x}L(x,y) \quad \text{for all } x \text{ and } y.$$

From the fact that increasing functions of independent random variables are associated, it follows that if $F^I \leq_{\text{assoc}} F$, then F is the distribution function of associated random variables, where F^I is defined in Section 9.A.

The following closure property is easy to prove.

Theorem 9.E.1. *Let* (X_1, X_2) *and* (Y_1, Y_2) *be two random vectors. If* $(X_1, X_2) \leq_{\text{assoc}} (Y_1, Y_2)$*, then* $(\phi(X_1), \psi(X_2)) \leq_{\text{assoc}} (\phi(Y_1), \psi(Y_2))$ *for all strictly increasing functions* ϕ *and* ψ.

The relationship between the association and the PQD orders is described in the next result.

Theorem 9.E.2. *Let* (X_1, X_2) *and* (Y_1, Y_2) *be two random vectors with distribution functions* $F, G \in \mathcal{M}(F_1, F_2)$. *Then*

$$(X_1, X_2) \leq_{\text{assoc}} (Y_1, Y_2) \Longrightarrow (X_1, X_2) \leq_{\text{PQD}} (Y_1, Y_2).$$

Proof. Denote by F and G the distribution functions of (X_1, X_2) and (Y_1, Y_2), respectively. By assumption, $(Y_1, Y_2) =_{\text{st}} (K(X_1, X_2), L(X_1, X_2))$ where K and L are increasing and satisfy (9.E.2). Fix a pair (x_1, x_2). First suppose that $K(x_1, x_2) \le x_1$ and that $L(x_1, x_2) \le x_2$. Then

$$\begin{aligned} P\{Y_1 \le x_1, Y_2 \le x_2\} &\ge P\{Y_1 \le K(x_1, x_2), Y_2 \le L(x_1, x_2)\} \\ &= P\{K(X_1, X_2) \le K(x_1, x_2), L(X_1, X_2) \le L(x_1, x_2)\} \\ &\ge P\{X_1 \le x_1, X_2 \le x_2\}, \end{aligned}$$

where the second inequality follows from the increasingness of K and of L. Thus (9.A.3) holds in this case. Next suppose that $K(x_1, x_2) \le x_1$ and that $L(x_1, x_2) > x_2$. Then

$$\begin{aligned} P\{Y_1 > x_1, Y_2 < x_2\} &\le P\{Y_1 > K(x_1, x_2), Y_2 < L(x_1, x_2)\} \\ &= P\{K(X_1, X_2) > K(x_1, x_2), L(X_1, X_2) < L(x_1, x_2)\} \\ &\le P\{X_1 > x_1, X_2 < x_2\}, \end{aligned}$$

where the second inequality follows from (9.E.2). From the fact that (X_1, X_2) and (Y_1, Y_2) have the same univariate marginals it is seen that (9.A.3) holds in this case too. For the remaining two cases the inequality (9.A.3) follows in a similar way. □

The relationship between the association and the PRD orders is described next.

Theorem 9.E.3. *Let (X_1, X_2) and (Y_1, Y_2) be two random vectors with distribution functions $F, G \in \mathcal{M}(F_1, F_2)$ such that $F_{X_2|X_1}(x_2|x_1)$ and $G_{Y_2|Y_1}(x_2|x_1)$ are continuous in x_2 for all x_1. Then*

$$(X_1, X_2) \le_{\text{PRD}} (Y_1, Y_2) \Longrightarrow (X_1, X_2) \le_{\text{assoc}} (Y_1, Y_2).$$

Proof. Suppose that $(X_1, X_2) \le_{\text{PRD}} (Y_1, Y_2)$. Define K and L by $K(x_1, x_2) \equiv x_1$ and $L(x_1, x_2) \equiv G^{-1}_{Y_2|Y_1}\left[F_{X_2|X_1}(x_2|x_1)\Big|x_1\right]$. Obviously K is an increasing function. Also, obviously $L(x_1, x_2)$ is increasing in x_2. Furthermore, from (9.C.21) it is seen that $L(x_1, x_2)$ is also increasing in x_1, and that (9.E.2) holds. Now note that since $X_1 =_{\text{st}} Y_1$, we have, using the continuity assumptions stated, that

$$(Y_1, Y_2) =_{\text{st}} L(X_1, X_2).$$

That is, (X_1, X_2) and (Y_1, Y_2) satisfy (9.E.1) and (9.E.2). □

Example 9.E.4. Let U and V be any independent random variables. Define

$$X_\alpha = (1 - \alpha)U + \alpha V, \quad Y = U, \quad \text{for } \alpha \in [0, 1].$$

Then $(X_{\alpha_1}, Y) \le_{\text{assoc}} (X_{\alpha_2}, Y)$ whenever $\alpha_1 \le \alpha_2$.

Example 9.E.5. Let U and V be any independent random variables. Define

$$X_\alpha = (1-\alpha)U + \alpha V, \quad Y = \alpha U + (1-\alpha)V, \quad \text{for } \alpha \in [0, \tfrac{1}{2}].$$

Then $(X_{\alpha_1}, Y) \leq_{\text{assoc}} (X_{\alpha_2}, Y)$ whenever $\alpha_1 \leq \alpha_2$.

Example 9.E.6. Let (X_1, X_2) and (Y_1, Y_2) have bivariate normal distributions with correlation coefficients ρ_1 and ρ_2, respectively. Then $(X_1, X_2) \leq_{\text{assoc}} (Y_1, Y_2)$ if, and only if, $-1 \leq \rho_1 \leq \rho_2 \leq 1$.

Capéraà, Fougères, and Genest [119] introduced an order that is related to the association order. In order to define it we need first to introduce some notation. Let (X_1, X_2) be a random vector with a continuous distribution function $F \in \mathcal{M}(F_1, F_2)$. Define

$$V_F \equiv F(X_1, X_2),$$

and let K_F denote the distribution function of V_F. For example, if the distribution function of (X_1, X_2) is the Fréchet upper bound $F_U \in \mathcal{M}(F_1, F_2)$ (see page 387), then $K_{F_U}(v) = v$, $v \in [0, 1]$. If the distribution function of (X_1, X_2) is the Fréchet lower bound $F_L \in \mathcal{M}(F_1, F_2)$, then $K_{F_L}(v) = 1$, $v \in [0, 1]$. Finally, if X_1 and X_2 are independent, with distribution function $F^I \in \mathcal{M}(F_1, F_2)$, then $K_{F^I}(v) = v - v \log v$, $v \in [0, 1]$. These facts suggest the following order. Let (X_1, X_2) and (Y_1, Y_2) be two random vectors with continuous distribution functions $F, G \in \mathcal{M}(F_1, F_2)$. Suppose that

$$K_F(v) \geq K_G(v), \quad \text{for all } v \in [0, 1].$$

Then we say that (X_1, X_2) is *smaller than* (Y_1, Y_2) *in the Capéraà-Fougères-Genest order* (denoted by $(X_1, X_2) \leq_{\text{CFG}} (Y_1, Y_2)$ or $F \leq_{\text{CFG}} G$).

Capéraà, Fougères, and Genest [119] showed that for every continuous distribution function $F \in \mathcal{M}(F_1, F_2)$ we have

$$F_L \leq_{\text{CFG}} F \leq_{\text{CFG}} F_U.$$

They also proved, under some regularity conditions, that

$$(X_1, X_2) \leq_{\text{assoc}} (Y_1, Y_2) \Longrightarrow (X_1, X_2) \leq_{\text{CFG}} (Y_1, Y_2).$$

However, Capéraà, Fougères, and Genest [119] showed that $\leq_{\text{CFG}} \not\Rightarrow \leq_{\text{PQD}}$, whereas Nelsen, Quesada-Molina, Rodríguez-Lallena, and Úbeda-Flores [433] showed that $\leq_{\text{PQD}} \not\Rightarrow \leq_{\text{CFG}}$.

Nelsen, Quesada-Molina, Rodríguez-Lallena, and Úbeda-Flores [432] introduced some generalizations of the order $\leq_{\text{CFG}}$.

Another related order of interest is based on the notion of weak positive association which is defined in (9.A.22). Let $\boldsymbol{X} = (X_1, X_2, \ldots, X_n)$ and $\boldsymbol{Y} = (Y_1, Y_2, \ldots, Y_n)$ be two random vectors that have the same univariate marginals, and that satisfy

$$\begin{aligned}\operatorname{Cov}(h_1(X_{i_1}, X_{i_2}, \dots, X_{i_k}), h_2(X_{j_1}, X_{j_2}, \dots, X_{j_{n-k}}))\\ \le \operatorname{Cov}(h_1(Y_{i_1}, Y_{i_2}, \dots, Y_{i_k}), h_2(Y_{j_1}, Y_{j_2}, \dots, Y_{j_{n-k}}))\end{aligned}$$

for all choices of disjoint subsets $\{i_1, i_2, \dots, i_k\}$ and $\{j_1, j_2, \dots, j_{n-k}\}$ of $\{1, 2, \dots, n\}$, and for all increasing functions h_1 and h_2 for which the above covariances are defined. Then $\boldsymbol{X}$ is said to be *smaller than* $\boldsymbol{Y}$ *in the weak association order* (denoted by $\boldsymbol{X} \le_{\text{w-assoc}} \boldsymbol{Y}$).

Some closure properties of the weak association order are described in the next theorem.

Theorem 9.E.7. (a) *Let* $(X_1, X_2, \dots, X_n)$ *and* $(Y_1, Y_2, \dots, Y_n)$ *be two* n*-dimensional random vectors. If* $(X_1, X_2, \dots, X_n) \le_{\text{w-assoc}} (Y_1, Y_2, \dots, Y_n)$*, then*

$$(g_1(X_1), g_2(X_2), \dots, g_n(X_n)) \le_{\text{w-assoc}} (g_1(Y_1), g_2(Y_2), \dots, g_n(Y_n))$$

whenever $g_i : \mathbb{R} \to \mathbb{R}$, $i = 1, 2, \dots, n$, *are all increasing.*

(b) *Let* $\boldsymbol{X} = (X_1, X_2, \dots, X_n)$ *and* $\boldsymbol{Y} = (Y_1, Y_2, \dots, Y_n)$ *be two* n*-dimensional random vectors. If* $\boldsymbol{X} \le_{\text{w-assoc}} \boldsymbol{Y}$*, then* $\boldsymbol{X}_I \le_{\text{w-assoc}} \boldsymbol{Y}_I$ *for each* $I \subseteq \{1, 2, \dots, n\}$*. That is, the weak association order is closed under marginalization.*

An important useful property of the weak association order is the following.

Theorem 9.E.8. *Let* $\boldsymbol{X}$ *and* $\boldsymbol{Y}$ *be two random vectors with the same univariate marginals. Then*

$$\boldsymbol{X} \le_{\text{w-assoc}} \boldsymbol{Y} \Longrightarrow \boldsymbol{X} \le_{\text{sm}} \boldsymbol{Y}.$$

Remark 9.E.9. Note that if $\boldsymbol{X} = (X_1, X_2, \dots, X_n)$ is a vector of weakly positively associated random variables, as defined in (9.A.22), and if $\boldsymbol{Y} = (Y_1, Y_2, \dots, Y_n)$ is a vector of independent random variables such that, marginally, $X_i =_{\text{st}} Y_i$, $i = 1, 2, \dots, n$, then $\boldsymbol{X} \ge_{\text{w-assoc}} \boldsymbol{Y}$. Similarly, if $\boldsymbol{X}$ is a vector of negatively associated random variables, as defined in (3.A.54), and if $\boldsymbol{Y}$ is a vector of independent random variables such that, marginally, $X_i =_{\text{st}} Y_i$, $i = 1, 2, \dots, n$, then $\boldsymbol{X} \le_{\text{w-assoc}} \boldsymbol{Y}$. Thus it is seen that Theorem 9.E.7 is a stronger result than Theorem 9.A.23.

9.F The PDD Order

Let the random variables X_1 and X_2 have the symmetric (or exchangeable, or interchangeable) joint distribution F. Shaked [501] defines F (or X_1 and X_2) to be positive definite dependent (PDD) if F is a positive definite kernel on $\mathbb{S} \times \mathbb{S}$, where $\mathbb{S}$ is the support of X_1 (and therefore, by symmetry, $\mathbb{S}$ is also the support of X_2). Shaked [501] has shown that X_1 and X_2 are PDD if, and only if,

$$\mathrm{Cov}(\phi(X_1), \phi(X_2)) \geq 0 \quad \text{for every real function } \phi, \tag{9.F.1}$$

provided the covariance is well defined. This notion naturally leads to the order that is defined below.

Let (X_1, X_2) be a bivariate random vector with distribution function $F \in \mathcal{M}^{(s)}(\hat{F})$, where $\mathcal{M}^{(s)}(\hat{F})$ is the class of all the bivariate symmetric distributions with univariate marginals $\hat{F}$. Let (Y_1, Y_2) be another bivariate random vector with distribution function $G \in \mathcal{M}^{(s)}(\hat{F})$. Suppose that

$$\mathrm{Cov}(\phi(X_1), \phi(X_2)) \leq \mathrm{Cov}(\phi(Y_1), \phi(Y_2)) \quad \text{for every real function } \phi, \tag{9.F.2}$$

provided the covariances are well defined. Then we say that (X_1, X_2) is *smaller than* (Y_1, Y_2) *in the* PDD *order* (denoted by $(X_1, X_2) \leq_{\mathrm{PDD}} (Y_1, Y_2)$ or $F \leq_{\mathrm{PDD}} G$). Since only symmetric random vectors with the same univariate marginals are compared in the PDD order, we will implicitly assume this fact throughout this section.

Since $E\phi(X_1) = E\phi(X_2) = E\phi(Y_1) = E\phi(Y_2)$ for every real function ϕ, it follows that $(X_1, X_2) \leq_{\mathrm{PDD}} (Y_1, Y_2)$ if, and only if,

$$E\phi(X_1)\phi(X_2) \leq E\phi(Y_1)\phi(Y_2) \quad \text{for every real function } \phi, \tag{9.F.3}$$

provided the expectations exist. Thus, if $(X_1, X_2) \leq_{\mathrm{PDD}} (Y_1, Y_2)$, then

$$P\{X_1 \in A, X_2 \in A\} \leq P\{Y_1 \in A, Y_2 \in A\}$$

for all Borel-measurable sets A in $\mathbb{R}$.

Another characterization of the PDD order is given in the next theorem.

Theorem 9.F.1. *Let F and G be two symmetric bivariate distributions in $\mathcal{M}^{(s)}(\hat{F})$. Then $F \leq_{\mathrm{PDD}} G$ if, and only if, $G(x, y) - F(x, y)$ is a positive definite kernel.*

From (9.F.1) and (9.F.3) it is easily seen that F is PDD if, and only if,

$$F^I \leq_{\mathrm{PDD}} F,$$

where F^I is defined in Section 9.A.

A powerful closure property of the PDD order is described in the next theorem.

Theorem 9.F.2. *Suppose that the four random vectors (X_1, X_2), (Y_1, Y_2), (U_1, U_2) and (V_1, V_2) satisfy*

$$(X_1, X_2) \leq_{\mathrm{PDD}} (Y_1, Y_2) \quad \text{and} \quad (U_1, U_2) \leq_{\mathrm{PDD}} (V_1, V_2), \tag{9.F.4}$$

and suppose that (X_1, X_2) and (U_1, U_2) are independent, and also that (Y_1, Y_2) and (V_1, V_2) are independent. Then

$$(\phi(X_1, U_1), \phi(X_2, U_2)) \leq_{\mathrm{PDD}} (\phi(Y_1, V_1), \phi(Y_2, V_2)),$$
$$\text{for every increasing function } \phi.$$

In particular, if (9.F.4) holds, then the PDD order is closed under convolutions, that is,

$$(X_1 + U_1, X_2 + U_2) \le_{\rm PDD} (Y_1 + V_1, Y_2 + V_2).$$

Using (9.F.3) it is easy to verify the following closure properties.

Theorem 9.F.3. (a) *Let* $\{(X_1^{(j)}, X_2^{(j)}),\ j = 1, 2, \ldots\}$ *and* $\{(Y_1^{(j)}, Y_2^{(j)}),\ j = 1, 2, \ldots\}$ *be two sequences of random vectors such that* $(X_1^{(j)}, X_2^{(j)}) \to_{\rm st} (X_1, X_2)$ *and* $(Y_1^{(j)}, Y_2^{(j)}) \to_{\rm st} (Y_1, Y_2)$ *as* $j \to \infty$*, where* $\to_{\rm st}$ *denotes convergence in distribution. If* $(X_1^{(j)}, X_2^{(j)}) \le_{\rm PDD} (Y_1^{(j)}, Y_2^{(j)})$*,* $j = 1, 2, \ldots$*, then* $(X_1, X_2) \le_{\rm PDD} (Y_1, Y_2)$.

(b) *Let* (X_1, X_2)*,* (Y_1, Y_2)*, and* $\boldsymbol{\Theta}$ *be random vectors such that* $[(X_1, X_2)\big|\boldsymbol{\Theta} = \boldsymbol{\theta}] \le_{\rm PDD} [(Y_1, Y_2)\big|\boldsymbol{\Theta} = \boldsymbol{\theta}]$ *for all* $\boldsymbol{\theta}$ *in the support of* $\boldsymbol{\Theta}$*. Then* $(X_1, X_2) \le_{\rm PDD} (Y_1, Y_2)$*. That is, the* PDD *order is closed under mixtures.*

Example 9.F.4. Let (X_1, X_2) and (Y_1, Y_2) have exchangeable bivariate normal distributions with common marginals and correlation coefficients ρ_1 and ρ_2, respectively. If $0 \le \rho_1 \le \rho_2 \le 1$, then $(X_1, X_2) \le_{\rm PDD} (Y_1, Y_2)$.

If (X_1, X_2) and (Y_1, Y_2) have distributions F and G which are not symmetric, but still have the same common marginals (that is, X_1, X_2, Y_1, and Y_2 are all identically distributed), then the PDD order can still be defined on the symmetrizations $\tilde{F}(x, y) = \frac{1}{2}[F(x, y) + F(y, x)]$ and $\tilde{G}(x, y) = \frac{1}{2}[G(x, y) + G(y, x)]$ of F and G.

Hu and Joe [234] applied the idea of the PDD order to stationary reversible Markov chains $\{X_1, X_2, \ldots\}$. They showed for such chains that, if X_1 and X_2 are PDD (in the sense (9.F.1)), then dependence (in the sense of the PDD order) is decreasing with the lag, namely,

$$F_{12} \ge_{\rm PDD} F_{13} \ge_{\rm PDD} \cdots \ge_{\rm PDD} F_{1n} \ge_{\rm PDD} \cdots \ge_{\rm PDD} F^{(2)},$$

where the F_{1j}'s and $F^{(2)}$ are as defined in (9.A.12).

An n-variate extension of the PDD order for the case when $n \ge 2$ is suggested by (9.F.3). Explicitly, let $\boldsymbol{X} = (X_1, X_2, \ldots, X_n)$ and $\boldsymbol{Y} = (Y_1, Y_2, \ldots, Y_n)$ have distribution functions with common marginals. Then we can say that $\boldsymbol{X}$ is less positively dependent than $\boldsymbol{Y}$ if

$$E \prod_{i=1}^{n} \phi(X_i) \le E \prod_{i=1}^{n} \phi(Y_i) \quad \text{for every nonnegative real function } \phi. \qquad (9.F.5)$$

Note that for this definition it is not required that $\boldsymbol{X}$ and $\boldsymbol{Y}$ have exchangeable distribution functions; it is only required that $\boldsymbol{X}$ and $\boldsymbol{Y}$ have the same common marginals.

One reason for the usefulness of inequality (9.F.5) is that it implies that

$$P\{X_1 \in A, X_2 \in A, \ldots, X_n \in A\} \le P\{Y_1 \in A, Y_2 \in A, \ldots, Y_n \in A\}$$

for all Borel-measurable sets A in $\mathbb{R}$.

9.G Ordering Exchangeable Distributions

Let $\boldsymbol{X} = (X_1, X_2, \ldots, X_n)$ and $\boldsymbol{Y} = (Y_1, Y_2, \ldots, Y_n)$ be two random vectors with exchangeable distributions. Let $X_{(1)} \le X_{(2)} \le \cdots \le X_{(n)}$ and $Y_{(1)} \le Y_{(2)} \le \cdots \le Y_{(n)}$ be the corresponding order statistics. Intuitively, if $\boldsymbol{Y}$ is "more positively dependent" than $\boldsymbol{X}$ (or, alternatively, $\boldsymbol{Y}$ is "less dispersed" than $\boldsymbol{X}$), then we can expect the Y_i's to "hang together" more than the X_i's. For example, we can expect quantities such as $X_{(n)} - X_{(1)}$ or $X_{(n)} + X_{(n-1)} - X_{(2)} - X_{(1)}$ to be stochastically larger than $Y_{(n)} - Y_{(1)}$ or $Y_{(n)} + Y_{(n-1)} - Y_{(2)} - Y_{(1)}$. This observation naturally leads to the following definitions.

Let $\boldsymbol{X}$ and $\boldsymbol{Y}$ be two n-dimensional random vectors with exchangeable distribution functions and with the same common marginals. We will write $\boldsymbol{X} \le_{\text{pd-1}} \boldsymbol{Y}$ if

$$\left|\sum_{i=1}^{n} c_i X_{(i)}\right| \ge_{\text{st}} \left|\sum_{i=1}^{n} c_i Y_{(i)}\right| \quad \text{whenever} \ \sum_{i=1}^{n} c_i = 0. \tag{9.G.1}$$

When the interest is in the unordered components of the random vectors, then the following definition is useful. We will write $\boldsymbol{X} \le_{\text{pd-2}} \boldsymbol{Y}$ if

$$\left|\sum_{i=1}^{n} c_i X_i\right| \ge_{\text{st}} \left|\sum_{i=1}^{n} c_i Y_i\right| \quad \text{whenever} \ \sum_{i=1}^{n} c_i = 0. \tag{9.G.2}$$

Recall from page 2 the definition of the majorization order $\boldsymbol{a} \prec \boldsymbol{b}$ among n-dimensional vectors. For any random variable W, let F_W denote the distribution function of W. We will write $\boldsymbol{X} \le_{\text{pd-3}} \boldsymbol{Y}$ if

$$\begin{aligned}&(F_{X_{(1)}}(x), F_{X_{(2)}}(x), \ldots, F_{X_{(n)}}(x))\\ &\qquad \succ (F_{Y_{(1)}}(x), F_{Y_{(2)}}(x), \ldots, F_{Y_{(n)}}(x)) \quad \text{for all } x.\end{aligned} \tag{9.G.3}$$

It is easy to verify that (9.G.3) is equivalent to

$$(E\phi(X_{(1)}), E\phi(X_{(2)}), \ldots, E\phi(X_{(n)})) \succ (E\phi(Y_{(1)}), E\phi(Y_{(2)}), \ldots, E\phi(Y_{(n)}))$$

for all monotone functions ϕ for which the expectations exist. A further insight into the meaning of (9.G.3) can be obtained by rewriting it as the set of inequalities

$$E\left[\sum_{i=1}^{j} I_{(-\infty,x]}(X_{(i)})\right] \ge E\left[\sum_{i=1}^{j} I_{(-\infty,x]}(Y_{(i)})\right], \quad \text{for } j = 1, 2, \ldots, n, \text{ and all } x, \tag{9.G.4}$$

with equality holding for $j = n$. That is, for each j, the expected value of the number of order statistics which are less than or equal to x among the first k

ordered X_i's is at least as large as the corresponding expected value based on the ordered Y_i's.

When one is concerned only with the expectations of the order statistics, then the following stochastic order is useful. We will write $\boldsymbol{X} \le_{\text{pd-4}} \boldsymbol{Y}$ if

$$(EX_{(1)}, EX_{(2)}, \ldots, EX_{(n)}) \succ (EY_{(1)}, EY_{(2)}, \ldots, EY_{(n)}). \tag{9.G.5}$$

The next result describes some interrelationships among the orders $\le_{\text{pd-}k}$, $k = 1, 2, 3, 4$.

Theorem 9.G.1. *Let $\boldsymbol{X}$ and $\boldsymbol{Y}$ be two n-dimensional random vectors with exchangeable distribution functions and with the same common marginals. Then*

$$\begin{array}{ccc} & & \boldsymbol{X} \le_{\text{pd-1}} \boldsymbol{Y} \Rightarrow \boldsymbol{X} \le_{\text{pd-2}} \boldsymbol{Y} \\ & & \Downarrow \\ \boldsymbol{X} \le_{\text{pd-3}} \boldsymbol{Y} \Rightarrow \boldsymbol{X} \le_{\text{pd-4}} \boldsymbol{Y} & & \end{array}$$

Proof. First suppose that $\boldsymbol{X} \le_{\text{pd-1}} \boldsymbol{Y}$. Let $\boldsymbol{\pi} = (\pi_1, \pi_2, \ldots, \pi_n)$ denote a permutation of $\{1, 2, \ldots, n\}$, and let $\sum_{\boldsymbol{\pi}}$ denote a summation over all such permutations. Then, by exchangeability, for any real z, and whenever $\sum_{i=1}^n c_i = 0$, we have

$$\begin{aligned} P\Big\{\Big|\sum_{i=1}^n c_i X_i\Big| > z\Big\} &= \sum_{\boldsymbol{\pi}} \frac{1}{n!} P\Big\{\Big|\sum_{i=1}^n c_i X_i\Big| > z \Big| X_{\pi_1} \le X_{\pi_2} \le \cdots \le X_{\pi_n}\Big\} \\ &= \sum_{\boldsymbol{\pi}} \frac{1}{n!} P\Big\{\Big|\sum_{i=1}^n c_{\pi_i} X_{(i)} i\Big| > z\Big\} \\ &\ge \sum_{\boldsymbol{\pi}} \frac{1}{n!} P\Big\{\Big|\sum_{i=1}^n c_{\pi_i} Y_{(i)} i\Big| > z\Big\} \\ &= P\Big\{\Big|\sum_{i=1}^n c_i Y_i\Big| > z\Big\}, \end{aligned}$$

and (9.G.2) follows.

If we denote $a_i = EX_{(i)}$ and $b_i = EY_{(i)}$, $i = 1, 2, \ldots, n$, then from (9.G.1) it follows that $a_i - a_{i-1} \ge b_i - b_{i-1}$, $i = 1, 2, \ldots, n-1$. Also, $\sum_{i=1}^n a_i = \sum_{i=1}^n b_i$. Now it is easily seen that $\boldsymbol{a} \succ \boldsymbol{b}$, and thus (9.G.5) holds.

The proof of $\boldsymbol{X} \le_{\text{pd-3}} \boldsymbol{Y} \Rightarrow \boldsymbol{X} \le_{\text{pd-4}} \boldsymbol{Y}$ is easy (see, for example, Marshall and Olkin [383, page 350]). □

Some closure properties of the above orders are described in the following theorem.

Theorem 9.G.2. (a) *For $j = 1, 2, \ldots,$ let $\boldsymbol{X}^{(j)}$ and $\boldsymbol{Y}^{(j)}$ be two random vectors with exchangeable distribution functions and with the same common marginals such that $\boldsymbol{X}^{(j)} \to_{\text{st}} \boldsymbol{X}$ and $\boldsymbol{Y}^{(j)} \to_{\text{st}} \boldsymbol{Y}$ as $j \to \infty$, where $\to_{\text{st}}$ denotes convergence in distribution. If $\boldsymbol{X}^{(j)} \le_{\text{pd-}k} \boldsymbol{Y}^{(j)}$, $j = 1, 2, \ldots,$ then $\boldsymbol{X} \le_{\text{pd-}k} \boldsymbol{Y}$, $k = 1, 2, 3$.*

(b) *Let* $\boldsymbol{X} = (X_1, X_2, \ldots, X_n)$ *and* $\boldsymbol{Y} = (Y_1, Y_2, \ldots, Y_n)$ *be two* n*-dimensional random vectors with exchangeable distribution functions and with the same common marginals. If* $\boldsymbol{X} \le_{\text{pd-}k} \boldsymbol{Y}$, *then* $\boldsymbol{X}_I \le_{\text{pd-}k} \boldsymbol{Y}_I$ *for each* $I \subseteq \{1, 2, \ldots, n\}$. *That is, the* $\le_{\text{pd-}k}$ *order is closed under marginalization,* $k = 1, 2, 3, 4$.

(c) *Let* $(X_1, X_2, \ldots, X_n)$ *and* $(Y_1, Y_2, \ldots, Y_n)$ *be as in part* (b). *If* $(X_1, X_2, \ldots, X_n) \le_{\text{pd-}k} (Y_1, Y_2, \ldots, Y_n)$, *then* $(aX_1 + b, aX_2 + b, \ldots, aX_n + b) \le_{\text{pd-}k} (aY_1 + b, aY_2 + b, \ldots, aY_n + b)$ *for any constants* a *and* b, $k = 1, 2, 3, 4$.

(d) *Let* $\boldsymbol{X}$ *and* $\boldsymbol{Y}$ *be as in part* (b), *and let* $\boldsymbol{\Theta}$ *be another random vector. If* $[\boldsymbol{X} | \boldsymbol{\Theta} = \boldsymbol{\theta}] \le_{\text{pd-}k} [\boldsymbol{Y} | \boldsymbol{\Theta} = \boldsymbol{\theta}]$ *for all* $\boldsymbol{\theta}$ *in the support of* $\boldsymbol{\Theta}$, *then* $\boldsymbol{X} \le_{\text{pd-}k} \boldsymbol{Y}$. *That is, the* $\le_{\text{pd-}k}$ *order is closed under mixtures,* $k = 1, 2, 3, 4$.

In the bivariate case we have the following relationship.

Theorem 9.G.3. *Let* (X_1, X_2) *and* (Y_1, Y_2) *be two random vectors with exchangeable distribution functions with common marginals. Then*

$$(X_1, X_2) \le_{\text{PDD}} (Y_1, Y_2) \Longrightarrow (X_1, X_2) \le_{\text{pd-3}} (Y_1, Y_2).$$

Proof. Suppose that $(X_1, X_2) \le_{\text{PDD}} (Y_1, Y_2)$. Then, for any real z we have

$$\begin{aligned} F_{X_{(1)}}(z) &= 1 - P\{\min(X_1, X_2) > z\} \\ &= 1 - EI_{(z,\infty)}(X_1) I_{(z,\infty)}(X_2) \\ &\ge 1 - EI_{(z,\infty)}(Y_1) I_{(z,\infty)}(Y_2) \\ &= F_{Y_{(1)}}(z), \end{aligned}$$

where the inequality follows from (9.F.3). Now, since $F_{X_{(1)}}(z) + F_{X_{(2)}}(z) = F_{Y_{(1)}}(z) + F_{Y_{(2)}}(z)$, it follows that $(F_{X_{(1)}}(z), F_{X_{(2)}}(z)) \succ (F_{X_{(1)}}(z), F_{Y_{(2)}}(z))$ which is (9.G.3). □

A relationship between the star order (see Section 4.B) and the order $\le_{\text{pd-4}}$ is described next.

Theorem 9.G.4. *Let* $\boldsymbol{X} = (X_1, X_2, \ldots, X_n)$ *and* $\boldsymbol{Y} = (Y_1, Y_2, \ldots, Y_n)$ *be two vectors, each consisting of independent and identically distributed nonnegative random variables. If* $X_1 \le_* Y_1$, *and if* $EX_1 = EY_1$, *then* $\boldsymbol{X} \le_{\text{pd-4}} \boldsymbol{Y}$.

Example 9.G.5. If $X_1, X_2, \ldots, X_n$ are conditionally independent and identically distributed (then they are exchangeable), and if $Y_1, Y_2, \ldots, Y_n$ are independent and identically distributed, and if all the X_i's and Y_i's have the same marginal distribution, then $\boldsymbol{X} = (X_1, X_2, \ldots, X_n)$ and $\boldsymbol{Y} = (Y_1, Y_2, \ldots, Y_n)$ satisfy $\boldsymbol{X} \le_{\text{pd-3}} \boldsymbol{Y}$ and, of course, also $\boldsymbol{X} \le_{\text{pd-4}} \boldsymbol{Y}$; this is shown in Shaked and Tong [523]. Hu and Hu [233] have shown that if $X_1, X_2, \ldots, X_n$ have some other properties of positive or negative dependence, and if $Y_1, Y_2, \ldots, Y_n$ are independent, and if $X_i =_{\text{st}} Y_i$ for $i = 1, 2, \ldots, n$, then the above (that is, $\boldsymbol{X} \le_{\text{pd-3}} \boldsymbol{Y}$ and $\boldsymbol{X} \le_{\text{pd-4}} \boldsymbol{Y}$) also hold. Ebrahimi and Spizzichino [178] obtained conditions on the expected values of the order statistics that are associated with $\boldsymbol{X} = (X_1, X_2, \ldots, X_n)$ and $\boldsymbol{Y} = (Y_1, Y_2, \ldots, Y_n)$, under which $\boldsymbol{X} \le_{\text{pd-4}} \boldsymbol{Y}$.

Paul [442] gave conditions under which $X_i \leq_{\rm cx} Y_i$, $i = 1, 2$ (where (X_1, X_2) and (Y_1, Y_2) are some bivariate random vectors) imply $(Y_1, Y_2) \leq_{\text{pd-4}} (X_1, X_2)$ (in fact the conclusion of Paul [442] is stated as $E\max\{X_1, X_2\} \leq E\max\{Y_1, Y_2\}$, but since $EX_i = EY_i$, $i = 1, 2$, the stated conclusion is the same as $(Y_1, Y_2) \leq_{\text{pd-4}} (X_1, X_2)$). Müller [414], however, noticed that in Paul [442] there was a subtle mistake which invalidated his Theorem 1. Müller [414] provided other conditions under which the conclusion above is valid.

9.H Complements

A good review of the theory of positive dependence orders is the survey by Scarsini and Shaked [496]. Section 2.2 in Joe [262] contains many of the results that are mentioned in Sections 9.A–9.F, as well as many examples and counterexamples.

Section 9.A: The PQD order is first defined in Yanagimoto and Okamoto [570]; it also can be found in Tchen [547]. The general closure property of the PQD order (Theorem 9.A.1) is taken from Kimeldorf and Sampson [295]. The definition of the PQD order for general n-dimensional vectors ($n > 2$) can be found in Joe [260]. The conditions under which Archimedean copulas are ordered in the PQD sense (Example 9.A.3) can be found in Joe [262]. Brown and Rinott [110] showed that some pairs of multivariate infinitely divisible distributions are PQD-ordered. The PQD comparisons of convolutions and of mixtures results (Theorems 9.A.6 and 9.A.7) are special cases of results of Belzunce and Semeraro [77]. The PQD ordering of random vectors with elliptically contoured densities (Example 9.A.8) follows from Theorem 5.1 of Das Gupta, Eaton, Olkin, Perlman, Savage, and Sobel [139]; see also Landsman and Tsanakas [331]. The results about the supermodular order (Section 9.A.4) are mostly taken from Meester and Shanthikumar [387] and from Shaked and Shanthikumar [517]; see also Joe [260] and Szekli, Disney, and Hur [545]. The closure results of the supermodular order given in Theorem 9.A.12, and the application to Markov chains given in Example 9.A.13, are taken from Li and Xu [350]. An extension of the result in Example 9.A.13 can be found in Kulik and Szekli [325]. The closure property of the order $\leq_{\rm sm}$ under random sums (Theorem 9.A.14) can be found in Denuit, Genest, and Marceau [145]; it generalizes some results of Hu and Pan [238]. Extensions of Theorem 9.A.14 are given in Lillo, Pellerey, Semeraro, and Shaked [363], and in Kulik and Szekli [325]. The supermodular comparison of mixtures result (Theorem 9.A.15) is taken from Denuit and Müller [157]. The property that is described in Theorem 9.A.16 can be found in Bäuerle [58] or in Bäuerle and Rieder [61], and the property that is described in Theorem 9.A.18 can be found in Müller [411]. The inequality that is described in Example 9.A.17 is taken from Vanichpun and Makowski [554, 555]; they credit it to Bäuerle [58]. The fact that sums of components of supermodular ordered vectors are ordered according to $\leq_{\text{icx}}$, described in (9.A.19), is taken from Müller [409]. The convex order comparison of random sums in Example 9.A.19 is a generalization of a result of O'Cinneide [439]. The result about the ordering of multivariate normal random vectors according to the $\leq_{\text{sm}}$ order (Example 9.A.20) can be found in Huffer [250]; see also Müller and Scarsini [416] and Block and Sampson [94, Section 2], though in the latter paper there is a mistake which is corrected in Müller and Scarsini [416]. An extension of the result in Example 9.A.20 to Kotz-type distributions is given in Ding and Zhang [168]. The bound on $\boldsymbol{X}$, which is described in Theorem 9.A.21, can be found in Tchen [547]. A geometric proof of (9.A.20) is given in Kaas, Dhaene, Vyncke, Goovaerts, and Denuit [268] and in Hoedemakers, Beirlant, Goovaerts, and Dhaene [224]. The convex comparison of sums (Proposition 9.A.22) is taken from Kaas, Dhaene, and Goovaerts [267]; some related results and extensions can be found in Goovaerts and Kaas [213] and in Hoedemakers, Beirlant, Goovaerts, and Dhaene [224]. The comparison of a vector of associated random variables with its independence version (Theorem 9.A.23) can be found in Christofides and Vaggelatou [130]; the first part of this theorem strengthens a result in Shaked and Shanthikumar [517] which states the same conclusion, but under the CIS condition (defined in (6.B.11)) which is stronger than the weak positive association condition. The lower bound on $\boldsymbol{X}$ by the so-called "mutually exclusive" random variables (that is, that satisfy (9.A.23)), given in Theorem 9.A.24, is taken from Dhaene and Denuit [162]; see related results in Frostig [207] and in references therein. The sufficient condition by means of copulas, which imply the $\leq_{\text{dir-cx}}$ order (Theorem 9.A.25), can be found in Juri [266]. Theorem 3.1 and Corollaries 3.2 and 4.1 in Rüschendorf [486] are variants of Theorem 9.A.25. The model that is described in Example 9.A.26 is a special case of a model discussed in Bäuerle [57]; in fact, her Theorem 3.1 can be obtained from the stochastic inequality of Example 9.A.26 and the closure of the supermodular order under mixtures (Theorem 9.A.9(d)). Rüschendorf [486] studied various extensions of Example 9.A.26. The comparison of sampling plans which is given in Example 9.A.27 was obtained in Karlin [276], and noted by Frostig [206]. The comparison of multivariate Archimedean copulas (Example 9.A.28), as well as further similar comparisons, can be found in Wei and Hu [559].

If F and G of (9.A.3) are the distribution functions of bivariate vectors with integer-valued components, then the comparison $F \leq_{\text{PQD}} G$ is the same as a comparison of the partial sums of two matrices with non-negative entries (which sum up to 1). Nguyen and Sampson [434] studied the geometry of such matrices.

The PQD comparison can be used also to compare contingency tables that have the same row and column sums. Nguyen and Sampson [435] obtained some results regarding the number of such contingency tables that are more PQD than a given contingency table.

Block, Chhetry, Fang, and Sampson [92] found necessary and sufficient conditions (by means of orders of permutations) for two bivariate empirical distributions to be ordered according to the PQD order. Further results in this vein are given in Metry and Sampson [392]. Examples of pairs of bivariate distributions that are PQD-ordered can be found in de la Horra and Ruiz-Rivas [227] and in Joe [261]. Bassan and Scarsini [55] characterized the PQD order by means of the usual stochastic ordering of some related stopping times.

Ebrahimi [175] discussed negatively dependent distributions that are ordered according to the PQD order.

Some positive dependence orders that are weaker than the PQD order were introduced in Rodríguez-Lallena and Úbeda-Flores [470].

Lu and Yi [366] gave a definition of an order that generalizes the bivariate PQD order to higher dimensions. However, this order does not have the desirable properties of being closed under mixtures and concatenations (this follows from the fact that parts (c) and (e) of Theorem 2.4 in Lu and Yi [366] may be incorrect).

Section 9.B: Most of the results in this section can be found in Colangelo, Scarsini, and Shaked [133].

Section 9.C: Most of the results in this section, about the LTD and RTI orders, are taken from Averous and Dortet-Bernadet [25]. The relationship between the strong orthant ratio orders and the LTD and RTI orders (Theorem 9.C.7) can be found in Colangelo, Scarsini, and Shaked [133]; the counterexamples that are mentioned after Theorem 9.C.7 can also be found in that paper. The results about the PRD order are taken from Yanagimito and Okamoto [570] and from Fang and Joe [192]. In addition to the characterizations (9.C.19)–(9.C.21) of the PRD order, the reader may find another characterization in Rüschendorf [484]. In addition to Examples 9.C.10–9.C.12, many other examples of pairs of random vectors that are PRD-ordered can be found in Fang and Joe [192].

Hollander, Proschan, and Sconing [225] briefly considered some LTD and RTI orders that are different than the ones in Section 9.C. Colangelo [132] studied the relationships among these orders and the LTD and RTI orders in Section 9.C, and Colangelo, Scarsini, and Shaked [133] studied the relationships among these orders and the orthant ratio orders.

Block, Chhetry, Fang, and Sampson [92] found necessary and sufficient conditions (by means of orders of permutations) for two bivariate empirical distributions to be ordered according to the PRD order. Some variations of the PRD order are discussed in Capéraà and Genest [120] and in Fang and Joe [192].

Avérous, Genest, and Kochar [26] introduced an extension of the PRD order which compares bivariate random vectors that need not have the same univariate marginals. Their order is equivalent to the requirement that the corresponding copulas are ordered in the PRD order.

Hollander, Proschan, and Sconing [225] briefly discussed the order according to which (X_1, X_2) is smaller than (Y_1, Y_2) if

$$G_{Y_2|Y_1}(x_2|x_1) - F_{X_2|X_1}(x_2|x_1) \quad \text{is increasing in } x_1 \text{ for all } x_2.$$

Section 9.D: Most of the material in this section is taken from Kimeldorf and Sampson [295]. The conditions under which Archimedean copulas are ordered in the PLRD sense (Example 9.D.5) can be found in Joe [262]. The comparison of two bivariate normal random vectors in the PLRD sense (Example 9.D.6) is taken from Genest and Verret [208].

Yanagimoto [569] introduced a collection of 16 orders based on the idea of (9.D.2). He did it by requiring (9.D.2) to hold for special choices of intervals I_1, I_2, J_1, and J_2. The PQD order is one of the 16 orders in the collection of Yanagimoto. Metry and Sampson [391] extended Yanagimoto's idea and presented a more general approach for generating positive dependence orderings. That approach makes it fairly easy to study the properties of the resulting orders and the interrelationships among them. Yanagimoto [569] also introduced an order that is similar to the PLRD order, and which applies to random vectors of dimension $n \geq 2$.

Kemperman [284] and Karlin and Rinott [278] suggested an order according to which the bivariate distribution F (with density f) is smaller than the bivariate distribution G (with density g) if

$$f(x_1, y_1)g(x_2, y_2) \geq f(x_1, y_2)g(x_2, y_1) \quad \text{whenever } x_1 \leq x_2 \text{ and } y_1 \leq y_2.$$

This order has not been studied in the literature as a positive dependence order. In fact, Kimeldorf and Sampson [295] have noticed that it does not satisfy some of the basic axioms that they introduced.

Section 9.E: The definition and many properties of associated random variables can be found in Esary, Proschan, and Walkup [184]. Most of the results described in this section are taken from Schriever [498] and from Fang and Joe [192]. In addition to Examples 9.E.4–9.E.6, many other examples of pairs of random vectors that are ordered by association can be found in Fang and Joe [192]. Some variations of the association order are also discussed in that paper.

Block, Chhetry, Fang, and Sampson [92] found necessary and sufficient conditions (by means of orders of permutations) for two bivariate empirical distributions to be ordered according to the association order.

The main result about the weak association order (Theorem 9.E.8) is extracted from Rüschendorf [486]; see also Yi and Tongyu [574].

Kimeldorf and Sampson [296] and Hollander, Proschan, and Sconing [225] discuss briefly an order according to which (X_1, X_2) is smaller than (Y_1, Y_2) if

$$\text{Cov}(K(X_1, X_2), L(X_1, X_2)) \le \text{Cov}(K(Y_1, Y_2), L(Y_1, Y_2)),$$

for all increasing functions K and L for which the covariance is well defined. Kimeldorf and Sampson [296] showed that this order does not satisfy one of their axioms. This order can clearly be extended to the case in which the dimension is $n \ge 2$.

Section 9.F: Most of the results in this section are taken from Shaked [501] and from Rinott and Pollak [467]. One can prove Theorem 9.F.1 using the method of proof of Theorem 3.1 in Shaked [501]. Tong [550] has listed some examples of vectors $\boldsymbol{X}$ and $\boldsymbol{Y}$ that satisfy (9.F.5), and has shown some applications of this order.

Rinott and Pollak [467] have essentially shown that if $(X_1, X_2) \le_{\text{PDD}} (Y_1, Y_2)$, then some of the first-passage times of related Gaussian processes are ordered in the usual stochastic order.

Section 9.G: The results in this section are mostly taken from Shaked and Tong [523]. Many examples of pairs of exchangeable vectors that satisfy the orders $\le_{\text{pd-}k}$, $k = 1, 2, 3, 4$, are listed in that paper. Further examples can be found in Shaked and Tong [522]. The relationship between the star order and the order $\le_{\text{pd-4}}$ (Theorem 9.G.4) is taken from Barlow and Proschan [35]; a slightly stronger result can be found in Shaked [502]. Gupta and Richards [218] have given examples of pairs of multivariate Liouville distributions that are ordered according to $\le_{\text{pd-1}}$ and therefore also according to $\le_{\text{pd-2}}$ and $\le_{\text{pd-4}}$.

Shaked and Tong [523] have noted that, intuitively, exchangeable random vectors are "more positively dependent" if, and only if, they are "less dispersed." Thus they suggested to define orderings according to which $(X_1, X_2, \ldots, X_n)$ is smaller than $(Y_1, Y_2, \ldots, Y_n)$ if

$$E\phi(X_1, X_2, \ldots, X_n) \ge E\phi(Y_1, Y_2, \ldots, Y_n),$$

for every ϕ which belongs to some properly chosen class of permutation symmetric functions. In addition to the classes defined in (9.G.1), (9.G.2) and (9.G.4) [there exists also a class under which the above inequality gives (9.G.5)], a natural choice of such a class is the class of all Schur-convex functions. Chang [124] considered some orders that are defined by the above inequality for several classes of permutation symmetric functions. His paper contains a rich bibliography regarding several stochastic majorization orders.

Mosler [399, Section 7.6] introduced some notions of positive dependence orders that are based on volumes of central regions.

References

1. Aalen, O.O., Hoem, J.M.: Random time changes for multivariate counting processes. Scandinavian Actuarial Journal, 81–101 (1978)
2. Adell, J.A., Badía, F.G., de la Cal, J.: Beta-type operators preserve shape properties. Stochastic Processes and Their Applications **48**, 1–8 (1993)
3. Adell, J.A., de la Cal, J.: Optimal Poisson approximation of uniform empirical processes. Stochastic Processes and Their Applications **64**, 135–142 (1996)
4. Adell, J.A., Lekuona, A.: Taylor's formula and preservation of generalized convexity for positive linear operators. Journal of Applied Probability **37**, 765–777 (2000)
5. Adell, J.A., Perez-Palomares, A.: Stochastic orders in preservation properties by Bernstein-type operators. Advances in Applied Probability **31**, 492–507 (1999)
6. Ahmadi, J., Arghami, N.R.: Some univariate stochastic orders on record values. Communications in Statistics—Theory and Methods **30**, 69–74 (2001)
7. Ahmed, A.-H. N.: Preservation properties for the mean residual life ordering. Statistical Papers **29**, 143–150 (1988)
8. Ahmed, A.N., Alzaid, A., Bartoszewicz, J., Kochar, S.C.: Dispersive and superadditive ordering. Advances in Applied Probability **18**, 1019–1022 (1986)
9. Ahmed, A.N., Soliman, A.A., Khider, S.E.: On some partial ordering of interest in reliability. Microelectronics Reliability **36**, 1337–1346 (1996)
10. Ahmed, A.N., Soliman, A.A., Khider, S.E.: Preservation results for ordered random variables, with applications to reliability theory. Microelectronics Reliability **37**, 277–287 (1997)
11. Alzaid, A., Kim, J.S., Proschan, F.: Laplace ordering and its applications. Journal of Applied Probability **28**, 116–130 (1991)
12. Alzaid, A.A.: Mean residual life ordering. Statistical Papers **29**, 35–43 (1988)
13. Alzaid, A.A.: Length-biased orderings with applications. Probability in the Engineering and Informational Sciences **2**, 329–341 (1988)
14. Alzaid, A.A., Proschan, F.: Dispersivity and stochastic majorization. Statistics and Probability Letters **13**, 275–278 (1992)
15. Arcones, M.A., Kvam, P.H., Samaniego, F.J.: Nonparametric estimation of a distribution subject to a stochastic precedence constraint. Journal of the American Statistical Association **97**, 170–182 (2002)

16. Argon, N.T., Andradóttir, S.: Partial pooling in tandem lines with cooperation and blocking. Queueing Systems **52**, 5–30 (2006)
17. Arias-Nicolás, J.P., Fernández-Ponce, J.M., Luque-Calvo, P., Suárez-Llorens, A.: Multivariate dispersion order and the notion of copula applied to the multivariate t-distribution. Probability in the Engineering and Informational Sciences **19**, 363–375 (2005)
18. Arjas, E.: A stochastic process approach to multivariate reliability systems: Notions based on conditional stochastic order. Mathematics of Operations Research **6**, 263–276 (1981)
19. Arnold, B.C.: Majorization and the Lorenz Order: A Brief Introduction. Springer-Verlag, New York (1987)
20. Arnold, B.C.: Inequality measures for multivariate distributions. Metron **63**, 317–327 (2005)
21. Arnold, B.C., Villasenor, J.A.: Lorenz ordering of order statistics and record values. In: Balakrishnan, N., Rao, C.R. (ed) Handbook of Statistics 16: Order Statistics: Theory and Methods. Elsevier, Amsterdam, 75–87 (1998)
22. Arrow, K.J.: Essays in the Theory of Risk-Bearing. North-Holland, New York (1974)
23. Asadi, M., Shanbhag, D.N.: Hazard measure and mean residual life orderings: A unified approach. In: Balakrishnan, N., Rao, C.R. (ed) Handbook of Statistics 20: Advances in Reliability. Elsevier, Amsterdam, 199–214 (2001)
24. Atakan, A.E.: Stochastic convexity in dynamic programming. Economic Theory **22**, 447–455 (2003)
25. Averous, J., Dortet-Bernadet, J.-L.: LTD and RTI dependence orderings. Canadian Journal of Statistics **28**, 151–157 (2000)
26. Avérous, J., Genest, C., Kochar, S.C.: On the dependence structure of order statistics. Journal of Multivariate Analysis **94**, 159–171 (2005)
27. Baccelli, F., Makowski, A.M.: Multi-dimensional stochastic ordering and associated random variables. Operations Research **37**, 478–487 (1989)
28. Baccelli, F., Makowski, A.M.: Stochastic orders associated with the forward recurrence time of a renewal process. Technical Report. Department of Electrical Engineering, University of Maryland, College Park (1992)
29. Bagai, I., Kochar, S.C.: On tail-ordering and comparison of failure rates. Communications in Statistics—Theory and Methods **15**, 1377–1388 (1986)
30. Baker, E.: Increasing risk and increasing informativeness: Equivalence theorems. Operations Research **54**, 26–36 (2006)
31. Bapat, R.B., Kochar, S.C.: On likelihood-ratio ordering of order statistics. Linear Algebra and Its Applications **199**, 281–291 (1994)
32. Barlow, R.E., Bartholomew, D.J., Bremner, J.M., Brunk, H.D.: Statistical Inference under Order Restrictions. Wiley, New York (1972)
33. Barlow, R.E., Campo, R.: Total time on test processes and applications to failure data analysis. In: Barlow, R.E., Fussel, R., Singpurwalla, N.D. (ed) Reliability and Fault Tree Analysis. SIAM, Philadelphia, 451–481 (1975)
34. Barlow, R.E., Doksum, K.A.: Isotonic tests for convex ordering. In: Proceedings of the Sixth Berkeley Symposium on Mathematical Statistics and Probability **1**. University of California Press, Berkeley, 293–323 (1972)
35. Barlow, R.E., Proschan, F.: Inequalities for linear combinations of order statistics from restricted families. Annals of Mathematical Statistics **37**, 1574–1592 (1966)

36. Barlow, R.E., Proschan, F.: Statistical Theory of Reliability and Life Testing, Probability Models. Holt, Rinehart, and Winston, New York (1975)
37. Bartoszewicz, J.: Moment inequalities for order statistics from ordered families of distributions. Metrika **32**, 383–389 (1985)
38. Bartoszewicz, J.: Dispersive ordering and monotone failure rate distributions. Advances in Applied Probability **17**, 472–474 (1985)
39. Bartoszewicz, J.: Dispersive ordering and the total time on test transformation. Statistics and Probability Letters **4**, 285–288 (1986)
40. Bartoszewicz, J.: A note on dispersive ordering defined by hazard functions. Statistics and Probability Letters **6**, 13–16 (1987)
41. Bartoszewicz, J.: Quantile inequalities for linear combinations of order statistics from ordered families of distributions. Applicationes Mathematicae **21**, 575–589 (1993)
42. Bartoszewicz, J.: Stochastic order relations and the total time on test transform. Statistics and Probability Letters **22**, 103–110 (1995)
43. Bartoszewicz, J.: Tail orderings and the total time on test transform. Applicationes Mathematicae **24**, 77–86 (1996)
44. Bartoszewicz, J.: Dispersive functions and stochastic orders. Applicationes Mathematicae **24**, 429–444 (1997)
45. Bartoszewicz, J.: Applications of a general composition theorem to the star order of distributions. Statistics and Probability Letters **38**, 1–9 (1998)
46. Bartoszewicz, J.: Characterizations of the dispersive order of distributions by the Laplace transform. Statistics and Probability Letters **40**, 23–29 (1998)
47. Bartoszewicz, J.: Characterizations of stochastic orders based on ratios of Laplace transforms. Statistics and Probability Letters **42**, 207–212 (1999)
48. Bartoszewicz, J.: Stochastic orders based on the Laplace transform and infinitely divisible distributions. Statistics and Probability Letters **50**, 121–129 (2000)
49. Bartoszewicz, J.: Stochastic comparisons of random minima and maxima from life distributions. Statistics and Probability Letters **55**, 107–112 (2001)
50. Bartoszewicz, J.: Mixtures of exponential distributions and stochastic orders. Statistics and Probability Letters **57**, 23–31 (2002)
51. Bartoszewicz, J., Skolimowska, M.: Preservation of classes of life distributions and stochastic orders under weighting. Statistics and Probability Letters **76**, 587–596 (2006)
52. Bassan, B., Denuit, M., Scarsini, M.: Variability orders and mean differences. Statistics and Probability Letters **45**, 121–130 (1999)
53. Bassan, B., Kochar, S., Spizzichino, F.: Some bivariate notions of IFR and DMRL and related properties. Journal of Applied Probability **39**, 533–544 (2002)
54. Bassan, B., Scarsini, M.: Convex ordering for stochastic processes. Commentationes Mathematicae Universitatis Carolinae **32**, 115–118 (1991)
55. Bassan, B., Scarsini, M.: Positive dependence orderings and stopping times. Annals of the Institute of Statistical Mathematics **46**, 333–342 (1994)
56. Bassan, B., Spizzichino, F.: Stochastic comparisons for residual lifetimes and Bayesian notions of multivariate ageing. Advances in Applied Probability **31**, 1078–1094 (1999)
57. Bäuerle, N.: Inequalities for stochastic models via supermodular orderings. Communications in Statistics—Stochastic Models **13**, 181–201 (1997)

58. Bäuerle, N.: Monotonicity results for MR/GI/1 queues. Journal of Applied Probability **34**, 514–524 (1997)
59. Bäuerle, N.: The advantage of small machines in a stochastic fluid production process. Mathematical Methods of Operations Research **47**, 83–97 (1998)
60. Bäuerle, N.: Asymptotic optimality of tracking policies in stochastic networks. Annals of Applied Probability **10**, 1065–1083 (2000)
61. Bäuerle, N., Rieder, U.: Comparison results for Markov-modulated recursive models. Probability in the Engineering and Informational Sciences **11**, 203–217 (1997)
62. Baxter, L.A.: Reliability applications of the relevation transform. Naval Research Logistics Quarterly **29**, 323–330 (1982)
63. Belzunce, F.: On a characterization of right spread order by the increasing convex order. Statistics and Probability Letters **45**, 103–110 (1999)
64. Belzunce, F., Candel, J., Ruiz, J.M.: Dispersive ordering and characterizations of ageing classes. Statistics and Probability Letters **28**, 321–327 (1996)
65. Belzunce, F., Candel, J., Ruiz, J.M.: The ageing curve and partial orderings of life distributions. Statistical Papers **37**, 141–152 (1996)
66. Belzunce, F., Franco, M., Ruiz, J.-M., Ruiz, M.C.: On partial orderings between coherent systems with different structures. Probability in the Engineering and Informational Sciences **15**, 273–293 (2001)
67. Belzunce, F., Gao, X., Hu, T., Pellerey, F.: Characterizations of the hazard rate order and IFR aging notion. Statistics and Probability Letters **70**, 235–242 (2004)
68. Belzunce, F., Hu, T., Khaledi, B.-E.: Dispersion-type variability orders. Probability in the Engineering and Informational Sciences **17**, 305–334 (2003)
69. Belzunce, F., Lillo, R.E., Ruiz, J.M., Shaked, M.: Stochastic comparisons of nonhomogeneous processes. Probability in the Engineering and Informational Sciences **15**, 199–224 (2001)
70. Belzunce, F., Mercader, J.-A., Ruiz, J.-M.: Stochastic comparisons of generalized order statistics. Probability in the Engineering and Informational Sciences **19**, 99–120 (2005)
71. Belzunce, F., Ortega, E.-M., Ruiz, J.M.: A note on replacement policy comparisons from NBUC lifetime of the unit. Statistical Papers **46**, 509–522 (2005)
72. Belzunce, F., Pellerey, F., Ruiz, J.M., Shaked, M.: The dilation order, the dispersion order, and orderings of residual lives. Statistics and Probability Letters **33**, 263–275 (1997)
73. Belzunce, F., Ruiz, J.M.: Multivariate dispersive ordering of epoch times of nonhomogeneous Poisson processes. Journal of Applied Probability **39**, 637–643 (2002)
74. Belzunce, F., Ruiz, J.M., Ruiz, M.C.: On preservation of some shifted and proportional orders by systems. Statistics and Probability Letters **60**, 141–154 (2002)
75. Belzunce, F., Ruiz, J.-M., Ruiz, M.-d.-C.: Multivariate properties of random vectors of order statistics. Journal of Statistical Planning and Inference **115**, 413–424 (2003)
76. Belzunce, F., Ruiz, J.M., Suárez-Llorens, A.: Multivariate conditional dispersive order. Technical Report, Dpto. Estadística e Investigación Operativa, Universidad de Cádiz, Spain (2005)

77. Belzunce, F., Semeraro, P.: Preservation of positive and negative orthant dependence concepts under mixtures and applications. Journal of Applied Probability **41**, 961–974 (2004)
78. Belzunce, F., Shaked, M.: Failure profiles of coherent systems. Naval Research Logistics **51**, 477–490 (2004)
79. Berger, E.: Majorization, exponential inequalities and almost sure behavior of vector-valued random variables. Annals of Probability **19**, 1206–1226 (1991)
80. Bergmann, R.: Some classes of semi-ordering relations for random vectors and their use for comparing covariances. Mathematische Nachrichten **82**, 103–114 (1978)
81. Bergmann, R.: Some classes of distributions and their application in queueing. Mathematische Operationsforschung und Statistik. Series Statistics **10**, 583–600 (1979)
82. Bergmann, R.: Stochastic orders and their applications to a unified approach to various concepts of dependence and association. In: Mosler, K., Scarsini, M. (ed) Stochastic Orders and Decision under Risk, IMS Lecture Notes—Monograph Series 19. Hayward, California, 48–73 (1991)
83. Bhattacharjee, M.C.: Some generalized variability orderings among life distributions with reliability applications. Journal of Applied Probability **28**, 374–383 (1991)
84. Bhattacharjee, M.C.: Exponentiality within class $\mathcal{L}$ and stochastic equivalence of Laplace ordered survival times. Probability in the Engineering and Informational Sciences **13**, 201–207 (1999)
85. Bhattacharjee, M.C.: Discrete convex ordered lifetimes: Characterization, equivalence and applications. Sankhyā **65**, 292–306 (2003)
86. Bhattacharjee, M.C.: Comparisons of random sums in some integral orderings and applications. Sankhyā **66**, 450–465 (2004)
87. Bhattacharjee, M.C., Bhattacharya, R.N.: Stochastic equivalence of convex ordered distributions and applications. Probability in the Engineering and Informational Sciences **14**, 33–48 (2000)
88. Bhattacharjee, M.C., Sethuraman, J.: Families of life distributions characterized by two moments. Journal of Applied Probability **27**, 720–725 (1990)
89. Bickel, P.J., Lehmann, E.L.: Dispersive statistics for nonparametric models. III. Dispersion. Annals of Statistics **4**, 1139–1158 (1976)
90. Birnbaum, Z.W.: On random variables with comparable peakedness. Annals of Mathematical Statistics **19**, 76–81 (1948)
91. Block, H.W., Bueno, V., Savits, T.H., Shaked, M.: Probability inequalities via negative dependence for random variables conditioned on order statistics. Naval Research Logistics **34**, 547–554 (1987)
92. Block, H.W., Chhetry, D., Fang, Z., Sampson, A.R.: Partial orders on permutations and dependence orderings on bivariate empirical distributions. Annals of Statistics **18**, 1840–1850 (1990)
93. Block, H.W., Langberg, N.A., Savits, T.H.: Comparisons for maintenance policies involving complete and minimal repair. In: Block, H.W., Sampson, A.R., Savits, T.H. (ed) Topics in Statistical Dependence. IMS Lecture Notes—Monograph Series 16. Hayward, California, 57–68 (1990)
94. Block, H.W., Sampson, A.R.: Conditionally ordered distributions. Journal of Multivariate Analysis **27**, 91–104 (1988)
95. Block, H.W., Savits, T.H., Shaked, M.: Some concepts of negative dependence. Annals of Probability **10**, 765–772 (1982)

96. Block, H.W., Savits, T.H., Singh, H.: The reversed hazard rate function. Probability in the Engineering and Informational Sciences **12**, 69–90 (1998)
97. Boland, P.J., El-Neweihi, E., Proschan, F.: Applications of the hazard rate ordering in reliability and order statistics. Journal of Applied Probability **31**, 180–192 (1994)
98. Boland, P.J., El-Neweihi, E., Proschan, F.: Schur properties of convolutions of exponential and geometric random variables. Journal of Multivariate Analysis **48**, 157–167 (1994)
99. Boland, P.J., Hu, T., Shaked, M., Shanthikumar, J.G.: Stochastic ordering of order statistics II. In: Dror, M., L'Ecuyer, P., Szidarovszky, F. (ed) Modeling Uncertainty: An Examination of Stochastic Theory, Methods, and Applications. Kluwer, Boston, 607–623 (2002)
100. Boland, P.J., Proschan, F., Tong, Y.L.: A stochastic ordering of partial sums of independent random variables and of some stochastic processes. Journal of Applied Probability **29**, 645–654 (1992)
101. Boland, P.J., Shaked, M., Shanthikumar, J.G.: Stochastic ordering of order statistics. In: Balakrishnan, N., Rao, C.R. (ed) Handbook of Statistics 16: Order Statistics: Theory and Methods. Elsevier, Amsterdam, 89–103 (1998)
102. Boland, P.J., Singh, H., Cukic, B.: Stochastic order in partition and random testing of software. Journal of Applied Probability **39**, 555–565 (2002)
103. Boland, P.J., Singh, H., Cukic, B.: The stochastic precedence ordering with applications in sampling and testing. Journal of Applied Probability **41**, 73–82 (2004)
104. Bon, J.-L., Illayk, A.: Ageing properties and series systems. Journal of Applied Probability **42**, 279–286 (2005)
105. Bon, J.-L., Păltănea, E.: Ordering properties of convolutions of exponential random variables. Lifetime Data Analysis **5**, 185–192 (1999)
106. Borglin, A., Keiding, H.: Stochastic dominance and conditional expectation—an insurance theoretical approach. Geneva Papers on Risk and Insurance Theory **27**, 31–48 (2002)
107. Boutsikas, M.V., Vaggelatou, E.: On the distance between convex-ordered random variables, with applications. Advances in Applied Probability **34**, 349–374 (2002)
108. Braverman, M.: A remark on stochastic monotonicity. Statistics and Probability Letters **26**, 259–262 (1996)
109. Brockett, P.L., Kahane, Y.: Risk, return, skewness and preference. Management Science **38**, 851–866 (1992)
110. Brown, L.D., Rinott, Y.: Inequalities for multivariate infinitely divisible processes. Annals of Probability **16**, 642–657 (1988)
111. Brown, M.: Bounds, inequalities, and monotonicity properties for some specialized renewal processes. Annals of Probability **8**, 227–240 (1980)
112. Brown, M., Shanthikumar, J.G.: Comparing the variability of random variables and point processes. Probability in the Engineering and Informational Sciences **12**, 425–444 (1998)
113. Bueno, V.: Models in Negative Dependence through Stochastic Ordering for Random Variables Conditioned on Order Statistics. Ph.D. Thesis. Department of Statistics, University of Pittsburgh, Pittsburgh, Pennsylvania (1985)
114. Bulinski, A., Suquet, C.: Normal approximation for quasi-associated random fields. Statistics and Probability Letters **54**, 215–226 (2001)

115. Burger, H.U.: Dispersion orderings with applications to nonparametric tests. Statistics and Probability Letters **16**, 1–9 (1993)
116. Cai, J., Wu, Y.: Characterization of life distributions under some generalized stochastic orderings. Journal of Applied Probability **34**, 711–719 (1997)
117. Cao, J., Wang, Y.: The NBUC and NWUC classes of life distributions. Journal of Applied Probability **28**, 473–479 (1991)
118. Capéraà, P.: Tail ordering and asymptotic efficiency of rank tests. Annals of Statistics **16**, 470–478 (1988)
119. Capéraà, P., Fougères, A.-L., Genest, C.: A stochastic ordering based on a decomposition of Kendall's Tau. In: Beneš, V., Štěpán, J. (ed) Distributions with given Marginals and Moment Problems. Kluwer, Boston, 81–86 (1997)
120. Capéraà, P., Genest, C.: Concepts de dépendance et ordres stochastiques pour des lois bidimensionnelles. Canadian Journal of Statistics **18**, 315–326 (1990)
121. Carletti, M., Pellerey, F.: A new necessary condition for higher order stochastic dominances with applications. Ricerche di Matematica **47**, 373–381 (1998)
122. Chacon, R.V., Walsh, J.B.: One-dimensional potential embedding. In: Meyer, P.A. (ed) Séminaire de Probabilités X, Université de Strasbourg, Lecture Notes in Mathematics 511. Springer Verlag, Berlin, 19–23 (1976)
123. Chan, W., Proschan, F., Sethuraman, J.: Convex-ordering among functions, with applications to reliability and mathematical statistics. In Block, H.W., Sampson, A.R., Savits, T.H. (ed) Topics in Statistical Dependence. IMS Lecture Notes—Monograph Series 16. Hayward, California, 121–134 (1990)
124. Chang, C.-S.: A new ordering for stochastic majorization: Theory and applications. Advances in Applied Probability **24**, 604–634 (1992)
125. Chang, C.-S., Chao, X.L., Pinedo, M., Shanthikumar, J.G.: Stochastic convexity for multidimensional processes and its applications. IEEE Transactions on Automatic Control **36**, 1347–1355 (1991)
126. Chang, K.-H.: Stochastic orders of the sums of two exponential random variables. Statistics and Probability Letters **51**, 389–396 (2001)
127. Chateauneuf, A., Cohen, M., Meilijson, I.: Four notions of mean-preserving increase in risk, risk attitudes and applications to the rank-dependent expected utility model. Journal of Mathematical Economics **40**, 547–571 (2004)
128. Cheng, D.W., Righter, R.: On the order of tandem queues. Queueing Systems **21**, 143–160 (1995)
129. Cheng, Y., Pai, J.F.: On the nth stop-loss transform order of ruin probability. Insurance: Mathematics and Economics **32**, 51–60 (2003)
130. Christofides, T.C., Vaggelatou, E.: A connection between supermodular ordering and positive/negative association. Journal of Multivariate Analysis **88**, 138–151 (2004)
131. Cohen, A., Sackrowitz, H.B.: On stochastic ordering of random vectors. Journal of Applied Probability **32**, 960–965 (1995)
132. Colangelo, A.: Some positive dependence orderings involving tail dependence. Technical Report. Università dell'Insubria, Varese, Italy (2005)
133. Colangelo, A., Scarsini, M., Shaked, M.: Some positive dependence stochastic orders. Journal of Multivariate Analysis **97**, 46–78 (2006)
134. Collet, P., López, F.J., Martínez, S.: Order relations of measures when avoiding decreasing sets. Statistics and Probability Letters **65**, 165–175 (2003)
135. Costantini, C., Pasqualucci, D.: Monotonicity of Bayes sequential tests for multidimensional and censored observations. Journal of Statistical Planning and Inference **75**, 117-131 (1998)

136. Cramer, E., Kamps, U.: Sequential k-out-of-n systems. In: Balakrishnan, N., Rao, C.R. (ed) Handbook of Statistics 20: Advances in Reliability. Elsevier, Amsterdam, 301–372 (2001)
137. Daduna, H., Szekli, R.: A queueing theoretical proof of increasing property of Polya frequency functions. Statistics and Probability Letters **26**, 233–242 (1996)
138. Daduna, H., Szekli, R.: Dependence structure of sojourn times via partition separated ordering. Operations Research Letters **31**, 462–472 (2003)
139. Das Gupta, S., Eaton, M.L., Olkin, I., Perlman, M., Savage, L.J., Sobel, M.: Inequalities on the probability content of convex regions for elliptically contoured distributions. In: LeCam, L.M., Neyman, J., Scott, E.L. (ed) Proceedings of the Sixth Berkeley Symposium on Mathematical Statistics and Probability, Volume II. University of California Press, Berkeley, 241–265 (1972)
140. Denuit, M.: Time stochastic s-convexity of claim processes. Insurance: Mathematics and Economics **26**, 203–211 (2000)
141. Denuit, M.: Laplace transform ordering of actuarial quantities. Insurance: Mathematics and Economics **29**, 83–102 (2001)
142. Denuit, M., De Vylder, E., Lefèvre, C.: Extremal generators and extremal distributions for the continuous s-convex stochastic orderings. Insurance: Mathematics and Economics **24**, 201–217 (1999)
143. Denuit, M., Dhaene, J., Ribas, C.: Does positive dependence between individual risks increase stop-loss premiums? Insurance: Mathematics and Economics **28**, 305–308 (2001)
144. Denuit, M., Frostig, E.: Heterogeneity and the need for capital in the individual model. Scandinavian Actuarial Journal, 42–66 (2006)
145. Denuit, M., Genest, C., Marceau, É.: Criteria for the stochastic ordering of random sums, with actuarial applications. Scandinavian Actuarial Journal, 3–16 (2002)
146. Denuit, M., Lefèvre, C.: Some new classes of stochastic order relations among arithmetic random variables, with applications in actuarial sciences. Insurance: Mathematics and Economics **20**, 197–213 (1997)
147. Denuit, M., Lefèvre, C.: Stochastic s-increasing convexity. In: Hadjisavvas, N., Martínez-Legaz, J., Penot, J.-P. (ed) Generalized Convexity and Generalized Monotonicity. Proceedings of the 6th International Symposium, Samos, Greece, September 1999. Springer, Berlin, 167–182 (2001)
148. Denuit, M., Lefèvre, C., Mesfioui, M.: A class of bivariate stochastic orderings, with applications in actuarial sciences. Insurance: Mathematics and Economics **24**, 31–50 (1999)
149. Denuit, M., Lefèvre, C., Mesfioui, M.: On s-convex stochastic extrema for arithmetic risks. Insurance: Mathematics and Economics **25**, 143–155 (1999)
150. Denuit, M., Lefèvre, C., Mesfioui, M.: Stochastic orderings of convex-type for discrete bivariate risks. Scandinavian Actuarial Journal, 32–51 (1999)
151. Denuit, M., Lefèvre, C., Shaked, M.: The s-convex orders among real random variables, with applications. Mathematical Inequalities and Applications **1**, 585–613 (1998)
152. Denuit, M., Lefèvre, C., Shaked, M.: On the theory of high convexity stochastic orders. Statistics and Probability Letters **47**, 287–293 (2000)
153. Denuit, M., Lefèvre, C., Shaked, M.: Stochastic convexity of the Poisson mixture model. Methodology and Computing in Applied Probability **2**, 231–254 (2000)

154. Denuit, M., Lefèvre, C., Shaked, M.: On s-convex approximations. Advances in Applied Probability **32**, 994–1010 (2000)
155. Denuit, M., Lefèvre, C., Utev, S.: Generalized stochastic convexity and stochastic orderings of mixtures. Probability in the Engineering and Informational Sciences **13**, 275–291 (1999)
156. Denuit, M., Lefèvre, C., Utev, S.: Stochastic orderings of convex/concave-type on an arbitrary grid. Mathematics of Operations Research **24**, 835–846 (1999)
157. Denuit, M., Müller, A.: Smooth generators of integral stochastic orders. Annals of Applied Probability **12**, 1174–1184 (2002)
158. Denuit, M., Vermandele, C.: Optimal reinsurance and stop-loss order. Insurance: Mathematics and Economics **22**, 229–233 (1998)
159. Deshpande, J.V., Kochar, S.C.: Dispersive ordering is the same as tail-ordering. Advances in Applied Probability **15**, 686–687 (1983)
160. Deshpande, J.V., Kochar, S.C., Singh, H.: Aspects of positive ageing. Journal of Applied Probability **23**, 748–758 (1986)
161. Deshpande, J.V., Singh, H., Bagai, I., Jain, K.: Some partial orders describing positive ageing. Communications in Statistics—Stochastic Models **6**, 471–481 (1990)
162. Dhaene, J., Denuit, M.: The safest dependence structure among risks. Insurance: Mathematics and Economics **25**, 11–21 (1999)
163. Di Crescenzo, A.: Dual stochastic orderings describing ageing properties of devices of unknown age. Communications in Statistics—Stochastic Models **15**, 561–576 (1999)
164. Di Crescenzo, A.: A probabilistic analogue of the mean value theorem and its applications to reliability theory. Journal of Applied Probability **36**, 706–719 (1999)
165. Di Crescenzo, A., Longobardi, M.: The up reversed hazard rate stochastic order. Scientiae Mathematicae Japonicae **54**, 575–581 (2001)
166. Di Crescenzo, A., Pellerey, F.: On lifetimes in random environments. Naval Research Logistics **45**, 365–375 (1998)
167. Di Crescenzo, A., Shaked, M.: Some applications of the Laplace transform ratio order. Arab Journal of Mathematical Sciences **2**, 121–128 (1996)
168. Ding, Y., Zhang, X.: Some stochastic orders of Kotz-type distributions. Statistics and Probability Letters **69**, 389–396 (2004)
169. Doksum, K.: Star-shaped transformations and the power of rank tests. Annals of Mathematical Statistics **40**, 1167–1176 (1969)
170. Downey, P.J., Maier, R.S.: Stochastic ordering and the mean growth of extremes. Technical Report. Department of Mathematics, University of Arizona, Tucson (1991)
171. Droste, W., Wefelmeyer, W.: A note on strong unimodality and dispersivity. Journal of Applied Probability **22**, 235–239 (1985)
172. Dubra, J., Maccheroni, F., Ok, E.A.: Expected utility theory without the completeness axiom. Journal of Economic Theory **115**, 118–133 (2004)
173. Dyckerhoff, R., Mosler, K.: Orthant ordering of discrete random vectors. Journal of Statistical Planning and Inference **62**, 193–205 (1997)
174. Dykstra, R., Kochar, S., Rojo, J.: Stochastic comparisons of parallel systems of heterogeneous exponential components. Journal of Statistical Planning and Inference **65**, 203–211 (1997)
175. Ebrahimi, N.: The ordering of negative quadrant dependence. Communications in Statistics—Theory and Methods **11**, 2389–2399 (1982)

176. Ebrahimi, N.: Ordering repairable systems. Journal of Applied Probability **27**, 193–201 (1990)
177. Ebrahimi, N.B., Pellerey, F.: New partial ordering of survival functions based on notion of uncertainty. Journal of Applied Probability **32**, 202–211 (1995)
178. Ebrahimi, N.B., Spizzichino, F.: Some results on normalized total time on test and spacings. Statistics and Probability Letters **36**, 231–243 (1997)
179. Ebrahimi, N.B., Zahedi, H.: Memory ordering of survival functions. Statistics **23**, 337–345 (1992)
180. Eeckhoudt, L., Gollier, C.: Demand for risky assets and monotone probability ratio order. Journal of Risk and Uncertainty **11**, 113-122 (1995)
181. Efron, B.: Increasing properties of Pólya frequency functions. Annals of Mathematical Statistics **36**, 272–279 (1965)
182. Elton, J., Hill, T.P.: Fusions of probability distributions. Annals of Probability **20**, 421–454 (1992)
183. Elton, J., Hill, T.P.: On the basic representation theorem for convex domination of measures. Journal of Mathematical Analysis and Applications **228**, 449–466 (1998)
184. Esary, J.D., Proschan, F., Walkup, D.W.: Association of random variables with applications. Annals of Mathematical Statistics **38**, 1466–1474 (1967)
185. Fagiuoli, E., Pellerey, F.: New partial orderings and applications. Naval Research Logistics **40**, 829–842 (1993)
186. Fagiuoli, E., Pellerey, F.: Mean residual life and increasing convex comparison of shock models. Statistics and Probability Letters **20**, 337–345 (1994)
187. Fagiuoli, E., Pellerey, F.: Moment inequalities for sums of DMRL random variables. Journal of Applied Probability **34**, 525–535 (1997)
188. Fagiuoli, E., Pellerey, F., Shaked, M.: A characterization of the dilation order and its applications. Statistical Papers **40**, 393–406 (1999)
189. Fahmy, S., Pereira, A.d.B., Proschan, F., Shaked, M.: The influence of the sample on the posterior distribution. Communications in Statistics—Theory and Methods **11**, 1757–1768 (1982)
190. Fan, K., Lorentz, G.G.: An integral inequality. American Mathematical Monthly **61**, 626–631 (1954)
191. Fang, Z., Hu, T., Joe, H.: On the decrease in dependence with lag for stationary Markov chains. Probability in the Engineering and Informational Sciences **8**, 385–401 (1994)
192. Fang, Z., Joe, H.: Further developments on some dependence orderings for continuous bivariate distributions. Annals of the Institute of Statistical Mathematics **44**, 501–517 (1992)
193. Fellman, J.: The redistributive effect of tax policies. Sankhyā, Series A **64**, 90–100 (2002)
194. Fernández, I.C., Molchanov, I.: A stochastic order for random vectors and random sets based on the Aumann expectation. Statistics and Probability Letters **63**, 295–305 (2003)
195. Fernandez-Ponce, J.M., Kochar, S.C., Muñoz-Perez, J.: Partial orderings of distributions based on right-spread functions. Journal of Applied Probability **35**, 221–228 (1998)
196. Fernández-Ponce, J.M., Rodríguez-Griñolo, R.: Preserving multivariate dispersion: An application to the Wishart distribution. Journal of Multivariate Analysis **97**, 1208–1220 (2006)

197. Fernández-Ponce, J.M., Suárez-Llorens, A.: An aging concept based on majorization. Probability in the Engineering and Informational Sciences **17**, 107–117 (2003)
198. Fernandez-Ponce, J.M., Suarez-Llorens, A.: A multivariate dispersion ordering based on quantiles more widely separated. Journal of Multivariate Analysis **85**, 40–53 (2003)
199. Ferreira, F., Pacheco, A.: Level-crossing ordering of semi-Markov processes and Markov chains. Journal of Applied Probability **42**, 989–1002 (2005)
200. Fill, J.A., Machida, M.: Stochastic monotonicity and realizable monotonicity. Annals of Probability **29**, 938–978 (2001)
201. Fishburn, P.C.: Continua of stochastic dominance relations for bounded probability distributions. Journal of Mathematical Economics **3**, 295–311 (1976)
202. Fishburn, P.C. Continua of stochastic dominance relations for unbounded probability distributions. Journal of Mathematical Economics **7**, 271–285 (1980)
203. Fishburn, P.C.: Third-degree stochastic dominance and random variables. Economic Letters **19**, 113–117 (1985)
204. Fishburn, P.C., Lavalle, I.H.: Stochastic dominance on unidimensional grids. Mathematics of Operations Research **20**, 513–525 (1995)
205. Franco, M., Ruiz, J.M., Ruiz, M.C.: Stochastic orderings between spacings of generalized order statistics. Probability in the Engineering and Informational Sciences **16**, 471–484 (2002)
206. Frostig, E.: A comparison between homogeneous and heterogeneous portfolios. Insurance: Mathematics and Economics **29**, 59–71 (2001)
207. Frostig, E.: On risk dependence and mrl ordering. Statistics and Probability Letters **76**, 231–243 (2006)
208. Genest, C., Verret, F.: The TP_2 ordering of Kimeldorf and Sampson has the normal-agreeing property. Statistics and Probability Letters **57**, 387–391 (2002)
209. Gerchak, Y., Golani, B.: Hiring policies in an uncertain environment: Cost and productivity trade-offs. European Journal of Operational Research **125**, 195–204 (2000)
210. Gerchak, Y., He, Q.-M.: Personal communication (1994)
211. Giovagnoli, A., Wynn, H.: Multivariate dispersion orderings. Statistics and Probability Letters **22**, 325–332 (1995)
212. Goldstein, L., Rinott, Y.: Multivariate normal approximations by Stein's method and size bias couplings. Journal of Applied Probability **33**, 1–17 (1996)
213. Goovaerts, M.J., Kaas, R.: Some problems in actuarial finance involving sums of independent risks. Statistica Neerlandica **56**, 253–269 (2002)
214. Gupta, P.L., Gupta, R.C.: Ageing characteristics of the Weibull mixtures. Probability in the Engineering and Informational Sciences **10**, 591–600 (1996)
215. Gupta, R.C., Gupta, R.D.: Generalized skew normal model. Test **13**, 501–524 (2004)
216. Gupta, R.C., Kirmani, S.N.U.A.: On order relations between reliability measures. Communications in Statistics—Stochastic Models **3**, 149–156 (1987)
217. Gupta, R.C., Kirmani, S.N.U.A.: Closure and monotonicity properties of nonhomogeneous Poisson processes and record values. Probability in the Engineering and Informational Sciences **2**, 475–484 (1988)
218. Gupta, R.D., Richards, D.St.P.: Multivariate Liouville distributions, III. Journal of Multivariate Analysis **43**, 29–57 (1992)

219. Heilmann, W.-R., Schröter, K.J.: Orderings of risks and their actuarial applications. In: Mosler, K., Scarsini, M. (ed) Stochastic Orders and Decision under Risk. IMS Lecture Notes—Monograph Series 19. Hayward, California, 157–173 (1991)
220. Hennessy, D.A.: Preferences over valuation distributions in auctions. Economics Letters **68**, 55–59 (2000)
221. Hesselager, O.: A unification of some order relations. Insurance: Mathematics and Economics **17**, 223–224 (1996)
222. Hesselager, O.: Closure properties of some partial orderings under mixing. Insurance: Mathematics and Economics **22**, 163–170 (1998)
223. Hickey, R.J.: Concepts of dispersion in distributions: A comparative note. Journal of Applied Probability **23**, 914–921 (1986)
224. Hoedemakers, T., Beirlant, J., Goovaerts, M.J., Dhaene, J.: Confidence bounds for discounted loss reserves. Insurance: Mathematics and Economics **33**, 297–316 (2003)
225. Hollander, M., Proschan, F., Sconing, J.: Information, censoring, and dependence. In: Block, H.W., Sampson, A.R., Savits, T.H. (ed) Topics in Statistical Dependence. IMS Lecture Notes—Monograph Series 16. Hayward, California, 257–268 (1990)
226. Holley, R.: Remarks on the FKG inequalities. Communications in Mathematical Physics **36**, 227–231 (1974)
227. de la Horra, J., Ruiz-Rivas, C.: A Bayesian method for inferring the degree of dependence for a positively dependent distributions. Communications in Statistics—Theory and Methods **17**, 4357–4370 (1988)
228. Hu, C.-Y., Lin, G.D.: Sharp reliability bounds for the $\mathcal{L}$-like distributions, a class in which ageing is based on Laplace transform inequalities. Journal of Applied Probability **42**, 1204–1208 (2005).
229. Hu, T.: Statistical dependence of multivariate distributions and stationary Markov chains with applications. Ph.D. Thesis, Department of Mathematics, University of Science and Technology of China (1994)
230. Hu, T.: Monotone coupling and stochastic ordering of order statistics. System Science and Mathematical Sciences **8**, 209–214 (1995)
231. Hu, T., Chen, J., Yao, J.: Preservation of the location independent risk order under convolution. Insurance: Mathematics and Economics **38**, 406–412 (2006)
232. Hu, T., He, F.: A note on comparisons of k-out-of-n systems with respect to the hazard and reversed hazard rate orders. Probability in the Engineering and Informational Sciences **14**, 27–32 (2000)
233. Hu, T., Hu, J.: Comparison of order statistics between dependent and independent random variables. Statistics and Probability Letters **37**, 1–6 (1998)
234. Hu, T., Joe, H.: Monotonicity of positive dependence with time for stationary reversible Markov chains. Probability in the Engineering and Informational Sciences **9**, 227–237 (1995)
235. Hu, T., Khaledi, B.-E., Shaked, M.: Multivariate hazard rate orders. Journal of Multivariate Analysis **84**, 173–189 (2003)
236. Hu, T., Kundu, A., Nanda, A.K.: On generalized orderings and ageing properties with their implications. In: Hayakawa, Y., Irony T., Xie, M. (ed) System and Bayesian Reliability. World Scientific, Singapore, 199–228 (2001)
237. Hu, T., Nanda, A.K., Xie, H., Zhu, Z.: Properties of some stochastic orders: A unified study. Naval Research Logistics **51**, 193–216 (2004)

238. Hu, T., Pan, X.: Preservation of multivariate dependence under multivariate claim models. Insurance: Mathematics and Economics **25**, 171–179 (1999)
239. Hu, T., Pan, X.: Comparisons of dependence for stationary Markov processes. Probability in the Engineering and Informational Sciences **14**, 299–315 (2000)
240. Hu, T., Wei, Y.: Stochastic comparisons of spacings from restricted families of distributions. Statistics and Probability Letters **53**, 91–99 (2001)
241. Hu, T., Xie, C., Ruan, L.: Dependence structures of multivariate Bernoulli random vectors. Journal of Multivariate Analysis **94**, 172–195 (2005)
242. Hu, T., Zhu, Z.: An analytic proof of the preservation of the up-shifted likelihood ratio order under convolutions. Stochastic Processes and Their Applications **95**, 55–61 (2001)
243. Hu, T., Zhu, Z., Wei, Y.: Likelihood ratio and mean residual life orders for order statistics of heterogenous random variables. Probability in the Engineering and Informational Sciences **15**, 259–272 (2001)
244. Hu, T., Zhuang, W.: Likelihood ratio ordering of the inspection paradox. Probability in the Engineering and Informational Sciences **18**, 503–510 (2004)
245. Hu, T., Zhuang, W.: Stochastic orderings between p-spacings of generalized order statistics from two samples. Technical Report. Department of Statistics and Finance, University of Science and Technology of China (2004)
246. Hu, T., Zhuang, W.: Stochastic properties of p-spacings of generalized order statistics. Probability in the Engineering and Informational Sciences **19**, 257–276 (2005)
247. Hu, T., Zhuang, W.: A note on stochastic comparisons of generalized order statistics. Statistics and Probability Letters **72**, 163–170 (2005)
248. Hu, T., Zhuang, W.: Stochastic comparisons of m-spacings. Journal of Statistical Planning and Inference **136**, 33–42 (2006)
249. Huang, J.S., Lin, G.D.: Equality in distribution in a convex ordering family. Annals of the Institute of Statistical Mathematics **51**, 345–349 (1999)
250. Huffer, F.W.: Slepian's inequality via the central limit theorem. Canadian Journal of Statistics **14**, 367–370 (1986)
251. Hürlimann, W.: Analytical evaluation of economic risk capital for portfolios of Gamma risks. ASTIN Bulletin **31**, 107–122 (2001)
252. Jacobsen, M.: Statistical Analysis of Counting Processes. Lecture Notes in Statistics 12. Springer-Verlag, New York (1982)
253. Jain, K., Nanda, A.K.: On multivariate weighted distributions. Communications in Statistics—Theory and Methods **24**, 2517–2539 (1995)
254. Jean-Marie, A., Liu, Z.: Stochastic comparisons for queueing models via random sums and intervals. Advances in Applied Probability **24**, 960–985 (1992)
255. Jeon, J., Kochar, S., Park, C.G.: Dispersive ordering—Some applications and examples. Statistical Papers **47**, 227–247 (2006)
256. Jewitt, I.: Choosing between risky prospects: The characterization of comparative statics results, and location independent risk. Management Science **35**, 60–70 (1989)
257. Joag-Dev, K.: Conditional negative dependence in stochastic ordering and interchangeable random variables. In: Block, H.W., Sampson, A.R., Savits, T.H. (ed) Topics in Statistical Dependence. IMS Lecture Notes—Monograph Series 16. Hayward, California, 295–298 (1990)
258. Joag-Dev, K.: Personal communication (1995)

259. Joag-Dev, K., Kochar, S., Proschan, F.: A general composition theorem and its applications to certain partial orderings of distributions. Statistics and Probability Letters **22**, 111–119 (1995)
260. Joe, H.: Multivariate concordance. Journal of Multivariate Analysis **35**, 12–30 (1990)
261. Joe, H.: Parametric families of multivariate distributions with given marginals. Journal of Multivariate Analysis **46**, 262–282 (1993)
262. Joe, H.: Multivariate Models and Dependence Concepts. Chapman and Hall, London (1997)
263. Johnson, N.L., Kotz, S.: Distributions in Statistics, Discrete Univariate Distributions. Wiley, New York (1969)
264. Johnson, N.L., Kotz, S.: A vector multivariate hazard rate. Journal of Multivariate Analysis **5**, 53–66 (1975)
265. Jun, C.: Characterizations of life distributions by moments of extremes and sample mean. Journal of Applied Probability **31**, 148–155 (1994)
266. Juri, A.: Supermodular order and Lundberg exponents. Scandinavian Actuarial Journal, 17–36 (2002)
267. Kaas, R., Dhaene, J., Goovaerts, M.J.: Upper and lower bounds for sums of random variables. Insurance: Mathematics and Economics **27**, 151–168 (2000)
268. Kaas, R., Dhaene, J., Vyncke, D., Goovaerts, M.J., Denuit, M.: A simple geometric proof that comonotonic risks have the convex-largest sum. ASTIN Bulletin **32**, 71–80 (2002)
269. Kaas, R., Heerwaarden, A.E. van, Goovaerts, M.J.: Ordering of Actuarial Risks. CAIRE, Brussels (1994)
270. Kaas, R., Hesselager, O.: Ordering claim size distributions and mixed Poisson probabilities. Insurance: Mathematics and Economics **17**, 193–201 (1995)
271. Kalashnikov, V.V., Rachev, S.T.: Mathematical Methods for Construction of Queueing Models. Wadsworth and Brooks, Pacific Grove, California (1990)
272. Kamae, T., Krengel, U., O'Brien, G.L.: Stochastic inequalities on partially ordered spaces. Annals of Probability **5**, 899–912 (1977)
273. Kamps, U.: A Concept of Generalized Order Statistics. B.G. Taubner, Stuttgart (1995)
274. Kan, C., Yi, W.W.: Preservation of some partial orderings under nonhomogeneous Poisson shock model and Laplace transform, Microelectronics Reliability **37**, 451–455 (1997)
275. Karlin, S.: Total Positivity. Stanford University Press, Palo Alto, California (1968)
276. Karlin, S.: Inequalities for symmetric sampling plans I. Annals of Statistics **2**, 1065–1094 (1974)
277. Karlin, S., Novikoff, A.: Generalized convex inequalities. Pacific Journal of Mathematics **13**, 1251–1279 (1963)
278. Karlin, S., Rinott, Y.: Classes of orderings of measures and related correlation inequalities. I. Multivariate totally positive distributions. Journal of Multivariate Analysis **10**, 467–498 (1980)
279. Karlin, S., Rinott, Y.: Classes of orderings of measures and related correlation inequalities. II. Multivariate reverse rule distributions. Journal of Multivariate Analysis **10**, 499–510 (1980)
280. Kebir, Y.: On hazard rate processes. Naval Research Logistics **38**, 865–876 (1991)

281. Kebir, Y.: Laplace transform characterization of probabilistic orderings. Probability in the Engineering and Informational Sciences **8**, 69–77 (1994)
282. Kebir, Y.: Order-preserving shock models. Probability in the Engineering and Informational Sciences **8**, 125–134 (1994)
283. Keilson, J., Sumita, U.: Uniform stochastic ordering and related inequalities. Canadian Journal of Statistics **10**, 181–198 (1982)
284. Kemperman, J.H.B.: On the FKG inequality for measures on a partially ordered space. Indagationes Mathematicae **39**, 313–331 (1977)
285. Khaledi, B.-E., Kochar, S.: Stochastic orderings between distributions and their sample spacings—II. Statistics and Probability Letters **44**, 161–166 (1999)
286. Khaledi, B.-E., Kochar, S.: On dispersive ordering between order statistics in one-sample and two-sample problems. Statistics and Probability Letters **46**, 257–261 (2000)
287. Khaledi, B.-E., Kochar, S.: Some new results on stochastic comparisons of parallel systems. Journal of Applied Probability **37**, 1123–1128 (2000)
288. Khaledi, B.-E., Kochar, S.: Dispersive ordering among linear combinations of uniform random variables. Journal of Statistical Planning and Inference **100**, 13–21 (2002)
289. Khaledi, B.-E., Kochar, S.C.: Ordering convolutions of Gamma random variables. Sankhyā **66**, 466–473 (2004)
290. Khaledi, B.-E., Kochar, S.: Dependence, dispersiveness, and multivariate hazard rate ordering. Probability in the Engineering and Informational Sciences **19**, 427–446 (2005)
291. Kijima, M.: Uniform monotonicity of Markov processes and its related properties. Journal of the Operations Research Society of Japan **32**, 475–490 (1989)
292. Kijima, M., Ohnishi, M.: Portfolio selection problems via the bivariate characterization of stochastic dominance relations. Mathematical Finance **6**, 237–277 (1996)
293. Kijima, M., Ohnishi, M.: Stochastic orders and their applications in financial optimization. Mathematical Methods of Operations Research **50**, 351–372 (1999)
294. Kim, J.S., Proschan, F.: Laplace ordering and its reliability applications. Technical Report. Department of Statistics, Florida State University, Tallahassee (1988)
295. Kimeldorf, G., Sampson, A.R.: Positive dependence orderings. Annals of the Institute of Statistical Mathematics **39**, 113–128 (1987)
296. Kimeldorf, G., Sampson, A.R.: A framework for positive dependence. Annals of the Institute of Statistical Mathematics **41**, 31–45 (1989)
297. Kirmani, S.N.U.A.: On sample spacings from IMRL distributions. Statistics and Probability Letters **29**, 159–166 (1996)
298. Kirmani, S.N.U.A.: Erratum: On sample spacings from IMRL distributions. Statistics and Probability Letters **37**, 315 (1998)
299. Kirmani, S.N.U.A., Gupta, R.C.: Some moment inequalities for the minimal repair process. Probability in the Engineering and Informational Sciences **6**, 245–255 (1992)
300. Klar, B.: A note on the $\mathcal{L}$-class of life distributions. Journal of Applied Probability **39**, 11–19 (2002)
301. Klar, B., Müller, A.: Characterizations of classes of lifetime distributions generalizing the NBUE class. Journal of Applied Probability **40**, 20–32 (2003)

302. Klefsjö, B.: A useful ageing property based on the Laplace transform. Journal of Applied Probability **20**, 615–626 (1983)
303. Kleiber, C.: Variability ordering of heavy-tailed distributions with applications to order statistics. Statistics and Probability Letters **58**, 381–388 (2002)
304. Kleiber, C.: Lorenz ordering of order statistics from log-logistic and related distributions. Journal of Statistical Planning and Inference **120**, 13–19 (2004)
305. Kochar, S.C.: Distribution-free comparison of two probability distributions with reference to their hazard rates. Biometrika **66**, 437–441 (1979)
306. Kochar, S.C.: On extensions of DMRL and related partial orderings of life distributions. Communications in Statistics—Stochastic Models **5**, 235–245 (1989)
307. Kochar, S.C.: Some partial ordering results on record values. Communications in Statistics—Theory and Methods **19**, 299–306 (1990)
308. Kochar, S.C.: Dispersive ordering of order statistics. Statistics and Probability Letters **27**, 271–274 (1996)
309. Kochar, S.C.: Some results on interarrival times of nonhomogeneous Poisson processes. Probability in the Engineering and Informational Sciences **10**, 75–85 (1996)
310. Kochar, S.C.: A note on dispersive ordering of record values. Calcutta Statistical Association Bulletin **46**, 63–67 (1996)
311. Kochar, S.C.: On stochastic orderings between distributions and their sample spacings. Statistics and Probability Letters **42**, 345–352 (1999)
312. Kochar, S.C., Carrière, K.C.: Connections among various variability orderings. Statistics and Probability Letters **35**, 327–333 (1997)
313. Kochar, S.C., Kirmani, S.N.U.A.: Some results on normalized spacings from restricted families of distributions. Journal of Statistical Planning and Inference **46**, 47–57 (1995)
314. Kochar, S.C., Korwar, R.: Stochastic orders for spacings of heterogeneous exponential random variables. Journal of Multivariate Analysis **57**, 69–83 (1996)
315. Kochar, S.C., Korwar, R.: Stochastic properties of order statistics from finite populations. Sankhyā, Series A **59**, 102–116 (1997)
316. Kochar, S.C., Li, X., Shaked, M.: The total time on test transform and the excess wealth stochastic orders of distributions. Advances in Applied Probability **34**, 826–845 (2002)
317. Kochar, S., Ma, C.: Dispersive ordering of convolutions of exponential random variables. Statistics and Probability Letters **43**, 321–324 (1999)
318. Kochar, S., Rojo, J.: Some new results on stochastic comparisons of spacings from heterogeneous exponential distributions. Journal of Multivariate Analysis **59**, 272–281 (1996)
319. Kochar, S.C., Wiens, D.P.: Partial orderings of life distributions with respect to their aging properties. Naval Research Logistics **34**, 823–829 (1987)
320. Kopocinska, I., Kopocinski, B.: The DMRL closure problem. Bulletin of the Polish Academy of Sciences, Mathematics **33**, 425–429 (1985)
321. Korwar, R.: On stochastic orders for the lifetime of a k-out-of-n system. Probability in the Engineering and Informational Sciences **17**, 137–142 (2003)
322. Korwar, R.: On the likelihood ratio order for progressive type II censored order statistics. Sankhyā **65**, 793–798 (2003)
323. Kottas, A., Gelfand, A.E.: Modeling variability order: A semiparametric Bayesian approach. Methodology and Computing in Applied Probability **3**, 427–442 (2001)

324. Ku, P.-S., Niu, S.-C.: On Johnson's two-machine flow shop with random processing times. Operations Research **34**, 130–136 (1986)
325. Kulik, R., Szekli, R.: Dependence orderings for some functionals of multivariate point processes. Journal of Multivariate Analysis **92**, 145–173 (2005)
326. Kurtz, T.G.: Representations of Markov processes as multiparameter time changes. Annals of Probability **8**, 682–715 (1980)
327. Kusum, K., Kochar, S.C., Deshpande, J.V.: Testing for the star-equivalence of two probability distributions. Journal of the Indian Statistical Association **24**, 21–30 (1986)
328. Kwieciński, A., Szekli, R.: Compensator conditions for stochastic ordering of point processes. Journal of Applied Probability **28**, 751–761 (1991)
329. Landsberger, M., Meilijson, I.: Mean-preserving portfolio dominance. Review of Economic Studies **60**, 479–485 (1993)
330. Landsberger, M., Meilijson, I.: The generating process and an extension of Jewitt's location independent risk concept. Management Science **40**, 662–669 (1994)
331. Landsman, Z., Tsanakas, A.: Stochastic ordering of bivariate elliptical distributions. Statistics and Probability Letters **76**, 488–494 (2006)
332. Langberg, N.A.: Comparison of replacement policies. Journal of Applied Probability **25**, 780–788 (1988)
333. Lapidoth, A., Moser, S.M.: Capacity bounds via duality with applications to multiple-antenna systems on flat-fading channels. IEEE Transactions on Information Theory **49**, 2426–2467 (2003)
334. Lawrence, M.J.: Inequalities of s-ordered distributions. Annals of Statistics **3**, 413–428 (1975)
335. Lefèvre, C.: Personal communication (1992)
336. Lefèvre, C., Malice, M.-P.: On a system of components with joint lifetimes distributed as a mixture of independent exponential laws. Journal of Applied Probability **26**, 202–208 (1989)
337. Lefèvre, C., Picard, P.: A multivariate stochastic ordering by the mixed descending factorial moments with applications. In: Shaked, M., Tong, Y.L. (ed) Stochastic Inequalities. IMS Lecture Notes—Monograph Series 22. Hayward, California, 235–252 (1992)
338. Lefèvre, C., Picard, P.: An unusual stochastic order relation with some applications in sampling and epidemic theory. Advances in Applied Probability **25**, 63–81 (1993)
339. Lefèvre, C., Utev, S.: Comparing sums of exchangeable Bernoulli random variables. Journal of Applied Probability **33**, 285–310 (1996)
340. Lefèvre, C., Utev, S.: Comparison of individual risk models. Insurance: Mathematics and Economics **28**, 21–30 (2001)
341. Lehmann, E.L.: Ordered families of distributions. Annals of Mathematical Statistics **26**, 399–419 (1955)
342. Lehmann, E.L.: Testing Statistical Hypotheses. Wiley, New York (1959)
343. Lehmann, E.L.: Some concepts of dependence. Annals of Mathematical Statistics **37**, 1137–1153 (1966)
344. Lehmann, E.L.: Comparing location experiments. Annals of Statistics **16**, 521–533 (1988)
345. Lehmann, E.L., Rojo, J.: Invariant directional orderings. Annals of Statistics **20**, 2100–2110 (1992)

346. Levy, H., Kroll, Y.: Ordering uncertain options with borrowing and lending. Journal of Finance **33**, 553–574 (1978)
347. Lewis, T., Thompson, J.W.: Dispersive distributions and the connection between dispersivity and strong unimodality. Journal of Applied Probability **18**, 76–90 (1981)
348. Li, H., Scarsini, M., Shaked, M.: Linkages: a tool for the construction of multivariate distributions with given nonoverlapping multivariate marginals. Journal of Multivariate Analysis **56**, 20–41 (1996)
349. Li, H., Shaked, M.: Stochastic convexity and concavity of Markov processes. Mathematics of Operations Research **19**, 477–493 (1994)
350. Li, H., Xu, S.H.: On the dependence structure and bounds of correlated parallel queues and their applications to synchronized stochastic systems. Journal of Applied Probability **37**, 1020–1043 (2000)
351. Li, H., Zhu, H.: Stochastic equivalence of ordered random variables with applications in reliability theory. Statistics and Probability Letters **20**, 383–393 (1994)
352. Li, X.: Some properties of ageing notions based on the moment-generating-function order. Journal of Applied Probability **41**, 927–934 (2004)
353. Li, X.: A note on expected rent in auction theory. Operations Research Letters **33**, 531–534 (2005)
354. Li, X., Li, Z., Jing, B.-Y.: Some results about the NBUC class of life distributions. Statistics and Probability Letters **46**, 229–237 (2000)
355. Li, X., Lu, J.: Stochastic comparisons on residual life and inactivity time of series and parallel systems. Probability in the Engineering and Informational Sciences **17**, 267–275 (2003)
356. Li, X., Shaked, M.: The observed total time on test and the observed excess wealth. Statistics and Probability Letters **68**, 247–258 (2004)
357. Li, X., Shaked, M.: A general family of univariate stochastic orders. Journal of Statistical Planning and Inference, to appear (2006)
358. Li, X., Zuo, M.J.: Preservation of stochastic orders for random minima and maxima, with applications. Naval Research Logistics **51**, 332–344 (2004)
359. Li, X., Zuo, M.J.: Stochastic comparison of residual life and inactivity time at a random time. Stochastic Models **20**, 229–235 (2004)
360. Liggett, T.M.: Monotonicity of conditional distributions and growth models on trees. Annals of Probability **28**, 1645–1665 (2000)
361. Lillo, R.E., Nanda, A.K., Shaked, M.: Some shifted stochastic orders. In: Limnios, N., Nikulin, M. (ed) Recent Advances in Reliability Theory. Birkhäuser, Boston, 85–103 (2000)
362. Lillo, R.E., Nanda, A.K., Shaked, M.: Preservation of some likelihood ratio stochastic orders by order statistics. Statistics and Probability Letters **51**, 111–119 (2001)
363. Lillo, R.E., Pellerey, F., Semeraro, P., Shaked, M.: On the preservation of the supermodular order under multivariate claim models. Ricerche di Matematica **52**, 73–81 (2003)
364. Lim, J.-H., Lu, K.-L., Park, D.-H.: Bayesian imperfect repair model. Communications in Statistics—Theory and Methods **27**, 965–984 (1998)
365. Loh, W.-Y.: Bounds on AREs for restricted classes of distributions defined via tail-ordering. Annals of Statistics **12**, 685–701 (1984)
366. Lu, T.-Y., Yi, Z.: Generalized correlation order and stop-loss order. Insurance: Mathematics and Economics **35**, 69–76 (2004)

367. Lynch, J.: Mixtures, generalized convexity and balayages. Scandinavian Journal of Statistics **15**, 203–210 (1988)
368. Lynch, J., Mimmack, G., Proschan, F.: Dispersive ordering results. Advances in Applied Probability **15**, 889–891 (1983)
369. Lynch, J., Mimmack, G., Proschan, F.: Uniform stochastic orderings and total positivity. Canadian Journal of Statistics **15**, 63–69 (1987)
370. Lynch, J.D.: When is a renewal process convexly parameterized in its mean parameterization? Probability in the Engineering and Informational Sciences **11**, 43–48 (1997)
371. Ma, C.: A note on stochastic ordering of order statistics. Journal of Applied Probability **34**, 785–789 (1997)
372. Ma, C.: On peakedness of distributions of convex combinations. Journal of Statistical Planning and Inference **70**, 51–56 (1998)
373. Ma, C.: Likelihood ratio ordering of order statistics. Journal of Statistical Planning and Inference **70**, 255–261 (1998)
374. Ma, C.: Uniform stochastic ordering on a system of components with dependent lifetimes induced by a common environment. Sankhyā, Series A **61**, 218–228 (1999)
375. Ma, C.: Convex orders for linear combinations of random variables. Journal of Statistical Planning and Inference **84**, 11–25 (2000)
376. Maccheroni, F., Muliere, P., Zoli, C.: Inverse stochastic orders and generalized Gini functionals. Metron **63**, 529–559 (2005)
377. Mailhot, L.: Ordre de dispersion et lois tronquées. C. R. Acad. Sc. Paris, t. 304, Série I **16**, 499–501 (1987)
378. Makowski, A.M.: On an elementary characterization of the increasing convex ordering, with an application. Journal of Applied Probability **31**, 834–840 (1994)
379. Makowski, A.M.: Bounding superposed on-off sources — variability ordering and majorization to the rescue. IEEE International Conference on Communications **4**, 2158–2162 (2004)
380. Makowski, A.M., Philips, T.K.: Stochastic convexity of sums of i.i.d. non-negative random variables with applications. Journal of Applied Probability **29**, 156–167 (1992)
381. Marshall, A.W.: Some comments on the hazard gradient. Stochastic Processes and Their Applications **3**, 293–300 (1975)
382. Marshall, A.W.: Multivariate stochastic orderings and generating cones of functions. In: Mosler, K., Scarsini, M. (ed) Stochastic Orders and Decision under Risk. IMS Lecture Notes—Monograph Series 19. Hayward, California, 231–247 (1991)
383. Marshall, A.W., Olkin, I.: Inequalities: Theory of Majorization and Its Applications. Academic Press, New York (1979)
384. Marshall, A.W., Proschan, F.: An inequality for convex functions involving majorization. Journal of Mathematical Analysis and Applications **12**, 87–90 (1965)
385. Meester, L.E.: Contributions to the Theory and Applications of Stochastic Convexity. Ph.D. Thesis. Department of Statistics, University of California, Berkeley (1990)
386. Meester, L.E., Shanthikumar, J.G.: Concavity of the throughput of tandem queueing systems with finite buffer storage space. Advances in Applied Probability **22**, 764–767 (1990)

387. Meester, L.E., Shanthikumar, J.G.: Regularity of stochastic processes: A theory based on directional convexity. Probability in the Engineering and Informational Sciences **7**, 343–360 (1993)
388. Meester, L.E., Shanthikumar, J.G.: Stochastic convexity on general space. Mathematics of Operations Research **24**, 472–494 (1999)
389. Meilijson, I., Nádas, A.: Convex majorization with an application to the length of critical paths. Journal of Applied Probability **16**, 671–677 (1979)
390. Menezes, C., Geiss, C., Tressler, J.: Increasing downside risk. The American Economic Review **70**, 921–932 (1980)
391. Metry, M.H., Sampson, A.R.: A family of partial orderings for positive dependence among fixed marginal bivariate distributions. In: Dall'Aglio, G., Kotz, S., Salinetti, G. (ed) Advances in Probability Distributions with Given Marginals. Kluwer Academic Publishers, Boston, 129–138 (1991)
392. Metry, M.H., Sampson, A.R.: Orderings for positive dependence on multivariate empirical distributions. Annals of Applied Probability **3**, 1241–1251 (1993)
393. Metzger, C., Rüschendorf, L.: Conditional variability ordering of distributions. Annals of Operations Research **32**, 127–140 (1991)
394. Mi, J.: Some comparison results of system availability. Naval Research Logistics **45**, 205–218 (1998)
395. Mi, J., Shaked, M.: Stochastic dominance of random variables implies the dominance of their order statistics. Journal of the Indian Statistical Association **40**, 161–168 (2002)
396. Misra, N., van der Meulen, E.C.: On stochastic properties of m-spacings. Journal of Statistical Planning and Inference **115**, 683–697 (2003)
397. Miyoshi, N.: A note on bounds and monotonicity of spatial stationary Cox shot noises. Probability in the Engineering and Informational Sciences **18**, 561–571 (2004)
398. Miyoshi, N., Rolski, T.: Ross-type conjectures on monotonicity of queues. Australia and New Zealand Journal of Statistics **46**, 121–131 (2004)
399. Mosler, K.: Multivariate Dispersion, Central Regions and Depth: The Lift Zonoid Approach. Springer, New York (2002)
400. Mosler, K., Scarsini, M.: Some theory of stochastic dominance. In: Mosler, K., Scarsini, M. (ed) Stochastic Orders and Decision under Risk. IMS Lecture Notes—Monograph Series 19. Hayward, California, 261–284 (1991)
401. Mosler, K.C.: Stochastic dominance decision rules when the attributes are utility independent. Management Science **30**, 1311–1322 (1984)
402. Mugdadi, A.R., Ahmad, I.A.: Moment inequalities derived from comparing life with its equilibrium form. Journal of Statistical Planning and Inference **134**, 303–317 (2005)
403. Mukherjee, S.P., Chatterjee, A.: Closure under convolution of dominance relations. Calcutta Statistical Association Bulletin **42**, 251–254 (1992)
404. Mukherjee, S.P., Chatterjee, A.: Stochastic dominance of higher orders and its implications. Communications in Statistics—Theory and Methods **21**, 1977–1986 (1992)
405. Muliere, P., Petrone, S.: Generalized Lorenz curve and monotone dependence orderings. Metron **50**, 19–38 (1992)
406. Muliere, P., Scarsini, M.: A note on stochastic dominance and inequality measures. Journal of Economic Theory **49**, 314–323 (1989)
407. Müller, A.: Orderings of risks: A comparative study via stop-loss transforms. Insurance: Mathematics and Economics **17**, 215–222 (1996)

408. Müller, A.: Stochastic orders generated by integrals: A unified approach. Advances in Applied Probability **29**, 414–428 (1997)
409. Müller, A.: Stop-loss order for portfolios of dependent risks. Insurance: Mathematics and Economics **21**, 219–223 (1997)
410. Müller, A.: Comparing risks with unbounded distributions. Journal of Mathematical Economics **30**, 229–239 (1998)
411. Müller, A.: On the waiting times in queues with dependency between interarrivals and service times. Operations Research Letters **26**, 43–47 (2000)
412. Müller, A.: Stochastic orders generated by generalized convex functions. In: Hadjisavvas, H. et al. (ed) Generalized Convexity and Generalized Monotonicity. Springer-Verlag, Berlin, 264–278 (2001)
413. Müller, A.: Stochastic ordering of multivariate normal distributions. Annals of the Institute of Statistical Mathematics **53**, 567–575 (2001)
414. Müller, A.: On the interplay between variability and negative dependence for bivariate distributions. Operations Research Letters **31**, 90–94 (2003)
415. Müller, A., Rüschendorf, L.: On the optimal stopping values induced by general dependence structures. Journal of Applied Probability **38**, 672–684 (2001)
416. Müller, A., Scarsini, M.: Some remarks on the supermodular order. Journal of Multivariate Analysis **73**, 107–119 (2000)
417. Müller, A., Scarsini, M.: Stochastic comparison of random vectors with a common copula. Mathematics of Operations Research **26**, 723–740 (2001)
418. Müller, A., Scarsini, M.: Stochastic order relations and lattices of probability measures. SIAM Journal on Optimization **16**, 1024–1043 (2006)
419. Müller, A., Stoyan, D.: Comparison Methods for Stochastic Models and Risks. Wiley, New York (2002)
420. Muñoz-Perez, J.: Dispersive ordering by the spread function. Statistics and Probability Letters **10**, 407–410 (1990)
421. Muñoz-Perez, J., Sanches-Gomez, A.: Dispersive ordering by dilation. Journal of Applied Probability **27**, 440–444 (1990)
422. Nanda, A.K.: Stochastic orders in terms of Laplace transforms. Calcutta Statistical Association Bulletin **45**, 195–201 (1995)
423. Nanda, A.K., Jain, K.: Some weighted distribution results on univariate and bivariate cases. Journal of Statistical Planning and Inference **77**, 169–180 (1999)
424. Nanda, A.K., Jain, K., Singh, H.: On closure of some partial orderings under mixtures. Journal of Applied Probability **33**, 698–706 (1996)
425. Nanda, A.K., Jain, K., Singh, H.: Properties of moments for s-ordered equilibrium distributions. Journal of Applied Probability **33**, 1108–1111 (1996)
426. Nanda, A.K., Jain, K., Singh, H.: Preservation of some partial orderings under the formation of coherent systems. Statistics and Probability Letters **39**, 123–131 (1998)
427. Nanda, A.K., Misra, N., Paul, P., Singh, H.: Some properties of order statistics when the sample size is random. Communications in Statistics-Theory and Methods **34**, 2105–2113 (2005)
428. Nanda, A.K., Shaked, M.: The hazard rate and the reversed hazard rate orders, with applications to order statistics. Annals of the Institute of Statistical Mathematics **53**, 853–864 (2001)
429. Nanda, A.K., Singh, H., Misra, N., Paul, P.: Reliability properties of reversed residual lifetime. Communications in Statistics—Theory and Methods **32**, 2031–2042 (2003)

430. Navarro, J., Shaked, M.: Hazard rate ordering of order statistics and systems. Journal of Applied Probability **43**, 391–408 (2006)
431. Nelsen, R.B.: An Introduction to Copulas. Springer-Verlag, New York (1999)
432. Nelsen, R.B., Quesada-Molina, J.J., Rodríguez-Lallena, J.A., Úbeda-Flores, M.: Distribution functions of copulas: A class of bivariate probability integral transforms. Statistics and Probability Letters **54**, 277–282 (2001)
433. Nelsen, R.B., Quesada-Molina, J.J., Rodríguez-Lallena, J.A., Úbeda-Flores, M.: Kendall distribution functions. Statistics and Probability Letters **65**, 263–268 (2003)
434. Nguyen, T.T., Sampson, A.R.: The geometry of a certain fixed marginal probability distributions. Linear Algebra and Its Applications **70**, 73–87 (1985)
435. Nguyen, T.T., Sampson, A.R.: Counting the number of $p \times q$ integer matrices more concordant than a given matrix. Discrete Applied Mathematics **11**, 187–205 (1985)
436. Norros, I.: Systems weakened by failures. Stochastic Processes and Their Applications **20**, 181–196 (1985)
437. Norros, I.: A compensator representation of multivariate life length distributions, with applications. Scandinavian Journal of Statistics **13**, 99–112 (1986)
438. O'Brien, G.L., Scarsini, M.: Multivariate stochastic dominance and moments. Mathematics of Operations Research **16**, 382–389 (1991)
439. O'Cinneide, C.A.: Phase-type distributions and majorization. Annals of Applied Probability **1**, 219–227 (1991)
440. Oja, H.: On location, scale, skewness and kurtosis of univariate distributions. Scandinavian Journal of Statistics **8**, 154–168 (1981)
441. Oja, H.: Descriptive statistics for multivariate distributions. Statistics and Probability Letters **1**, 327–332 (1983)
442. Paul, A.: On bivariate dependence and the convex order. Operations Research Letters **30**, 181–188 (2002)
443. Paul, A., Gutierrez, G.: Mean sample spacings, sample size and variability in an auction-theoretic framework. Operations Research Letters **32**, 103–108 (2004)
444. Pellerey, F.: Personal communication (1993)
445. Pellerey, F.: Partial orderings under cumulative damage shock models. Advances in Applied Probability **25**, 939–946 (1993)
446. Pellerey, F.: Correction. Advances in Applied Probability **26**, 280 (1994)
447. Pellerey, F.: Personal communication (1994)
448. Pellerey, F.: On the preservation of some orderings of risks under convolution. Insurance: Mathematics and Economics **16**, 23–30 (1995)
449. Pellerey, F.: Correction note to "On the preservation of some orderings of risks under convolution". Insurance: Mathematics and Economics **19**, 81–83 (1996)
450. Pellerey, F.: Some new conditions for the increasing convex comparison of risks. Scandinavian Actuarial Journal, 38–47 (1997)
451. Pellerey, F.: Stochastic comparisons for multivariate shock models. Journal of Multivariate Analysis **71**, 42–55 (1999)
452. Pellerey, F.: Random vectors with HNBUE-type marginal distributions. Statistics and Probability Letters **50**, 265–271 (2000)
453. Pellerey, F., Petakos, K.I.: Closure property of the NBUC class under formation of parallel systems. IEEE Transactions on Reliability **51**, 452–454 (2002)
454. Pellerey, F., Semeraro, P.: Ageing and stochastic comparisons for a covariate failure model. Journal of Applied Probability **39**, 421–425 (2002)

455. Pellerey, F., Shaked, M.: Stochastic comparison of some wear processes. Probability in the Engineering and Informational Sciences **7**, 421–435 (1993)
456. Pellerey, F., Shaked, M.: Characterizations of the IFR and DFR aging notions by means of the dispersive order. Statistics and Probability Letters **33**, 389–393 (1997)
457. Perlman, M.D., Olkin, I.: Unbiasedness of invariant tests for MANOVA and other multivariate problems. Annals of Statistics **6**, 1326–1341 (1980)
458. Pledger, G., Proschan, F.: Comparisons of order statistics and spacings from heterogeneous distributions. In: Rustagi, J.S. (ed) Optimizing Methods in Statistics. Academic Press, New York, 89–113 (1971)
459. Pratt, J.W.: Risk aversion in the small and in the large. Econometrica **32**, 122–136 (1964)
460. Preston, C.J.: A generalization of the FKG inequalities. Communications in Mathematical Physics **36**, 233–241 (1974)
461. Proschan, F.: Peakedness of distributions of convex combinations. Annals of Mathematical Statistics **36**, 1703–1706 (1965)
462. Purcaru, O., Denuit, M.: Dependence in dynamic claim frequency credibility models. ASTIN Bulletin **33**, 23–40 (2003)
463. Ramos, H.M., Sordo, M.A.: Dispersive measures and dispersive orderings. Statistics and Probability Letters **61**, 123–131 (2003)
464. Ramos Romero, H.M., Sordo Díaz, M.A.: The proportional likelihood ratio order and applications. Qüestiió **25**, 211–223 (2001)
465. Raqab, M.Z., Amin, W.A.: Some ordering results on order statistics and record values. IAPQR Transactions **21**, 1–8 (1996)
466. Righter, R., Shanthikumar, J.G.: Extensions of the bivariate characterization for stochastic orders. Advances in Applied Probability **24**, 506–508 (1992)
467. Rinott, Y., Pollak, M.: A stochastic ordering induced by a concept of positive dependence and monotonicity of asymptotic test sizes. Annals of Statistics **8**, 190–198 (1980)
468. Rinott, Y., Scarsini, M.: Total positivity order and the normal distribution. Journal of Multivariate Analysis **97**, 1251–1261 (2006)
469. Rivest, L.-P.: Products of random variables and star-shaped ordering. Canadian Journal of Statistics **10**, 219–223 (1982)
470. Rodríguez-Lallena, J.A., Úbeda-Flores, M.: Distribution functions of multivariate copulas. Statistics and Probability Letters **64**, 41–50 (2003)
471. Rojo, J.: A pure-tail ordering based on the ratio of the quantile functions. Annals of Statistics **20**, 570–579 (1992)
472. Rojo, J., He, G.Z.: New properties and characterizations of the dispersive ordering. Statistics and Probability Letters **11**, 365–372 (1991)
473. Rolski, T.: Order relations in the set of probability distribution functions and their applications in queueing theory. Dissertationes Mathematicae **CXXXII**. Polska Akademia Nauk, Instytut Matematyczny, Warsaw, Poland (1976)
474. Rolski, T., Szekli, R.: Stochastic ordering and thinning of point processes. Stochastic Processes and Their Applications **37**, 299–312 (1991)
475. Ross, S.M.: Stochastic Processes (2nd Edition). Wiley, New York (1996)
476. Ross, S.M.: The inspection paradox. Probability in the Engineering and Informational Sciences **17**, 47–51 (2003)
477. Ross, S.M., Schechner, Z.: Some reliability applications of the variability ordering. Operations Research **32**, 679–687 (1984)

478. Ross, S.M., Shanthikumar, J.G., Zhu, Z.: On increasing-failure-rate random variables. Journal of Applied Probability **42**, 797–809 (2005)
479. Rowell, G., Siegrist, K.: Relative aging of distributions. Probability in the Engineering and Informational Sciences **12**, 469–478 (1998)
480. Roy, D.: Classification of multivariate life distributions based on partial ordering. Probability in the Engineering and Informational Sciences **16**, 129–137 (2002)
481. Rüschendorf, L.: Inequalities for the expectation of Δ-monotone functions. Zeitschrift für Wahrscheinlichkeitstheorie und Verwandte Gebiete **54**, 341–349 (1980)
482. Rüschendorf, L.: Ordering of distributions and rearrangement of functions. Annals of Probability **9**, 276–283 (1981)
483. Rüschendorf, L.: Solutions of a statistical optimization problem by rearrangement methods. Metrika **30**, 55–61 (1983)
484. Rüschendorf, L.: Monotonicity and unbiasedness of tests via a. s. construction. Statistics **17**, 221–230 (1986)
485. Rüschendorf, L.: On conditional stochastic ordering of distributions. Advances in Applied Probability **23**, 46–63 (1991)
486. Rüschendorf, L.: Comparison of multivariate risk and positive dependence. Journal of Applied Probability **41**, 391–406 (2004)
487. Samorodnitsky, G., Taqqu, M.: Stochastic monotonicity and Slepian-type inequalities for infinitely divisible and stable random vectors. Annals of Probability **21**, 143–160 (1993)
488. Samuels, S.M.: On the number of successes in independent trials. Annals of Mathematical Statistics **36**, 1272–1278 (1965)
489. Saunders, D.J.: Dispersive ordering of distributions. Advances in Applied Probability **16**, 693–694 (1984)
490. Scarsini, M.: Stochastic dominance with pair-wise risk aversion. Journal of Mathematical Economics **14**, 187–201 (1985)
491. Scarsini, M.: Multivariate stochastic dominance with fixed dependence structure. Operations Research Letters **7**, 237–240 (1988)
492. Scarsini, M.: Multivariate convex orderings, dependence, and stochastic equality. Journal of Applied Probability **35**, 93–103 (1998)
493. Scarsini, M., Shaked, M.: Ordering distributions by scaled order statistics. Zeitschrift für Operations Research **31**, A1–A13 (1987)
494. Scarsini, M., Shaked, M.: Some conditions for stochastic equality. Naval Research Logistics **37**, 617–625 (1990)
495. Scarsini, M., Shaked, M.: Stochastic ordering for permutation symmetric distributions. Statistics and Probability Letters **9**, 217–222 (1990)
496. Scarsini, M., Shaked, M.: Positive dependence orders: A survey. In: Heyde, C.C., Prohorov, Y.V., Pyke, R., Rachev, S.T. (ed) Athens Conference on Applied Probability and Time Series I: Applied Probability. Springer-Verlag, New York, 70–91 (1996)
497. Schöttl, A.: A new ordering for point processes. Mathematical Methods of Operations Research **43**, 373–387 (1996)
498. Schriever, B.F.: An ordering for positive dependence. Annals of Statistics **15**, 1208–1214 (1987)
499. Schweder, T.: On the dispersion of mixtures. Scandinavian Journal of Statistics **9**, 165–169 (1982)

500. Sengupta, D., Deshpande, J.V.: Some results on relative ageing of two life distributions. Journal of Applied Probability **31**, 991–1003 (1994)
501. Shaked, M.: Some concepts of positive dependence for bivariate interchangeable distributions. Annals of the Institute of Statistical Mathematics **31**, 67–84 (1979)
502. Shaked, M.: On mixtures from exponential families. Journal of the Royal Statistical Society B **42**, 192–198 (1980)
503. Shaked, M.: Dispersive ordering of distributions. Journal of Applied Probability **19**, 310–320 (1982)
504. Shaked, M., Shanthikumar, J.G.: The multivariate hazard construction. Stochastic Processes and Their Applications **24**, 241–258 (1987)
505. Shaked, M., Shanthikumar, J.G.: Multivariate hazard rate and stochastic ordering. Advances in Applied Probability **19**, 123–137 (1987)
506. Shaked, M., Shanthikumar, J.G.: Characterization of some first passage times using log-concavity and log-convexity as aging notions. Probability in the Engineering and Informational Sciences **1**, 279–291 (1987)
507. Shaked, M., Shanthikumar, J.G.: Temporal stochastic convexity and concavity. Stochastic Processes and Their Applications **27**, 1–20 (1988)
508. Shaked, M., Shanthikumar, J.G.: Stochastic convexity and its applications. Advances in Applied Probability **20**, 427–446 (1988)
509. Shaked, M., Shanthikumar, J.G.: Parametric stochastic convexity and concavity of stochastic processes. Annals of the Institute of Statistical Mathematics **42**, 509–531 (1990)
510. Shaked, M., Shanthikumar, J.G.: Convexity of a set of stochastically ordered random variables. Advances in Applied Probability **22**, 160–177 (1990)
511. Shaked, M., Shanthikumar, J.G.: Multivariate stochastic ordering and positive dependence in reliability theory. Mathematics of Operations Research **15**, 545–552 (1990)
512. Shaked, M., Shanthikumar, J.G.: Dynamic multivariate aging notions in reliability theory. Stochastic Processes and Their Applications **38**, 85–97 (1991)
513. Shaked, M., Shanthikumar, J.G.: Dynamic multivariate mean residual life functions. Journal of Applied Probability **28**, 613–629 (1991)
514. Shaked, M., Shanthikumar, J.G.: Regular, sample path and strong stochastic convexity: A review. In: Mosler, K., Scarsini, M. (ed) Stochastic Orders and Decision under Risk. IMS Lecture Notes—Monograph Series 19. Hayward, California, 320–333 (1991)
515. Shaked, M., Shanthikumar, J.G.: Stochastic Orders and Their Applications. Academic Press, Boston (1994)
516. Shaked, M., Shanthikumar, J.G.: Hazard rate ordering of k-out-of-n systems. Statistics and Probability Letters **23**, 1–8 (1995)
517. Shaked, M., Shanthikumar, J.G.: Supermodular stochastic orders and positive dependence of random vectors. Journal of Multivariate Analysis **61**, 86–101 (1997)
518. Shaked, M., Shanthikumar, J.G.: Two variability orders. Probability in the Engineering and Informational Sciences **12**, 1–23 (1998)
519. Shaked, M., Shanthikumar, J.G., Tong, Y.L.: Parametric Schur convexity and arrangement monotonicity properties of partial sums. Journal of Multivariate Analysis **53**, 293–310 (1995)

520. Shaked, M., Suarez-Llorens, A.: On the comparison of reliability experiments based on the convolution order. Journal of the American Statistical Association **98**, 693–702 (2003)
521. Shaked, M., Szekli, R.: Comparison of replacement policies via point processes. Advances in Applied Probability **27**, 1079–1103 (1995)
522. Shaked, M., Tong, Y.L.: Stochastic ordering of spacings from dependent random variables. In: Tong, Y.L. (ed) Inequalities in Statistics and Probability. IMS Lecture Notes—Monograph Series 5. Hayward, California, 141–149 (1984)
523. Shaked, M., Tong, Y.L.: Some partial orderings of exchangeable random variables by positive dependence. Journal of Multivariate Analysis **17**, 333–349 (1985)
524. Shaked, M., Wong, T.: Preservation of stochastic orderings under random mapping by point processes. Probability in the Engineering and Informational Sciences **9**, 563–580 (1995)
525. Shaked, M., Wong, T.: Stochastic orders based on ratios of Laplace transforms. Journal of Applied Probability **34**, 404–419 (1997)
526. Shaked, M., Wong, T.: Stochastic comparisons of random minima and maxima. Journal of Applied Probability **34**, 420–425 (1997)
527. Shanthikumar, J.G.: On stochastic comparison of random vectors. Journal of Applied Probability **24**, 123–136 (1987)
528. Shanthikumar, J.G., Koo, H.-W.: On uniform conditional stochastic order conditioned on planar regions. Journal of Applied Probability **27**, 115–123 (1990)
529. Shanthikumar, J.G., Yamazaki, G., Sakasegawa, H.: Characterization of optimal order of servers in a tandem queue with blocking. Operations Research Letters **10**, 17–22 (1991)
530. Shanthikumar, J.G., Yao, D.D.: The preservation of likelihood ratio ordering under convolution. Stochastic Processes and Their Applications **23**, 259–267 (1986)
531. Shanthikumar, J.G., Yao, D.D.: Second-order stochastic properties in queueing systems. Proceedings of the IEEE **77**, 162–170 (1989)
532. Shanthikumar, J.G., Yao, D.D.: Bivariate characterization of some stochastic order relations. Advances in Applied Probability **23**, 642–659 (1991)
533. Shanthikumar, J.G., Yao, D.D.: Strong stochastic convexity: Closure properties and applications. Journal of Applied Probability **28**, 131–145 (1991)
534. Shanthikumar, J.G., Yao, D.D.: Spatiotemporal convexity of stochastic processes and applications. Probability in the Engineering and Informational Sciences **6**, 1–16 (1992)
535. Shao, Q.-M.: A comparison theorem on moment inequalities between negatively associated and independent random variables. Journal of Theoretical Probability **13**, 343–356 (2000)
536. Singh, H.: On partial orderings of life distributions. Naval Research Logistics **36**, 103–110 (1989)
537. Singh, H., Vijayasree, G.: Preservation of partial orderings under the formation of k-out-of-n:G systems of i.i.d. components. IEEE Transactions on Reliability **40**, 273–276 (1991)
538. Sordo, M.A., Ramos, H.M.: Characterization of stochastic orders by L-functionals. Statistical Papers, to appear (2006)
539. Spizzichino, F.: Subjective Probability Models for Lifetimes. Chapman and Hall/CRC, Boca Raton (2001)

540. Stoyan, D.: Comparison Methods for Queues and Other Stochastic Models. Wiley, New York (1983)
541. Strassen, V.: The existence of probability measures with given marginals. Annals of Mathematical Statistics **36**, 423–439 (1965)
542. Suárez-Llorens, A.: Personal communication (2005)
543. Sun, L., Zhang, X.: Stochastic comparisons of order statistics from gamma distributions. Journal of Multivariate Analysis **93**, 112–121 (2005)
544. Szekli, R.: Stochastic Ordering and Dependence in Applied Probability. Springer-Verlag, New York (1995)
545. Szekli, R., Disney, R.L., Hur, S.: MR/GI/1 queues with positively correlated arrival stream. Journal of Applied Probability **31**, 497–514 (1994)
546. Taillie, C.: Lorenz ordering within the generalized gamma family of income distributions. In: Taillie, C., Patil, G.P., Baldessari, B.A. (ed) Statistical Distributions in Scientific Work, Volume 6. Reidel, Boston, 181–192 (1981)
547. Tchen, A.H.: Inequalities for distributions with given marginals. Annals of Probability **8**, 814–827 (1980)
548. Thistle, P.D.: Negative moments, risk aversion, and stochastic dominance. Journal of Financial and Quantitative Analysis **28**, 301–311 (1993)
549. Thorlund-Peterson, L.: Comparison of probability measures: Dominance of the third degree. Operations Research Letters **26**, 237–245 (2000)
550. Tong, Y.L.: Inequalities for a class of positively dependent random variables with a common marginal. Annals of Statistics **17**, 429–435 (1989)
551. Torgersen, E.: Stochastic orders and comparison of experiments. In: Mosler, K., Scarsini, M. (ed) Stochastic Orders and Decision under Risk. IMS Lecture Notes—Monograph Series 19. Hayward, California, 334–371 (1991)
552. Townsend, J.T., Colonius, H.: Variability of the max and min statistic: A theory of the quantile spread as a function of sample size. Psychometrika **70**, 759–772 (2005)
553. Valdés, J.E., Zequeira, R.I.: On the optimal allocation of two active redundancies in a two-component series system. Operations Research Letters **34**, 49–52 (2006)
554. Vanichpun, S., Makowski, A.M.: Positive correlations and buffer occupancy: Lower bounds via supermodular ordering. Proceedings of IEEE INFOCOM 2002, 1298–1306 (2002)
555. Vanichpun, S., Makowski, A.M.: When are on-off sources SIS? Probability in the Engineering and Informational Sciences **18**, 423–443 (2004)
556. Veinott, R.: Optimal policy in a dynamic, single product, non-stationary, inventory model with several demand classes. Operations Research **13**, 761–778 (1965)
557. Vyncke, D., Goovaerts, M.J., De Schepper, A., Kaas, R., Dhaene, J.: On the distribution of cash flows using the Esscher transform. Journal of Risk and Insurance **70**, 563–575 (2003)
558. Wang, S.S., Young, V.R.: Ordering risks: Expected utility theory versus Yaari's dual theory of risk. Insurance: Mathematics and Economics **22**, 145–161 (1998)
559. Wei, G., Hu, T.: Supermodular dependence ordering on a class of multivariate copulas. Statistics and Probability Letters **57**, 375–385 (2002)
560. Whitt, W.: A note on the influence of the sample on the posterior distribution. Journal of the American Statistical Association **74**, 424–426 (1979)
561. Whitt, W.: Uniform conditional stochastic order. Journal of Applied Probability **17**, 112–123 (1980)

562. Whitt, W.: The effect of variability in the $GI/G/s$ queue. Journal of Applied Probability **17**, 1062–1071 (1980)
563. Whitt, W.: Multivariate monotone likelihood ratio and uniform conditional stochastic order. Journal of Applied Probability **19**, 695–701 (1982)
564. Whitt, W.: Uniform conditional variability ordering of probability distributions. Journal of Applied Probability **22**, 619–633 (1985)
565. Whitt, W.: The renewal-process stationary-excess operator. Journal of Applied Probability **22**, 156–167 (1985)
566. Wilfling, B.: Lorenz ordering of power-function order statistics. Statistics and Probability Letters **30**, 313–319 (1996)
567. Wolff, R.W.: Stochastic Modeling and the Theory of Queues. Prentice Hall, Englewood Cliffs, New Jersey (1989)
568. Wong, T.: Preservation of multivariate stochastic orders under multivariate shock models. Journal of Applied Probability **34**, 1009–1020 (1997)
569. Yanagimoto, T.: Dependence ordering in statistical models and other notions. In: Block, H.W., Sampson, A.R., and Savits, T.H. (ed) Topics in Statistical Dependence. IMS Lecture Notes—Monograph Series 16. Hayward, California, 489–496 (1990)
570. Yanagimoto, T., Okamoto, M.: Partial orderings of permutations and monotonicity of a rank correlation statistic. Annals of the Institute of Statistical Mathematics **21**, 489–506 (1969)
571. Yanagimoto, T., Sibuya, M.: Stochastically larger component of a random vector. Annals of the Institute of Statistical Mathematics **24**, 259–269 (1972)
572. Yang, G.L.: Estimation of a biometric function. Annals of Statistics **6**, 112–116 (1978)
573. Yao, D.D.: Stochastic convexity and submodularity, with production applications. In: Borthakur, A.C., Choudhury, H. (ed) Probability Models and Statistics. New Age International Publishers Limited, New Delhi, 1–25 (1996)
574. Yi, Z., Tongyu, L.: A further study on correlation order. Applied Mathematics, a Journal of Chinese Universities, Series B **19**, 429–434 (2004)
575. Yue, D., Cao, J.: Some results on successive failure times of a system with minimal instantaneous repairs. Operations Research Letters **29**, 193–197 (2001)
576. Zenga, M.: Concentration curves and concentration indexes derived from them. In: Dagum, C. et al. (ed) Income and Wealth Distribution, Inequalities and Poverty, Studies in Contemporary Economics. Springer-Verlag, Berlin, 94–110 (1990)
577. Zijlstra, M., de Kroon, J.M.P.: Stochastic ordering related to ranking in an nth order record process. Journal of Statistical Planning and Inference **5**, 79–91 (1981)
578. van Zwet, W.: Convex Transforms of Random Variables. Mathematische Centrum, Amsterdam (1964)

Author Index

Subject Index

Springer Series in Statistics

(continued from p. ii)

Lahiri: Resampling Methods for Dependent Data.
Le/Zidek: Statistical Analysis of Environmental Space-time Processes
Le Cam: Asymptotic Methods in Statistical Decision Theory.
Le Cam/Yang: Asymptotics in Statistics: Some Basic Concepts, 2nd edition.
Liu: Monte Carlo Strategies in Scientific Computing.
Manski: Partial Identification of Probability Distributions.
Mielke/Berry: Permutation Methods: A Distance Function Approach.
Molenberghs/Verbeke: Models for Discrete Longitudinal Data.
Mukerjee/Wu: A Modern Theory of Factorial Designs.
Nelsen: An Introduction to Copulas, 2nd edition.
Pan/Fang: Growth Curve Models and Statistical Diagnostics.
Politis/Romano/Wolf: Subsampling.
Ramsay/Silverman: Applied Functional Data Analysis: Methods and Case Studies.
Ramsay/Silverman: Functional Data Analysis, 2nd edition.
Reinsel: Elements of Multivariate Time Series Analysis, 2nd edition.
Rosenbaum: Observational Studies, 2nd edition.
Rosenblatt: Gaussian and Non-Gaussian Linear Time Series and Random Fields.
Särndal/Swensson/Wretman: Model Assisted Survey Sampling.
Santner/Williams/Notz: The Design and Analysis of Computer Experiments.
Schervish: Theory of Statistics.
Shaked/Shanthikumar: Stochastic Orders.
Shao/Tu: The Jackknife and Bootstrap.
Simonoff: Smoothing Methods in Statistics.
Sprott: Statistical Inference in Science.
Stein: Interpolation of Spatial Data: Some Theory for Kriging.
Taniguchi/Kakizawa: Asymptotic Theory for Statistical Inference for Time Series.
Tanner: Tools for Statistical Inference: Methods for the Exploration of Posterior Distributions and Likelihood Functions, 3rd edition.
Tillé: Sampling Algorithms.
Tsaitis: Semiparametric Theory and Missing Data.
van der Laan/Robins: Unified Methods for Censored Longitudinal Data and Causality.
van der Vaart/Wellner: Weak Convergence and Empirical Processes: With Applications to Statistics.
Verbeke/Molenberghs: Linear Mixed Models for Longitudinal Data.
Weerahandi: Exact Statistical Methods for Data Analysis.

Printed in the United States of America.